essentials of environmental science

second edition

Bernard J. Nebel
Richard T. Wright

STUDY GUIDE MATERIALS
PREPARED BY CLARK E. ADAMS

D1450986

Pearson
Custom
Publishing

Printed in the United States of America

10 9 8 7 6 5 4 3 2

This manuscript was supplied camera-ready by the authors.

Please visit our web site at www.pearsoncustom.com

ISBN 0–536–60239–5

BA 990618

PEARSON CUSTOM PUBLISHING
75 Arlington Street, Boston, MA 02116
A Pearson Education Company

Bernard J. Nebel is Professor Emeritus of Biology at Catonsville Community College in Maryland. He earned his Bachelor of Arts from Earlham College and his Ph.D. from Duke University. Nebel was one of the first college professors to develop a comprehensive environmental science course and write a text for the subject. Nebel is now interested in developing and writing an elementary (K–5) science curriculum designed to help children develop an understanding of the world, their place in it, and their responsibility toward it. Nebel is a member of the American Association for the Advancement of Science, the Institute of Biological Sciences, the American Solar Energy Society, and the National Association of Science Teachers. He strives to make a difference in the environment in his personal life; his urban backyard is a small ecosystem complex of pond, fruit trees, and garden that is supported by composted wastes. He is an active supporter of Freedom From Hunger, Habitat for Humanity, the World Wildlife Fund, Conservation International, and other environmental organizations.

Richard T. Wright is Professor Emeritus of Biology at Gordon College in Massachusetts, where he taught an environmental science course for 28 years. He earned a Bachelor of Arts from Rutgers and a Master of Arts and Ph.D. from Harvard University. For many years Wright received research grants from the National Science Foundation for his work in marine microbiology and, in 1981, was a founding faculty member of Au Sable Institute of Environmental Studies in Michigan, where he also served as Academic Chairman for 11 years. He is a Fellow of the American Association for the Advancement of Science, and in 1996 was appointed a Fulbright Scholar to Kenya. He is a member of many environmental organizations, including the National Wildlife Federation, Bread for the World, Habitat for Humanity, the Union of Concerned Scientists, and others, and is a supporting member of the Trustees of Reservations. In his personal life, Wright is involved full time in speaking and writing about the environment, and strives to keep his environmental impact at a minimum by recycling, planting trees and vegetables in his yard, and working with numerous environmental organizations.

Contents

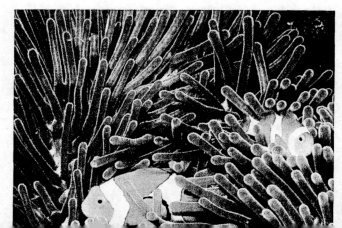

Acknowledgments

Completion of this edition of the text represents the fruition of devoted labors and contributions from many people. I wish to gratefully acknowledge and express my sincere thanks and appreciation to all those who have contributed to this work in so many ways.

I offer heartfelt, special thanks to all the dedicated people at Prentice Hall who had a hand in the production of this volume. In particular, I thank my Editor, Teresa Ryu, for her careful and creative supervision over the entire process; to my Production Editor, Joanne Hakim, for her careful attention to the seemingly endless details of putting the text together; to my Development Editor, Nancy García, who edited the entire manuscript and was in constant touch with me as we decided on the appropriate structure and wording of each chapter and selected each figure; to Barbara Salz, enterprising photo researcher who hunted down appropriate photos; to Karen Horton, Project Manager, who has fallen heir to all of the supplementary materials; to Andrew T. Stull, for effectively coordinating all the media components that are part of this text; to Tressa Smith, who worked on all of the end-of-chapter review and material; to Clark Adams (Texas A & M University) for writing the Study Guide; to Nancy Ostiguy (Pennsylvania State University) for a fine job with the Instructor's Guide; to Steven Ailstock (Ann Arundel Community College) and Shari Snitovsky (Skyline College) for revising the Test Bank, and finally but very sincerely to Isobel Heathcote (University of Guelph) for the new *Environment on the Web* essays, and for her and Steve Overmann's superb job with the text's home page on the World Wide Web.

Eight years ago Prentice Hall editor David Brake asked me if I would be interested in helping Bernard Nebel in writing the fourth edition of his environmental science text. Because of my long-time concern about environmental issues and my interest in writing, I accepted the offer. As the years passed, my commitment to environmental stewardship and deep concerns about our society's interactions with the environment have led me to direct more and more of my energy and ability to writing and speaking about environmental issues. As I have accepted more of the responsibility for writing this text, I have realized what an amazing job Bernie Nebel did in producing the first three editions alone while also teaching full-time. He did it because he was frustrated with existing environmental science texts, and was convinced he could produce a more readable and effective book. Bernie and I share very similar philosophical and educational values and have enjoyed collaborating over the years. I am deeply indebted to Bernie for his wonderful sense of organization and beautiful and clear prose, which still form the major part of the book. Both of us have offered this book in its successive editions as our contribution to the students of the '80s and '90s, in the hope that they will join us in helping to bring about the environmental revolution that must come—hopefully sooner rather than later.

I wish to offer some very personal thanks to my wife, Ann, who has been with me since the beginning of my work in biology and has provided the emotional base and companionship without which I would be far less of a person and a biologist. Her love and patience have sustained me in immeasurable ways. Finally, I offer my gratefulness to God, who is the author of the amazing Creation I love so much. I count it a privilege to be involved in the care of his Creation.

Richard T. Wright

Environmental Science

1

Introduction: Sustainability, Stewardship, and Sound Science

〰〰〰〰〰〰〰

Key Issues and Questions

1. Sharp contrasts exist in the way in which different peoples of the world interact with their environments. Give a range of contrasting "pictures" supporting this statement.

2. History is a saga of rises and falls of civilizations. What are the factors that brought about the collapse of the Easter Island civilization in the South Pacific? Are there any parallels to the present?

3. There is cause for concern about the general condition of the global environment today. What are four global trends that are of particular concern?

4. *Sustainability* is the practical goal toward which we should be working. What is meant by sustainability? For a sustainable society, what are the principal prerequisites?

5. *Sustainable development* is now a broadly accepted ideal. How do different disciplines address sustainable development?

6. *Stewardship* represents the ethical and moral framework that should inform our public and private actions. How can stewardship be applied to the natural world and to the concerns for justice?

7. The modern environmental movement has achieved much in recent years. How did this movement start, and what have been some recent reactions against it?

8. Sound science is the basis for understanding how the world works and how human systems interact with it. What is the essence of science and the scientific method?

9. Science occurs within a community of scientists. How does this community function to prevent the occurrence of poor, or "junk," science?

Imagine that you are assigned the task of traveling throughout the world to document the range of human interactions with the environment. Armed with a camera, you start your trip in South America. Boating the Amazon River through Peru's rain forest, you might photograph a village along the shore where there are a few small houses. Constructed of poles cut from the forest, lashed together with vines, and covered with a thatch of braided palm leaves, these primitive dwellings provide simple but adequate shelter in a climate where temperatures range from 75° to 85°F (24° to 29°C) year-round (Fig. 1–1).

People living here have no running water or sewers, no electricity or telephones, and the closest shops or markets are many miles downriver. The rain forest and the river provide all basic needs: fish, game, fruits, and even medicines for those who learn which plants to use. People here have lived this way for centuries, but their way of life is gradually changing as modern civilization makes its inroads, as forests are cleared for pasture or logging, or as mining and oil exploration bring technologies far upriver.

In Tanzania, in contrast, you might observe a much harsher life: Women and girls in rural villages may have to walk miles across a denuded, eroded landscape each day to collect the water—often polluted—that they will use for drinking, cooking, and washing. Similar treks of increasing length must be taken to collect the firewood for cooking (Fig. 1–2). Food is mostly coarse grains, such as sorghum and maize, raised in small landholdings by the women, and the amounts are often limited.

◀ **FIGURE 1–1** *People closely tied to their environment.* People living in the rain forest along the banks of the Amazon River in Peru depend on the forest and river for their needs. Canoes are made from hollowed-out logs. Food is the daily catch. The river serves as both water supply and sewer.

Both of these pictures are from what are called the **developing countries.** They are pictures of people closely tied to their environment. In the United States and other **industrialized countries,** people appear to be more detached from their environment. People live in well-built

▼ **FIGURE 1–2** *Impact on people's singular source of fuel as a result of landscape depletion.* For one-third of the world population, the only source of fuel for cooking is firewood. Many women in less developed countries must spend several hours each day gathering wood. Treks become longer and more difficult as the landscape is increasingly denuded.

homes surrounded by manicured lawns. They adjust the indoor temperature to their liking, and any amount of safe hot or cold water is available with the turning of a faucet handle. Travelers go nearly anywhere they wish in the air-conditioned comfort of private cars. Their greatest concern about food is that they not gain weight. News and entertainment from around the world are displayed on several television sets in each home, and many families own computers with rapidly growing interactive capabilities. There is an appearance of a civilization that is detached from its environment (Fig. 1–3). This is an illusion.

Human societies, whether they appear to be closely tied to their environment or not, are absolutely dependent on the natural environment, and their impact on the environment is crucial for their continued success. We can look to past civilizations for better understanding. On Easter Island, a remote spot of land in the South Pacific, we find giant stone heads (Maoi) standing as sentinels with their backs toward the sea (Fig. 1–4). These statues are evidence of a once sophisticated civilization. Yet, Easter Island natives encountered by eighteenth-century explorers were living at a primitive level, scratching out a meager existence on a desolate island. When asked about the great stone heads, the natives could say only, "Our legends say that our ancestors made them, but we don't know how or why." The past culture and civilization of the island were not sustained.

Working from the legends that natives told, and conducting excavations for evidence, archaeologists have pieced together the following chronology of events. The original inhabitants of Easter Island were Polynesians who probably arrived at the island as part of a deliberate colonization mission sometime between 400 and 800 A.D. The evidence from pollen grains found in the soil and in artifacts shows that these early arrivals found an island abundantly forested with a wide

◀ FIGURE 1–3 *Living in industrialized countries.* Life in industrialized countries appears to be detached from the environment.

variety of trees, including palms, conifers, and mahogany. As their population grew and flourished, they cut trees for agriculture, structural materials, and to move the huge stone heads from the quarries to their erection sites. By 1600 all the trees were gone. Without plant roots, the cleared land failed to hold water, and the soil washed into the sea, killing the fish and shellfish near the shore. The eroded soil baked hard and dry after rains, offering little support for agriculture.

As the forest was depleted and soil and water resources were degraded, the work necessary for existence became harder and the rewards fewer. The gap between the ruling and worker classes widened, apparently becoming intolerable. In 1678 there was a sudden revolt of the workers. In the

great war that ensued, virtually the entire ruling class was killed. Still, the situation worsened. Anarchy broke out among the workers, who splintered into groups and continued to fight among themselves. Starvation and disease became epidemic. Without any trees, no one could escape the island by boat. A population that had numbered about 8000 at the time of the revolt was down to a few hundred people by the mid-1800s. Many have puzzled over the reasons why the Easter Islanders, who could walk around their island in a day, were unable to foresee the consequences of their practices.

The lesson of Easter Island is all too clear: When a society fails to care for the environment that sustains it, when its population increases beyond the capacity of the land and

◀ FIGURE 1–4 *Easter Island.* The great stone heads and other artifacts found on Easter Island provide evidence of a once prosperous culture. The present barren, eroded landscape indicates that the civilization collapsed as a result of overexploitation of forest and soil resources. Is the story of Easter Island a parable for modern civilization?

water to provide adequate food for all, and when the disparity between haves and have-nots widens into a gulf of social injustice, the result is disaster. The civilization collapses. History is replete with the ruins of other civilizations, such as the Mayans, Greeks, Incas, and Romans, who each failed to recognize the constraints of their environment.

Are there some parallels between Easter Island and the start of the twenty-first century? Is there evidence that we are making some of the same mistakes made by the Easter Islanders? Like Easter Island, Earth has only limited resources to support human societies and their demands. There is no escape. The social unrest on Easter Island may parallel the tensions between the industrialized and the developing countries. Is there some possibility that we will so damage our environment that our civilization will collapse? Are we already in trouble? Can we take corrective action before it is too late?

These are important questions, and to answer them will take the rest of this text. The task before us—both in this text and as a twenty-first century civilization—can be laid out in four steps:

- To understand how the natural world works
- To understand how human systems are interacting with natural systems
- To accurately assess the status and trends of crucial natural systems
- To promote and follow a long-term, sustainable relationship with the natural world

1.1   The Global Environmental Picture

As we consider this task just outlined, it is enlightening to step back and take a panoramic view of the interaction of human systems and natural systems. As we do, four global trends are of particular concern: (a) population growth and increasing consumption per person, (b) degradation of soils, (c) global atmospheric changes, and (d) loss of biodiversity. Each of these issues will be explored in greater depth in later chapters.

Population Growth

The world's human population, over 6 billion persons in mid-1999, has grown by 2 billion in just the last 25 years. It is continuing to grow rapidly, adding nearly 83 million persons per year. Even though the growth rate is gradually slowing, the world population is projected to reach about 10 billion by the year 2050 (Fig. 1–5). Each person creates a certain demand on Earth's resources, and the demand tends to increase with greater affluence. Compare, for example, the resources required to support a typical American or Canadian lifestyle with those required to support indigenous people living along the banks of the Peruvian Amazon.

Vital resources are stressed by the dual demands of increasing population and increasing consumption per person. Around the world we see groundwater supplies being depleted, agricultural soils being degraded, oceans being overfished, oil reserves being drawn down, and forests being cut faster than they can regrow. It is significant that the global economy has quintupled since 1950; a substantial share of that growth can be attributed to the exploitation of natural resources. In spite of that economic growth, 841 million people in the developing world—one out of every five—remain undernourished. Over 1 billion people live in abject poverty, lacking sufficient income to meet their basic needs for food, clothing, and shelter.

▶ **FIGURE 1–5** *World population in the last 200 years.* World population started a rapid growth phase in the early 1800s and has grown sixfold in the last 200 years. It continues to grow by nearly 83 million people per year (see Chapter 6).

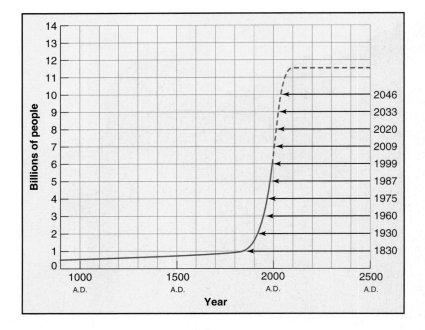

How can Earth support a near doubling of its human population over the next 50 years, as is projected, and still increase standards of living?

Degradation of Soils

Fertile soil is the foundation for plant growth and food production. Yet, around the world, soils are being degraded by erosion (Fig. 1–6a), grazing lands are turning into deserts, irrigated lands are becoming too salty to support crops, water supplies for irrigation are being depleted, and millions of acres of agricultural land are being sacrificed for development (Fig. 1–6b).

Global Atmospheric Changes

Historically, pollution has been a relatively local problem, affecting a given river, lake, bay, or the air in a city. Today, scientists are analyzing pollution on a global scale. A case in point is the danger of global climate change due to carbon dioxide (CO_2), an unavoidable by-product of burning gasoline and other liquid fuels from fossil fuels, such as crude oil, coal, and natural gas.

Carbon dioxide is a natural component of the lower atmosphere, along with nitrogen and oxygen. It is required by plants for photosynthesis and is important to the Earth-atmosphere energy budget. Carbon dioxide is transparent to incoming light from the Sun but absorbs infrared (heat) energy radiated from Earth's surface, thus delaying its loss to space. This process warms the lower atmosphere in a process known as the *greenhouse effect*. Although the concentration

of carbon dioxide is a small percentage of the atmosphere, even slight increases in its volume affects temperatures.

Because of the large amount of fossil fuels being burned today, carbon dioxide levels in the atmosphere have grown from about 280 parts per million (ppm), or 0.028%, in 1900 to over 370 ppm as we approach the end of the century. It is increasing at 0.4% per year and is expected to double during the next century. The latest conclusion of the Intergovernmental Panel on Climate Change (IPCC), published in 1996, stated that

> Human activities, including the burning of fossil fuels…are increasing the atmospheric concentrations of greenhouse gases. These changes…are projected to change regional and global climate and climate-related parameters such as temperature, precipitation, soil moisture, and sea level.…The balance of evidence suggests that there is a discernible human influence on global climate.

Figure 1–7 graphs air temperatures from 1880 to the present and illustrates an overall warming trend. Carbon dioxide is thought to be responsible for almost 60% of the global-warming trend. Concern about global climate change led representatives of 166 nations to meet in Kyoto, Japan, in December of 1997 to negotiate a treaty to reduce emissions of carbon dioxide and other greenhouse gases. Most of the industrialized nations agreed to reduce emissions to below 1990 levels, to be achieved by the year 2010. Even if the treaty is ratified and conformed to by all parties, greenhouse gases will continue to rise. At issue are the conflicting

▼ FIGURE 1–6 *Degradation of soils.* (a) Throughout much of the world, agricultural soils are degraded by erosion. (b) Around every metropolitan area in North America, as well as in other parts of the world, agricultural land is consumed by development, causing serious loss of millions of acres of soil from production every year.

(a)

(b)

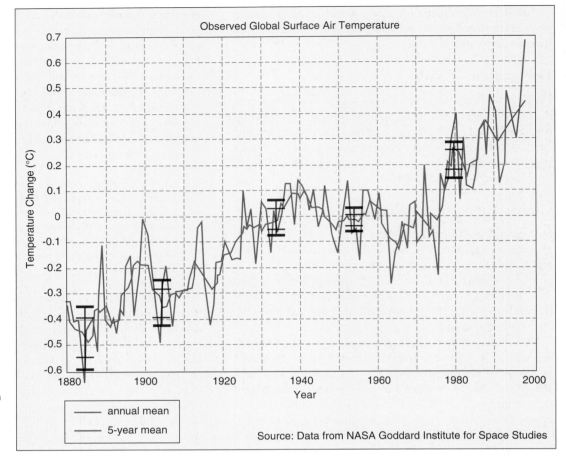

FIGURE 1–7
Global temperature trends from 1880 to 1998. The zero baseline represents the 1950–1980 global average. Note the cooling effect of the Mount Pinatubo volcanic eruption in the Philippines in 1991. Global temperatures quickly recovered, setting a new record in 1998.

concerns between the short-term economic impacts of reducing fossil fuel use and the long-term consequences of a climate change for the planet and all its inhabitants.

Loss of Biodiversity

Rapidly increasing human populations, along with increasing consumption, are accelerating the conversion of forests, grasslands, and wetlands to agriculture and urban development (Fig. 1–8). The inevitable result is the extermination of most of the wild plants and animals that occupied those natural habitats. If the species involved have no populations at other locations, they are doomed to extinction by such **habitat alteration**. Pollution alters additional habitats—particularly aquatic and marine habitats—destroying the species they house. Also, hundreds of species of mammals, reptiles, amphibians, fish, birds, butterflies, and innumerable plants are exploited for their commercial value. Even where species are protected by law, their hunting, killing, and marketing continues illegally.

Thus, Earth is rapidly losing many of its species—as many as 4000 per year by some estimates. The term used to refer to the total diversity of living things—plants, animals, and microbes—that inhabit the planet is **biodiversity**. About 1.75 million species have been described and classified, but scientists estimate that up to 100 million still remain unidentified. Because so many species remain unidentified, the exact number of species becoming extinct can only

FIGURE 1–8 ***Natural ecosystems giving way to development.*** Continuing growth requires a massive reorganization and exploitation of natural resources, bringing record levels of extinctions of plant and animal species. Here we see the stripping away of a mature forest in the northeastern United States.

be estimated. At present, loss of biodiversity is accelerating because of mounting habitat alteration, pollution, and pressures for exploitation.

Why is losing biodiversity so critical? For one thing, all domestic plants and animals used in agriculture are derived from wild species, and we still rely on introducing genes from wild species into our domestic species to keep them vigorous and able to adapt to different conditions. For another thing, between 1959 and 1980, 25% of all prescription drugs were originally derived from plants, even though only a few percent had been thoroughly studied from this medicinal viewpoint. Biodiversity is the mainstay of agricultural crops and of medicines. The loss of biodiversity can only curtail potential development in these areas. Biodiversity is a critical factor in maintaining the stability of natural systems and enabling them to recover after disturbances such as fires or volcanic eruptions.

There are also aesthetic and moral arguments for maintaining biodiversity. Forty or 50 years from now, do you want to show your grandchildren pictures of animals such as rhinoceroses, tigers, and orangutans and have to say, "These animals don't exist anymore. We killed them"? The question for society is: Do we have a moral responsibility to protect and preserve such animals and other species? More and more people are answering this question with a firm yes.

1.2 Three Unifying Themes

What will it take to move our civilization in the direction of "a long-term relationship with the natural world," as we have stated previously in our list of tasks? The answer to this question is not simple, but we would like to present three interrelated themes that are applicable to changing or giving direction to the interactions between human and natural systems (Fig. 1–9). These themes are **sustainability**—the practical goal that our interactions with the natural world should be working toward; **stewardship**—the ethical and moral framework that informs our public and private actions; and **sound science**—the basis for our understanding of how the world works and how human systems interact with it. These themes will be applied to public policy and individual responsibility throughout the text, and we will briefly explore them here.

Sustainability

To say that a system or process is sustainable is to say that it can be continued indefinitely without depleting any of the material or energy resources required to keep it running. The term was first applied to the idea of **sustainable yields** in human endeavors such as forestry and fisheries. Trees, fish, and other biological species normally grow and reproduce at rates faster than that required just to keep their populations stable. This built-in capacity allows every species to increase or replace a population following some natural disaster.

Thus, it is possible to harvest a certain percentage of trees or fish every year without depleting the forest or reducing the fish population below a certain base number. As long as the number harvested stays within the capacity of the population to grow and replace itself, the practice can be continued indefinitely. The harvest then represents a sustainable yield. It becomes nonsustainable only when the cutting of trees or catching of fish exceeds the capacity for their present population to reproduce and grow. The concept of sustainable yield can also be applied to freshwater supplies, soils, and the ability of natural systems to absorb pollutants without being damaged.

▼ **FIGURE 1–9** *Three unifying themes.* Sustainability, stewardship, and sound science represent three vital concepts that must be embraced by our society and employed in the development of environmental public policy and private environmental concern.

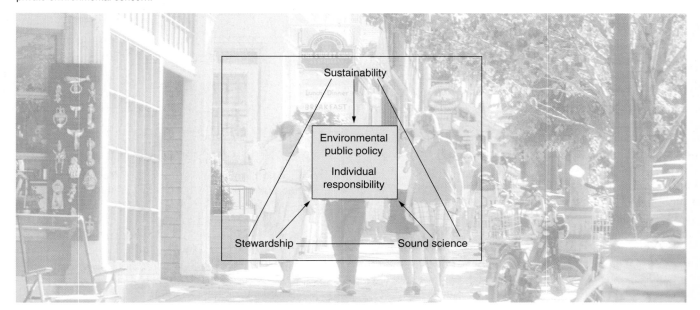

The concept of sustainability can be extended to include ecosystems. Sustainable ecosystems are entire natural systems that persist over time by recycling nutrients and maintaining a diversity of species in balance and by using the Sun as a sustainable energy source. We will investigate these systems in the next several chapters.

To illustrate the concept further, we can picture a sustainable society as a society that is in balance with the natural world, continuing generation after generation, neither depleting its resource base by exceeding sustainable yields nor producing pollutants in excess of nature's capacity to absorb them. Many primitive societies were sustainable in this sense, but population growth was restrained by high mortality rates, and life in these societies was usually harsh and short.

When we apply the concept of sustainability to modern societies, we immediately recognize that many of our interactions with the environment are *not sustainable*, as seen in the aforementioned global trends. Although population growth in the industrialized countries has almost halted, these countries are using energy and other resources at unsustainable rates, producing pollutants that are accumulating in the atmosphere, water, and land. In contrast, developing countries are experiencing a continued rapid population growth yet are often unable to meet the needs of many of their people in spite of heavy exploitation of their natural resources. Based on the expectations of continued economic growth and progress, the crux of the problem is modern society's inexperience with sustainability. No civilization on Earth has ever done it. How do we resolve this dilemma? One answer is the concept of **sustainable development**.

Sustainable Development. Sustainable development is a term that was first brought into common use by the World Commission on Environment and Development, a group appointed by the United Nations. The commission made sustainable development the theme of its final report, *Our Common Future*, published in 1987. They defined the term as a form of development or progress that "meets the needs of the present without compromising the ability of future generations to meet their own needs." The concept arose in the context of a debate between the *environmental* and *developmental* concerns of different groups of countries. **Development** refers to the continued improvement of living standards by economic growth, usually in the so-called developing countries. (The economically developed industrialized countries are often referred to as the *North*, since most are in the Northern hemisphere; those countries that are behind the level of economic development of the North are referred to as the developing countries, or the *South*.) Both groups of countries have embraced the concept of sustainable development, although the industrialized countries are usually more concerned about environmental sustainability, while the developing countries are more concerned about economic development.

The concept is now so well entrenched in international circles that it has become almost an article of faith. It sounds comforting, so people want to believe it is possible, and it appears to incorporate some ideals that are sorely needed, such

as **equity**—where the needs of the present are actually met, and where future generations are seen as equally deserving as those living now. However, sustainable development means different things to different people, and this is well illustrated by the viewpoints of three important disciplines traditionally concerned with the processes involved. The **economists** are mainly concerned with growth, efficiency, and maximum use of resources. **Sociologists** focus on human needs and concepts like equity, empowerment, social cohesion, and cultural identity. **Ecologists** show their greatest concern for preserving the integrity of natural systems, for living within the carrying capacity of the environment, and for dealing effectively with pollution. Yet, it can be argued that sustainable solutions will be found only where the concerns of these three groups intersect, as illustrated in Fig. 1–10.

There are many dimensions to sustainable development—environmental, social, economic, political—and it is clear that no societies today have achieved anything resembling it. Nevertheless, like justice, equality, and freedom, it is important to uphold sustainable development as an ideal, a goal toward which all human societies need to be moving, even if we have not achieved it completely anywhere.

The transition to a sustainable civilization is hard to picture at present. To achieve it will take a radical shift in our societies, involving deeply held ideologies both of nature and of human relationships (see "Ethics" essay, page 11). It will require a special level of dedication and commitment to care for the natural world and to act with justice and equity toward one another as citizens of a global community. As one recent report states, "Sustainability requires that society itself, within and among nations, become a steward of the planet." We now turn to stewardship as a concept that captures much of what is needed in the realm of ethics and ideals to achieve sustainability.

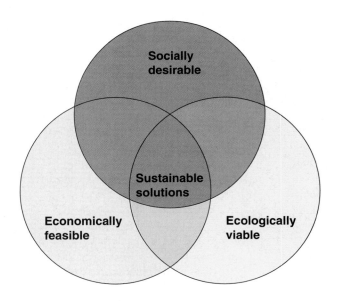

▲ **FIGURE 1–10** *Sustainable solutions.* The concerns of sociologists, economists, and ecologists must intersect in order to achieve sustainable solutions in a society.

ETHICS

ARE WE IN THE PROCESS OF A MAJOR PARADIGM SHIFT?

Very rare but significant events in human history are what scientists call *paradigm shifts*. They are major changes in the way humans view the world and their place and role in it. They can also be called major shifts in worldviews. They tend to be fraught with controversy and conflict at the time they are first presented, but then they usher in a whole new era in the advancement of knowledge and understanding. An example will help illustrate this concept.

Prior to the time of Nicolaus Copernicus (1473–1543), most people believed that Earth was the center of the universe. The Sun, Moon, planets, and stars were thought to all revolve about Earth. In 1512, Copernicus presented his theory that the Sun was the center of a system and that Earth and the other planets revolved around the Sun. Steeped in the old worldview, not only did people ignore the new theory, but anyone who suggested that it had merits was vigorously attacked by the existing power structure, which was dominated by the Catholic Church and which had a vested interest in maintaining the old beliefs. Indeed, about 100 years later, Galileo faced an inquisition

and was prohibited by the Church from doing further scientific work for daring to present hard evidence supporting the Copernican theory. However, other scientists extended Galileo's observations, thus opening up a whole new era of advancing understanding about the universe.

The question now is, Are we in the midst of another, even more significant paradigm shift regarding an old worldview and that of stewardship? The old worldview, dating from the earliest rise of Western culture, is that the world's plants, animals, minerals, etc., exist for the primary purpose of benefiting humans. Therefore, treating these things simply as resources to be exploited not only is acceptable but is the right and proper course of action. It is further implied and assumed in this view that the "bounty" awaiting exploitation is infinite, making conservation or preservation needless. Given this worldview, it is perhaps understandable that the entire economy and lifestyle of Western civilization came to be what it is.

The worldview presented and promoted by stewardship is the antithesis of this old worldview in virtually all respects. According to this worldview, the world is not

infinite. Continued exploitation is not sustainable. The continued well-being of humans will depend on the conservation of wild plants and animals and the protection of air and water. In short, the new view amounts to a paradigm shift from seeing humans as the center of things, free to ride over nature in any manner possible, to seeing humans as the caretakers of nature, intricately linked to it in life processes and global systems.

Of course, we want to consider the evidence presented by proponents of this new worldview. But if it stands up to scrutiny, can we afford to ignore it? There is more riding on this paradigm shift than on the shift in the old view of the solar system. Environmental science is telling us that the future of humanity in the years immediately ahead is at risk. Moreover, stewardship will demand more than just a theoretical awareness. To have any meaning, it must come down to our personal lives, our lifestyles, and how we personally affect the environment. Thus, this new worldview requires an ethical and moral commitment to the stewardship of Earth that will engender concrete actions.

Stewardship

Stewardship is a concept that emerged from slavery. A steward was a slave put in charge of the master's household, responsible for maintaining the welfare of the people and the property of the owner. Since a steward did not own the property himself, the ethic of stewardship is one that expands on the concept of caring for something on behalf of someone else. Modern-day stewardship, therefore, is an ethic that provides a guide to actions taken to benefit the natural world and other people. To whom is the steward responsible? To present and future generations of people who depend on the natural world as their life-support system, many would say. For people with religious convictions, stewardship stems from a recognition that the world and everything in it belongs to a higher being and, thus, they are stewards on behalf of God. For others, stewardship becomes a matter of concern that stems from a deep understanding and love of the natural world and the necessary limitations on human use of that world. The bottom line, however, is that stewards recognize that a trust has been given to them and that they are responsible to care for something that is not theirs—whether it be elements of the natural world or of human culture—which they will pass on to the next generation.

Very often those who are given responsibility for some part of the natural world, or whose living depends on making use of natural resources, end up really caring deeply for nature. Their contact with the natural world often leads them not only to love their work, but also to love the natural world they work in. It is common for foresters, hunters, or fishers—who began as exploiters—to become aware of the vulnerability and beauty of the resources they are using and to instead turn exploitation into responsible, stewardly care for the natural systems in which they are working. Sometimes stewardship leads people into battle to stop the destruction of the environment or to stop the pollution that is degrading human neighborhoods and health. There are many examples of this kind of stewardly action: Lois Gibbs and other homeowners at Love Canal, who drew attention to the chemical wastes buried in their neighborhood; Rachel Carson, who in her book *Silent Spring* alerted the public about pesticides; Sam LaBudde, who took on the tuna industry because of their killing of dolphins; Ken Saro-Wiwa, who was executed by the Nigerian dictatorship for defending the Ogoni people of Nigeria, where oil spills and toxic wastes were devastating their farming and fishing (Fig. 1–11).

▲ **FIGURE 1–11** *Stewardship at work.* Ken Saro-Wiwa was a Nigerian writer who lost his life defending the Ogoni people against government-backed interests that were degrading their environment. He is an example of unusual environmental stewardship at work.

Justice and Equity. The stewardship ethic is concerned not only for the care of the natural world but also for the establishment of just relationships among humans. This concern for justice has been applied to the United States in what is called the environmental justice movement. The major problem addressed by this movement is environmental racism—the placement of waste sites and other hazardous facilities in towns and neighborhoods where the majority of residents is nonwhite. People of color are seizing the initiative to correct these wrongs, creating citizen groups and watchdog agencies to bring effective action and to monitor progress. An example of this can be found in the neighborhoods of Piney Woods and Alton Park, suburbs of Chattanooga, Tennessee, largely populated by African Americans. In these neighborhoods, there are 42 known hazardous waste sites, 12 of which have been listed as Superfund sites. Cancer and asthma rates there were unusually high, so as a result, people organized the group

Stop TOxic Pollution (STOP) to draw attention to the toxic waste and health problems and to eventually acquire a Technical Assistance Grant from the Environmental Protection Agency (EPA) to help fence off the worst sites.

The flip side of this problem is seen when wealthier, more politically active, and often predominantly white communities receive disproportionately greater benefits for facilities such as new roads, public buildings, and water and sewer projects.

Justice is especially crucial for the developing world, where unjust relationships often leave people without land, adequate food, or health. Abject poverty is the condition of at least 1 billion people, and their poverty is often brought on by injustices within the societies where wealthy elites maintain political power and, through corruption and nepotism, steal money and create corporations that receive preferred treatment. The names Marcos (the Philippines), Mobutu (Zaire), and Sukarno (Indonesia) are reminders of the capacity of some leaders to divert billions of dollars from a country's resources for their private gain while ignoring the needs of the desperately poor and powerless majorities in their countries (Fig. 1–12). These leaders plundered their countries and enriched their families until they were finally driven out of office.

Some of the poverty of the developing countries can be attributed to unjust economic practices of the wealthy industrialized countries. The history of European colonialism demonstrates how a dominant culture can enrich itself by seizing power in an undeveloped region and exploiting the peoples and resources. Although this colonialism is now dead, its legacy is still seen in the dependency fostered in so many of the former colonies and the poorly developed technological, educational, and social conditions there. Other sources of the continued poverty of many of the developing countries are the current patterns of international trade and the problem of international debt. By imposing restrictive tariffs and import quotas, industrialized countries have maintained inequities that discriminate against the developing countries. The United Nations reports that "the effective rate of protection against exports from developing countries is considerably higher than the rate against exports from industrial countries." Although

▼ **FIGURE 1–12** *Leaders who stole from their people.* (a) Mobutu Sese Seku, former President of Zaire; (b) Sukarno, former President of Indonesia; and (c) Ferdinand Marcos, former President of the Philippines. These national leaders diverted billions of dollars from the people of their countries to enrich themselves and their families.

(a)  (b)  (c)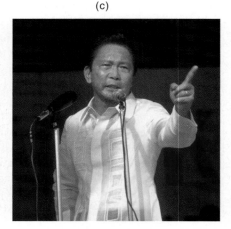

these barriers are falling, they are still in place for many manufactured goods coming from the developing world. Such barriers deprive people in developing countries of jobs and money that would go far to improve their living conditions.

The international debt problem is complex; debtor nations are most commonly those in the process of development, and the borrowed funds were intended to speed up that process. Unfortunately, much of the borrowed money went to the purchase of arms, to ill-planned projects, or to corruption. One example is the Bura Irrigation project in Kenya that cost over $100 million, displaced thousands of people, and was then never used. Today there are some 32 low-income countries that are basically unable to repay all their debts, according to the World Bank. These countries must continue to make scheduled repayments that are often greater than the amount they spend on basic education and health care. One can argue that the leaders of these countries should not have borrowed the funds, but such arguments do nothing to help the poor, who are most affected by this problem. Although the issue of international justice is a difficult one, it is clearly part of the mission of stewardly action to address it.

The degradation of the natural world and concerns about environmental justice have led many to become involved in efforts to resolve these problems; such people are called environmentalists, and the worldview they embrace has become known as environmentalism.

Environmentalism

A Brief History. Although what we now term the environmental movement began less than 50 years ago, it had its roots in the late nineteenth century, when some people realized that the unique, wild aspects of the United States were disappearing. The 1890 U.S. census demonstrated the "closing of the frontier," an event that was noted with some sadness. No longer could any area of the country be classified as totally uninhabited. Around that time, several groups devoted to conservation formed: the National Audubon Society and the National Wildlife Federation; the Sierra Club, founded in California by naturalist John Muir, who helped popularize the idea of wilderness. Also the national parks system was created. The national environmental consciousness was stirring.

Technological achievements following World War I, however, eventually helped create an environmental crisis—the Dust Bowl of the 1930s, when wheat failed to hold soil in place and the topsoil eroded (Fig. 1–13). During the Great Depression (1930–1936), conservation provided a means both of restoring the land and providing work for the unemployed. Many trails, erosion-control projects, and other improvements in national parks and forests were originally installed by the Civilian Conservation Corps (CCC). This program played a major role in pulling the country out of the Depression.

The two decades following World War II (1945–1965) were full of optimism. We had won the "greatest" of all wars, in part because of technology. The tremendous productive capacity built up during the war, and new developments—ranging from rocket science to computers and from pesticides to antibiotics—could be redirected to peacetime applications. Except for the tensions of the Cold War during those years and some apprehension about atomic energy, it seemed as if nothing but opportunity and prosperity lay ahead.

Although economic expansion enabled most families to have a home, a car, and other possessions, certain problems became obvious. The air in and around cities was becoming murky and irritating to people's eyes and respiratory systems. Streetlights from St. Louis to Pittsburgh were left on during the day because of the smoke from coal-burning industries. In winter, even freshly fallen snow soon turned gray as the soot fell from city chimneys. Rivers and beaches were increasingly fouled with raw sewage, garbage, and chemical wastes. The effects of air, land, and water pollution affected all living systems: Conspicuous declines occurred in many bird populations—including our national symbol, the bald eagle—aquatic species, and other animals.

It was easy to identify some culprits—belching industrial smokestacks, open burning dumps, and municipal and industrial sewers, discharging raw sewage and chemical wastes into waterways. The decline of the bald eagle and other bird populations was traced to the accumulation in

◀ **FIGURE 1–13** *The Dust Bowl.* In the 1930s, modern farming methods combined with an extended drought led to the loss of enormous amounts of topsoil due to wind erosion.

their bodies of DDT, the long-lasting pesticide that had been used in large amounts since the 1940s. In short, it was clear that we were seriously contaminating our environment.

The Modern Environmental Movement. In 1962 biologist Rachel Carson wrote *Silent Spring*, presenting her scenario of a future with no songbirds and with other dire consequences if pollution of the environment with DDT and other pesticides continued. Carson's voice was soon joined by others, many of whom formed organizations to focus and amplify the voices of thousands more in demanding a cleaner environment. This was the beginning of the modern **environmental movement**, in which a newly militant citizenry demanded the curtailment of pollution, the cleanup of polluted environments, and the protection of pristine areas. It is significant that the environmental movement began as a grassroots initiative, and it maintains its momentum and force today only by continuing to command public interest and support.

Members of the environmental movement of the 1960s joined forces with older organizations, such as the National Audubon Society and the National Parks Conservation Association, that had a considerable history of dedication to preserving wildlife and natural habitats. There is an obvious overlap between the goal of reducing pollution and that of protecting wildlife. Wildlife cannot survive in polluted surroundings, and preserving an uncontaminated environment presupposes that the plant and animal species within that environment will be protected. Therefore, established wildlife-preservation organizations and their members became significant players in the blossoming environmental movement, along with newly formed organizations such as the Environmental Defense Fund, the Natural Resources Defense Council, Greenpeace, and Zero Population Growth. This dual focus continues today. As the broad category of "environmentalists" includes everyone concerned with reducing pollution and/or protecting wildlife, it is no wonder that at least 70% of the population identify themselves as environmentalists.

By almost any measure, the environmental movement has been successful. Pressured by a concerned citizenry, Congress created the Environmental Protection Agency in 1970 and passed numerous laws for pollution control and wildlife protection—among them the Endangered Species Act of 1973, the Marine Mammals Protection Act of 1972, the Clean Air Act of 1970, the Clean Water Act of 1972, and the Safe Drinking Water Act of 1974. You will read more about these laws and their effects in subsequent chapters. For now, let us simply say that a number of species probably have been saved from extinction—at least for the present. In the area of pollution abatement alone, industry has spent hundreds of billions of dollars, both on installing pollution control devices and on redesigning procedures and products so that less pollution is created. The pollution control devices on our cars are an example. Governments have spent additional billions on upgrading sewage treatment, on refuse disposal, and on other measures to reduce pollution. As a result of these expenditures, the air in most cities and the water in numerous lakes and rivers are cleaner now than they were in the late 1960s.

Without a doubt, they are immensely cleaner than they would be had the environmental movement not come into existence.

Expenditures for pollution control are continuing. In its various aspects—ranging from law enforcement to the design, manufacture, and operation of equipment—pollution abatement has grown to be a major sector of our economy, providing jobs that did not even exist in 1960. Anyone contemplating becoming a professional environmentalist may do well to consider careers in areas such as environmental law, environmental economics, environmental engineering, or geographic information systems (GIS).

Environmentalism Acquires Its Critics. In the early stages of the environmental movement, the sources of the problems were specific and visible. That made it easy to identify certain polluters as the "bad guys" and take the side of the environmental movement. Likewise, the solutions seemed relatively straightforward—install waste treatment and pollution control equipment and ban the use of DDT, substituting safer pesticides. Importantly, it was possible in many cases to achieve significant improvements in air and water quality with only modest expenditures by addressing these **point sources** (specific producers of pollution).

Given this beginning at cleaning up major point sources of pollution, further improvements in air quality demand addressing **diffuse sources**—the innumerable small sources such as automobiles and transportation, home fireplaces, barbecue grills, gas lawn mowers, and the runoff from agricultural fields and every homeowner's lawn and garden. Although no one individual is responsible for much of this pollution, the sum total from everyone is significant. In other words, the "bad guys" are no longer "them," but "us," and many of us do not take well to this notion: We perceive restrictions on such things as lawn mowers as unacceptable infringements on our personal liberty. Many people are concerned about regulations that may threaten their well-being and even their economic survival. This makes them easily swayed by critics of the environmental movement who want to be free of regulations protecting the environment. A significant contingent of such critics is made up of the *Wise-Use Movement*, which we will discuss later in Chapter 12.

The bitterness of the conflict between environmentalists and those with special business interests is illustrated by the case of the spotted owl that inhabits old-growth forests of the Northwest (Fig. 1–14). Saving the spotted owl from extinction requires putting large tracts of old-growth forests off limits to logging. Indeed, the central interest of environmentalists is to save some of these old-growth forests in their natural state. However, curtailing logging of the remaining old-growth forests is portrayed as threatening the jobs of many loggers, some logging companies, and even whole communities whose economies are based solely on logging. Many loggers are hostile toward environmentalists who would effectively trade their jobs for "some bird no one ever heard of before the controversy started" (Fig. 1–15). Similarly, many landowners become hostile toward environmentalists as they find themselves prevented from

◀ **FIGURE 1–14** *Spotted owl and old-growth forests of the Pacific Northwest.* Listing of the spotted owl as a threatened species led to battles between environmentalists who wanted to preserve the old-growth forests and the logging interests who wanted to cut them.

developing their property because it is home to an endangered species or is subject to some other environmental restriction. They may also become hostile toward the government in general, since the government is responsible for enforcing these regulations.

Environmentalism to Stewardship. Has the environmental movement accomplished everything it can? Is environmentalism merely a passing fad? Has it served its purpose? Along with most other professionals in the environmental arena, we feel that our future will depend on redefining the role of environmentalism. Most important is that the ethic of stewardship be broadly communicated and practiced by whole societies—individuals, communities, corporations, and nations. How can this be accomplished? The President's Council on Sustainable Development addressed this concern in its initial report, *Sustainable America:*

> The answer is multifaceted, but it starts with understanding the dynamics at work in the environment and the connection among environmental protection, economic prosperity, and social equity and well-being. It depends on the processes by which individuals, institutions, and government at all levels can work together toward protecting and restoring the country's inherited natural resource base. Education, information, and communication are all important for developing a stewardship ethic. Also important is the widespread understanding that people, bonded by a shared purpose, can work together to make sustainable development a reality.

In their report, the Council identified a number of policy recommendations to accomplish the stewardship of natural resources, each of which is further developed in the report by proposing specific actions:

- Enhance, restore, and sustain the health, productivity, and biodiversity of terrestrial and aquatic ecosystems through cooperative efforts to use the best ecological, social, and economic information to manage natural resources.
- Create and promote incentives to stimulate and support the appropriate involvement of corporations, property

▲ **FIGURE 1–15** *Anti-environmentalism.* This scene is part of a crowd of 15 to 20 thousand demonstrators at a pro-logging rally in Victoria, B.C., Canada.

owners, resource users, and government at all levels in the individual and collective pursuit of stewardship of natural resources.

- Manage and protect agricultural resources to maintain and enhance long-term productivity, profitability, human health, and environmental quality.
- Establish a structured process involving a representative group of stakeholders to facilitate public and private efforts to define and achieve the national goal of sustainable management of forests by the year 2000.
- Restore habitat and eliminate overfishing to rebuild and sustain depleted wild stocks of fish in U.S. waters.
- Create voluntary partnerships among private landowners at the local and regional levels to foster environmentally responsible management and protection of biological diversity, with government agencies providing incentives, support, and information.

It should be evident that these recommendations are both present- and future-directed; where they are implemented, they represent actions taken that will lead in the direction of sustainability.

As we have suggested, sustainability encompasses the goals that we should work toward, while stewardship defines an ethic that should guide our practices. Still missing from our picture, however, is the crucial information that informs us how the natural world works and what is happening to it as a result of human activities. This is the domain of sound science and the scientist.

Sound Science

Many environmental issues are embroiled in controversies that are so polarized that no middle ground seems possible. On the one hand are persons who argue from presumably sound facts and proven theories. On the other are persons who dismiss the theories and mistrust scientists and their motives, particularly when findings and conclusions of scientists conflict with corporate interests and traditional ways of doing things. Many people are understandably left confused. It is our objective to give a brief overview of the nature of science and the scientific method so that you can evaluate for yourself the two sides of such controversies.

Science and the Scientific Method. In its essence, science is simply a way of gaining knowledge; the way is called the **scientific method**. The term science further refers to all the knowledge gained through that method. We employ the term *sound science* to distinguish legitimate science from what has been called "junk science," information that is presented as valid science but that does not conform to the rigors of the methods and practice of legitimate science. What is the scientific method?

First, the scientific method rests on certain basic assumptions that most of us accept without argument. The first assumption is that what we perceive with our basic five senses represents the existence of an objective reality; that is, our

perceptions are not some kind of mirage or dream. The second assumption is that this "objective reality" functions according to certain basic principles and natural laws that remain consistent through time and space.

The third assumption, which follows directly from the second, is that every result has a cause, and every event, in turn, will cause other events. In other words, we assume that events do not occur without reason and that there is an explainable cause behind every happening. The fourth and final assumption is that through our powers of observation, manipulation, and reason, we can discover and understand the basic principles and natural laws by which the universe functions.

Although assumptions, by definition, are premises that cannot be proved, the fact is that the assumptions underlying science have served us well and are borne out with everyday experience. For example, we suffer severe consequences if we do not accept our perception of fire as real. All our experience confirms that gravity is a predictable force acting throughout the universe and that it is not subject to unpredictable change. (Weightlessness in orbit is not a change in gravity; it is effectively a state of free fall.) The same can be said for any number of other phenomena that we observe. If our car fails to start, we know that there is a logical reason, and we call a mechanic to fix it.

Thus, whether we are conscious of the fact or not, we all accept the basic assumptions of science in the conduct and understanding of our everyday lives. Scientists and scientific investigations only extend the boundaries of everyday experience, deepen our understanding of cause-effect relationships, and provide greater appreciation for the principles and natural laws that seem to determine the behavior of all things, from the outcome of a chemical reaction to the functioning of the biosphere.

Observation. In previous schooling you may have learned that the scientific method consists of the following sequence: question, hypothesis, test (experiment), theory (Fig. 1–16). This sequence is an oversimplification in that it both fails to describe what is involved in these steps and it omits what is really the most fundamental aspect of science.

The foundation of all science and scientific discovery is **observation** (seeing, hearing, smelling, tasting, feeling). Indeed, many branches of science, such as natural history (where and how various plants and animals live, mate, reproduce, etc.), astronomy, anthropology, and evolutionary biology, are based almost entirely on observation because experimentation is either inappropriate or impossible. For example, experimentation is obviously counterproductive if you want to discover what plants and animals do under completely natural conditions, and it simply is impossible to conduct experiments on stars or past events.

Likewise, many of the data and conclusions of other sciences, such as zoology, botany, geology, comparative anatomy, and taxonomy (classification of organisms), are based on nothing more (or less) than the careful observing and chronicling of things and events by persons taking the pains and time to do so. Even experimentation, as we will discuss shortly, is conducted to gain another window of observation.

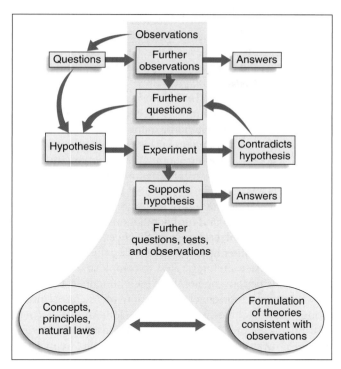

▲ **FIGURE 1–16** *Steps of the scientific method.* The process leading to theory formation, natural laws, and concepts is a continual interplay between observations, experimentation, hypothesis formulation, and further refinement.

Therefore, in all science, careful observation is the keystone. How can we be sure that observations are accurate?

Of course, not every reported observation is accurate, for reasons ranging from honest misperceptions to calculated mischief. Therefore, an important aspect of science and a trait of scientists is to be skeptical of any new report until it is confirmed or verified. Such confirmation usually involves other investigators' repeating and checking out the observations of the first investigator and validating (or invalidating) the accuracy. As observations are confirmed by additional investigators, they gain the status of factual data. In other words, **facts** are things or events that have been confirmed by more than one observer and remain open to be reconfirmed by additional people. Things or events that do not allow such confirmation—UFOs, for example—remain in the realm of speculation from a scientific standpoint.

Various observations by themselves, like the pieces of a puzzle, may be put together into a larger picture—a model of how a system works. To give a simple example, we observe that water evaporates and leads to moist air, that water from moist air condenses on a cool surface, and we observe clouds and precipitation. Putting these observations together logically, we derive the concept of the hydrologic cycle. Water evaporates and then condenses as air is cooled, condensation forms clouds, and precipitation follows. Water thus makes a cycle from the surface of Earth into the atmosphere and back to Earth (Fig. 9–3). Note how this simple example incorporates the four assumptions described above: there is an objective reality; it

operates according to principles; every result has a cause; we can discover the principles. Also note how the example broadens our everyday experiences of the evaporation of water and falling of rain into an understanding of a cycle involving both.

Thus, the essence of science and the scientific method may be seen as a process of making observations and logically integrating those observations into a model of how the world works. To be sure, many areas of science get more complex and difficult to comprehend. Still, the basic process—constructing a logically coherent picture of causes and effects from basic observations—is the same.

Where, then, does experimentation, that additional hallmark of science, fit in?

Experimentation. **Experimentation** is basically setting up situations to make more systematic observations regarding causes and effects. For example, the number of chemical reactions one can readily observe in nature is limited. However, in the laboratory, it is possible to purify elements or compounds, mix them together in desired combinations, and carefully observe and measure how they react (or fail to react). From the way chemicals react, chemists have constructed a coherent cause-effect picture, the **atomic theory**, and they have determined the attributes of each element. Similarly, biologists put plants or animals into determined situations where they can carefully observe and measure their responses to particular conditions or treatments. (Again, note that experimentation is necessarily limited to things that lend themselves to artificial manipulation. In many cases, such as stars, geological events, atmospheric events, and events that have occurred in the past, one has only observations to work with in the construction of the broader picture.)

Some experimentation may be more or less spontaneous and random—the childhood inclination to "do this and see what happens." Sometimes valuable information may be obtained in this way if careful accounts are kept so that one has an accurate record of causes and effects. However, to solve a particular problem—What is the cause of this event?—a more systematic line of experimentation is used. This is where the sequence question, hypothesis, test (experiment), and theory comes in. Let's consider the observation of the die-off of submerged aquatic vegetation in Chesapeake Bay (see Chapter 18).

The question is, What has caused the die-off? The next step is to make educated guesses as to the cause. Each educated guess as to the cause is a **hypothesis**. Each hypothesis is then tested by making further field observations or conducting experiments to determine if the hypothesis really accounts for the observed effect. In this particular case, the first hypothesis or presumed cause was industrial wastes. However, this hypothesis was disproved by the observations that the die-off occurred in locations where no industrial wastes were present. A second hypothesis, herbicides being used on farmlands, was rejected as the primary cause by laboratory testing and field measurement showing that herbicide levels were generally too low to cause damage.

However, the third hypothesis, reduction of light due to increased turbidity, stood up to tests of measuring photosynthesis at various light intensities. Additional observations concerning the causes for the increased turbidity filled in the picture and brought what started out as a hypothesis to the status of the proven cause, which we might express as a *theory* or *concept*.

Theories. We have already noted in the case of the hydrologic cycle how various specific observations (evaporation, precipitation, etc.) may fit together to give a logically coherent cause-effect picture of a broader phenomenon, i.e., the hydrologic cycle. At first, the broader picture is termed a hypothesis because it is really a tentative explanation as to how various observations are related. A hypothesis generally is a tentative explanation for a specific event, whereas a **theory** is a more well-established understanding of how things happen and behave in the real world.

In any case, theories may be tested and confirmed (or denied) and perfected so they are logically consistent with all observations (Fig. 1–16). Further, through cause-effect reasoning, theories will generally suggest additional aspects: "If this is so, it should follow that …" If the outcome predicted by the theory is indeed found, it provides strong evidence for the correctness of the theory. Predictions require experiments and testing and more data gathering and observation. When theories reach a state of providing a logically consistent framework for all relevant observations (facts) and can be used to reliably predict outcomes, they represent a valid interpretation. For example, we have never seen atoms as such, but innumerable observations and experiments are coherently explainable by the theory that all gases, liquids, and solids consist of various combinations of only slightly more than a hundred kinds of atoms. Hence, we fully accept the atomic theory of matter.

Sometimes people will argue that because a theory is not proven fact, one theory is as good as another. That notion is false! One theory may have overwhelming supporting evidence, whereas much evidence may contradict another theory. In evaluating a theory, ask: What is the supporting evidence? Is there more evidence, or less, supporting an alternative theory?

Natural Laws and Concepts. The second assumption underlying science—that the universe functions according to certain basic principles and natural laws that remain consistent through time and space—cannot be proved, but every observation and test bear it out. All our observations, whether direct or through experimentation, demonstrate that matter and energy do not behave randomly or even inconsistently but in precise and predictable ways. We refer to these principles by which we can define and precisely predict the behavior of matter and energy as **natural laws**. Examples are the *law of gravity*, the *law of conservation of matter*, and the various *laws of thermodynamics*. Our technological success in space exploration and many other fields is in no small part due to our recognition of these principles and precise calculations based on them. Conversely, trying to make something work in a manner contrary to a natural law invariably results in failure.

In many situations, the outcome of scientific work results in well-established theories that do not have the inevitability of natural laws; they are probabilistic in nature.

This is especially true for biological phenomena such as predator-prey relationships or the effects of pesticides. Here it is appropriate to speak of *concepts* rather than laws. **Concepts** are perfectly valid explanations of data gathered from the natural world and can also be predictive. They model for us the way we believe the natural world works and enable us to make qualified predictions of future outcomes. Thus, we might say that based on our understanding of the effects of DDT (a potent pesticide) on mosquitoes and the development of resistance to DDT in mosquito populations, spraying the local salt marshes with DDT to eradicate unwelcome mosquitoes is *likely* to result in the development of a more resistant mosquito population in the future (not to mention the other effects of DDT on fish and bird populations).

The Role of Instruments in Science. Complex instrumentation is another hallmark of science, which often gives it an aura of mystery. Yet, regardless of complexity, all scientific **instruments** perform one of three basic functions. First, they may extend our powers of observation—telescopes, microscopes, x-ray machines, and CAT scans, for example. Second, instruments are used to quantify observations; that is, they enable us to measure exact quantities. For example, we may feel cold, but a thermometer enables us to measure and quantify exactly how cold it is. Comparisons, communication, and verification of different observations and events would be impossible if it were not for such measurement and quantification. Third, instruments such as growth chambers and robots help us achieve conditions and perform manipulations required to make certain observations or to perform experiments.

All instruments used in science are themselves subjected to testing and verification to be sure that they are giving an accurate representation of what is there as opposed to creating illusions.

Scientific Controversies. With the scientific method capable of coming to objective conclusions, why does so much controversy still prevail? There are four main reasons. First, we are continually confronted by new observations—the hole in the ozone layer, for instance, or the dieback of certain forests. There is considerable time before all the hypotheses regarding the cause can be adequately tested. During this time, there may be honest disagreement as to which hypothesis is most likely. Such controversies are gradually settled by further observations and testing, but this process leads into the second reason for continuing controversy.

Phenomena such as the hole in the ozone layer or the loss of forests do not lend themselves to simple tests or experiments. Therefore, it is difficult and time-consuming to prove the causative role of one factor or to rule out the involvement of another. Gradually, different lines of evidence will support one hypothesis and exclude another and enable the issue to be resolved.

When is there enough evidence to say unequivocally that one hypothesis is right and another wrong? Deciding that there is enough evidence to be convincing involves subjective judgment. The biases or vested interests of a person may affect the amount of information that person requires to be convinced. The Tobacco Institute, a lobbying association for the tobacco industry, provides a prime example. It continually makes the point that the connection between smoking and illness has not been proved and that more studies are necessary. By harping on the lack of absolute proof and simply ignoring the overwhelming amount of evidence supporting the connection between smoking and illness, the tobacco lobby has succeeded in keeping the issue controversial and thereby has delayed restrictions on smoking. Thus, the third reason for controversy is that there are many vested interests that wish to maintain and promote disagreement, because they stand to profit by doing so. The need to keep a watch for this kind of behavior in evaluating the two sides of a controversy is self-evident. (Internal tobacco-industry memoranda now show that they have perjured themselves in the public forum.)

The fourth reason for controversy, which may be seen as a generalization of the third reason, is that subjective value judgments, as well as subjective judgments of facts, may be involved. This is particularly true in environmental science because it deals with the human response to environmental issues. For example, there is virtually no controversy regarding nuclear power as long as it is considered at the purely scientific level of physics. However, when it comes to the environmental level of deciding whether to promote the further use of nuclear energy to generate electrical power, controversy arises, and it stems from the fact that different people have different subjective feelings regarding the relative risks and benefits involved.

The Scientific Community. We have seen that the scientific method is a way of gaining understanding that starts with basic observations which are developed into a logically coherent picture of broader phenomena through logical cause-effect reasoning. The value in this approach is seen in the scientific and technological progress it has made possible. However, what is often left out of the usual presentation of the scientific method is the fact that science and its outcomes take place in the context of a **scientific community** and a larger society. There is no single authoritative source that makes judgments on the soundness of scientific theories. Instead, it is the collective body of scientists working in a given field who, because of their competence and experience, establish what is sound science and what is not. They do so by communicating their findings to each other and to the public as they publish their work in peer-reviewed journals. The process of *peer review* is crucial; here, experts in a given field review the analysis and results of their colleagues' work. Careful scrutiny is given to grant proposals and published research with the objective of rooting out poor or sloppy "science" and affirming work that is clearly meritorious.

As we have just seen, controversy often arises in the context of environmental science, proof that science is a thoroughly human enterprise. Some controversy is the inevitable outcome of the scientific process itself, but much of it is attributable to less noble causes. Such is the case for "junk science."

Junk Science. **Junk science** is information that is presented as valid science but that does not conform to the rigors of the scientific community process. According to the Union of Concerned Scientists (UCS), junk science can take many forms: presentation of selective results (picking and choosing the results that agree with your preconceived ideas), politically motivated distortions of scientifically sound papers, or publication in quasi-scientific, unreviewed journals and books. Quite often, junk science is generated by special-interest groups who are trying to influence the public debate about science-related concerns. Information presented by the Tobacco Institute, for example, is clearly suspect. Unfortunately, the media and the public may not be aware of the true nature of the science—sound or junk—and will often give equal credibility to, say, opposing views on an issue. The result is that the junk science is given equal respectability with the sound science. The UCS is so concerned about this problem that they have launched the **Sound Science Initiative**, which is intended to help people distinguish between sound science and junk science and to help legislators apply sound science to the formulation of public policy.

Evaluating Science. The human mind (and its inclination to gain advantage as opposed to the truth) is extremely adept at starting with a desired conclusion and working in the opposite direction. The danger is that any action or policy based on such unfounded conclusions will almost certainly end in failure, if not disaster. Too often, technological optimism, fueled by profit motive, runs ahead of evidence and experience. Therefore, it is important that we distinguish between such dubious work and conclusions that are soundly based. Whether we are scientists or not, we can use facets of the scientific method to judge the relative validity of alternative viewpoints and also to develop our own capacity for logical reasoning. The basic questions to ask are:

- What are the basic observations (facts) underlying the conclusion (theory)?
- Can the observations be satisfactorily verified?
- Do the conclusions follow logically from the observations?
- Does the conclusion account for all observations? (If the conclusion is logically inconsistent with any observations, it must be judged as questionable at best.)
- Is the conclusion or predicted outcome supported by the community of scientists with the greatest competence to judge the work? If not, it is highly suspect.

In summary, sound science is absolutely essential to the project of guiding us into the twenty-first century with any success for preserving our life-support system. Our planet is human-dominated—our activities have reached such intensity and scale that we are now one of the major forces affecting nature. The natural ecosystems that support most of the world economy and process our wastes are heavily impacted by what we do, and we need to know how to manage the planet so as to maintain a sustainable relationship with it. It is obvious that the information gathered by scientists needs to be accurate, credible, and communicated clearly to decision makers.

EARTH WATCH

AGENDA 21

Agenda 21 is the official document signed by world leaders representing 98% of the global population at the United Nations Earth Summit in Rio de Janeiro, Brazil, in June of 1992. The following is an excerpt from an abridged version edited by Daniel Sitarz:[1]

Agenda 21 is, first and foremost, a document of hope…. [I]t is the principal global plan to confront and overcome the economic and ecological problems of the late twentieth century. It provides a comprehensive blueprint for humanity to use to forge its way into the next century by proceeding more gently upon the Earth….

Humanity is at a crossroads of enormous consequence. Never before has civilization faced an array of problems as critical as the ones now faced. As forbidding and portentous as it may sound, what is at stake is nothing less than the global survival of humankind.

… Where once nature seemed forever the dominant force on Earth, evidence is rapidly accumulating that human influence over nature has reached a point where natural forces may soon be overwhelmed. Only very recently have the citizens of Earth begun to appreciate the depth of the

potential danger of the human impact on our planet…. Scientists around the world, in every country on Earth, are documenting the hazard of ignoring our dependence upon the natural world…. For the first time in history, humanity must face the risk of unintentionally destroying the foundation of life on Earth…. To prevent such a collapse is an awesome challenge for the global community….

Agenda 21 is not a static document. It is a plan of action. It is meant to be a hands-on instrument to guide the development of the Earth in a sustainable manner…. It is based on the premise that sustainable development of the Earth is not simply an option: it is a requirement—a requirement increasingly imposed by the limits of nature to absorb the punishment which humanity has inflicted upon it. *Agenda 21* is also based on the premise that sustainable development of the Earth is entirely feasible.

The bold goal of *Agenda 21* is to halt and reverse the environmental damage to our planet and to promote environmentally sound and sustainable development in all countries on Earth. It is a blueprint for action in all areas relating to the sustainable development of our planet into the

twenty-first century…. It includes concrete measures and incentives to reduce the environmental impact of the industrialized nations, revitalize development in developing nations, eliminate poverty world-wide and stabilize the level of human population.

Agenda 21 provides a myriad of opportunities. Suggestions are furnished to develop new industries, pioneer innovative technologies, evolve fresh techniques and institute novel trade arrangements.

Various meetings involving leaders of both governments and nongovernmental organizations are continuing to take place around the world to develop, refine, and implement strategies to confront the world's environmental problems. In particular, the newly created UN Commission on Sustainable Development is charged with the responsibility of overseeing *Agenda 21*. "Gradually it is being understood that the issues of poverty, population growth, industrial development, depletion of natural resources and destruction of the environment are all very closely interrelated."

---

[1]Daniel Sitarz, Agenda 21 (Boulder, CO: Earth Press, 1993), pp. 1–5.

A New Commitment

People in all walks of life—scientists, sociologists, workers and executives, economists, government leaders, and clergy, as well as traditional environmentalists—are recognizing that "business as usual" is not sustainable. Many global trends are on a collision course not only with basic human needs but also with fundamental systems that maintain our planet as a tolerable place to live. A finite planet cannot go on adding more than 80 million additional persons annually, nor can we tolerate the current losses of soil, atmospheric changes, losses of species, and depletion of water resources without arriving at a point where resources are no longer adequate to support the human population and where, consequently, civil order breaks down. As one observer put it, "If we don't change direction, we will end up where we are heading."

The news is not all bad, however. Food production has improved the nutrition of millions in the developing world, and the percentage of undernourished has declined from 35% to 20% over the past 30 years. Population growth rates

continue to decline in many of the developing countries. A rising tide of environmental awareness in the industrialized countries has led to the establishment of policies, laws, and treaties that protect environmental resources. It is clear that environmental degradation can be slowed down and reversed, that people can be freed from hunger and poverty, and that peoples' behavior toward the environment can be transformed from an exploitive mode to a conserving one.

Many caring people are beginning to play an important role in changing society's treatment of Earth. For example, many are now adding their voices to the literally hundreds of traditional environmental and professional organizations devoted to controlling pollution and protecting wildlife. People in business have formed the Business Council for Sustainable Development, economists have formed the International Society for Ecological Economics, religious leaders have formed the National Religious Partnership for the Environment, and philosophers are speaking out for a new ethic of "caring for creation."

Sustainable development was the primary focus of a 1992 world summit meeting of leaders and representatives from 180 nations—the United Nations Conference on Environment and Development (UNCED)—held in Rio de Janeiro, Brazil. The outcome of this conference, a "blueprint" intended to guide development in sustainable directions into and through the next century, was published in book form as *Agenda 21* (see "Earth Watch" essay, page 20). Five years later, in June 1997, the U.N. General Assembly held a special session to review progress during the five years since UNCED. Many observers concluded that there were more instances of failure than progress, citing a continued deterioration in the global environment, a continued decline in aid to the developing countries, and a failure to set targets for reducing CO_2 emissions. The conference did, however, recognize some achievements and served to revitalize commitments made at UNCED to sustainable development.

Thus, we are seeing a continuing concern on the part of many individuals and groups that appreciate the problems jeopardizing sustainability. Together they are working to bring about corrective measures. In this light, the textbook you are holding is our own "best effort" to contribute

to the cause from our perspective as scientists and educators in biology and ecology. Given the vastness of what is now the environmental arena, we are the first to say that this text is far from exhaustive, nor should it be taken as the last word. Of course, viewpoints are subject to change as human experience and understanding increase. Our basic premise, however, is that sound environmental public policy in the long run must be based on sustainability, stewardship, and sound science.

In the first part of the text (Chapters 2 through 5) we describe the basics of how natural systems function and perpetuate themselves, and their limits in terms of adaptability to changing conditions. In addition to giving us some appreciation for natural systems, this study will reveal certain basic principles underlying sustainability. We contend that to achieve sustainability, we must incorporate these principles into the functioning of our own society. Subsequent chapters will address issues of population and development, issues relating to resources, energy and land, and pollution in its various forms. In each case, we will attempt to give you a deeper understanding of just what the problems are, how far we have come, and a view of the path ahead toward sustainability.

ENVIRONMENT ON THE WEB

THE MYTH OF OBJECTIVE SCIENCE

The image of the white-coated scientist is a familiar and reassuring one. Scientific evidence now provides the foundation for many important public debates, including those in environmental management. But how accurate is the image of the scientist as an objective seeker of truth?

Scientific analysis is not just a straightforward process of observation and reporting but, rather, a complex series of personal decisions, value judgments, and guesses, influenced by the scientist's unique combination of personal experiences, fears, hopes, desires, and values. Like the rest of us, scientists worry about their careers, their families, and their finances. Ultimately, these human qualities influence the ways that individuals see and interpret scientific information. For this reason, it is not uncommon for two scientists to examine the same data set but reach very different conclusions.

Beth Savan, in her book *Science Under Siege*, cites the example of Stephen Jay Gould, a professor of geology at Harvard University. Gould reanalyzed data compiled by Samuel Morton, a nineteenth-century physician, on the physical and intellectual differences among human races. Gould's analysis showed that Morton had consciously or unconsciously manipulated his data to arrive at the conclusion—widely held when Morton was alive—that white people are a superior race. Yet in his own analysis, Gould misread one of Morton's figures, leading him to underestimate racial differences in the data and thus to arrive at a conclusion more in keeping with his own preconceptions—that the differences among races are small.

This example demonstrates another feature of scientific analysis—that we tend to favor familiar, widely accepted views, while demanding a higher standard of proof for new ideas. Sometimes these biases can create obstacles to sound decision making. For example, a group of Western scientists planned to conserve Peary caribou in

the High Arctic by protecting females and juveniles but allowing some hunting of adult males. Inuit hunters, knowledgeable about the social structure of caribou herds, warned that this practice would instead speed the decline of the population. Subsequent monitoring has confirmed the validity of their position.

Human emotions and values underlie most of the environmental disputes of this century. Divergent scientific analyses are often seen in the development of environmental standards. Environmental managers can reveal these subjective influences and make them explicit in decision making by including a wide range of viewpoints in their analysis and by recognizing and, where possible, compensating for their own unique values and biases.

Web Explorations

The Environment on the Web activity for this essay describes some of the debate—much of it driven by the human cultural and economic context of the decision—surrounding U.S. EPA's recent reevaluation of the standard for the cancer-causing agent dioxin. Go to the Environment on the Web activity (select Chapter 1 at http://www.prenhall.com/nebel) and learn for yourself:

1. how human biases may influence applications of the scientific method;
2. how science can be used to persuade and influence political and social decisions; and
3. how individual interpretations of scientific data can vary greatly.

A suggested time frame for each of these exercises is 10–30 minutes.

REVIEW QUESTIONS

1. Describe the range of ways that people in different parts of the world live and interact with their environments.
2. What factors brought about the collapse of the Easter Island civilization? Draw parallels and counterparallels between the current global situation and the prelude to the collapse of the Easter Island civilization.
3. Cite four global trends that indicate we are "still losing the environmental war."
4. List some indications that things are not all bad and that progress toward sustainability is possible.
5. Define sustainability and a sustainable society.
6. Define and give features of sustainable development.
7. Describe the origins of the stewardship ethic and its modern usage.
8. What are the concerns of the environmental justice movement?

9. How does justice become an issue between the industrialized countries and the developing countries?
10. How and when did environmentalism arise?
11. What are the successes of the environmental movement?
12. What are the grounds and concerns of present critics of environmentalism?
13. Explain the role of observation, experimentation, and theory formation in the operation of scientific work.
14. Give several reasons for the existence of scientific controversies.
15. Describe the major distinction between sound science and junk science.
16. What are the new directions, new focus, and new adherents of modern environmental concern?

THINKING ENVIRONMENTALLY

1. Have a class debate between environmental critics and people representing environmentalists. Characterize the two sides in terms of long-term vs. short-term viewpoint, personal interests vs. interests of society in general, and local vs. global perspective.
2. Some people say that the concept of sustainable development either is an oxymoron or represents going back to some kind of primitive living. Present an argument demonstrating that neither is the case but that sustainable development is the only course that will allow the continued advancement of civilization.

3. List all the prerequisites for a sustainable society that can continue to gain further understanding of the universe and our place in it. Why is it necessary that people from all walks of life be involved? Give the roles (in general) that each needs to play in achieving a sustainable society.
4. How can you contribute to a sustainable society?
5. Set up a debate between proponents and opponents of smoking. Use the concepts of sound science and junk science to conduct the debate, and show how the two kinds of science will lead to different conclusions.

WEB REFERENCES

On-line resources for this chapter can be found on the World Wide Web at: **http://www.prenhall.com/nebel**.
Click on Chapter 1 on the chapter selector.

PART ONE

Ecosystems
and How They Work

Tropical rain forests—humid, warm, dense, full of unusual plants and animals. These are the impressions that our senses bring to us when we take the first few steps into a rain forest. However, our senses will not tell us that the tropical rain forests are home to a greater diversity of living things than anywhere else on Earth, that they are the result of millions of years of adaptive evolution, or that they are the site of storage of more carbon than the entire atmosphere. Nor can we sense how energy flows through these forests or how nitrogen and phosphorus are cycled and recycled. This information comes from the work of many scientists who have been asking the basic questions of how such natural systems work, how they came to be, and how they relate to the rest of the world.

How important is the information that comes from the work of such scientists? We hope in this first part to convince you that information about how the natural world works is absolutely crucial to the human enterprise of living on our planet. Ecosystems—the name we give to natural units like the tropical rain forest—are not just the backdrop for human activities, they are the basic context of life on Earth, including human life. They are self-sustaining systems, and if we accept the notion that the human enterprise should also be self-sustaining (and currently is not), then perhaps we can learn how to construct a sustainable society from our study of ecosystems. Environmental science begins by understanding how ecosystems work.

Tropical rain forest.

2

Ecosystems: Units of Sustainability

Key Issues and Questions

1. Natural communities are organized into units we call ecosystems. What are ecosystems? How are they organized into larger units?

2. The organisms in every ecosystem can be assigned to trophic categories. What are the categories? How do they function together in food webs to make a sustainable system?

3. Many non-trophic relationships also exist in ecosystems. What role do mutualism and competition play in ecosystems?

4. Environmental factors can be categorized into conditions and resources. How do these factors act as limiting factors in the distribution of different species in ecosystems?

5. Precipitation and temperature are the predominant determinants of climate. How do these factors interact to produce the different biomes around the globe?

6. Revolutionary changes have occurred in human culture that have greatly changed the relationship between humans and the environment. How have the Neolithic and Industrial Revolutions impacted the natural environment, and what is meant by the Environmental Revolution?

Barrier islands are found along the East and Gulf Coasts of the United States. Plum Island, in Massachusetts, is a nine-mile long barrier island in the northern part of the state (see opposite page). The southern two-thirds of the island is part of the Parker River National Wildlife Refuge, while the northern third is occupied by a host of summer cottages and year-round houses. On the refuge part of the island, sand dunes emerge at the upper edge of the ocean beach and extend back several hundred meters. Occupying the sand dunes is a mosaic of vegetation: the primary dunes, closest to the ocean, are colonized by beach grass (*Ammophila brevigulata*). Behind these dunes, a heathlike low growth occurs, dominated by false heather (*Hudsonia tomentosa*). Further back, a shrub community can be found, with poison ivy and bayberry as dominants. Even farther back, we encounter the maritime forest, in which we find pitch pine, poplar, wild cherry, and a few other deciduous trees. Here and there is a depression that reaches the water table, occupied by cranberry plants.

Animals on the barrier island—mice, rabbits, deer, skunk, red foxes—are seldom seen by visitors but leave their tracks on the sand. A diverse bird community can be found here, drawing birders who often find rarities like the peregrine falcon. Life for the plants—especially close to the ocean—is harsh; the sand holds little water or nutrients, salt spray from the ocean stresses leaf surfaces, and wind either blows loose sand away from plants' roots or piles it up around the plants and buries them. Yet the barrier island community is a functioning system that has endured for centuries.

On the northern third of the island, scores of streets were laid out in the early 1900s, and hundreds of small cottages were built. The dunes were either removed or built upon. Now, the only vegetation is a few beach grass plants that lie between the beach and the first houses. There are no dunes to speak of, no shrubs or maritime forest, no deer or foxes. It is not hard to imagine the entire island like this; only the presence of the protected refuge prevents further development. The contrast between these "communities" is unforgettable.

A barrier island. Plum Island in northern Massachusetts is a nine-mile-long sandy island separating the ocean from a bay, salt marsh, and the upland.

From time to time, the "environment" tests this barrier island, as it tests all barrier islands, with storms. On the southern section, the dunes absorb the powerful waves that pound the beach and wash upward toward the land. After a storm, the beach grass and primary dunes are intact and the back dunes untouched. The dune communities continue to thrive, forming a true barrier to the ocean that protects the fragile salt marshes and land behind the island. However, on the northern section, storms often wash beach sand by the houses and onto the streets, and occasionally a house or two are swept into the ocean. No one visits the northern Plum Island dunes to view birds or animals; in fact, there are no dunes, birds, or animals. In comparison, thousands of visitors enjoy the Parker River National Wildlife Reserve for its wildlife and beauty each year.

This situation is a microcosm of our interaction with the natural environment. Where we have managed to save the natural plants and animals occupying the land, they repay us by providing many goods and services. Where we have replaced or damaged them, we lose those benefits and often suffer the consequences. Can we learn to interact with the natural world and conduct our own endeavors in ways that are sustainable—that is, meet the needs of the present without compromising the ability of future generations to meet their own needs?

The study of natural ecosystems—those groupings of plants, animals, and other organisms that we think of as unspoiled forests, grasslands, coral reefs, ponds, barrier islands—can serve several objectives.

First, our agriculture and technology notwithstanding, we remain dependent on the natural world and its biodiversity for such basic needs as clean air, clean water, a suitable climate for growing food, and a host of other things. Our study of natural ecosystems will help us understand the relationships between the environment and living things and between the natural world and ourselves. From this we can more clearly understand the impact of humans on the natural world and the consequences this may have. We can then use our knowledge to improve our management of ecosystems.

Second, natural ecosystems are models of sustainability. The individual kinds of trees and other plants, animals, and microbes making up a forest, for example, are known to have propagated themselves over many tens of thousands, or even millions, of years. Our premise is that bringing out the basic principles that enable the perpetuation or sustainability of natural ecosystems will provide insights into the ways in which we need to direct human development in order to achieve a sustainable future.

Finally, it is our hope that the study of natural ecosystems will lead you to a greater appreciation of the amazing diversity and beauty of our planet home and a heightened desire to care for it as stewards. Thus, our study begins with an examination of natural ecosystems.

2.1 What Are Ecosystems?

The grouping or assemblage of plants, animals, and microbes we observe when we study a natural forest, grassland, pond, coral reef, or other undisturbed area is referred to as the area's **biota** (*bio*, living) or **biotic community**. The plant portion of the biotic community includes all vegetation, from large trees down through microscopic algae. Likewise, the animal portion includes everything from large mammals, birds, reptiles, and amphibians through earthworms, tiny insects, and mites. Microbes encompass a large array of microscopic bacteria, fungi, and protozoans. Thus, one may speak of the biotic community as comprising a plant community, an animal community, and a microbial community.

The particular kind of biotic community that we witness in a given area is, in large part, determined by **abiotic** (nonliving, chemical, and physical) factors, such as the amount of water or moisture present, temperature, salinity, and soil type. These abiotic factors both support and limit the particular community. For example, a relative lack of available moisture prevents the growth of most species of plants but supports certain species, such as cacti; we recognize such areas as deserts. Land with plenty of available moisture and suitable temperature supports forests. Obviously, the presence of water is the major factor that sustains aquatic communities.

The first step in investigating a biotic community may be to simply catalogue all the *species* present. **Species** are the different kinds of plants, animals, and microbes. Each species includes all those individuals that have a very strong similarity in appearance to one another and which are distinct in appearance from other such groups (robins vs. redwing blackbirds, for example). Similarity in appearance implies a close genetic relationship. Indeed, the biological definition of a species is the entirety of a population that can interbreed and produce fertile offspring, whereas members of different species generally do not interbreed, or if they do, fertile offspring are not produced. Breeding is often impractical or impossible to observe, however, so for purposes of identification, the aspect of appearance suffices.

In cataloguing the species of a community, one will observe that each species is represented by a certain **population**—that is, by a certain number of individuals that make up the interbreeding, reproducing group. The distinction between *population* and *species* is that the term *population* is used to refer only to those individuals of a certain species that live within a given area. The term *species* is all inclusive—it refers to all the individuals of a certain kind, even though they may exist in different populations in widely separated areas.

Continuing our study, along with the incredible variety of species and communities, it is impressive that the species within a community depend on and support one another in many ways. Most evident, certain animals will not be present unless particular plants that provide their necessary food and shelter are also present. Thus, the plant community supports (or limits by its absence) the animal community. Additionally, every plant and animal species is adapted to cope with the abiotic factors of the region. For example, every species that lives in temperate regions is adapted in one way

▲ FIGURE 2–1 *Winter in the forest.* Many trees and other plants of temperate forests are so adapted to the winter season that they actually require a period of freezing temperature in order to recommence growth in the spring.

or another to survive the winter season, which includes a period of freezing temperatures (Fig. 2–1). We shall explore these interactions among organisms and their environments later. For now, the point is that the populations of different species within a biotic community are constantly interacting with each other and the abiotic environment.

This brings us to the concept of an *ecosystem.* The ecosystem concept brings together the biotic community *and*

the abiotic conditions that it lives in. It includes considerations of the ways populations interact with each other and the abiotic environment to reproduce and perpetuate the entire grouping. In one sentence, an **ecosystem** is a grouping of plants, animals, and microbes occupying an explicit unit of space and interacting with each other and with their environment. For study purposes, an ecosystem may be taken to be any more or less distinctive biotic community living in a certain environment. Thus, a forest, a grassland, a wetland, a marsh, a pond, a barrier island, and a coral reef, each with its respective species in a particular environment, can be studied as distinct ecosystems.

Since no organism can live apart from its environment or from interactions with other species, ecosystems are the functional units of sustainable life on Earth. The study of ecosystems and the interactions that occur among organisms and between organisms and their environment is the science of **ecology**, and the investigators who conduct such studies are called **ecologists**.

While it is convenient to divide the living world into different ecosystems, any investigation soon reveals that there are seldom distinct boundaries between ecosystems, and they are never totally isolated from one another. Many species will occupy (and thus be a part of) two or more ecosystems at the same time. Or, they may move from one ecosystem to another at different times, as in the case of migrating birds. In passing from one ecosystem to another, one may observe only a gradual decrease in the populations of one biotic community and an increase in the populations representing another. Thus, one ecosystem may gradate into the next through a transitional region, known as an **ecotone**, that shares many of the species and characteristics of the two adjacent ecosystems (Fig. 2–2).

◄ FIGURE 2–2 *Ecotones on land.* Ecosystems are not isolated from one another. One ecosystem blends into the next through a transitional region, an ecotone, that contains many species common to the two adjacent systems.

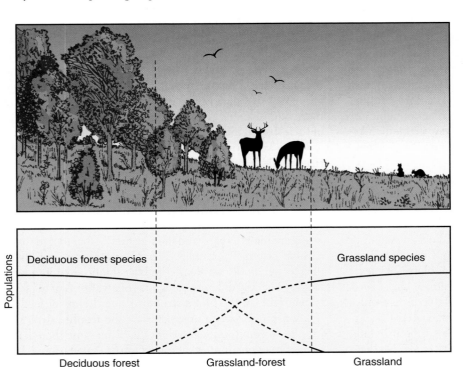

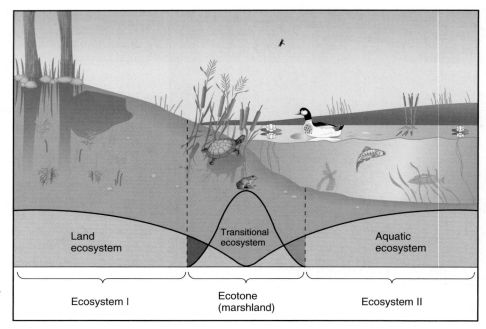

▶ **FIGURE 2–3** *Terrestrial to aquatic system ecotone.* An ecotone may create a unique habitat that harbors specialized species not found in either of the ecosystems bordering it. Typically, cattails, reeds, and lily pads grow here, along with several species of frogs and turtles as well as egrets and herons.

The ecotone between adjacent systems may also include *unique* conditions that support distinctive plant and animal species. Consider, for example, the marshy area that often occurs between the open water of a lake and dry land (Fig. 2–3). Ecotones may be studied as distinct ecosystems in their own right.

Furthermore, what happens in one ecosystem will definitely affect other ecosystems. For this reason, ecologists have begun using the concept of **landscapes**, defined as a group of interacting ecosystems. Thus, a barrier island, a saltwater bay, and the salt marsh behind it could be considered a landscape. Landscape ecology is then the science that studies the interactions between ecosystems.

Similar or related ecosystems or landscapes are often grouped together to form major kinds of ecosystems called **biomes**. Tropical rain forests, grasslands, and deserts are examples. While more extensive than an ecosystem in its breadth and complexity, a biome is still basically a certain biotic community supported and limited by certain abiotic environmental factors. Like ecosystems, there are generally no distinct boundaries between biomes, but one grades into the next through transitional regions. Table 2-1 presents six major terrestrial biomes and their major characteristics.

Likewise, there are major categories of aquatic and wetland ecosystems that are determined primarily by the depth, salinity, and permanence of water. These include lakes, marshes, streams, rivers, estuaries, bays, and ocean systems. As units of study, these aquatic systems may be viewed as ecosystems, as parts of landscapes, or as major biome-like features such as seas or oceans. (The biome category has been exclusively reserved for terrestrial systems.) Table 2-2 presents six major aquatic systems and their major characteristics.

Regardless of how we choose to divide (or group) and name different ecosystems, it is important to recognize that they all remain interconnected and interdependent. Terrestrial biomes are connected by the flow of rivers between them and by migrating animals. Sediments and nutrients washing from the land may nourish or pollute the ocean. Seabirds and mammals connect the oceans with the land, and all biomes share a common atmosphere and water cycle.

Therefore, all the species on Earth, along with all their environments, can be seen as one vast ecosystem, often called the **biosphere**. Although the separate local ecosystems are the individual units of sustainability, they are all interconnected to form the biosphere. The concept is analogous to the idea that the cells of our bodies are the units of living systems but are all interconnected to form the whole body. Carrying the analogy further, to what degree may individual ecosystems be degraded or destroyed before the entire biosphere is affected? And to what degree can basic global parameters such as atmosphere and temperature be altered before all ecosystems on Earth are affected? Let us now begin a more intensive study of ecosystems to discover the principles of their sustainability.

The key terms introduced in this section are summarized in Table 2-3.

2.2 The Structure of Ecosystems

Structure refers to parts and the way they fit together to make a whole system. There are two key aspects to every ecosystem: the biota or biotic community and the abiotic environmental factors. The way different categories of organisms fit together is referred to as the **biotic structure**, and the major feeding relationships between organisms are the **trophic structure** (*trophic*, feeding). All ecosystems have the same three basic categories of organisms that interact in the same ways.

TABLE 2-1	Major Terrestrial Biomes			
Biome	Climate and Soils	Dominant Vegetation	Dominant Animal Life	Geographic Distribution
Deserts	Very dry; hot days and cold nights; rainfall less than 10 in./yr.; soils thin and porous	Widely scattered thorny bushes and shrubs, cacti	Rodents, lizards, snakes, numerous insects, owls, hawks, small birds	N. and S.W. Africa; parts of Middle East and Asia; s.w. United States, northern Mexico
Grasslands	Seasonal rainfall, 10 to 60 in./yr; fires frequent; soils rich and often deep	Grass species, from tall grasses in higher rainfall areas to short grasses where drier; bushes and woodlands in some areas	Large grazing mammals: bison, goats; wild horses; kangaroos; antelopes, rhinos, warthogs, prairie dogs, coyotes, jackals, lions, hyenas; termites important	Central North America, central Asia, subequatorial Africa and South America, much of southern India, northern Australia
Tropical Rain Forests	Nonseasonal; annual temperature average 28°C; rainfall frequent and heavy, average over 95 in./yr.; soils thin and nutrient-poor	High diversity of broad-leafed evergreen trees, dense canopy, abundant epiphytes and vines; little understory	Enormous biodiversity; exotic colorful insects, amphibians, birds, snakes; monkeys, small mammals, tigers, jaguars	Northern South America, Central America, and western central Africa, islands in Indian and Pacific Oceans, s.e. Asia
Temperate Forests	Seasonal; temperature below freezing in winter, summers warm, humid; rainfall from 30–80 in./yr.; soils well developed	Broad-leafed deciduous trees, some conifers; shrubby undergrowth, ferns, lichens, mosses	Squirrels, raccoons, opossums, skunks, deer, foxes, black bear; snakes amphibians, rich soil microbiota, birds	Western and central Europe, eastern Asia, eastern North America
Coniferous Forests	Seasonal; winters long and cold; precipitation light in winter, heavier in summer; soils acidic, much humus and litter	Coniferous trees (spruce, fir, pine, hemlock), some deciduous trees (birch, maple); poor understory	Large herbivores as mule deer, moose, elk; mice, hares, squirrels; lynx, bears, foxes, fisher, marten; important nesting area for neotropical birds	Northern portions of North America, Europe, Asia, extending south-ward at high elevations
Tundra	Bitter cold except for an 8 to 10-week growing season with long days and moderate temperatures; precipitation low, soils thin and underlain with permafrost	Low-growing sedges, dwarf shrubs, lichens, mosses, and grasses	Year-round: lemmings, arctic hares, arctic foxes, lynx, caribou, musk, ox; summers: abundant insects, many migrant sandpipers, geese, and ducks	North of the coniferous forest in northern hemi-sphere, extending southward at elevations above the coniferous forest

Trophic Categories

The major categories of organisms are (1) *producers*, (2) *consumers*, and (3) *detritus feeders* and *decomposers*. Together, these groups produce food, pass it along food chains, and return the starting materials to the abiotic parts of the environment.

Producers. Producers are organisms that capture energy from the Sun or from chemical reactions to convert carbon dioxide (CO_2) to organic matter. Most producers are green plants, which use light energy to convert CO_2 and water to a sugar called glucose and then release oxygen as a byproduct. This chemical conversion, which is driven by light energy, is called **photosynthesis**. Plants are able to manufacture all the complex molecules that make up their bodies from the glucose produced in photosynthesis, plus a few additional *mineral nutrients* such as nitrogen, phosphorus, potassium, and sulfur, which they absorb from the soil or from water (Fig. 2–4).

The molecule that plants use to capture light energy for photosynthesis is **chlorophyll**, a green pigment. Hence, plants that carry on photosynthesis are easily identified by their green color. (In some cases, additional red or brown pigments may overshadow the green. Thus, red algae and brown algae also carry on photosynthesis.) Producers range in diversity from microscopic photosynthetic bacteria and

TABLE 2-2   Major Aquatic Systems

Aquatic Systems	Major Environmental Parameters	Dominant Vegetation	Dominant Animal Life	Distribution
Lakes & Ponds (fresh water)	Standing water bodies; low concentration of dissolved solids; seasonal vertical stratification of water	Rooted and floating plants, phytoplankton	Zooplankton, fish, insect larvae, ducks, geese, herons	Physical depressions in the landscape where precipitation and groundwater accumulate
Streams & Rivers (fresh water)	Flowing water; dissolved solids low; high dissolved oxygen, often turbid	Attached algae, rooted plants	Insect larvae, fish, amphibians; otters, raccoons, wading birds, duck, geese, swans	Landscapes where precipitation and groundwater flow by gravity toward oceans or lakes
Inland Wetlands (fresh water)	Standing water, at times seasonally dry; thick organic sediments; high nutrients	Marshes: grasses, reeds, cattails; swamps: water-tolerant trees; bogs: sphagnum moss, low shrubs	Amphibians, snakes, numerous invertebrates, wading birds, ducks, geese; alligators, turtles	Shallow depressions, poorly drained, often occupy sites of lakes and ponds that have filled in
Estuaries	Variable salinity; tides create two-way currents, often nutrient-rich, turbid	Phytoplankton in water column, rooted grasses like saltmarsh grass, mangrove swamps in tropics with salt-tolerant trees and shrubs	Zooplankton, rich shellfish, worms, crustaceans, fish; wading birds, sandpipers, ducks, geese	Coastal regions where rivers meet the ocean; may form bays behind sandy barrier islands
Coastal Ocean (saltwater)	Tidal currents promote mixing; nutrients high	Phytoplankton, large benthic algae; turtle grass, symbiotic algae in corals	Zooplankton, rich bottom fauna of worms, shellfish, crustaceans, echinoderms; coral colonies, jellyfish, fish, turtles, gulls, terns, ducks, sea lions, seals, dolphins, penguins, whales	From coastline outward over continental shelf; coral reefs abundant in tropics
Open Ocean	Great depths (to 11,000 meters); all but upper 200 m. dark and cold; nutrient-poor except in upwelling regions	Exclusively phytoplankton	Diverse zooplankton and fish adapted to different depths; seabirds, whales, tuna, sharks, squid, flying fish	Covering 70% of Earth, from edge of continental shelf outward

single-celled algae through medium-sized plants such as grass, daisies, and cacti to gigantic trees. Every major ecosystem, both aquatic and terrestrial, has its particular producers carrying on photosynthesis.

The term **organic** is used to refer to all those materials that make up the bodies of living *organ*isms—molecules such as proteins, fats or lipids, and carbohydrates. Likewise, materials that are specific products of living organisms, such as dead leaves, leather, sugar, and wood, are considered *organic*. On the other hand, materials and chemicals of air, water, rocks and minerals, which exist apart from the activity of living organisms, are considered **inorganic** (Fig. 2–5). Interestingly, there are bacteria that are able to use the energy in some inorganic chemicals to form organic matter from CO_2 and water. This process is called **chemosynthesis**, and these organisms clearly are producers.

The key feature of *organic* materials and molecules is that they are in large part constructed from bonded carbon and hydrogen atoms, a structure that is not found among *inorganic* materials. This carbon-hydrogen structure has its origins in the process of photosynthesis. Hydrogen atoms taken from water molecules and carbon atoms taken from carbon dioxide are joined together to form organic compounds in the process of photosynthesis. Green plants use light as the energy source to produce all the complex organic molecules their bodies need from the simple inorganic chemicals (carbon dioxide, water, mineral nutrients) present in the environment. As this conversion from inorganic to organic occurs, some of the energy from light is stored in the organic compounds.

Now, all organisms in the ecosystem *other than the producers* feed on organic matter as their source of both energy and nutrients. These include not only all animals

Heterotrophs may be divided into numerous subcategories, the two major categories being **consumers** (which eat living prey) and **detritus feeders** and **decomposers,** both of which feed on dead organisms or their products.

Consumers. Consumers encompass a wide variety of organisms ranging in size from microscopic bacteria to blue whales that includes such diverse groups as protozoans, worms, fish and shellfish, insects, reptiles, amphibians, birds, and mammals (including humans).

For the purpose of understanding ecosystem structure, consumers are divided into various subgroups according to their food source. Animals, as large as elephants or as small as mites, that feed directly on producers are called **primary consumers**. They are also called **herbivores** (*herb*, grass).

Animals that feed on primary consumers are called **secondary consumers**. Thus, elk, which feed on vegetation, are primary consumers, whereas wolves, because they feed on elk, are secondary consumers (Fig. 2–7). There may also be third, fourth, or even higher levels of consumers, and certain animals may occupy more than one position on the consumer scale. For instance, humans are primary consumers when they eat vegetables, secondary consumers when they eat beef, and third-level consumers

TABLE 2-3	Important Terms
Species	All the members of a specific kind of plant, animal, or microbe; a kind given by similarity of appearance and/or capacity for interbreeding and producing fertile offspring.
Population	All the members of a particular species occupying a given area.
Biotic community	All the populations of different plants, animals, and microbes occupying a given area.
Abiotic factors	All the factors of the physical environment: moisture, temperature, light, wind, pH, soil type, salinity, etc.
Ecosystem	The biotic community together with the abiotic factors; all the interactions among the members of the biotic community and between the biotic community and the abiotic factors within an explicit unit of space.
Landscape	A group of interacting ecosystems in a particular area.
Biome	A grouping of all the ecosystems of a similar type; e.g., tropical forests, grasslands, etc.
Biosphere	All species and physical factors on Earth functioning as one mammoth ecosystem.

but also **fungi** (mushrooms, molds, and similar organisms), most bacteria, and even a few higher plants, such as Indian Pipe (Fig. 2–6), that do not have chlorophyll and that therefore cannot carry on photosynthesis.

Thus, green plants, which carry on photosynthesis, are absolutely essential to every ecosystem (with the exception of a few ecosystems like the deep-sea hydrothermal vents, which depend on chemosynthesis). Their photosynthesis and growth constitute the production of organic matter, which sustains all other organisms in the ecosystem.

Indeed, all organisms in the biosphere can be divided into two categories, *autotrophs* and *heterotrophs*, on the basis of whether they do or do not produce the organic compounds they need to survive and grow. Those organisms such as green plants and chemosynthetic bacteria, which produce their own organic material from inorganic constituents in the environment using an external energy source, are **autotrophs** (*auto*, self; *troph*, feeding). The most important and common autotrophs by far are green plants, which use chlorophyll to capture light energy for photosynthesis. All other organisms, which must *consume* organic material to obtain energy and nutrients, are **heterotrophs** (*hetero*, other).

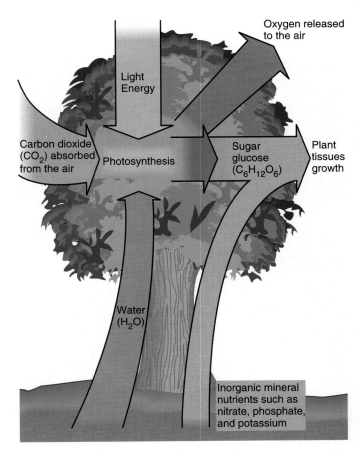

▲ FIGURE 2–4 *Green plant photosynthesis.* The producers in all major ecosystems are green plants.

Inorganic

Air:
Oxygen

Carbon
dioxide

Nitrogen

Other gases

Water

Rock and
soil minerals

Organic

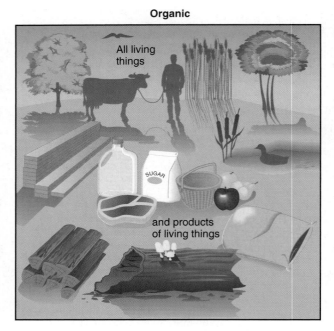

All living
things

and products
of living things

▲ **FIGURE 2–5** *Organic and inorganic.* Water and the simple molecules found in air and in rocks and soils are *inorganic*. The complex molecules that make up plant and animal tissues are *organic*.

when they eat fish that feed on smaller fish that feed on algae. Secondary and higher-order consumers are also called **carnivores** (*carni*, meat). Consumers that feed on both plants and animals are called **omnivores** (*omni*, all).

▲ **FIGURE 2–6** *Indian Pipe.* A flowering plant that is not a producer, the Indian Pipe does not carry on photosynthesis, but derives its energy from other organic matter, as do animals and other consumers (heterotrophs).

In a relationship in which one animal attacks, kills, and feeds on another, the animal that attacks and kills is called the **predator**; the animal that is killed is called the **prey**. Together, the two animals are said to have a **predator-prey** relationship.

Parasites are another important category of consumers. Parasites are organisms—either plants or animals—that become intimately associated with their "prey" and feed on it over an extended period of time, typically without killing it, but sometimes weakening it so that it becomes more prone to being killed by predators or adverse conditions. The plant or animal that is fed upon is called the **host**; thus, we speak of a **host-parasite** relationship.

A tremendous variety of organisms may be parasitic. Various worms are well-known examples, but certain protozoans, insects, and even certain mammals (vampire bats) and plants (dodder) (Fig. 2–8a) are also parasites. Many serious plant diseases and some animal diseases (such as athlete's foot) are caused by parasitic fungi. Indeed, virtually every major group of organisms has at least some members that are parasitic. Parasites may live inside or outside their host, as the examples shown in Fig. 2–8 illustrate.

In medical circles, a distinction is generally made between bacteria and viruses that cause disease, and parasites that are usually larger organisms. Ecologically, however, there is no real distinction. Bacteria are foreign organisms, and viruses are organism-like entities feeding on and multiplying in their hosts over a period of time and doing the same damage as do other parasites. Therefore, disease-causing bacteria and viruses can be considered highly specialized parasites. Representative examples of producers and consumers, and feeding relationships among them, are shown in Fig. 2–9.

◀ **FIGURE 2–7** *Secondary consumers.* Gray wolves have brought down an elk.

Detritus Feeders and Decomposers. Dead plant material such as fallen leaves, branches and trunks of dead trees, dead grass, the fecal wastes of animals, and occasional dead animal bodies is called **detritus**. Many organisms are specialized to feed on detritus, and we refer to such consumers as detritus feeders, or *detritivores*. Examples include earthworms, millipedes, fiddler crabs, termites, ants, and wood beetles. As with regular consumers, one can identify *primary* **detritus feeders** (those that feed directly on detritus), *secondary* **detritus feeders** (those that feed on primary detritus feeders), and so on.

An extremely important group of primary detritus feeders is the decomposers, namely, fungi and bacteria. Much of the detritus in an ecosystem—particularly dead leaves and the wood of dead trees or branches—does not appear to be eaten as such, but rots away. Rotting is the result of the metabolic activity of fungi and bacteria. These organisms secrete digestive enzymes that cause the breakdown of wood, for example, into simple sugars that the fungi or bacteria then absorb for their nourishment. Thus, the rotting we observe is really the result of material being consumed by fungi and bacteria. Even though fungi and bacteria are called decomposers because of their unique behavior, we group them with detritus feeders because their function in the ecosystem is the same. In turn, such secondary detritus feeders as protozoans, mites, insects, and worms (Fig. 2–10) feed upon decomposers. When a fungus or other decomposer dies, its body becomes part of the detritus and the source of energy and nutrients for still more detritus feeders and decomposers.

▼ **FIGURE 2–8** *Diversity of parasites.* Nearly every major biological group of organisms has at least some members that are parasitic on others. Shown here are (a) dodder, a plant parasite that has no leaves or chlorophyll. The orange "strings" are the dodder stems, which suck sap from the host plant. (b) Nematode worms (*Ascaris lumbricoides*), the largest of the human parasites, reach a length of 14 inches (35 cm). (c) Lamprey attached to a salmon. Lampreys are parasitic on fish.

(a)   (b)   (c)

Producers	Consumers		
	Primary consumers	Secondary consumers	Third-order consumers

Herbivores

Carnivores

Prey

Predator

Prey

Prey

Predator-prey relationships

Host

Parasite

Host-parasite relationship

Omnivore

◄ **FIGURE 2–9** *Trophic relationships among producers and consumers.*

▼ **FIGURE 2–10** *Detritus food web.* The feeding (trophic) relationships among primary detritus feeders, secondary detritus feeders, and consumers.

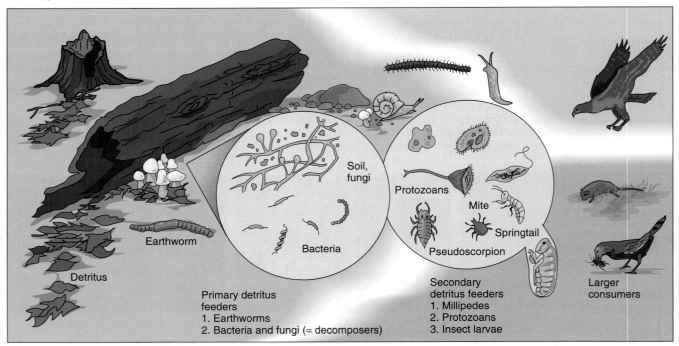

Detritus

Earthworm

Soil, fungi

Bacteria

Protozoans

Mite

Springtail

Pseudoscorpion

Primary detritus feeders
1. Earthworms
2. Bacteria and fungi (= decomposers)

Secondary detritus feeders
1. Millipedes
2. Protozoans
3. Insect larvae

Larger consumers

In summary, despite the apparent diversity of ecosystems, they all have a similar *biotic structure*. They can all be described in terms of autotrophs, or producers, which produce organic matter that becomes the source of energy and nutrients for heterotrophs, which are various categories of consumers and detritus feeders and decomposers (Fig. 2–11).

Trophic Relationships: Food Chains, Food Webs, and Trophic Levels

In describing the trophic structure of ecosystems, we can identify innumerable pathways where one organism is eaten by a second, which is eaten by a third, and so on. Each such pathway is called a **food chain**. While it is interesting to trace such pathways, it is important to recognize that food chains seldom exist as isolated entities. A herbivore population feeds on several kinds of plants and is preyed upon by several secondary consumers or omnivores. Consequently, virtually all food chains are interconnected and form a complex *web* of feeding relationships—the **food web**.

Despite the number of theoretical food chains and the complexity of food webs, there is a simple overall pattern: They all basically lead through a series of steps or levels—namely, from producers to primary consumers (or primary detritus feeders) to secondary consumers (or secondary detritus feeders), and so on. These *feeding levels* are called **trophic levels**. All producers belong to the first trophic level; all primary consumers (in other words, all herbivores) belong to the second trophic level; organisms feeding on these herbivores belong to the third level, and so on.

Whether we visualize the biotic structure of an ecosystem in terms of food chains, food webs, or trophic levels, we should see, through each feeding step, that there is a fundamental movement of the chemical nutrients and stored energy they contain from one organism or level to the next. These movements of energy and nutrients will be described in more detail later. A diagrammatic comparison of a food chain, a food web, and trophic levels is shown in Fig. 2–12a. A marine food web is shown in Fig. 2–12b.

How many trophic levels are there? Usually, no more than three or four in any ecosystem. This answer comes from straightforward observations. The **biomass**, or total combined (net dry) weight, of all the organisms at each trophic level can be estimated by collecting (or trapping) and weighing suitable samples. In terrestrial ecosystems, the biomass is about 90–99% less at each higher trophic level. If the biomass of producers in a grassland is 10 tons (20,000 lb) per acre, the biomass of herbivores will be about 2000 pounds and that of carnivores about 200 pounds. Clearly, you can't go through very many trophic levels before the biomass approaches zero. Depicting this graphically gives rise to what is commonly called a **biomass pyramid** (Fig. 2–13).

The biomass decreases so much at each trophic level for two reasons. First, much of the food that is consumed by a heterotroph is not converted to the body tissues of the heterotroph; rather, it is broken down so that the stored energy

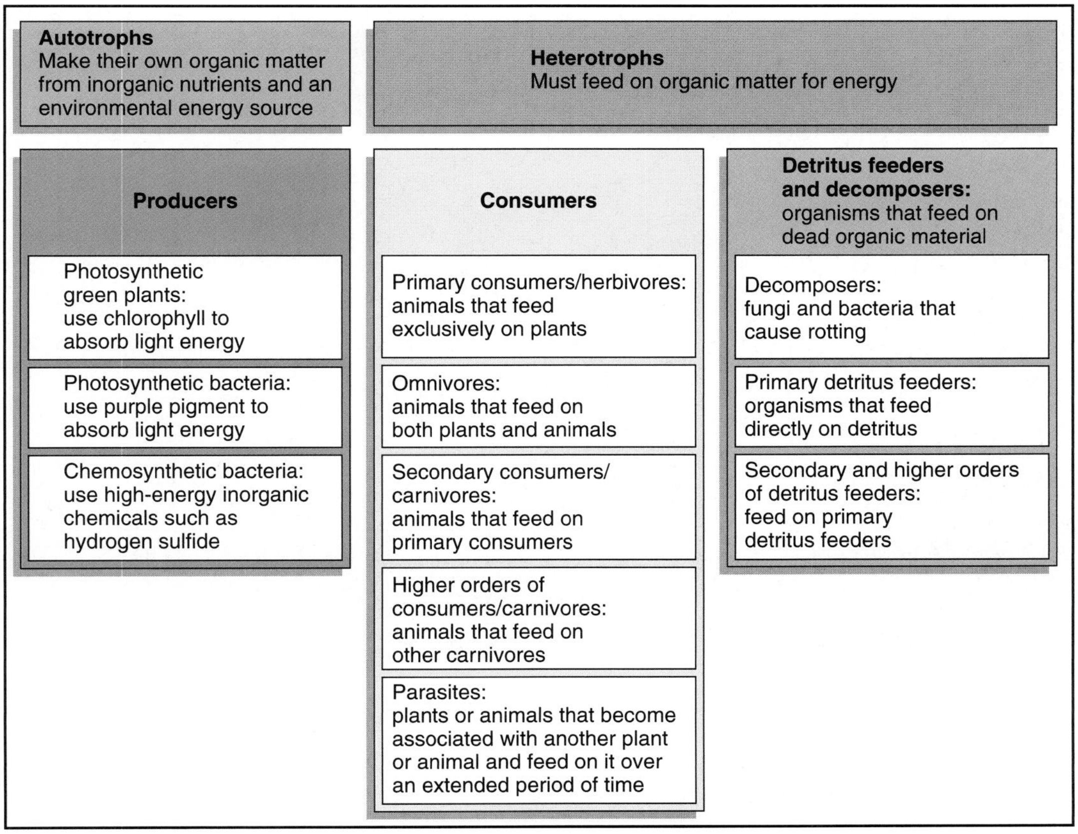

◀ FIGURE 2–11
Trophic categories. A summary of how living organisms are ecologically categorized according to feeding attributes.

Third
trophic
level:
all
primary
carnivores

Second
trophic
level:
all
herbivores

First
trophic
level:
all
producers

(a)

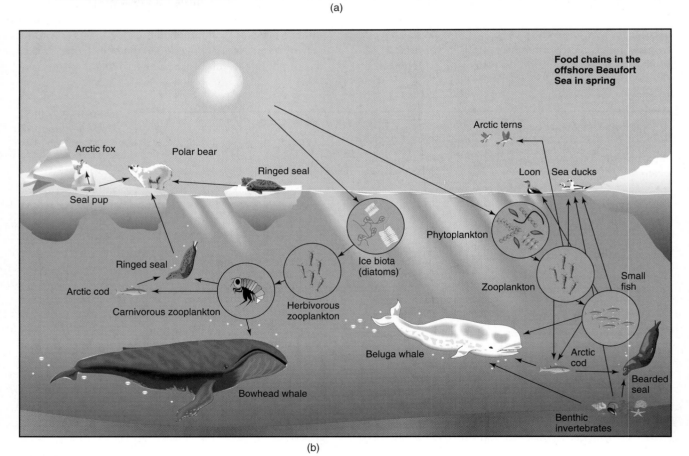

Food chains in the
offshore Beaufort
Sea in spring

Arctic terns

Loon   Sea ducks

Phytoplankton

Arctic fox   Polar bear

Ringed seal

Seal pup

Zooplankton

Small
fish

Ringed seal

Arctic cod

Carnivorous zooplankton

Ice biota
(diatoms)

Herbivorous
zooplankton

Beluga whale

Arctic
cod

Bearded
seal

Bowhead whale

Benthic
invertebrates

(b)

▲ **FIGURE 2–12** *Food webs.* (a) Specific pathways, such as from nuts to squirrels to foxes (shown by green arrows),
are referred to as *food chains.* A *food web* refers to the collective consideration of all food chains, which are invariably
interconnected (all arrows). Trophic levels, indicated by shading at the left, stresses the general pattern that food always
flows from producers to herbivores to carnivores. (b) A marine food web.

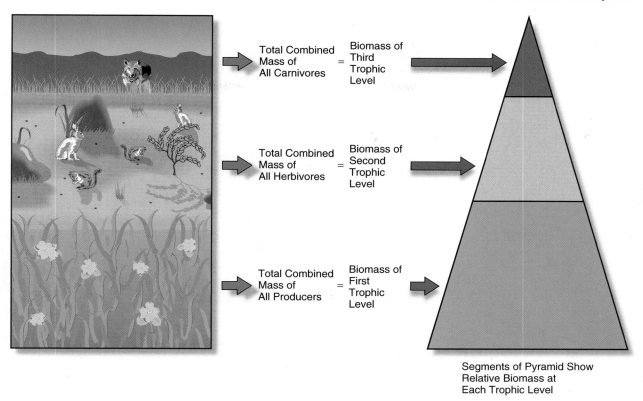

Total Combined Mass of All Carnivores = Biomass of Third Trophic Level

Total Combined Mass of All Herbivores = Biomass of Second Trophic Level

Total Combined Mass of All Producers = Biomass of First Trophic Level

Segments of Pyramid Show Relative Biomass at Each Trophic Level

▲ **FIGURE 2–13** *Biomass pyramid.* A graphic representation of the biomass (the total combined mass of organisms) at successive trophic levels has the form of a pyramid.

it contains can be released and used by the heterotroph. Second, much of the biomass—especially at the producer level—is never eaten by herbivores and goes directly to the decomposers. Hence, there is an inevitable loss of biomass with the movement to higher trophic levels. It is obvious that all heterotrophs depend on the continual input of fresh organic matter produced by the autotrophs (green plants). Without such input, the heterotrophs would all run out of food and starve.

As the breakdown of organic matter occurs, the chemical elements are released back to the environment in the inorganic state, where they may be reabsorbed by autotrophs (producers). Thus, there is a continuous cycle of nutrients from the environment through organisms and back to the environment. The spent energy, on the other hand, is lost as heat is given off from bodies (Fig. 2–14). These concepts of recycling chemical nutrients and the flow of energy will be discussed in greater detail in Chapter 3.

In summary, all food chains, food webs, and trophic levels *must start with producers,* and producers must have suitable environmental conditions to support their growth. Populations of all heterotrophs, including humans, are ultimately limited by what plants produce, in accordance with the concept of the biomass pyramid. Should any factor cause the productive capacity of green plants to be diminished, all other organisms at higher trophic levels will be diminished accordingly.

Nonfeeding Relationships

Mutually Supportive Relationships. The overall structure of ecosystems is characterized by feeding relationships, as we have just seen. In any feeding relationship, we generally think of one species benefiting and the other being harmed to a greater or lesser extent. However, there are many relationships that provide a mutual benefit to both species. This phenomenon is called **mutualism.** A common example is the relationship between flowers and insects: The insects benefit by obtaining nectar from the flowers, and the plants benefit by being pollinated in the process. Another example is observed in tropical seas: Clownfish are immune to the toxin in the tentacles of sea anemones, which the anemones use to immobilize their prey. Thus, these fish are able to feed on detritus around the anemones, at the same time receiving protection from would-be predators that are *not* immune. The anemones benefit by being cleaned (Fig. 2–15).

In some cases, the mutualistic relationship has become so close that the species involved are no longer capable of living alone. A classic example is the group of plants known as *lichens* (Fig. 2–16). Lichen is actually comprised of two organisms: a fungus and an alga. The fungus provides protection for the alga, enabling it to survive in dry habitats where it could not live by itself, and the alga, which is a producer, provides food for the fungus, which is a heterotroph. These two species living together in close union are said to have a

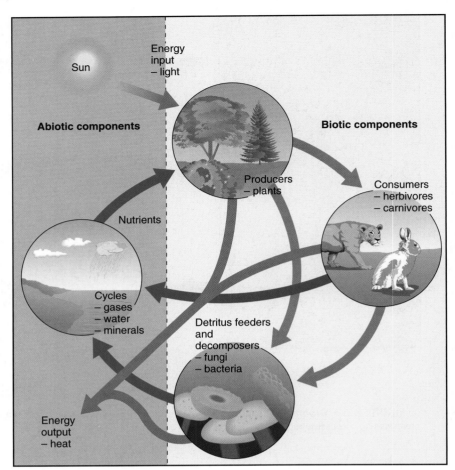

▶ **FIGURE 2–14** *Nutrient cycles and energy flow.* The movement of nutrients (blue arrows) and energy (red arrows) and both (brown arrows) through the ecosystem. Nutrients follow a cycle, being used over and over. Light energy absorbed by producers is released and lost as heat energy as it is "spent."

▶ **FIGURE 2–15** *A mutualistic relationship.* The clownfish, protected by the anemones, can forage without risk of predation; the anemones are cleaned of detritus.

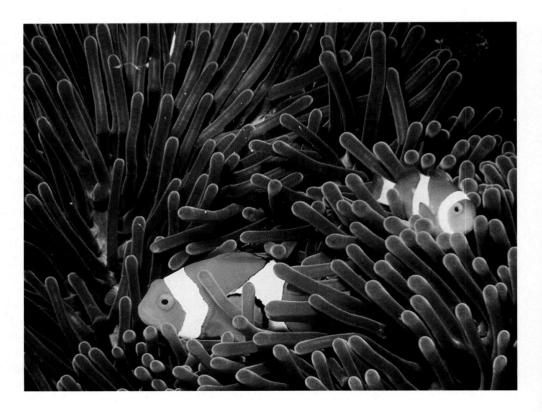

▲ **FIGURE 2–16** *Lichens.* The crusty-appearing "plants" commonly seen growing on rocks or the bark of trees are actually composed of a fungus and an alga growing in a symbiotic relationship.

symbiotic relationship. However, **symbiosis** by itself simply refers to the fact of "living together" in close union (*sym*, together; *bio*, living); it does not specify a mutual benefit or harm. Therefore, symbiotic relationships may include parasitic relationships as well as mutualistic relationships.

While not categorized as mutualistic, countless relationships in an ecosystem may be seen as aiding its overall sustainability. For example, plant detritus provides most of the food for decomposers and soil-dwelling detritus feeders such as earthworms. Thus, these organisms benefit from plants, but the plants also benefit because the activity of the organisms is instrumental in releasing nutrients from the detritus and in returning them to the soil where they can be reused by the plants. In another example, insect-eating birds benefit from vegetation by finding nesting materials and places among trees, while the plant community benefits because the birds feed on and reduce the populations of many herbivorous insects. Even in predator-prey relationships, some mutual advantage may exist. The killing of individual prey that is weak or diseased may benefit the population at large by keeping it healthy. Predators and parasites may also prevent herbivore populations from becoming so abundant that they overgraze their environment.

Competitive Relationships. Given the concept of food webs, it might seem that species of animals would be in a great "free-for-all" competition with each other. In fact, fierce competition rarely occurs, because each species tends to be specialized and adapted to its own *habitat* and/or *niche*.

Habitat refers to the kind of place—defined by the plant community and the physical environment—where a species is biologically adapted to live. For example, a deciduous forest, a swamp, and an open, grassy field denote types of habitats. Types of forests (e.g., conifer vs. deciduous) provide markedly different habitats and support a variety of wildlife.

Even when different species occupy the same habitat, competition may be slight or nonexistent because each species has its own *niche*. An animal's **niche** refers to what it feeds on, where it feeds, when it feeds, where it finds shelter, how it responds to abiotic factors, and where it nests. Seeming competitors can coexist in the same habitat but have separate niches. Competition is minimized because potential competitors are using different resources. For example, woodpeckers, which feed on insects in dead wood, are not in competition with birds that feed on seeds. Bats and swallows both feed on flying insects, but they are not in competition, because bats feed on night-flying insects and swallows feed during the day. Sometimes the "resource" can be the space used by different species as they forage for food, as in the case of five species of warblers that coexist in the spruce forests of Maine (Fig. 2–17). This is a well-known case of what is called *resource partitioning*. It is assumed that by adapting to each other's presence over time, competition is avoided and all species benefit (see Chapter 5).

Depending on how a set of resources is "divided up" between species, there may be overlap between the niches of the species. If two species compete directly in many respects, as sometimes occurs when a species is introduced from another continent, one of the two generally perishes in the competition. This is the *competitive exclusion principle*.

All green plants require water, nutrients, and light, and where they are growing in the same location, one species may eliminate others through competition. (Hence, maintaining flowers and vegetables against the advance of weeds is a constant struggle.) However, different plant species are also adapted and specialized to particular conditions. Thus, each species is able to hold its own against competition where conditions are well suited to it. The same concepts hold true for species in aquatic and marine ecosystems.

Abiotic Factors

We now turn to the *abiotic* side of the ecosystem. As noted before, the environment involves the interplay of many physical and chemical factors that different species respond to, or **abiotic factors**. It is helpful to make a distinction between two types of abiotic factors: **conditions** and **resources**. *Conditions* are abiotic factors that vary in space and time but are not used up or made unavailable to other species. Such factors include temperature (extremes of heat and cold, as well as average), wind, pH (acidity), salinity (saltiness), and fire. For example, within aquatic systems, the key conditions are salinity (fresh water vs. saltwater), temperature, texture of the bottom (rocky vs. silty), depth and turbidity (cloudiness) of water (determining how much, if any, light reaches the bottom), and currents.

Resources are any factors—biotic or abiotic—that are consumed by organisms. Abiotic resources include water, chemical nutrients (like nitrogen, phosphorus, and carbon dioxide), light (for plants), and oxygen. The concept also includes spatial needs,

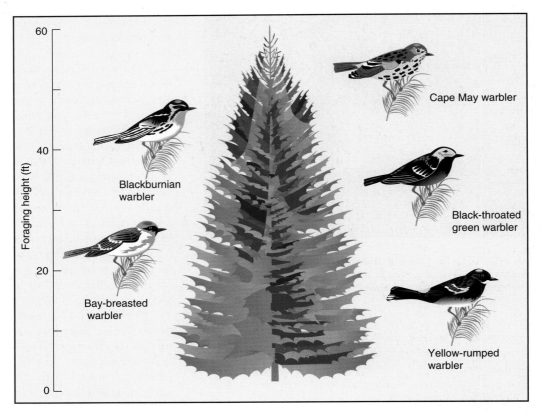

▶ **FIGURE 2–17**
Resource partitioning.
Five species of North
American warblers reduce
the competition among
themselves by feeding at
different levels and on
different parts of the trees.

such as a place on the intertidal rocks or a hole in a tree. Resources, unlike conditions, can be the objects of competition between individuals or species.

The degree to which each abiotic factor is present or absent and high or low profoundly affects the ability of organisms to survive. However, each species may be affected differently by each factor. We shall find that this difference in response to environmental factors determines which species may or may not occupy a given region or particular area within a region. In turn, which organisms do or do not survive determines the nature of a given ecosystem.

Optimum, Zones of Stress, and Limits of Tolerance. In any study of ecology, a primary observation is that *different species thrive under different environmental regimes.* This principle applies to all living things, both plants and animals. Some survive where it is very wet; others where it is relatively dry. Some thrive in warmth; others do best in cooler situations. Some tolerate freezing; others do not. Some require bright sun; others do best in shade. Aquatic systems are divided into fresh water and saltwater, each with its respective fish and other organisms.

Laboratory experiments clearly bear out the fact that different species are best adapted to different factors. Organisms can be grown under controlled conditions where one factor is varied while other factors are held constant. Such experiments demonstrate that for every factor there is an **optimum**, a certain level at which the organisms do best. At higher or lower levels the organisms do less well, and at further extremes they may

not be able to survive at all. This concept is shown graphically in Fig. 2–18. Temperature is shown as the variable in the figure, but the idea pertains to any abiotic factor that might be tested.

The point at which the best response occurs is called the optimum, but since this often occurs over a range of several degrees, it is common to speak of an *optimal range.* The entire span that allows any growth at all is called the **range of tolerance**. The points at the high and low ends of the range of tolerance are called the **limits of tolerance**. Between the optimal range and the high or low limit of tolerance, there are **zones of stress**. That is, as the factor is raised or lowered from the optimal range, the organisms experience increasing stress, until, at either limit of tolerance, they cannot survive.

Of course, not every species has been tested for every factor; however, the consistency of such observations leads us to conclude that the following is a fundamental biological principle: *Every species (both plant and animal) has an optimum range, zones of stress, and limits of tolerance with respect to every abiotic factor.*

This line of experimentation also demonstrates that different species vary in characteristics with respect to the values at which the optimum and limits of tolerance occur. For instance, what may be an optimal amount of water for one species may stress a second and result in the death of a third. Some plants cannot tolerate any freezing temperatures, others can tolerate slight but not intense freezing, and some actually require several weeks of freezing temperatures in order to complete their life cycles. Also, some species have a very

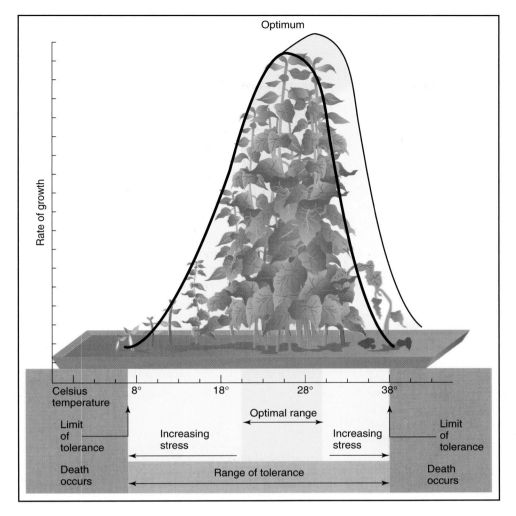

Optimum

Rate of growth

Celsius
temperature

8°   18°   28°   38°

Optimal range

Limit
of
tolerance

Increasing
stress

Increasing
stress

Limit
of
tolerance

Death
occurs

Range of tolerance

Death
occurs

◀ **FIGURE 2–18** *Survival
curve.* For every factor influencing
growth, reproduction, and survival,
there is an optimum level. Above and
below the optimum, there is increasing
stress, until survival becomes
impossible at the limits of tolerance.
The total range between the high and
low limits is the range of tolerance.

broad range of tolerance, whereas others have a much nar-rower range. While optimums and limits of tolerance may differ from one species to another, there may be great over-lap in their ranges of tolerance.

The concept of a range of tolerance does not just affect the growth of individuals; insofar as the health and vigor of indi-viduals affect reproduction and survival of the next genera-tion, the population is also influenced. That is, the population density (individuals per unit area) of a species will be greatest where all conditions are optimal, and it will decrease as any one or more conditions depart from the optimum. Different ranges of tolerance for different factors make an important contribu-tion to the identity of an ecological niche for a given species.

Law of Limiting Factors. In 1840 Justus von Liebig studied the effects of chemical nutrients on plant growth. He ob-served that restricting any one of the many different nutrients at any given time had the same effect: It limited growth. A fac-tor that limits growth is called the **limiting factor**. Therefore, it follows that *any one factor* being outside the optimal range will cause stress and limit the growth, reproduction, or even the survival of a population. This observation is referred to as the **law of limiting factors**, or Liebig's law of minimums.

The limiting factor may be a problem of "too much," as well as a problem of "too little." For example, plants may be stressed or killed not only by underwatering or underfertiliz-ing but also by overwatering or overfertilizing, which are common pitfalls for beginning gardeners. Note also that the limiting factor may change from one time to another. For in-stance, in a single growing season, temperature may be lim-iting in the early spring, nutrients may be limiting later, and then water may be limiting if a drought occurs. Also, if one limiting factor is corrected, growth will increase only until an-other factor comes into play. Of course, the organism's ge-netic potential is an ultimate limiting factor: A daisy will never grow to be the height of a tree, nor a mouse to the bulk of an elephant, regardless of optimal environmental factors.

Observations made since Liebig's time show that his law has a much broader application: Growth may be limited not only by abiotic factors but also by biotic factors. Thus, the limiting factor for a population may be competition or pre-dation from another species. This is certainly the case with our agricultural crops, where it is a constant struggle to keep them from being limited or even eliminated by weeds and "pests."

Finally, while one factor may be determined to be lim-iting at a given time, several factors outside the optimum may combine to cause additional stress or even death. Particularly, pollutants may act in a way that causes organisms to become more vulnerable to disease or drought. Such cases are examples of **synergistic effects**, or **synergisms**, which are defined as two or more factors interacting in a way that causes an effect much greater than one would anticipate from the effects of each of the two acting separately.

2.3 Global Biomes

We can now use the concepts of optimums and limiting factors to gain a better understanding of why different regions or even localized areas may have distinct biotic communities, creating a variety of ecosystems, landscapes, and biomes.

The Role of Climate

The **climate** of a given region is a description of the average temperature and precipitation—the weather—that may be expected on each day throughout the entire year (see Chapter 21). Climates in different parts of the world vary widely. In general, the temperature of equatorial regions is continuously warm, with high rainfall and no discernible seasons. Above and below the equator, the temperatures become increasingly seasonal (warm or hot summers and cool or cold winters); the farther we go toward the poles, the longer and colder the winters become, until at the poles it is perpetually winter. Likewise, colder temperatures are found at higher elevations, so that there are even snowcapped mountains on or near the equator.

Annual precipitation in any area also may vary greatly, from virtually zero to well over 100 inches (250 cm) per year. It may be evenly distributed throughout the year or occur in certain months, dividing the year into wet and dry seasons.

Different temperature and rainfall conditions may occur in almost any combination to give a wide variety of climates. In turn, a given climate will support only those species that find the temperature and precipitation levels optimal or at least within the range of tolerance. As indicated in Fig. 2–18, population densities will be greatest where conditions are optimal and will decrease as any condition departs from the optimum. A species will be excluded from a region (or local areas) where any condition is beyond its limit of tolerance. How will this affect the biotic community?

To illustrate, let us consider six major types of biomes and their global distribution. (Fig. 2–19 shows these six types and subdivisions of several.) Within the temperate zone (between 30° and 50° latitude) the amount of rainfall is the key limiting factor. The **temperate deciduous forest biome** will be found where annual precipitation is 30–80 inches (75–200 cm). Where rainfall tapers off or is highly seasonal (10–60 inches, or 25–150 cm per year), **grassland biomes** are found, and regions receiving an average of less than 10 inches (25 cm) per year are occupied by a **desert biome** (Table 2-1).

The effect of temperature, the other dominant parameter of climate, is largely superimposed on that of rainfall. That is, 30

inches (75 cm) or more of rainfall per year will usually support a forest, but temperature will determine the *kind* of forest. For example, broadleaf evergreen species, which are extremely vigorous and fast growing but cannot tolerate freezing temperatures, predominate in the **tropical rain forest**. By dropping their leaves and going into dormancy each autumn, deciduous trees are well adapted to freezing temperatures. Therefore, wherever rainfall is sufficient, deciduous forests predominate in temperate latitudes. Most deciduous trees, however, cannot tolerate the extremely harsh winters and short summers that occur at higher latitudes and higher elevations. Therefore, northern regions and high elevations are occupied by the **coniferous forest biome**, as conifers are better adapted to these conditions.

Temperature by itself limits forests only when it becomes low enough to cause **permafrost** (permanently frozen subsoil). Permafrost prevents the growth of trees, because roots cannot penetrate deeply enough to provide adequate support. However, a number of grasses, clovers, and other small flowering plants can grow in the topsoil above permafrost. Consequently, where permafrost sets in, the coniferous forest biome gives way to the **tundra biome** (Table 2-1). Of course, at still colder temperatures, the tundra gives way to permanent snow and ice cover.

The same relationship of rainfall effects being primary, and temperature effects secondary, applies in deserts. Any

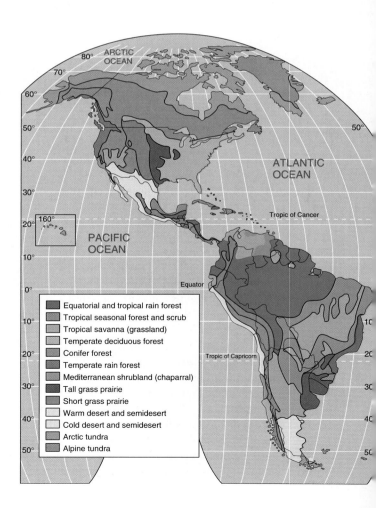

Equatorial and tropical rain forest
Tropical seasonal forest and scrub
Tropical savanna (grassland)
Temperate deciduous forest
Conifer forest
Temperate rain forest
Mediterranean shrubland (chaparral)
Tall grass prairie
Short grass prairie
Warm desert and semidesert
Cold desert and semidesert
Arctic tundra
Alpine tundra

region receiving less than about 10 inches (25 cm) of rain per year will be a desert, but the plant and animal species found in hot deserts are different from those found in cold deserts.

Temperature also exerts considerable influence on an ecosystem by its effect on the rate of evaporation of water. Higher temperatures effectively reduce the amount of available water because more is lost through evaporation. As a result, the transitions from deserts to grasslands and from grasslands to forests are found at higher precipitation levels in hot regions than in cold regions.

A summary of the relationship between biomes and the temperature and rainfall conditions is given in Fig. 2–20. The average temperature for a region varies with both latitude and altitude, as shown in Fig. 2–21.

Microclimate and Other Abiotic Factors

A specific site may have temperature and moisture conditions that are significantly different from the overall climate of the region in which it is located, which is necessarily an average. For example, a south-facing slope, which receives more direct sunlight, will be relatively warmer and hence also drier than a north-facing slope. Similarly, temperature range in a sheltered ravine will be narrower than that in a more exposed location, and so on.

The conditions found in a specific localized area are referred to as the **microclimate** of that location. In the same way that different climates determine the major biome of the region, different microclimates result in variations of the biotic community within the biome.

Soil type and topography may also contribute to the diversity found in a biome because these two factors affect the availability of moisture. For example, in the eastern United States, oaks and hickories generally predominate on rocky, sandy soils and on hilltops, which retain little moisture, whereas beeches and maples are found on richer soils, which hold more moisture, and red maples and cedars inhabit low, swampy areas. In the transitional region between desert and grassland [10–20 inches (25–50 cm) of rainfall per year], a soil with good water-holding capacity will support grass, but a sandy soil with little ability to hold water will support only desert species (Fig. 2–22).

In certain cases, an abiotic factor other than rainfall or temperature may be the primary limiting factor. For example, the strip of land adjacent to a coast frequently receives a salty spray from the ocean, a factor that relatively few plants can tolerate. Consequently, a community of salt-tolerant plants frequently occupies this strip, as we saw on the barrier island. Relative acidity or alkalinity (pH) may also have an overriding effect on a plant or animal community.

◀ **FIGURE 2–19** *World distribution of the major terrestrial biomes.* *(After Robert Christopherson, Geosystems, 4th ed. ©2000. Reprinted by permission of Prentice Hall, Upper Saddle River, NJ.)*

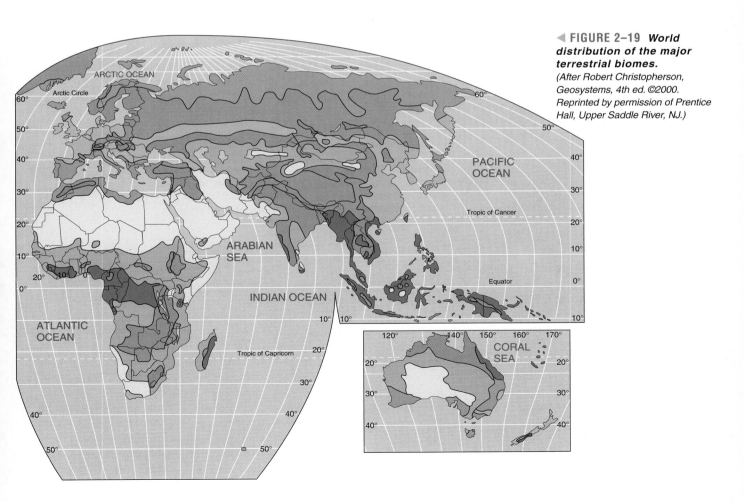

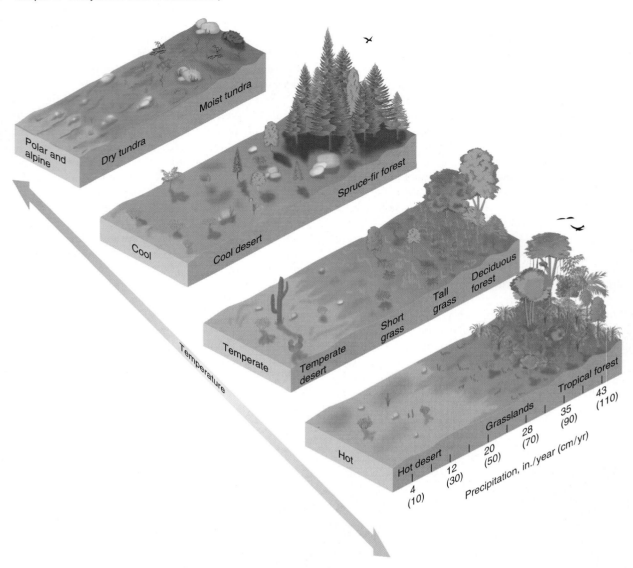

▲ **FIGURE 2–20** *Climate and major biomes.* Moisture is generally the overriding factor determining the type of biome that may be supported in a region. Given adequate moisture, an area will generally support a forest. Temperature, however, determines the *kind* of forest. The situation is similar for grasslands and deserts. At cooler temperatures, there is a shift toward less precipitation because lower temperatures reduce evaporative water loss. Temperature becomes the overriding factor only when it is low enough to sustain permafrost.

Biotic Factors

Limiting factors may also be biotic—that is, caused by other species. Grasses thrive when rainfall is more than 30 inches (75 cm). However, when the rainfall is great enough to support trees, increased shade may limit grasses. Thus, the factor that limits grasses from taking over high-rainfall regions is biotic: overwhelming competition from taller species. The distribution of plants may also be limited by the presence of certain herbivores; elephants are notorious destroyers of woodlands, and their presence leads predominantly to grasslands.

The concept of limiting factors also applies to animals. As with plants, the limiting factor may be abiotic—cold temperatures or lack of open water, for example—but it is more frequently biotic, in the form of a lack of a plant community that provides suitable food or habitat or both.

Physical Barriers

A final factor that may limit species to a particular region is the existence of a physical barrier, such as an ocean, desert, or mountain range, which species are unable to cross. Thus, species making up the communities on separate continents or remote islands are usually quite different despite having similar climates.

When such barriers are overcome—for example, by humans transporting a species from one continent to another—the introduced species may make a successful "invasion." However, a successful invasion by a foreign species may cause an ecological disaster, because the invader often displaces existing species through competition. Examples of such problems with imported species will be explored further in Chapter 4. Note also that humans erect barriers—dams, roadways,

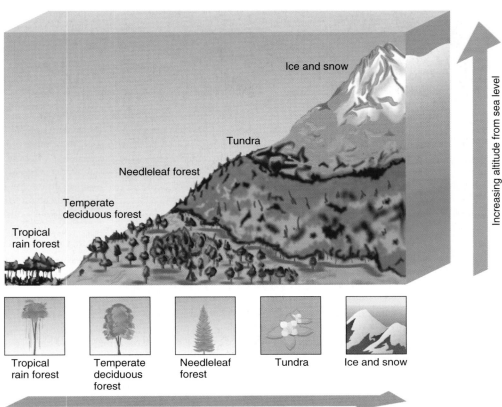

Ice and snow

Tundra

Needleleaf forest

Temperate
deciduous forest

Tropical
rain forest

Increasing altitude from sea level

| Tropical rain forest | Temperate deciduous forest | Needleleaf forest | Tundra | Ice and snow |

Increasing latitude from the equator

◀ **FIGURE 2–21** *Effects of latitude and altitude.* Decreasing temperatures that result in the biome shifts noted in Fig. 2–20 occur both with increasing latitude (distance from the equator) and increasing altitude.

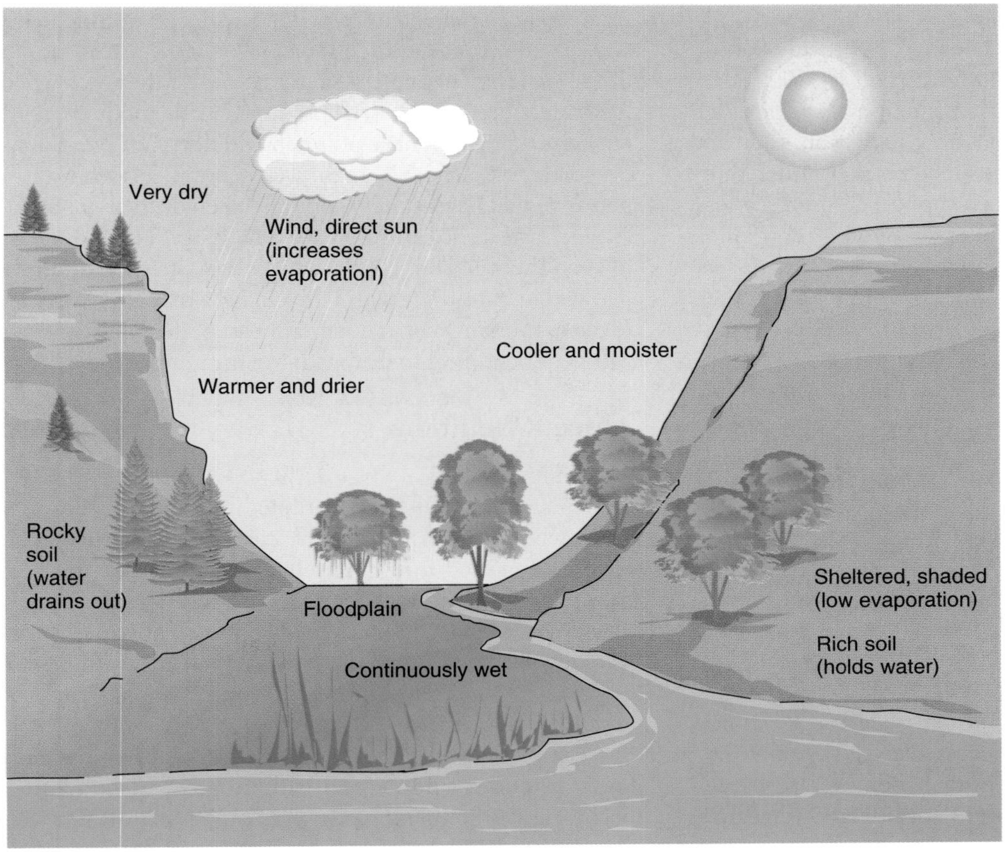

◀ **FIGURE 2–22**
Microclimates. Abiotic factors such as terrain, wind, and type of soil create different microclimates by influencing temperature and moisture in localized areas.

Very dry

Wind, direct sun
(increases
evaporation)

Cooler and moister

Warmer and drier

Rocky
soil
(water
drains out)

Floodplain

Continuously wet

Sheltered, shaded
(low evaporation)

Rich soil
(holds water)

EARTH WATCH

A DOSE OF LIMITING FACTOR

Environmental factors—particularly temperature and rainfall—are always changing. What does it mean to say that rainfall, for example, is a limiting factor when weather is almost always changing?

Consideration of a limiting factor should always include the concept of dose. Dose is defined as the level of exposure multiplied by the length of time over which the exposure occurs. You have probably experienced this kind of thing for yourself. For example, you can enter a walk-in freezer or expose yourself to the intense heat from an oven for a few seconds without particular discomfort, whereas prolonged exposure to such heat or cold would be extremely damaging, if not fatal. You might not be hurt by breathing noxious fumes for a short time, but longer exposures might well kill you.

Similarly, a limiting factor that causes the die-off of a species involves both the intensity of the factor and the duration of exposure to it. For example, most plants can tolerate drought to some degree. It is more a matter of how long they can tolerate it. Cacti and other desert plants can tolerate much longer periods of drought than non-desert species. Likewise, it is not a question of whether plants can tolerate flooding. Rather, it is a question of how long they can tolerate being flooded. Marsh plants, of course, thrive on perpetually flooded land; in contrast, a few days of flooding will kill many other terrestrial species.

Thus, the balance between ecosystems is not a perfect steady-state balance with a difference in average rainfall as a sharp dividing line. What actually occurs, where forest meets grassland, for example, is that trees encroach into grasslands during years

of normal rainfall, but then a severe drought occurs, and trees in the grassland are killed back, and then the cycle starts again. Hence, there is a seesaw effect between adjoining ecosystems that depends on the occurrence of limiting doses of one or another factor. This seesaw effect is a particularly important consideration in the face of global warming.

There is no question that most species could probably tolerate the average temperature being a few degrees higher if this were the only factor involved. The problem is, how will the slightly higher average temperature affect the occurrence of heat waves and the duration of droughts and floods? Such warming could create limiting doses that have far-reaching and unpredictable effects. Scientists are already tracking changing habitats attributable to global warming.

cities, and farms—that may block the normal movement of populations and cause their demise.

In summary, the biosphere consists of a great variety of environments, both aquatic and terrestrial. In each environment we find plant, animal, and microbial species that are adapted to all the abiotic factors. In addition, they are adapted to each other, in various feeding and nonfeeding relationships. Each environment supports a more or less unique grouping of organisms interacting with each other and with the environment in a way that perpetuates or sustains the entire group. That is, each environment, together with the species it supports, is an ecosystem. Every ecosystem is tied with others through species that migrate from one system to another and through exchanges of air, water, and minerals common to the whole planet. At the same time, each species and, as a result, each ecosystem, is kept within certain bounds by limiting factors. The spread of each species is at some point limited by its not being able to tolerate particular conditions, to compete with other species, or to cross some physical barrier. Species distribution is always due to one or more limiting factors.

2.4 Implications for Humans

So far, we have been looking at ecosystems as natural functioning systems, unaffected by human impacts. In reality, human impacts must be taken into consideration because

we have become such a dominant presence on Earth. We have replaced many natural systems with agriculture, urban and suburban developments; we make heavy use of most of the remaining "natural" systems as we use them for wood, food, and other commercial products; the byproducts of our economic activities have polluted and degraded ecosystems everywhere. Our involvement with natural ecosystems is so pervasive that if we want to continue to enjoy their goods and services, we must learn how to *manage* natural ecosystems to keep them healthy and productive. A brief look into the past will help us understand the changes that have occurred to make humans such a dominant part of the landscape and will better equip us to deal with the future.

Three Revolutions

Neolithic Revolution. Natural ecosystems have existed and perpetuated themselves on Earth for hundreds of millions of years, while humans are relative newcomers on this scene. Evidence gained through archaeology and anthropology shows that hominid ancestry goes back several million years. The evidence indicates that several different hominid species were involved on the evolutionary pathway from primate ancestors to our present-day human species, which emerged about 100,000 years ago.

Early hominids survived in small tribes as hunter-gatherers, catching wildlife and gathering seeds, nuts, roots, berries, and other plant foods (Fig. 2–23). Settlements were

◀ **FIGURE 2–23** *Hunter-gatherer culture.* Before the advent of agriculture, all human societies had to forage for their food as these bushmen from Namibia are doing.

never large and were of relatively short duration because, as one area was "picked over," the tribe was forced to move on. As hunter-gatherers, hominids were much like other omnivorous consumers in natural ecosystems. Populations could not expand beyond the sizes that natural food sources supported, life was harsh, and deaths from predators, disease, and famine were common.

About 10,000 years ago, however, a highly significant change in human culture occurred: Humans in the Middle East began to develop agriculture—the domestication of wild species. Animal husbandry and agriculture are processes of taking particular animal and plant species out of the wild, clearing space, and providing other conditions to grow them preferentially. Plants are protected from competition (weeds) and other would-be consumers, and additional nutrients (fertilizer) and water may be provided. Animals are protected from predators and given food for optimal growth.

The development of agriculture provided a more abundant and reliable food supply, but it created a turning point in human history for other reasons as well. Because of this, it is referred to as a *revolution*—specifically, the **Neolithic Revolution**. Conducting agriculture does not just allow, but *requires*, permanent (or at least long-term) settlements and the specialization of labor. Some members of the settlement specialize in tending crops and producing food, freeing others to specialize in other endeavors. With this specialization of labor in permanent settlements, there is more incentive and potential for technological development: better tools, better dwellings, and better means of transporting water and other materials. Trade with other settlements begins, and thus commerce is born. Also, living in settlements enables better care and protection for everyone, and therefore the number of early deaths is reduced. This reduced mortality rate, coupled with more reliable food production, supports population growth, which in turn is supported by expanding agriculture. In short, civilization had its origins in the invention of agriculture about 10,000 years ago.

Industrial Revolution. For another 9000-plus years the human population increased and spread throughout the Earth. Agriculture and natural ecosystems provided the support for the growth of a civilization and culture that increased in knowledge and mastery over the natural world. With the birth of modern science and technology in the seventeenth and eighteenth centuries, the human population—by 1800 almost a billion strong—was on the threshold of another revolution: the **Industrial Revolution** (Fig. 2–24). It is fair to say that this revolution created the modern world, with its worldwide commerce, factories, large cities, and pollution, on a massive scale. The Industrial Revolution and its technological marvels were energized by fossil fuels—first coal, then oil and gas. Pollution and exploitation took on new dimensions as the industrial world turned to the extraction of raw materials from all over the world (hence the desire for colonies). In time, every part of Earth was impacted by this revolution and continues to be even today. As a result, we are now living in a time of great population and economic expansion, with all of the environmental problems we outlined in Chapter 1 and will be dealing with in the rest of this text.

In this historical progression, we see natural systems being displaced by the **human system**, a term we will use to refer to our total system, including animal husbandry, agriculture, and all human developments. If the human system functioned as a true ecosystem, there might be an argument for its sustainability. We could see the process of human domination of the environment as one set of ecosystems displacing others, a phenomenon that does occur occasionally in nature.

Indeed, the human system does have some features in common with natural ecosystems; the series of trophic levels from crop producers to human consumers is an example. But, in other respects, it is far off the mark—failing to break down and recycle its "detritus" (trash, chemical wastes) and other byproducts and suffering the consequences of pollution is one example. Remaining highly

ETHICS

CAN ECOSYSTEMS BE RESTORED?

The human capacity for destroying ecosystems is well established. To some degree, however, we also have the capacity to restore them. In many cases, restoration simply involves stopping the abuse. For example, it has been found that water quality improves, and fish and shellfish gradually return to previously polluted lakes, rivers, and bays, after the input of pollutants is curtailed. Similarly, forests may gradually return to areas that have been cleared. Humans can speed up this process by seeding, planting seedling trees, and reintroducing populations of fish and animals that have been eliminated.

In some cases, however, specific ecosystems have been eliminated or disturbed to such an extent that they require the efforts of a new breed of scientist, the restoration ecologist. Two ecosystem types that have suffered most from removal or conversion are the prairie and wetland ecosystems. The potential for restoration of any ecosystem rests on three assumptions: (a) abiotic factors must have remained unaltered or can be returned to their original state; (b) viable populations of the species involved remain in existence, and (c) the ecosystem has not been upset by introduction of one or more foreign species that cannot be eliminated, and that may preclude the survival of reintroduced native species. If these conditions are met, restoration efforts have the potential of successful restoration to some semblance of the original ecosystem.

For example, let us assume that the Nature Conservancy, a land-trust organization, has acquired land in the Great Plains and wishes to restore the prairie that once flourished there. The problems are many: Lack of grazing and regular fires have led to much woody vegetation; exotic species abound in the region and can continuously rain seeds on the experimental prairie; there may be no remnants of the original prairie grasses and herbs on the site. How to proceed? First, do an inventory of what is on the site and, from historical records, what used to be present. Then, develop a hypothetical model of the desired ecosystem in structural and functional terms. Set goals for the restoration efforts by defining the desired future condition of the site. Then, design an implementation plan that can convert the goals into specific actions. For example, it may be necessary to completely remove all herbaceous vegetation with herbicides and plow and plant the land with native grasses. Then, to maintain the prairie, it may be necessary to conduct regular burnings of the prairie or, in the case of larger landholdings, to introduce bison. Finally, it is necessary to monitor the results and frequently make midcourse corrections to the implementation plan.

Why should we be involved in restoring ecosystems? Restoration ecologists Steven Apfelbaum and Kim Chapman cite several compelling reasons in a recent article on ecological restoration[1]: First, for aesthetic reasons: Natural ecosystems are often beautiful, and the restoration of something beautiful and pleasing to the eye is a worthy project that can be uplifting to many people; second, for the benefit of human use: The ecosystem services of a wetland, for example, can be restored and enjoyed by present and future generations; third, for the benefit of the species and ecosystems themselves: Nature has value and a right to continued existence. It can be argued that people should act to preserve and restore ecosystems and species in order to preserve that right. Do you find these reasons compelling?

---

[1] "Ecological Restoration: A Practical Approach," Ch. 15 in *Ecosystem Management: Applications for Sustainable Forest and Wildlife Resources*, ed. by Mark S. Boyce and Alan Haney (New Haven: Yale University Press, 1997).

▶ **FIGURE 2–24** *Industrial Revolution.* The Industrial Revolution began in England in the 1800s. Coal was the energy source, and economic growth and pollution were the consequences.

dependent on fossil fuels and thereby suffering the buildup of carbon dioxide in the atmosphere is another. Another revolution is needed.

Environmental Revolution. In Chapter 1 we referred to the fact that a "paradigm shift" needs to occur, one that leads to a worldview characterized by stewardship as its leading ethic, and sustainability as its goal. Some observers have referred to this shift as the **Environmental Revolution**, recognizing that all of the changes that will be required to move the human system from its present state to one that is sustainable are indeed revolutionary. We shall be exploring those changes in succeeding chapters. Revolutions suggest overthrow, and indeed what is involved is an overthrow of reigning worldviews that perpetuate a "business-as-usual" attitude toward the environment. This does not have to be a violent revolution; it could take place so gradually and quietly that it would take a future generation to look back

and realize that a major revolution has taken place. Yet it appears that the options are limited as we look to the future: We can choose to undergo the necessary changes to achieve sustainability by planning properly and learning as we go, or we can ignore the signs of unsustainability and move toward a gradual disintegration of biological systems and the global atmosphere. At some point the environmental revolution will be thrust upon us by the realities of limits of the environment to support an irresponsible human population. Preventing this from happening is the subject of Vice President Al Gore's book *Earth in the Balance.* He makes the case for "… bold and unequivocal action: *we must make the rescue of the environment the central organizing principle for civilization.* Whether we realize it or not, we are now engaged in an epic battle to right the balance of our earth, and the tide of this battle will turn only when the majority of people in the world become sufficiently aroused by a shared sense of urgent danger to join an all-out effort."

ENVIRONMENT ON THE WEB

CHANGING TROPHIC STRUCTURE IN ESTUARIES AFFECTED BY HUMAN ACTIVITIES

Estuaries—the zones where rivers empty into the ocean—are among the most productive ecosystems on Earth, with the most complex food webs. The combination of terrestrial habitat, fresh water and saltwater, the tidal flux of nutrients and other materials, and the flushing action of ocean currents create a system in which many habitat types can coexist. This diversity of habitat in turn allows the coexistence of a wide range of species that would not normally be found together.

Estuaries are also a focus for impacts from human activities, like dam building, in upstream areas. Such practices can fundamentally alter downstream ecosystems through habitat degradation, blockage of migratory routes, and increased pollution levels.

Think, for example, about the food web of the harp seal. Although it is not strictly speaking an estuarine species, the harp seal is tied to a wide range of organisms at several trophic levels, some of them estuarine in adult or juvenile forms. (A migratory species, the harp seal spends time both in near-shore waters and in offshore waters and thus is tied to different marine ecosystems.) This food web is complex, and perhaps because of its complexity is able to withstand minor stresses imposed by fluctuating water-quality conditions, temperature, and flows. However, when major estuarine flow and sedimentation changes occur, as would be the case when a dam is built in a river that flows to the coast, habitat alteration may be so significant that some species no longer thrive and may die out locally. In particular, as the reduction of sediment loads to downstream estuaries decline, so do the number of organisms that depend on it for habitat through some or all of their life cycles. These can include sediment-dwelling worms and other invertebrates, but also higher level consumers, such as fish, which require a particular type of substrate for effective spawning and rearing of young.

Even modest changes in the number and abundance of sediment-dwelling polychaete worms, for example, can have a "snowball" effect, reducing the number of flatfish and sand lance (a small fish). As a species at higher trophic levels fail to find enough prey, their own numbers will decline, while competing species, such as capelin and redfish, flourish. The net result is a simplified food web, now dependent on a few "keystone" species. And although the harp seal itself, which feeds on a variety of prey, may be largely unaffected in such a system, the food web overall is now much more vulnerable to the effects of other temperature, climatic, or chemical stressors.

Web Explorations

The Environment on the Web activity for this essay describes how human activities like dam building can influence the ecology of rivers and estuaries. Go to the Environment on the Web activity (select Chapter 2 at http://www.prenhall.com/nebel) and learn for yourself:

1. how infrared images can reveal the nature and extent of habitat alteration in estuaries;
2. about human influences on the development of hydrologic systems in Everglades National Park, Florida; and
3. how human activities affect food webs and aquatic habitat.

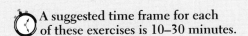 **A suggested time frame for each of these exercises is 10–30 minutes.**

REVIEW QUESTIONS

1. What is the difference between the biotic community and the abiotic environmental factors of an ecosystem?
2. Define and compare the terms species, population, and ecosystem.
3. In relation to the definition of an ecosystem, what is an ecotone, landscape, biome, and biosphere?
4. Identify and describe the biotic and the abiotic components of the biome of your region.
5. Name and describe the roles of the three main trophic categories that comprise the biotic structure of every ecosystem. Give examples of organisms from each category.
6. How do the terms "organic" and "inorganic" relate to the biotic and abiotic components of an ecosystem?
7. Name and describe the attributes of the two categories into which all organisms can be divided.
8. Give four categories of consumers in an ecosystem and the role that each plays.
9. Give similarities and differences between detritus feeders and decomposers in terms of what they do, how they do it, and the kinds of organisms involved in each category.

10. Differentiate between the concepts food chain, food web, and trophic levels.
11. Relate the concept of the biomass pyramid to the fact that all heterotrophs are dependent upon autotrophic production.
12. Describe three nonfeeding relationships that exist between organisms.
13. How is competition among different species of an ecosystem reduced?
14. Differentiate between the two types of abiotic factors. What is the effect on a population when any abiotic factor shifts from the optimum to the limit of tolerance and beyond? What things in addition to abiotic factors may act as limiting factors?
15. Describe how differences in climate cause the Earth to be distributed into six major biomes.
16. What are three situations that might cause microclimates to develop within an ecosystem?
17. What is significant about each of the following revolutions: Neolithic, Industrial, and Environmental?

THINKING ENVIRONMENTALLY

1. From local, national, and international news, compile a list of the many ways humans are altering abiotic and biotic factors on a local, regional, and global scale. Analyze ways that local changes may affect ecosystems on larger levels and ways that global changes may affect local levels.
2. Write a scenario of what would happen to an ecosystem or to the human system in the event of one of the following: (a) all producers were killed through losses of soil fertility or toxic contamination; (b) all parasites were eliminated; (c) decomposers/detritus feeders were eliminated. Support all of your statements with reasons drawn from your understanding of the way ecosystems function.

3. Consider the various kinds of relationships humans have with other species, both natural and domestic. Give examples of relationships that benefit humans but harm other species; that benefit both humans and the other species; that benefit the other species but harm humans. Give examples where the relationship may be changing—for instance, from exploitation to protection. Discuss the ethical issues involved in changing relationships.
4. Can the human system be modified into a sustainable ecosystem in balance with (i.e., preserving) other natural ecosystems without losing the benefits of modern civilization?

WEB REFERENCES

On-line resources for this chapter can be found on the World Wide Web at: **http://www.prenhall.com/nebel**. Click on Chapter 2 on the chapter selector.

3

Ecosystems:
How They Work

∿∿∿∿∿∿∿

Key Issues and Questions

1. All the elements that comprise living things come from the environment. What are these key elements? Where is each found?

2. All chemical reactions taking place in living things involve energy. What are the different forms of energy, and what are the laws that govern energy exchanges?

3. Photosynthesis and cell respiration are the two fundamental biological processes. What matter and energy changes occur in these two processes? Relate them to the dynamics of ecosystems.

4. Detritivores and decomposers promote the breakdown of organic matter. What are they, and how important is their role?

5. The flow of energy is one of the vital processes that occurs in all ecosystems. Describe how energy flows in terms of trophic levels. How efficient are the transfers of energy?

6. The recycling of elements is another vital functional process occurring in all ecosystems. Describe the biogeochemical cycles for carbon, phosphorus, and nitrogen. How have humans impacted these three cycles?

7. The goods and services performed by natural ecosystems are essential to human survival. What is their overall value, and of what significance is it to measure this value?

"There is a place on earth where it is still the morning of life and the great herds still run free." With these words, James Earl Jones begins the narration of the stirring IMAX/OMNIMAX production *Africa: The Serengeti*. Pictured in the opening photo, the Serengeti is a vast tropical savannah ecosystem of 25,000 km^2 in northern Tanzania and southern Kenya. The rain falls bimodally: The short rains are normally in November and December, and the long rains are during March through May. There is a rainfall gradient from the drier southeastern plains (50 cm/year) to the wet northwest in Kenya (120 cm/year). The southeastern plains are a treeless grassland, and as the volcanic soils thicken to the north and west, the grasses gradually shift to woodlands—mixed woods and grassy patches. Large herbivores (wildebeest, zebra, and Thompson's gazelle) dominate the ecosystem. Large groups of over 1.5 million animals are common on the plains during the rainy seasons.

As the rain fails, the large herds move to the woodlands in the north, where forage is available year-round. There they face heavier predation from lions and hyenas, which are more tied to the woodlands because of the need for cover for raising young and for successful stalking. This migration is energetically costly; the animals travel 200 or more kilometers and back again each year. Why do they do it if, as it seems, there is sufficient forage in the woodlands? The answer is still being investigated, but evidence points to two possible factors: 1) The vegetation in the plains is high in phosphorus, which the herbivores need for successful growth and lactation. Their annual presence there maintains the high phosphorus content as their wastes are broken down and nutrients are returned to the soils. 2) The presence of high numbers of predators in the woodlands may force the flocks to migrate to the grasslands, where they are less vulnerable, especially when giving birth to their young.

◀ ***The Serengeti.*** A female lion looks over a herd of impala in Masai Mara National Park, Kenya, part of the Serengeti ecosystem.

TABLE 3-1 Principles of Ecosystem Sustainability

For Sustainability:

- Ecosystems use sunlight as their source of energy.
- Ecosystems dispose of wastes and replenish nutrients by recycling all elements.
- The size of consumer populations is maintained such that overgrazing and other forms of overuse do not occur.
- Biodiversity is maintained.

In this picture of the Serengeti it is clear that producers, herbivores, carnivores, and scavengers/detritus feeders are interacting in this ecosystem in a sustainable set of relationships—although the system is not without its human impacts. We explore in this chapter how ecosystems like the Serengeti work at the fundamental level of chemicals and energy. We will also gain some understanding of how natural ecosystems sustain human life, look at an estimate of what ecosystem goods and services are worth to humankind, and consider how best to manage such vital resources. Our look at how ecosystems work will reveal two underlying principles that enable natural systems to be sustainable, and it will provide insight into the pathways that we must take to make our human system sustainable (Table 3-1). Two principles of sustainability will be discussed in this chapter, and the remaining two principles will be covered in Chapter 4.

3.1 Matter, Energy, and Life

The basic building blocks of all **matter** (all gases, liquids, and solids in both living and nonliving systems) are **atoms**. Only 92 different kinds of atoms occur in nature, and these are known as the naturally occurring **elements**. In addition, physicists have created 14 more in the laboratory, but these are unstable and break down into simpler elements (see Table C-1, p. 609).

How can these relatively few building blocks make up the innumerable materials of our world, including the tissues of living things? Like blocks, the elements can be put together to build a great variety of things. Also, like blocks, nature's materials can be taken apart into their separate constituent atoms, and the atoms can then be reassembled into different materials. All chemical reactions, whether they occur in a test tube, in the environment, or inside living things, and whether they occur very slowly or very fast, involve rearrangements of atoms to form different kinds of matter.

Atoms do not change during the disassembly and reassembly of different materials. A carbon atom, for instance, will always remain a carbon atom. Furthermore, atoms are not created or destroyed during any chemical reactions. This constancy of atoms is regarded as a fundamental natural law, the *law of conservation of matter*.

On the chemical level, then, the cycle of growth, reproduction, death, and decay of organisms can be seen as a continuous process of taking various atoms from the environment, assembling them into living organisms (growth) and then disassembling them (decay) and repeating the process. Driving the process is the irresistible, genetically predisposed urge of living things to grow and reproduce.

Which atoms make up living organisms? Where are they found in the environment? How do they become part of living organisms? We answer these questions next.

Matter in Living and Nonliving Systems

A more detailed discussion of atoms—how they differ from one another, how they bond to form various gases, liquids, and solids, and how we use chemical formulas to describe different chemicals—is given in Appendix C (p. 608). Studying that appendix first may give you a better comprehension of the material we are about to cover. At the very least, the definitions of two terms are essential: *molecule* and *compound*.

A **molecule** refers to *any* two or more atoms bonded together in a specific way. The properties of a material are dependent on the specific way in which atoms are bonded to form molecules as well as on the atoms themselves. Similarly, a **compound** refers to any *two or more different kinds* of atoms bonded together. (Note the distinction that a molecule may consist of two or more of the *same kind*, as well as different kinds, of atoms bonded together.) A compound always implies that at least two different kinds of atoms are involved. For example, the fundamental units of oxygen gas, which consist of two oxygen atoms bonded together, are molecules but not a compound. Water, on the other hand, can be referred to as either a molecule or a compound, since the fundamental units are two hydrogen atoms bonded to an oxygen atom. Some further distinctions are given in Appendix C.

The key elements in living systems (and their chemical symbols) are carbon (C), hydrogen (H), oxygen (O), nitrogen (N), phosphorus (P), and sulfur (S). You can remember them by the acronym N. CHOPS. These six elements are the building blocks of all the organic molecules that make up the tissues of plants, animals, and microbes. We have said that growth and decay can be seen as a process of atoms moving from the environment into living things and returning to the environment. By looking at the chemical nature of air, water, and minerals, we shall see where our six key elements and others occur in the environment (Table 3-2).

TABLE 3-2 Elements Found in Living Organisms and Locations in the Environment

Element (Kind of Atom)	Symbol	Biologically Important Molecule or Ion in Which the Element Occurs[a]			Location in the Environment		
		Name	Formula	Air	Dissolved in Water	Some Rock and Soil Minerals	
Carbon	C	Carbon dioxide	CO_2	X	X	$X(CO_3^-)$	
Hydrogen	H	Water	H_2O		(Water itself)		
Atomic oxygen (required in respiration)	O	Oxygen gas	O_2	X	X		
Molecular oxygen (released in photosynthesis)	O_2	Water	H_2O		(Water itself)		
Nitrogen	N	Nitrogen gas	N_2	X	X	Via fixation	
		Ammonium ion	NH_4^+		X	X	
		Nitrate ion	NO_3^-		X	X	
Sulfur	S	Sulfate ion	SO_4^{2-}		X	X	
Phosphorus	P	Phosphate ion	PO_4^{3-}		X	X	
Potassium	K	Potassium ion	K^+		X	X	
Calcium	Ca	Calcium ion	Ca^{2+}		X	X	
Magnesium	Mg	Magnesium ion	Mg^{2+}		X	X	
Trace Elements[b]							
Iron	Fe	Iron ion	Fe^{2+}, Fe^{3+}		X Fe^{2+} only	X	
Manganese	Mn	Manganese ion	Mn^{2+}		X	X	
Boron	B	Boron ion	B^{3+}		X	X	
Zinc	Zn	Zinc ion	Zn^{2+}		X	X	
Copper	Cu	Copper ion	Cu^{2+}		X	X	
Molybdenum	Mo	Molybdenum ion	Mo^{2+}		X	X	
Chlorine	Cl	Chloride ion	Cl^-		X	X	

NOTE: These elements are found in *all* living organisms—plants, animals, and microbes. Some organisms require certain elements in addition to the ones given. For example, humans require sodium and iodine.

[a] A molecule is a chemical unit of two or more atoms bonded together. An ion is a single atom or group of bonded atoms that has acquired a positive or negative charge as indicated.
[b] Only small or trace amounts of these elements are required.

The lower atmosphere is a mixture of molecules of three important gases—oxygen (O_2), nitrogen (N_2), and carbon dioxide (CO_2)—along with water vapor and trace amounts of several other gases that have no immediate biological importance (Fig. 3–1). Saying that air is a **mixture** means that there is no chemical bonding between the molecules involved. Indeed, it is this lack of connection between molecules that results in air being gaseous. Attraction, or bonding, between molecules results in liquid or solid states.

While air is a potential source of carbon, oxygen, and nitrogen for all organisms, the source of the key element hydrogen is water. Each molecule of water consists of two hydrogen atoms bonded to an oxygen atom, as indicated by the formula for water: H_2O. A weak attraction between water molecules is known as *hydrogen bonding*. At temperatures below freezing, hydrogen bonding holds the molecules in position with respect to one another, and the result is a solid (ice or snow). At temperatures above freezing, but below vaporization (evaporation), hydrogen bonding still holds the molecules close, but allows them to move around one another, producing the liquid state. Vaporization occurs as hydrogen bonds break and water molecules move into the air independently. With a lowering of temperature, all these changes in state go in the reverse direction (Fig. 3–2). We reemphasize that regardless of the changes in state, the water molecules themselves retain their basic structure of two hydrogen atoms bonded to an oxygen atom. It is only the relationship between the molecules that changes.

All the other elements required by living organisms, as well as the 72 or so elements that are not required, are found in various rock and soil minerals. A **mineral** refers to any hard, crystalline, inorganic material of a given chemical composition. Most rocks are made up of relatively small crystals of two or more minerals, and soil generally consists of particles of many different minerals. Each mineral is made up of dense clusters of two or more kinds of atoms bonded together by an attraction between positive and negative charges on the atoms, as explained in Appendix C and Fig. 3–3.

There are simple but significant interactions between air, water, and minerals. Gases from the air and ions (charged atoms) from minerals may dissolve in water. Therefore, natural water is inevitably a *solution* containing variable amounts of dissolved gases and minerals. This solution is constantly subject to change, as any dissolved substances in it may be removed by various processes, or additional materials may dissolve in it. Molecules of water enter the air by evaporation and leave it by

Clean, dry air is a mixture of molecules of three important gases.

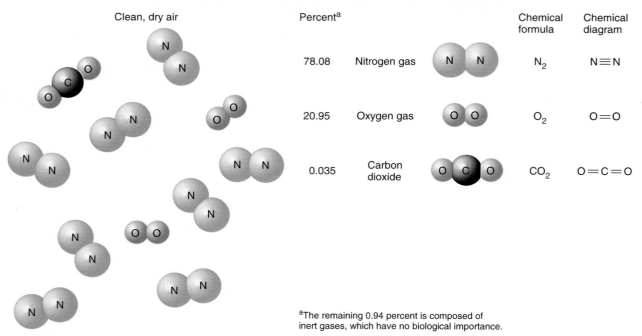

^aThe remaining 0.94 percent is composed of
inert gases, which have no biological importance.

▲ **FIGURE 3–1** *The major gases of clean, dry air.* From a biological point of view, the three most important
gases of the lower atmosphere are nitrogen, oxygen, and carbon dioxide.

means of condensation and precipitation (see the hydrologic cycle, Chapter 9). Thus, the amount of moisture in air is constantly fluctuating. Wind may carry a certain amount of dust or mineral particles, and this amount is also changing constantly,

since the particles gradually settle out from the air. The various interactions are summarized in Fig. 3–4.

By contrast to the relatively simple molecules that occur in the environment (for example, CO_2, H_2O, N_2),

▶ **FIGURE 3–2** *Water and its three states.*
(a) Water consists of molecules, each of which is formed by two hydrogen atoms bonded to an oxygen atom (H_2O). (b) In water vapor, the molecules are separate and independent. (c) In liquid water, the weak attraction between water molecules known as hydrogen bonding gives the water its liquid property. (d) At freezing temperatures, hydrogen bonding holds the molecules firmly, giving the solid state—ice.

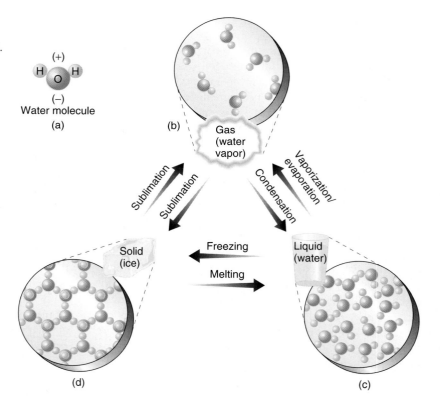

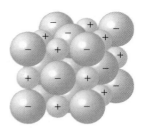

▲ **FIGURE 3–3** *Minerals.* Minerals (hard crystalline compounds) are composed of dense clusters of atoms of two or more elements. The atoms of most elements gain or lose one or more electrons, becoming negative (-) or positive (+) ions. The ions are held together by an attraction between positive and negative charges.

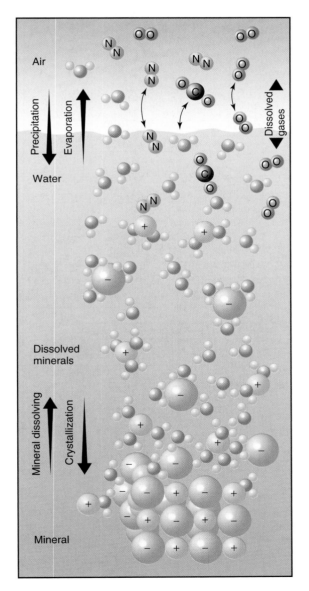

▲ **FIGURE 3–4** *Interrelationship among air, water, and minerals.* Minerals and gases dissolve in water, forming solutions. Water evaporates into air, causing humidity. These processes are all reversible: Minerals in solution recrystallize, and water vapor in the air condenses to form liquid water.

the key atoms in living organisms (C, H, O, N, P, S) bond into very large, complex molecules known as proteins, carbohydrates (sugars and starches), lipids (fatty substances), and nucleic acids. Some of these molecules may contain millions of atoms, and their potential diversity is infinite. Indeed, the diversity of living things is a reflection of the diversity of such molecules.

The molecules that make up the tissues of living things are constructed mainly from carbon atoms bonded together into chains with hydrogen atoms attached. Oxygen, nitrogen, phosphorus, and sulfur may be present also, but the key common denominator is carbon-carbon and/or carbon-hydrogen bonds (Fig. 3–5). Recall (Chapter 1) that material making up the tissues of living organisms is referred to as *organic*. Hence, these carbon-based molecules, which make up the tissues of living organisms, are called **organic molecules**. (Don't miss the connection between the words *organic* and *organism*.) **Inorganic**, then, refers to molecules or compounds with neither carbon-carbon nor carbon-hydrogen bonds.

Causing some confusion is the fact that all plastics and countless other human-made compounds are based on carbon-carbon bonding and are, chemically speaking, organic compounds. We resolve this confusion by referring to the compounds of living organisms as **natural organic compounds** and the human-made ones as **synthetic organic compounds**.

In conclusion, we can see that the elements essential to life (C, H, O, and so on) are present in air, water, or minerals in relatively simple molecules. In living *organ*isms, on the other hand, they are *organized* into very complex *organ*ic molecules. These organic compounds in turn make up the various parts of cells, which make up the tissues and organs of the body (Fig. 3–6). Growth and reproduction, then, may be seen as using the atoms from simple molecules in the environment to construct the complex organic molecules of an organism. Decomposition and decay may be seen as the reverse process. We shall look at each of these processes in more detail later in the chapter; first, however, we must consider another factor: *energy.*

▼ **FIGURE 3–5** *Organic molecules.* The organic molecules making up living organisms are larger and more complex than the inorganic molecules found in the environment. Glucose and cystine show this relative complexity.

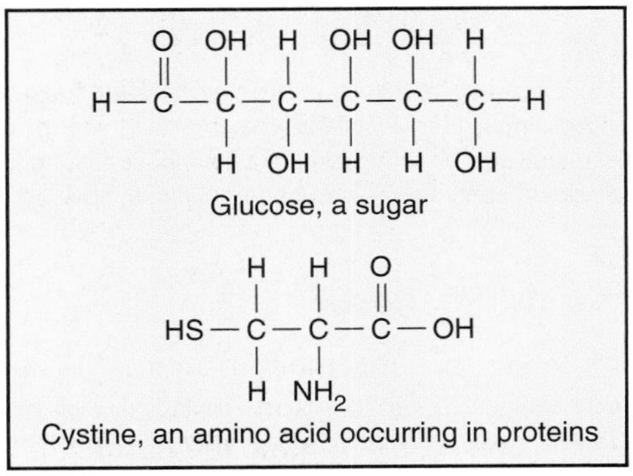

Glucose, a sugar

Cystine, an amino acid occurring in proteins

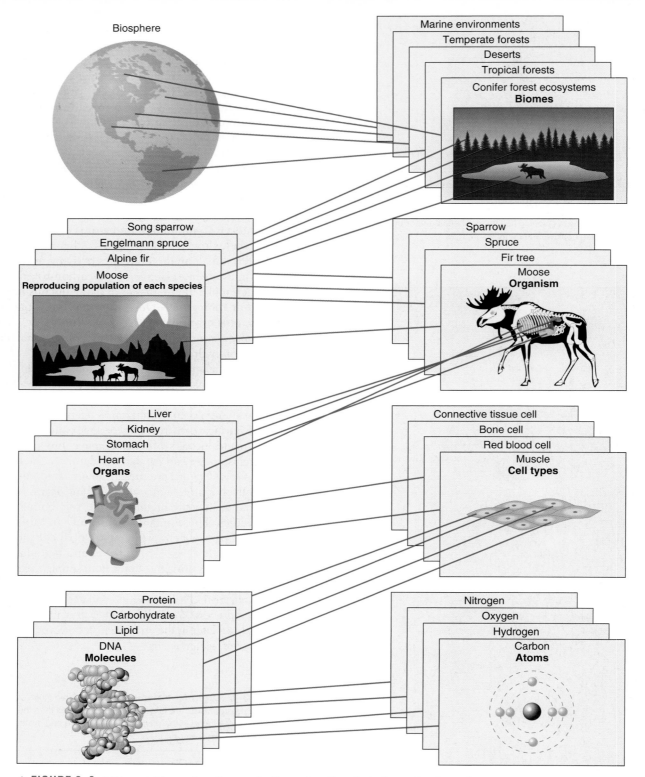

▲ FIGURE 3–6 *Life as a hierarchy of organization of matter.* In the inorganic sphere, elements are arranged simply in molecules of the air, water, and minerals. In living organisms, they are arranged in very complex organic molecules, which in turn make up cells that constitute tissues, organs, and, thus, the whole organism. Levels of organization continue up through populations, species, ecosystems, and, finally, the whole biosphere.

Energy Considerations

In addition to the rearrangement of atoms, chemical reactions also involve the absorption or release of energy. To grasp this concept, let us examine the distinction between matter and energy.

Matter and Energy. The universe is made up of *matter* and *energy*. A more technical definition of **matter** than the one given earlier in this chapter is anything that occupies space and has mass—that is, anything that can be weighed when gravity is present. This definition obviously covers all solids, liquids, and gases, and living as well as nonliving things.

Atoms are made up of protons, neutrons, and electrons, which in turn are made of still smaller particles. Thus, physicists debate about what is the most basic unit of matter. However, since atoms are the basic units of all elements and remain unchanged during chemical reactions, it is practical to consider them as the basic units of matter.

Light, heat, movement, and *electricity,* on the other hand, do not have mass, nor do they occupy space. (Note that heat, as used here, refers not to a hot object but to the heat energy we can feel radiating from the hot object.) These are the common forms of energy that we are familiar with. What do forms of energy have in common? They *affect* matter, causing changes in its *position* or its *state.* For example, the release of energy in an explosion causes things to go flying, a change in position. Heating water causes it to boil and change to steam, a change in state. On a molecular level, changes in state may be seen as movements of atoms or molecules. For instance, the degree of heat energy is actually a measure of the relative vibrational motion of the atoms and molecules of the substance. Therefore, **energy** is the ability to move matter.

Energy is commonly divided into two major categories: *kinetic* and *potential* (Fig. 3–7). **Kinetic energy** is energy in action or motion. Light, heat energy, physical motion, and electrical current are all forms of kinetic energy. **Potential energy** is energy in storage. A substance or system with potential energy has the capacity, or *potential,* to release one or more forms of kinetic energy. A stretched rubber band, for

example, has potential energy; it can send a paper clip flying. Numerous chemicals, such as gasoline and other fuels, release kinetic energy—heat energy, light, and movement—when ignited. The potential energy contained in such chemicals and fuels is called **chemical energy**.

Energy may be changed from one form to another in innumerable ways. How many examples can you think of in addition to those shown in Fig. 3–8? Besides seeing that potential energy may be converted to kinetic energy, it is especially important to recognize that kinetic energy may be converted to potential energy, as in charging a battery or pumping water into a high-elevation reservoir. We shall see shortly that photosynthesis does just that.

Because energy does not have mass or occupy space, it cannot be measured in units of weight or volume, but it can be measured in other kinds of units. One of the most common units is the **calorie**, which is defined as the amount of heat required to raise the temperature of 1 gram (1 milliliter) of water 1 degree Celsius. Since this is a very small unit, it is frequently more convenient to speak in terms of kilocalories (1 kilocalorie = 1000 calories), the amount of heat required to raise 1 liter (1000 milliliters) of water 1 degree Celsius. (Kilocalories are sometimes denoted as "Calories" with a capital "C." Food Calories, which are a measure of the energy in given foods, are actually kilocalories.) Any form of energy can be measured in calories by converting it to heat energy and measuring that heat in terms of a rise in the temperature of water. Temperature is a measurement of the molecular motion in a substance caused by the kinetic energy present.

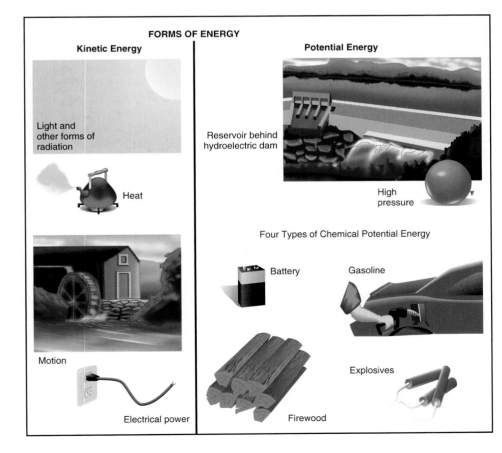

FORMS OF ENERGY

Kinetic Energy

Light and other forms of radiation

Heat

Motion

Electrical power

Potential Energy

Reservoir behind hydroelectric dam

High pressure

Four Types of Chemical Potential Energy

Battery

Gasoline

Firewood

Explosives

◀ **FIGURE 3–7** *Forms of energy.*
Energy is distinct from matter in that it neither has mass nor occupies space. It has the ability to act on matter, changing the position of the matter and/or its state. Kinetic energy is energy in one of its active forms. Potential energy refers to systems or materials that have the potential to release kinetic energy.

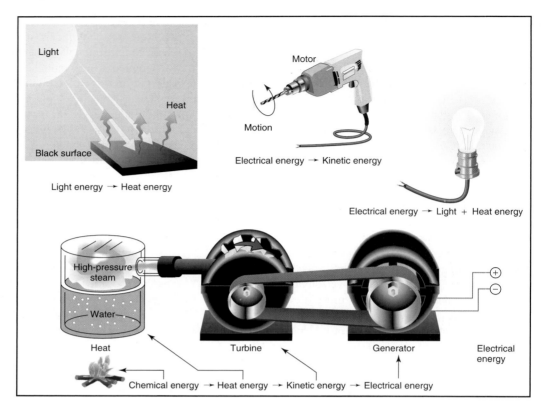

▶ **FIGURE 3–8** *Energy conversions.* Any form of energy can be converted to any other form, except heat energy. Heat is a form of energy that flows from one system or object to another because the two are at different temperatures; therefore, it can be transferred only to something cooler.

We define energy as the ability to move matter. Conversely, no change in the movement of matter can occur *without* the absorption or release of energy. Indeed, no change in matter—from a few atoms coming together or apart in a chemical reaction to a major volcanic eruption—can be separated from respective changes in energy.

Energy Laws: Laws of Thermodynamics. Knowing that energy can be converted from one form to another has led numerous would-be inventors over the years to try to build machines or devices that would produce more energy than they consumed. A common idea is to use the output from a generator to drive a motor that, in turn, drives the generator to keep the cycle going and yields additional power in the bargain. Unfortunately, all such devices have one feature in common: They don't work. When all the inputs and outputs of energy are carefully measured, they are found to be *equal.* There is no net gain or loss in total energy. This observation is now accepted as a fundamental natural law, **the law of conservation of energy**, also called the **first law of thermodynamics**: *Energy is neither created nor destroyed, but may be converted from one form to another.* The law is also commonly stated as "You can't get something for nothing."

Fanciful "energy generators" fail for two reasons: First, in every energy conversion, a portion of the energy is converted to heat energy (thermal infrared). Second, heat always flows toward cooler surroundings. There is no way of trapping and recycling heat energy, since it can flow only "downhill" toward a cooler place. Consequently, in the absence of energy inputs, any and every system will sooner or later come to a stop as its

energy is converted to heat and lost. This is now accepted as another natural law, the **second law of thermodynamics**. Basically, the second law says that *in any energy conversion, you will end up with less usable energy than you started with.* So, not only can you not get something for nothing (the first law), you can't even break even.

A principle that underlies the loss of heat is the principle of increasing *entropy*. **Entropy** refers to the degree of disorder: Increasing entropy means increasing disorder. The principle is that without energy inputs, everything goes in one direction only: toward increasing entropy. This principle of ever-increasing entropy is most readily apparent in the fact that all human-made things tend to deteriorate. We never observe the reverse—a run-down building renovating itself, for example. Students often like to speak of the increasing disorder of their dormitory rooms as the semester wears on as an example of entropy.

The conversion of energy and the loss of heat are both aspects of increasing entropy. Heat energy is the result of the random vibrational motion of atoms and molecules. Thus, it is the lowest (most disordered) form of energy, and its flow to cooler surroundings is a way for that disorder to spread. Therefore, the second law of thermodynamics is nowadays more generally stated as: *Systems will go spontaneously in one direction only—toward increasing entropy.* The second law also says that systems will go spontaneously only toward *lower* potential energy, a direction that releases heat from the systems (Fig. 3–9).

Very important in the statement of the second law is the word *spontaneously.* It is possible to pump water uphill, charge

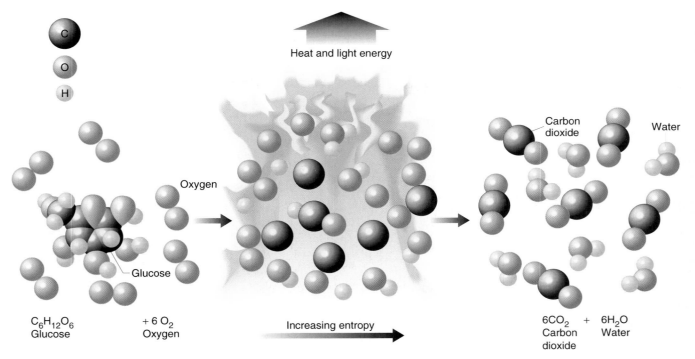

▲ **FIGURE 3–9** *Entropy.* Systems go spontaneously only in the direction of increasing entropy. When glucose, the building-block molecule of wood, is burned, heat is released, and the atoms become more and more disordered, both aspects of increasing entropy. The fact that wood will burn but not form spontaneously is an example of the second law of thermodynamics.

a battery, stretch a rubber band, compress air, or otherwise increase the potential energy of some system. However, inherent in such words as *pump, charge, stretch,* and *compress* is the fact that energy is being put into the system; in contrast, the opposite direction, which releases energy, occurs spontaneously.

The conclusion is that whenever you see something gaining potential energy, you should realize that that energy is being obtained from somewhere else (the first law). Moreover, the amount of energy lost from that somewhere else is greater than the amount gained (the second law). Let us now relate these concepts of matter and energy to organic molecules, organisms, ecosystems, and the biosphere.

Energy Changes in Organisms and Ecosystems

All organic molecules, which make up the tissues of living organisms, contain *high potential energy.* This is evident from the simple fact that they burn: The heat and light of the flame are the potential energy being released as kinetic energy. On the other hand, try as you might, you will not be able to get energy by burning inorganic molecules, such as carbon dioxide, water, or mineral compounds that occur in nature. Indeed, many of these materials are used as fire extinguishers. This extreme nonflammability is evidence that such materials have very *low potential energy.* Thus, the production of organic material from inorganic material involves a *gain* in potential energy. Conversely, the breakdown of organic matter involves a *release* of energy.

In this relationship between the formation and breakdown of organic matter and the gain and release of energy,

we can see the energy dynamics of ecosystems. Producers (green plants) play the role of making high-potential-energy organic molecules for their bodies from low-potential-energy raw materials in the environment—namely, carbon dioxide, water, and a few dissolved compounds of nitrogen, phosphorus, and other elements. Such "uphill" conversion is made possible by the light energy absorbed by chlorophyll. On the other hand, all consumers, detritus feeders, and decomposers obtain their energy requirement for movement and other body functions from feeding on and breaking down organic matter made by producers (Fig. 3–10). Let us now look at this *energy flow* in somewhat more detail for each category of organisms.

Producers. Recall from Chapter 2 that producers are green plants, which use light energy in the process of *photosynthesis* to make sugar (glucose, stored chemical energy) from carbon dioxide and water and release oxygen gas as a byproduct. The process is expressed by the following formula:

PHOTOSYNTHESIS

$$6\,CO_2 + 12\,H_2O \rightarrow C_6H_{12}O_6 + 6\,O_2 + 6\,H_2O$$

| carbon dioxide (gas) | water | light energy input | glucose | oxygen (gas) | water |

(low potential energy)   (high potential energy)

The kinetic energy of light is absorbed by chlorophyll in the cells of the plant and used to remove the hydrogen

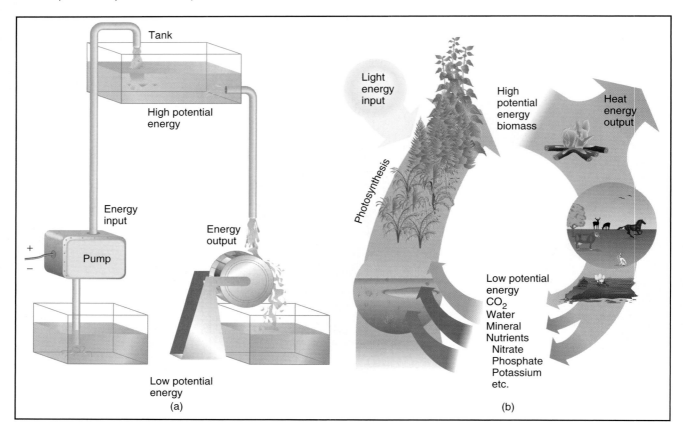

▲ **FIGURE 3–10** *Storage and release of potential energy.* (a) A simple physical example of the storage and release of potential energy. (b) The same principle of storage and release of potential energy shown in ecosystems.

atoms from water (H$_2$O) molecules. The hydrogen atoms join carbon atoms coming from carbon dioxide as the carbons join a chain to begin forming a glucose molecule. After the removal of hydrogen from water, the oxygen atoms that remain combine with each other to form oxygen gas, which is released into the air. Water is seen on both sides of the equation because twelve molecules are consumed and six molecules are newly formed during photosynthesis.

The key energy steps in photosynthesis are removing the hydrogen from water molecules and joining carbon atoms together to form the high-potential-energy carbon-carbon and carbon-hydrogen bonds of glucose in place of the low-potential-energy bonds in water and carbon dioxide molecules. But the laws of thermodynamics are not violated or even strained in this process. Careful measurements show that the rate of photosynthesis (which determines the amount of glucose formed) is proportional to the intensity of light, and, at most, 2 calories of sugar are formed for each 100 calories of light energy falling on the plant. Thus, plants are not particularly efficient "machines" in performing this conversion of light energy to chemical energy.

The glucose produced in photosynthesis plays three roles in the plant. First, either by itself or along with nitrogen, phosphorus, sulfur, and other mineral nutrients absorbed by the plant's roots, glucose is the raw material used for making all the other organic molecules (proteins, carbohydrates, and

so on) that make up the stems, roots, leaves, flowers, and fruits of the plant. Second, the synthesis of all these organic molecules requires additional energy, as does the plant's absorption of nutrients from the soil and certain other functions. This energy is provided when the plant breaks down a portion of the glucose to release its stored energy in a process called *cell respiration*, which will be discussed shortly. Third, a portion of the glucose produced may be stored for future use. For storage, the glucose is generally converted to starch, as in potatoes, or to oils, as in seeds. These conversions are summarized in Fig. 3–11.

As the plants in ecosystems convert sunlight into new organic matter, they are "setting the table" for the components of other living ecosystems—the herbivores, carnivores, and decomposers we met in the food webs in the last chapter. Since they are creating new organic matter for the ecosystem, we refer to them as *primary* producers. Given suitable conditions and resources, the producers of an ecosystem will maintain their photosynthetic activity over time in the process we call *primary production*. The total amount of photosynthetic activity in producers is called *gross primary production*; subtracting the energy consumed by the plants themselves yields the *net primary production*. Thus, net primary production refers to the *rate* at which new organic matter is made available to consumers in an ecosystem. In the example of the Serengeti, the grasses in the plains had to produce at the

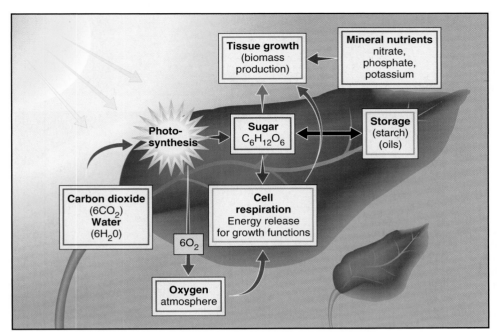

◀ **FIGURE 3–11** *Producers as chemical factories.* Using light energy, producers make glucose from carbon dioxide and water, releasing oxygen as a byproduct. Breaking down some of the glucose to provide additional chemical energy, they combine the remaining glucose with certain nutrients from the soil to form other complex organic molecules that the plant then uses for growth.

rate of 560 kg dry weight per km² per day in order to keep up with the rate at which they were being grazed.

Consumers—*Energy of Food.* Obviously, consumers need energy to move about and to perform such bodily functions as pumping blood. In addition, consumers need energy to synthesize all the molecules required for growth, maintenance, and repair of the body. Where does this energy come from? It comes from the breakdown of organic molecules of food (or of the body's own tissues if food is not available). About 60–90 percent of the food that we or other consumers eat and digest acts as "fuel" to provide energy.

First, the starches, fats, and proteins that you eat are digested in the stomach and/or intestine, which means that they are broken into simpler molecules—starches into sugar (glucose), for example. These simpler molecules are then absorbed from the intestine into the bloodstream and transported to individual cells of the body.

Inside each cell, organic molecules may be broken down through a process called **cell respiration** to release the energy required for the work done by that cell. Most commonly, cell respiration involves the breakdown of glucose, and the overall chemical equation is basically the reverse of that for photosynthesis:

CELL RESPIRATION

(An energy-releasing process)

$$C_6H_{12}O_6 \ + \ 6\,O_2 \quad \rightarrow \quad 6\,CO_2 \ + \ 6\,H_2O$$

Glucose	oxygen	energy released	carbon dioxide	water

(high potential energy) (low potential energy)

Again, the key point of cell respiration is to release the potential energy contained in organic molecules to perform

the activities of the organism. However, other aspects of the chemistry are also significant. Note that *oxygen* is released in photosynthesis, but in cell respiration it is *used* to complete the breakdown of glucose to carbon dioxide and water. Oxygen is absorbed through the lungs with every inhalation (or through the gills, in the case of fish) and is transported to all cells via the circulatory system. Carbon dioxide, which is formed as a waste product, moves from the cells into the circulatory system and is eliminated through the lungs (or gills) with every exhalation. The other byproduct, water, serves any of the body's needs for water, which reduces the need to drink water. A number of desert animals, which are adapted to conserve water, do not need to drink any water, because that produced by cell respiration is sufficient. However, the bodies of most animals, including ourselves, are less conserving of water; therefore, drinking additional water is necessary.

Again, in keeping with the laws of thermodynamics, the energy conversions involved in the body's use of the potential energy from glucose to do work are not 100% efficient. Considerable waste heat is produced, and this is the source of body heat. This heat output can be measured in cold-blooded animals and in plants, as well as in warm-blooded animals. It is more noticeable in warm-blooded animals only because they produce extra heat, via cell respiration, to maintain their warm body temperature.

The basis of weight gain or loss should become evident here also. Organic matter is broken down in cell respiration only as it is needed to meet the energy demands of the body; this is why your breathing rate, the outer reflection of cell respiration, varies with changes in your level of exercise and activity. If you consume more calories from food than your body needs, the excess is converted to fat and stored, and the result is a gain in weight. Conversely, the principle of dieting is to eat less and exercise more, to create an energy demand that exceeds the amount of energy

GLOBAL PERSPECTIVE

LIGHT AND NUTRIENTS: THE CONTROLLING FACTORS IN MARINE ECOSYSTEMS

Even though running on solar energy and recycling nutrients are basic principles of sustainability, light and nutrients are limiting factors in marine ecosystems. First, light is diminished as water depth increases, because even clear water absorbs light. The layer of water from the surface down to the greatest depth at which there is adequate light for photosynthesis is known as the **euphotic zone**. Below the euphotic zone, by definition, photosynthesis does not occur. In clear water, the euphotic zone may be as deep as 600 feet (200 m), but in very turbid (cloudy) water, it may be a matter of only a few centimeters. In coastal waters where the euphotic zone extends to the bottom, the bottom may support abundant plant life—that is, aquatic vegetation attached to or rooted in the bottom sediments. If the euphotic zone does not extend to the bottom, however, the bottom will be barren of plant life.

That the euphotic zone does not extend to the bottom does not preclude an ecosystem from existing in it. Phytoplankton—algae and photosynthetic bacteria that grow as single cells or in small groups of cells—can maintain themselves close to the surface in the euphotic zone. Phytoplankton support a diverse food web, from the zooplankton (small crustaceans, protozoans) that feed on them to many species of fish and sea mammals (whales and porpoises) at the higher trophic levels.

Also, an entire ecosystem operates in the cold, dark depths below the euphotic layer nourished by detritus raining down from above, and closer to the ocean floor, by vents and fissures that produce mineral-rich water and warmth.

In a phytoplankton-based system, nutrients dissolved in the water become critically important. If the water contains too few

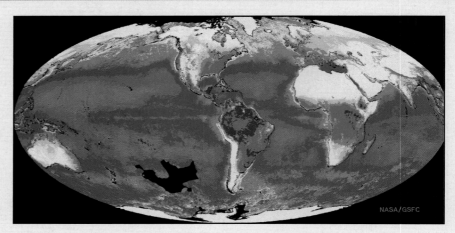

NASA/GSFC

Magenta—mid oceans: lowest productivity (0.1 mg chlorophyll/m^3 or less).
Red/orange—along coasts: highest productivity (10 mg chlorophyll/m^3 or more).

dissolved nutrients such as phosphorus or nitrogen compounds, the growth of phytoplankton and, hence, the rest of the ecosystem will be limited. If the bottom receives light, it may support vegetation despite nutrient-poor water because such vegetation draws nutrients from the bottom materials. Indeed, in some estuaries, nutrient-rich water is counterproductive to bottom vegetation, because the dissolved nutrients support the growth of phytoplankton, which makes the water turbid and shades out the bottom vegetation.

Let us put these concepts together to understand particular marine environments. The most productive areas of the ocean—the areas supporting the most abundant marine life of all sorts—are mostly found within 200 miles (300 km) of shorelines. This is true because either the bottom is within the euphotic zone and thus supports abundant bottom vegetation, or nutrients washing in from the land support an abundant primary production of phytoplankton.

In the open ocean, there is less and less marine life as one moves farther from shore. Indeed, marine biologists speak of most of the open ocean as being a "biological desert." The scarcity of life occurs both because the bottom is well below the euphotic zone and because the water is nutrient-poor.

The nutrients carried to the bottom with the settling detritus are released into solution by decomposers, thus making the bottom water nutrient-rich. This nutrient-rich bottom water is carried along by deep-running ocean currents. Where the currents hit underwater mountains or continental rims, the nutrient-rich water is brought to the surface. Phytoplankton flourish in these areas of **upwelling** (rising) nutrient-rich water and support a rich diversity of fish and marine mammals.

In sum, the world's oceans are far from being uniformly stocked with fish. By far, the richest marine fishing areas are continental shelves and regions of upwelling as shown on the accompanying map. Unfortunately, however, many of these areas are now being depleted by overfishing.

---

contained in the food being consumed. This imbalance forces the body to break down its own tissues to make up the difference, and the result is a weight loss. Of course, carried to an extreme, this imbalance leads to *starvation* and death when the body runs out of anything expendable to break down for its energy needs.

The overall reaction for cell respiration is the same as that for simply burning glucose. Thus, it is not uncommon to speak of "burning" our food for energy. Such

a breakdown of molecules is also called **oxidation**. The distinction between burning and cell respiration is that in cell respiration the oxidation takes place in about 20 small steps, so that the energy is released in small "packets" suitable for driving the functions of each cell. If all the energy from glucose molecules were released in a single "bang," as occurs in burning, it would be like heating and lighting a room with large firecrackers—energy, yes, but hardly useful energy.

▲ **FIGURE 3–12** *Animal wastes are plant fertilizer.* When consumers burn food to obtain energy, the waste products are the inorganic nutrients needed by plants. Here, dog urine has been deposited on a lawn. The ring of dark green grass is where the urine has been diluted to optimal concentration; the grass in the center has been killed by overfertilization with concentrated urine.

The Fate of Food. Whereas 60–90% of the food that consumers eat, digest, and absorb is oxidized for energy, the remaining 10–40%, which is converted to the body tissues of the consumer, is no less important. This is the fraction that enables the body to grow, as well as to maintain and repair itself. A portion of what is ingested by consumers is not digested (broken down so that it can be absorbed), but simply passes through the digestive system and out as fecal wastes. For consumers that eat plants, such waste is largely the material of plant cell walls, or **cellulose.** We often refer to it as

fiber, bulk, or *roughage.* Some fiber is a necessary part of the diet in order for the intestines to have something to push through to keep clean and open. Waste products can also include compounds of nitrogen, phosphorus, and any other elements present, in addition to the usual carbon dioxide and water. These byproducts are excreted in the urine (or as similar waste in other kinds of animals) and returned to the environment, where they may be reabsorbed by plants (Fig. 3–12). Here you can see the movement of elements in a cycle between the environment and living organisms. We expand on these cycles shortly.

In summary, organic material (food) eaten by any consumer follows one of three pathways: (1) More than 60% of what is digested and absorbed is oxidized to provide energy, and waste products are released back to the environment; (2) the remainder of what is digested and absorbed goes into body growth, maintenance and repair, or storage (fat); and (3) the portion that is not digested or absorbed passes out as fecal waste (Fig. 3–13). Recognize that in an ecosystem, it is only that portion of the food that becomes body tissue of the consumer that becomes food for the next organism in the food chain. This process is often referred to as *secondary* production, and like primary production, it also can be expressed as a rate (amount of growth of the consumer or consumer trophic level) over time.

Detritus Feeders and Decomposers—The Detritivores. Recall that detritus is mostly dead leaves, the woody parts of plants, and animal fecal wastes. As such, it is largely cellulose. Nevertheless, it is still organic and high in potential energy for those organisms that can digest it—namely, the decomposers we learned about in Chapter 2. Beyond having this ability to digest cellulose, decomposers (various species of fungi and bacteria, and a few other microbes) act as any other consumer, using the cellulose as a source of both energy and nutrients.

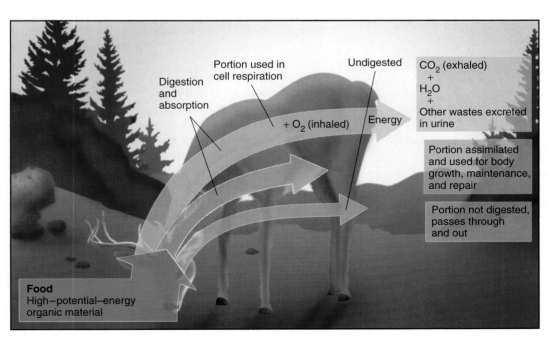

Digestion and absorption

Portion used in cell respiration

Undigested

+ O_2 (inhaled)

Energy

CO_2 (exhaled) + H_2O + Other wastes excreted in urine

Portion assimilated and used for body growth, maintenance, and repair

Portion not digested, passes through and out

Food High–potential–energy organic material

◄ **FIGURE 3–13**
Consumers. Only a small portion of the food ingested by a consumer is assimilated into body growth, maintenance, and repair. A larger amount is used in cell respiration to provide energy; waste products are carbon dioxide, water, and various mineral nutrients. A third portion is not digested and becomes fecal waste.

Termites and some other detritus feeders can digest woody material by virtue of maintaining decomposer microorganisms in their guts in a mutualistic symbiotic relationship. The termite (a detritus feeder) provides a cozy home for the microbes (decomposers) and takes in the cellulose, which the microbes digest for both their own and the termites' benefit.

Most decomposers make use of cell respiration. Thus, the detritus is broken down to carbon dioxide, water, and mineral nutrients. Likewise, there is a release of waste heat, which you may observe as the "steaming" of a manure or compost pile on a cold day. The release of nutrients by decomposers is highly important to the primary producers, as it is the major source of nutrients in most ecosystems.

Some decomposers (certain bacteria and yeasts) can meet their energy needs through the partial oxidation of glucose that can occur in the absence of oxygen. This modified form of cell respiration is called **fermentation**. It results in such end products as ethyl alcohol (C_2H_6O), methane gas (CH_4), and acetic acid ($C_2H_4O_2$). The commercial production of these compounds is achieved by growing the particular organism on suitable organic matter in a vessel without oxygen. In nature, **anaerobic**, or *oxygen-free*, environments commonly exist in the sediments of lakes, marshes, or swamps, buried deep in the Earth, and in the guts of animals, where oxygen does not penetrate readily. Methane gas is commonly produced in such locations. A number of large grazing animals, including cattle, maintain fermenting bacteria in their digestive systems in a mutualistic, symbiotic relationship similar to that just described for termites. Both cattle and termites produce methane as a result.

For simplicity, our orientation in this chapter is directed toward terrestrial ecosystems. It is important to realize that exactly the same processes occur in aquatic ecosystems. As aquatic plants and algae absorb dissolved carbon dioxide and mineral nutrients from the water, their photosynthetic production becomes the food and dissolved oxygen that sustain consumers and other heterotrophs. Likewise, aquatic heterotrophs return carbon dioxide and mineral nutrients to the aquatic environment. Of course, aquatic and terrestrial systems are never entirely isolated from one another, and exchanges between them go on all the time.

3.2   Principles of Ecosystem Function

The preceding examination of how the different biotic components of ecosystems function reveals that two common processes underlie them all: (a) the flow of energy, using sunlight as the basic energy source, and (b) the recycling of nutrients. In turn, these common features reveal basic principles underlying the sustainability of ecosystems. Let us examine these processes further at the ecosystem level.

Energy Flow in Ecosystems

Primary Production.   In most ecosystems, sunlight, or solar energy, is the initial source of energy absorbed by producers through the process of photosynthesis (the only exceptions are

ecosystems near the ocean floor or in dark caves, where the producers are chemosynthetic bacteria). As we saw, the process of primary production is capable of capturing at best only about 2% of incoming solar energy. Even though this seems like a small fraction, it is enough to fuel all of life on Earth—estimated at some 170 billion tons of organic matter per year produced! In a given ecosystem, the actual biomass of primary producers at any given time is referred to as the *standing crop biomass*. Both biomass and primary production vary greatly in different ecosystems. For example, a forested ecosystem maintains a very large biomass in comparison with a tropical grassland, yet the rate of primary production could be higher in the grassland, where animals continually graze newly produced organic matter.

It is instructive to examine the productivity of different ecosystem types (biomes and aquatic ecosystems) to evaluate their contribution to global productivity and to investigate why some are more productive than others. Fig. 3–14 presents (a) the percentage of different ecosystems over Earth's surface, (b) the average net primary productivity, and subsequently, (c) the percentage of global net primary productivity attributed to 19 of the most important ecosystem types. Some key relationships between abiotic factors and specific ecosystems can be seen in the data. Tropical rain forests are both very productive and contribute highly to global productivity; they cover a large area of the land and are blessed with ideal climatic conditions for photosynthesis—warm temperature and abundant rainfall. The open oceans, because they cover 65% of Earth's surface, account for a large portion of global productivity, yet the actual rate of production is low enough to refer to them as veritable biological deserts. Although light, temperature, and water are abundant, primary production in the oceans is limited by the scarcity of nutrients—a good lesson in the significance of limiting factors (Chapter 2). The seasonal effects of differences in latitude can also be seen in the comparison between productivity in tropical, temperate, and boreal (coniferous) forests.

Energy Flow and Efficiency.   As primary producers are consumed by herbivores, a transfer of energy is actually occurring. Recall that we refer to each of these components as a trophic level. Thus, we can examine energy flow in an ecosystem by considering how it moves from one trophic level to another. Fig. 3–15 presents a scheme of energy flow through three trophic levels of a grazing food web. Notice that at each trophic level, some energy goes into growth (production), some to heat (respiration), and some is given off as waste or is not consumed. It is quite clear that as energy flows from one trophic level, only a small fraction is actually passed on to the next trophic level. This is a consequence of three things: (1) Much of the preceding trophic level is standing biomass and is not consumed; (2) much of what is consumed is used for energy; and (3) some of what is consumed is undigested and passes through the organism.

Typically, a very large proportion of the primary-producer trophic level is not consumed in the grazing food web. As this material dies (leaves drop, grasses wither and die, etc.), it is joined by the fecal wastes and dead bodies of higher

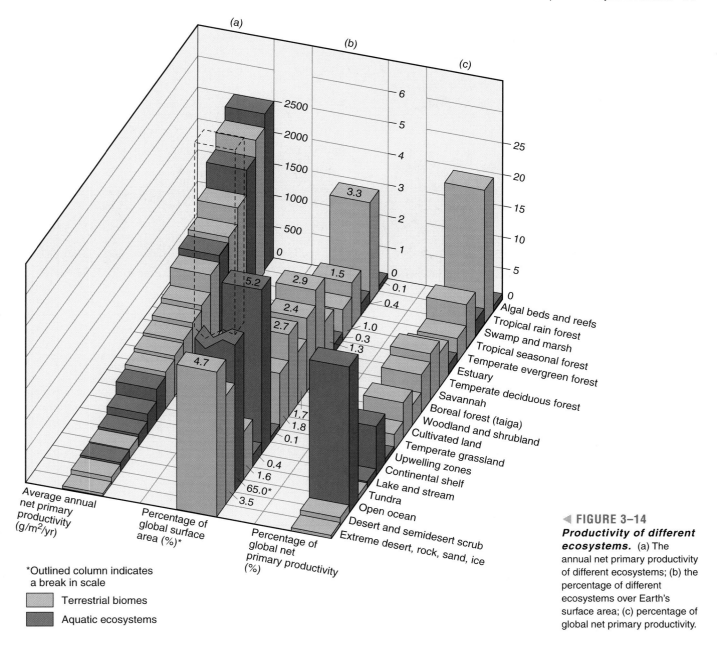

◀ FIGURE 3–14
Productivity of different ecosystems. (a) The annual net primary productivity of different ecosystems; (b) the percentage of different ecosystems over Earth's surface area; (c) percentage of global net primary productivity.

trophic levels and represents the starting point for a separate food web, the detritus food web, pictured earlier in Fig. 2–11. It is often the case that the majority of energy in an ecosystem flows through the detritus food web.

Because of the losses of energy in transfer at each trophic level, it is clear that each successive trophic level will capture only a fraction of the energy that entered the previous trophic level and usually will be represented by a much smaller biomass. Calculations have been made of the efficiency of transfer in a number of ecosystems, and the efficiencies of transfer range from 5 to 20%, 10% on average. Thus there is an estimated 90% loss of energy as it moves from one trophic level to the next. This loss gets quite critical at increasingly higher trophic levels and is the reason why carnivores are much less abundant than herbivores; carnivores that eat other carnivores are even less abundant, and so forth. This helps to explain why in any given ecosystem there are usually only three to five trophic levels. There simply isn't enough energy left to pass along to "super carnivores."

What happens to all the energy entering ecosystems? High potential energy in living plants, synthesized by using high-energy solar radiation, is either passed along to the next trophic level or degraded into the lowest and most disordered form of energy—heat—as it undergoes decomposition. As Fig. 3–15 indicates, eventually all of the energy escapes as heat energy. The laws of thermodynamics require that no energy is actually lost, but because so many energy conversions are taking place in ecosystem trophic activities, entropy is increased and all the energy is degraded to a form unavailable to do further work. The ultimate result is that *energy flows in a one-way direction through ecosystems*; it is not recycled, so it must be resupplied by sunlight.

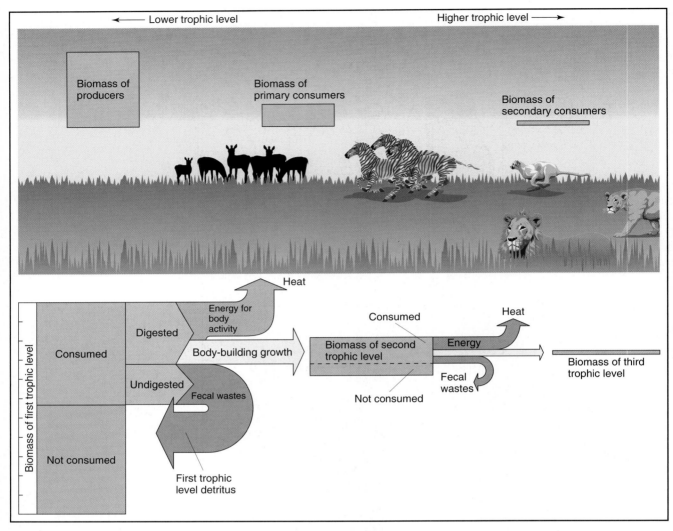

▲ **FIGURE 3–15** *Energy flow through trophic levels in a grazing food web.* Each trophic level is represented as biomass boxes, and the pathways taken by the energy flow are indicated with arrows.

Running on Solar Energy. We have seen that no system can run without an input of energy, and living systems are no exception. For all major ecosystems, both terrestrial and aquatic, the initial source of energy is *sunlight.* Using sunlight as the basic energy source is fundamental to sustainability for two reasons: It is both *nonpolluting* and *nondepletable.*

Nonpolluting. Light from the Sun is a form of pure energy; it contains no substance that can pollute the environment. All the matter and pollution involved in the production of light energy are conveniently left behind on the Sun some 93 million miles (150 million kilometers) away in space.

Nondepletable. The Sun's energy output is constant. How much or how little of this energy is used on Earth will not influence, much less deplete, the Sun's output. For all practical purposes, the Sun is an everlasting source of energy. True, astronomers tell us that the Sun will burn out in another 3–5 billion years, but we need to put this figure in perspective. One thousand is only 0.0001% of a billion. Thus, even the passing of millennia is hardly noticeable on this time scale.

Hence, we uncover the **first basic principle of ecosystem sustainability**:

For sustainability, ecosystems use sunlight as their source of energy.

Energy flow is one of the two fundamental processes that make ecosystems work. The second process is the recycling of nutrients in what are known as biogeochemical cycles.

Biogeochemical Cycles

Looking at the various inputs and outputs of producers, consumers, detritus feeders, and decomposers, you should be impressed by how they fit together. The products and byproducts of each group are the food and/or essential nutrients for the other. Specifically, the organic material and oxygen produced by green plants are the food and oxygen required by consumers and other heterotrophs. In turn, the carbon dioxide and other wastes generated when heterotrophs break down their food are exactly the nutrients needed by green plants. Such recycling is fundamental, for two reasons: (a) It prevents the accumulation of wastes that would cause problems; and (b) it assures that the ecosystem will not run out of essential elements. Thus, we encounter the **second basic principle of ecosystem sustainability**:

> *For sustainability, ecosystems dispose of wastes and replenish nutrients by recycling all elements.*

If we reconsider the natural law of conservation of matter, which says that atoms cannot be created, destroyed, or changed, we can see that recycling is the only possible way to maintain a dynamic system, and the biosphere has mastered this to a profound degree. We can see this even more clearly by focusing on the pathways of three key elements heavily impacted by human activities: carbon, phosphorus, and nitrogen. Because these pathways all lead in circles and they involve biological, geological, and chemical processes, they are known as biogeochemical cycles (recall that energy is not recycled).

The Carbon Cycle. For descriptive purposes, it is convenient to start the carbon cycle (Fig. 3–16) with the "reservoir" of carbon dioxide (CO_2) molecules present in the air, and bicarbonate (HCO_3^-) molecules present in water. Through photosynthesis and further metabolism, carbon atoms from CO_2 become the carbon atoms of the organic molecules making up the plant's body. Through food webs, the carbon atoms then move into and become part of the tissues of all the other organisms in the ecosystem. At any point, a given atom may be respired and returned to the atmosphere; in aquatic systems, it will be returned to the inorganic carbonate in solution. Processes other than direct involvement in trophic transfer are significant. Fig. 3–16 indicates two in particular: 1) geological sedimentation of carbon in the oceans, which removes carbon from solution; 2) the combustion of fossil-fuel carbon that was laid down millions of years ago by biological systems.

By calculating the total amount of carbon dioxide in the atmosphere and the amount of primary production (photosynthesis) occurring in the biosphere, scientists have concluded that about a third of the total atmospheric carbon dioxide is taken up in photosynthesis in a year, but an equal amount is returned to the atmosphere through cell respiration. This means that on the average, a carbon atom makes a cycle from the atmosphere through one or more living things and back to the atmosphere every three years.

▼ **FIGURE 3–16** *The carbon cycle.* Boxes in the figure refer to *pools* of carbon, and arrows refer to the movement, or *fluxes*, of carbon from one pool to another.

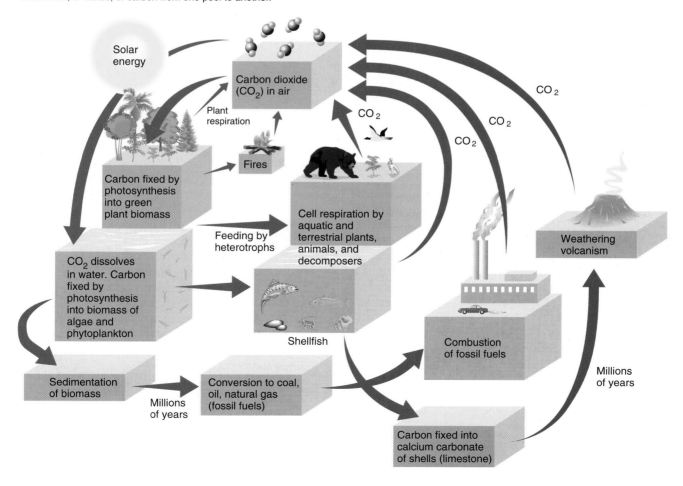

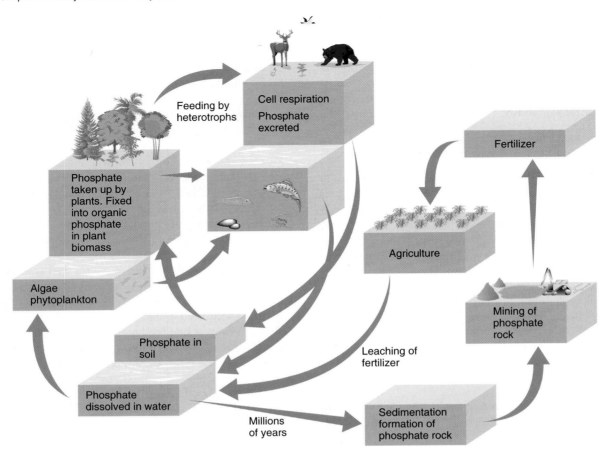

▲ **FIGURE 3–17** *The phosphorus cycle.*

The Phosphorus Cycle. The phosphorus cycle is representative of the cycles for all the mineral nutrients—those required elements that have their origin in rock and soil minerals (see Table 3-2). We focus on phosphorus because its shortage tends to be a limiting factor in a number of ecosytems and its excess can be a serious stimulant for unwanted algal growth in freshwater systems.

Phosphorus exists in various rock and soil minerals as the inorganic ion *phosphate* (PO_4^{3-}). As rock gradually breaks down, phosphate and other ions are released. PO_4^{3-} dissolves in water, but does not enter the air. Plants absorb phosphate from the soil or from a water solution, and when it is bonded into organic compounds by the plant, it is referred to as **organic phosphate**. Moving through food chains, organic phosphate is transferred from producers to the rest of the ecosystem. As with carbon, at each step there is a high likelihood that the organic compounds containing phosphate will be broken down in cell respiration, releasing PO_4^{3-} in urine or other waste. The phosphate may then be reabsorbed by plants to start another cycle.

The phosphorus cycle is illustrated in Fig. 3–17, again set up as a set of pools and fluxes to indicate key processes. Phosphorus enters into complex chemical reactions with other substances that are not shown in this simplified version of the cycle. For example, PO_4^{3-} forms insoluble chemical precipitates with a number of cations (positively charged ions), such

as iron (Fe^{3+}), aluminum (Al^{3+}) and calcium (Ca^{2+}). If these cations are in sufficiently high concentration in soil or aquatic systems, the phosphorus can be tied up in chemical precipitates and rendered largely unavailable to plants. This precipitated phosphorus can slowly release PO_4^{3-} as plants withdraw naturally occurring PO_4^{3-} from soil water or sediments.

There is an important difference between the carbon cycle and the phosphorus cycle. No matter where CO_2 is released, it will mix into and maintain the concentration of CO_2 in the air. Phosphate, however, which does not have a gas phase, is recycled only if the wastes containing it are deposited on the soil *from which it came.* The same holds true for other mineral nutrients. Of course, in natural ecosystems, wastes (urine, detritus) are deposited in the same area so that recycling occurs efficiently. As mentioned, the rich growth of grasses on the plains of the Serengeti is traced to phosphorus brought and maintained there by the dense herbivore herds. Humans have been extremely prone to interrupt this cycle, however.

A very serious case of disruption of the phosphorus cycle is the cutting of tropical rain forests. This type of ecosystem is supported by a virtually 100% efficient recycling of nutrients. There are little or no reserves of nutrients in the soil. When the forest is cut and burned, the nutrients that were stored in the organisms and detritus are readily washed away by the heavy rains, and the land is thus rendered unproductive.

Another human effect on the cycle is that much phosphate from agricultural croplands makes its way into waterways—either directly, in runoff from the croplands, or indirectly, in sewage effluents. When humans use manure, compost (rotted plant wastes), or sewage sludge (see Chapter 18) on crops, lawns, or gardens, the foregoing natural cycle is duplicated. But in too many cases it is not, and the applied chemical fertilizers end up being leached (carried by water seepage) into waterways. Because there is essentially no return of phosphorus from water to soil, this addition results in overfertilization of bodies of water, which in turn leads to a severe pollution problem known as eutrophication (see Chapter 18).

The Nitrogen Cycle. The nitrogen cycle (Fig. 3–18) is unique; it has aspects of both the carbon cycle and the phosphorus cycle. It is also unique in that many of the steps of the cycle are performed by bacteria in soils, water, and sediments. Like phosphorus, nitrogen is in high demand by plants in both aquatic and terrestrial systems and is a major limiting factor in many. Let's look at some of the details of the cycle as illustrated in Fig. 3–18.

The main reservoir of nitrogen is the air, which is about 78% nitrogen gas (N_2). Plants and animals cannot use nitrogen gas directly from the air. Instead, the nitrogen must be in mineral form, such as ammonium ions (NH_4^+) or nitrate ions (NO_3^-). We can begin with the uptake of nitrate-nitrogen by green plants, which incorporate the nitrogen into essential organic compounds such as proteins and nucleic acids. The nitrogen then follows the classic energy-flow pattern from producers to herbivores to carnivores and finally to decomposers (referred to as heterotrophs in the figure). At various points, nitrogen wastes are released, primarily in the form of ammonium compounds. A group of soil bacteria, the nitrifying bacteria, then convert the ammonium to nitrate. These are oxidations that yield energy to the bacteria (this is a chemosynthetic process). At this point, the nitrate is once again available for uptake by green plants—an internal ecosystem cycle within the larger cycle. However, in most ecosystems the supply of nitrate or ammonium nitrogen is quite limited, yet there is an abundance of nitrogen gas—if it can be accessed.

A number of bacteria and cyanobacteria (bacteria that contain chlorophyll, formerly referred to as blue-green algae) can convert nitrogen gas to the ammonium form, a process called biological **nitrogen fixation**. For terrestrial ecosystems, the most important among these nitrogen-fixing organisms is a bacterium called *Rhizobium*, which lives in nodules on the roots of legumes, the plant family that includes peas and beans (Fig. 3–19). (This is another example of mutualistic symbiosis:

▼ **FIGURE 3–18** *The nitrogen cycle.*

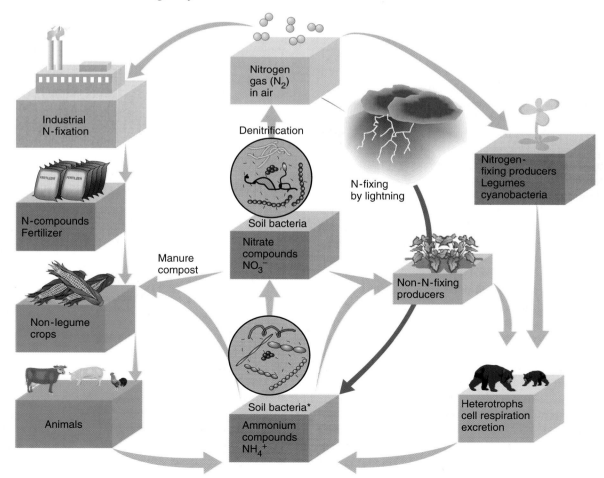

▲ **FIGURE 3–19** *Nitrogen fixation.* Bacteria in root nodules of legumes convert nitrogen gas in the atmosphere to forms that can be used by plants.

The legume provides the bacteria with a place to live and with food (sugar) and gains a source of nitrogen in return.) From the legumes, nitrogen enters the food web as described above. Many ecosystems, then, are "fertilized" by nitrogen-fixing organisms; legumes, with their symbiotic bacteria are by far the most important. The legume family includes a huge diversity of plants, ranging from clovers (common in grasslands) to desert shrubs and many trees. Every major terrestrial ecosystem, from tropical rain forest to desert and tundra, has its representative legume species, and legumes are generally the first plants to recolonize a burned-over area. Without them, all production would be sharply impaired because of lack of available nitrogen—precluding the formation of proteins, nucleic acids, and other building blocks of life. The nitrogen cycle in aquatic ecosystems is similar. There, cyanobacteria are the most significant nitrogen fixers.

Three other important processes also "fix" nitrogen. One is the conversion of nitrogen gas to the ammonium form by discharges of lightning in a process known as *atmospheric nitrogen fixation*; the ammonium then comes down with rainfall. The second is the *industrial fixation* of nitrogen in the manufacture of fertilizer, and the third is the consequence of *combustion of fossil fuels*, where nitrogen from coal and oil is oxidized; some nitrogen gas is also oxidized during high-temperature combustion. Both of these lead to nitrogen oxides in the atmosphere, which are converted to nitric acid and then brought down to Earth as acid precipitation.

One final microbial process must be introduced: **denitrification**. This process occurs in soils and sediments, where oxygen is unavailable for normal bacterial decomposition. A number of microbes can take nitrate (which is highly oxidized) and use it as an oxygen substitute. In so doing, the nitrogen is reduced (gains electrons) to the state of nitrogen gas

and released back to the atmosphere; much organic matter is decomposed. This is a process farmers want to avoid, as it results in a loss of soil fertility. They do so by plowing as early as possible in the spring to restore oxygen to the soil. In sewage treatment systems, this is a desirable process and is promoted in order to remove nitrogen from the wastewater before it is released to the environment (Chapter 18).

Human involvement in the nitrogen cycle is significant and a major cause for concern. Many agricultural crops are legumes (peas, beans, soybeans, alfalfa), and because they draw nitrogen from the air, they increase the normal rate of nitrogen fixation on land. Crops that are nonleguminous (corn, wheat, potatoes, cotton, etc.) are heavily fertilized with nitrogen derived from industrial fixation. And fossil-fuel combustion fixes nitrogen from the air. All told, these processes are estimated to add some 140 billion metric tons of nitrogen to terrestrial ecosystems annually. This is approximately the same amount of nitrogen fixation as that which occurs naturally. In effect, we are doubling the rate at which nitrogen is moved from the atmosphere to the land.

The consequences of this doubling for natural ecosystems are serious. Acid deposition has destroyed thousands of lakes and ponds and caused extensive forest damage (Chapter 22). The surplus nitrogen has led to "nitrogen saturation" of many natural areas, where the nitrogen can no longer be incorporated into living matter and is released into the soil. There it leaches cations like calcium and magnesium from the soil and leads to mineral deficiencies in forest trees and other vegetation. Washed into surface waters, the nitrogen makes its way to estuaries and coastal oceans, where it promotes rich "blooms" of algae. Some of these algae are toxic to fish and shellfish; then, when algal blooms die, they sink to deeper water or sediments, where they reduce the oxygen supply and kill the bottom-dwelling organisms like crabs, oysters, and clams. These are just the observable effects of nitrogen enrichment; there may be other effects that have not yet been documented, such as a loss of biodiversity by encouraging luxuriant growth of a few dominant species.

While we have focused on the cycles of carbon, phosphorus, and nitrogen, it should be evident that cycles exist for oxygen, hydrogen, and all the other elements that play a role in living things. Also, while the routes taken by distinct elements may differ, it should be evident that all are going on simultaneously and that all come together in the tissues of living things. And as the elements move in these cyclical patterns in ecosystems, energy is flowing in from the Sun and through the living members of the ecosystems. The links between these two fundamental processes of ecosystem function are shown in Fig. 3–20.

3.3 Implications for Humans

Sustainability

We have said that a major part of our purpose in studying natural ecosystems lies in the fact that they are models of sustainability. We have also said that if we can elucidate the

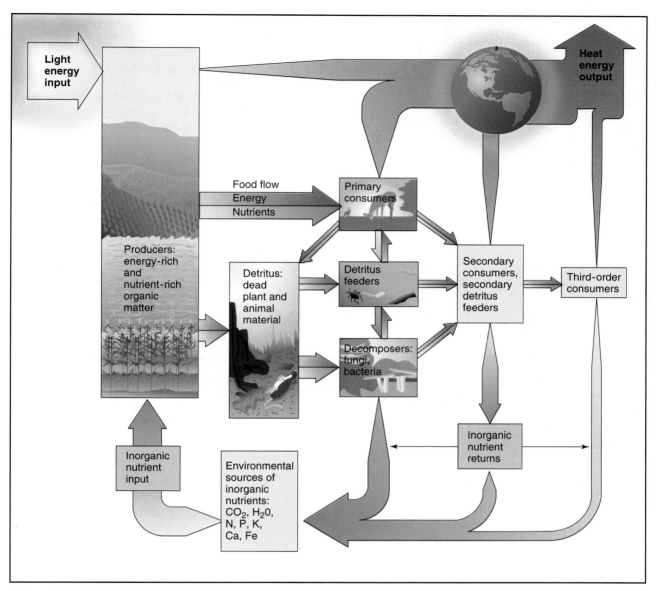

▲ FIGURE 3–20 *Nutrient recycling and energy flow through an ecosystem.* Arranging organisms by feeding relationships and depicting the energy and nutrient inputs and outputs of each relationship show a continuous recycling of nutrients (blue) in the ecosystem, a continuous flow of energy through it (red), and a decrease in biomass (thickness of arrows).

principles that underlie their sustainability, we may be able to apply those principles toward our own efforts to achieve a sustainable society. Let us look at the first principle of sustainability: *For sustainability, ecosystems use sunlight as their source of energy.* How are we making use of solar energy?

Significance of Energy Flow. The human system makes heavy use of the energy flowing through natural and agricultural ecosystems. These systems provide all of our food. Consider the use of grasslands to provide animals for labor, meat, wool, leather, and milk. Forest biomes provide us with 3.3 billion cubic meters of wood annually for fuel, building material, and paper. Some 15% of the world's energy consumption is derived directly from plant material. We have converted almost 11% of Earth's land area from forest and grassland biomes to agricultural systems. Calculations have indicated

that because of our use of primary production in terrestrial systems, both natural and cultivated, we are diverting 27% of primary production to support human enterprises. And because of our conversion of many natural and agricultural lands to urban and suburban housing, highways, dumps, factories, and the like, we cancel out an additional 13% of potential primary production. Thus, we divert 40% of the land's primary production to support human needs. In so doing, we have become the dominant biological force on Earth.

One practical application of our insights into energy flow involves the human fondness for meat. A trend that parallels increasing affluence in every country observed is increasing meat consumption. Because of the principles involved in the biomass pyramid, it takes about 10 pounds of grain to grow a pound of meat [more for beef, less for chicken (Fig. 3–21)]. Therefore, for every increase in meat consumption, there is a

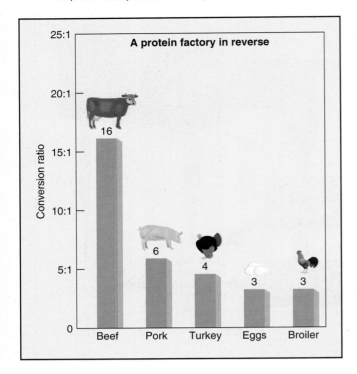

▲ **FIGURE 3–21** *Conversion of grain to protein.* To obtain one pound of meat, poultry, or eggs, farmers must invest these many pounds of grains and soy feeds. To get one pound of beef requires an expenditure of 16 pounds of feed. Said another way, the grain consumed to support one person eating meat could support 16 persons eating the grains directly.

tenfold increase in the demand on plant production and, consequently, on the land, fertilizer, pesticides, and energy used and pollution produced. Of course, the reverse is also true. Dropping down a trophic level alleviates the demand proportionally. The implication of this is profound when you consider that half the cultivated acreage in the United States (roughly the size of Texas and Oklahoma) produces animal feed!

Another energy source. In addition to running on solar energy, which is nonpolluting and nondepletable, we have constructed a human system that is heavily dependent on fossil fuels—coal, natural gas, and crude oil. Crude oil is the base for refinement of all liquid fuels: gasoline, diesel fuel, fuel oil, and so on. Even in the production of food, which is fundamentally supported by sunlight and photosynthesis, it is estimated that we use about 10 calories of fossil fuel for every calorie of food consumed. This additional energy is used in the course of field preparation, fertilizing, controlling pests, harvesting, processing, preserving, transportation, and, finally, cooking.

The most pressing problem in connection with consuming these fuels is the limited capacity of the biosphere to absorb the waste byproducts produced from burning them. Air pollution problems, including urban smog, acid rain, and the potential for global warming, are the result of these byproducts. Also, problems stemming from depletion, particularly that of crude oil, are on the horizon. You see why most people concerned about sustainability are also

solar-energy advocates. Solar energy is extremely abundant. Just as important, we do have the technology to obtain most, if not all, of our energy needs from sunlight and the forces it causes, such as wind (see Chapter 15).

Sustainability and nutrient cycling. Our second sustainability principle is very simple: *For sustainability, ecosystems dispose of wastes and replenish nutrients by recycling all elements.* In contrast to the remarkable recycling seen in natural ecosystems, we have constructed our human system, in large part, on the basis of a *one-directional flow* of elements (Fig. 3–22). We have already noted that the fertilizer-nutrient phosphate, which is mined from deposits, ends up going into waterways with effluents from sewage treatment. The same one-way flow can be seen in such metals as aluminum, mercury, lead, and cadmium, which are the "nutrients" of our industry. At one end, they are mined from the Earth; at the other, they end up in dumps and landfills as items containing them are discarded. Is it any wonder that there are depletion problems at one end and pollution problems at the other? The Earth has vast deposits of most minerals; however, the capacity of ecosystems (even the whole biosphere) to absorb wastes without being disturbed is comparatively limited. This limitation is aggravated even further by the fact that many of the products we use are nonbiodegradable. Conversely, can you see the rationale for expanding the concept of recycling to include not just paper, bottles, and cans, but everything from sewage to industrial wastes as well?

Value

How much are natural ecosystems worth to us? Answering this question requires that we introduce a concept that will be more fully developed in Chapter 23: **natural capital**. Use of the term "capital" suggests that there is monetary value involved, which can be converted into income. One component of natural capital is the natural assets of a country in the form of nonrenewable minerals and renewable land, timber, rangelands, and the like—the so-called natural resources of a country, which are exploited for meeting human needs and generating economic profit. The second component, however, is the services performed by natural systems that benefit human life and enterprises. This natural capital is represented in natural ecosystems, which provide *goods* (like lumber, fiber, and food) and vital *services* (like waste assimilation and nutrient cycling). In a first-ever attempt of its kind, a team of 13 natural scientists and economists[1] recently collaborated to produce a report entitled "The value of the world's ecosystem services and natural capital." Their reason for making such an effort was that the goods and services provided by natural ecosystems is not easily seen in the market (meaning the market economy that normally allows us to place value on things) and in fact may not be in the market

---

[1] Robert Costanza et al., "The value of the world's ecosystem services and natural capital," *Nature* 387 (1997): 253–260.

THE HUMAN SYSTEM
How can we make it into a sustainable cycle?

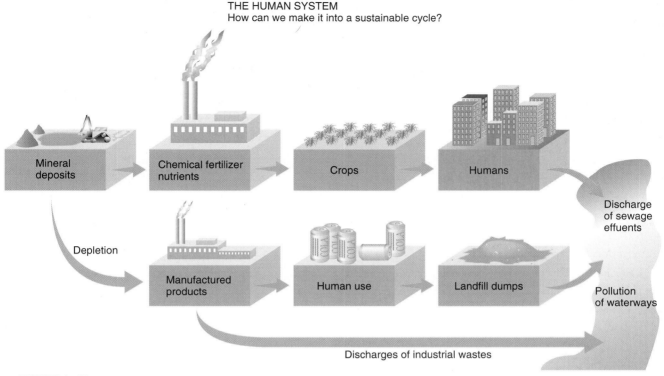

▲ FIGURE 3–22 *One-directional nutrient flow in human society.*

at all. Thus, things like clean air to breathe, soil formation, the breakdown of pollutants, and the like never pass through the market economy. People are often not even aware of their importance. Thus, they are undervalued or not valued at all.

The team identified 17 major ecosystem goods and services, which provide vital functions we depend on. They also identified the ecosystem functions that actually carry out the vital human support and gave examples of each (Table 3-3). The team made the point that it is useless to consider human welfare without these ecosystem services, so in one sense, their value is infinite as a whole. However, one can calculate the **incremental value** of each type of service—how changes in the quantities or quality of various types of services may impact human welfare. For example, removing a given forest will impact the ability of the forest to provide lumber in the future as well as other services, like soil formation and promotion of the hydrologic cycle. This value can be calculated in economic terms. So, by performing this kind of analysis and making many approximations of value and collecting information from other researchers who have worked on individual processes, they produced a table of annual global value of the ecosystem services performed. The total value to human welfare of a year's services amounts to $33 trillion, and this is considered to be a conservative estimate! This is far more than the $18 trillion for the global gross national product as calculated for the world economy.

The team pointed out that the real power of the analysis is in its use for local decisions; thus, the value of a wetlands cannot just be represented by the amount of soybeans that could be grown on the land if it were drained. Wetlands provide other

vital ecosystem services, and these should be added to the value of the soybeans in calculating costs and benefits of a given proposed change in land use. The bottom line of their analysis is, in their words, "...that ecosystem services provide an important portion of the total contribution to human welfare on this planet." For this reason, we need to give the natural capital stock (the ecosystems and populations in them, the lakes and wetlands) adequate weight in public policy decisions involving changes to them. Because these services are outside the market and uncertain, they are too often ignored or undervalued, and the net result is human changes to natural systems whose social costs far outweigh their benefits.

This remarkable analysis is a good example of *sound science* being employed to generate vital information for public policy. Our great dependence on natural systems should lead, then, to proper management of those systems.

Managing Ecosystems

Practically everything we do affects natural ecosystems. It has long been recognized that ecosystem health is essential to the provision of the goods and services, so valuable to us. By using ecosystems for these goods and services, we actively "manage" them. Management simply may be a consequence of reckless exploitation (poor management!) or it may be intentional, as in the case of silviculture practice in our national forests. Ecosystems have continued on in time past without us, and if we were to pass from the scene, they would undoubtedly continue into the future. This is to remind us that it is in our own interest to manage ecosystems well.

TABLE 3-3 Ecosystem Services and Functions

Ecosystem Service	Ecosystem Functions	Examples
Gas regulation	Regulation of atmospheric chemical composition	CO_2/O_2 balance, O_3 for UVB protection, and SO_x levels
Climate regulation	Regulation of global temperature, precipitation, and other biologically mediated climatic processes at global or local levels	Greenhouse gas regulation, dimethylsulfoxide production affecting cloud formation
Disturbance regulation	Capacitance, damping and integrity of ecosystem response toenvironmental fluctuations	Storm protection, flood control, drought recovery, and other aspects of habitat response to environmental variability mainly controlled by vegetation structure
Water regulation	Regulation of hydrological flows	Provisioning of water for agricultural (such as irrigation) or industrial (such as milling) processes or transportation
Water supply	Storage and retention of water	Provisioning of water by watersheds, reservoirs, and aquifers
Erosion control and sediment retention	Retention of soil within an ecosystem	Prevention of loss of soil by wind, runoff, or other removal processes; storage of silt in lakes and wetlands
Soil formation	Soil-formation processes	Weathering of rock and the accumulation of organic material
Nutrient cycling	Storage, internal cycling, processing, and acquisition of nutrients	Nitrogen fixation, N, P, and other elemental or nutrient cycles
Waste treatment	Recovery of mobile nutrients and removal or breakdown of excess or xenic nutrients and compounds	Waste treatment, pollution control, detoxification
Pollination	Movement of floral gametes	Provisioning of pollinators for the reproduction of plant populations
Biological control	Trophic-dynamic regulations of populations	Keystone predator control of prey species, reduction of herbivory by top predators
Refugia	Habitat for resident and transient populations	Nurseries, habitat for migratory species, regional habitats for locally harvested species, or overwintering grounds
Food production	That portion of primary production extractable as food	Production of fish, game, crops, nuts, and fruits by hunting, gathering, subsistence farming, or fishing
Raw materials	That portion of primary production extractable as raw materials	The production of lumber, fuel, or fodder
Genetic resources	Sources of unique biological materials and products	Medicine, products for materials science, genes for resistance to plant pathogens and crop pests, ornamental species (pets and horticultural varieties of plants)
Recreation	Providing opportunities for recreational activities	Ecotourism, sport fishing, and other outdoor recreational activities
Cultural	Providing opportunities for noncommercial uses	Aesthetic, artistic, educational, spiritual, and/or scientific values of ecosystems

Source: Reprinted, by permission, from Robert Costanza, Ralph d'Arge, Rudolf de Groot, Stephen Farber, Monica Grasso, Bruce Hannon, Karin Limburg, Shahid Naeem, Robert V. O'Neill, Jose Paruelo, Robert G. Raskin, Paul Sutton, and Marjan van den Belt, "The value of the world's ecosystem services and natural capital," *Nature* 387 (1997): 253–260.

EARTH WATCH

BIOSPHERE 2

The proof of a theory lies in testing it. If the biosphere functions as we have described—running on solar energy and recycling all the elements from the environment through living organisms and back to the environment—then it might be possible to create an artificial biosphere that functions similarly. Indeed, students commonly conduct an exercise of creating a "biosphere in a bottle": Some photosynthetic and compatible consumer organisms are sealed in a bottle and kept in the light. Varying degrees of success are achieved, however; such systems usually do not sustain themselves beyond a few weeks for various reasons.

The largest such experiment to date is Biosphere 2, constructed in Arizona 30 miles north of Tucson. Biosphere 2 was developed entirely with private venture capital, with a view toward gaining information and experience that might be used in creating permanent space stations on the Moon or other planets or in long-distance space travel. Additionally, it was hoped that Biosphere 2 would yield information that would further our understanding of our own biosphere—Biosphere 1.

Biosphere 2 is a supersealed "greenhouse," including seals underneath, enclosing an area of 2.5 acres (1 ha). Entry and exit is through a double air lock. Different environmental conditions within the containment support several ecosystems. Accordingly, there is an area of tropical rain forest, savannah, scrub forest, desert, fresh- and saltwater marshes, and a mini-ocean complete with a coral reef, each stocked with respective species—over 4000 in all. There is an agricultural area and living quarters for a crew of up to 10 "Biospherians."

Water vapor from evaporation is condensed to create high rainfall over the tropical rain forest. The water trickles back toward the marshes and the ocean through soil filters, providing a continuous supply of fresh water for both humans and ecosystems. Carbon dioxide from respiration is reabsorbed, and oxygen is replenished through photosynthesis. Thus, these "natural ecosystems" provide the basic ecological stability for the atmosphere and hydrosphere of Biosphere 2. All

wastes, including human and animal excrements, are treated, decomposed, and recycled to support the growth of plants.

Biosphere 2 is not self-sufficient with regard to solar energy's falling on the structure. Biosphere 2's energy demands for machinery are such that an additional 30 acres (12 ha) of solar collectors would be required. (Actually, external natural gas-driven generators are used.)

The first crew of four men and four women—a crew with a variety of skills and academic backgrounds—completed the first two-year mission in September 1993. In addition to monitoring and collecting data on the natural systems, their main occupation was intensive organic agriculture (no chemical pesticides) to produce plant foods both for themselves and for feeding a few goats and chickens, which produced eggs, milk, and a little meat. Some fish farming supplied additional protein. The living quarters included the comforts and conveniences of modern living, but all communication with the outside world was via electronics. Their environment with other plants and animals was totally sealed from that of the outside world. In such a closed system, the water soil and nutrients they had when they began were the same when they finished, having cycled innumerable times within the system.

Not everything went perfectly; in fact, quite a few things went wrong. At one point additional oxygen had to be introduced, because oxygen was absorbed by the huge concrete structure that supports the greenhouse. Nitrous oxide levels rose to dangerous levels and had to be controlled in order to preserve the health of the inhabitants. Nearly all the birds and animals thought to be able to tolerate the enclosure died off, with the exception of cockroaches and ants. It was reported that the food got so poor that the occupants smuggled some in.

At the end of their two-year experiment sojourn, the "Biospherians" emerged somewhat thinner but all in good health. Grave doubts were cast on the scientific value of the experiment, however, because of the necessary interventions and the loss of so many other species. The owners of the facility were at a loss to come up with a useful future for their $200 million establishment.

Biosphere 2 received new life, however, when Columbia University's Lamont-Doherty Earth Observatory agreed to take it over. Researchers are being invited to make use of it—the world's largest closed ecological laboratory presents some unique opportunities for climate and ecosystem research. For example, experiments are being conducted with tree growth within a doubled carbon dioxide atmosphere—conditions thought to be not many decades ahead, because of the high rate of fossil-fuel combustion of today's civilization. Next to Earth, it is the world's largest greenhouse—a suitable place to explore the greenhouse effect.

Some writers have suggested that if we trash our own planet, we may end up living in Biosphere 2–like enclosures, although the costs for such escape from a polluted environment would be prohibitive. An important lesson from Biosphere 2 is an appreciation of the operational complexity of Biosphere 1. If we fail to maintain our natural biosphere, there will be no alternative for survival.

The **goal** of ecosystem management is *managing ecosystems so as to assure their sustainability*, where sustainability is defined as maintaining ecosystems such that they continue to provide the same level of goods and services indefinitely into the future. This, in very simple terms, means preserving the productivity of ecosystems and maintaining the biodiversity of ecosystems.

How can this best be accomplished? The Ecological Society of America's Committee on Ecosystem Management has identified the following necessary elements: (1) clear operational goals, (2) sound ecological models and understanding, (3) an understanding of complexity and interconnectedness, (4) recognition of the dynamic character of ecosystems, (5) attention to context and scale, (6) acknowledgment of ignorance and uncertainty, (7) commitment to adaptability and accountability, and (8) acknowledgment of humans as ecosystem components. It is highly significant that since 1992 the U.S. Forest Service, the U.S. Bureau of Land Management, the U.S. Fish and Wildlife Service, the U.S. National Park Service, and the U.S. Environmental Protection Agency have all officially adopted Ecosystem Management as their management paradigm. This is, in the words of Jerry Franklin, a leading forester, "a paradigm shift of massive proportions." It should be evident that there is no way to accomplish this kind of ecosystem management without the involvement of sound science. Every one of these agencies recognizes the need to use the best available science as a guiding policy.

This commitment of our federal agencies is in fact a commitment to stewardship, an acceptance of responsibility "…for protecting the integrity of natural resources and their underlying ecosystems, and, in so doing, safeguarding the interests of future generations," as the President's Council on Sustainable Development put it. However, there are many pitfalls on the way to these good intentions, and we will explore the recent record of our federal agencies in subsequent chapters as we look at how our natural resources are in fact being managed. For the moment, let us be glad that there is now clear recognition, at the highest levels of resource management, of the goal of sustainability and the ethic of stewardship as guiding principles.

We turn our attention in the next chapter to the question of how ecosystems are balanced and how they change over time. As we do so, we will encounter the remaining principles of sustainability.

ENVIRONMENT ON THE WEB

HUMAN IMPACTS ON THE BIOGEOCHEMICAL CYCLES

Biogeochemical cycles describe the movement and transformation of chemical substances through the global environment. Humans impact these cycles in many ways. Sometimes the change arises because of the ways that humans have changed the face of the planet, through the building of cities and widespread agriculture. Other changes arise because of human additions to or removal of substances from the environment. In a sense, biogeochemical cycles provide simple conceptual models that help us understand the sources, pathways, and sinks of natural materials—and the implications of human intervention in those cycles. The following examples illustrate this point.

Sometimes humans impact biogeochemical cycles by changing the rates at which materials move from one stage to another. An example of this can be found in the carbon cycle. In undisturbed ecosystems, large quantities of carbon are tied up in biological tissues and are released again into the atmosphere only when the organism dies and its tissues decay. However, when human activities affect the way the system operates—for instance, through the cutting or the burning of forests—the rate at which organisms die and carbon is released into the atmosphere increases greatly.

Another example comes from the phosphorus cycle. In nature, phosphorus enters lakes and streams gradually as phosphorus is released from rock weathering and decay of biological tissue. Rock weathering in particular is a very slow process. Human activities such as phosphate extraction for fertilizer manufacture greatly increase the rate at which mineral phosphorus becomes available for biological processes. In agricultural systems, clearance of the natural vegetative cover and planting of simpler plant ecosystems contributes to the smoothness of the landscape and increases the rate at which water and dissolved nutrients like phosphorus can move through and over soils to reach waterways.

In other cases, human activities actually add new sources to a biogeochemical cycle. The best example of this can be found in the nitrogen cycle, where human manufacture of fertilizer creates new sources of some nitrogen compounds through the manufacture of synthetic fertilizers. These fertilizers are not naturally occurring substances (as in the case of phosphate fertilizers) but rather are new products formed from raw materials containing nitrogen. So although these processes do not create new nitrogen, they do fundamentally change the proportions of nitrogen compounds in the cycle.

Web Explorations

The Environment on the Web activity for this essay describes how people use biogeochemical cycles to understand the impact of human activities on ecosystems. Go to the Environment on the Web activity (select Chapter 3 at http://www.prenhall.com/nebel) and learn for yourself

1. about your own "carbon budget;"
2. how you can use simple knowledge of the nitrogen cycle to identify sources of nitrate pollution; and
3. about the use of the phosphorus cycle in the management of cottage development around lakes.

A suggested time frame for each of these exercises is 10–30 minutes.

REVIEW QUESTIONS

1. What are the six key elements of living organisms, and where does each occur in the environment? In four cases, identify the specific molecule from which the element comes.
2. What is the "common denominator" that distinguishes between organic and inorganic molecules?
3. In one sentence, define matter and energy, as well as demonstrate how they are related. What are the three categories of matter?
4. Give five examples of potential energy. How can these be converted into kinetic energy?
5. State the two energy laws. How do they relate to entropy?
6. What is the chemical equation for photosynthesis? Examine the origin and destination of each molecule of the equation. Do the same for cell respiration.
7. Food ingested by a consumer follows three different pathways. Describe each pathway in terms of what happens to the food involved and what products and byproducts are produced in each case.
8. Compare and contrast the decomposers with other consumers in terms of matter and energy changes that they perform.
9. What are factors that affect the rate and amount of primary production?
10. What three factors account for decreasing biomass at higher trophic levels—that is, the food pyramid?
11. What are the two principles of ecosystem sustainability stated in this chapter?
12. Where do carbon, phosphorus, and nitrogen exist in the environment, and how do they move into and through organisms and back to the environment?
13. Refer to Table 3-3: Ecosystem Services and Functions. These have been defined by scholars as the natural resources vital for human life. How can we improve our actions to become better stewards of these goods and services?
14. Name some of the elements necessary to achieve sustainable management of ecosystems.

THINKING ENVIRONMENTALLY

1. Describe the consumption of fuel by a car in terms of the laws of conservation of matter and energy. That is, what are the inputs and outputs of matter and energy? (*Note:* Gasoline is an organic carbon-hydrogen compound.)
2. Relate your level of exercise—breathing hard and "working up an appetite"—to cell respiration in your body. What materials are consumed, and what products and byproducts are produced?
3. Using your knowledge of photosynthesis and cell respiration, create an illustration showing the hydrogen cycle and the oxygen cycle.
4. Tundra and desert ecosystems support a much smaller biomass of animals than do tropical rainforests. Give two reasons for this fact.
5. Evaluate the sustainability of parts of the human system, such as transportation, manufacturing, agriculture, and waste disposal, by relating them to the two principles of sustainability in this chapter. How can such things be modified to make them more sustainable?

WEB REFERENCES

On-line resources for this chapter can be found on the World Wide Web at: **http://www.prenhall.com/nebel**. Click on Chapter 3 on the chapter selector.

4

Ecosystems:
Populations and Succession

∿∿∿∿∿∿∿∿

Key Issues and Questions

1. Population growth is the result of a balance between biotic potential and environmental resistance. What do these terms mean, and what are the basic patterns of growth of natural populations?

2. Density-dependent mechanisms help to regulate natural populations. Explain density-dependence and density-independence, and apply the concept of critical number.

3. Predators, parasites, and grazers are important in controlling populations. Explain how they can operate in a density-dependent manner.

4. The introduction of a foreign species frequently has disruptive ecological results. Explain why this is the case. Give examples.

5. Competition between species can be an important check on populations. How is competition between plant species minimized by different plant adaptations and balanced herbivory?

6. Natural ecosystems may undergo gradual succession until they reach a climax, or a more stable state of ongoing adaptation. What is meant by succession, and what factors are responsible?

7. Fire is a major form of disturbance to terrestrial ecosystems. Of what significance are disturbances, and what role can they play in maintaining high levels of biodiversity?

8. The major ecological imbalance on Earth is between the human species and the rest of the Earth's biota. What will be the probable consequences of not achieving a balance? What directions can be taken toward achieving a balance?

In May and June of 1988, following an extended drought, lightning started a large number of fires in Yellowstone National Park, a 2.2-million-acre gem in the U.S. National Park System. The fires burned slowly for a while, and then, fanned by high winds behind a series of cold fronts with no rain, they broke out and swept across hundreds of thousands of acres. Prior to the 1988 fires, the Park Service policy on fires was to allow them to burn their course unless they were near human habitations. This itself was a reversal of the Smokey Bear policy of earlier years of suppressing and preventing all possible forest fires. The Yellowstone fires touched off a great political controversy over Park Service policy and led to fruitless efforts to put the fires out. In spite of the largest fire-fighting effort in U.S. history, the fires were finally put out by a snowfall, leading to a standing joke at the time: "How do you put out the Yellowstone fires? Pour a hundred million dollars on it and wait for it to snow."

The fires burned nearly 10% of the park area, leaving behind a "patchy" landscape, where areas heavily or lightly burned were interspersed among areas untouched by the fires. However, within two weeks, grasses and herbaceous vegetation began sprouting from the ashes, and a year later, abundant vegetation covered the burned areas (Fig. 4–1a). Herbivores like bison and elk fed on the lush new growth. Ten years later, lodgepole pines several feet tall carpeted much of the ground, competing with aspen sprouts (Fig. 4–1b). It is expected that some 25 years after the fires, the diversity of plants and animals in the burned areas will have completely recovered; in the meantime, the total diversity of the park has been unaffected.

Yellowstone National Park. This scene illustrates the patchy landscape of the park as it appeared shortly after the devastating fires of 1988.

(a)   (b)

▲ FIGURE 4–1 *Recovery from fire.* In 1988, a fire swept through Yellowstone National Park. (a) One year after the fire, ground vegetation was well established. (b) Eight years after the fires, lodgepole pines were replacing the herbaceous vegetation.

It is hard to imagine a more devastating disaster to a forest ecosystem than a roaring forest fire. Yet, in the long run, as events at Yellowstone proved, the total landscape was recovering, populations of herbivores and predators reacted positively to the disturbance, and biodiversity was, if anything, enhanced. Many populations of plants revealed remarkable adaptations to fire, flourishing far more than in the unburned areas. The natural landscape demonstrated the capacity to recover from a profound disturbance that was looked on at the time as a disaster.

In our previous two chapters, our focus has been on the structure and function of ecosystems—viewing them as units of sustainability in the natural landscape. We considered the environmental conditions and resources that act as limiting factors in the broad distribution of ecosystems on Earth. We saw how important ecosystems are in human affairs, we sought to assign a value to the goods and services we derive from them, and we maintained that such importance should be reflected in the manner that we manage them, for manage them we must. In addition, we are recommending that ecosystems serve as models of sustainability, and we are in the process of deriving principles of sustainability extracted from our understanding of how ecosystems work.

In this chapter, we will depart from our look at processes in ecosystems and will instead take a closer look at the populations of different species that make up the living community of ecosystems. We will address the very important question of how it is that populations are sustained over time and what kinds of interactions between populations we can expect in ecosystems. In particular, we address the question of balance, or equilibrium, in ecosystems. What does it mean that an ecosystem is in balance? Are populations always in a state of equilibrium? What happens when a major disturbance like fire interrupts the "balance" that we observe in an ecosystem? Can ecosystems and the natural populations in them change over time and yet maintain the important processes that make them sustainable units?

Far from retaining the notion that nature is stable and will take care of itself, we will see that an ecosystem is a dynamic system in which changes are constantly occurring. The notion of a "balanced" ecosystem, then, must be questioned. Our objective in this chapter is to examine more of the basic mechanisms that underlie the sustainability of all ecosystems, including human systems and the rest of the biosphere. We do so by first considering the perspective of populations and how they are controlled, and then we will consider ways in which changes occur over time in ecosystems and the reasons for those changes.

4.1 Population Dynamics

Each species in an ecosystem exists as a population—that is to say, a reproducing group. Studies have indicated that over a long period of time, the populations of many species in an ecosystem tend to remain more or less constant in size and geographic distribution. In short, on average, deaths equal births; otherwise the population would shrink or grow accordingly. We speak of a balance between births and deaths as **population equilibrium**. Yet we know that populations are capable of growth under the right conditions. How can we relate both growth and equilibrium?

Population Growth Curves

When the size of a population is plotted over time, two basic kinds of curve can be seen: S-curves and J-curves (Fig. 4–2). For example, suppose some abnormally severe years have reduced a population to a low level, but then conditions return to normal. Once normal conditions return, the population may increase exponentially for a time, but then either of two things may occur: 1) Natural mechanisms come into play and cause the population to level off and continue in a dynamic equilibrium. This pattern is known as the S-curve; 2) In the absence of natural enemies, the population keeps growing until it exhausts essential resources—usually food—and then dies off precipitously due to famine and, perhaps, diseases related to malnutrition. This pattern is known as a J-curve.

An important characteristic of J-curve growth is the rapidity with which a population can go from modest levels to the peak and then crash. Consider, for example, that an insect population is doubling (hence doubling the amount it eats) *each week*. Suppose it has taken a given insect population eight weeks to devour one half of a crop. How long will it take to devour the second half? The answer is *1 week!* In any doubling sequence, the last doubling necessarily includes one half of the total. You may test this for yourself by taking a sheet of graph paper and blackening first one square, then two, then four, then eight, and so on, doubling the number each time. You will find that regardless of the size of the paper or the size of the squares, the last turn will involve blackening half the paper. If you double the size of the paper, how many more turns will that provide?

What follows the J-curve crash? Any one of three scenarios may unfold. First, if the ecosystem has not been too seriously damaged, the producers may recover, allowing a recovery of the herbivore population, and the J-curve may be repeated. This scenario is seen in periodic outbreaks of certain pest insects, even in natural ecosystems. Second, after the initial J, natural mechanisms may come into the picture as the ecosystem recovers and bring the population into an S-balance. There is evidence that such a balance has been established in the eastern United States for the introduced gypsy moth (Fig. 4–3). Stands of oak trees that were devastated by the initial invasion of gypsy moths a few years ago have recovered, and the gypsy moth is remaining at low levels. In the third scenario, damage to the ecosystem may be so severe that recovery does not occur, but small surviving populations eke out an existence in a badly degraded environment.

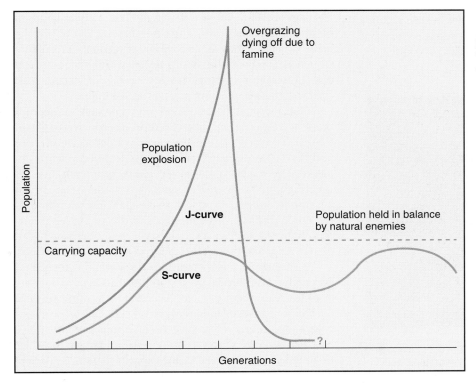

◀ FIGURE 4–2 *Two types of growth curves.* The *J*-curve (blue) demonstrates population growth under optimal conditions, with no restraints. The *S*-curve (green) shows a population at equilibrium.

EARTH WATCH

MAXIMUM VERSUS OPTIMUM POPULATION

Later in this chapter we define *carrying capacity* as the *maximum* population a habitat will support without being degraded over the long term. This population level, however, is not a simple fixed number that can be arrived at easily, and it will vary from year to year in any given ecosystem. Carrying capacity may be considerably higher in wet years than in dry years, which inhibit the growth of vegetation. In addition, the line between degrading and not degrading the habitat over the long term is anything but clear. Degradation may take place slowly and be hardly perceptible from year to year, yet it does occur.

Therefore, managers nowadays focus more on what is the *optimal* rather than the maximum population. The optimal population is large enough to ensure a healthy breeding stock, yet it is considerably less than the theoretical maximum. It allows flexibility between good years and bad years without upsetting the ecosystem.

The optimum population is not simple or fixed, either. It too can be influenced by habitat management. For example, planting certain species, thinning a forest, and starting new trees will increase the amount of browse available and enhance the carrying capacity for deer. Creation of ponds and marshes will attract and support waterfowl.

What is the minimum size of habitat that is required to support a breeding population? Natural areas isolated from surrounding similar areas proceed to lose species.

The smaller the isolated area, the greater the number of species lost. Thus, fragmentation of forests by intervening development is leading to many species declines because the fragments are insufficient to support critical numbers even though the ecosystems are otherwise preserved. It is being found that the loss of species may be reduced to some extent by keeping the remaining fragments connected by "green corridors."

Over 350 formal *biosphere reserves* are recognized worldwide. These areas contain a core in which genetic material is protected from outside disturbances by a surrounding buffer zone, which is in turn surrounded by a transition area. In an ideal sense, these reserves are meant to slow extinction rates and the fragmenting of natural sites.

The outstanding feature of natural ecosystems—ecosystems that are more or less undisturbed by human activities—is that they are made up of populations that are usually in the dynamic equilibrium represented by S-curves. J-curves come about when there are unusual disturbances, such as the introduction of a foreign species, the elimination of a predator, or the sudden alteration of a habitat. A J-curve is a picture of nonsustainability—a population growing out of control. Such increases cannot be sustained in animal populations; there is an inevitable dying off as resources are exhausted. Nevertheless, under the right conditions, population growth is always a possibility for species.

▲ FIGURE 4–3 *Gypsy moth caterpillar.* The Gypsy moth, an introduced species that has often caused massive defoliation of trees, now seems to have been brought under natural control in forests.

Biotic Potential Versus Environmental Resistance

We refer to the ability of populations to increase as the **biotic potential**, the number of offspring (live births, eggs laid, or seeds or spores set in plants) that a species may produce under ideal conditions. Looking at different species, you can readily see that biotic potential varies tremendously, averaging from less than one birth per year in certain mammals and birds to many millions per year in the case of many plants, invertebrates, and fish. However, to have any effect on the size of subsequent generations, the young must survive and reproduce in turn. Survival through the early growth stages to become part of the breeding population is called **recruitment**.

Considering differences in biotic potential and recruitment, we observe two different **reproductive strategies** in the natural world. The first strategy is to produce massive numbers of young but then leave survival to the whims of nature. This strategy results in very low recruitment. Thus, despite high biotic potential, population increase may be nil because recruitment is so low. (Note that "low recruitment" is a euphemism for high mortality of the young.) The second strategy is to have a much lower reproductive rate but then care for and protect the young to enhance recruitment.

Additional factors that influence population growth and geographic distribution are: the ability of animals to migrate, or of seeds to disperse, to similar habitats in other regions; the ability to adapt to and invade new habitats; defense mechanisms; and resistance to adverse conditions and disease.

Taking all these factors for population growth together, we find that every species has the capacity to increase its population when conditions are ideal. Furthermore, growth of a population under ideal conditions will be *exponential*. For example, a pair of rabbits producing 20 offspring, 10 of which are female, may grow by a factor of 10 each generation: 10, 100, 1000, 10,000, (10^1, 10^2, 10^3, 10^4) and so on. Such a series is called an **exponential increase**. When this occurs in a population, it is commonly called a **population explosion**. A basic feature of such an exponential increase is that the numbers increase faster and faster as the population doubles and redoubles, with each doubling occurring in the same amount of time. If we plot numbers over time during exponential increase, the pattern produced looks like a J (Fig. 4–2).

Population explosions are seldom seen in natural ecosystems, however, because a large number of both biotic and abiotic factors tends to decrease population. Among the biotic factors are predators, parasites, competitors, and lack of food. Among abiotic factors are unusual temperatures, moisture, light, salinity, pH, lack of nutrients, and fire. The combination of all the abiotic and biotic factors that may limit population increase is referred to as **environmental resistance**.

You may already foresee the result of the interplay between the factors promoting population growth and those leading to population decline. Conditions are always changing. When they are favorable, populations may increase. When they are unfavorable, populations decrease. If we plot numbers over time in this case, the pattern looks more like an S (Fig. 4–2).

In general, the reproductive ability of a species, or its biotic potential, remains fairly constant, because that ability is part of the genetic endowment of the species. What varies substantially is recruitment. It is in the early stages of growth

that individuals (plants or animals) are most vulnerable to predation, disease, lack of food (or nutrients) or water, and other adverse conditions. Consequently, environmental resistance effectively reduces recruitment. Of course, some adults also perish, particularly the old or weak. If recruitment is at the **replacement level**, that is, just enough to replace these adults, then the population will remain constant—a population equilibrium. If recruitment is not sufficient to replace losses in the breeding population, at that point the population will decline.

However, there is a definite upper limit to the population of any particular animal that an ecosystem can support. This limit is known as the *carrying capacity* (Fig. 4–2). More precisely, **carrying capacity** is defined as the maximum population of an animal that a given habitat will support without the habitat being degraded over the long term. If a population greatly exceeds the carrying capacity, it will undergo a J-curve crash, as Fig. 4–2 shows.

In certain situations, environmental resistance may affect reproduction as well as causing mortality directly. For example, the loss of suitable habitat often prevents animals from breeding. Also, certain pollutants adversely affect reproduction. However, we can still view these situations as environmental resistance that either blocks a population's growth or causes its decline.

In sum, whether a population grows, remains stable, or decreases is the result of an *interplay between its biotic potential and environmental resistance* (Fig. 4–4). In general, biotic potential remains constant; it is changes in environmental resistance that allow populations to increase or cause

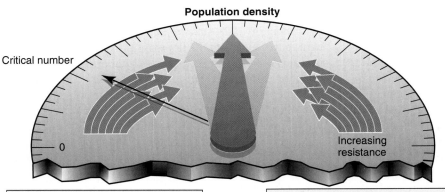

◄ FIGURE 4–4 *Biotic potential and environmental resistance.* A stable population in nature is the result of the interaction between factors tending to increase population (biotic potential) and factors tending to decrease population (environmental resistance).

Population density

Critical number

0

Increasing resistance

Biotic Potential
• Reproductive rate
• Ability to migrate (animals) or disperse (seeds)
• Ability to invade new habitats
• Defense mechanisms
• Ability to cope with adverse conditions

Environmental Resistance
• Lack of food or nutrients
• Lack of water
• Lack of suitable habitat
• Adverse weather conditions
• Predators
• Disease
• Parasites
• Competitors

ETHICS

HUNTING VERSUS ANIMAL RIGHTS

Hunting and animal rights are increasingly the topic of moral and ethical debates, with hunting enthusiasts and animal rights activists at the extremes of the issue. Certainly, each individual has a right to choose for himself or herself whether or not to engage in legal hunting, eating meat, wearing leather, or otherwise being a party to killing animals. But which point of view will best serve the long-term interests of society and the biosphere?

Proponents on both sides of the question could profit from a greater ecological understanding and awareness. If maintaining the biodiversity and ecological balances in the biosphere is taken as the primary aim, then there is a rationale for both sides. If we are looking at an endangered or threatened species, condoning its further hunting for any reason seems unconscionable because such hunting can only hasten its extinction,

causing a permanent loss for both society and the biosphere. This reasoning can be extended beyond hunting to include habitat destruction for development or other purposes, a process that is also causing the extinction of species. Where animal rights projects lend support to saving endangered species and their habitats, the efforts of the people involved are commendable.

On the other hand, if we are looking at a species that is overpopulating and overgrazing because its natural enemies have been removed (e.g., deer) or because it is an introduced species without natural enemies, then some would argue for human intervention. Continuing to allow such animals to overpopulate and overgraze can lead only to widespread damage to ecosystems and crops, damage that will include the death and even extinction of other animals dependent on the same vegetation for food and habitat. It will lead also to the death of

the animals that are overpopulating, because they will eventually deplete their food supply and die of starvation or disease.

Is keeping their population in check by hunting more cruel? Does allowing animals to overgraze in any way serve the value of preserving the integrity of the biosphere? Does one particular animal have rights that exceed those of others or of the ecosystem as a whole?

In conclusion, it makes little sense to argue, much less act on, the ethics of animal rights or hunting as issues in and of themselves. Meaningful resolution can be reached, however, when the debate or action is put in terms of particular species and the ecosystems in which they exist, and when the determining value is based on the preservation of that ecosystem. Preserving the ecosystem is the only way to preserve biodiversity. As for the actual act of killing an animal, this remains a personal decision based on individual values.

---

them to decrease. For example, a number of favorable years (low environmental resistance) will allow a population to increase; then a drought or other unfavorable conditions may cause it to die back, and the cycle may be repeated.

We emphasize that population balance is a **dynamic balance**, which implies that additions (births) and subtractions (deaths) are occurring continually and the population may fluctuate around a median (Fig. 4–2). Some fluctuate very little; others fluctuate widely, but as long as decreased populations restore their numbers, and the ecosystem's carrying capacity is not exceeded, the population is considered to be at an equilibrium. Still, the questions remain: What maintains the equilibrium within a certain range? What prevents a population from "exploding" or, conversely, becoming extinct?

Density Dependence and Critical Numbers

In general, the size of a population remains within a certain range because most factors of environmental resistance are *density-dependent*. That is, as **population density** (the number of individuals per unit area) increases, environmental resistance becomes more intense and causes such an increase in mortality that population growth ceases or declines. Conversely, as population density decreases, environmental resistance is generally lessened, allowing the population to recover. This balancing act will become clearer as we discuss

specific mechanisms of population equilibrium in the next section. Factors in the environment that cause mortality can also be *density-independent*. That is, their effect is independent of the density of the population. A sudden deep freeze in spring, for example, can cause high mortality in early germinating plants, independent of their density. Although density-independent factors can be important sources of mortality, they are not responsive to population density and thus are not involved in maintaining population equilibria.

There are no guarantees that a population *will* recover from low numbers. Extinctions can and do occur in nature. For example, where are the dinosaurs? The survival and recovery of a population depends on a certain minimum population base, which is referred to as the **critical number**. You can see the idea of critical number at work in terms of a herd of deer, a pack of wolves, a flock of birds, or a school of fish. Often, the group is necessary to provide protection and support for its members. In some cases the critical number is larger than a single pack or flock, because interactions between groups may be necessary as well. In any case, if a population is depleted to below the critical number needed to provide such supporting interactions, the surviving members actually become more vulnerable, breeding fails, and extinction is virtually inevitable. Thus, we must recognize that the density-dependent decline and recovery of a population occurs well above the critical number for the population.

As we will see subsequently, human activities are responsible for the decline and even extinction of many plants and animals (Chapter 11). Why is this? It is simply that human impacts, such as altering habitats, pollution, hunting, and other forms of exploitation are not density-dependent; they can even intensify as populations decline toward extinction. Concern for these declining populations eventually led to the Endangered Species Act, which calls for recovery of two categories of species. Species whose populations are declining rapidly because of human impacts are classified as **threatened**. If the population is near what scientists believe to be its critical number, the species may be classified as **endangered**. These definitions, when officially assigned by the U.S. Fish and Wildlife Service, set into motion a number of actions aimed at stemming the negative impacts on, and providing protection for and even artificial breeding of, the species in question (see Chapter 11). Nongovernmental organizations, such as the World Wildlife Fund and the National Wildlife Federation are also playing a great role in protecting threatened and endangered species.

4.2 Mechanisms of Population Equilibrium

With our general understanding of population equilibrium as a dynamic interplay between biotic potential and environmental resistance, we now turn our attention to some specific kinds of population interactions. (Keep in mind, however, that in the natural world, a population is subjected to the total array of all the biotic and abiotic environmental factors around it. Single factors are seldom totally responsible for the regulation of a given population; rather, regulation results from many factors acting together.)

Predator-Prey and Host-Parasite Dynamics

The best known mechanism of population control is regulation of a population by a predator—that is, a *predator-prey* relationship. An example well documented in nature is the interaction between wolves and moose on Isle Royale, a 45-mile-long island in Lake Superior not far from the Canadian shore.

During a hard winter early in this century, a small group of moose crossed the ice to the island and stayed. Their population grew considerably in the absence of predators. Then, in 1949, a pack of wolves also managed to reach the island. Nine years later, in 1958, wildlife biologists began carefully tracking the populations of the two species (Fig. 4–5). As seen in the figure, the rise in the moose population is followed by a rise in the wolf population, and the decline in the moose population is followed by a decline in the wolf population. The data can be interpreted as follows: A paucity of wolves represents low environmental resistance for the moose, so the moose population increases. Then, the abundance of moose represents optimal conditions (low environmental resistance) for the wolves, so the wolf population increases. The growing wolf population means higher predation on the moose, so the moose population falls. The decline in the moose population is followed by a decline in the wolf population, because now there are fewer prey. We can see how this cycle can be repeated indefinitely, providing a dynamic balance between the populations of moose and wolves.

The most recent data from Isle Royale shows another cycle in progress, but the wolf population has not increased as rapidly as expected. Apparently, the wolves suffered the decline seen in the 1980s due to a canine virus introduced by human visitors and their dogs. Also, the most recent dramatic fall in the moose population cannot be attributed entirely to predation by the small number of wolves. Deep snow and

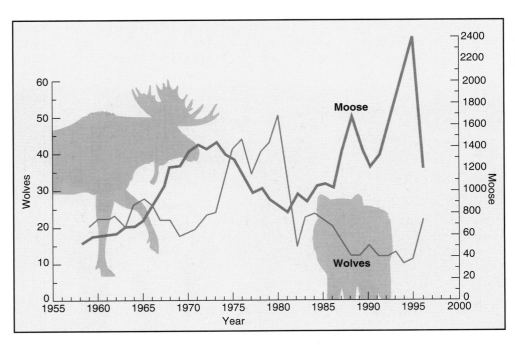

◀ **FIGURE 4–5** *Predator-prey relationship.* Wolves and moose populations on Isle Royale from 1955 to 1996.

an increase in ticks are thought to have caused additional mortality. Therefore, the situation may be less straightforward than implied above. In both predators and prey species, other factors may come into play to cause the observed fluctuations in population densities. For example, the shortage of vegetation that occurs as a herbivore population grows may stress the animals, especially the old, sick, and young, and make them more vulnerable to predation and disease. It has been observed, for instance, that wolves are incapable of bringing down an adult moose in good physical condition. The animals they kill are the young and those weakened by another factor—which is not the case when the predators are human hunters.

The observation that predators are often incapable of killing individuals of their prey that are mature and in good physical condition is extremely significant. This is what prevents predators from eliminating their prey. As the prey population is culled down to those healthy individuals that can escape attack, the predator population will necessarily decline, unless it can switch to other prey. Meanwhile, the survivors of the prey population are healthy and can readily reproduce the next generation.

Much more abundant and equally important as predators in population control is a huge diversity of *parasitic organisms*. Recall from Chapter 2 that these organisms range from tapeworms, which may be a foot or more in length , to microscopic disease-causing bacteria, viruses, protozoans, and fungi. All species of plants and animals (including the predators), and even microbes themselves, may be infected with parasites.

In terms of population impacts, parasitic organisms often act in the same way as large predators do—in a density-dependent manner. As the population density of the host organism increases, parasites and their *vectors* (agents that carry the parasites from one host to another), such as disease-carrying insects, have little trouble finding new hosts, and infection rates increase, causing higher mortality. Conversely, when the population density of the host is low, transfer of infection is less efficient, and there is a great reduction in levels of infection, a condition that allows the population to recover.

You can readily see how a parasite can work in conjunction with a large predator to control a given herbivore population. Parasitic infection breaks out in a dense population. Individuals weakened by infection are more easily removed by predators, leaving a smaller but healthier population.

The wide swings observed in the populations of moose and wolves on Isle Royale are seen to occur in very simple ecosystems involving relatively few species. However, most food webs are more complicated than that, and a population of any given organism is affected by a number of predators and parasites simultaneously. As a result, population density of a species can be thought of more broadly as a consequence of its relationships with all of its **natural enemies**. Relationships between a prey population and its natural enemies are generally much more stable and less prone to wide fluctuations than when only a single predator or parasite is involved, because different predators or parasites come into play at different population densities. Also, when the preferred prey is at a low density, the population of the predator may be supported by feeding on something else. Thus, the lag time between an increase in the prey population and that of the predator is diminished. These factors have a great damping effect on the rise and fall of the prey population.

Introduced Species

Another way of seeing that relationships among species are indeed delicate is to observe how vulnerable they are to the introduction of species from a foreign ecosystem. Consider the following examples.

In 1859, rabbits were introduced to Australia from England to be used for sport shooting. The Australian environment proved favorable to the rabbits and contained no carnivore or other natural enemies capable of controlling them. The result was that the rabbit population exploded and devastated vast areas of rangeland by overgrazing (Fig. 4–6). The devastation was extremely damaging to both native kangaroos and ranchers' sheep. It was finally brought under control by the introduction of a rabbit-disease virus. Over time,

▶ **FIGURE 4–6** *Rabbits overgrazing in Australia.* On one side of a rabbit-proof fence, there is lush pasture; on the other side, it is barren.

(a)

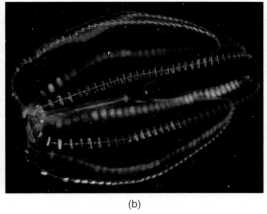

(b)

▲ **FIGURE 4–7** *Introduced species.* (a) Zebra mussels introduced from Europe are now proliferating throughout the Great Lakes and the Mississippi valley. (b) Ctenophores, originating on the South Atlantic Coast of the United States and introduced in Europe, have destroyed fishing in the Black and Azov Seas.

however, the rabbits adapted to the virus and began to repeat their explosive growth. After a recent release of a different virus rabbits have undergone a 95% decline in some regions, and kangaroos and rare plants are thriving once again.

Prior to 1900, the dominant tree in the Eastern deciduous forests of the United States was the American chestnut, which was highly valued for both its high-quality wood and its prolific production of chestnuts, which were eaten by wildlife as well as people. In 1904, however, a fungal disease called the chestnut blight was accidentally introduced when some Chinese chestnut trees carrying the disease were planted in New York. The fungus spread through the forests, killing nearly every American chestnut tree by 1950. Although oaks filled in where the chestnuts died, the ecological and commercial loss was incalculable. There is hope, however. Researchers have recently cross-bred the American and Chinese chestnut, resulting in a hybrid that is 94% native and resistant to the blight. It will be several decades before enough of the hybrids are available for sale to nurseries and wider distribution.

Most of the important insect pests in croplands and forests—Japanese beetles, fire ants, and gypsy moths, for example—are species introduced from other ecosystems. Domestic cats introduced into island ecosystems have often proved to be effective predators and have exterminated many species of wildlife unique to the islands. They are also responsible for greatly diminished songbird populations in urban and suburban areas, including parks (Chapter 11). Goats introduced onto islands, because of their voracious appetites, have been devastating to both native plants and animals dependent on that vegetation.

Such unfortunate introductions are not all in the past. With expanding world trade and travel, the problem is increasing. In 1986, the zebra mussel was introduced to the Great Lakes with the discharge of ballast water from European ships (Fig. 4–7a). The mussel is now spreading through the Mississippi River basin and may go much farther, causing untold ecological and commercial damage as it displaces other species and clogs water-intake pipes. Of

course, such problems may be exported as well as imported. In 1982, several species of jellyfish-like animals known as ctenophores were similarly transported from the east coast of the United States to the Black and Azov Seas in Eastern Europe. The ctenophores have cost Black Sea fisheries an estimated $250 million and have totally shut down those in the Azov, for they both kill larval fish directly and deprive larger fish of food (Fig. 4–7b).

Plant species have also been moved all over the world and, in some cases, have raised havoc. For example, in 1884 the water hyacinth, a plant originally from South and Central America, was introduced into Florida as an ornamental flower. It soon escaped into waterways, where it had little competition and few natural enemies. By now it has proliferated to the extent of making navigation difficult or impossible on the waters in which it thrives (Fig. 4–8a). Millions of dollars have been spent attempting to get rid of this weed, but with little success.

Kudzu, a vigorous vine introduced from Japan in 1876, was widely planted on farms throughout the southeastern United States with the idea of using it for cattle fodder and also for erosion control. From wherever it is planted, however, kudzu invades and climbs over adjacent forests, smothering everything (Fig. 4–8b). Considerable efforts are now being exerted by the Forest Service to eradicate it.

Spotted knapweed (Fig. 4–8c), which was probably unwittingly introduced into this country with alfalfa seed imported from Europe in the 1920s, is taking over vast areas of rangelands in the northwestern United States and southwestern Canada. Totally inedible, knapweed is endangering wildlife such as elk and rendering the lands worthless for grazing by domestic cattle as it displaces native plants and grasses. It has also been found to promote greater soil erosion where it is abundant.

Wetlands throughout temperate regions of the United States and Canada, already reduced by over 50% by development pressures, are now being further degraded by the invasion of purple loosestrife, which was introduced from Europe in the 1800s as an ornamental and medicinal plant

(a)

(b)

(c)

(d)

▲ FIGURE 4–8 *Introduced plant species.* (a) Water hyacinth overgrowing waterways. (b) Kudzu overgrowing forests. (c) Spotted knapweed taking over rangeland in the northwestern United States and southwestern Canada. (d) Purple loosestrife, a native of Eurasia, now is abundant in wetlands of the northeastern United States and Canada.

(Fig. 4–8d). By outcompeting and eliminating native wetland vegetation and being inedible itself, purple loosestrife is threatening many wildlife species, including waterfowl. Control of such weeds by the introduction of plant-eating insects is being investigated. However, it is hard to find an insect (or other organism) that will control the target weed but not attack desired species (see Chapter 17).

The balance among competing plant species may also be upset by the introduction of a herbivore, which invariably attacks some plant species but not others. For example, millions of acres of pasture and rangelands are now dominated by inedible weeds and shrubs as cattle have eaten back the grass and allowed the competing weeds to flourish (Fig. 4–9). (Shortly, we will consider the importance of maintaining a balance among competing species of plants.)

The list of introductions goes on and on, totaling in the tens of thousands. Still, not all of them have turned out badly. For instance, two species that lack the ability to become dominants, the ring-necked pheasant and the day lily, have found places in new ecosystems, apparently occupying vacant niches.

The ecological lesson to be learned from our experience of unfortunate introductions is twofold. First, we should emphasize again that regulation of populations is a matter of complex interactions among the members of the biotic community. Second, and just as important, the relationships are specific to the organisms in each particular ecosystem. Therefore, when we transport a species over a physical barrier from one ecosystem to another, it may find the environment favorable, but it is unlikely to fit into the framework of relationships in the new biotic community. Actually, in the best case, it finds the environmental resistance of the new system too severe and dies out. No harm is then done: The receiving system continues unaltered. In the worst case, the transported species finds physical conditions and a food supply that are hospitable, together with an insufficient number of natural enemies to stop its population growth. The damage is done as its population explodes, and it drives out native species by outcompeting them for space, food, or other resources, if not by predation. Such disruptions may be caused by any category of organism—plant, herbivore, carnivore, or parasite, large or small.

▲ **FIGURE 4–9** *Impact of a herbivore on competing plant species.* Grazing by a herbivore (left side of fence) can unbalance the equilibrium between competing plant species, as grazing will tip the scale in favor of the plants (inedible weeds and shrubs) resistant to grazing.

You may be wondering why the species interactions in different ecosystems should differ. It is because ecosystems on different continents or remote islands have been isolated by physical barriers for millions of years. Consequently, the species within each system have developed adaptations to other species *within their own ecosystem, and these are independent of those that have developed in other ecosystems.* We shall address the question of how species adapt to and develop such balances with each other in Chapter 5.

The seemingly obvious solution to the takeover by a "foreign invader" is to introduce a natural enemy. Indeed, this approach has been used in a number of cases; rabbit control in Australia is one. Others are discussed in connection with the biological control of pests in Chapter 17. However, such control is more easily said than done. Recall that control is achieved as a result of the interplay among *all* the factors of environmental resistance, which often include several natural enemies, as well as *all* the abiotic factors. Thus, a single natural enemy that will control a pest simply may not exist. There is no guarantee that the natural enemy, when introduced into the new ecosystem, will focus its attention on the target pest. Control of the rabbit population in Australia was initially attempted by introducing foxes. However, the foxes soon learned that they could catch other Australian wildlife more easily than rabbits and thus went their own way. In short, to prevent doing more harm than good, a great amount of research needs to be done before introducing a natural enemy.

Territoriality

In discussing predator-prey balances, we said that in lean times the carnivore population—the wolf, for instance—either had to switch to other prey or starve. Actually, another factor is often involved in the control of carnivore and some

herbivore populations of a great number of species, ranging from fish to birds and mammals: *territoriality*. **Territoriality** refers to individuals or groups such as a pack of wolves defending a territory against the encroachment of others of the *same species*. For example, the males of many species of songbirds claim a territory at the time of nesting. Their song has the function of warning other males to keep away (Fig. 4–10). The males of many carnivorous mammals, including dogs, "stake out" a territory by spotting it with urine, the smell of which warns others to stay away. If there is encroachment, there may be a fight, but in natural species a large part of the battle is intimidation—an actual fight rarely results in death.

In territoriality, what is really being defended or sought after by the "invader" is claim to an area from which adequate food resources can be obtained in order to rear a brood successfully. Hence, the territory is only defended against others that would cause the most direct competition for those resources—usually, members of the same species. As a consequence of territoriality, some members of the population are able to gain access to sufficient food resources to rear a well-fed next generation. Thus, a healthy population of the species survives. If, instead, there was an even rationing of inadequate resources to all the members, all of them trying to raise broods, the entire population would become malnourished and might perish. By territoriality, breeding is restricted to only those individuals that are capable of claiming and defending territory, and thus population growth is curtailed.

Individuals unable to claim a territory are in large part the young of the previous generation(s). Some may hang out on the fringes and seize their opportunity as they become more mature, and older members with territories weaken. Some fall prey to one or another factors of environmental resistance as they are continually chased out of one territory after another. Finally, some may be driven to migrate. Of course, it is an open question whether such migration will lead them to another region where they can successfully

▲ **FIGURE 4–10** *Territoriality.* A sedge wren is announcing its claim on a territory.

breed, or whether it will lead them to perish in conditions beyond their limit of tolerance along the way. In any case, territoriality may be seen as a powerful force behind migration as well as population stabilization.

Plant-Herbivore Dynamics

In previous discussions, we observed that herbivore populations are commonly held in check by various natural enemies, (and vice versa). In turn, keeping the herbivore population in check prevents it from growing to the extent that overgrazing occurs. Let us examine this relationship further.

In a grazing situation, it is readily apparent that if the herbivores eat the grass faster than the grass can grow, sooner or later the grass will be destroyed, and many of the animals will starve. This situation is known as **overgrazing**. The best way to appreciate the role of natural enemies in preventing a herbivore population from overgrazing is to observe what occurs when natural enemies are *not* present. A classic example with good documentation is the case of reindeer on St. Matthew Island, a 128-square-mile island in the Bering Sea midway between Alaska and Russia. In 1944 a herd of 29 reindeer (5 males and 24 females) was introduced onto the island, where they had no predators. From these 29 animals, the herd multiplied some two-hundred-fold over the next 19 years. Early in the cycle, the animals were observed to be healthy and well nourished, as supporting vegetation was abundant. However, by 1963, when the size of the herd had reached an estimated 6000, the animals were obviously malnourished. Lichens, an important winter food source, had been virtually eliminated and replaced by unpalatable sedges and grasses. During the winter of 1963–64, this factor, combined with harsh weather, resulted in death by starvation of nearly the entire herd; there were only 42 surviving animals in 1966 (Fig. 4–11).

The lesson is that no population can escape ultimate limitation by environmental resistance. But the form of environmental resistance and the consequences may differ. If a population is not held in a reasonable balance, it may explode, overgraze, and then crash as a result of starvation. It is also crucial to note that the consequence is not just to the herbivore in question. One or more types of vegetation may be eliminated and replaced by other forms or not replaced at all, leaving behind a "desert." Other herbivores that were dependent on the original vegetation, and secondary and higher levels of consumers dependent on them, also are eliminated as food chains are severed. Innumerable extinctions have occurred among the unique flora and fauna of islands because goats were introduced by sailors to create a convenient food supply for return trips.

Eliminating predators or other natural enemies upsets basic plant-herbivore relationships in the same way as introducing an animal without natural enemies does. Examples of this type of folly abound as well. For example, in much of the United States, deer populations were originally controlled by wolves, mountain lions, and black bear, most of which have been killed because they were felt to be a threat to livestock and even humans. Now, if it were not for human hunting in place of these natural predators, deer populations in most areas would increase to the point of overgrazing. Indeed, drastic population increases do occur where hunting is prevented. Similarly, prairie dogs and other small rodents are becoming an increasing problem in the western United States as a result of a reduction in the number of their predators, such as coyotes. In this case, hunting is obviously not practical—who wants to go hunting for mice?

A second factor influencing plant-herbivore balance is that large herbivores—bison in the American West or elephants in Africa, for example—were originally able to roam vast regions. As forage was reduced in one area, a herd would simply migrate before overgrazing could occur, as with the large Serengeti herbivore populations (Chapter 3). Because humans have fenced such regions for agriculture and cattle ranching, however, wild herbivores are increasingly confined

▼ FIGURE 4–11 *Plant-herbivore interaction.* In 1944, a population of 29 reindeer was introduced onto St. Matthew Island, where they increased exponentially to about 6000 and then died off due to overgrazing.

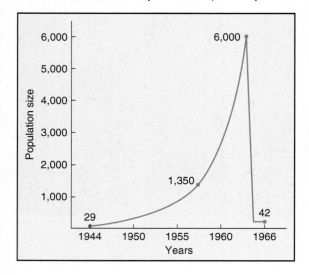

to areas such as parks and reserves. Overgrazing in such presumably "protected" areas has become an increasing danger.

We see, then, that in undisturbed ecosystems, herbivore populations are held in check by a number of important interactions—especially by predators and parasites. They rarely increase to populations numerous enough to overgraze their food. The result is that primary producers (plants) are able to maintain a substantial standing biomass and sustain their production of the organic matter that is so important to the entire ecosystem. We can readily see that this is another feature that is fundamental to sustainability. Hence, here is the **third basic principle of ecosystem sustainability**:

For sustainability, the size of consumer populations is controlled so that overgrazing or other overuse does not occur.

Competition Between Plant Species

We discussed in Chapter 2 the significance of competition between species and introduced the concept of the ecological niche. If two species are competing for some scarce resource, we say that their niches overlap. One outcome of such competition may be the elimination of one of the species, or competitive exclusion. However, competitive exclusion requires a stable and homogeneous environment and may take decades or longer. Consider the situation for plants. A natural ecosystem may contain hundreds or even thousands of species of green plants, all competing for nutrients, water, and light. What prevents one plant species from driving out others in competition—again, an event that may occur in the case of introduced species?

First, we observed in Chapter 2 that because of differences in topography, soil type, and so on, the landscape is far from uniform, even within a single ecosystem. It is actually comprised of numerous microclimates or microhabitats. That is to say, the specific conditions of moisture, temperature, light, and so on differ from location to location. Thus, the adaptation of a species to specific conditions enables it to thrive and overcome its competitors in one location but not in another. An example is the distribution of trees along streams and rivers. In the Great Plains states, trees only grow along waterways, because elsewhere the environment is too dry. This creates what are called **riparian** woodlands (Fig. 4–12). In the eastern United States, sycamore and/or red maple, which can thrive in water-saturated soil, grow along river banks. Oaks, which require well-drained soil, occupy higher elevations. In the West, white alder, willow, and cottonwoods can survive in water-saturated soils.

A second factor affecting the competition between plant species is the fact that a single species generally cannot utilize all of the resources in a given area. Therefore, any resources that remain may be gathered by other species having different adaptations (Chapter 2). For example, grasslands contain both grasses that have a fibrous root system and plants that have tap roots (Fig. 4–13). These different root systems enable the plants to coexist because they get their water and nutrients from different layers of the soil. Also, trees in a forest, while competing with each other for light in the canopy (the "layer" of treetops),

▲ **FIGURE 4–12** *Riparian woodlands.* Competition between plant communities is often maintained by differing amounts of available moisture. Riparian woodlands are shown growing only along a river in this prairie region.

leave lots of space near the ground, and this space may be occupied by plants (ferns and mosses, for example) that can tolerate the reduced light intensity. Another adaptation is the plethora of spring wildflowers that inhabit temperate deciduous forests. Sprouting from perennial roots or bulbs in the early part of that season, these plants take advantage of the light that can reach the forest floor before the trees grow leaves.

▼ **FIGURE 4–13** *Coexistence in plants.* Plants with fibrous roots may coexist with plants having tap roots because each is drawing water and nutrients from a different part of the soil.

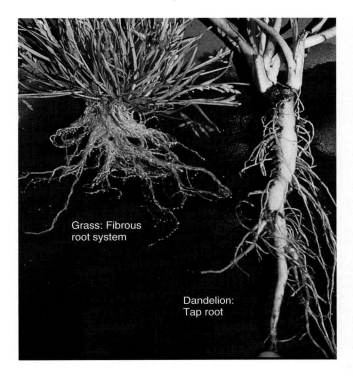

Grass: Fibrous root system

Dandelion: Tap root

▲ **FIGURE 4–14** *Epiphytes in the Amazon region.* The plants (bromeliads) growing on the tree branches are epiphytes. They are not parasitic, but perch on the branches of trees to gain access to light.

Another option is a form of mutualistic symbiosis. For example, in warm, humid climates, the branches of trees are often covered with **epiphytes**, or air plants (Fig. 4–14). Such plants are not parasitic. There is some evidence that the epiphytes help to gather the minute amounts of nutrients that come with rainfall (e.g., nitrogen compounds fixed by lightning, page 72) and make them accessible to the tree on which the epiphytes are located.

A third and very important factor in multiple-plant balance is called *balanced herbivory*. It is easiest to understand if we start from the point of view of a **monoculture**—the growth of a single species over a wide area, a practice commonly followed in agriculture and forestry for economic efficiency.

Experience shows that monocultures are exceedingly vulnerable to insects, fungal diseases, or other pests, while diverse ecosystems are much more resistant to them. To understand why this should be, consider the following. First, insects, fungal diseases, and other parasites are, for the most part, **host specific**. That is, they will attack only one species and, possibly, its close relatives. They are unable to attack species unrelated to their specific host. Second, such organisms have an enormous biotic potential. An individual often produces thousands of offspring—even millions, in the case of fungal spores—and they have a generation time of only a few days or weeks.

Now, a monoculture may be seen as a continuous, lush food supply for its particular host-specific attacker, a situation highly conducive to supporting a population explosion. Indeed, the pest population may explode so fast that its natural enemies cannot keep up with it even if they are present. Only virtual elimination of the monoculture halts the multiplication of the pest—a scenario not unlike that of the reindeer described earlier. (It is for this reason that many farmers and forest managers feel obliged to use chemical pest controls despite recognizing that they may present certain environmental hazards. See Chapter 17.)

On the other hand, in a diverse ecosystem—one consisting of a mixture of many different species of plants growing together—the host-specific attacker has trouble reaching its next host. With this limitation, most of the pest's offspring may perish, and the surviving population may be held in check by its natural enemies.

To conclude, visualize a monoculture developing in a natural situation. Its being largely wiped out by an outbreak of its host-specific pest would leave space that might be invaded by another plant species, which in turn might be largely wiped out by an outbreak of *its* pest, leaving space that might be occupied by a third plant species, and so on. The end result of this process would be a diversified plant community, with each species held down in density by its specific herbivore(s) and the herbivores held in check by their natural enemies (Fig. 4–15a). Thus, a **balanced herbivory** may be defined as a balance among competing plant populations being maintained by herbivores feeding on the respective populations.

A prime example of a balanced herbivory is seen in the tropical rain forests of the Amazon River Basin in Brazil and Peru (Fig. 4–15b). A single acre may contain a hundred or more species of trees, but often no more than a single individual of each. The next individual of the same species may be as much as a half mile away. Evidence that this diversity is maintained by a balanced herbivory is seen in that attempts to create plantations of single species—rubber plantations, for example—met with failure because outbreaks of various pests proved uncontrollable in the monoculture situation. (However, rubber plantations proved successful in Malaysia, where climatic and soil conditions are different enough to limit pests while still supporting the rubber trees.)

4.3 Disturbance and Succession

Equilibrium theory is the view that ecosystems are maintained by equilibria or balances between species. In the previous section, we considered a variety of biotic mechanisms that contribute to the regulation of species in ecosystems. Equilibrium theory suggests that ecosystems are stable environments in which species are able to interact constantly in predator-prey and competitive relationships. Thus, biotic interactions determine the structure of living communities within ecosystems. This approach has led to the popular idea of "the balance of nature"—that natural systems maintain a delicate balance over time that lends great stability to them. However, in our previous analysis of ecosystems, we have largely ignored the dimension of *time*. Are species and ecosystems at equilibrium

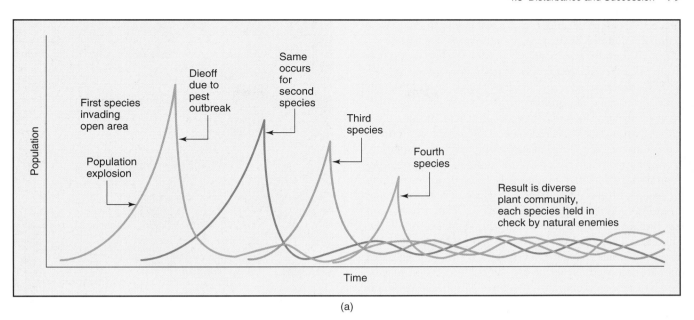

(a)

(b)

◀ **FIGURE 4–15** *Balanced herbivory.*
(a) A plant species may experience a population explosion as it invades an open area. If it dies back and is held down by a herbivore, space is opened up for a second invader, which may experience the same fate, etc. The result is a plant community of many species, each held in check by its specific herbivores. (b) View from canopy walkway overlooking the Amazonian rain forest near Iquitos, Peru. Nearly every tree in view is a different species.

all the time? If the environment were entirely stable and uniform, we might expect the answer to be yes. But this is not the case. How do ecosystems respond to disturbances?

When disturbed, as by human intervention, ecosystems may lose that balance and decline or disappear. Or, they may undergo a recovery that eventually leads them back to a final, stable state called the *climax*. So we must examine the possibility that time brings disturbances to ecosystems that are highly important to their sustainability. Indeed, many studies of ecosystems have shown that they are in fact very patchy environments; conditions and resources within ecosystems vary in temperature, moisture, exposure to sunlight, soil conditions, and the like. The distribution of species is similarly patchy, reflecting the patchiness in conditions. In a forest, for example, it is common to find groups of tree species varying independently in space, such that it is difficult to predict what

species will tend to be found associated together in a given stand at a given time. If neither conditions nor species distributions are uniform even within ecosystems, then perhaps the concept of a stable equilibrium should be questioned. This has caused many ecologists to think of ecosystems as *non-equilibrium systems* that seldom exhibit the characteristics of a true equilibrium. This controversy is well illustrated in the phenomenon known as ecological succession.

Ecological Succession

There are situations in nature where, over the course of years, we observe one biotic community gradually giving way to a second, the second perhaps to a third, and even the third to a fourth. This phenomenon of *transition* from one biotic community to another is called **ecological**, or **natural, succession**.

Succession occurs because the physical environment may be gradually modified by the growth of the biotic community itself, such that the area becomes more favorable to another group of species and less favorable to the present occupants.

The succession of species does not go on indefinitely, however. A stage of equilibrium is reached during which there is a dynamic balance between all species and the physical environment. This final state is referred to as a **climax ecosystem**, or assemblage of species in constant adaptation, each striving to achieve an optimal range and low environmental stress. The major biomes discussed up to this point in the text are climax ecosystems. Of course, it remains important to keep in mind that all balances are relative to the *current* biotic community and the *existing* climatic conditions. Therefore, even climax systems are subject to change if climatic conditions change, if new species are introduced, or old ones are removed. Nevertheless, natural succession may be seen as a progression toward a relatively more stable climax situation, one that no longer changes over time. Note: There may be several "final" stages, or a polyclimax condition, with adjoining ecosystems in the same environment at different stages. The following are three classic examples.

Primary Succession. If the area has not been occupied previously, the process of initial invasion and then progression from one biotic community to the next is termed "primary succession." An example is the gradual invasion of a bare rock surface by what eventually becomes a climax forest ecosystem. Bare rock is an inhospitable environment. There are few places for seeds to lodge and germinate, and if they do, the seedlings are killed by lack of water or by exposure to wind and sun on the rock surface. However, certain species of moss are uniquely adapted to this environment. Their tiny spores, specialized cells that function reproductively, can lodge and germinate in minute cracks, and moss can withstand severe drying simply by becoming dormant. With each bit of moisture, it grows and gradually forms a mat that acts as a sieve, catching and holding soil particles as they are broken from the rock or as they blow or wash by. Thus, a layer of soil, held in place by the moss, gradually accumulates (Fig. 4–16). The mat of moss and soil provides a suitable place for seeds of larger plants to lodge, and the greater amount of water held by the mat supports their germination and growth. The larger plants in turn collect and build additional soil, and eventually there is enough soil to support shrubs and trees. In the process, the fallen leaves and other litter from the larger plants smother and eliminate the moss and most of the smaller plants that initiated the process. Thus, there is a gradual succession from moss through small plants and finally to trees that form a climax forest ecosystem. The nature of the climax ecosystem, of course, differs according to the prevailing abiotic factors of the region, giving us the biomes typical of different climatic regions, as described in Chapter 2.

Because bare rock substrate can be reexposed by glaciation, earthquakes, landslides, and volcanic eruptions, there are always places for primary succession to start anew.

▲ **FIGURE 4–16** *Primary succession on bare rock.* Moss invades bare rock and acts as a collector, accumulating a layer of soil sufficient for additional plants to become established.

Secondary Succession. When an area has been cleared by fire or artificial means and then left alone, the surrounding ecosystem may gradually reinvade the area—not at once, but through a series of distinct stages termed *secondary succession.* The major difference between primary and secondary succession is that secondary succession starts with the preexisting soil substrate. Therefore, the early, prolonged stages of soil building are bypassed. Still, as you can readily experience, a clear area has a microclimate quite the opposite from the cool, moist, shaded conditions beneath a forest canopy. Those plant species that propagate themselves in the microclimate of the forest floor cannot tolerate the harsh conditions of the clearing. Hence, the process of reinvasion necessarily begins with different species and proceeds accordingly. The steps leading from abandoned agricultural fields in the eastern United States back to deciduous forests provide a classic example of secondary succession (Fig. 4–17a).

On an abandoned agricultural field, crabgrass is predominant among the initial invaders. Crabgrass is particularly well adapted to invading bare soil. Its seeds germinate in the spring, and it grows and spreads rapidly by means of runners; moreover, it is exceptionally resistant to drought. In spite of its vigor on bare soil, crabgrass is easily shaded out by taller plants. Consequently, taller weeds and grasses, which take a year or more to develop, eventually take over from the crabgrass. Next, young pine trees, which are well adapted to thrive in the direct sunlight and heat of open fields, gradually develop and shade out the

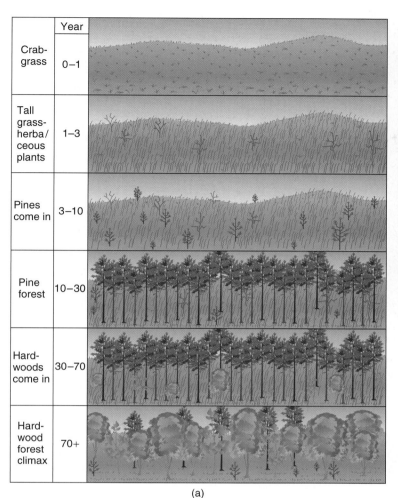

	Year	
Crab-grass	0–1	
Tall grass-herba/ceous plants	1–3	
Pines come in	3–10	
Pine forest	10–30	
Hard-woods come in	30–70	
Hard-wood forest climax	70+	

(a)

(b)

◀ **FIGURE 4–17** *Secondary succession.*
(a) Reinvasion of an agricultural field by a forest ecosystem occurs in the stages shown. (b) Hardwoods (species of oak) growing up underneath and displacing pines in eastern Maryland.

smaller, sun-loving weeds and grasses, eventually forming a pine forest. But pine trees also shade out their own seedlings, which need bright sun to grow. Thus, the seedlings of deciduous trees, not pines, develop in the cool shade beneath the pine trees (Fig. 4–17b). Consequently, as the pines die off (their life span is 40 to 100 years), they are replaced by oaks, hickories, beeches, maples, and other species of hardwoods that characterize eastern deciduous forests. The seedlings of the latter continue to flourish beneath the cover of their parents, providing a stable balance—the climax deciduous forest ecosystem.

We emphasize again that secondary succession requires a suitable soil base to start. If this base is lost by erosion, even the initial plants cannot get a start. Therefore, the bare subsoil may continue to erode and degrade indefinitely, preventing even the possibility of reforestation.

Aquatic Succession. Another example of natural succession is seen as lakes or ponds are gradually filled and taken over by the surrounding terrestrial ecosystem. This process occurs because a certain quantity of soil particles inevitably erodes from the land and settles out in ponds or lakes, gradually filling them. Aquatic vegetation produces detritus that also contributes to

the filling process. As the buildup occurs, terrestrial species can advance, and aquatic species move farther out into the lake. In short, the shoreline gradually advances toward the center of the lake until, finally, the lake disappears altogether (Fig. 4–18).

Disturbance and Biodiversity

In order for natural succession to occur, the spores and seeds of the various invading plants and the breeding populations of the various invading animals must already be present in the vicinity. Ecological succession is not a matter of new species developing, or even old species adapting, to new conditions. It is a matter of populations of existing species taking advantage of a new area as conditions become favorable. Where do these early stage species come from if their usual fate is to be replaced by late-stage or climax species? The answer is a key to the non-equilibrium theory: They come from other surrounding ecosystems in early stages of succession. Similarly, the late-stage species are recruited from ecosystems in later stages of succession. Thus, in any given landscape, there are likely to be all stages of succession represented in the ecosystems present. The reason for this is that disturbances are constantly creating gaps or patches in

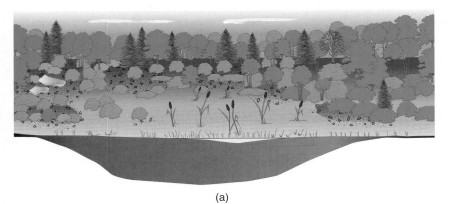

◀ **FIGURE 4–18** *Aquatic succession.*
(a) Ponds and lakes are gradually filled and invaded by the surrounding land ecosystem. (b) In this photograph, taken in Banff National Park in the Canadian Rockies, you can visualize the lake that used to exist in the low-level area. It is now filled with sediment and covered by scrub willow. Spruce and fir forest is gradually encroaching.

(a)

(b)

▲ **FIGURE 4–19** *Iceland.* Forests that originally covered much of this island nation were totally stripped for fuel in the eighteenth and nineteenth centuries. With natural succession impossible, Iceland has remained barren and tundralike, as seen here.

the landscape. It will be readily seen that when a variety of successional stages is present in a landscape, versus one single climax stage, a greater diversity of species can be expected; in other words, biodiversity is enhanced by disturbance! And consequently, natural succession is affected.

Of course, if certain species have been eliminated, natural succession will be blocked or modified. For example, beginning with the early colonization by Norsemen in the eleventh and twelfth centuries, the forests of Iceland were cut for fuel, a process that accelerated with further European colonization in the eighteenth and nineteenth centuries. By 1850 not a tree was left standing, and Iceland remained a barren, tundralike habitat, because there was no remaining source of seeds to foster natural regeneration (Fig. 4–19). Tree seedlings are now being imported and planted in Iceland in the hope that a natural succession may be reestablished.

Fire and Succession

Fire is an abiotic factor that has particular relevance to succession. It is a major form of disturbance common to terrestrial ecosystems. About 75 years ago, forest managers interpreted the potential destructiveness of fire to mean that all fire is bad and embarked on fire-prevention programs that eliminated fires from many areas. Unexpectedly, fire prevention did not preserve all ecosystems in their existing state. In pine forests of the southeastern United States, for instance, economically worthless scrub oaks and other broadleaf species began to displace the more valuable pines. Grasslands were gradually taken over by scrubby, woody species that hindered grazing. Pine forests of the western United States that were once clear and open became cluttered with the trunks and branches of trees that had died in the normal aging process. This deadwood became the breeding ground for

wood-boring insects that proceeded to attack live trees. In California, the regeneration of redwood seedlings began to be blocked by the proliferation of broadleaf species.

Scientists now recognize that fire, which is often started by lightning, is a natural abiotic factor. As with all abiotic factors, different species have different degrees of tolerance to fire. In particular, grasses and pines have their growing buds located deep among the leaves or needles, where they are protected from fire; broadleaf species, such as oaks, have their buds exposed, where they are sensitive to damage from fire. Consequently, where these species exist in competition, periodic fires are instrumental in maintaining a balance in favor of pines, grasses, or redwood trees. In relatively dry ecosystems, where natural decomposition is slow, fire may also play a role in releasing nutrients from dead organic matter. Some plant species actually depend on fire. The cones of lodgepole pine, for example, will not release their seeds until they have been scorched by fire.

Ecosystems that depend on the recurrence of fire to maintain the existing balance are now referred to as **fire climax ecosystems**. This category includes various grasslands and pine forests. Fire is now increasingly used as a tool in the management of such ecosystems. In pine forests, if ground fires occur every few years, there is relatively little accumulation of deadwood. With only small amounts of fuel, fires usually just burn along the ground, harming neither pines nor wildlife significantly (Fig. 4–20). In forests where fire has not occurred for more than 60 years, however, so much deadwood has accumulated that if a fire does break out, it will almost certainly become a crown fire. That is, so much heat is generated that entire living trees are ignited and destroyed. This long-term lack of fire was a major factor in the fires in Yellowstone National Park in the summer of 1988.

Crown fires do occur in nature, since not every area is burned on a regular basis, and exceedingly dry conditions can make a forest vulnerable to crown fires even when large amounts of deadwood are not present. Thus, humans have only actually promoted the potential for crown fires by fire prevention programs. Even crown fires, however, serve to clear the deadwood and sickly trees that provide a breeding ground for pests, to release nutrients, and to provide for a fresh ecological start. Burned areas soon become productive meadows as secondary succession starts anew. We have seen this in Yellowstone National Park. Thus, periodic crown fires create a patchwork of meadows and forests at different stages of succession that lead to a more varied, healthier habitat that supports a greater diversity of wildlife than does a uniform, aging conifer forest (Fig. 4–1).

The devastating fire that spread from brush land in Oakland, California, and destroyed some 1000 homes in the fall of 1991 is another example of lack of fire management. The grasslands on the hills east of San Francisco are a fire climax ecosystem. Preventing fire in this area for many years allowed a superabundance of woody shrubs to grow, not the least of which was eucalyptus that had been introduced, which has a particularly flammable wood. Homeowners created additional fuel by planting a lush, forested landscape,

▶ **FIGURE 4–20** *Ground fire.* Periodic ground fires are necessary to preserve the balance of pine forests. Such fires remove excessive fuel and kill competing species.

unnatural for this dry-summer climatic regime. The result, finally, was a fire that could not be controlled (Fig. 4–21).

Non-equilibrium Systems. In summary, the concept most important to recognize is that disturbances such as fires, floods, windstorms, and droughts are important in structuring ecosystems. The disturbances remove organisms, reduce populations, and create opportunities for other species to colonize. Much of the patchiness observed in natural landscapes is evidence of periodic disturbances. Some ecosystems are more stable than others and experience disturbances very infrequently. In these, biotic relationships become important for maintaining ecosystem stability. Thus, the sustainability of ecosystems is dependent upon maintaining equilibria among populations of the species in the biotic community, and it is also dependent upon maintaining existing relationships between the biotic community and abiotic factors of the environment like disturbances. We have seen that the natural biotic community itself may induce changes in abiotic factors that in turn result in changes in the biotic community (succession). Given this dynamic succession, it should be clear that if any one or more physical factors of the environment are shifted, portions of the biotic community may again be pushed into a state of flux in which certain species that are stressed by the new conditions die out and other species that are better suited to the new conditions thrive and become more abundant. This is the essence of the non-equilibrium theory of ecosystem structure.

In this light, a so-called "climax ecosystem" is perhaps only a figment of our limited perspective. Ecosystems are constantly experiencing disturbances that occur within the life span of the dominant organisms. Even tropical forests, long thought of as exceedingly stable environments, are known to suffer damage from windstorms and insects, opening up gaps in the canopy that allow different species to colonize. And pollen records inform us that the climate itself has been changing over time spans of hundreds and thousands of years, as glaciation and changes in the Earth's orbit have occurred (Chapter 21). In response to past climate changes, the forests of the temperate zone have shifted from coniferous to deciduous, with many changes in the dominant species of trees.

▲ **FIGURE 4–21** *Fire management.* The Oakland, California, fire in 1991 destroyed about 1000 homes. This tragedy could have been prevented with proper fire management of surrounding natural areas.

EARTH WATCH

AN ENDANGERED ECOSYSTEMS ACT?

The Endangered Species Act is a formal recognition of the importance of saving species that are in serious trouble. The Act has enjoyed some successes (Chapter 11) but is considered seriously flawed by many environmentalists because it does not directly address the major reason why species become endangered, namely, the loss of crucial ecosystems that are the habitats of the endangered species. Thus, there is a higher level of biodiversity that is important to maintain: ecosystem diversity.

Ecosystem diversity is the variety of ecosystem types in a given area, be it local, national, or global. Protecting many different ecosystem types consequently means protecting many endangered species and preventing the bulk of other species from becoming endangered. Maintaining the goods and services provided to us by ecosystems requires going beyond preserving basic *types* of ecosystems and protecting landscapes and regions of ecosystems large enough to sustain the functions we value. Thus, we are talking about qualitative as well as quantitative preservation of ecosystems.

Is there a need for this kind of preservation in the United States? Wildlife ecologists

Reed Noss and Michael Scott performed an extensive review of ecosystem declines and presented their results in terms of critically endangered, endangered, and threatened ecosystems of the United States.[1] Critically endangered ecosystems are considered to be those that have suffered greater than a 98% decline in the area formerly occupied. Twenty-eight ecosystems fit this description, including such ecosystems as native grasslands in California, spruce-fir forest in the southern Appalachians, and tallgrass prairie east of the Missouri River. Over 40 ecosystems were considered endangered (85–98% decline), including all tallgrass prairies, coastal redwood forests, and large streams and rivers in all major regions. Many more made the category of threatened ecosystems (70–84% decline).

What has happened to these ecosystems? They have largely been highly altered by urban development, agriculture, resource exploitation, and pollution. Even though these transformations have served many legitimate human needs, Noss and Scott maintain that much could have been done to reduce the damage. Further, there are pressures—largely economic—to continue the pattern of transformation for many of

the endangered ecosystems. In response to such real and continuing losses, many scientists and environmentalists have called for a national Endangered Ecosystems Act.

Such an act would involve taking steps similar to those taken to protect endangered species, i.e., taking an inventory of the status of ecosystems to identify those in trouble, protecting those that are indeed endangered from further activities that would damage them, establishing recovery plans for many of the most critical ecosystems, and promoting research and monitoring to maintain surveillance and to gain knowledge for better management and protection. What do you think about this suggestion? Moreover, what could be done to make it less controversial than the Endangered Species Act?

---

[1]Reed F. Noss and J. Michael Scott, "Ecosystem Protection and Restoration: The Core of Ecosystem Management," Ch. 12 in *Ecosystem Management: Applications for Sustainable Forest and Wildlife Resources*, eds. Mark S. Boyce and Alan Haney (New Haven: Yale University Press, 1997.) See also Peters, Robert L. and Reed F. Noss, "America's Endangered Ecosystems," *Defenders Magazine*, Fall 1995.

Note, however, that these changes have taken place over many hundreds or thousands of years.

What happens if the climate changes more rapidly? Can forests adapt to climate warming that takes place over decades? Currently, one of our most serious concerns is global warming. Because of our burning of fossil fuels, we are increasing the carbon dioxide level of the atmosphere to the point where, by the mid-twenty-first century, it will have doubled since the beginning of the Industrial Revolution. Since carbon dioxide is a powerful greenhouse gas (Chapter 21), a general warming of the atmosphere is expected to occur, with many other effects on the climate. Many ecosystems will be impacted by the changes, and it is questionable whether they can maintain their functional sustainability in the face of rapid changes. This threat of disturbance on such an enormous scale is a major reason why the nations of the world have begun to take steps to reduce the emissions of greenhouse gases.

The Fourth Principle of Ecosystem Sustainability

Throughout this chapter, you may have noted a great importance on maintaining a diversity of species—that is, biodiversity. To reiterate some of the major points: The most stable population equilibria are achieved by a diversity of natural enemies. Simple systems, especially monocultures, are inherently unstable. Most or all succession depends on a preservation of biodiversity, and succession underlies the ability of an ecosystem to recover from damage (e.g., fire, the loss of the American chestnut due to blight) and its ability to accommodate to changing conditions. Thus, we begin to see another basic principle of sustainability, the fourth. Very simply, the **fourth principle of ecosystem sustainability** may be stated:

For sustainability, biodiversity is maintained.

In Chapter 5 we will find that biodiversity is also required in order to sustain a much longer-term, evolutionary adaptation of species to changing conditions that are inevitable, even without human interference.

Our four principles of ecosystem sustainability are listed in Table 3–1 for review purposes.

4.4   Implications for Humans

How can we use the information presented in this chapter to create a sustainable future? The answer to this question has two aspects. One is protecting or managing the natural environment to maintain the beauty, interest, biodiversity, and other intrinsic values of the natural world. The second aspect is establishing a balance between our own species and the rest of the biosphere. The two aspects are interrelated.

Focusing on the first aspect, the basic rule for anyone interested in wildlife management and ecosystem protection is to think in terms of the factors that contribute to maintenance of ecosystems. In ecosystems that have not been degraded, simple protection from adverse human impacts and avoidance of the introduction of foreign species may be adequate for protection. However, if an ecosystem has suffered damage, restorative measures may involve a number of options, depending on the damage. Observe that restoration may involve anything from the painstaking reintroduction of a predator or parasite to the hunting of a species that is multiplying beyond the carrying capacity of the system.

Likewise, preventing alteration of the environment generally is necessary for the protection of ecosystems, but in some cases, alterations may be done to encourage certain kinds of wildlife. This is called **ecological restoration**: How do you restore some semblance of a natural ecosystem after an area has been impacted or totally altered by such activities as development, agriculture, or mining? Restoration begins with the creation of the desired physical environment and introduction of appropriate plants to support the desired animals. An example is the creation of ponds and wetlands to encourage waterfowl.

In the long run, any efforts to protect natural ecosystems will be overwhelmed and thwarted by growing pressures from the human system if current trends of population growth and exploitation of natural ecosystems continue. You may observe that a graph of the human population over the last 200 years (see page 6) has a distinct similarity to the upsweeping portion of the J-curve for the reindeer population on St. Matthew Island. We can also draw parallels in that humans have overcome, for the most part, those factors that generally keep natural populations in balance—disease, parasites, and predators. Therefore, we may observe that the human population is behaving much like any population in the absence of natural enemies.

But what lies ahead? If we were being viewed by an alien who made the assumption that we were just "dumb" animals like the reindeer, the alien could predict the peak and crash portions of the J-curve with total confidence. After all, the alien would see that we are "overgrazing" our environment as we cut forests, overfish the seas, deplete water resources, and cause the extinction of ever more species even as our population continues to grow. It is only a matter of time, the alien would reason, before essential resources are exhausted and the human population crashes.

The glaring fallacy in the alien's scenario, of course, is that we are not dumb animals. We have options not available to the reindeer. But this, by itself, does not rule out the scenario: Smart people are still capable of making dumb decisions and doing dumb things, particularly en masse. We need to be not only smart enough to *recognize* the critical need to establish a balance between our human system and the rest of the biosphere but also wise enough to *do so*. Once again, we need to see the absolutely essential need for becoming stewards of the planet.

What is the urgency? In discussing the reindeer population on St. Matthew Island, we noted the rapidity with which an exponentially growing population can deplete its resource base. Half of the original resource base is consumed in the last doubling. This is to say that half of the original resource base is needed just to accomplish the last doubling, and then the crash will occur anyhow, because no resources remain to sustain the high population, much less to allow it to grow further, as was the case with many ancient civilizations that perished (Fig. 4–22).

Now, the human population doubled from 2.5 billion in 1950 to 5 billion in 1987 (37 years). Reaching 6 billion in 1999 and still growing at the rate of 1.52%, nearly 83 million persons per year, it is expected to reach 10 billion in the 2040s. But in many populous regions of the world, forests, water, fish, fertile soil, and other resources have already been exploited well beyond the 50% level needed for the final doubling. Are there enough resources to support another doubling? Are there enough to sustain a population of 10 billion, should that level be reached? What level of resource consumption per person will see the *human* carrying capacity exceeded? If the answer to either of the first two questions is no, and the exploitation or consumption of resources continues as in the past, we will find ourselves facing an impoverished future.

We need to ask the question: What is the carrying capacity of our planet for human beings? Many experts have considered and debated this question. Unfortunately, there is no uniform conclusion because humans are capable of such diverse behavior, but let us consider the major points. Most significantly, we observe that major environmental parameters such as forest, fisheries, soils, water resources, and atmosphere are in a state of decline under the pressures of the present population of 6 billion. Therefore, by definition we are already over the Earth's carrying capacity. How much over?

Human pressures on the environment vary with affluence as higher incomes both enable higher levels of consumption and produce higher levels of pollution as waste or byproducts from consumption are inevitably disposed. Only 20% of the world's population, or 1.2 billion people, live at a level of affluence typical of the United States. The rest live appreciate ecological principles and in using our techno-logical skills to bring our human system into compliance with those principles.

In relation to the concepts presented in this chapter, anything that we are doing, or that we may do, toward preserving or reestablishing natural equilibria among biota and protecting natural biota from destructive human impacts may be seen as progress toward sustainability. Likewise, protecting ecosystems, to say nothing of the total biosphere, from unnatural changes in abiotic factors such as pollution or atmospheric alterations is equally important. We must work toward stabilizing the human population in morally acceptable ways and husbanding resources such that they remain available to future generations. A concern for posterity is one of the basic elements of the stewardship ethic.

Chapters 6 and 7 will focus on the problems of stabilizing the human population, and later chapters will focus on husbanding resources. Before proceeding with those issues, however, we will examine in Chapter 5 the question of how species become adapted to one another and their abiotic environment in such a way as to form balanced, sustainable ecosystems — the perspective of evolutionary change.

▲ **FIGURE 4–22** *An unsustainable civilization.* On the Yucatán Peninsula of Mexico, the ruins of Chichen-Itza are only one example of civilizations that arose, prospered, and then passed into oblivion due to a failure to maintain a suitable balance between their human population and environmental resources.

at considerably lower levels. It is calculated that the Earth's carrying capacity (sustainability) for people living a typical American lifestyle would be about 2 billion. On the other hand, the Earth's carrying capacity for people living at minimal levels of resource consumption, as most do in China for example, is estimated to be in the order of 12 billion.

Neither of these end figures, however, considers the potential of implementing new sustainable technologies, most particularly in the areas of solar energy, total product and waste recycling, and improvements in crop production and soil conservation. If ideal states could be reached in all these areas, it is conceivable that the Earth might sustain 8–10 billion people at a reasonably comfortable lifestyle. Thus, we do not have to imagine either reducing the world's population by 70% or all having to live in abject poverty. There is realistic hope that the overall condition of humanity and the environment can be improved through a combination of stabilizing population growth and implementing appropriate sustainable technologies. But this cannot be expected to happen automatically through "business as usual." It will require using our intelligence and wisdom to understand and

ENVIRONMENT ON THE WEB

FIRE IN PROTECTED ECOSYSTEMS: FRIEND OR FOE?

In 1994, the Smokey Bear campaign—"Only *you* can prevent forest fires!"—celebrated its fiftieth anniversary, testimony to the long fight humans have waged against the natural force of fire. The battle has been successful: Forest fires are now suppressed throughout North America, saving millions of dollars in damages every year. But what is the cost to ecosystems of removing a natural cycle of disturbance and succession?

Ecologists now know that many ecosystems have evolved in the presence of fire and need periodic fire to develop normally. In ecosystems like the Indiana Dunes National Lakeshore, maintenance of the natural parklike oak savannah depends on periodic fire for the maintenance of its characteristic open habitat. Also, species such as jack pine need the heat of an intense fire to burn off the pitch that protects seeds, permitting normal germination. When fire is suppressed, jack pine populations decline, and species that depend on jack pine habitat (such as the endangered Kirtland's Warbler) are jeopardized. Yet fire, however natural, also has real costs in managed ecosystems. Historically, national parks have suppressed fire in an attempt to protect valued aesthetic characteristics—and associated tourism revenues. This chapter describes the 1988 fire in Yellowstone Park, one of the most visited parks in the U.S. National Park system. In a controversial policy decision, park managers allowed the fire to burn freely, destroying thousands of acres of forest. Was this the right decision?

Evidence from other ecosystems shows that letting the fire burn was probably the right thing to do. In nature, forests are subjected to occasional high winds and lightning-induced fire, forces that create openings in the canopy of trees above. Opportunistic species colonize these spaces, creating a mosaic of species of different ages,

shapes, and reproductive strategies. Periodic fires are usually moderate in intensity and are confined to dry litter and understory plants, with relatively little damage to older trees. In a sense, regular fires remove accumulated "fuel," reducing the potential for catastrophic fire in a future blaze. Fire also provides a mechanism for the remobilization of nutrients stored in plant and animal tissue. When fire is suppressed, however, the result may be major changes in habitat structure and species richness. In commercially valuable forests, research suggests that fire suppression results in more, but smaller, trees. And because dense vegetation is more vulnerable to drought and insect pests, the system may be less able to withstand normal stresses than a fire-controlled forest. Letting the fire burn may therefore be a case of "short-term pain, long-term gain."

Web Explorations

The Environment on the Web activity for this essay compares clearance by natural forces with clearcutting and observes the impacts these different forces have on an ecosystem. Go to the Environment on the Web activity (select Chapter 4 at http://www.prenhall. com/nebel) and learn for yourself.

1. about trends in human control of fires in Florida;
2. how natural clearance by fire or wind differs from clearcutting; and
3. about the importance of the scale of disturbance in its impact on an ecosystem.

A suggested time frame for each of these exercises is 10–30 minutes.

REVIEW QUESTIONS

1. What are the two fundamental kinds of population growth curves? What are the causes and consequences of each?

2. Define biotic potential and environmental resistance and give factors of each. Which generally remains constant, and which controls a population's size?

3. Differentiate between the terms critical number and carrying capacity.

4. Give two factors of environmental resistance that would act density-dependently on a field mouse population. Give two of the same that would act density-independently.

5. What does it mean when a species is threatened? Endangered? What piece of U.S. legislature is responsible for defining these terms? What department of U.S. government is responsible for protecting threatened and endangered organisms?

6. Describe the predator-prey relationship between the moose and wolves of Isle Royale. What other factors are involved with these two populations? How are the dynamics of this ecosystem different from most others?

7. What are some of the problems when a species is introduced from a foreign ecosystem? What is the basic explanation for why these problems occur?

8. What is meant by territoriality, and how does it control certain populations in nature?

9. What is the third basic principle of ecosystem sustainability? How does it relate to the current U.S. white-tailed deer population?

10. State the three main ways that enable different plant species to coexist in the same region.

11. Define the terms ecological succession and climax ecosystem. How do disturbances allow for ecological succession?

12. What role may fire play in ecological succession, and how may fire be used in the management of certain ecosystems?

13. What is the fourth principle of ecosystem sustainability? What are some potential consequences if this principle is not upheld?

14. Do humans need to control their population? If so, what methods are available that are not available to all other animals?

THINKING ENVIRONMENTALLY

1. Describe, in terms of biotic potential and environmental resistance, how the human population is affecting natural ecosystems.

2. Consider the plants, animals, and other organisms present in a natural area near you, and then do the following: (a) think of how the area has experienced ecological succession; (b) analyze the population-balancing mechanisms that are operating among the various organisms; and (c) choose one species and predict what will happen to it if two or three other species native to the area are removed and, again, if two or three foreign species are introduced into the area.

3. Evaluate such practices as legal hunting, controlling pests with chemical sprays, using or preventing fires, and poaching of endangered species in terms of supporting sustainable balances.

4. Make an argument, pro or con, regarding sustainability of the human system in terms of concepts you have learned in this chapter. What new directions, if any, do humans need to take to achieve sustainability?

WEB REFERENCES

On-line resources for this chapter can be found on the World Wide Web at: **http://www.prenhall.com/nebel**. Click on Chapter 4 on the chapter selector.

PART TWO

The Human Population

Human Diversity.

Six billion is an enormous number, impossible to imagine in any context. Yet sometime during 1999 the human population on Earth reached and began surpassing 6 billion people. The United Nations projects continued population growth well into the twenty-first century, and there is no reason to doubt that our numbers may well reach 9 billion before the middle of the century—a 50% increase in 50 years. Virtually all of the increase will be in the developing countries, which are already densely populated and straining to meet the needs of their people for food, water, health, shelter, and employment.

In the industrialized world, where populations have almost stopped growing, there is a tendency to become complacent about population. Yet it is a known fact that the developed countries, with only 20% of the world's people, account for two-thirds of the world's energy use and create pressures on water, forests, and fisheries that are leading to the depletion of these renewable resources. In other words, even though our population is stable and thus at an equilibrium, our interactions with the natural world are highly unsustainable. When we picture a world of 9 billion, with a greater proportion of people hoping to improve their economic standards and catch up to the developed world, we may be staring into the face of disaster.

In Chapter 6, we first look at the dynamics of the human population, examine the pressures we put on natural systems as we continue to grow, and focus especially on the demographic transition—the shift from high birth and death rates to low birth and death rates that has brought stable populations to the industrialized world. We then consider, in Chapter 7, what needs to be done to bring the developing countries through this transition and the steps that are now being taken on the part of the international community to address this need. It is appropriate that we take up population concerns before other significant issues like resource use and pollution, because population affects every environmental issue, and we have no chance of achieving sustainability in these other areas until we achieve population stability.

6

The Human Population: Demographics

〜〜〜〜〜〜〜

Key Issues and Questions

1. The human population is undergoing an explosion. When did it start? What are its causes? What is the current growth rate?

2. The world comprises high-, middle-, and low-income nations. Identify the nations or regions, and describe representative lifestyles in these three groups.

3. The most rapid population growth is occurring in developing countries. What are the social and environmental consequences of such growth for developing countries? For developed countries?

4. Population profiles give the age structure of populations. How are these profiles used to project future populations?

5. Population profiles for developed and developing countries are fundamentally different. What are the differences?

6. Populations in the developed countries have experienced great reductions in birth and death rates over time. What is the demographic transition, and what are the different phases of this transition?

7. Different regions of the world are at different phases of the demographic transition. What are the consequences of remaining at earlier transitional phases?

The Mombasa Highway leads out from the center of Nairobi, Kenya, and is lined with factories for several miles. Outside every factory gate every morning is a crowd of people—mostly men—waiting for the gates to open. Dressed in tattered clothes and worn-out shoes, they are the "casual workers," hired for periods up to three months but never under any employment contract. Many are there just for the hope that they will be hired for the day. One such worker—we will call him Charles—has worked at one of these factories for almost a year. He is paid 100 Kenya shillings a day, about $1.70. He walks 10 km to work from his home in one of the shantytowns on the edge of the city. He can't afford transport and can only occasionally buy lunch, which is a bowl of githeri—a 10-shilling dish of corn and beans sold in the kiosks that cluster around every factory. Charles has no work contract, nor do any of his co-workers. Virtually all of the factories pay workers 100 shillings a day; working six days a week for 10 hours a day, they earn, at most, $50 a month. Any worker who becomes a union

representative is fired. Any worker who complains about an injury on the job is fired. For every job, there are scores of applicants, so no one dares attract attention. They work in dusty, dangerous conditions. Cement-factory workers have no masks to prevent their breathing the dust. Metal workers lack protective gear and often lose limbs or their lives due to the nature of the jobs, which are often monotonous, making them lose concentration. There are 14,000 such factories in cities like Nairobi, Mombasa, Kisumu, Nakuru, and Eldoret, in Kenya.

Scenes like the Mombasa highway can be found around most cities in the developing world. The incredible growth of cities is one of the realities of the population explosion in the developing world. It is no secret why. The rural countryside provides only subsistence farming and livestock herding for most of its people—and with population growth rates of 3% or more per year, there are continually more people than there is land to till and cattle to herd. It is the tragic story of *Cry the Beloved Country* (a classic novel by Alan Paton, set in South

---

Casual workers in Kenya. High population growth and slow economic development combine to produce a large unemployed and underemployed work force. Competition for the scarce, low-paying factory jobs leaves many men like these without employment.

137

Africa) multiplied by hundreds of thousands. So the young people and the men migrate to the cities to find work. Such is the case in Nairobi. With one of the most rapidly expanding populations in the world, Kenya has an enormous unemployed and underemployed workforce.

The global human population has recently undergone an explosion. It has grown almost fourfold in the last 100 years, and while the rate of growth is slowing somewhat, the increase in absolute numbers continues to be great—over 80 million people per year (births minus deaths). Remarkable changes in technology and substantial improvements in living standards have accompanied this growth. It comes as no surprise that this great growth in the human population and the accompanying demands for resources have impacted the natural environment and will continue to do so. Much of this text is dedicated to documenting concerns over the resource use and pollution stemming from the human system in all of its extensions.

What are the implications of these trends for sustainability? Another way to put this question is to ask, What is the carrying capacity of Earth for humans? How many people can Earth support, and at what quality of life? Some observers are convinced that there are already too many people and that the global environment is being stressed beyond its limits even now. No one argues about the fact that the human population cannot continue to increase indefinitely. There is broad consensus that continued population increase makes solving other problems more difficult, and thus there are many calls for bringing human population growth to a halt. The focus is especially on the developing world, where 98% of world population growth is occurring.

Yet, specific efforts to reduce population growth often collide with ethical and moral values, creating controversy. Moreover, not everyone agrees that the population issue is serious. The major counterarguments are as follows:

- Some claim that population growth is beneficial in that more people provide more ideas, creativity, and work. This belief is supported by the fact that the greatest technological advances and improvements in living standards have occurred in parallel with the population explosion of the past 100 years.
- Because of their religious convictions, some take the position that any artificial interference in the reproductive process (including the use of sex education, contraceptives, and, especially, abortions) is immoral. Therefore, for them, any thought of altering the course of population growth, except by abstinence, is not debatable.
- Some argue that population growth is not the issue as much as consumption is, at least for the present. What needs to be achieved more than reducing growth in numbers, they say, is adopting conservation measures that will reduce consumption.
- Others take the position that population growth will level off by itself well within the capacity to support the population. This view is supported by the facts that we have been able to expand agricultural production even faster than population growth and that the average number of births per woman is decreasing.
- Still others are disturbed by the fact that population programs often seem to have the trappings of social engineering—the rich trying to hold down the poor or minorities by preventing them from having children.

Our objective in this chapter is to gain an understanding of the dynamics of population growth and its social and environmental consequences. We will see that a continually growing population is unsustainable, and so our focus in this and the next chapter will be population stability and what it will take to get there.

6.1 The Population Explosion and Its Cause

Considering all the thousands of years of human history, the explosion of the global human population is a recent and unique event—a phenomenon of just the past 100 years (Fig. 6–1). Let us look more closely at this event and why it occurred.

The Explosion

From the dawn of human history until the beginning of the 1800s, population increased slowly and variably with periodic setbacks. It was roughly 1830 before world population reached the 1 billion mark. But by 1930, just 100 years later, the population had doubled to 2 billion. Barely 30 years later, in 1960, it reached 3 billion, and in only 15 more years,

by 1975, it had climbed to 4 billion. Thus, the population doubled in just 45 years, from 1930 to 1975. Then, 12 years later, in 1987, it crossed the 5 billion mark! In 1999, world population passed 6 billion, and it is currently growing at the rate of nearly 80 million people per year. This rate is equivalent to fitting into the world every year the combined populations of New York, Los Angeles, Chicago, Philadelphia, Detroit, Dallas, Boston, and 10 other U.S. metropolitan areas.

Based on current trends (which assume declining fertility rates in the future), the Population Reference Bureau (a private educational organization) projects that world population will pass the 7 billion mark in 2009, the 8 billion mark in 2020, the 9 billion mark in 2033, and the 10 billion mark in 2046, before population finally levels off by the end of the next century (Fig. 6–1).

EARTH WATCH

ARE WE LIVING LONGER?

It is commonly said that with the introduction of modern medical technology and disease control, average longevity increased from about 40 to 65 years of age. While mathematically correct, this statement may be misleading. Many people take it to mean that nearly everyone used to die around age 40 and now everyone lives to around 65. However, the years between 35 and 45 are generally the healthiest period of human life. With or without modern medicine, a relatively small portion of the population dies in this age range. The "average longevity of 40" before modern techniques of disease control was a function of a large fraction of the population's dying in childhood, thus counterbalancing another fraction that lived into their sixties and beyond.

Through disease control, most of those who once would have died in childhood now live past 50. Extending the life span of this group raises the average age of death of the population. But the basic life span of the human species has changed little, if at all, as a result of modern medicine. All modern medicine has done is to increase the proportion of people who get to or near the maximum age.

Reasons for the Explosion

The main reason for slow and fluctuating population growth prior to the early 1800s was the prevalence of diseases that were often fatal, such as smallpox, diphtheria, measles, and scarlet fever. These diseases hit infants and children particularly hard. It was not uncommon for a woman who had seven or eight live births to have only one or two children reach adulthood. In addition, epidemics of diseases such as the black plague of the fourteenth century, typhus, and cholera would eliminate large numbers of adults. Famines also were not unusual.

Biologically speaking, prior to the 1800s, the population was essentially in a dynamic balance with natural enemies—mainly diseases—and other aspects of environmental resistance. High reproductive rates were largely balanced by high mortality, especially among infants and children. With high birth and death rates, the population growth rate was low in these preindustrial societies.

In the 1800s, Louis Pasteur and others made the major discovery that diseases were caused by infectious agents (now identified as various bacteria, viruses, and parasites) and that these were transmitted via water and food, insects, and other vermin. With these discoveries came major improvements in sanitation and personal hygiene. Then, techniques of providing protection by means of vaccinations came into play. Later, in the 1930s, the discovery of penicillin, the first in a long line of antibiotics, resulted in cures for otherwise often-fatal diseases such as pneumonia. Improvements in nutrition began to be significant as well. In short, better sanitation, medicine, and nutrition brought about spectacular reductions in mortality, especially among infants and children, while birth rates remained high.

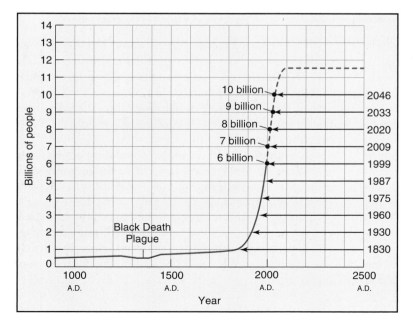

◀ **FIGURE 6–1** *The world population explosion.* For most of human history, population grew slowly, but in modern times it has suddenly "exploded." (Source: Data from Joesph A. McFalls, Jr., "Population: A Lively Introduction," *Population Bulletin* 46, no.2 [1991]: 4.)

TABLE 6-1 Demographic Terms Used in This Chapter

Term	Definition
Growth Rate (annual rate of increase)	The rate of growth of a population, as a percentage. Multiplied by the existing population, this rate gives the net yearly increase for a population.
Total Fertility Rate	The average number of children each woman has over her lifetime, expressed as a yearly rate based on fertility occurring during a particular year.
Replacement-Level Fertility	A fertility rate that will just replace a woman and her partner, theoretically 2.0 but adjusted slightly higher because of mortality and failure to reproduce.
Infant Mortality	Infant deaths per thousand live births.
Population Profile (age structure)	A bar graph plotting numbers of males and females for successive ages in the population, starting with youngest at the bottom.
Population Momentum	The tendency of a population to continue growing even after replacement-level fertility has been reached, due to continued reproduction by already existing age groups.
Crude Birth Rate	The number of live births per thousand in a population in a given year.
Crude Death Rate	The number of deaths per thousand in a population in a given year.
Doubling Time	The time it takes for a population growing at a given growth rate to double in size.
Epidemiologic Transition	The shift from high death rates to low death rates in a population as a result of modern medical and sanitary developments.
Fertility Transition	The decline of birth rates from high to low levels in a population.
Demographic Transition	The tendency of a population to shift from high birth and death rates to low birth and death rates as a result of the epidemiologic and fertility transitions. The consequence is a population that grows very slowly, if at all.

From the biological point of view, the human population entered into exponential growth, as does any natural population on being freed from natural enemies and other environmental restraints. Note that the rapidly declining death rates must be compared to the birth rates that remained high.

In the last few decades, the average *fertility rate*—that is, the average number of babies born to a woman over her lifetime—has declined, resulting in a decreasing *rate of growth* of population. (Table 6-1 provides a list of definitions of technical terms used in this chapter.) Still, as the number of people of reproductive age has expanded with the increasing numbers of children growing up, we continued until 1990 to add absolute numbers faster than at any other time in history (Fig. 6–2). Extrapolating the trend of lower fertility rates leads to the projection by the United Nations Population Division that the global human population will level off at around 11 billion toward the end of the next century.

The projected leveling off at around 11 billion is what leads some to believe that the population situation will take care of itself. However, such complacency fails to recognize that the currently declining fertility rates are a reflection of already significant efforts at reducing births, including what many consider an extremely controversial one-child-per-couple policy exercised in China.

Even more, the future global population projection of 11 billion in no way considers any of the looming environmental questions of how Earth can sustain such numbers. Where are the additional billions of people going to live, and how are they going to be fed, clothed, housed, educated, and otherwise cared for? Will enough energy and material resources be available for them to fulfill a satisfying lifestyle? What will the natural environment look like by then?

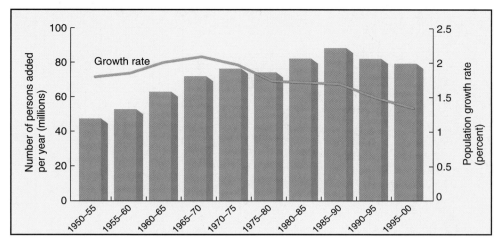

◀ **FIGURE 6–2** *World population growth rate and absolute growth.* Declining fertility rates in the last three decades have resulted in a decreasing rate of growth. However, absolute numbers are still adding 80 million per year. (Source : Data from Shiro Horiuchi, "World Population Growth Rate," *Population Today* (June 1993): 7 and June/July 1996): 1; updated data from U.S. Bureau of the Census, International Data Base.)

6.2 Different Worlds

To begin to answer the foregoing questions, we must first recognize the tremendous disparities among nations. In fact, people in wealthy and poor countries live almost in separate worlds, isolated by radically different economic and demographic conditions.

Rich Nations and Poor Nations

The World Bank, an arm of the United Nations, divides the countries of the world into three main economic categories according to average per capita gross national product (Fig. 6–3).

1. **High-income, highly developed, industrialized countries.** This group includes the United States, Canada, Japan, Australia, New Zealand, and the countries of Western Europe and Scandinavia. (1995 gross national product per capita = $24,930 average; Switzerland first at $40,630, United States sixth at $26,980, and South Korea at the bottom of this list at $9,700.)

2. **Middle-income, moderately developed countries:** These are mainly the countries of Latin America (Mexico, Central America, and South America), northern and western Africa, eastern Asia, Eastern Europe, and countries of the former U.S.S.R. (1995 gross national product per capita range from $770–$8200.)

3. **Low-income countries.** This group comprises the countries of eastern and central Africa, India, and other countries of central Asia. (1995 gross national product per capita = less than $730.)

The high-income nations are commonly referred to as **developed countries**, whereas the middle- and low-income countries are often grouped together and referred to as **developing countries**. The terms *more developed countries* (MDCs), *less developed countries* (LDCs), and *Third World countries* are being phased out, although you will still hear them used. (The Second World was the former Communist bloc, which no longer exists. Therefore, referring to the developing countries as Third World countries is obsolete.)

The disparity in distribution of wealth among the countries of the world is mind boggling. The highly developed countries hold just 20% of the world's population, yet they control about 80% of the world's wealth. Thus, developing countries, which have 80% of the world's population, have only about 20% of the world's wealth. Of course, the distribution of wealth *within* each country is also disproportionate. Between 10% and 15% of the people in more developed countries are recognized as poor (unable to afford adequate food, shelter, and/or clothing), while about 90% of those in developing countries are.

The disparity of wealth is difficult to understand just by looking at monetary figures. Therefore, Allen Durning of the WorldWatch Institute describes the relative wealth for different peoples of the world in terms of their access to different kinds of food, drink, and transportation, as is shown in Table 6-2. It is sobering that although some people in developed nations such as the United States may feel poor, they are still well within the rich category in terms of representative living standards for the world. The people in the lower-middle and poor categories of consumption, predominantly in low- and middle-income countries, live along a bare "subsistence margin," with little beyond the minimum requirements for survival. Roughly half of these—somewhat over a billion people—live in a condition of "absolute poverty," defined by Robert McNamara, former president of the World Bank, as "A condition of life so limited by malnutrition, illiteracy, disease, squalid surroundings, high infant mortality, and low-life expectancy as to be beneath any reasonable definition of human decency." Yet, it is among the poorest people that population growth is usually the highest.

Population Growth in Rich and Poor Nations

The population growth shown in Figs. 6–1 and 6–2 is for the world as a whole. If we look at population growth in developed versus developing countries, we find a discrepancy that parallels the great differences in wealth of these two groups

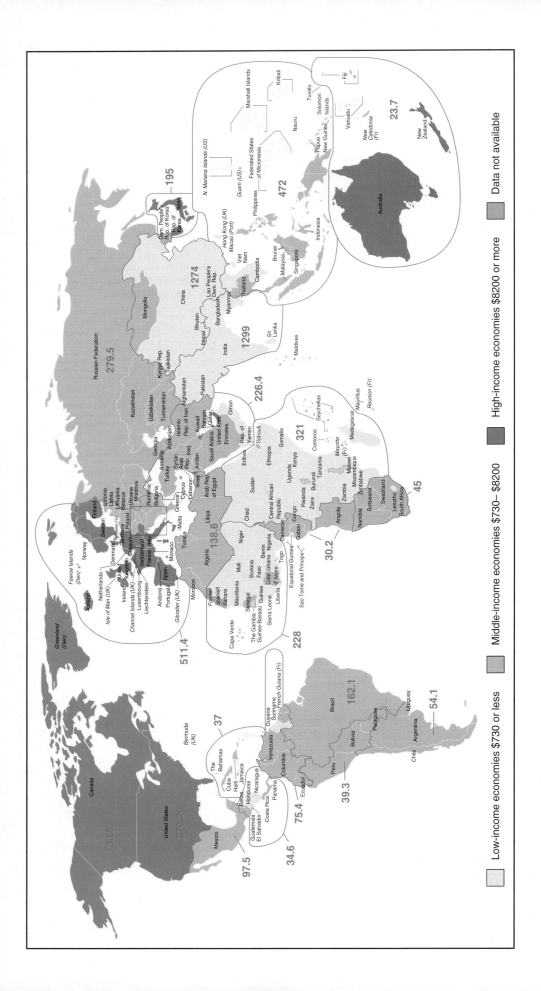

▲ FIGURE 6–3 *Major economic divisions of the world.* Nations of the world are grouped according to gros national product per capita. the population of various regions is also shown by magenta lines and numbers. (Source: Reprinted, by permission of the publisher, from the World Development Report, [New York: Oxford University Press, Inc. 1994]. Copyright © 1994 by the International Bank for Reconstruction and Development/The World Bank. Population and per capita gross national product data from the Population Reference Bureau, *World Population Sheet.* Updated data from The World Bank, *World Development Report 1997.*)

Low-income economies $730 or less

Middle-income economies $730– $8200

High-income economies $8200 or more

Data not available

TABLE 6-2 World Consumption Classes

Category of consumption	Consumers (1.1 billion)	Middle (3.5 billion)	Poor (1.4 billion)
Diet	Meat, packaged food, soft drinks	Grain, clean water	Insufficient grain, unsafe water
Transport	Private cars	Bicycles buses	Walking
Materials	Throwaways	Durables	Local biomass

Source: Data from Alan Durning, *How Much Is Enough*, (Washington, DC: Worldwatch Inst., 1992) 27; updated Population Reference Bureau, Inc. 1998.

of countries. The developed world, with a population of 1.18 billion in mid-1999, is growing at a rate of 0.1% per year. The remaining countries, whose mid-1999 population was 4.82 billion, are increasing at a rate of 1.7% per year. Consequently, *98% of world population growth is occurring in the developing countries.* Why is this so?

In the absence of high mortality, the major determining factor for population growth is **total fertility rate**—the average number of children each woman has over her lifetime. It follows that a total fertility rate of 2.0 will give a stable population, since two children per woman will just replace the parents when they eventually die. Fertility rates greater than 2.0 will give a growing population because each generation is replaced by a larger one. Conversely, barring immigration, a total fertility rate less than 2.0 will lead to a declining population because each generation will eventually be replaced by a smaller one. Because infant and childhood mortality are not in fact zero, and some women do not reproduce, **replacement-level fertility**—the fertility rate that will just replace the population of parents—is slightly higher: 2.03 for developed countries and 2.16 for developing countries, which have higher infant and childhood mortality.

For reasons we shall discuss later, total fertility rates in developed countries have declined over the past several decades to the point where they now average 1.6. The one major exception is the United States, with a total fertility rate of 2.0 (1998). In developing countries, on the other hand, although fertility rates have come down considerably in recent years, they still average 3.3, with some as high as 6 or more, rates that will cause the populations of those countries to double in just 20 to 40 years (Table 6-3). Thus, the populations of poor countries will continue growing, while the populations of more developed countries will stabilize or even decline. As a consequence, the percentage of the world's population living in developing countries—already 80%—is expected to climb steadily to over 90% by 2075 (Fig. 6–4). However, this is not to say that only the developing countries have a population problem.

Different Populations Present Different Problems

Today, increasing numbers of people put increasing demands on the environment, both through demands for resources, including food, energy, and water, and through the production of wastes. However, it should also be clear that the demand

TABLE 6-3 Population Data for Selected Countries

Country	Total Fertility	Doubling Time (Years)
World	2.9	49
Developing Countries		
Average (excluding China)	3.8	35
Egypt	3.6	32
Kenya	4.5	35
Madagascar	6.0	23
India	3.4	37
Iraq	5.7	25
Vietnam	2.3	57
Haiti	4.8	33
Brazil	2.5	48
Mexico	3.1	32
Developed Countries		
Average	1.6	548
United States	2.0	116
Canada	1.6	136
Japan	1.4	330
Denmark	1.8	431
Germany	1.3	—*
Italy	1.2	—*
Spain	1.2	1733

Source: Data from *1998 World Population Data Sheet* (Washington, D.C.: Population Reference Bureau, 1998).

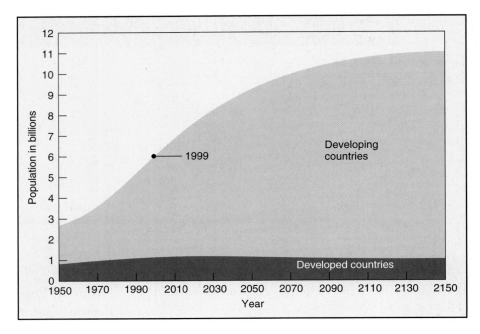

► **FIGURE 6–4** *Population increase in developed and developing countries.* Because of higher populations and higher birth rates, developing countries represent a larger and larger share of the world's population.

each individual makes on the environment depends on how much and what that individual consumes. Each additional thing purchased represents a certain additional demand on resources for its production, as well as additional wastes produced in the course of its production, use, and, finally, disposal. Therefore, negative effects on the environment also increase dramatically as consumption increases.

For example, it is estimated that because of differences in consumption, the average American causes at least 20 times the demand on Earth's resources, including its ability to absorb pollutants, as does the average person in Bangladesh, a poor Asian country. Major world pollution problems, including depletion of the ozone layer, the implications of global climate change, and the accumulation of toxic wastes in the environment, are almost exclusively the consequence of the high consumption associated with affluent lifestyles. For instance, it is people who drive cars and heat and cook with fossil fuels that contribute most significantly to rising levels of carbon dioxide in the atmosphere. Likewise, much of the global deforestation and loss of biodiversity is a consequence of consumer demands in developed countries.

Environmental impacts of affluent lifestyles may be moderated to a large extent by a concern for environmental stewardship. For example, suitable attention to wildlife conservation, pollution control, energy conservation and efficiency, and recycling may offset, to some extent, the negative impact of a consumer lifestyle. One might make a good argument that a life devoted to conservation or other aspects of environmental stewardship might entirely offset the negatives and have a very positive effect overall. The relationship between these factors may be expressed as

$$\text{Negative environmental impact} \propto \frac{\text{Population size} \times \text{Affluence of lifestyle}}{\text{Stewardship}}.$$

This proportionality should be read as "Negative environmental impact is proportional to the population multiplied by the affluence of the population's lifestyle, moderated by the stewardly regard of the population." Note that it is not strictly a mathematical proportionality, because meaningful numbers cannot be assigned to any of the factors except population.

You may hear debates among people, one side arguing that population growth is the main problem, the other arguing that our highly consumption-oriented lifestyle is the main problem. Our contention in this text is that in order to reach sustainability, all *three* areas must be addressed: Stabilize population, decrease consumption, and increase stewardly care. For the present, we wish to maintain our focus on the factors and consequences of population growth, mainly in developing countries.

6.3 Environmental and Social Impacts of Growing Populations and Affluence

Both growing populations and increasing affluence have numerous environmental and social implications for the whole world.

The Growing Populations of Developing Countries

Prior to the Industrial Revolution, most of the human population survived through subsistence agriculture. That is, families lived on the land and produced enough food for their own consumption and perhaps enough extra to barter for other essentials. Natural forests provided firewood, structural materials for housing, and wild game for meat. With a small, stable population, this system was basically sustainable. As the older generation passed away, the land and natural systems could still support the next generation. Indeed, many cultures sustained themselves in this way over thousands of years, and a few areas of Latin America, Africa, and Asia maintain this tradition today.

After World War II, modern medicines—chiefly vaccines and antibiotics—were introduced to developing nations, whereupon death rates plummeted, and population growth increased. What are the impacts of rapid growth on a population that is largely engaged in subsistence agriculture? Five basic alternatives are possible, all of which are being played out to various degrees by people in these societies.

1. Subdivide farms among the children of the next generation and/or intensify cultivation of existing land to increase production per unit area.
2. Open up new land to farm.
3. Move to cities and seek employment.
4. Engage in illicit activities for income.
5. Emigrate to other countries legally or illegally.

In addition, rapid population growth especially affects women and children.

Let us look at each of the preceding alternatives and their consequences in a little more detail.

Subdividing Farms and Intensifying Cultivation. Over wide areas of Asia and Africa, plots of land have been divided and redivided to the point that the United Nations Food and Agriculture Organization (FAO) estimates that over a billion rural people live in households that have too little land to meet even their own meager needs for food and fuel, much less producing extra for income to barter.

Adding to the problem is the quest for water and firewood. Some 3 billion people, or 60% of the world's population, do not have gas, electric, or even kerosene stoves; hence, they depend on firewood in the preparation of their daily food. As a consequence, local woodlands are being cut faster than they can regenerate. Much of East Africa, Nepal, and Tibet, as well as many slopes of the Andes in South America, have been deforested as a result. Already, women in many developing countries spend a large portion of each day on increasingly long treks to gather firewood and water (Fig. 6–5). As many as 3 billion poor people face acute shortages of firewood. Worse, in the long term, as the trees are removed, soil erosion occurs.

The consequences of soil erosion are manifold. The loss of topsoil diminishes the future productivity of the land. It also results in more water running off rather than soaking into the ground, a situation that aggravates flooding in the lowlands and diminishes groundwater. Soil washing into streams and rivers destroys fisheries, clogs channels, and also aggravates flooding. (These effects are explained further in Chapter 8.)

With regard to the intensification of cultivation, the introduction of more highly productive varieties of basic food grains has had a dramatic positive effect in supporting the

▲ **FIGURE 6–5** *Gathering firewood.* Like the Nepalese woman seen here, some 60% of the world's population still depends on gathering firewood for cooking and other fuel needs. The resulting deforestation is causing both ecological and human tragedy.

growing population, but is not without some concerns, as will be described in Chapter 10. However, intensification of cultivation also means simply working the land harder. For example, traditional subsistence farming in Africa involved rotating cultivation among three plots. In that way, after being cultivated for one year, the soil in each plot had two years to regenerate. With pressures for increasing productivity, plots have been put into continuous production with no time off. The results have been deterioration of soil, decreased productivity (ironically), and erosion.

In addition, the increasing intensity of grazing is damaging the land. (Desertification, the extremely serious consequence of overgrazing, will be considered further in Chapter 8.) Given the countertrends of rapidly increasing population and deterioration of land from overcultivation, food production per capita in Africa, for example, is currently on a downward course.

Opening Up New Lands for Agriculture. With respect to opening up new lands for agriculture, first consider that there really is no such thing as "new land." Instead, it is always a matter of converting the land of natural ecosystems to agricultural production, which means increasing pressure on wildlife and an unavoidable loss of biodiversity. Even then, converted land may not be well suited for agriculture. (Most good agricultural land is already in production.) For example, it is estimated that two thirds of the tropical deforestation that is occurring in Brazil and Central America is for the purpose of increasing agricultural production (Fig. 6–6). Much of this deforestation is done by poor, young people who are seeking an opportunity to get

▶ **FIGURE 6–6** *Deforestation in the tropics.* Millions of acres of rain forest in Central and South America are being cut down each year to make room for agriculture, as shown in this photograph from Peru.

ahead but are unskilled and untrained in the unique requirements of maintaining tropical soils. Consequently, beyond the loss of biodiversity, we have the additional problem that between a third and a half of the cleared land becomes unproductive within three to five years, again leaving the people in absolute poverty.

Additional "new lands" include steep slopes, which suffer horrendous erosion when plowed, and areas with minimal rainfall that turn to deserts under cultivation.

Migration to Cities. Faced with the poverty and hardship of the countryside, many hundreds of millions of people in developing nations continue to migrate to cities in search of employment and a better life. The result is that a number of cities of the developing world are now among the world's largest and are still growing rapidly (Fig. 6–7a). Opportunities in many cities have not expanded fast enough to handle the influx. People are forced to live in sprawling, wretched squatter settlements and slums that do not even provide adequate water and sewers, much less other services (Fig. 6–7b, Table 6-4). Consider the following description of São Paulo, the industrial center of Brazil:

> Too many people in the wrong places. The shacks and shanty towns surround São Paulo in concentric circles—called the rings of misery—of millions of people living below the poverty level, trying to earn, beg or steal a living with virtually no hope of aid from a government that feels it has a long way to go before it can consider social welfare programs.

And most of those people are young; each year three million Brazilians enter the job market. They come to the cities because, bad as life is there, it is better than in the desolate rural area where many were born. It is expected that the population of São Paulo—as well as of Rio de Janeiro, with its five million people—will double by the end of the century. Brazil's economy must grow rapidly if the country is to keep from eating itself alive.[1]

Consider also this, said of Bombay, India:

> More than a million people crouch cheek by jowl along a maze of suffocating narrow footpaths. Naked children play in black puddles of stagnant sewage beside tents of rags and corrugated tin huts. A choking stench, unending noise and festering disease are everywhere.[2]

It is not coincidental that three epidemics of cholera have occurred in the developing world in recent years. Cholera is caused by a bacterium spread via sewage that gets into drinking water. The disease causes extreme vomiting and diarrhea, which result in great loss of body fluids; it is frequently fatal if not properly treated.

Worse, these cities do not provide the jobs people seek. Indeed, the high numbers of rural immigrants in the cities

[1] Brian Kelly and Mark Landon. *Amazon.* (New York: Harcourt Brace Jovanovich, 1983) 19.

[2] Robert Benjamin, Baltimore *Sun,* June 16, 1991.

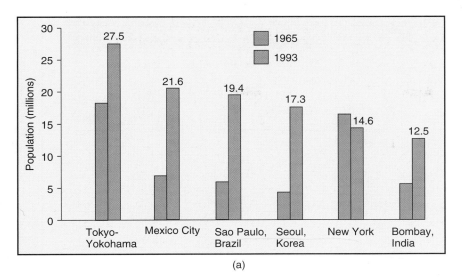

(a)

(b)

◀ **FIGURE 6–7** *Growing cities.*
(a) Growth of some major world metropolitan areas. Since 1965, cities in the developing world have grown phenomenally, and a number of them are now among the world's largest. (b) Slums on the outskirts of São Paulo, Brazil, where 32% of the city's population live. (Source: Data for [a] from *Christian Science Monitor* [Dec. 1990]: 12; updated from U.S. Bureau of the Census.)

dilute the value of the one thing they have to sell—their labor. As we saw earlier, a common wage for a day's unskilled work is often equivalent to no more than a dollar or two, not enough for food, much less housing, clothing, and other amenities. Thousands, including many children, make their living by scavenging in dumps to find items they can salvage, repair, and sell. Many survive by begging—or worse.

Illicit Activities. Anyone who doesn't have a way to grow sufficient food must gain enough income to buy it—and sometimes, desperate people break the law to do this. Of course, it is difficult to draw the line between the need and the greed that also draws people into illicit activities. However, it is undeniable that the shortage of adequately paying employment exacerbates the problem. Besides the rampant petty thievery and corruption that pervade many developing countries, income is also obtained from the following illegal activities.

Illicit drugs. A small peasant farmer with too little land to make a living growing food can make a decent income growing the various crops from which illicit drugs are made. Of course, those who synthesize and sell the drugs make even greater profits.

Poaching of wildlife. A person in a tropical developing country can make a considerable income hunting or trapping various fish, birds, reptiles, and other animals and selling them into the "pet" trade. Illegal trade involving endangered species has grown steadily over the years; in

TABLE 6-4 Incidence of Household Environmental Problems in Accra, Jakarta, and São Paulo

Environmental Indicator	Incidence of Problem (percentage of all households surveyed)		
	Accra, Ghana	Jakarta, Indonesia	São Paulo, Brazil
Water			
No water source at residence	46	13	5
No drinking water source at residence	46	33	5
Sanitation			
Toilets shared with > 10 households	48	14–20	13
Solid waste			
No home garbage collection	89	37	5
Waste stored indoors in open container	40	27	14
Indoor air			
Wood or charcoal is main cooking fuel	76	2	0
Mosquito coils used	45	28	8
Pests			
Flies observed in kitchen	82	38	17
Rats/mice often seen in home	61	82	25

Note: Sample sizes were as follows: Accra, N = 1000; Jakarta, N t=1055; and São Paulo, N = 1000. From: McGranahan & Songsore, "Health, Wealth, and the Urban Household," *Environment* (July–Aug. 1994): 9.

black-market activity, it is second only to drugs. Since few of the animals collected survive, and fewer yet are put into situations in which they will breed, poaching is a major factor in pushing many species toward extinction.

Given the poverty in developing countries, can you see why these activities persist despite efforts to stop them? Of course, it shouldn't escape your notice that the affluent consumer at the end of the chain provides the basic incentive for such activities.

Emigration and Immigration. Facing the stresses and limited opportunities in their own countries, many people of developing countries see emigration to developed countries as their best hope for a brighter future. Historically, the New World was colonized by the overflow population from Europe, which experienced its population explosion early in the Industrial Revolution, and certainly, immigrants have contributed immensely to the development and economic growth of the United States, Canada, and other countries. However, is it feasible for the United States or any other developed nation simply to open its doors to all who would flee from poverty and lack of opportunity in the developing world?

The Worldwatch Institute estimates that there are already over 23 million environmental refugees—people living outside their homeland because they can't make a living there—and the number is growing rapidly. The gates for these immigrants are rapidly closing, however. Concerned over already high unemployment and strains on state welfare systems, the European Union countries are taking steps to severely limit immigration from the developing world.

The one developed country that accepts more immigrants than all others combined is the United States (see the "Ethics" essay, p. 149). The current (1999) population of the United States—273 million—is growing at about 2.7 million per year. About one third of this growth is new immigrants.

Impoverished Women and Children. The hardships and deprivation of poverty fall most heavily on women and children. Men are more free to roam and pick up whatever work is available, and they may keep their wages for themselves. Some men take no responsibility at all for the women they make pregnant, much less the children that are produced. Even many married men, under the stress of poverty, abandon wives and children. Few developing countries have a welfare system that will provide care in such situations. Too often, the women cannot cope; children are abandoned, and women turn to begging, stealing, or prostitution.

And what happens to their children? If they survive at all, it is by begging, scrounging through garbage, stealing, and finding shelter in any hole or crevice they can find (Fig. 6–8). The problem is great: Nearly every sizable developing world city has thousands of these "stray" children, on the order of 20 million in all by some estimates, and their numbers are growing. Forced child labor, child prostitution, and selling children for adoption are additional problems that exist in no small measure. One can speculate as to the

◀ **FIGURE 6–8** *Foraging in trash.* In cities of the developing world, many poor people, including mothers and children, subsist only by scrounging through refuse for bits of food and items they can resell.

kinds of adults these children become as they grow up. At the very least, all of the factors tend to lock the poor into the vicious cycle of illiteracy and squalid conditions that defines absolute poverty.

A summary of the consequences of rapid population growth in developing countries is given in Fig. 6–9. It is clear that population growth, poverty, and environmental degradation are not separate issues. They are very much interrelated.

◀ **FIGURE 6–9** *Consequences of exploding populations in the developing world.* The diagram shows the numerous connections between unchecked population growth and social and environmental problems.

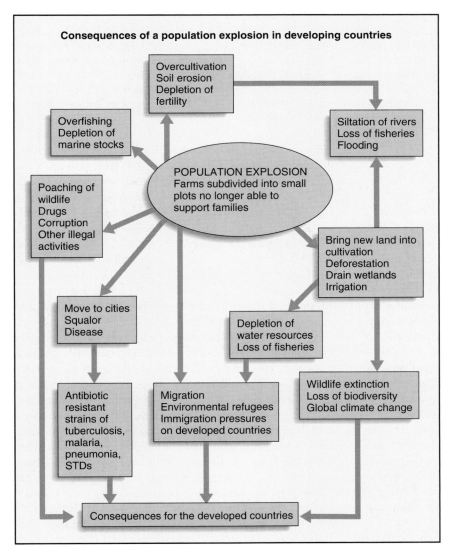

Consequences of a population explosion in developing countries

Overcultivation
Soil erosion
Depletion of fertility

Overfishing
Depletion of marine stocks

Siltation of rivers
Loss of fisheries
Flooding

POPULATION EXPLOSION
Farms subdivided into small plots no longer able to support families

Poaching of wildlife
Drugs
Corruption
Other illegal activities

Bring new land into cultivation
Deforestation
Drain wetlands
Irrigation

Move to cities
Squalor
Disease

Depletion of water resources
Loss of fisheries

Antibiotic resistant strains of tuberculosis, malaria, pneumonia, STDs

Migration
Environmental refugees
Immigration pressures on developed countries

Wildlife extinction
Loss of biodiversity
Global climate change

Consequences for the developed countries

ETHICS

THE DILEMMA OF IMMIGRATION

For people trapped by poverty or lack of opportunity in their homeland, emigration to another country has always seemed a way to achieve a better life. As the New World opened up, many millions of people recognized this dream by emigrating to the United States and other countries. The United States and Canada are countries largely composed of immigrants and their descendants. Until 1875, all immigration into the United States was legal; all who could manage to arrive could stay and become citizens. This openness was inscribed on the Statue of Liberty, which reads, in part, "Give me your tired, your poor, your huddled masses yearning to breathe free, the wretched refuse of your teeming shore....Send these, the homeless, tempest-tossed to me."

Emigration from the Old World created a flood of migrants immigrating to the New World. This relieved population pressures in European countries and aided in the development of the New World. It is apparent,

however, that a totally open policy toward immigration would be untenable today. The United States, with its current population of 273 million, is no longer a vast, open land awaiting development; yet hundreds of millions of people would immigrate if they could. How much immigration should be permitted, and should some groups be favored over others? For example, in 1882, the U.S. Congress passed the Chinese Exclusion Act, which barred the immigration of Chinese laborers but not Chinese teachers, diplomats, students, merchants, or tourists. This act remained in effect until 1943, when China and the United States became allies in World War II. However, the current immigration policy still makes it easier for trained people to gain citizenship and relatively difficult for untrained people to do so. This policy created what is commonly referred to as a "brain drain." Brain power, many pointed out, is the "export" that developing nations can least afford.

Under the current immigration laws, the

United States now accepts around 800,000 new immigrants per year, a number larger than we have received at any time since the 1920s and larger than is accepted by all other countries combined. Further, a lottery system was adopted for selection. Under the act, immigration accounts for about 33% of U.S. population growth, which is presently 2.4 million per year. The remainder of the population growth is called natural increase, births exceeding deaths. If the fertility rate remains low, immigration will account for a growing proportion of population growth.

The preceding deals, of course, with legal immigration; illegal immigration is another matter. Hundreds of thousands, unable to gain access through legal channels, seek ways to illegally enter the country. The United States maintains an active border patrol, especially along the border with Mexico, that several thousand people try to cross each night. Most are caught and returned, but an undetermined number slip through—estimated at 200,000 to 300,000 per year. The Illegal Immigration

Effects of Increasing Affluence

Increasing the average wealth of a population affects the environment both positively and negatively. An affluent country certainly can and does provide such things as safe drinking water, sanitary sewage systems and sewage treatment, and collection and disposal of refuse. Thus, many forms of pollution decrease, and the environment we live in improves with increasing affluence. In addition, if we can afford gas and electricity, we are not destroying our parks and woodlands for firewood. In short, we are able to afford conservation and management, better agricultural practices, and pollution control and thereby improve our environment.

But affluence has its negative aspects as well. For example, by using such large quantities of fossil fuel (coal, oil, and natural gas) to drive our cars, heat and cool our homes, and generate electricity, the United States is responsible for a large share of the production of carbon dioxide. Specifically, with about 4.5% of the world's population, the United States is responsible for about 25% of the emissions of carbon dioxide that may be changing global climate. Similarly, emissions of chlorofluorocarbons (CFCs) that degrade the ozone ...missions of chemicals that cause acid rain, emissions ...micals and production of nuclear wastes are ...roducts of affluent societies.

...factors place further demands on the environment, ...xample, many rain forests in Brazil have been

cut in order to convert the land into cattle pasture, serving the appetites of the affluent for meat. A large portion of the oceans have been overfished for the same reason. Currently, old-growth forests in southern South America are being clearcut and turned into chips to make fax paper. Oil spills are a "byproduct" of our appetite for energy. We have already mentioned the pressure being put on endangered species from people willing to pay exorbitant prices for exotic "pets" or "medicines" made from animal parts. As increasing numbers of people strive for and achieve greater affluence, it seems more than likely that these and similar pressures will mount.

One way of generalizing the effect of affluence is to say that it enables humans to clean up their immediate environment by disposing of their wastes to more distant locations. It also allows them to obtain resources from more distant locations such that they do not see or feel the impacts of obtaining those resources. Thus, in many respects, the affluent isolate themselves and may become totally unaware of the environmental stresses they cause with their consumption-oriented lifestyles. On the other hand, affluence also provides people with opportunities to exercise lifestyle choices that are consistent with the concerns for stewardship and sustainability.

With this picture of population growth and its impacts in view, we turn now to some additional dimensions of population growth to provide a more thorough understanding of the population problem.

Reform Act of 1996 addressed this problem by increasing the border patrol and stepping up efforts to locate and deport illegal aliens, actions which are supported by most observers.

In 1991 and 1992 Haitian people made headlines as they fled their country in crowded, unseaworthy boats and headed for the United States. These people presented a different issue—political asylum. U.S. immigration policy permits admission to people who are fleeing their country because of political oppression. Although political oppression was certainly present in Haiti, the evidence—as reflected in a Supreme Court decision—indicated that the boat people were economic refugees. That is, they were simply seeking a better life. They were deported back to Haiti, and subsequently, the United States intervened in Haiti in an attempt to restore democracy and economic progress to that impoverished country.

A recent report by the National Research Council examined the economic impacts of our immigration policies and concluded that legal immigration basically benefits the U.S. economy and has little negative impact on native-born Americans. A few areas of the country where immigrants are especially numerous (California, Texas, Florida) experience some challenges in assimilating new immigrants, but the overall impact is positive, according to the National Research Council.

In its report to the nation, the President's Council on Sustainable Development addressed the question of immigration in the context of achieving a sustainable society, one where population growth has come to a halt. The Council decided not to take a definite stand on immigration policy and instead deferred to a coming report from the U.S. Commission on Immigration Reform, established by Congress in 1990, and directed to make its report at the end of 1997. The Commission recommended cutting back immigration to a core level of 550,000 per year and to allow 150,000 additional visas annually for spouses and minor children of legal permanent residents. This policy is to be phased out when the existing backlog is eliminated—judged to take up to eight years. Interestingly, the Commission recommended discontinuing entirely the immigration of unskilled laborers, while continuing to welcome (under quotas) more highly skilled immigrants. The Commission has now disbanded, and it is up to Congress to take up the report and act on it—or not. It should be pointed out that these recommendations are a small reduction of the current rate of legal immigration. It looks like immigration will continue to be a large and increasing part of growth in the United States well into the future.

As population pressures in developing countries continue to mount, the question of how many immigrants to accept, from what countries, and where to draw the line regarding asylum seems certain to become more and more pressing. In addition to compassion, the social, economic, and environmental consequences—both national and global—of the alternatives must be weighed in making the final decision. Where do you stand?

6.4 Dynamics of Population Growth

In considering population growth, we consider more than just the increase in numbers, which is simply births minus deaths; we also consider how the numbers of births ultimately affect the entire population over the **longevity**, or lifetimes, of the individuals.

Population Profiles

A **population profile** is a bar graph showing the number of people (males and females separately) at each age for a given population. The population profile of the United States in 1991 is shown in Fig. 6–10. The data are collected through a census of the entire population, a process in which each household is asked to fill out a questionnaire concerning the status of each of its members. Various estimates are made for those who do not maintain regular households. For the United States and most other countries, a detailed census is taken every 10 years. Between censuses, the population profile may be adjusted by using data regarding births, deaths, immigration, and, of course, the aging of the population. The field of collecting, compiling, and presenting information about populations is called **demography**; the people engaged in this work are **demographers**.

A population profile shows the **age structure** of the population—that is, the proportion of people in each age group at a given date. However, leaving out the complication of emigration and immigration for the moment, recognize that each bar in the profile started out as a cohort of babies at a given point in the past, and that cohort has only been diminished by deaths as it has aged. In developed countries such as the United States, the proportion of people who die before age 60 is relatively small. Therefore, the population profile below age 60 is an "echo" of past events insofar as those events affected birth rates. For example, in Fig. 6–10 you can observe that smaller numbers of people were born between 1931 and 1936. This is a reflection of lower birth rates during the Great Depression. The dramatic incre... people born in 1946 and in following years is a... returning veterans and others starting fami... to have relatively large numbers of... following World War II—the "... in numbers of people b... of sharply declin... have significan'... in numbers ... the "baby boo... ducing a simila... the actual total fe... tapering off of num... of the increasing nu...

More than a vie... profile provides govern...

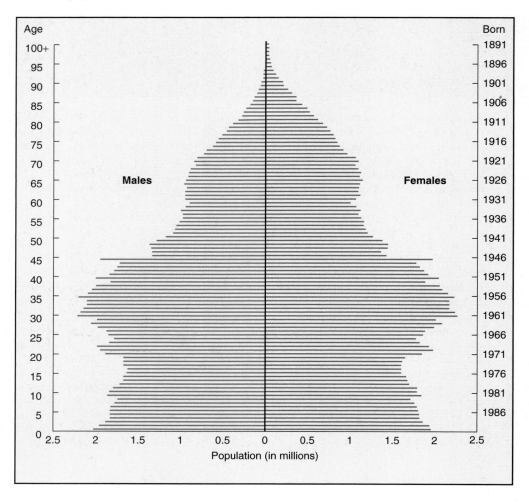

FIGURE 6–10 *Population profile of the United States, 1991.* The age structure of the U.S. population, showing the effects of major shifts in fertility. (Source: *Population Today,* Sept. 1993, p. 10. Washington, DC: Population Reference Bureau, Inc.)

of realistic planning for future demand for various goods and services, ranging from elementary schools to retirement homes. Consumer demands are largely age specific; that is, what children need and want differs from what teenagers, or young adults, or older adults, or finally people entering retirement want and need. Using a population profile, one can literally see the projected populations of particular age cohorts and plan to expand or retrench accordingly. A number of industries expanded and then contracted as the baby boom generation moved through a particular age range, and [th]is phenomenon is not yet past. You can readily see that [con]cerns about Social Security are well founded, because [Socia]l Security is paid not from funds contributed by re-[tirees b]ut from contributions being made by current work-[ers. Th]e large baby boom generation, now in middle age, [...] the population profile, there will be a huge addi-[tional deman]d for Social Security outlays to retirees, and [...] relatively small population of workers to sup-[port them. ...] In contrast, any business or profession in [...] goods or services to seniors is looking [at ... o]f growth.

[...] [impo]rtance of fertility rate cannot be un-[derstated. ...] [ra]tes differing from replacement will

affect the economy, and hence the environment, in one way or another over the entire lifetime of the individuals born.

Population Projections

Most significantly, we can use a population profile and certain other data to estimate the future overall growth of a population.

The Forecasting Technique. The technique of estimating future population growth (or decline) is one of estimating numbers of future births and deaths. Simple subtraction of deaths from births gives the absolute change in size of the population.

Estimating births: Using the population profile, you can see the numbers of young people moving into reproductive ages, and from other statistics, you know what percentage of women at each age have babies. From here, it is a process of multiplying the number of women coming into each age by the percent who have babies at that age, then totaling, and then adding that total number of babies as a new bar at the bottom of the profile.

Estimating deaths: From statistics regarding deaths, you know the percentage of people who die at each age. These are the actuarial tables used by insurance companies to determine life insurance rates. Using the population profile, you can see the number of people moving into each age. Particularly significant are the numbers of people moving into older ages, where death rates are highest. Again, multiplying the number of people moving into each age by the percentage of people who die at that age, then totaling, gives the overall deaths for the year, and all the bars on the profile are adjusted accordingly.

Repeating this process over and over allows demographers to project births and deaths and, by subtraction, change in population as far into the future as desired. You can see the need for many calculations, but computers make the process relatively straightforward. Of course, such projections become increasingly subject to error as they are extended further into the future, because one has only current statistics and trends to work from, and many things may occur to either increase or decrease longevity. Just as troublesome in making long-term projections are changes in fertility rates, which are basically a function of how many babies women choose to have at any given age. As we noted at the beginning of this chapter, the projection that the world population will level off at around 11 billion is based on the assumption that fertility rates will continue a gradually declining trend. However, fertility rates tend to rise and fall, for reasons that are not always well understood or predictable. We have already seen how fertility rates in the United States rose after World War II, producing the baby boom generation, and then how the rates sharply declined between 1961 and 1976.

Therefore, different projections are made on the basis of different assumptions regarding future fertility rates. Thus, the United Nations gives three different projections of future world population (Fig. 6–11). The medium-fertility scenario assumes that replacement-level fertility will be reached by 2050; the high-fertility scenario assumes a total fertility rate of 2.5 by 2050 and continuing at that rate; the low-fertility scenario assumes

that the total fertility rate will reach 1.5 by 2050 and be maintained. Note how each fertility assumption generates profoundly different world populations. Still imponderable, of course, are conditions that may drastically increase death rates, something we wish to make every effort to avoid.

Simplified Population Projections. Demographers use statistics and computers to make projections, because they need those projections to be as accurate as possible. However, we may gain a reasonable picture of future population by a more simplified procedure in which population profiles are presented in a "condensed" form, lumping people into five-year age groups (Fig. 6–12). In this version, each one of the bars moves up one space every five years. Since, in modern human societies, relatively few deaths occur before old age, the population loss from deaths can be approximated simply by removing the uppermost bar. To determine the number of births, observe that the average childbearing age range is 20–24. Therefore, we simply multiply the number of *women* (only the left side of the bar) by the total fertility rate and obtain an approximation of the number of babies born in the relevant five-year period. Again, the process can be repeated over and over to project the population into the future as far as desired, and we can make different projections based on different total fertility rates.

Population Projections for a More Developed Country. The 1990 population profile of Denmark, a developed country in northern Europe, shown in Fig. 6–12a, is more or less columnar in shape because Danish women have had a low fertility rate for some time. If we assume that the 1991 total fertility rate of 1.6 remains constant for the next 20 years and we follow the technique just described, we

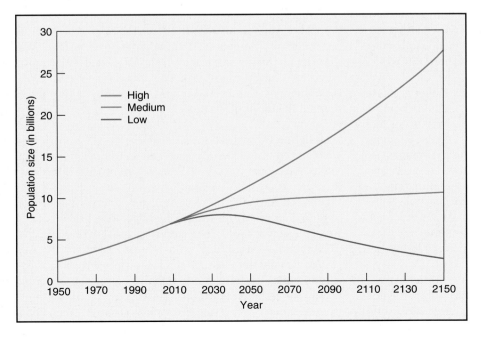

◀ **FIGURE 6–11** *Projected world population according to three different fertility scenarios.* U.N. projections of the future world population, using different total fertility rates. (Source: Population Division of the Dept. of Economic and Social Affairs at the United Nations Secretariat, *World Population Projections to 2150*, (New York: United Nations, 1998).

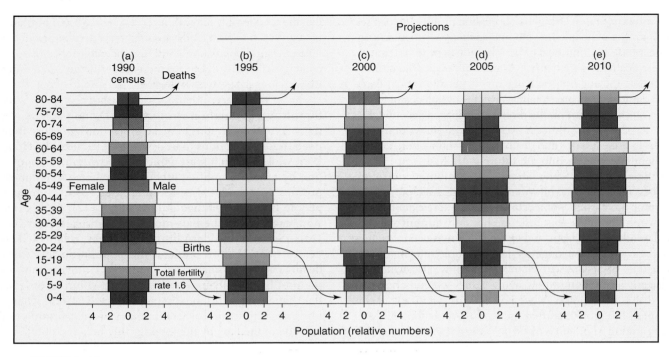

▲ **FIGURE 6–12** *Projecting future populations: developed country.* (a) A population profile representative of a highly developed country, Denmark, in 1990. (b–e) Projections of Denmark's population made as described in the text, assuming that the total fertility rate remained at its 1991 value of 1.6. Note how larger numbers of persons are moving into older age groups and the number of children is diminishing. (Profile shown in [a] from Joseph A. Mcfalls, Jr., "Population: A Lively Introduction," *Population Bulletin* 46, no 2 [1991]: 4.

will obtain the profiles presented in Figs. 6–12b–e. The profiles show an increase in the number of older people and, with one exception, a marked decline in the number of children and young people.

The profiles also show that for the next 20 years, Denmark's total population will remain relatively constant because the number of deaths (removal of top bars) will be nearly the same as the number of births (addition of bottom bars). But the population will be **graying**, a term used to indicate that the proportion of elderly people is increasing. We can also predict a considerable population decline following 2010, as the large elderly population suffers mortality while fewer and fewer children are born.

What opportunities and risks does the changing population profile imply for Denmark? If you were an adviser to the Danish government, what would your advice be for the short term? For the longer term? Unless a smaller population is the goal, our advice here might well be to encourage and to provide incentives for Danish couples to bear more children. We might also advocate allowing more immigration. But allowing the declining numbers of Danish people to be replaced by an immigrant population requires one to reflect on the effect that would have on the Danish culture, religion, etc. (In fact, Denmark's total fertility rate in 1998 has increased to 1.8 births since 1991.)

The very low fertility rate and prospective declining population seen in Denmark are typical of nearly all highly developed nations. Therefore, the preceding analysis and questions can be applied to any of them. The one exception

is the United States: In contrast to other developed countries, the U.S. fertility rate reversed directions in the late 1980s and started back up. Based on the lower fertility rate, the U.S. population had been projected to stabilize at between 290 and 300 million toward the middle of the next century. With a higher fertility rate of 2.0, the U.S. population is projected to be 392 million by 2050 and to continue growing indefinitely (Fig. 6–13). For this projection, immigration is assumed to remain constant at current levels—880,000 per year.

These projections for the United States show what a profound effect on population slight differences in the total fertility rate make when they are extrapolated 50 or more years forward. In light of the concerns for sustainable development, what do you think the population policy of the United States should be?

Population Projections for a Less Developed Country. Developing countries are in a vastly different situation from that of developed countries. In developing countries, while fertility rates are generally declining, they are still well above 2.1. (The average is currently 3.8, excluding China.) Because of even higher past fertility rates, the population profiles of developing countries have a pyramidal shape. For example, the 1990 population profile for the African country of Kenya, which had a fertility rate of 6.7, is shown in Fig. 6–14a. Assuming that this fertility rate remains constant, projecting Kenya's population ahead just 10 years gives us the profile shown in Fig. 6–14b. Note that even the relatively high child mortality rate (currently 62 per thousand), comes nowhere

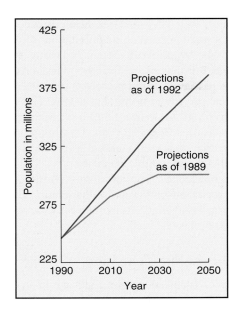

▲ **FIGURE 6–13** *Population projections for the United States.* Projections shift drastically with changes in fertility. Contrast the 1988 projection, based on a fertility rate of 1.8, with the 1993 projection, based on the increased fertility rate of 2.13. (Source: Data from Kelvin M. Pollard, *Population Stabilization No Longer in Sight for U.S. Population Today* [Washington, D.C. Population Reference Bureau, Inc.,1994]: 1.)

close to offsetting the high fertility rate. Thus, the pyramidal form of the age structure remains the same because each rising generation of young adults produces an even larger generation of children. The pyramid just gets wider and wider, leading to a doubling and redoubling of the population.

Kenya's total fertility has been declining during the '90s—to 4.5 by 1998—but even at the lower rate the population will still double in just 35 years.

While highly developed countries are facing the problems of a graying population, the high fertility rates in developing countries maintain an exceedingly young population. An "ideal" population structure with equal numbers of persons in each age group and a longevity of 75 years would have one fifth, or 20%, of the population in each 15-year age group. By comparison, in many developing countries, 40%–50% of the population is below 15 years of age. In contrast, in most developed countries, less than 20% of the population is below the age of 15 (Table 6-5).

Consider what all this means in terms of the need for new schools, housing units, hospitals, roads, sewage collection and treatment facilities, telephones, etc. If a country such as Kenya is simply to maintain its current standard of living, the amount of housing and all other facilities (to say nothing of food production) must be doubled in as little as 25 years. Accordingly, it is not difficult for a developing country's efforts to get ahead to be more than nullified by its own population growth.

A comparison of present and projected population profiles for developed and developing countries is shown in Fig. 6–15. The figure shows that while little growth will occur in developed countries over the next 25 years, enormous growth is in store for the developing world. What is worse, this is assuming that fertility rates in the developing world continue their current downward trend.

Again, these or any other population projections should not be confused with predicting the future. They are intended only to show where we will end up if we continue the present

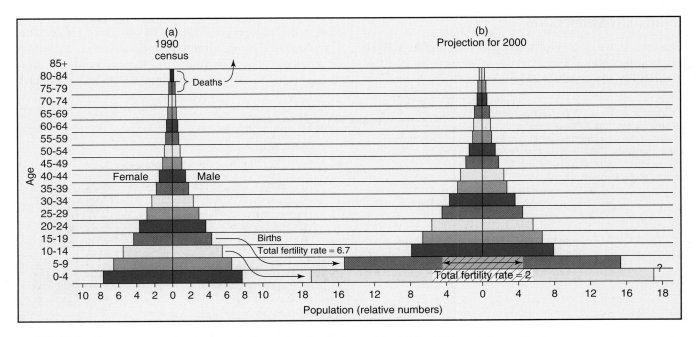

▲ **FIGURE 6–14** *Projecting future populations: developing country.* (a) The 1990 population profile of Kenya, a developing country. (b) Projection for the year 2000, based on the assumption that the total fertility rate will continue at 6.7. Even if the total fertility rate immediately dropped to 2, the number of births (hatched inner portion of the bottom two bars) would still greatly exceed the number of deaths because so few persons are in the upper age groups. (Profile shown in [a] from Joseph A. Mcfalls, Jr., "Population: A Lively Introduction," *Population Bulletin* 46, no. 2 [1991]: 4.

TABLE 6-5 Populations by Age Group

Region/country	Percent of Population in Specific Age Groups		
	<15	15 to 65	>65
Africa	44	53	3
Kenya	46	51	3
Latin America	34	61	5
Asia	32	62	6
Europe	19	67	14
Germany	16	69	15
China	26	68	6
United States	22	65	13

Source: Data from *1998 World Population Data Sheet*, (Washington, D.C.: Population Reference Bureau, 1998).

course. As the old saying goes, "If you don't like where you are going, change direction." In other words, if we feel that the projected population growth is not desirable, we can try to bring fertility rates down faster. However, even bringing the fertility rates of developing countries down to 2.0 will not stop their growth immediately. This is because of a phenomenon known as *population momentum*, which we examine next.

Population Momentum

Countries with a pyramid-shaped population profile, such as Kenya, will continue to grow for 50–60 years, even after the total fertility rate is reduced to the replacement level. This phenomenon, called **population momentum**, occurs because such a small portion of the population is in the upper age groups, where most deaths occur, and many children are entering their reproductive years. Even if these rising generations have only two children per woman, the number of births will far exceed the number of deaths. The imbalance will continue until the current children reach the Kenyan limits of longevity—50 to 60 years. In other words, only a population at or below replacement-level fertility for many decades will achieve a stable population.

To acknowledge population momentum is not, of course, to say that efforts to stabilize population are fruitless. It is, however, to say that fertility rates *must* be reduced before a crisis point is reached.

The Demographic Transition

The concept of a stable, nongrowing global human population based on people freely choosing to exercise a lower fertility rate is possible because it is already the fact in developed countries. If we can understand the factors that have brought this about in developed countries, then perhaps we can make those factors operative in developing countries.

Early demographers observed that modernization of a nation brings about more than just a lower death rate resulting from better health care; a decline in fertility rate also occurs as people choose to limit the size of their families. Thus, as economic development occurs, human societies move from a primitive population stability, in which high birth rates are offset by high infant and childhood mortality, to a modern population stability, in which low infant and childhood mortality are balanced by low birth rates. This gradual shift in birth and death rates from the primitive to the modern condition in the industrialized societies is called the **demographic transition**. The *basic premise* of the demographic transition is that there is a causal link between modernization and a decline in birth and death rates.

▶ **FIGURE 6–15 *Comparing projected populations.*** Population profiles for developed and developing countries, projected to the year 2025. (Source: Data from *Population and the Environment: The Challenges Ahead,* United Nations Population Fund, 1991.)

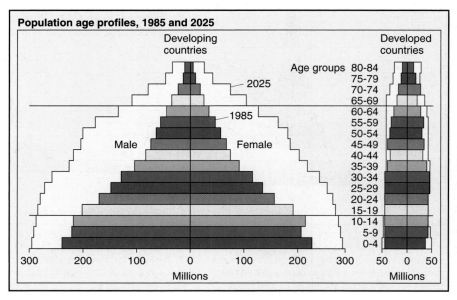

Birth Rates and Death Rates. To understand the demographic transition, we need to introduce two new terms: the **crude birth rate (CBR)** and **crude death rate (CDR)**. These terms are defined as the number of births or deaths per 1000 of the population per year. By giving the data in terms of "per 1000," populations of different countries can be compared regardless of their total size. The term *crude* is used because no consideration is given to what proportion of the population is old or young, male or female. Subtracting the CDR from the CBR gives the increase (or decrease) per 1000 per year. Dividing this result by 10 then puts it in terms of "per 100," or percent.

CBR		CDR		Natural increase
Number of		Number of		(or decrease)
births per	−	*deaths* per	=	in population
1000		1000		per 1000
per year		per year		per year
				÷ 10 = %

Of course, a stable population is achieved if, and only if, the CBR and CDR are equal.

The **doubling time**, or the number of years it will take a population growing at a constant percent per year to double, is calculated by dividing the percentage rate of growth into 70.

The answer is the number of years it will take the population to double. (The 70 has nothing to do with population; it is derived from an equation for population growth.) CBR, CDR, and doubling times of various countries are shown in Table 6-6.

Epidemiologic Transition. Throughout most of human history, crude death rates were high—in the 40s and above per thousand for most societies. As we have seen, by the middle of the nineteenth century, the epidemics and other social conditions responsible for high death rates began to recede, and death rates in Europe and North America underwent a decline. It is significant that this decline was gradual in the now developed countries, lasting for many decades and finally stabilizing at a CDR of about 10 per thousand. At present, cancer and cardiovascular disease and other degenerative diseases account for most mortality, as many people survive to old age. This pattern of change in mortality factors has been called the *epidemiologic* transition (Fig. 6–16) and represents one element of the demographic transition (epidemiology is the study of diseases in human societies).

Fertility Transition. Another pattern of change over time can be seen in crude birth rates. Again, in the now developed countries, the birth rates have undergone a decline from a high in the 40s to 50s down to 10 to 12 per thousand—a *fertility* transition. As Fig. 6–16 shows, this does not happen

TABLE 6-6 Crude Birth and Death Rates for Selected Countries

Country	Crude Birth Rate	Crude Death Rate	Annual Rate of Increase (%)	Doubling Time (Years)
World	23	9	1.4	49
Developing Nations				
Average (excluding China)	29	10	2.0	35
Egypt	28	6	2.2	32
Kenya	33	13	2.0	35
Madagascar	44	14	3.0	23
India	27	9	1.9	37
Iraq	38	10	2.8	25
Vietnam	19	7	1.2	57
Haiti	34	13	2.1	33
Brazil	22	8	1.4	48
Mexico	27	5	2.2	32
Developed Nations				
Average	11	10	0.1	548
United States	15	9	0.6	116
Canada	12	7	0.5	136
Japan	10	7	0.2	330
Denmark	13	11	0.2	431
Germany	10	10	-0.1	—*
Italy	9	9	-0.0	—*
Spain	9	9	0.0	1733

Source: Data from 1998 *World Population Data Sheet* (Washington, D.C.: Population Reference Bureau, 1998).

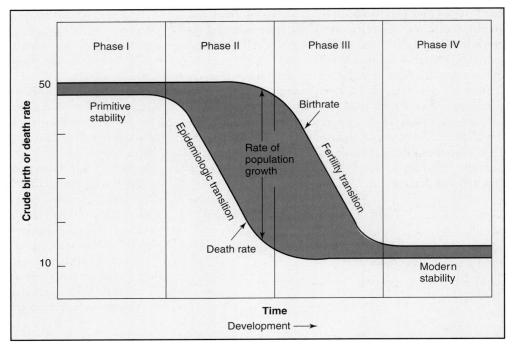

▶ **FIGURE 6–16** *The demographic transition.* The epidemiologic transition and the fertility transition combined to produce the demographic transition in the developed countries over many decades. (Adapted from Joseph A. McFalls, Jr., "Population: A Lively Introduction," *Population Bulletin* 46, no. 2 [1991]: 33.)

at the same time as the epidemiologic transition; it is delayed by decades or more. Since net growth is the difference between CBR and CDR, the time during which these two patterns are out of phase is a time of rapid population growth. The developed countries underwent such a time during the nineteenth and early twentieth centuries, and one result of this was the massive emigration from the Old World to the less populated New World.

Phases of the Demographic Transition. The demographic transition is typically presented as occurring in four phases, as shown in Fig. 6–16. **Phase I** is the primitive stability resulting from a high CBR being offset by an equally high CDR. **Phase II** is marked by a declining CDR—the epidemiologic transition. Because fertility and, hence, the CBR remain high during Phase II, this is a phase of accelerating population growth. **Phase III** is a phase of declining CBR resulting from a declining fertility rate; population growth is still significant. Finally, **Phase IV** is reached, where modern stability is achieved by a continuing low CDR but an equally low CBR.

Basically, developed countries have completed the demographic transition. Developing countries, on the other hand, are still in Phases II and III. Death rates have declined markedly, and fertility and birth rates are declining but are still considerably above replacement levels. Therefore, populations in developing countries are still growing rapidly. Using the concept of the demographic transition, we can plot the major regions of the world in terms of current birth and death rates (Fig. 6–17). We can draw a dividing line through this plot; nations to the right of the line are on a fast track to completing the demographic transition or are already there. Nations to the left of the line (about half the world) are in the middle of the demographic transition, most

in the fourth decade of rapid population increase. They appear to be trapped there, with serious consequences.

It is on the basis of the demographic transition that some argue that we not worry about population; it will stabilize by itself as developing countries reach Phase IV. Therefore, the argument goes, we need only focus on free world trade and other factors that will speed economic growth in developing countries. But there are major flaws in this argument. First, it must be emphasized that the demographic transition occurred in the developed countries over a period of many decades; modernization did not happen overnight. Second, many of the most populous developing countries are still very far behind the developed countries economically (Fig. 6–3) and are making very slow progress toward modernization. If they must modernize before population growth comes under control, their population growth and its demands for resources and services will undercut the very economic growth that is so necessary—a catch-22 with profound consequences. Third, we emphasize again that present stresses on the biosphere and loss of biodiversity are largely a consequence of the consumption-oriented lifestyles of the current 1.2 billion people in developed nations. Recognizing that severe stresses are being caused by the lifestyles of 1.2 billion people makes any notion of a world with 11 billion people living with the same lifestyle utterly absurd. Finally, and most important, the demographic transition really only shows a *correlation* between development and changing birth and death rates; it does not prove that development is necessary for the demographic transition to occur. Other factors may be much more significant.

In Chapter 7, we will investigate the factors that influence birth rates and explore the relative contributions of economic development and family planning for bringing about the fertility transition.

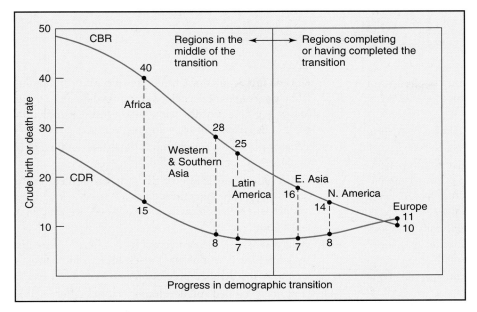

◀ **FIGURE 6–17** *World regions in the process of demographic transition.* Crude birth rates and crude death rates are shown for major regions of the world. A dividing line separates countries at or well along in the demographic transition from those apparently stuck in the middle of the transition. (Source: Data from *1998 World Population Data Sheet* [Washington, D.C.: Population Reference Bureau, 1998].)

ENVIRONMENT ON THE WEB

TRACKING THE DEMOGRAPHIC TRANSITION

This chapter describes how a demographic transition is likely to occur as a population (or economy) modernizes. When we read about population projections for this or that country, it is easy to think of ourselves as somehow removed from the system. In fact, we are all a part of some demographic trend, whether rising or falling, and our own actions will affect the size and composition of future populations. Like us, people in other countries are faced with lifestyle decisions—for instance, about education, work, and family size. These decisions are influenced by our personal histories and the patterns we have observed in our own families and the society around us.

For example, in more industrialized societies, a larger proportion of the population stays in school longer, is engaged in high paying, complex work, and defers marriage and childbearing. With fewer childbearing years, the individual has fewer children than someone with less education and a menial job—or no job at all. Conversely, people in less developed societies may not go to school at all, are less likely to work at complex jobs, and tend to marry and bear children earlier. In these societies, fertility rates are generally higher than in developed nations.

Therefore, it is possible to observe some general patterns in birth and death rates as societies modernize and to be able to predict these demographics for managerial purposes. Yet, in order to predict demographic change, we must understand the human behaviors that cause it. We must also understand that it is much more difficult to predict how these behaviors may or may not change over time.

A key challenge in predicting demographics relates to uncertainties in birth rates. While medical advancements, and thus changes in death rates, can occur over a matter of years, birth rates may take generations to respond to societal change. People like to have children, especially if they have come from a large family or a culture that values children highly. Even when a society has made good progress on industrialization, it can be hard to encourage a move to a smaller family size.

Web Explorations

The Environment on the Web activity for this essay describes the challenges involved in predicting and managing demographic change and the human behaviors that underlie these factors. Go to the Environment on the Web activity (select Chapter 6 at http://www.prenhall.com/nebel) and learn for yourself:

1. about the relationship between population and birth rate;
2. about optimistic and pessimistic models for population growth;
3. how population pyramids can help forecast health care and educational needs: and
4. how decision makers can use population growth models for long-term resource management planning.

 A suggested time frame for each of these exercises is 10–30 minutes.

REVIEW QUESTIONS

1. How has the global human population changed from early times to 1800? From 1800 until the present? What is projected over the next 50 years?

2. How is the world divided in terms of relative per capita incomes? Fertility rates? Population growth rates?

3. What two factors are multiplied to give total environmental impact? Are developed nations exempt from environmental impact? Why not? What third factor affects environmental impact, and how does it do so?

4. What are the social and environmental consequences of rapid population growth in developing countries? How may developed countries be affected?

5. What information is given by a population profile? How is the information presented?

6. How are numbers of future births and deaths predicted from a population profile? What simplifying assumptions can be made to enable you to approximately project the future size and structure of a population?

7. How do the population profiles, fertility rates, and population projections of developed countries differ from those of developing countries? How might future population goals of developed and developing countries contrast?

8. What is meant by population momentum, and what is its cause?

9. Define crude birth rate (CBR) and crude death rate (CDR). Describe how these rates are used to calculate the percent rate of growth and the doubling time of a population.

10. What is meant by the demographic transition? Relate the epidemiologic transition and the fertility transition, two elements of the demographic transition, to its four phases.

11. How do developed and developing nations differ regarding their current positions in the demographic transition?

12. Can future sustainability be attained through development alone? Why or why not?

THINKING ENVIRONMENTALLY

1. Make a cause-and-effect "map" showing the many social and environmental consequences linked to unabated population growth. Include crossovers between the developed and developing worlds.

2. It has been proposed that excess human populations be accommodated by building orbiting space stations that would house about 10,000 persons each. Each station would be able to produce its own food. How many space stations would be required to accommodate the world's projected population growth over the next 10 years? What kind of population policy would have to be enforced on the space stations? Are space stations a logical solution to the population problem?

3. Starting with a hypothetical population of 14,000 people and an even age distribution (1000 in each five-year age group from 1–5 to 65–70), assume that this population initially has a total fertility rate of 2 and an average longevity of 70 years. Project how the population will change over the next 60 years under each of the following conditions:

a. Total fertility rate and longevity remain constant.
b. Total fertility rate changes to 4; longevity remains constant.
c. Total fertility rate changes to 1; longevity remains constant.
d. Total fertility rate remains at 2; longevity increases to 100.
e. Total fertility rate remains at 2; longevity decreases to 50.

4. From the 1998 crude birth and crude death rates, calculate the rate of population growth and the population doubling time for each of the following countries:

	CBR	CDR
Algeria	31	7
Ethiopia	46	21
Argentina	19	8
Iran	24	6
Russia	9	14
France	12	9
Australia	14	7

WEB REFERENCES

On-line references for this chapter can be found on the World Wide Web at: **http://www.prenhall.com/nebel**. Click on Chapter 6 on the chapter selector.

7

Addressing the
Population Problem

〰〰〰〰〰

Key Issues and Questions

1. Many regions of the developing world are stuck in the middle of the demographic transition. Must they modernize before fertility will decline, or must they bring fertility down before they can modernize?

2. The factors that influence fertility rates are more specific than development in general. What are the factors that actually influence the number of children desired?

3. The vicious cycle of poverty, high fertility, and environmental degradation continues over much of the globe despite past efforts at development. Why have such efforts missed the mark? What is being done to change this record of failure?

4. The shift from high to low fertility in developing countries is being accomplished by social modernization. How is this different from industrialization, and what five areas must development efforts focus on for successful delivery of social modernization?

5. In 1994, world leaders met in Cairo, Egypt, at the United Nations Conference on Population and Development. What was the significance of this meeting? What are the agreed-upon strategies for addressing the problems of poverty, excessive population, and environmental degradation?

In September 1994, some 15,000 leaders and representatives from 179 nations and nearly 1000 nongovernmental organizations (NGOs) met in Cairo, Egypt, at the United Nations Conference on Population and Development (Fig. 7–1). Before the delegates was a draft document of a "Program of Action" to address the world's persistent problems of poverty and population. In her keynote address to the Convention (Fig. 7–2), Norway's Prime Minister, Gro Harlem Brundtland, stated:

> Population growth is one of the most serious obstacles to world prosperity and sustainable development....
>
> We may soon be facing new famine, mass migration, destabilization, and even armed struggle as people compete for ever more scarce land and water resources....
>
> Today's newborns will be facing the ultimate collapse of vital resource bases....Only when people have the right to take part in the shaping of society by participating in democratic

political processes will changes be politically sustainable. Only then can we fulfill the hopes and aspirations of generations yet unborn.

Some objections were raised to the draft document by the Vatican and by Muslim countries regarding the wording of certain sections alluding to birth control methods and abortions. However, differences were resolved, and in the end, all 179 nations—large and small, rich and poor—signed the final document, committing themselves to achieve basic goals set forth therein by the year 2015.

The historic significance of this event is that for the first time in history, the political, religious, and scientific communities of the world reached a consensus on the population issue. Some outspoken individuals are still voicing arguments to the effect that population is not a critical issue. However, such voices are now overwhelmed by worldwide agreement that the intertwined issues of exces-

◄ **Family planning in India**. Realizing that increasing population is causing a large segment of the population to become locked in poverty, India is actively promoting family planning. The poster is extolling the virtues of the two-child family.

▲ FIGURE 7–1 *1994 ICPD Conference.* World leaders at the U.N. Conference on Population and Development congratulate each other as they reach a consensus regarding a "Program of Action."

▲ FIGURE 7–2 *Keynote conference speaker.* Prime Minister Gro Harlem Brundtland of Norway, the keynote speaker at the Cairo Conference.

sive population, poverty, and environmental degradation are jeopardizing the future of the planet. If they are not mitigated over the next 20 years, the world is more than likely to see a future of unprecedented biologic and human impoverishment.

This is not to say that the world is just awakening to the problems of population, poverty, and environmental degradation. Indeed, these problems have been long recognized, and many billions of dollars have been directed toward their solution through governmental, charitable, and U.N. organizations, and marked accomplishments have occurred. However,

the persistence of the problems speaks to the requirement for improvement.

In this chapter we follow up on the demographic information presented in Chapter 6. We begin with a reassessment of the demographic transition, examining the factors that promote movement of a country through the transition. Then, we examine various policies and programs that have been used in the past to improve development and lower fertility. We then look at evidence for a new direction in development, and finally we examine the international efforts being made to address population issues in the developing world.

7.1 Reassessing the Demographic Transition

At the end of Chapter 6, we presented evidence that many regions of the developing world are stuck in the middle phases of the demographic transition (Fig. 6–17), resulting in continuing rapid population growth. The countries in these regions would all like to experience the economic growth that has brought South Korea, Indonesia, Malaysia, Brazil, and others into the middle- and even high-income nation groups. If they did so, it is likely that they would then move through the demographic transition. Yet there are great disparities in economic growth among developing countries; some 100 countries have been experiencing economic stagnation or decline and remain mired in their poverty. These same countries have the most rapid growth in population.

The **key question** is: Must these countries bring down population growth before they can grow economically, or must they make serious progress toward modernization before their population growth declines? This question has been debated for some time.

In 1798, a British economist, Thomas Malthus, pointed out that populations tend to grow exponentially, but there are ultimate limits to the expansion of agriculture. Thus, at the very beginning of the population explosion, Malthus foresaw a world headed toward calamity if something was not done to control population. But Malthus could foresee neither the tremendous expansion of agriculture that would come with the Industrial Revolution nor the demographic transition—the fact that fertility rates would decline with the Industrial Revolution.

Therefore, from early on, there were two basic schools of thought regarding the growth of population: (a) We need to focus on controlling population, and (b) a focus on development will take care of the situation and bring about a balance "automatically." These two schools of thought were reflected at two previous U.N. population conferences, the first in Bucharest, Romania, in 1974, and the second in Mexico City in 1984. (The 1994 conference in Cairo became the third.) At the Bucharest conference, the United States was a strong advocate of population control through family planning, while the developing nations argued that "development was the best contraceptive." Their resistance to family planning was also bolstered by feelings that the developed world's promotion of population control was another form of economic imperialism or even genocide.

At the second conference in Mexico City, the sides were somewhat reversed. Developing nations facing real problems of excessive population growth were asking for more assistance with family planning, whereas the United States, under pressure from "right-to-life" advocates, took the position that development was the answer and terminated all contributions to international family planning efforts, a policy that remained in effect until 1993. The other developed countries, however, remained convinced that family planning was essential and continued to support international efforts to aid the developing countries in their efforts to implement policies designed to bring fertility rates down.

To try to resolve this debate, it is instructive to examine the factors that influence people to have more or less children. It should be self-evident that no one makes a decision

whether to have a child based on the average GNP of the home country. Interestingly, plotting fertility rate against GNP per capita shows a weak correlation (Fig. 7–3). Yet there is no question that fertility rates in industrialized countries have declined with development. Therefore, we need to determine the specific aspects of development that influence childbearing.

Factors Influencing Family Size

Many students in developed countries find it difficult to understand why poor women in developing countries have large numbers of children. It is obvious from our perspective that more children spread a family's income more thinly and handicap efforts to get ahead economically. "Why," many ask, "do poor people behave so irrationally?"

What we fail to recognize is that the poor in developing countries live in a very different sociocultural situation. When we understand that situation, we find that their choices for larger families are quite logical. Numerous studies and surveys reveal the following as primary reasons that the poor in developing countries desire large numbers of children:

1. **Security in one's old age.** The traditional custom and expectation in developing countries is that old people will be cared for by their children. Social Security, welfare, Medicare, retirement, and nursing homes are all relatively new developments, found in high-income nations

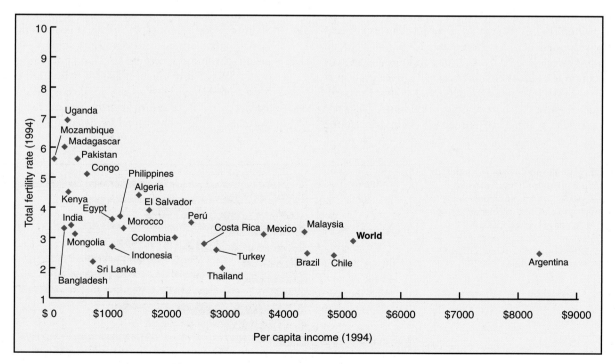

▲ FIGURE 7–3 *Fertility rate and income for selected developing countries.* There is a weak correlation between income and lower total fertility. Factors that affect fertility more directly are health care, education for women, and the availability of contraceptive information and services. (Source: Data from *1998 World Population Data Sheet* [Washington, DC: Population Reference Bureau, 1998].)

alone. Such things are not available to the poor of developing countries. Therefore, a primary reason given by poor women in developing nations for desiring many children is "to assure my care in old age."

2. **Infant and childhood mortality.** Closely coupled with security in one's old age is a high infant and childhood mortality. According to the Population Reference Bureau, 20,000 children below one year of age die every day in the developing world. This high infant mortality is the most profound indicator of the conditions of squalor and poverty in which people are living; it is unacceptable on any moral, ethical, or religious grounds. Nor does it serve to stabilize population, for the following reason: The common and often personal experience of children dying leads people to desire additional children as an "insurance policy" for security in their old age. Therefore, high infant mortality actually leads to higher fertility rates. It is only when there is a very high likelihood that children will survive (low childhood mortality) that people feel secure in having just two children or even one child.

3. **Children: an economic asset or a liability?** A third reason given by women of developing nations for desiring many children is "to help me with my work." In the subsistence-agriculture societies of the developing world, it has been and remains traditional for women to do most of the work relating to the direct care and support of the family. Clearing fields and turning the soil in preparation for planting is done largely by men, but all of the rest of the work—from planting, weeding, and harvesting to going to the market and gathering firewood and water—falls to the women (Fig. 7–4). A child as young as 5 can begin to help with many of these chores, and 12-year-olds can do an adult's work (Fig. 7–5). In short, children are seen as an economic asset. As environmental degradation occurs and more time is required to fetch water, collect firewood, and tend crops, the asset of many little hands becomes even greater. It is only in an urban setting in a developed nation that opportunities for children to contribute to the economic welfare of the family become extremely limited, that the costs of feeding, clothing, and educating children are prolonged, and that the economic burden of children is acutely felt.

4. **Importance of education.** The importance given by a society to education is closely allied to whether children are seen as an economic asset or a liability. If it is felt that children do not need to be educated, then it is easy to cast them in the role of simply being the "many little hands" to help with the chores of everyday survival, and the more the better. If, on the other hand, it is required that they go to school, they not only are removed from the labor force but also need additional economic support for suitable clothing, school supplies, and so on. In short, the requirement that children be educated changes their position from being seen as an economic asset to being viewed as an economic liability and influences fertility rates. Again, in traditional, subsistence-agriculture societies, education has been deemed unnecessary, and this remains the case for many children in the developing world, especially girls. Thus, the continued high fertility of parents is supported by a belief that educating girls is unnecessary. Of course, education does much more than keep children from working; it opens up any number of additional opportunities that also affect fertility rates (see the "Global Perspective" essay, p. 170).

5. **Status of women: opportunities for women's education and careers.** The traditional social structure in many developing countries still discourages and, in many cases, bars women from obtaining higher education, owning businesses, owning land, and pursuing many careers. Such discrimination against women forces them into doing what only they can do: bear children. Worse, in many of these countries, the male's respect for a woman is only proportional to the number of children

▶ **FIGURE 7–4** *Gender-related work in developing countries.* In developing countries—especially Africa—women do most of the work relating to care and maintenance of the family, including heavy farming tasks. (Source: Data from Jodi Jacobson, "Gender Bias: Roadblock to Sustainable Development," *Worldwatch Paper* 110 [1992].)

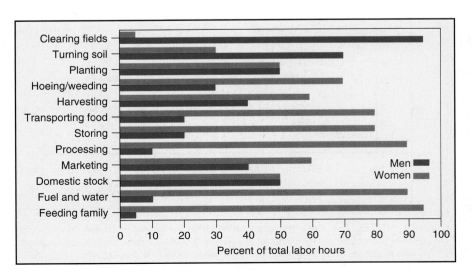

◀ **FIGURE 7–5** *Children as an economic asset.* Children working with adults in the fields in Bali, Indonesia. In most developing countries, children perform adult work and thus contribute significantly to the income of the family.

she bears. Breaking down such barriers of discrimination so that girls are educated and gain status outside the context of raising children has probably contributed more than anything toward the very low fertility rates seen in developed countries today. Indeed, studies show that women with just an eighth-grade education have, on average, only half the number of children as their uneducated counterparts.

6. **Availability of contraceptives.** There can be no doubt regarding the importance of contraceptives in achieving a lower fertility rate. Studies show a strong correlation between lower fertility rates and the percentage of couples using contraception (Fig. 7–6); each 13% increase in contraceptive use translates into one less child. In the developed world, we take the availability of contraceptives almost for granted. Perhaps the most profound finding in surveys of women in the developing world is that large numbers state that they want to delay having their next child or that they do not want any more children. Yet many of these women are not using contraceptives. Women in rural areas report that contraceptives are frequently not available. Women in cities also have trouble getting them in spite of free clinics; the clinics may be too far away or crowded, or they may run out of contraceptives. Providing contraceptives to women is a major facet of family planning.

Students frequently raise the point that religious beliefs play a role in determining family size. To some extent, this may be true, but it becomes less of a factor as educational and other barriers to women are broken down. For example, Italy, primarily a Catholic country, has the lowest fertility rate of any country. Obviously, most Italian Catholics are not taking the Pope's admonitions seriously. The same is true for Mexico, host country for the 1984 World U.N. population conference.

Conclusions

Reflecting on the foregoing items, we can see that the factors supporting large families are common to preindustrialized, agrarian societies, while those conducive to raising small families (or no children) generally appear with industrialization and development. Those factors include the following: the costs of raising children, the existence of pensions and a Social Security system, opportunities for women to join the work force, free access to inexpensive contraceptives, adequate health care, high educational opportunity and achievement, and higher age at marriage. Fertility rates in developing countries remain high not because people in those countries are behaving irrationally, but because the sociocultural climate in which they live favors high fertility and, often, contraceptives are not available. Furthermore, we should be able to understand how poverty, environmental degradation, and high fertility drive one another in a vicious cycle (Fig. 7–7). Increasing population density leads to greater depletion of rural community resources like firewood, water, and land, which encourages couples to have more children to help gather resources and so on.

One may ask how the now industrialized nations came through the demographic transition without getting caught in the poverty-population trap. Two points are significant in this regard. First, the improvements in disease control that lowered death rates occurred gradually through the 1800s and early 1900s. Industrialization, which introduced factors

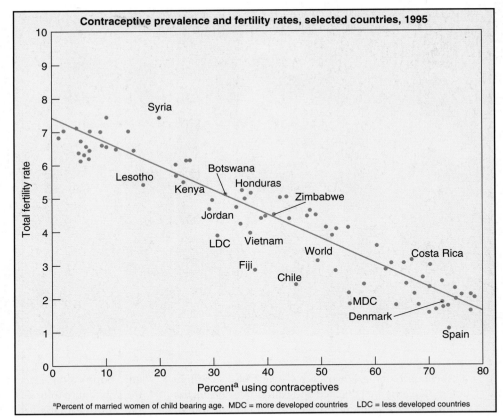

Contraceptive prevalence and fertility rates, selected countries, 1995

(Graph: *Total fertility rate* on y-axis, 0 to 10; *Percent^a using contraceptives* on x-axis, 0 to 80. Labeled data points: Syria, Lesotho, Kenya, Botswana, Honduras, Zimbabwe, Jordan, LDC, Vietnam, World, Fiji, Chile, Costa Rica, MDC, Denmark, Spain.)

^aPercent of married women of child bearing age.   MDC = more developed countries   LDC = less developed countries

▶ **FIGURE 7–6 *Contraceptive prevalence and fertility rates.*** More than any other single factor, lower fertility rates are correlated with the percent of the population using contraceptives. (Source: Data from *1996 World Population Data Sheet* [Washington, DC: Population Reference Bureau, 1996].)

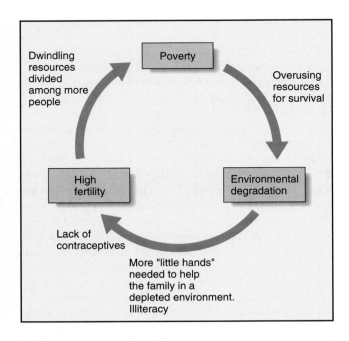

▲ **FIGURE 7–7 *The poverty cycle.*** Poverty, environmental degradation, and high fertility rates become locked in a self-perpetuating vicious cycle.

that lowered fertility, occurred over the same period. Therefore, there was never a huge discrepancy between birth and death rates (Fig. 7–8a). Second, and perhaps even more significant, surplus population from European nations could and did readily emigrate to the United States, Canada, Latin America, New Zealand, and Australia.

In contrast, modern medicine was introduced to the developing world relatively suddenly, bringing about a precipitous decline in death rates, while the fertility-lowering effects of development were delayed (Fig. 7–8b). All these observations point to the fact that it is not economic development by itself that leads to declining fertility rates. Rather, fertility rates decline insofar as development provides (1) security in one's old age apart from ministrations of children, (2) lower infant and childhood mortality, (3) mandatory education for children, (4) opportunities for higher education and careers for women, and (5) unrestricted access to contraceptives.

We turn our attention now to efforts that have been made to bring about development in the poor nations.

7.2 Development

The absence of development and the impoverished societies in what we now call developing countries is largely a legacy of eighteenth and nineteenth century colonialism, which persisted well into this century. Thus, reversing colonial policy

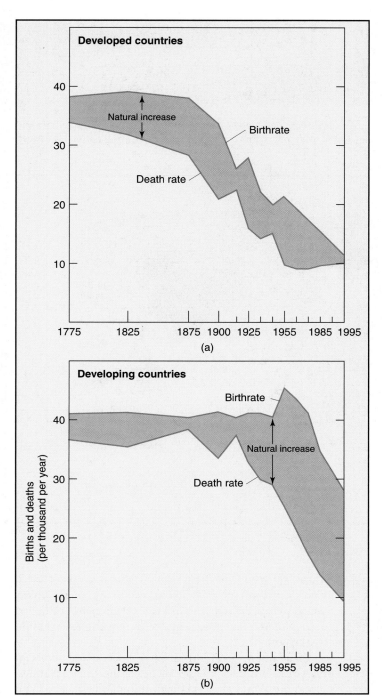

◀ **FIGURE 7–8** *Demographic transition in developed and developing countries.* (a) In developed countries, the decrease in birth rates proceeded soon after and along with the decrease in death rates, so very rapid population growth never occurred. (b) In developing countries, both birth and death rates remained high until the mid 1900s. Then a precipitous decline in death rates was caused by the rapid introduction of modern medicine, whereas birth rates remained high, causing very rapid population growth. (Redrawn with permission of Population Reference Bureau, Inc., Washington, D.C.)

and fostering the development of these countries can be amply justified on all grounds: humanitarian, political, and economic. By providing better jobs and incomes, development fosters improved standards of living, which in turn creates expanded markets for the developed world. Finally, development fosters trade, cooperation, and peace among nations. If fertility rates decline to replacement levels in the process, then, in the perspective of sustainable development, we can look forward to a future in which a stable world population lives in a state of relative affluence.

Promoting the Development of Low-Income Countries

In 1944, during World War II, delegates from around the world met in Bretton Woods, New Hampshire, and conceived a vision of development for poor countries. They established the International Bank for Reconstruction and Development (or, as it is more commonly known now, the **World Bank**). The World Bank now functions as a special agency within the United Nations. With deposits from governments and

GLOBAL PERSPECTIVE

FERTILITY AND LITERACY

Illiteracy, particularly among women, is one of the prime indicators of poverty and high fertility. The following map depicts countries according to female illiteracy rate and fertility rate. The strong correlation is evi-

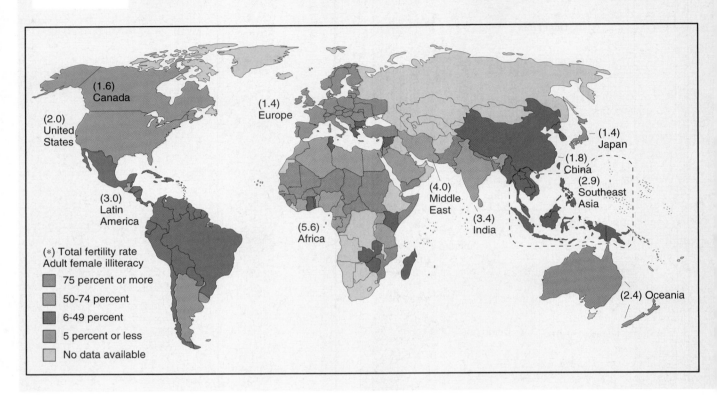

(1.6) Canada

(1.4) Europe

(2.0) United States

(1.4) Japan

(1.8) China

(2.9) Southeast Asia

(3.0) Latin America

(4.0) Middle East

(3.4) India

(5.6) Africa

(2.4) Oceania

(*) Total fertility rate
Adrulf female illiteracy

- 75 percent or more
- 50-74 percent
- 6-49 percent
- 5 percent or less
- No data available

---

commercial banks in the developed world, the World Bank lends money to developing nations for a variety of projects at interest rates somewhat below the going market rates. Effectively, the World Bank helps governments of developing countries (the Bank loans only to governments) borrow large sums of money for projects they otherwise could not afford.

Annual loans from the World Bank have climbed steadily, from $1.3 billion in 1949 to $29.5 billion in 1997. With the power to approve or disapprove loans, and through the amount of money it lends, the World Bank has been the major instrument in providing aid to developing countries for the past 50 years. How does the World Bank's record look?

Past Successes and Failures of the World Bank

Many developing countries have made remarkable *economic* progress. The gross national products of some countries have increased as much as fivefold, bringing them from the low- to the medium-income countries, and some medium-income nations have achieved high-income status. Although the world economy is still strongly dominated by the industrialized countries,

the developing countries have become more and more involved in what is now an integrated global economy. Foreign investment is playing a large role in this development; in the last decade, foreign investment in developing countries has increased fourfold.

In addition, great strides have been made in *social* progress. Efforts from other branches of the United Nations, such as the World Health Organization (WHO), Food and Agricultural Organization (FAO), U.N. Educational, Scientific, and Cultural Organization (UNESCO), and United Nations Children's Fund (UNICEF), have augmented the work of various government programs. Private charitable organizations have also played a large role. Literacy rates, the percentage of the population with access to clean drinking water and sanitary sewers, and other social indicators of development, generally speaking, have improved (Table 7-1). Further, in keeping with the concept of the demographic transition, the fertility rates of most developing countries have declined, although not to the replacement level (Table 7-2).

However, these successes are dulled by the facts described in Chapter 6: A fifth of the world population, 1.3

TABLE 7-1 Improvement in Development Indicators 1970–1995

	Low-Income Countries		Upper-Middle-Income Countries	
	1970	1995	1970	1995
Population with access to:				
Safe drinking water (percent)	22	51	59	84
Sanitary sewers (percent)	29	38	61	79
Age group enrolled in education:[2]				
Primary (percent of total)	55	105[1]	94	99
% females per 100 males	61	88	95	99
Secondary (percent of total)	13	27	32	70
% females per 100 males	44	83	95	100
Infant mortality:				
(per 1000 live births)	114	69	70	35
Life expectancy at birth:				
(years)	47	63	62	69

[1]Adults in addition to children enrolled.
[2]Data from 1993.
Source: Data from World Bank, *World Development Report*, 1997.

billion people (an increase of 300 million over 1990), remains in absolute poverty, largely illiterate, and without access to clean water or adequate nutrition; environmental degradation is rampant; and fertility rates remain unacceptably high. In short, the vicious cycle of high fertility, poverty, and environmental degradation seems about to overwhelm development efforts and commence a cycle of decline. Indeed, more than 1 billion people are worse off economically than they were in 1980. The gap between the rich and poor countries is growing; the difference between average per capita income in the industrialized vs. the developing countries tripled between 1960 and 1993.

It is not accurate to either credit the World Bank for all the progress made or to blame them for all areas where progress has been lacking. However, critics point to many examples where the Bank's projects have actually exacerbated the cycle of poverty and environmental decline. For example, the World Bank loaned India nearly a billion dollars to create a huge electric-generating facility consisting of five coal-burning power plants at Singraali and to develop openpit coal mines to support the plants. However, the increased power does little for the poor, who cannot afford electrical hookups. The project displaced over 200,000 rural poor people, who had farmed the fertile soil of the region for generations, and moved them to a much less fertile area without allowing them any say in the matter and giving them little, if any, compensation. In addition, the project has caused extensive air and water pollution. Hydroelectric dams in a number of countries have similarly displaced people and heightened their poverty.

Nowhere are the failures and environmental destructiveness of large-scale projects more evident than in agriculture. The World Bank funneled $1.5 billion into Latin America from 1963 to 1985 for clearing millions of acres of tropical forests. Most of the cleared land was given over to large cattle operations for producing beef for export. However, no type of agriculture requires less labor per acre than ranching. Spreads of more than a million acres are run by millionaire cattle barons checking their herds by aircraft (Fig. 7–9). Meanwhile, the poor were pushed into more marginal lands or into cities, as described in Chapter 6. Because the soil of cleared tropical forests is so poor, some of the ranches have already been abandoned, and most of the remaining are only marginally profitable. Projects in other countries have emphasized growing cash crops for export, fostering huge mechanized plantations while leaving the poor likewise marginalized.

TABLE 7-2 Decline in Total Fertility Rate

	1981	1985	1990	1995	1998
Africa	6.4	6.3	6.2	5.8	5.6
Latin America and Caribbean	4.4	4.2	3.5	3.1	3.0
Asia (excluding China)	5.5	4.6	4.1	3.5	3.3
China	2.3	2.1	2.3	1.9	1.8
Developed Countries	2.0	2.0	2.0	1.6	1.6

Source: Data from *1998 World Population Data Sheet* (Washington, DC: Population Reference Bureau, 1998).

▶ **FIGURE 7–9** *Cattle ranch in eastern Brazil.*
Large World Bank loans went to clear rain forest and
convert the land into rangeland, as seen here.

The Debt Crisis

Another consequence of promoting development through
World Bank loans is similar to enticing people to buy on credit. Borrowers become overwhelmed by interest payments.
Theoretically, development projects were intended to generate additional revenues that would be sufficient for the recipient to pay back the loan with interest. However, a number
of things have gone wrong with this theory, such as corruption, mismanagement, and, perhaps, honest miscalculations,
not the least of which are the responsibility of the recipient
countries. In their eagerness to obtain a loan for a billion-
dollar project, government officials often overestimate the
virtues of and expected revenues from the project.

In any case, far from paying off loans, developing countries as a group have become increasingly indebted. Their
total debt reached $1.4 trillion in 1995 and is still climbing.
Of course, interest obligations climb accordingly, and any

failure to pay interest gets added to debt, increasing the interest owed—the typical credit-debt trap. Many developing
countries are now paying a substantial amount of their export earnings in interest, and the situation in Africa is worsening (Fig. 7–10).

Because of the rising interest being paid by developing
countries, there is now a net flow of capital from developing
to developed countries in the amount of some $50 billion
per year, and there is no question that income disparities between rich and poor countries are widening. Ironically, then,
who is aiding the development of whom?

This debt situation continues to be an economic, social,
and ecological disaster for many developing countries. In
order to keep up even partial interest payments, poor countries have done one or more of the following:

1. Focused agriculture on large-scale growing of cash crops
   for export. This has occurred at the expense of peasant

▶ **FIGURE 7–10** *The debt trap.* Loans aimed at
promoting development have caught many countries in a debt
trap. (Source: Data from Gary Gardener, "The Third World
Debt Is Still Growing," *Worldwatch* [Jan/Feb 1995]: 38.)

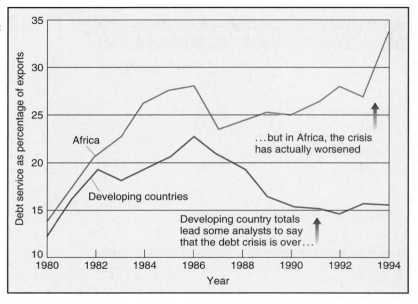

farmers growing food. Thus, hunger and malnutrition have increased, and so has poverty, as peasants have been pushed from the land.

2. Adopted austerity measures. Government expenditures have been drastically reduced so that income can go to pay interest. But what is cut? Usually, it is schools, health clinics, police protection in poor areas, building and maintenance of roads in rural areas, and other goods and services that benefit not only the poor but the country as a whole.

3. Invited the rapid exploitation of natural resources (e.g., logging of forests and extraction of minerals) for quick cash. With the emphasis on quick cash, few if any environmental restrictions are imposed. Thus, the debt crisis has meant disaster for the environment. Ironically, this forced exploitation of material resources has placed such oversupplies of commodities on the market that prices have been severely depressed, so that actual earnings are low.

In essence, these are examples of liquidating natural capital assets to raise cash for short-term needs. Clearly, they do not represent sustainability. Also, it is clear that the brunt of these measures falls on the poor. Thus, many observers point out that it is the poor, who gained nothing from the development, who are now being required to pay for it. As is typical with the credit trap, many countries have paid back in interest many times what they originally borrowed, yet the debt remains. Is there any point—humanitarian, ecological, or economic—to keeping such debts in place?

In sum, the concept of fostering the development of poor countries through massive loans for large-scale projects—whatever the advantages in enhancing gross national products—has not broken the cycle of excess population, poverty, and environmental degradation. For decades, policymakers at the World Bank saw development in terms of the hallmarks of the industrialized world: energy—in particular, electrification through centralized plants; transportation—in particular, building roads and cars; and mechanization of agriculture. Thus, about three fourths of the money loaned by the World Bank has been for projects in these three areas. Moreover, the policymakers at the Bank were mainly economists and engineers, and the traditional training in these two professions often involves viewing Earth as yielding endless resources and seeing humans as able to manipulate anything to their benefit. Thus, such professionals, lacking any ecological training, were notoriously slow to appreciate the environmental ramifications of the projects they proposed.

World Bank Reform

It is fair to say that the World Bank has undergone major reform in recent years in how they do business with the developing world. The Bank now has a well-organized and influential Environment Department that provides "intellectual and practical leadership and support in fulfilling the World Bank's ... environmental and social agenda," as the department's mission statement reads. The Environment Department has established

policies and procedures to ensure that World Bank–financed projects are not environmentally damaging. The Department is also engaged in helping developing countries to strengthen their own environmental institutions and policies. Interestingly, an independent global forum of NGOs (nongovernmental organizations) has been created—the NGO Working Group on the World Bank (NGOWG)—to dialogue with and monitor the World Bank's work with developing countries.

The Bank is also more directly addressing the problems of poverty through two new initiatives: the Consultative Group to Assist the Poorest (CGAP) and the Heavily Indebted Poor Country (HIPC) initiative. The first of these is designed to increase access to financial services for very poor households through what is called "microfinancing." The HIPC initiative addresses the debt problem of the poorest developing countries by providing direct debt relief to a level deemed sustainable. To qualify, the countries have to demonstrate a track record of successfully carrying out economic and social reforms that lead to greater stability. By 1998, six countries (Bolivia, Burkina Faso, Côte d'Ivoire, Guyana, Mozambique, and Uganda) had qualified for debt relief, totaling $3 billion. In addition to these initiatives, the Bank has produced a major document laying out their "Poverty Reduction Strategy," designed to focus attention on all of the Bank's efforts to address poverty in the developing world.

Undoubtedly, the work of the World Bank and other international agencies is sorely needed to provide assistance for economic development and represents a vital element in the response of the industrialized world to the needs of the developing world. Meeting the objectives of *Agenda 21* for sustainable development (see the "Earth Watch" essay, Chapter 1, p. 20) will require continued large-scale funding from these sources. However, we need to return to our assessment of the relative roles played by development and family planning in an attempt to resolve this decades-long debate.

7.3 A New Direction for Development—Social Modernization

Kerala is the southernmost state of India, with a population of 32 million occupying an area of 39.9 thousand square miles, the second most densely populated state in India and close to the highest in the world. Situated only 10° north of the equator, Kerala is tropical, with lush plantations of coffee, tea, rubber, and spices in the highlands, and rice, coconut, sugarcane, tapioca, ginger, and bananas typically grown in the lowlands. With the Arabian Sea on its coastline, fishing is an important industry. Kerala is very much like the rest of India in some ways: It is crowded, per capita income is low—less than $300 per year—food intake is around 2200 calories/day, considered adequate but on the low end. Here the comparison ends, however.

The people of Kerala have a life expectancy of 71 years vs. 59 for all of India; infant mortality is 17 per thousand vs. 92 per thousand for India; the fertility rate is 1.8 (below replacement level) vs. 3.4 for India. Literacy is over 95%, almost all villages have access to school and modern health services, and women have achieved high offices in the land and are as well educated as the men. Even though Kerala must be considered a poor country in every economic measurement, its people are well on the way to achieving a stable population.

What is different about Kerala is a strong public policy commitment to health development and education. Land distribution is relatively equitable, food distribution is efficient, and the old caste system of India has all but disappeared. These investments in social policy have paid off. Kerala's total fertility rate dropped from 3.7 to 1.8 in just two decades; in the last few years, Kerala has been making twice the progress of India in per capita income growth. In short, Kerala is an example of the possibility of bringing a developing region from the midphase of the demographic transition to the threshold of its completion without the thorough economic development that characterized the industrial countries as they underwent their demographic transition.

As the case of Kerala demonstrates, demographic experts are recognizing that the shift from high to low fertility rates in the poorer developing countries is brought about by what is called **social modernization**. In order to be achieved, social modernization does not require the economic trappings of a developed country. Instead, what is needed are efforts made on behalf of the poor, with particular emphasis on the following:

1. Education—especially improving literacy and educating girls and women equally with boys and men.
2. Improving health—especially lowering infant mortality.
3. Making family planning accessible.
4. Enhancing income through employment opportunities.
5. Improving resource management (reversing environmental degradation).

In all of these areas, the focus should be on women, because they not only bear the children but are also the primary providers of nutrition, child care, hygiene, and early education. In short, it is women who are most relevant in determining the number and welfare of subsequent generations.

Fortunately, the world is by no means starting from scratch in any of the preceding areas. Numerous programs, both private and government funded, have been going on for many years, and a wealth of experience and knowledge has been gained. We shall look at each area in somewhat more detail, and then we shall return to the program set forth at the Cairo Conference, which aims to put them all together.

Education

The education we are speaking of in this context is not college or graduate school nor even advanced high school. It is basic literacy—learning to read, write, and do simple calculations (Fig. 7–11). Illiteracy rates among poor women in developing countries are commonly between 50 and 70%, in part because the education of women is not considered important and in part because the population explosion has overwhelmed school systems and transportation systems. Providing basic literacy will empower people to glean information from pamphlets on everything from treating diarrhea to conditioning soils with compost and baking bread. Consider how much more efficiently information can be distributed with written materials as opposed to one-on-one oral instruction, which is required if people cannot read.

An educated populace is an important component of the "wealth" of a nation. Investing in education of children represents a key element of the public policy options of a developing country, one that returns great dividends. For example, Pakistan and South Korea both had similar incomes and population growth rates (2.6%) in 1960, but very different school enrollments—94% in Korea vs. 30% in Pakistan. Within 25 years, Korea's economic growth was three times that of Pakistan's, and now the rate of population growth in Korea has declined to 0.6% per year, while Pakistan's is still 2.4% per year. Kerala is another case in point.

Improving Health

Like education, the health care required most by poor developing-world communities is not high-tech bypass surgery or chemotherapy; rather, it is the basics of good nutrition and hygiene—steps such as boiling water to avoid the spread of disease, and proper treatment of infections and common ailments such as diarrhea. (In developing countries, diarrhea is a major killer of young children but is easily treated by giving suitable liquids, a technique called oral rehydration therapy.) Health care in the developing world must emphasize pre- and postnatal care of the mother, as well as that of the children. Many governmental, charitable, and religious organizations are involved in providing basic health care, and when this is extended to rural countrysides in the form of clinics, it is one of the most effective ways of delivering family planning information and contraceptives to women.

AIDS. One of the greatest challenges to health care in the developing countries is the sexually transmitted disease AIDS (acquired immune deficiency syndrome). It is unfortunately the case that the global epidemic of AIDS is most severe in many of the poorest developing countries, which are least able to cope with the consequences (Fig. 7–12). More than 90% of all HIV-infected people (30 million) live in the developing countries, and most of these people are not aware of their infection—thus guaranteeing that the epidemic will continue. In sub-Saharan Africa the virus is spread by heterosexual contact, resulting in high incidences of female infections. The incidence of infected adults is over 20% in some countries (Zimbabwe, Botswana).

The impacts of this epidemic are horrendous for the developing world. The mortality rate is climbing in many countries; life expectancy in Botswana, for example, has declined from 61 years in the late 1980s to the present 50 years. In Kenya, the rate of population growth is expected to drop

◀ FIGURE 7–11 *Education in developing countries.* Providing education in developing countries can be very cost-effective, since any open space can suffice as a classroom, and few materials are required. This is a class in Bombay, India.

a net 0.8% by 2010 due to the death rate from AIDS alone. There may be as many as 40 million AIDS orphans by 2010 in the developing world. Already inadequate health care systems are being swamped by the victims. Presently, the only hope is to convince people to change their sexual behavior—to practice "safe sex," limit their partners, avoid prostitution, and delay sexual activity until marriage. Again, education and literacy are important components of combating this social problem.

Family Planning

For those who can pay, information on contraceptives and related materials and treatments are readily available from private doctors and health-care institutions. The poor, however, must depend on family planning agencies, which are supported by a combination of private donations, government funding, and small amounts the clients may be able to afford. The stated policy of family planning agencies (or

◀ FIGURE 7–12 *AIDS.* Four members of a Central African family are victims of AIDS. The disease is often acquired from other family members who are unaware of their infection.

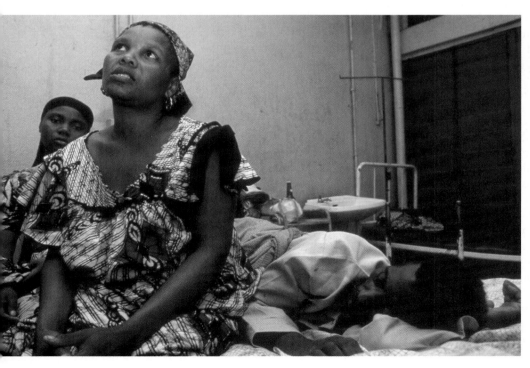

176 Chapter 7 Addressing the Population Problem

private services) is, as the name implies, to enable people to plan their own family size—that is, to have children only if and when they want them. In addition to helping people avoid unwanted pregnancies, planning often involves determining and overcoming fertility problems for those couples who are having reproductive difficulties. More specifically, family planning services include the following:

- Counseling and education for singles, couples, and groups regarding the reproductive process, the hazards of sexually transmitted diseases (AIDS, in particular), and the benefits and risks of various contraceptive techniques.
- Counseling and education on achieving the best possible pre- and postnatal health for mother and child. The emphasis is on good nutrition, sanitation, and hygiene.
- Counseling and education to avoid high-risk pregnancies. Pregnancies that occur when a woman is too young or too old, and pregnancies that follow too closely on a previous pregnancy, are considered high risk; they seriously jeopardize the health and even the life of the mother. Any existing children are also at risk if the mother suffers injury or death.
- Provision of contraceptive materials and/or treatments after people have been properly instructed about all alternatives.

The vigorous promotion and provision of contraceptives has proved all by itself to have a decided effect in lowering fertility rates, as seen in Fig. 7–6. Those countries that have implemented effective family planning programs have experienced the most rapid decline in fertility. For example, Thailand initiated a vigorous family planning program as part of a national population policy in 1971. Population growth declined from 3.1% per year to its present 1.1% per year, and the Thai economy has shown one of the most rapid rates of increase over the years. Encouraging and implementing family planning is the first and most important step a country can take to improve its chances to develop economically. Unfortunately, to many people, family planning conjures up images of the abortion clinic.

Abortion. Abortions, by definition, are terminations of unwanted pregnancies. Nearly everyone agrees that an abortion is the least desirable way to avoid having an unwanted child—especially in view of the other alternatives available. Indeed, the document resulting from the Cairo Conference explicitly states that abortions should *never* be used as a means of family planning. Therefore, it is particularly important to observe that the primary functions of family planning are education and services directed at avoiding unwanted or high-risk pregnancies. If family planning services were universally available and people availed themselves of them, there virtually would be no unwanted pregnancies and, hence, few abortions. The recourse to abortions should be seen as a consequence of the lack of family planning education and services. All studies show that cutbacks in family planning services result in *more* unwanted pregnancies and *more* demand for abortions, not *less*. This occurred following the cutoff of family planning monies by the Reagan administration in 1984—abortions increased an estimated 70,000 a year.

Furthermore, if abortions are not legal in a particular country, they tend to be provided illegally, or women will attempt to induce them on their own. Both of these practices have tremendous adverse health consequences. It is estimated that 500,000 women die annually from pregnancy-related problems. Most of these deaths result from illegal and self-induced abortions.

Planned Parenthood, which operates clinics throughout the world, is probably the best known family planning agency. Another significant player is the United Nations Population Fund (UNFPA), which provides financial and technical assistance to developing countries at their request. The emphasis of UNFPA is on combining family planning services with maternal and child health care and expanding the delivery of such services in rural and marginal urban areas. Support of this U.N. agency and other family planning agencies has become a political football in the United States, where the administration—through the Agency for International Development—is trying to continue funding for the agencies, and many Republican members of the House are opposing it. The argument, used since the Reagan years, is that these agencies promote abortions or, in the case of China, condone government efforts that force couples to have no more than one child (see the "Ethics" essay, p. 177). One far-reaching consequence of this battle is that Congress has continued to reject bills authorizing payment of our commitment to the United Nations, placing us in arrears for over $1 billion. This means that UNICEF and all other U.N. agencies are receiving less than they need to carry out their mission, with tragic consequences for the developing world.

Enhancing Income

The bottom line of any economic system is the exchange of goods and services. At its simplest level, this entails a barter economy in which people agree on direct exchanges of certain things and/or services. Barter economies are still widespread in the developing world.

The introduction of a cash economy facilitates the exchange of a wider variety of goods and services, and everyone may prosper, as they have a wider market for what they can provide and a wider choice of what they can get in return. In a poor community, the ironic twist is that everyone may have the potential to provide certain goods or services and may want other things in return, but there is no money to get the system off and running. In a going economy, people who wish to start a new business venture generally begin by obtaining a bank loan to "set up shop." However, the poor are considered high credit risks, they may want a smaller loan than what a commercial bank wants to deal with, and many are women who are denied credit because of gender discrimination alone. For these three reasons, poor communities are handicapped in getting start-up capital.

In 1976, Muhammad Yunus, an economics professor in Bangladesh, conceived and created a new kind of bank (now known as the Grameen Bank) that would engage in **microlending** to the poor. As the name implies, microloans are small—they average just $67—and they are short term,

ETHICS

ADDITIONAL INCENTIVES FOR REDUCING FERTILITY

What are our options as we foresee the limits of Earth's carrying capacity for the human species? What are the ethical and moral implications of each option?

1. We can argue, as some optimists do, that the problems do not really exist—that technology will always have a solution to make life better and better for more and more people. Does global environmental evidence support this position, or is this blinding oneself to reality? Is this simply delaying making tough choices? Is it creating an even greater crisis for the future as more people will have to support themselves with ever more depleted or degraded resources?

2. Perhaps a disease, famine, or a natural disaster—or even war—will come along and take care of the situation for us in "nature's way." Again, this argument seems to lack an appreciation for the magnitude of the numbers involved. For example, the current world death toll from AIDS, tragic as it is, is about 2.3 million—less than 3% of what would actually stabilize population. Therefore, looking for a "death solution" is looking for untold human suffering. It is doubtful that anyone would escape the ramifications of that suffering. The entire human endeavor has been to try to avoid such calamities. Are we going to change that?

3. Some consider forced sterilizations or abortions as a solution to the population problem. This option immediately implies that some person or group is deciding who is going to reproduce and who is not. The backlash from this kind of endeavor in the few cases where it has been tried has been so severe that the agencies or governments

that have tried it have been summarily thrown out of power.

4. The most acceptable option—the one that has been the theme of this chapter—is to create an economic and social climate in which people of their own volition will desire to have fewer children and then provide the means (family planning) to enable them to meet that choice. This, of course, has happened "unconsciously" in developed countries, although we are confronted by dilemmas. How far can one go in manipulating the social or economic environment before choice becomes coercion?

With its current population of 1.25 billion (a fifth of the world's people), China provides the most comprehensive example of a country that offers extensive economic incentives and disincentives for reducing population growth. Some years ago, China's leaders recognized that unless population growth was stemmed, the country would be unable to live within the limits of its resources. Because of inevitable population momentum, the leaders felt that the country could not even afford a total fertility rate of 2.0; they set a goal of a one-child family, and to achieve that goal, they instituted an elaborate array of incentives and deterrents. The prime *incentives* were as follows:

- Paid leave to women who have fertility-related operations—namely, sterilization or abortion procedures.
- A monthly subsidy to one-child families.
- Job priority for only children.
- Additional food rations for only children.
- Housing preferences for single-child families.
- Preferential medical care to parents whose only child is a girl. (There is a

strong preference for sons in China, and parents generally wish to have children until at least one son is born.)

Penalties for an excessive number of children in China included the following:

- Repayment to the government of bonuses received for the first child if a second is born.
- Payment of a tax for having a second child.
- Payment of higher prices for food for a second child.
- Maternity leave and paid medical expenses only for the first child.

Along with improving economic opportunities, these incentives and deterrents have helped China achieve a precipitous drop in its fertility rate, from about 4.5 in the mid-1970s to 1.8 in 1998 (although recent work by Chinese demographers have pegged the fertility rate at 2.1 in spite of the official government figure of 1.8). The one-child policy has been pursued only in urban areas, some 29% of the population; a higher fertility rate lies in rural areas and with minorities (who are exempted from the policy). However, the population of China is still growing because of momentum (a large percentage of the population is still at, or below, reproduction age). One disturbing consequence of China's policy is the skewed ratio of males to females; males are preferred in Chinese families, and if only one child is permitted, there is a tendency on the part of some parents to make sure that their one child is a male.

Is China's population policy morally just or unjust? Is it possible that some other countries will have to resort to a China-like policy in the near future as they face severe limits on resources? At what point are such policies necessary? What do you think?

usually just four to six months. Nevertheless, they provide such basic things as seed and fertilizer for a peasant farmer to start growing tomatoes, some pans for a baker to start baking bread, a supply of yarn for a weaver, some tools for an auto mechanic, and so on.

Yunus secured his loans by having the recipients form **credit associations**, groups of several people who agreed to be responsible for each other's loans. With this arrangement, the Grameen Bank experienced an exceptional rate of payback:

Less than 3% of the people defaulted on their loans. Loans from the Grameen Bank have had outstanding results when applied to small-scale agriculture. In a rural area of Bangladesh, small loans, along with horticultural advice, are now enabling peasant farmers to raise tomatoes and other vegetables for sale to the cities. These people have doubled their incomes in three years.

Microlending has been found to have the greatest social benefits when focused on women because, as Yunus observed,

"When women borrow, the beneficiaries are the children and the household. In the case of a man, too often the beneficiaries are himself and his friends." Note that the credit associations also create another level of cooperation and mutual support within the community, particularly when the loans are directed toward women.

The unqualified success of microlending in stimulating the economic activity and enhancing the incomes of people within poor communities has been so remarkable that the concept has been adopted with various modifications by a considerable number of private organizations dedicated to alleviating hunger and poverty—Oxfam and Freedom from Hunger among them. Freedom from Hunger, which has projects in six countries around the world, combines its lending with "problem-solving education" in a program called Credit with Education (see the "Earth Watch" essay, page 179).

Recently, the U.S. Agency for International Development (AID) has entered the picture by providing grants to other organizations that wish to do microlending. Even the World Bank is expressing the need to be more sensitive to local communities in its lending practices and is making $200 million available for microlending.

Improving Resource Management

The world's poor, almost by definition, are dependent on local resources, particularly water, soil for growing food, and forests for firewood. We have amply described how the pressures of excessive population are degrading these resources. However, a considerable part of the problem lies in poor utilization of resources—failing to replant trees and prevent erosion of the soil, for example. Conversely, a major factor in enhancing income is providing the technical skills and information necessary to manage those basic resources more effectively (Fig. 7–13). This in itself can greatly slow or even reverse the tide of environmental degradation. Allowing impoverishment of the environment can only lead to further impoverishment of the people.

▼ **FIGURE 7–13 Improving resource management.** A major step in enhancing incomes and protecting the environment is to encourage better resource management. Here local people are learning the skills to raise tree seedlings that will later be transplanted in a reforestation project. This is part of the Greenbelt Movement in Kenya.

Putting It All Together

Each of the five components just described both depends on and supports the other components. For example, better health and nutrition support better economic productivity, better economic productivity supports obtaining a better education, and a better education leads to a delay in marriage and the desire to have fewer children. The availability of family planning services is essential to realizing the desire of parents to have fewer children. In short, all the components work together in harmony to alleviate the conditions of poverty, reverse environmental degradation, and stabilize population (Fig. 7–14). Conversely, the lack of any component—especially family planning services—will undercut the ability to achieve all other components.

It is not too hard to imagine putting these components together without undue expense. Most important, doing so involves the local people's uplifting themselves. For example, Zimbabwe is training 5000 women to be preventive healthcare workers in their own communities. Already, a number of developing countries have itinerant "nurses" who make the rounds of rural villages, treating illnesses and injuries, giving advice regarding better health maintenance, and dispensing contraceptive supplies. Another one or two persons traveling with the nurses might conduct classes to improve literacy, provide technical skills, and dispense microloans.

The importance of putting the five components together into a single "package" may seem self-evident. However, probably the biggest reason that U.N., governmental, and

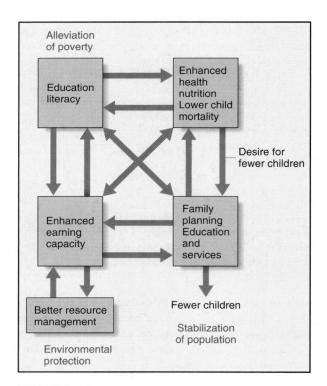

▲ **FIGURE 7–14 Social development.** Five main aspects of enhancing the well-being of the poor are mutually supporting and dependent on one another as illustrated.

EARTH WATCH

AN INTEGRATED APPROACH TO ALLEVIATING THE CONDITIONS OF POVERTY

Freedom from Hunger is a nongovernmental organization that is pioneering an integrated approach toward improving the lives of the poor described in the text. Freedom from Hunger's **Credit with Education** program provides opportunities for women in extreme poverty to invest in their own small businesses and to save for emergency needs. These financial services are linked to education for better health, nutrition, and family planning—always respecting local beliefs and culture.

Women interested in receiving loans come together to form credit associations, composed of about 20–30 members, the great majority of whom are very poor. The members guarantee repayment of each other's loans, so they must agree that each woman is capable of making a sufficient profit from her proposed income-earning activity. After training the new members to manage their own association within specified rules, Freedom from Hunger makes a four- to six-month loan to the credit association. The members then break the large loan into small loans averaging $64 per individual. Women invest in activities in which they are already skilled and need no technical assistance, such as baking and selling food, raising chickens, operating a small shop, and making or buying and selling clothing.

Initially attracted by the offer of credit, the women of the credit association become engaged in weekly "learning sessions" to discuss and plan how to provide more and better

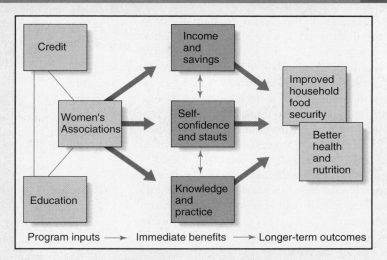

food for their children. In addition, learning about family planning—with natural or artificial methods that prevent or postpone conception (rather than with postconception methods)—is critically important, because multiple, closely spaced births often lead to maternal and child malnutrition, poor health, and even death.

Freedom from Hunger has found that poor women who enter the Credit with Education program are good credit risks; their repayment currently stands at 99% over three or more years. The impact of Credit with Education goes beyond augmenting income and gaining some knowledge. The women increase their confidence and self-esteem, and they go on to provide more active leadership in their communities.

Freedom from Hunger's desire is to have its program adopted by others. The organization is actively enlisting the direct support of banks and other financial and nonfinancial institutions in each country and training them to deliver Credit with Education. Thus, the program is designed to become self-financing and therefore can be expanded to reach very large numbers of women and families in each country. Started in 1989, Credit with Education programs now exist in seven countries: Bolivia, Burkina Faso, Ghana, Honduras, Mali, Togo, and Uganda. As of 1998, over 85,000 women have been helped by the program.

From Freedom from Hunger, Davis, CA 95617

---

charitable programs have not achieved greater success is because they have often focused on only a single component in their approach. This brings us back to the significance of the Cairo Conference.

7.4 The Cairo Conference

The decades since World War II must be recognized as a time when a substantial portion of world leaders held views to the effect that technological progress could always support more people and prevent environmental degradation or that, with more general development, the population problem would take care of itself. They were also decades in which we were gaining knowledge and experience regarding the

effectiveness of different kinds of programs. Therefore, it is perhaps understandable why the world has not reached a consensus before now.

Recognizing this diversity of opinions makes the consensus reached at the Cairo Conference in 1994 all the more significant. Effectively, *all nations agreed that population is an issue of crisis proportions that must be confronted forthrightly.* This sentiment was summed up in the words of Lewis Preston, then president of the World Bank: "Putting it bluntly, if we do not deal with rapid population growth, we will not reduce poverty—and development will not be sustainable." These words and hearing them emanate from the World Bank, which traditionally had been a proponent of the "development will take care of it" view, reflects a fundamental change in attitudes on the part of numerous leaders and organizations.

The document that nations have signed and, thus, given their commitment to achieve, is technically known as the *1994 International Conference on Population and Development Program of Action* (1994 ICPD Program of Action). It sets forth various principles and goals to be achieved by 2015.

Importantly, the goals of the 20-year Program of Action are not cast simply in terms of reducing fertility or population growth per se. Instead, the 1994 ICPD document asks that "interrelationships between population, resources, the environment and development should be fully recognized, properly managed and brought into a harmonious, dynamic balance." Thus, the goals are set in terms of creating an economic, social, and cultural environment in which *all people*, regardless of race, gender, or age, can equitably share a state of well-being. In this context of providing opportunities such that people can improve their quality of life, it is assumed that they will choose to have fewer children as a result, and population growth will level off to the medium projection shown in Fig. 6–11. This means that no more than 7.5 billion persons will have to share the world by 2015, and 9.8 billion by the year 2050.

The basic premises that underlie the Program of Action affirm the following rights:

Every child should be a wanted child.
Every person should have access to family planning services.
Every person should have access to basic education.
Every person should have access to good nutrition.
Every person should have access to basic health care.
Every person should have rights to own and manage property.
Every person should have access to employment.
No person should fear for his or her care and support in old age.

We have observed that the burdens of poverty fall unduly on women and children because women are often denied access to education and opportunities for business and professional careers. Therefore, the phrase "every person" in the foregoing premises has special significance to women—so much so, that the program has been widely reported as a document for the empowerment of women.

By signing the Program of Action, the governments of 179 nations have indicated their commitment to achieving, over the next 20 years, a host of objectives, including the following, in seven different categories:

General

1. Implementing strategies for income generation and employment, especially for the rural poor and those living within or on the edge of fragile ecosystems. Strategies should include maintaining and enhancing the productivity of natural resources.

Empowerment of Women

2. Eliminating gender discrimination in hiring, wages, benefits, training, and job security, with a view toward eradicating gender-biased disparities in income.

3. Changing customs and laws, where necessary, such that women can buy, hold, and sell property and land, obtain credit, and negotiate contracts equally with men.

4. Promoting the full involvement of women in community life, with speaking and voting rights equal to those of men.

Family

5. Promoting the full involvement of men in family life and creating policies to ensure men's responsibility to, and financial support for, their children and families.

6. Assuring that programs are created and administered in ways that encourage keeping family members together.

Reproductive and Basic Health

7. Making access to basic health care and health maintenance central strategies for reducing mortality and morbidity, especially among infants, children, and childbearing women.

8. Ensuring community participation in planning health policy.

9. Strengthening education and communication regarding health and nutrition.

10. Making reproductive health counseling and information accessible to all.

11. Placing special emphasis on controlling the spread of AIDS.

12. Integrating basic health and reproductive health services.

Education

13. Ensuring complete access to primary school education for both girls and boys.

14. Removing any gender-based barriers that prevent girls from going on to reach their full potential in education.

15. Sensitizing parents to the value of education for girls.

Migration

16. Addressing the root causes of migration into cities and emigration to other areas.

International Cooperation

17. The developed countries are called upon to cooperate in transferring technology such that the need for contraceptives and other basic items can be met by local production.

18. Further, each member country of the developed world is called upon to set aside 0.7% of its GNP for the achievement of the Program of Action's objectives throughout the world.

What is to prevent the 1994 ICPD Program of Action document from simply being forgotten? It is significant that the program does not spell out exactly how the objectives should be met, and it does not have any power of enforcement. It is left to each nation to decide how it will meet the objectives. National governments and multilateral agencies like the World Bank and UNFPA will implement actions. NGOs like the Population Institute and a number of women's groups are expected to play the role of watchdog, becoming involved at the grassroots level.

A five-year review and appraisal of the ICPD is now in progress (ICPD+5) that will culminate in a Special Session of the U.N. General Assembly in July 1999, five years after the original ICPD Cairo conference. The process involves several preliminary forums and roundtables held in different parts of the world, focusing on assessments of progress and various implementation strategies and problems of the Program of Action.

The developing countries have agreed to cover two thirds of the costs and are keeping their commitment to date. Unfortunately, the industrialized donor countries have not kept their promises to fund the rest—$5.7 billion a year until 2000, and even more after that. Donor contributions were $2 billion in 1995 and have been declining ever since. With the 0.7% GNP recommendation, the United States, with by far the world's largest GNP ($7.1 trillion in 1995), could be a major donor (or a major holdout). It is significant that two U.S. billionaires—Ted Turner and Bill Gates—have stepped into the gap with multimillion-dollar donations to support U.N. population activities! Regardless of what happens in the United States, it is clear that the international community is taking the ICPD Program of Action seriously and is, on the whole, engaged in implementing the recommendations. This is good news for everyone—even us—as it means, in the end, greater world security and a much greater chance to achieve sustainability in our interactions with the environment that supports us.

ENVIRONMENT ON THE WEB

WOMEN AS THE "KEY TO DEVELOPMENT"

In 1994, the Cairo Conference helped to focus the world's attention on the need to satisfy women's rights and needs (education, reproductive health care, economic and political equality) as part of attaining the world's goals of development and population stabilization. Indeed, women, as bearers of children, have a central role to play in demographic change. However, rather than focus on women solely as the targets of increasing birth control policies, the conference saw women as a central factor in development. Some authors have gone so far as calling women the "key to development."

In India, there is a long tradition of empowerment of women through community groups. Sometimes these groups are initiated by aid organizations such as Oxfam or the Canadian International Development Agency. In other cases, a single individual within the community can begin activities that eventually draw in many women. Often, a woman is attracted to a community group because of its social opportunities—especially the chance to get away from the home and the burden of routine work, and to interact with other women with similar interests. Once a participant, the woman may feel able to share painful experiences of infant death, family violence, or simple poverty with the others. Sometimes these women have few other opportunities for support and advice, and the community group becomes a highly valued part of their lives. Some travel dozens of kilometers to attend a meeting, often bringing with them their children and friends.

Oxfam, through its Community Aid Abroad (CAA) program, sponsors a number of women's groups in India. Its RUCHI (Rural and Urban Center for Human Interest) program began as an attempt to halt the rapid deforestation in the northern Indian state of Himachel Pradesh. Women had found that adequate fuel for cremation was rarely available, and excessive tree harvesting had also encouraged erosion, which resulted in landslide risks and poor productivity on hillside agricultural fields. Working through women, RUCHI and similar programs have made tremendous progress on the environmental problems they set out to solve, while raising women's awareness of issues and options for themselves and their communities.

Chaitanya, another (CAA) program in India, began as a simple community nonprofit savings-and-loan organization, encouraging women to save a few rupees a month (ten to twenty cents). As the program developed, participants came to realize that the more they can save—even if only a rupee or two—the more they will earn in interest. Women have begun to plan their savings strategically and to look for other ways of earning income, such as through cottage crafts.

Through women's advocacy groups, women are becoming more autonomous and confident. They are taking control of land and household improvements as well as making economic and political decisions. Also, as they gain control over their futures, more of these women will be equipped to make sound decisions about their reproductive health, family size, and timing. As their knowledge and confidence grow, so does their awareness of their own abilities. Whereas most used to spend all of their time around home and children, they are now traveling to distant meetings, making community decisions—even playing cricket.

Web Explorations

The Environment on the Web activity for this essay describes some of the ways in which governments have intervened in women's reproductive health. Go to the Environment on the Web activity (select Chapter 7 at http://www.prenhall.com/nebel) and learn for yourself:

1. how education can affect a woman's childbearing and childrearing behavior;
2. about the availability of birth control in U.S. schools;
3. about family planning programs; and
4. about the ethics of forced birth control.

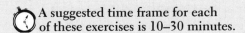 A suggested time frame for each of these exercises is 10–30 minutes.

REVIEW QUESTIONS

1. What have been the two basic schools of thought regarding population growth?
2. Discuss the six specific factors that influence the number of children desired.
3. Describe how poverty, environmental degradation, and high fertility rates drive one another in a vicious cycle.
4. What has been the major agency and mechanism for promoting development in poor nations over the past 45 years? Discuss its past successes and failures.
5. What is meant by the developing-world debt crisis? How has it been disastrous economically, socially, and ecologically?
6. How has the World Bank undergone major reform in recent years?
7. What are the five interdependent components that must be addressed to bring about social modernization?
8. What are the key aspects of family planning, and why is family planning of critical importance to all other aspects of development?
9. What is meant by microlending?
10. How can each of the following be addressed in a cost-effective way: education; reduction of infant mortality; income enhancement; environmental degradation?
11. What was the significance of the 1994 Cairo conference? What are the agreed-upon strategies for addressing the problems of poverty, excessive population, and environmental degradation?

THINKING ENVIRONMENTALLY

1. Is the world population below, at, or above the optimum? Defend your answer by pointing out things that may improve and things that may worsen by increasing population.
2. Suppose you are the head of an island nation with a poor, growing population, and the natural resources of the island are being degraded. What kinds of programs would you initiate, and what help would you ask for to try to provide a better, sustainable future for your nation's people?
3. List and discuss the benefits and harms of writing off debts owed by developing nations.
4. Describe what you think will be long-term results (for the people, the society, and the environment) of limiting family planning services in developing nations.
5. What priority do you give to the cyclic factors—population, poverty, and environmental decline? Give reasons pro and con as to your ranking. What proportion of national security expenditures would you allocate toward addressing the population-poverty issue?

MAKING A DIFFERENCE

PART TWO: Chapters 6 and 7

1. Think carefully about your own reproduction. What concerns will you and your mate weigh as you plan your family?
2. Become involved in the abortion debate, pro or con. Whether or not the United States supports international family planning, whether or not legal abortions remain available in your state, and other issues will be determined by votes cast by you or your representatives.
3. Become involved in and support programs promoting effective sex education and responsible sexual behavior.

Consider the advantages of abstinence and monogamy as ways to avoid needing an abortion or contracting a sexually transmitted disease.

4. Share your knowledge and energy by joining the Peace Corps or another organization engaged in appropriate technology in a developing country.
5. Encourage sustainability by buying products that originate from appropriate technology in the developing world.

WEB REFERENCES

On-line resources for this chapter can be found on the World Wide Web at: **http://www.prenhall.com/nebel**.
Click on Chapter 7 on the chapter selector.

PART THREE

Renewable Resources

A mountainside on the island of Bali is terraced for rice cultivation; tall-grass prairies of Illinois and Iowa are plowed under to raise corn; tropical rain forests in Brazil are converted to cattle pasture; desert in Israel is irrigated to raise vegetables. Forests are harvested for wood and paper pulp; coastal oceans are fished; grasslands are grazed by sheep. We get food and fiber from a host of natural ecosystems—some thoroughly managed and some not. All of these activities depend on water, nutrients, and sunlight—the basis for productivity in natural systems. They are renewable resources—that is, they are replenished as energy flows and water and nutrients cycle in ways that have long sustained life on Earth.

At the same time, we embrace the beauty of wild ecosystems and turn to them for enjoyment and rest. Can we have it both ways? Can we hope to preserve nature while we also make use of it, especially in light of continuing population growth and therefore, continuing pressure on all resources? In this part, it is our task to dig more deeply into the science of soil, water, food production, forest growth, and fisheries. We will examine all of these renewable resources and ways of managing them while again keeping our eyes on sustainability.

Delaware River Valley, New Jersey and Pennsylvania.

9

Water: Hydrologic Cycle and Human Use

〜〜〜〜〜〜

Key Issues and Questions

1. All water on Earth is constantly recycled, repurified, and reused. How does the hydrologic cycle inform us on how recycling and repurification occur?
2. Humans have three major impacts on the hydrologic cycle. What are they, and what are their effects?
3. All the water humans use must come out of the hydrologic cycle. What are the major uses, points of withdrawal, and limitations and consequences of overdrawing water?
4. Historically, humans have addressed water problems by obtaining more water. To what degree is this not a viable option for the future?
5. Humans can reduce their water demands in numerous ways. How can demands be reduced in agriculture, industry, and domestic use?
6. Urbanization seals surfaces with pavement, increasing stormwater runoff and quickening concentration times. Discuss related problems caused by paving over soil. How should these concepts influence development?
7. There is potential for all parties' getting together to work out compromises for water usage between agriculture, cities, and natural ecosystems. What are some policy options that would encourage this process?

Mono Lake is a 63-square-mile (163-km^2) saline lake in east-central California. The lake has no outlet but is fed by tributary streams of fresh water from snowmelt off the Sierra Nevada. Like other saline lakes around the world, Mono Lake supported numerous species of wildlife, including huge flocks of aquatic birds that nested there and fed on the unique species of brine shrimp in the lake.

Beginning in 1941, however, much of the freshwater inflow was diverted to support water-hungry Los Angeles, 350 miles to the south. Soon Mono Lake was losing more water to evaporation than it was receiving, and gradually it began to shrink. As the lake shrank, the salt in its water became more concentrated, threatening to turn the lake into a sterile, hypersaline system. The streams feeding the lake lost much of their water and failed to support once abundant trout populations. In the meantime, the city of Los Angeles continued to grow and become increasingly dependent on imported water.

Concerned about what was happening to the lake, a group of students formed the Mono Lake Committee in 1978, which brought legal action under the Public Trust Doctrine, an old law originally intended to protect navigation. The idea that destruction of the integrity of an ecosystem constituted a violation of the public trust was a new concept that had to be tried in the courts, and it was bitterly opposed by the city of Los Angeles. Nevertheless, the citizens of the Mono Lake Committee persevered through the many appeals and won the final decision in September 1994. The State Water Resources Control Board ruled that Los Angeles must sufficiently reduce diversions of water to allow Mono Lake, some 40 feet below its original level, to recover by at least 15 feet (5 m) above the 1995 level. The lake is now on its way back up, having gained almost 9 feet in height since the Water Board's decision. Recovery of the 15 feet will require about 20 years, but at least a major ecological collapse of wildlife populations supported by the lake and the feeder streams will seemingly have been averted.

◄ **Mono Lake.** With its freshwater input diverted to Los Angeles, Mono Lake was drying up, leaving unique sculptured forms of tufa (precipitated calcium carbonate) and threatening ecological collapse.

And Los Angeles? The decision forces L.A. to initially reduce its takings from the Mono Lake supply by 80%, which amounts to a 12% cut in the city's total water supply. When the lake reaches its prescribed height, L.A. will be able to divert about one-third of their historical take of water. This controversy and frequent droughts have had the effect of forcing citizens to develop water-saving habits and

additional conservation and recycling schemes, which have been more than adequate to make up for the loss. In recent years, Mono Lake has been joined by the Aral Sea, the Colorado River, the Nile, the Jordan, the Danube, and countless other water bodies that have become the focus of national and international battles over access to fresh water.

9.1 Water—a Vital Resource

Water is absolutely fundamental to life as we know it. It is difficult even to imagine a form of life that might exist without water. Happily, Earth is virtually flooded with water; a total volume of some 325 million cubic miles (1.4 billion cubic kilometers) covers 71% of Earth's surface. Yet it is still difficult in many locations to obtain desired amounts of water of suitable purity (see the "Global Perspective" essay, p. 213).

All major terrestrial biota, ecosystems, and humans depend on **fresh water**, water that has a salt content of less than 0.1% (1000 ppm). Over 97% of Earth's water is the salt water of oceans and seas. Then, of the 2.5% that is fresh water, two-thirds is bound up in the polar ice caps and glaciers. Only 0.77% of all water is found in lakes, wetlands, rivers, groundwater, biota, soil, and the atmosphere (Fig. 9–1). To be sure, evaporation from the seas and precipitation continually resupply that small percentage through the solar-powered hydrologic cycle, as we shall describe in detail shortly. Thus, fresh water is a continually renewable resource.

As we know, precipitation patterns around the globe are far from even. Regions with abundant precipitation support

lush forest ecosystems; other regions have minimal rainfall and are deserts as a result. Thus, we can visualize different volumes of flow through different natural regions (over 1 million gallons of water per acre per year in a temperate forest region; 2500 gallons or less per acre per year in desert regions). Human societies must draw on the same water for drinking, irrigating crops, and supplying industries.

In high-rainfall regions, there is plenty of water for both human demands and natural biota. However, in dryer regions with growing human populations, there are growing conflicts between human needs and those of the natural ecosystems. Around the world, there are countless examples of ecosystems under stress or already dead because of diversions of water for human uses. Moreover, within the human arena, there is growing contention between agricultural, urban, and industrial demands, and between countries that share a common water source.

Hydrologists (water experts) estimate that water shortages place a severe constraint on food production, economic development, and protection of natural ecosystems as available water drops below about 1000 cubic meters per person per year (725 gallons per day). So defined, water-supply shortages loom in at least 26 countries as of 1997. In many other countries, seasonal shortages of water occur. Water shortages increase conflicts and public health problems, reduce food production, and endanger the environment. As populations grow, more countries will be joining the list. And as shortages become more severe, already contentious relations between nations sharing common supplies will become more so (Fig. 9–2). Similar contentions are growing between water-rich and water-poor areas within national boundaries as well. Finally, a considerable number of countries and regions are now satisfying their water needs only by withdrawing groundwater faster than it is replenished, thereby depleting their supply for future generations. Obviously, these trends are not sustainable.

A sustainable future will depend on learning stewardship of water resources. There are abundant opportunities for stewardly management in this arena. Our objective in this chapter is threefold: (1) to understand the natural water cycle, its capacities, and its limitations; (2) to understand how we are overdrawing certain water resources and the consequences of this action; and (3) to understand how water must be managed if we are to achieve sustainable supplies.

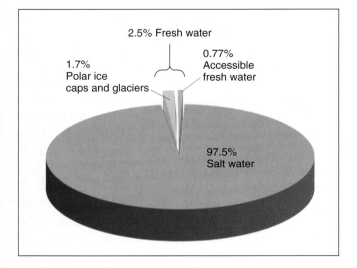

▲ **FIGURE 9–1** *Earth's water.* The Earth has an abundance of water, but terrestrial ecosystems, humans, and agriculture depend on accessible fresh water, which constitutes only 0.77% of the total.

(a)

(b)

◀ **FIGURE 9–2**
Differences in availability. Human conditions range from (a) being able to luxuriate in water to (b) having to walk long distances to obtain enough simply to survive. In many cases, however, the abundance of water is illusory, since water supplies are being overdrawn to supply apparent abundance.

9.2 The Hydrologic Cycle

Earth's **water cycle**, also called the **hydrologic cycle**, is represented in Fig. 9–3. The basic cycle consists of water rising to the atmosphere through either evaporation or transpiration and returning to the land and oceans through condensation and precipitation. However, these and additional aspects bear more consideration.

Evaporation, Condensation, and Purification

As we discussed in Chapter 3, a weak attraction known as hydrogen bonding tends to hold water molecules (H_2O) together. Below 32°F (0°C), the kinetic energy of the molecules is so low that the hydrogen bonding is enough to hold the molecules in place with respect to one another, and the result is ice. At temperatures above freezing but below boiling, the kinetic energy of the molecules is such that hydrogen bonds keep breaking and re-forming with different molecules. The result is liquid water. As water molecules absorb energy from sunlight or an artificial source, the kinetic energy they gain may be enough to allow them to break away from other water molecules entirely and enter the atmosphere. This is the process we know as **evaporation**, and the water molecules are said to be in the gaseous state.

We speak of water molecules in the air as **water vapor**; the amount of water vapor in the air is **humidity**. Humidity is generally measured as **relative humidity**, the amount of water vapor as a percentage of what the air can hold *at that temperature*. For example, a relative humidity of 60% means that the air contains 60% of the maximum amount of water vapor it could hold at that temperature. The amount of water vapor air can hold increases with rising temperature and decreases with falling temperature. Consequently, relative humidity will decrease as air warms and increase as air cools quite apart from any change in the actual amount of water vapor present. The key point is that when warm, moist air is cooled, its relative humidity rises until it reaches 100%; further cooling causes the excess vapor to *condense*, because the air can no longer hold as much vapor (Fig. 9–4).

Condensation simply is water molecules rejoining by hydrogen bonding to form liquid water or ice. If the droplets form in the atmosphere, the result is fog and clouds (fog simply is a very low cloud). If the droplets of condensing vapor form on the cool surfaces of vegetation, the result is dew.

One very important aspect of evaporation and condensation is that these processes result in natural *water purification*. When water evaporates, only the water molecules leave the surface; salts and other solids in solution remain behind. (We noted this process when we discussed the problem of salinization in Chapter 8.) The condensed water is thus purified water—except as it picks up pollutants in the air. (The most chemically pure water for use in laboratories is obtained by distillation, a process of boiling water and recondensing the vapor.) Thus, evaporation and condensation of water vapor are the source of all natural fresh water on Earth. Fresh water from precipitation falling on the land gradually makes its way through aquifers, streams, rivers,

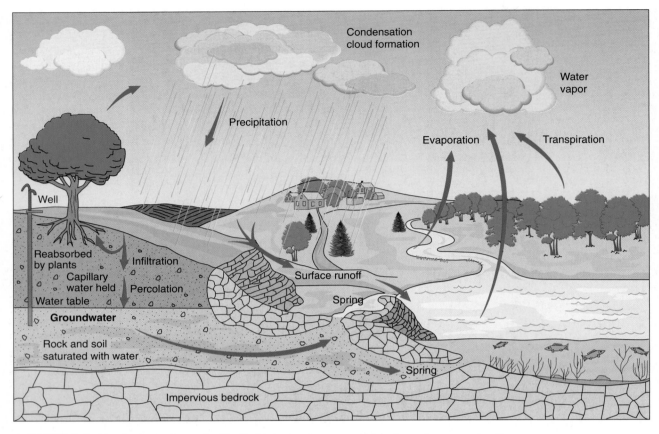

▲ **FIGURE 9-3** *The hydrologic cycle.* The Earth's fresh waters are replenished as water vapor enters the atmosphere by evaporation or transpiration from vegetation, leaving salts and other impurities behind. As precipitation hits the ground, note that three additional pathways are possible.

▶ **FIGURE 9-4** *Condensation.* The amount of water vapor that air can hold increases and decreases with corresponding changes in temperature. Therefore, as warm, moist air is cooled, the amount of water it can hold decreases. Cooling air beyond the point where relative humidity (RH) reaches 100% forces excess moisture to condense, forming clouds. Further cooling and condensation results in precipitation.

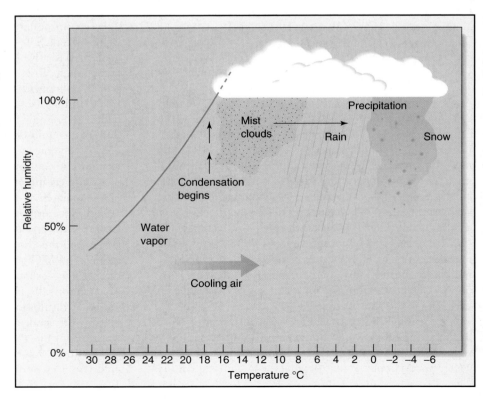

GLOBAL PERSPECTIVE

PEOPLE AND WATER

Safe drinking water is fundamental to human health and well-being. Yet much of the world's population does not have access to this essential resource, as indicated by the color coding on the accompanying map. Furthermore, water scarcity will become a severe constraint on food pro-duction, economic development, and protection of natural ecosystems if total annual water supplies diminish below 1000 cubic meters per person. Twenty-six countries, indicated by cross-hatching on the map, are already below this threshold, and with rapid population growth, many more will cross into this category in the near future. Finally, a number of regions, including some major cities (Beijing, New Delhi, and Mexico City), are meeting current demands only by depleting groundwater reserves (black dots), a nonsustainable solution.

_____

Sources: "Percent of population with access to safe

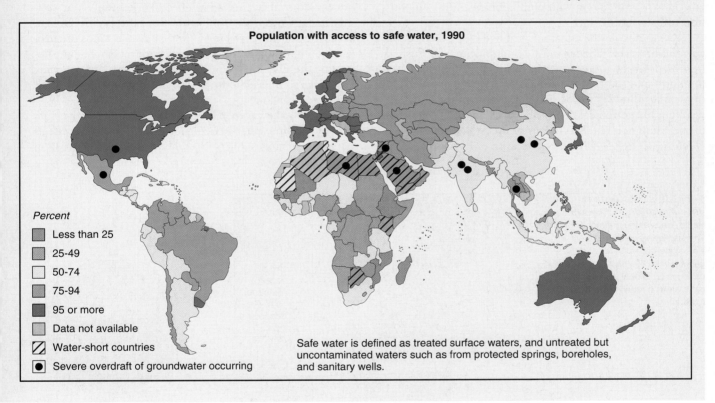

Population with access to safe water, 1990

Percent
- Less than 25
- 25-49
- 50-74
- 75-94
- 95 or more
- Data not available
- Water-short countries
- Severe overdraft of groundwater occurring

Safe water is defined as treated surface waters, and untreated but uncontaminated waters such as from protected springs, boreholes, and sanitary wells.

and lakes to oceans or seas. In the process, salts from the land are constantly flushed toward locations where the only exit is by evaporation, and the salts accumulate at those points. Oceans are the prime example, but there are notable inland salt seas or lakes, such as the Great Salt Lake in Utah. Salinization of irrigated croplands is a notable human-made example.

Precipitation

Recall that as evaporation occurs, air readily picks up water vapor from any wet or moist surface and as transpiration occurs from vegetation. Warm air rises because it is less dense than the cooler air above, and as it encounters lower atmospheric pressure, it gradually cools as it expands—a process known as _adiabatic cooling_. When relative humidity reaches 100% and cooling continues, condensation occurs and clouds form. With intensifying condensation, water droplets become large enough to fall as precipitation. _Adiabatic warming_ occurs as air descends and is compressed by the higher air pressure in the lower atmosphere.

The distribution of precipitation over Earth, which ranges from near zero in some areas to more than 100 inches (2.5 m) per year in others, basically depends on patterns of rising or falling air currents. As air rises, cooling and condensation occur, and precipitation results. As air descends, it tends to become warmer, causing an increase in evaporation and dryness. A rain-causing event that you see in almost every television weather report is the movement of a cold front. As a cold front moves into an area, the existing warm, moist air is forced upward because the cold air of the advancing front is denser. The rising warm air cools, causing condensation and precipitation along the leading edge of the cold front.

Two factors may cause more or less continuously rising or falling air currents over particular regions, with major effects on precipitation. First are global convection currents. Solar heating of Earth is most intense over and near the equator, where rays of sunlight are almost perpendicular to Earth's surface. As the air at the equator is heated, it expands, rises, and cools; condensation and precipitation occur. The constant intense heat in these equatorial areas ensures that this process is repeated often, thus causing high amounts of rainfall. This rainfall, along with continuous warmth, supports tropical rain forests.

Rising air over the equator is just half of the convection current, however. The air must come down again. Pushed from beneath by more rising air, it literally "spills over" to the north and south of the equator and descends over subtropical regions (25° to 35° north and south of the equator), resulting in subtropical deserts. The Sahara of Africa is the prime example. The two halves of this system make up a *Hadley cell* (Fig. 9–5a and b). Because of Earth's rotation,

winds are deflected from the strictly vertical and horizontal paths indicated by a Hadley cell and tend to flow in easterly and westerly directions—the *trade winds* (Fig. 9–5c), which blow almost continuously from the same direction.

The second situation that causes continually rising and falling air occurs where trade winds hit mountain ranges. As the moisture-laden air in the trade winds encounters a mountain range, the air is deflected upward, causing cooling and high precipitation on the windward side of the range. As the air crosses the range and descends on the other side, it becomes warmer and increases its capacity to pick up moisture. Hence, deserts occur on the leeward side of mountain ranges. The dry region downwind of a mountain range is referred to as a **rain shadow** (Fig. 9–6). The severest deserts in the world are caused by the rainshadow effect. For example, the westerly trade winds, full of moisture from the Pacific Ocean, strike the Sierra Nevada in California. As the winds rise over the mountains, large amounts of water precipitate out, supporting the lush forests on the western slopes. Immediately east of the southern Sierra Nevada lies Death Valley, a result of the rain shadow.

Fig. 9–7 shows general precipitation patterns for the land; the general atmospheric circulation (Fig. 9–5) and the effects of mountain ranges (Fig. 9–6) combine to give great variation

▶ **FIGURE 9–5 *Global air circulation.*** (a) The two Hadley cells at the equator. (b) The six Hadley cells on either side of the equator, indicating general vertical airflow patterns. (c) Global trade-wind patterns as a result of Earth's rotation.

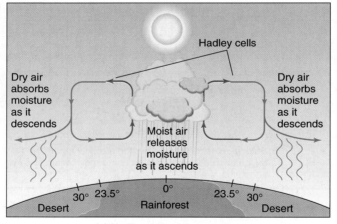

(a) Hadley cells at the equator

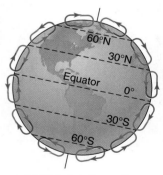

(b) Global air flow patterns

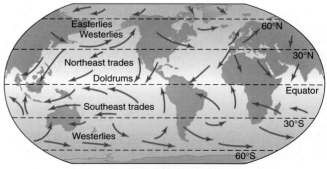

(c) Global trade winds

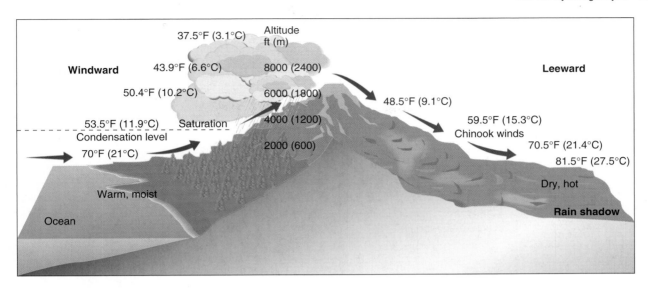

▲ **FIGURE 9–6** *Rain shadow.* Moisture-laden air in a trade wind cools as it rises over a mountain range, resulting in high precipitation on the windward slopes. Desert conditions arise on the leeward side as the descending air warms and tends to evaporate water from the soil.

in the precipitation reaching the land. This in turn is a major determinant of the ecosystems found in a given region.

Water over and through the Ground

As precipitation hits the ground, it may follow either of two pathways. It may soak into the ground, **infiltration**, or it may run off the surface, **runoff**. We speak of the amount that soaks in compared with the amount that runs off as the **infiltration-runoff ratio**.

Runoff flows over the ground surface into streams and rivers, which make their way to the ocean or to inland seas. All the land area that contributes water to a particular stream or river is referred to as the **watershed** for that stream or river. All ponds, lakes, streams, rivers, and other waters on the surface of Earth are referred to as **surface waters**.

For water that infiltrates, there are two more alternatives (Fig. 9–3). The water may be held in the soil; the amount held depends on the water-holding capacity of the soil, as was discussed in Chapter 8. This water, called **capillary water**, returns to the atmosphere either by way of evaporation from the soil or by transpiration through plants. The combination of evaporation and transpiration is referred to as **evapotranspiration**.

The second alternative is **percolation**. Infiltrating water that is not held in the soil is called **gravitational water** because it trickles, or percolates, down through pores or cracks under the pull of gravity. Sooner or later, however, gravitational water comes to an impervious layer of rock or dense clay. There it accumulates, completely filling all the spaces above the impervious layer. This accumulated water is called **groundwater**, and its upper surface is the **water table** (Fig. 9–3). Gravitational

water becomes groundwater as it hits the water table in the same way that rainwater is defined as lake water as it hits the surface of the lake. Wells must be dug to below the water table; then groundwater, which is free to move, seeps into the well and fills it to the level of the water table.

Groundwater will seep laterally as it seeks its lowest level. Where a highway has been cut through rock layers, you can frequently observe groundwater seeping out. Layers of porous material through which groundwater moves are called **aquifers**. It is often difficult to determine the location of an aquifer. Many times layers of porous rock are found between layers of impervious material, and the entire formation may be folded or fractured in various ways. Thus, groundwater in aquifers may be found at various depths between layers of impervious rock. Also, the **recharge area**—the area where water enters an aquifer—may be many miles away from where the water leaves the aquifer.

As water percolates through the soil, debris and bacteria from the surface are generally filtered out. However, water may dissolve and leach certain minerals. Underground caverns, for example, are the result of the gradual leaching away of limestone (calcium carbonate). In most natural situations, the minerals that leach into groundwater are not harmful. Indeed, calcium from limestone is considered beneficial to health. Thus, groundwater is generally high-quality fresh water that is safe for drinking. A few exceptions occur where there is leaching of minerals containing sulfide, arsenic, or other poisonous elements that make groundwater unsafe to drink.

Drawn by gravity, groundwater may move through aquifers until it finds some opening to the surface. We observe such natural exits as springs or seeps. In a **seep**, water flows out over a relatively wide area; in a **spring**, water exits

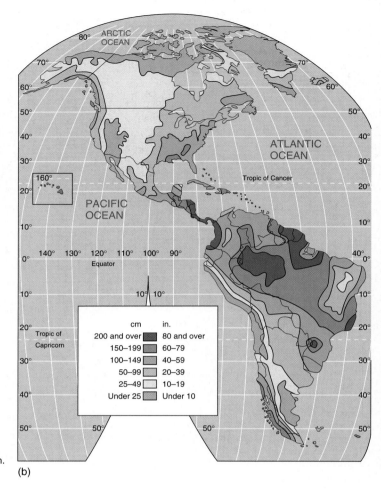

▶ **FIGURE 9–7** *Global precipitation.* Note the high rainfall in equatorial regions and the regions of low rainfall to the north and south.

(b)

the ground as a significant flow from a relatively small opening. Since springs and seeps feed streams, lakes, and rivers, groundwater joins and becomes part of surface water. A spring will flow, however, only if the water table is higher than the spring. Whenever the water table drops below the level of the spring, the spring will dry up.

Summary of the Hydrologic Cycle

The hydrologic cycle consists of evaporation, condensation, precipitation, and gravitational flow. There are three principal "loops" in the cycle: (1) the *surface runoff loop*, in which water runs across the ground surface and becomes part of the surface water system; (2) the *evapotranspiration loop*, in which water infiltrates, is held as capillary water and then returns to the atmosphere by way of evapotranspiration; and (3) the *groundwater loop*, in which water infiltrates, percolates down to join the groundwater, and then moves through aquifers, finally exiting through springs, seeps, or wells, where it rejoins the surface water. In addition, there are substantial exchanges of water between the land, the atmosphere, and the oceans. Estimates of these loops and exchanges provide a quantitative overview of the

cycle (Fig. 9–8) and establish a basis for calculating the human appropriation of water in the cycle.

Terms commonly used to describe water are given in Table 9-1.

9.3 Human Impacts on the Hydrologic Cycle

A large share of the environmental problems we face today stem from direct or indirect impacts on the water cycle. These impacts can be categorized into three areas: (1) changing Earth's surface, (2) pollution, and (3) withdrawals for use.

Changing the Surface of Earth

We are most concerned, perhaps, by the loss of forests and other ecosystems to various human enterprises as this loss diminishes biodiversity. However, the indirect effects of the loss on the water cycle are also profound. In most natural ecosystems, there is relatively little runoff; rather, precipitation is intercepted by vegetation and infiltrates into a porous topsoil, as we describe in Chapter 8, and goes on to recharge the groundwater reservoir.

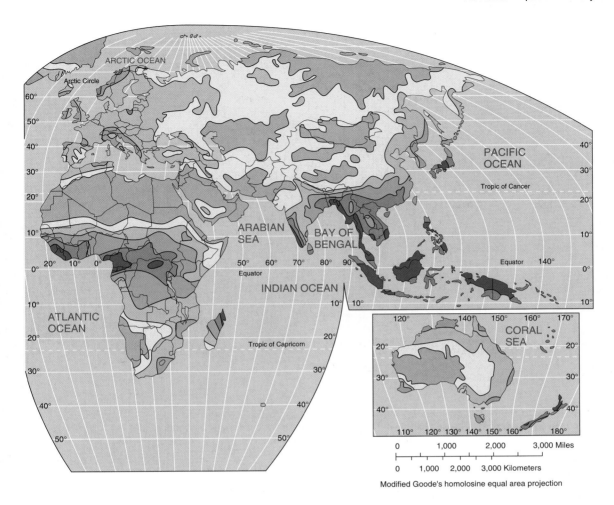

Modified Goode's homolosine equal area projection

Then its gradual release through springs and seeps maintains the flow of streams and rivers at relatively uniform rates. The reservoir of groundwater may be sufficient to maintain a flow even during a prolonged drought. In addition, dirt, detritus, and microorganisms are filtered out as the water percolates through soil and porous rock, resulting in groundwater that is drinkable in most cases. Similarly, streams and rivers fed by springs contain high-quality water.

As forests are cleared or land is overgrazed, the pathway of the water cycle is shifted from infiltration and groundwater recharge to runoff. With runoff, the water runs into the stream or river almost immediately. This sudden influx of water into the waterway not only is likely to cause a flood but also brings along sediments and other pollutants from surface erosion.

Of course, floods are not unknown in nature. However, in many parts of the world, the frequency and severity of flooding are increasing—not because precipitation is greater, but because both deforestation and cultivation cause erosion and reduce infiltration. For example, extreme flooding in Bangladesh (a very flat country only a few feet above sea level) is now common because Himalayan foothills in India and Nepal have been deforested. Because of sediment deposition from upriver erosion, the Ganges river basin has risen 15–22 feet (5–7 m) in recent years! The 1988 flood inundated 60% of the country and caused an estimated 2000 deaths (Fig. 9–9).

Also profound and far-reaching is the fact that increased runoff necessarily means less infiltration and groundwater recharge. Thus, there may be insufficient groundwater to keep springs flowing during dry periods. Typical of deforested regions are dry, barren, and lifeless stream beds—a tragedy for both the ecosystems and the humans dependent on the flow. Wetlands function to store and release water in a manner similar to the groundwater reservoir. Therefore, the destruction of wetlands has the same impacts as deforestation: Flooding is exacerbated, and waterways are polluted during wet periods and dry up during drought periods.

Urban and suburban development provides an extreme case of altering Earth's surface as it replaces porous soil with asphalt. This problem and remedial measures will be discussed later in this chapter.

Polluting the Water Cycle

You can see that the water cycle involves the entire biosphere. Therefore, wherever wastes are put, they are

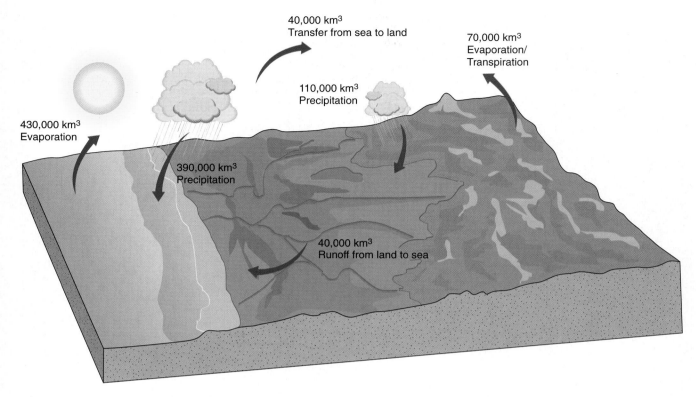

▲ **FIGURE 9–8** *Water balance in the hydrologic cycle.* The data show: (1) The contribution of water from the oceans to the land via evaporation and then precipitation; (2) Movement of water from the land to the oceans via runoff and seepage; and (3) The net balance of water movement between terrestrial and oceanic regions of Earth.

TABLE 9-1 Terms Commonly Used to Describe Water

Term	Definition
Water quantity	The amount of water available to meet desired demands.
Water quality	The degree to which water is pure enough to fulfill the requirements of various uses.
Fresh water	Water having a salt concentration below 0.1%. As a result of purification by evaporation, all forms of precipitation are fresh water, as are lakes, rivers, groundwater, and other bodies of water that have a throughflow of water from precipitation.
Salt water	Water, typical of oceans and seas, that contains at least 3% salt (30 parts salt per 1000 parts water).
Brackish water	A mixture of fresh and salt water, typically found where rivers enter the ocean.
Hard water	Water that contains minerals, especially calcium or magnesium, that cause soap to precipitate, producing a scum, curd, or scale in boilers.
Soft water	Water that is relatively free of minerals.
Polluted water	Water that contains one or more impurities, making the water unsuitable for a desired use.
Purified water	Water that has had pollutants removed or is rendered harmless.
Storm water	Water from precipitation events that runs off of land surfaces in surges.

EARTH WATCH

WATER PURIFICATION

Polluted water is defined as water that contains one or more materials that make the water unsuitable for a given use. Water purification is any method that will remove one or more such materials. Several methods may be used in combination to obtain water that is sufficiently pure for a given use. Methods that are commonly used in water purification are:

1. *Settling.* Soil particles and other solid material carried by flowing water may be removed by holding the water still and allowing the solids to settle. Clarified water is removed from the top. Settling may be aided by the addition of alum (aluminum sulfate). The +3 charge on aluminum ions pulls clay and other particles, which are negatively charged, into clumps that settle more readily than do the individual particles.

2. *Filtration.* Filtration is the passage of water through a porous material. Any materials larger than the pores will be filtered out. A bed of sand is often used for this purpose.

3. *Adsorption.* Passing water through an adsorbing material that binds and holds other materials on its surface will remove certain pollutants. Activated carbon is a material commonly used in this way to remove organic contaminants from water or air.

4. *Biological oxidation.* Organic material (detritus and organisms) is fed upon by detritus feeders and decomposers, broken down in cell respiration and thus removed. Passage of water through systems supporting the growth of such organisms accomplishes removal of organic material (see Chapter 18).

5. *Distillation.* Distillation is the evaporation and condensation of water. All materials present in the water before the evaporation step remain behind in the holding tank and are therefore not present when the water vapor is condensed.

6. *Disinfection.* Water is treated with chlorine or other agents that kill disease-causing organisms.

The natural water cycle includes all of these purification methods except disinfection. Sitting in lakes, ponds, or the oceans, water is subject to settling. As it percolates through soil or porous rock, it is filtered. Soil and humus are also good chemical adsorbents. As water flows down streams and rivers, detritus is removed by biological oxidation. As water evaporates and condenses, it is distilled.

Thus, numerous sources of fresh water might be safe to drink were it not for human pollution. The most serious threat to human health is contamination with disease-causing organisms and parasites, which come from the excrements of humans and their domestic animals. In human settlements, you can see how these organisms can get into water and be passed on to people before any of the natural purification processes can work.

▼ **FIGURE 9–9** *Flooding in Bangladesh.* A flooded street in Dhaka, the capital of Bangladesh. Floods in 1998 alone claimed more than 850 lives and displaced millions of people.

inevitably introduced into the water cycle. Any smoke or fumes exhausted or evaporated into the air will come back down as contaminated precipitation. Acid rain (see Chapter 22) is a case in point. Chemicals that we use on the soil surface, such as fertilizers, pesticides, and road salt, may either leach into groundwater or be carried into streams by runoff. The same is true of any oil, grease, or other materials we drop or spill on the ground. Any waste we bury in the ground (landfills) may eventually leach into groundwater (Fig. 9–10). (It should be added that modern landfills are constructed so as to minimize this problem; see Chapter 19.) Of course, all the water we use in the course of washing or flushing away wastes directly adds the pollutants into surface waters unless there is intervening water treatment (Chapter 18).

Withdrawing Water Supplies

Finally are the many problems centering around withdrawals of water for human use, not the least of which is having insufficient water to sustain human needs. We shall expand on this issue in the following sections.

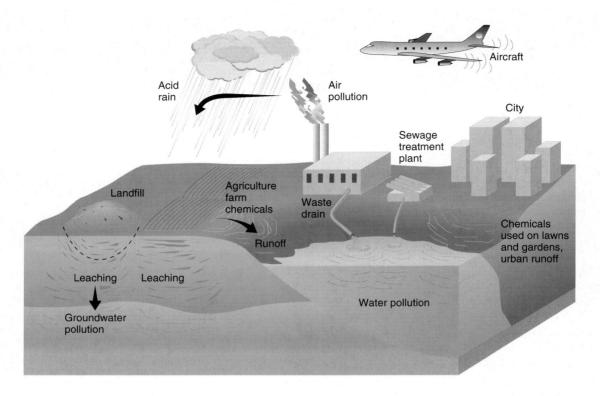

▲ FIGURE 9–10 *Pollution of the hydrologic cycle.* Human activities introduce pollution into the water cycle at numerous points as shown.

9.4 Sources and Uses of Fresh Water

Human concerns regarding water can be divided into two categories: *quantitative* and *qualitative*. **Quantitative** refers to such issues as: Is there enough water to meet our needs? What are the impacts of diverting water from one point of the cycle to another? **Qualitative** refers to such issues as: Is the water of sufficient purity so as not to harm human or environmental health? In the remainder of this chapter, we shall focus primarily on the quantitative aspects, although you should bear in mind that quality is an ever-present concern. In following chapters, we shall focus on the qualitative aspects—that is, pollution.

The major uses of fresh water are given in Table 9-2. Most of the water used in homes and industries is for washing and flushing away undesired materials, and the water used in electrical power production is used for taking away waste heat and increasing the efficiency of the process. Such uses are termed **nonconsumptive** because the water, though now contaminated with the wastes, remains available to humans for the same or other uses if its quality is adequate or if it can be treated to remove undesired materials. In contrast, irrigation is called a **consumptive** use because the applied water does not return to the water resource. It can only percolate into the ground or return to the atmosphere through evapotranspiration. Of course, in either case, the water does reenter the overall water cycle, but it is gone from human control.

Worldwide, the largest use of water is for irrigation (69%); second is for industry (23%); and third is for direct human consumption (8%). These percentages vary greatly from one region to another, depending on natural precipitation and the degree of development (Fig. 9–11).

Humans take fresh water from whatever source they can. In some cases, this means capturing precipitation directly in a rain barrel under a downspout. The major sources of fresh water, however, are surface water (rivers and lakes) and groundwater. In the United States, about half of domestic

TABLE 9-2 U.S. Demands on Fresh Water	
Use	Gallons (liters) used per person per day
Consumptive	
Irrigation and other agricultural use	700 (2800)
Nonconsumptive	
Electrical power production	600 (2400)
Industrial use	370 (1500)
Residential use	100 (400)

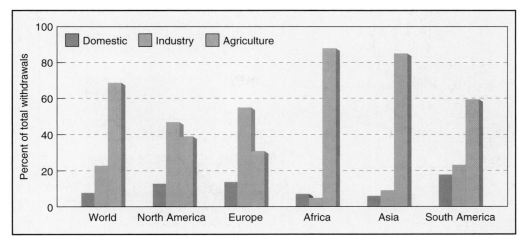

◀ **FIGURE 9-11** *Human usage of water.* The percentage used in each category varies with climate and relative development of the country. A dry-climate, less developed region uses most of its water for irrigation (e.g., Africa), whereas moist-climate, industrialized countries (e.g., Europe) require the largest percentage for industry.

water comes from each of these sources. Before municipal water supplies were made available, each family dipped into a local stream or shallow well for their own use. This method is still used to a considerable extent, and women in many developing countries walk long distances each day to fetch water. Because surface water and shallow wells often receive runoff, they frequently are polluted with various wastes, including animal excrements and human sewage likely to contain pathogens (disease-causing organisms). Yet, unsafe as it is, this polluted water is the only available water for an estimated 1.4 billion poor people in less-developed countries (Fig. 9–12). It is commonly consumed without treatment, but not without consequences. According to U.N. estimates,

▼ **FIGURE 9-12** *Water in the developing world.* In many villages and cities of the developing world, people withdraw water from rivers, streams, and ponds. Such sources are often contaminated with pathogens and other pollutants. This woman in Burkina Faso is collecting water from a pond used by people and animals.

contaminated water is responsible for 80% of the diseases in the developing world and for the deaths of 4–5 million children per year, over 1 million from simple diarrhea (see the "Global Perspective" essay, p. 213).

In the developed countries, also, the major freshwater sources are rivers and lakes, but methods for collection, treatment, and distribution are more sophisticated. Dams are built across rivers to create reservoirs, which hold water in times of excess flow and can be drawn down at times of lower flow. In addition, dams and reservoirs may provide for power generation, recreation, and flood control. Water for municipal use is piped from the reservoir to a treatment plant, where it is treated to kill pathogens and remove undesirable materials, as shown in Fig. 9–13. After treatment, water is distributed through the water system to homes, schools, and industries. The wastewater, collected by the sewage system, is carried to a sewage-treatment plant, where it is treated before being discharged into a natural waterway (see Chapter 18). Often it is discharged into the same river from which it was withdrawn, but farther downstream.

So far, where possible, both water and sewage systems are laid out so that gravity maintains the flow through the system. This arrangement minimizes pumping costs and increases reliability. On major rivers, such as the Mississippi, water is reused many times. Each city along the river takes water, treats it, uses it, and then returns it to the river. In developing nations, the wastewater is discharged often with minimal or no treatment. Thus, as the water moves downstream, each city has a higher load of pollutants to contend with than the previous city had, and ecosystems at the end of the line may be severely affected by the pollution. Pollutants include industrial wastes as well as pollutants from households, since industries and residences generally utilize the same water and sewer systems.

Reservoirs created by dams on rivers are also major sources of water for irrigation. In this case, no treatment is required. As noted above, the croplands are the end point for this water except insofar as it reenters the water cycle through evaporation and infiltration into groundwater.

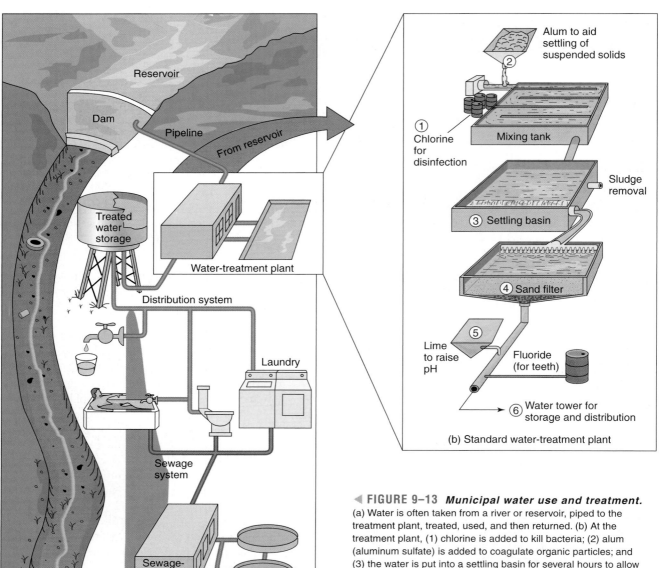

(a) Municipal water use

FIGURE 9–13 *Municipal water use and treatment.*
(a) Water is often taken from a river or reservoir, piped to the treatment plant, treated, used, and then returned. (b) At the treatment plant, (1) chlorine is added to kill bacteria; (2) alum (aluminum sulfate) is added to coagulate organic particles; and (3) the water is put into a settling basin for several hours to allow the coagulated particles to settle. It is then (4) filtered through sand; (5) treated with lime to adjust pH; and (6) put into a storage water tower or reservoir for distribution to your home.

Both to augment surface water supplies and to obtain water of higher quality, in the past few decades there has been an increasing trend of drilling wells and tapping groundwater. Since 1950, hundreds of cities have drilled huge wells for municipal supplies, and millions of wells have been drilled for individual households in suburbs beyond municipal supply systems. In addition, farmers have turned to the center-pivot irrigation system (see Fig. 9–16c). The use of such wells has risen tremendously in the last 25 years, increasing agricultural production, as noted in Chapter 10, but also consuming huge amounts of groundwater. One system may use as much as 10,000 gallons (40,000 liters) per minute.

From your understanding of the water cycle, you can see that both surface water and groundwater are replenished through the cycle. Therefore, in theory at least, such water represents a sustainable or renewable (self-replenishing) resource, but it is not inexhaustible. If humans attempt to extract amounts that exceed those of natural flow, shortages will occur. Furthermore, as water is diverted from its natural pathway for human use, there will be ecological consequences. Using and diverting water in volumes that lead to shortages or other undesirable consequences are spoken of as *overdrawing*, or *overdraft of*, water resources. Let's turn our attention now to its consequences.

9.5 Overdrawing Water Resources

Because the consequences of overdrawing surface waters differ somewhat from those of overdrawing groundwater, we shall consider these two categories separately. However, because all water is tied together in the same overall cycle, you should be aware of the similarities and interconnections between the two discussions.

Consequences of Overdrawing Surface Waters

Inevitable Shortages. There are wet years and dry years, and surface-water flows vary accordingly. On the average of once every 20 years, surface-water flow may drop to only 30% of its annual average. Therefore, the rule of thumb is that no more than 30% of a river's average flow can be taken out each year without risking a shortfall every 20 years. This rule has not always been heeded, however. In some river systems in the United States, for instance, water demand has grown to 100% (and even *more than* 100%!) of average flow, making water shortages of increasing length and severity inevitable (Fig. 9–14).

Southern California is a case in point. Because of a seven-year drought, much of that part of the state entered the spring of 1992 with reservoirs down to 20%–40% of capacity. If the drought had continued another year, there would have been a severe crisis, but fortunately, rains came and mitigated the situation—a narrow escape. Of course, people blamed the water shortage on the "terrible drought," but was the drought the real cause? Southern California has a desert climate, and prolonged droughts are not abnormal. Their likelihood must be taken into account in long-term planning for sustainability. California's water future must be at the forefront of growth and development planning.

Demands on the Colorado River by dry southwestern states are another example, setting the stage for inevitable shortages. Around the world, there are many cases where tensions are mounting as different nations share common rivers and, hence, compete for the same water (see the "Ethics" essay, p. 224).

Ecological Effects. When a river is dammed and its flow is diverted to cities or croplands, the waterway below the diversion is deprived of that much water. The impact on fish and other aquatic organisms is obvious, but the ecological ramifications go far beyond the river. Wildlife that depends on the water or on food chains involving aquatic organisms—virtually all wildlife—are also adversely affected. Wetlands along many rivers, no longer nourished by occasional overflows, dry up, resulting in a tremendous die-off of waterfowl and other wildlife that depended on these habitats (Fig. 9–15). Fish such as salmon, which swim from the ocean far upriver to spawn, are seriously affected by the reduced water level and have problems getting around the dam, even one equipped with fish ladders (stepwise series of pools on the side of a dam, where the water flows in small falls that fish can negotiate). If the fish do get upriver, the hatchlings have similar problems getting back to the ocean.

The problems extend to **estuaries**, bays in which fresh water from a river mixes with seawater. Estuaries are among the most productive ecosystems on Earth; they are rich breeding grounds for many species of fish, shellfish, and waterfowl. As a river's flow is diverted to other locations, there is less fresh water entering and flushing the estuary. Consequently, the salt concentration increases, profoundly upsetting the ecology. The San Francisco Bay is a prime example. Over 60% of the fresh water that once flowed from rivers into the bay has been diverted for irrigation in the Central Valley and for municipal use in southern California.

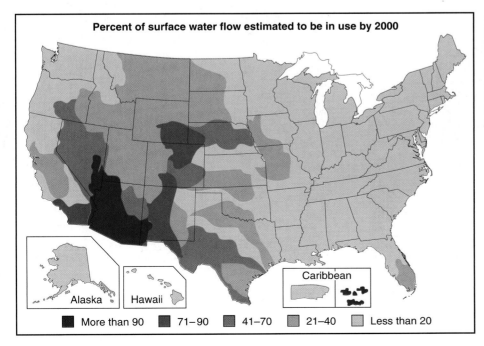

Percent of surface water flow estimated to be in use by 2000

Alaska   Hawaii   Caribbean

More than 90   71–90   41–70   21–40   Less than 20

◀ **FIGURE 9–14** *Water shortage in the U.S.* Droughts occur on an average of every 20 years and may reduce normal water flows by 70%. Therefore, no more than 30% of the average surface-water flow can be counted on to be continuously available. By the year 2000, large areas of the United States will be above the 30% level, making severe, recurring water shortages inevitable.

ETHICS

WATER: WHO SHOULD GET IT?

Many rivers form borders between or flow through a number of states or countries, and, of course, having different governments involved complicates the situation. As water demands rise with expanding populations, the issue of who has the right to the water is leading to ever increasing acrimony between states and between nations.

For example, in a 1922 agreement, the states of the U.S. Southwest divided the flow of the Colorado River among themselves: 7.5 million acre-feet to be shared by Colorado, Wyoming, Utah, and New Mexico—all states in which tributaries of the river originate—and 7.5 million acre-feet to be shared by Arizona, Nevada, and California—states through which the lower river flows. (An acre-foot is the amount of water that will cover 1 acre to a depth of 1 foot, about 325,000 gallons.) Another agreement gave Mexico, where the mouth of the Colorado River empties into the Gulf of California, the right to 1.5 million acre-feet and gave Native Americans whatever amounts they needed. (Their use was almost none at the time.) All went well as long as demands were below the allocated rights. As the population grew, however, each party began demanding its full share. The only problem is that the *total* average flow of the Colorado is only 13.0 million acre-feet (1930 to 1985 average). Even worse, for the past few years, flow has dropped to just 9 million acre-feet.

Another water-allocation problem takes us to the other side of the world. About 90% of the tributaries of the Nile River, which flows through Egypt into the Mediterranean Sea, arise in Ethiopia, and Ethiopia is now considering ways to divert more of the water to ease its drought and famine. Egypt, which depends on the Nile water, has blocked a loan for the Ethiopian project.

The conflict between Israel and Jordan over the West Bank, a piece of territory on the western bank of the Jordan River, is as much about access to water as it is about territory. Many people speculate that the next war(s) in the Mideast may well be over water.

What moral and ethical principles are involved in such disputes? What are the moral dilemmas? How do you think they should be resolved? Who should be the final authorities to enforce agreements?

Without the freshwater flows, salt water from the Pacific has intruded into the bay, with devastating consequences. Chinook salmon have become all but nonexistent; sturgeon, Dungeness crab, and striped bass populations are greatly reduced; and water quality for drinking and irrigation in the region is at risk. New standards will require freshwater flows to be restored to the Delta (the convergence of rivers leading into the bay), but central and southern California will have to give up some water.

The problem is not limited to the United States. The southeastern end of the Mediterranean Sea was formerly flushed by water from the Nile River. Because this water is now held back and diverted for irrigation by the Aswan High

(a)

(b)

▲ FIGURE 9–15 *Ecological effects of diversion* (a) Ducks and geese at Klamath, Oregon. Vast flocks of such waterfowl are supported by wetlands. (b) What happens to these birds when wetlands dry up because of water diversion or a falling water table?

Dam in Egypt, this part of the Mediterranean is suffering the ecological consequences.

The world's most dramatic example of water mismanagement is the Aral Sea, an inland sea in south-central Russia (see the "Earth Watch" essay, p. 226).

Consequences of Overdrawing Groundwater

To augment supplies of high-quality fresh water, humans have increasingly turned to groundwater; it is the largest reservoir of fresh water now available to us. In tapping groundwater, we are tapping a large but not unlimited natural reservoir. Its sustainability ultimately depends on balancing withdrawals with rates of recharge.

In some dry regions, the groundwater found is actually water that accumulated millennia ago, when the climate in the region was wetter. Current rates of recharge are nil. The practice of tapping such pockets of ancient water is frequently spoken of as "mining fossil water" to emphasize that the resource will ultimately be depleted regardless of rates of withdrawal.

Falling Water Tables and Depletion. Rates of groundwater recharge aside, however, the simple indication that groundwater withdrawals are exceeding recharge is a falling water table, a situation that is common throughout the world (see the "Global Perspective" essay, p. 213).

Since irrigation consumes far and away the largest amount of fresh water, depleting water resources will ultimately have its most significant impact on crop production. A prime example is the Great Plains region of Texas, Oklahoma, New Mexico, Colorado, Kansas, Wyoming, and Nebraska. Within this arid region, some 150,000 wells tap the Ogallala aquifer to supply irrigation water to 140 million acres (56 million hectares) (Fig. 9–16). The irrigated farming in

(a)

(b)

(c)

▲ **FIGURE 9–16** *Exploitation of an aquifer.* (a) Pumping up water from the Ogallala aquifer has made this arid region of the United States into some of the most productive farmland in the country. (b) Water is applied by means of center-pivot irrigation, in which water is pumped from a central well to a self-powered boom that rotates around the well, spraying water as it goes. (c) Aerial photograph shows the extent of center-pivot irrigation throughout the region. Groundwater depletion will bring an end to this kind of farming.

EARTH WATCH

THE DEATH OF THE ARAL SEA

In the 1930s, economic planners sitting in thick-walled stone buildings in Moscow set in motion a chain of events that has almost killed an entire sea.

In order to create vast cottonfields in the drylands of Soviet Central Asia, the planners had long irrigation canals dug, fed by the waters of two rivers, the Amu Daria and the Syr Daria, that flow into the inland Aral Sea. In the statistics of the central planners, the project was a huge success. The cotton harvests grew until the Soviet Union became the world's second largest cotton exporter, after China.

But the statistics did not show the effect of diverting most of the river flow away from the Aral Sea. By 1989, the sea was receiving only one-eighth the level of water as in 1960. Its water level had dropped by 47 feet (16 meters), and its volume had shrunk by two-thirds. Once the size of North America's Lake Huron, its total area had diminished by 44%.

The shores of the sea receded, leaving fishing villages tens of miles from the shore. A new desert was created around the sea, with salt strewn in massive dust storms across a vast area.

At the same time, Moscow's demands for cotton cultivation were met with saturation use of pesticides, fertilizers, and herbicides that flowed into the rivers and canals. The population around the sea, deprived of clean drinking water and living on poisoned soil, experienced rising rates of disease and infant mortality.

"The problem of the Aral Sea is very simple," said Igor Zonn, a specialist on the subject, "the Aral Sea will be dead, not soon and not completely, but it will be dead."

Dr. Zonn and fellow Russian scientist Nikita Glazorsky, head of the Institute of Geology and until recently deputy environment minister of Russia, have been working for a long time to try to save the Aral Sea. The Russian scientists say a technical solution for the sea is already well formulated. Much of the water is now wasted because of evaporation and drainage out of unlined irrigation canals and primitive irrigation technology.

At least half the (120-km3) flow of the two rivers could be saved by rebuilding the canals and introducing new irrigation systems. It would be enough to begin to stabilize the Aral Sea at its present level, though not enough to restore it. At the same time there must be a program of health care and a concerted effort to shift the economy away from cotton cultivation, they say.

But such measures require resources and a political will that is not present. The breakup of the Soviet Union and its replacement by a loose commonwealth have given the Russian government an opportunity to dump the problem into the laps of the five former Soviet Central Asian states that form the Aral Sea's water basin. This is the source of anger for Central Asians who see the cotton monoculture imposed by Moscow as a classic example of colonial-style exploitation.

"We are left face to face with the Aral Sea problem," Kazakhstan president Nursultan Nazarbayev said, "meanwhile 97% of the cotton of the Central Asia and Kazakhstan is taken out to the European part of the Commonwealth of Independent States where the employment of 10 million workers depends on cotton use."

Russia has an ethical responsibility to help, agrees Dr. Glazorsky. "All of us live in this system, and we must solve this problem together."

---

this region, which has developed since World War II, now supplies 15% of the nation's total value of wheat, corn, sorghum, and cotton, and about 40% overall of the grain fed to cattle. However, the withdrawal rate is about 24 million acre-feet per year, whereas natural recharge is only about 3 million acre-feet, and water tables are dropping rapidly. In the past 40 years, the water table has dropped 30 m (100 ft) and is lowering at 2 m per year. Irrigated farming has already come to a halt in some sections, and it is predicted that over the next 10 years another 3.5 million acres (1.4 million hectares) in this region will be abandoned or converted to dryland farming (ranching and production of forage crops) because of water depletion.

Although running out of water is the obvious eventual conclusion of overdrawing groundwater, falling water tables have other consequences. Let us now examine some of them.

Diminishing Surface Water. Surface waters are also affected by falling water tables. In various wetlands, for instance, the water table is essentially at or slightly above the ground surface. Dropping water tables result in such wetlands drying up, with the ecological results described earlier. Further, as water tables drop, springs and seeps dry up, diminishing streams and rivers even to the point of dryness. Thus, excessive groundwater removal leads to the same effects as diversion of surface water.

Land Subsidence. Over the ages, groundwater has leached cavities in Earth. Where these spaces are filled with water, the water helps support the overlying rock and soil, but as the water table drops, this support is lost. Then there may be a gradual settling of the land, a phenomenon known as **land subsidence**. The rate of sinking may be 6–12 inches (10–15 cm) per year. In some areas of the San Joaquin Valley in California, land has settled as much as 27 feet (9 m) because of groundwater removal. Land subsidence causes building foundations, roadways, and water and sewer lines to crack. In coastal areas, subsidence causes flooding unless levees are built for protection. For example, where a 4000-square-mile

▲ **FIGURE 9–17** *Sinkhole.* Removal of groundwater may drain an underground cavern until the roof, no longer supported by water pressure, collapses. The result is the sudden development of a sinkhole, such as this one, which consumed a home in Frostproof, Florida, July 12, 1991.

Another kind of land subsidence, the occurrence of a **sinkhole**, may be sudden and dramatic (Fig. 9–17). A sinkhole results when an underground cavern, drained of its supporting groundwater, suddenly collapses. Sinkholes may be at least 300 feet (91 m) across and as much as 150 feet deep. Formation of sinkholes is particularly severe in the southeastern United States, where groundwater has leached numerous passageways and caverns through ancient beds of underlying limestone. An estimated 4000 sinkholes have occurred in Alabama alone, some of which have "consumed" buildings, livestock, and sections of highways.

Saltwater Intrusion. Another problem resulting from dropping water tables is **saltwater intrusion**. In coastal regions, springs of outflowing groundwater may lie under the ocean. As long as a high water table maintains a sufficient head of pressure in the aquifer, there is a flow of fresh water into the ocean. Thus, wells near the ocean yield fresh water (Fig. 9–18a). However, a lowering of the water table or a rapid rate of groundwater removal may reduce the pressure in the aquifer, permitting salt water to flow back into the aquifer and hence into wells (Fig. 9–18b). Saltwater intrusion is a problem at many locations along U.S. coasts.

(10,000-km²) area in the Houston-Galveston Bay region of Texas is gradually sinking because of groundwater removal, coastal properties are being abandoned as they are gradually being inundated by the sea. Land subsidence is also a serious problem in New Orleans, sections of Arizona, Mexico City, and many other places throughout the world.

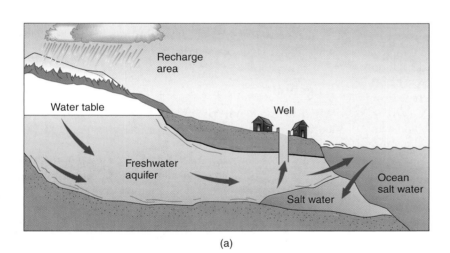

(a)

◄ **FIGURE 9–18** *Saltwater intrusion.* (a) Where aquifers open into the ocean, fresh water is maintained in the aquifer by the head of fresh water inland. (b) Excessive removal of water may reduce the pressure, so that salt water moves into the aquifer.

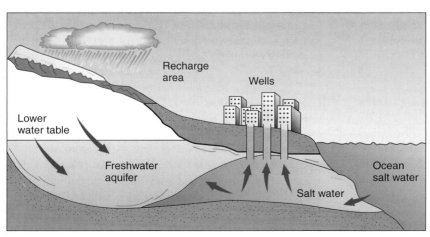

(b)

9.6   Obtaining More Water

Despite the obvious and growing negative impacts of over-drawing water resources, growing populations create an ever-increasing demand for additional water for both irrigation and municipal use. In the United States, 75,000 dams tame wild rivers, provide irrigation water, and harness energy. The Snake River, once home to great salmon runs, now runs from dam to dam—four have been built—and as a consequence, the salmon have all but disappeared.

Increasingly, however, people are recognizing the inevitable trade-offs that occur with such projects and are considering them to be unacceptable. Intense counterlobbying by various environmental organizations has, in recent years, led to the rejection of a number of dam proposals. Existing dams are now being challenged in several states; some have already been dismantled in Vermont and North Carolina, with others under consideration. The Army Corps of Engineers is even studying the possibility of dismantling the Snake River dams in spite of the fact that the dams provide power for 300,000 homes and irrigate 35,000 acres of farmland.

Protection has been accorded to some rivers with passage of the Wild and Scenic Rivers Act of 1968, which protects rivers designated as "wild and scenic" from damming and other harmful operations. Some 9300 miles of rivers have been protected, but more than 60,000 miles qualify for inclusion in the Act. These designated wild and scenic rivers are in many ways equivalent to national parks. Like national parks, they need public supporters and defenders. The primary public supporter and defender of wild and scenic rivers is the association American Rivers.

Bitter controversies between environmentalists and dam-building interests continue and are not limited to the industrialized world. An example currently unfolding is the construction of the Three Gorges Dam across the Yangtze River in China. For the Chinese government, this project represents a centerpiece of their effort to industrialize and join the modern age. But many Chinese scientists have opposed the dam, calling it a major ecological and social disaster. Concerns over sediment, endangered species, and the loss of fertile farmlands in the Yangtze delta downstream have, in the words of one official, "compelled us to take every step with double or triple caution." Over 1.2 million people—including entire cities, farms, homes, and factories—will be displaced and relocated to make way for the 430-mile-long (690 km) reservoir.

A lesson might be taken from the Aswan High Dam, which was constructed across the Nile River in the 1970s to bring Egypt into the modern industrial age. The loss of fisheries and productive wetlands below the dam, the loss of land flooded by the reservoir above, and the unbridled population growth have largely canceled out any gains. The people in general are as bad off as they were before the dam was built.

In conclusion, does it appear that a sustainable future can be found in large-scale water exploitation and diversion projects? Fortunately, there are alternatives. They are based on the conservation and recycling of water.

9.7   Using Less Water

In the past, water has been treated as an inexhaustible resource that can be taken for granted. This viewpoint has led to extravagant and wasteful use of water. A developing-nation family living where water must be carried several miles from a well finds that one gallon per day per person is sufficient to provide for all essential needs, including cooking and washing. Yet, a typical household in America consumes an average of 106 gallons (400 L) per day per person. If all indirect uses are added (including irrigation), this increases to a per capita use of 1350 gallons (5100 L). Similarly, a peasant farmer may irrigate by carefully ladling water onto each plant with a dipper, while typical modern irrigation floods the whole field. This is not to say that we in the developed world should take up developing-world habits, but it does point out that water consumption can be cut back greatly without people suffering hardship. It is through such cutbacks that we have an opportunity to meet our needs without undercutting needs of future generations or natural ecosystems—the essence of sustainable development.

Let's consider some specific measures that are being implemented to reduce water demands.

Irrigation

Where irrigation water is applied by traditional flood or center-pivot systems, about 60% is wasted in evaporation, percolation, or runoff. Several strategies have been employed recently to cut down this waste. One is the *surge flow* method, where microprocessor control allows periodic release of water instead of the continuous flood method; this can cut water use in half. Another is the *drip irrigation* system, a network of plastic pipes with pinholes that literally drip water at the base of each plant (Fig. 9–19). With this system, less than 5% of the water is wasted, although such systems are costly. They have the added benefit of retarding salinization (see Chapter 8). Although drip irrigation is spreading worldwide, especially in arid lands like Israel and Australia, 97% of the irrigation in the United States and 99% throughout the world is still done by traditional flood or center-pivot methods.

The reason for the low changeover is the cost to farmers; it costs about $1000 per acre to install a drip system. In comparison, water for irrigation is heavily subsidized by the government; the farmer pays next to nothing for it. Therefore, it makes financial sense to use the cheapest system for distributing water, even if it is wasteful. Calculations of the construction cost and energy subsidies indicate an annual subsidy of $4.4 billion for the 11 million acres of irrigated land in the western United States, an average of $400 an acre. Eliminating this subsidy would greatly encourage water conservation through the use of more efficient irrigation technologies.

▲ **FIGURE 9–19** *Drip irrigation.* Irrigation is the most consumptive water use. Drip irrigation offers a conservative method of applying water.

▲ **FIGURE 9–20** *The 1.6-gallon commode.* Now required for all new installations, this device saves 10 gallons of water or more a day per person.

Municipal Systems

The water consumption of 105 gallons (400 liters) per day per person in modern homes is mostly used for washing and flushing away wastes: flushing toilets (3–5 gallons per flush), taking showers (2–3 gallons per minute), doing laundry (20–30 gallons per wash), and so on. Watering lawns, filling swimming pools, and other indirect consumption add to this use.

Water conservation has long been promoted as a "save the environment" measure without, it may be added, much effect. Now numerous cities are facing the stark reality that it will be extremely expensive and in many cases impossible to increase supplies by the traditional means of building more reservoirs or drilling more wells. The only practical alternative, they are discovering, is to take real steps toward reducing water consumption and wastage. A considerable number of cities have programs whereby leaky faucets will be repaired and low-flow shower heads and water-displacement devices in toilets will be installed free of charge. Phoenix is paying homeowners to replace their lawns with **xeroscaping**—landscaping with desert species that require no additional watering—and it is becoming a thriving business.

In 1997, the last phase of a regulation authorized by the 1992 National Energy Act took effect, and it became illegal to sell the 3.5-gallon commodes. In their place is the new wonder of the flush world—the 1.6-gallon toilet (Fig. 9–20). As the new toilets came into use, homeowners found that they could no longer depend on an easy flush; in fact, plumbers would usually offer a free plunger with installation of the early 1.6-gallon models. Newer versions, however, work perfectly well, and the new models save 10 gallons or more a day per person. New York City and Los Angeles are providing rebates of $100 to $250 dollars to people who replace their old toilets with the new models.

Also, *gray-water* recycling systems are being adopted in some water-short areas. **Gray water**, the slightly dirtied water from sinks, showers, bathtubs, and laundry tubs, is collected in a holding tank and used for such things as flushing toilets, watering lawns, and washing cars. Going further, a number of cities are using treated wastewater (sewage water) for irrigation, both to conserve water and to reduce pollution of receiving waters (see Chapter 18). If the idea of reusing sewage water turns you off, recall that *all* water is recycled by nature. There is hardly a molecule of water you drink that has not moved through organisms—including humans—numerous times. A number of communities are already treating their wastewater to such a degree that its quality surpasses what many cities take in from lakes and rivers.

9.8 Desalting Seawater

The world's oceans are an inexhaustible source of water, not only because they are vast but because any water removed will ultimately flow back in. Traditionally, seawater has not been used, because of both the cost required to remove the salt and pump it uphill to where it is used, and the availability of adequate sources of natural fresh water. However, with increasing water shortages and most of the world's population living near coasts, there is a growing trend toward **desalination** (desalting) of seawater. More than 7500 desalination plants already exist, primarily in Saudi Arabia, Israel, and other countries of the Middle East.

Two technologies are commonly used for desalination: microfiltration, or reverse osmosis, and distillation.

Small desalination plants generally use the microfiltration process in which great pressure forces seawater through a membrane filter fine enough to remove the salt. Large plants, particularly where a source of waste heat is available (as from electrical power plants, for instance), generally use distillation (evaporation and recondensation of vapor). Additional efficiency is gained by using the heat given off by condensing water to heat the incoming water. Even where waste heat is used, however, the costs of building and maintaining the plant, which is subject to corrosion from seawater, are considerable. Under the best of circumstances, the production of desalinized water costs about $3 per 1000 gallons (4000 L). This is three to six times what most city dwellers in the United States currently pay, but it is still not a high price to pay for drinking water. The high cost might cause some people to cut back on watering lawns and to implement other conservation measures, but most people in the United States could afford it without undue strain.

Irrigation, however, commonly consumes 500,000 gallons (20,000 L) or more per acre for producing one crop. Farmers currently pay as little as 2 cents per 1000 gallons ($10 per crop for water). Thus, it is cost-effective to irrigate crops that bring in only a few hundred dollars per acre. However, paying $3 per 1000 gallons of desalinated water would raise the cost to $1500 per acre of crop. Then there would be the additional pumping costs to get the water from the coast to farmlands. As these costs are added up, to say nothing of future constraints on energy use (see Chapter 13), a future of irrigating croplands with desalinized seawater seems out of the question.

9.9 Storm Water

You have already seen the general aspects of how changing the soil surface changes the infiltration-runoff ratio and in turn alters the rest of the hydrologic cycle. This problem is particularly acute where development results in exchanging porous, water-receiving topsoil for asphalt pavements and rooftops that shed nearly 100% of the water falling on them. Even the soil in lawns is much more compacted than in a natural ecosystem and sheds a high percentage of rainwater.

On the other side of the infiltration-runoff ratio, decreased infiltration also exacerbates saltwater intrusion, land subsidence, and other problems related to falling water tables. The traditional practice was, and in many cases still is, to channel the storm runoff down storm drains. These led to the nearest convenient off-site location to discharge the water, usually the side of a valley or a natural stream bed. The ramifications of this practice are many.

Mismanagement and Its Consequences

The increase in stream flow during rains because of the sudden influx of runoff is well documented, as is the decrease in flow between rains as springs have gone dry because of reduced groundwater recharge (Fig. 9–21).

Most streams in urban and suburban areas that used to flow quietly year-round and supported a diverse biota are now little more than open storm drains alternating between surges of water when it rains and dry, dead beds when it doesn't (Fig. 9–22). Indeed, many such streams have been covered and simply incorporated into the storm-drain system. In

▶ **FIGURE 9–21** *Effect of development on stormwater discharge.* Curves are for similar storms on Brays Bayou in Houston, Texas, before, during, and after development in the early 1950s. Note the increasing height of the surge occurring with the storm and the decreasing volume of flow that occurs later in the cycle. Because of increasing runoff from new development in the watershed, many areas experience flooding where none occurred before.

23-25 September 1941
12-19 May 1953
23-27 June 1960

(a)

(b)

▲ **FIGURE 9–22** *Effect of development on streamflow.*
Before development, this stream maintained a continuous, generally
modest flow of water throughout the year. Now, after development of the
upstream watershed, the stream is dry most of the time. (a) During rains,
there are high surges of water because of increased runoff. (b) A few
hours later the stream is dry again because of depletion of groundwater.

urban areas, great numbers of streams and their tributaries no
longer exist except as underground storm drains.

The surges from increasing runoff have a number of
other effects as well.

Flooding. The increased potential for flooding is self-evident. Countless communities, many of them expensive, new
suburban developments, have experienced flooding with increasing frequency and severity as development has led to
paving more and more of the upstream watershed.

Streambank Erosion. The erosion resulting from poorly
placed storm drain outlets can be horrendous (Fig. 9–23).

The effects continue downstream. Even short of flooding,
the surges of water greatly accelerate erosion of the streambanks, undercutting trees and causing them to topple into
the stream; hence, diverting water against and over the
banks and causing further erosion. While finer soil particles (clay and silt) are carried on to finally settle out in lakes
and bays, coarser materials (sand, stones, and rocks) eroded from gullies below poorly placed storm drains and
streambanks are deposited in the bottom of the stream channel itself. This filling of the stream channel causes the water
to eat away even more at the sides, exacerbating the process.
The result is that the stream channel gets wider and shal-
lower. Indeed, a stream channel may become completely
filled such that the water is diverted and simply floods the
valley floor (Fig. 9–24). The results are topsoil erosion and
the death of many trees because of waterlogged soil. Gradually, a narrow, tree-lined stream is converted into a broad
wash of sand and gravel.

Increased Pollution. Consider all the materials used, carelessly dropped or spilled on the ground, or even worse, disposed of down storm drains. All may wash away with runoff
directly into adjacent natural waterways. Indeed, urban runoff
is now recognized as a major nonpoint source of pollution for
many rivers and estuaries (Chapter 18). Major categories of
contaminants include:

- Nutrients from lawn and garden fertilizer
- Insecticides and herbicides used on lawns and gardens
- Bacteria from fecal wastes of pets

▼ **FIGURE 9–23** *Erosion gully.* The gully seen in this photograph
is the result of erosion caused by water exiting from a storm drain, just out
of the picture in the background, funneling the runoff from a parking lot
onto the hillside. The soil and stones eroded from such gullies add
sediments to, and clog the channels of, waterways below (see Fig. 9–24).

▶ **FIGURE 9–24** *Erosion deposits.* Sediment from upstream erosion fills and clogs stream channels, causing flooding and further erosion of the stream valley as the water is forced to find new pathways. In this photo, you can see how the stream channel has been nearly filled with fine gravel sediment, which eroded from upstream areas.

- Road salt and other chemicals from surface treatments or spills
- Grime and toxic chemicals from settled vehicle exhaust and other air pollution
- Oil and grease picked up from road surfaces or disposed of in storm drains
- Trash and litter carelessly discarded on the ground

How many other things can you think of?

Channelization. What is to be done about the problems arising from increasing stormwater runoff? The old approach to streambank erosion and flooding was **channelization**. The stream bed was dredged, straightened, or gently curved, and lined with rock or concrete to prevent bank erosion (Fig. 9–25). To be sure, a channelized stream carries both water and sediment efficiently and reduces flooding and erosion in the area of the channel. However, you may also observe that it obliterates any semblance of a natural stream ecosystem. The channelized stream is simply an open storm drain. Indeed, it may be covered and made into a storm drain proper, as indicated previously.

The real defeat of channelization, however, came from recognizing that channelized sections simply served to transfer water farther downstream and to create flooding there. The flooding on the Mississippi River in 1993 provides an extreme example. Millions of acres of upstream tributaries were channelized to drain wetlands that normally would have held the water, and extensive levees were constructed downriver to protect floodplains. Heavy spring and summer rains, together with the channelization and levee construction, led to a disastrous flood that swept over levees, engulfed more than 9.3 million acres, and took 50 lives. More than 70,000 people were made homeless by the flood.

A major impetus behind preservation of wetlands, besides protecting their natural ecology, is to maintain their capacity to store storm water. In contrast, development of such wetlands invariably translates into increasing runoff, lowering water tables, and degrading water quality. It is this understanding that is behind the pressure of environmentalists to preserve wetlands, an effort that is threatened by "takings legislation"—legislation that basically says a landowner is free to develop wetlands or any other ecotype unless he or she is compensated for not developing it.

Improving Storm Water Management

A number of states now require stormwater management to be a part of any new development. The management technique most commonly used is to construct a *stormwater retention reservoir* at the low extremity of the site being developed. The **stormwater retention reservoir** simply is a "pond" that receives and holds runoff from the site during storms. From the pond, the water may gradually infiltrate into the soil, or it may trickle out slowly through a standpipe mounted in the pond. Thus, the pond plays a role imitating groundwater storage, and it may also create a pocket of natural wetland habitat supporting wildlife (Fig. 9–26). Large stormwater retention–flood control reservoirs may serve additionally as recreational areas for boating, fishing, and so on.

Techniques of stormwater management on small sites include such things as "wells" and trenches filled with rock

▲ **FIGURE 9–25** *A channelized stream.* To reduce flooding and streambank erosion, many streams and rivers have been channelized; that is, the water course has been straightened and lined with concrete or stone.

FIGURE 9-26 *Stormwater management.* Rather than letting excessive runoff from developed areas cause flooding and other environmental damage, runoff can be funneled into a retention pond, as shown here. Then it can drain away slowly, maintaining natural streamflow, or recharge groundwater. The retention pond may be designed to retain a certain amount of water and thus create a pocket wildlife habitat in an otherwise urban or suburban setting.

that receive water and allow it to percolate; terraces that receive and hold water on the "steps" and allow it to infiltrate; and rooftops and parking lots designed to "pond" water and let it trickle away slowly. Of course, storm water collected in any sort of reservoir can act as a source of water for watering lawns, washing cars, and other nondrinking purposes.

Water Stewardship

Our study of the hydrologic cycle shows us that nature provides a finite flow of water through each region. When humans come on the scene, this flow of water is inevitably divided between the needs of the existing natural ecosystems and the agricultural, industrial, and domestic needs of humans, with human needs usually being met first. In fact, recent calculations indicate that humans now use 26% of total terrestrial evapotranspiration and 54% of the accessible precipitation runoff. We are major players in the hydrologic cycle, and as we have seen, many facets of our use of water are unsustainable.

We have also seen many opportunities for conserving and recycling water. The question is, How can we get ourselves and others to implement these measures short of social or ecological disaster? Laws, litigation, and the courts have been and are being used to settle disputes, but we can see that this is probably the most costly and time-consuming method; parties are left embittered and antagonistic, and wins are questionable. For example, we saw that the recovery of Mono Lake will be long and only partial despite the win by environmentalists.

We might at least imagine a better way. With suitable leadership, the contending parties might be brought together and provided with information regarding such questions as: How much water may be removed from a given source without disrupting natural ecosystems? If that amount is less than desired, what are the conservation or recycling measures available to reduce needs for irrigation and cities? Is irrigating crops really essential? (In a number of cases, cities are buying farm lands simply for the water rights.) How can revenues be raised and costs be shared to implement such measures? In short, how can we best balance our direct needs for water with the broader objectives of Earth stewardship and creating a sustainable future?

The report of the President's Council on Sustainable Development addresses these concerns with some helpful policy recommendations:

- Executive orders should be issued by the President and state governors directing federal agencies to promote voluntary multi-stakeholder collaborative approaches toward managing and restoring natural resources.
- Public and private leaders, community institutions, nongovernmental organizations, and individual citizens should take collective responsibility for practicing environmental stewardship.
- The federal government should play a more active role in building consensus on difficult issues and identifying actions that would allow stakeholders to work together toward common goals.
- Government agencies, conservation groups, and the private sector should expand the use of ecosystem approaches to natural-resource management by using collaborative partnerships, developing compatible information databases, and carrying out appropriate incentives for responsible stewardship.

As a case study example of how this could be accomplished, the council reported on the protection of coastal wetlands in Louisiana. Early in the process of addressing erosion of the wetlands, the Louisiana Coastal Wetlands Interfaith Stewardship Group began getting involved with state, business, and environmental agencies. The religious community helped keep up front the message of environmental stewardship and social justice for the coastal people as different solutions to the problem were being considered. As a result of the collaboration of all these groups, the people of Louisiana voted in 1989 to establish the Louisiana Wetlands Conservation and Restoration Trust Fund. Subsequently, $1.5 billion in funds was appropriated by Congress to help restore Louisiana's wetlands.

In conclusion, the choice is either to find similar, sustainable solutions that meet the economic needs of people and the ecological needs of the land that supports us, or to continue disputes; in which case, all parties ultimately stand to lose.

ENVIRONMENT ON THE WEB

WATER CONSERVATION: HOW FAR CAN WE GO?

Arid regions have long recognized the need for careful use of precious water resources. In recent years, even moist temperate zones have come to realize the value of water conservation. As climatologists debate the likelihood of global warming, water managers are confronting the need for careful planning for water supplies into the twenty-first century. In Southern Canada, for instance, water managers are facing the possibility that river levels may drop as much as 1 m under worst-case climate projections. In the United Kingdom, water companies are analyzing expected changes in the distribution of precipitation and associated impacts on water supply. Faced with the potential for shortages, water managers must either develop new water sources or find ways to encourage water users to conserve.

In many parts of North America, domestic water use now exceeds 600 L per person per day, yet in dry African nations people consume only 8–20 L per person per day. Clearly, North American water usage is often wasteful compared to other water-poor nations. But how much reduction is feasible as a resource conservation measure?

Associations such as the American Water Works Association and individual municipalities suggest that 25% of domestic water use could be saved through simple measures such as water-saving toilets, shower heads, and leak controls, without noticeable impact on current lifestyles. Some estimates place potential savings as high as 50%. But despite its obvious advantages for resource protection, water conservation has been a hard sell in many municipalities, especially those in water-rich areas. And some uses, such as fire-fighting and electric power generation, require immense volumes of water that cannot easily be reduced without jeopardizing the benefits of such water use. Industrial uses of water can also be massive, particularly in older operations.

One of the simplest and most effective measures available for water conservation is water metering. In a metered system, a small device measures usage, and the consumer is billed for each unit of water used. So-called "increasing block" pricing, in which the consumer pays a proportionately higher rate with higher use, is a particularly effective way to encourage water efficiency. Most new residential developments in North America have water metering, but in some areas of the world, such as the United Kingdom, metering has met with strong resistance from politicians and the public alike. To be effective, water-conservation strategies must therefore be developed in the context of local conditions — including the needs and priorities of the users themselves.

Web Explorations

The Environment on Web activity for this essay describes some of the debate around residential water conservation and the tools for achieving efficient water use. Go to the Environment on the Web activity (select Chapter 9 at http://www.prenhall.com/nebel) and learn for yourself:

1. about the debate surrounding residential water conservation;
2. what tools are available to achieve efficient water use; and
3. how your local water use compares to that of neighboring areas.

A suggested time frame for each of these exercises is 10–30 minutes.

REVIEW QUESTIONS

1. How does water change its state with changes in energy (heat)?

2. What are the two processes that result in natural water purification? State the difference between them.

3. Describe how a Hadley cell works, and explain how Earth's rotation creates the trade winds.

4. Why do different regions receive different amounts of precipitation?

5. Define precipitation, infiltration, runoff, capillary water, transpiration, evapotranspiration, percolation, gravitational water, groundwater, water table, aquifer, recharge area, seep, and spring.

6. Use the terms defined in Question 5 to give a full description of the hydrologic cycle, including each of its three "loops." What is the water quality (purity) at different points of the cycle? Explain the reasons for the differences.

7. How does changing Earth's surface by deforestation, for example, change the pathway of water? How does it affect streams and rivers? Humans? Natural ecology?

8. Describe how groundwater and wetlands act as natural reservoirs to moderate the flow of streams and rivers in wet periods and dry periods. What impact will the development of wetlands have on water quality and quantity?

9. What are the three major categories of water use? From what point(s) in the cycle is water generally withdrawn? How is it withdrawn and distributed?

10. What are the indicators of overdrawing water supplies? What are the human, ecological, and environmental consequences of overdrawing surface waters? Of overdrawing groundwater?

11. Describe how water demands might be reduced in agriculture, industry, and households.

12. What is meant by stormwater runoff, and what causes it? What are the consequences of stormwater runoff for groundwater? For water quality in streams and rivers? For streamflow during rains? For streamflow between rains? For the natural environment and ecology?

13. Describe techniques now being implemented to manage stormwater runoff.

14. Are new policies called for to achieve sustainable water supplies? How is the government addressing this need? What policies would you support or promote?

THINKING ENVIRONMENTALLY

1. Pretend you are a water molecule, and describe your travels through the many places you have been and might go in the future as you make your way around the cycle time after time. Include travels through organisms.

2. Commercial interests wish to create a large new development on what is presently wetlands. Have a debate between those representing commercial interests and environmentalists who will bring out the environmental and economic costs of development. Work toward negotiating a compromise.

3. Describe how many of your everyday activities, including your demands for food and other materials, add pollution to the water cycle or alter it in other ways. How can you be more stewardly with your water consumption?

4. Describe the natural system that maintains uniform streamflow despite fluctuations in weather. How do humans upset this regulation? What are the consequences?

5. An increasing number of people are moving to the arid Southwest despite the fact that water supplies are already being overdrawn. If you were the governor of one of these states, what policies would you advocate regarding this situation?

6. Commercial interests wish to develop a golf course on presently forested land next to a reservoir used for city water. Describe the impacts this development might have on water quality in the reservoir.

7. Divide the class into five groups representing these various interest groups: environmentalists, farmers, city officials responsible for domestic water, industrialists, and developers. The actual supply of water will support only 50% of prospective demands. Negotiate a compromise that will honor the interests of all parties.

WEB REFERENCES

On-line resources for this chapter can be found on the World Wide Web at: **http://www.prenhall.com/nebel**. Click on Chapter 9 on the chapter selector.

10

The Production and Distribution of Food

∿∿∿∿∿∿

Key Issues and Questions

1. In industrialized societies, an agricultural revolution has taken place that has radically affected the practice of farming and its environmental impact. What is industrialized agriculture, how did it develop, and what are its environmental costs?

2. The agricultural revolution has been transferred to the developing world in a process called the Green Revolution. What are the origins and impacts of the Green Revolution?

3. Agriculture in most of the developing world is still practiced in traditional ways, called subsistence agriculture. How important is subsistence agriculture?

4. Continued population growth puts pressure on agricultural practice to keep producing more food. What are the prospects for increasing food production?

5. There is a lively and important world trade in foodstuffs. What are the global patterns of food trade, and what are its consequences?

6. Hunger and malnutrition still plague human societies. What is the extent of hunger, malnutrition, and undernutrition in the world?

7. Famines continue to occur. What are the causes of famine, and which geographical areas are affected?

8. Food aid is distributed to countries all over the world. Is food aid necessary? How is aid distributed?

9. Agricultural sustainability is a desirable goal. How can the four principles of ecosystem sustainability be applied to agriculture—both in developed and developing countries?

There is a land where hills and valleys were once covered with green forests and fields. Now the hillsides are cut by erosion gullies that have long since lost their soil. Trees are almost totally gone—less than 2% of the forests remain. Present vegetation on the arid, mountainous landscape is continually grazed by more than a million goats and sheep. Over half of the original cropland has been lost to erosion. To produce charcoal for fuel, people cut the mangroves along the coast and the mango trees in the hilly uplands; even the tree stumps remaining on the deforested land are dug up for charcoal.

Even with substantial food and economic aid ($400 million in 1997), the average person here receives only 80% of daily food needs. One in three children is chronically malnourished. Unemployment is at 70%; the per capita gross domestic product is $330, and the economic growth rate is negative. Until recently, a corrupt military regime maintained control over a human population now numbered at 7.5 million with an annual growth rate of 2.1%. The people fled their country by the thousands in makeshift boats, refugees from political and economic oppression. Behind them, they left a land that can only be described as an environmental wasteland. That land is Haiti (Fig. 10–1).

The sad plight of the Haitians is reflected in other areas throughout the world, where three decades of rapid population growth and declining food harvests have left hundreds of millions dependent on food aid. One in five in the developing world remains undernourished. As the world population continues its relentless rise (Fig. 6–1), no resource is more vital than food. Can the world's farmers and herders produce food

◀ **Marketplace in Bolivia, South America.** Subsistence farmers from the countryside bring their produce to these open air markets in the cities to earn much-needed cash.

fast enough to keep up with population growth? After all, our planet holds only a finite amount of the resources needed for food production: suitable land, water, energy, and fertilizer.

By many measures, human societies have done very well at putting food on the table. More people are being fed than ever before, with more nutritional food. In the last 25 years, world food production has more than doubled, rising more rapidly than population. A lively world trade in foodstuffs forms the bulk of economic production for many nations. Optimists would say that although regions of chronic hunger and malnutrition—and occasional famines—exist, these are the exceptions to an otherwise remarkable accomplishment. They are convinced that when the world population levels off (at 11 billion?), enough food will still be produced to sustain it. Others question whether that is at all possible, considering that we are not doing very well at feeding *half* that number.

This chapter takes a look at food-production systems, problems surrounding the lack of food, and the ways in which to build sustainability, as well as food security and justice, into the food arena.

▲ **FIGURE 10–1** *The Haitian landscape.* Decades of rapid population growth and intense poverty have left the Haitian landscape in a state of ruin, as forests have been cut down, grasslands overgrazed, and croplands eroded.

10.1 Crops and Animals: Major Patterns of Food Production

The Neolithic Revolution saw the introduction of agriculture and animal husbandry some 10,000 years ago, which probably did more than anything else to foster the development of human civilization. Virtually all of the major crop plants and domestic animals were established in the first thousand years of agriculture. Between 1450 and 1700, world exploration and discovery led to an exchange of foods that greatly influenced agriculture and nutrition of the Western world. From the New World came potatoes, maize (corn), beans, squash, tomatoes, pineapples, and cocoa. Rice came from the Orient. In exchange, the Europeans brought to the New World wheat, onions, sugar cane, and a host of domestic animals—horses, pigs, cattle, sheep, and goats. With the advancing capabilities of science and technology by the 1800s, the stage was set for a remarkable change in agriculture.

The Development of Modern Industrialized Agriculture

Until 150 years ago, the majority of people in the United States lived and worked on small farms. Human and animal labor turned former forests and grasslands into systems that produced enough food to supply a robust and growing nation (Fig. 10–2a). Farmers used traditional approaches to combat pests and soil erosion: Crops were rotated regularly, many different crops were grown, and animal wastes were returned to the soil. The land was good, and farming was efficient enough to allow a substantial segment of the population to leave the farm and join the growing ranks of merchants

and workers living in cities and towns. Then in the mid-1800s the Industrial Revolution came to the United States, and it had a major impact on farming.

The Transformation of Traditional Agriculture. The Industrial Revolution contributed to a revolution in agriculture so profound that today less than 3% of the U.S. work force produces enough food for all the nation's needs plus a substantial amount for trade on world markets (Fig. 10–2b). Indeed, this revolution has achieved such gains in production that the United States has had to formulate policies to cope with surpluses of many crops.

Virtually every industrialized nation has experienced this agricultural revolution. The pattern of developments in U.S. agriculture could just as well describe that of France, Australia, or Japan. Crop production has been raised to new heights, doubling or tripling yields per acre (Fig. 10–3). However, there are environmental costs. According to many agricultural experts, expanding production has reached, or even exceeded, sustainable limits. Let us examine the components of the agricultural revolution.

Machinery. The shift from animal labor to machinery has created a dependency on fossil-fuel energy that adds significantly to the energy demands of the industrial societies (Chapter 13). However, as long as oil continues to flow at its unusually low costs to consumers, this is an affordable expense. But as we have seen, continued use of farm machines for plowing, planting, and harvesting causes some soil compaction (Chapter 8).

Land Under Cultivation. Before 1960, much of the increased production in the United States came from bringing new land into production. Since then, attempts have been made to increase the land used to raise grain, but these

(a)

(b)

▲ **FIGURE 10–2** *Traditional vs. modern farming.* (a) Traditional farming practiced in Arkansas. For hundreds of years, traditional practices on American farms supported the growing U.S. population. (b) Modern agricultural practice illustrated by a combine harvesting wheat in America's Midwest.

new land areas were not well suited for agriculture and have been abandoned because erosion or depletion of water resources has rendered them no longer productive. Current farm policy—the Conservation Reserve Program (CRP)—reimburses farmers for "retiring" erosion-prone land and planting it to produce trees or grasses. The law allows up to 36 million acres to be placed in this program; land in the program currently totals 28 million acres. CRP acres may be returned to cultivation. In 1998, some 5 million acres was returned to production of corn, soybeans, and wheat.

Essentially all of the good cropland in the United States is now under cultivation or held in short-term reserve. Worldwide, cropland on a *per capita* basis is on the decline as population continues to rise. Any significant expansion in cropland will come at the expense of forests and wetlands, which are both economically important and ecologically fragile.

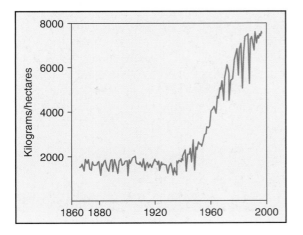

▲ **FIGURE 10–3** *U.S. corn yield.* This graph demonstrates two phenomena: the long-term rise in yields and the effects of droughts in 1970, 1973, 1980, 1983, and 1988 (8000 kg/hectare = 3.5 tons/acre).

Fertilizers and Pesticides. When fertilizers were first employed, 15 to 20 additional tons of grain were gained from each ton of fertilizer used. Between 1950 and 1990, worldwide fertilizer use rose tenfold. Now, however, farmers in many countries are applying near-optimal levels of fertilizers, and the gain is less than 2 tons of grain per additional ton of fertilizer applied. When levels of fertilizer are too high, plants become more vulnerable to attack by insects and other pests; when fertilizer is washed away, the result is water pollution (Chapter 18). Worldwide use of fertilizers has resumed its rise, following a six-year decline traced to economic conditions in the countries of the former Soviet Union. Most of the current increase is in "underfertilized" countries like China and India.

Chemical pesticides have provided significant control over insect and plant pests, but the pests have become resistant to most of the pesticides as a result of natural selection. Also, there are efforts to reduce the use of pesticides because of side effects to human and environmental health. Many of the chemicals in use have not been adequately tested in terms of human concerns and their potential to cause genetic defects. As we shall see in Chapter 17, progress is being made toward developing natural means of control that will be environmentally safe, but these new methods will be unlikely to increase yields.

Irrigation. Worldwide, irrigated acreage increased about 2.6 times from 1950 to 1980 and by 1995 represented 16% of all cropland—some 255 million hectares—and produced 36% of the world's food. It is still expanding, but at a much slower pace because of limits on water resources (irrigation is by far the greatest consumer of water). More ominous, much present irrigation is not sustainable, because groundwater resources are being depleted. In addition, production is being adversely affected on as much as one-third of the world's irrigated land, because of waterlogging and accumulation of salts in the soil, consequences of irrigating where there is poor drainage (Chapter 8).

High-yielding Plant Varieties. Several decades ago, plant geneticists developed new varieties of wheat, corn, and rice that gave yields of double to triple those of traditional varieties. This was accomplished by selecting strains that diverted more of the plants photosynthate (photosynthetic product) to the seed and away from the stems, leaves, and roots. As these new varieties were introduced throughout the world, production soared. However, most of their potential has now been realized, as the seeds of modern wheat, rice, and corn now receive more than 50% of the photosynthate, close to the calculated physiological limit of 60%. On the minus side, the widespread use of genetically identical crops has given rise to major pest damage, as pests have become adapted to the new varieties and resistant to pesticides.

The Green Revolution. The same technologies that gave rise to the agricultural revolution in the industrialized countries were eventually introduced into the developing world. There they gave birth to the remarkable increase in rice and wheat production called the **Green Revolution**.

In 1943, the Rockefeller Foundation sent agricultural expert Norman Borlaug and three other U.S. agricultural scientists to Mexico, with the objective of exporting U.S. agricultural technology to a less developed nation that had serious food problems. Their aim was to improve the traditional crops grown in Mexico, especially wheat. Mexican wheat was well adapted to the subtropical climate, but it gave low yields and responded to fertilization by growing very tall stalks that were easily blown over. Using wheat from other areas of the world, Borlaug and his coworkers bred a dwarf hybrid with a large head and a thick stalk; the hybrid did well in warm weather when provided with fertilizer and sufficient water (Fig. 10–4). The program was highly successful. By

the 1960s, Mexico had closed the gap between food production and food needs, wheat production had tripled, and Mexican wheat appeared on the export market.

Research workers with the Consultative Group on International Agricultural Research (CGIAR) extended the work done in Mexico, introducing both high-yielding wheat and high-yielding rice to other developing countries. To cite one success, Borlaug induced India to import hybrid Mexican wheat seed in the mid-1960s, and in six years, India's wheat production tripled. Within a few years, many of the world's most populous countries turned the corner from being grain importers to achieving stability and, in some cases, even becoming grain exporters. Thus, while the world population was increasing at its highest rate (2% per year), rice and wheat production underwent increases in yields of 4% or more per year. This Green Revolution has probably done more than any other single scientific or other achievement to prevent hunger and malnutrition. Norman Borlaug was awarded the Nobel Peace Prize in 1970 in recognition of his contribution.

The high-yielding grains are now cultivated throughout the world and have become the basis of food production in China, Latin America, the Middle East, southern Asia, and, of course, the industrialized nations. Because the technology raises yields without requiring new agricultural lands, the Green Revolution has also held back a significant amount of deforestation in the developing world. But as remarkable as it was, the Green Revolution is not a panacea for all of the world's food-population difficulties, for the following reasons:

1. Most of its potential has been realized; many of the most populous countries are reaching a plateau in their grain production and in acreage planted to high-yielding varieties, while their populations continue to increase.

(a)   (b)

▲ **FIGURE 10–4** *Traditional vs. high-yielding wheat.* Comparison of (a) an old variety of wheat, shown growing in Rwanda, with (b) a new, high-yielding variety of dwarf wheat growing in Mexico.

◄ **FIGURE 10–5** *Subsistence farming.* Mexican farmer plowing marginally productive land with a team of oxen. Subsistence farming feeds more than 1.4 billion people in the developing world.

2. Without irrigation, it does not work in drought-prone lands. It also requires constant inputs of fertilizers, pesticides, and energy-using mechanized labor, all of which are often in short supply in developing countries.

3. Because it is patterned after agriculture in the developed world, Green Revolution agriculture tends to benefit larger landholders. More food is raised by a smaller farm work force, causing many farm laborers and small landholders to become displaced and migrate to the cities, joining the ranks of the unemployed.

4. The most important African food crops (sorghum, millet, and yams) are not commonly used in the developed world, so they have not benefited from the Green Revolution technology. The fact is, the Green Revolution has had little impact on the large part of the developing world, where another kind of agriculture—subsistence agriculture—is practiced.

Subsistence Agriculture in the Developing World

In most of the developing world, plants and animals continue to be raised for food by *subsistence farmers*, using traditional agricultural methods. These farmers represent the great majority of the rural populations. **Subsistence farmers** live on small parcels of land that provide them with the food for their households and, it is hoped, a small cash crop. From the point of view of the modern world, such farmers are very poor, although some do not consider themselves to be so. Like past agricultural practice in the United States, subsistence farming is labor intensive and lacks practically all of the inputs of industrialized agriculture. Also, it is often practiced on marginally productive land (Fig. 10–5).

Typically, a family owns a small parcel of land for growing food and maintains a few goats, chickens, or cattle. Such a family is making the best use of very limited resources, and very often the people are adapted well enough to the prevailing social and environmental conditions to provide a livelihood for a household. An important fact to remember, however, is that subsistence agriculture is practiced in regions experiencing the most rapid population growth, even though this kind of agriculture is best suited for low population densities. An estimated 1.4 billion people in Latin America, Asia, and Africa—over one-third of the people there—are sustained by subsistence agriculture.

The pressures of population and the diversion of better land to industrialized agriculture often lead to practices that are at best nonsustainable and at worst ecologically devastating. In many regions in developing countries, woodlands and forests are cleared for agriculture or removed for firewood and animal fodder, leaving the soil susceptible to erosion and forcing the gatherers to travel farther and farther from their homes. The scarcity of firewood leads the residents to burn animal dung for cooking and heat, thus diverting nutrients from the land. Erosion-prone land suited only to growing grass or trees is planted to produce annual crops. Good land is forced to produce multiple crops instead of being left fallow to recover nutrients. Growing populations force the continued subdivision of existing land, which diminishes the land's ability to support each household. All these factors tend to increase the poverty characteristic of populations supported by subsistence agriculture, and, in a relentless cycle (see Fig. 7–7), the added poverty in turn puts increased pressures on the land to produce food and income.

Because subsistence agricultural practice varies with the local climate and with local knowledge, it is difficult to draw

▲ **FIGURE 10–6** *Slash-and-burn agriculture.* Countryside in Mindanao, Philippine Islands, showing the results of sustainable slash-and-burn agriculture. Cultivated fields are interspersed with trees and natural areas in a diverse ecosystem.

sweeping generalities. In some areas, subsistence agriculture involves shifting cultivation within tropical forests—often called slash-and-burn agriculture. Research has shown that such a practice can be sustainable (Fig. 10–6). The cultivators create highly diverse ecosystems, where the cleared land supports a few years of crops and gradually shifts into agroforestry—a system of tree plantations with different ground crops arising as the trees grow.

In other areas, subsistence farmers are showing remarkable success in adapting to the changing needs of local societies as they are forced to support expanding populations on the same land. For example, in Kenya, land in the Machakos region was seriously degraded during the 1930s. Now the land has recovered and supports a population six times larger, as the 1.5 million people there have diversified their agriculture and practiced soil and water conservation.

Animal Farming and Its Consequences

Raising livestock—sheep, goats, cattle, buffalo, and poultry—has many parallels to raising crops (Table 10-1), and there are also direct connections between the two. Fully one-fourth of the world's croplands are used to feed domestic animals; in the United States alone, 70% of the grain crop goes to animals—

Table 10-1 Parallels Between Plant and Animal Farming		
	Plant	**Animal**
Major products	Grains, fruits, and vegetables for food	Meat, dairy products, and eggs for food
Other important products	Oils, fabrics, rubber, specialty crops (spices, nuts, etc.)	Labor, leather, wool, manure, lanolin
Modern practices	Industrialized agriculture on former grasslands and forests	Ranching, dairy farming, and stall feeding
Traditional practices	Subsistence agriculture on marginally productive lands	Pastoral herding on nonagricultural land
Current global land use	3.7 billion acres (1.5 billion ha, or 11% of land surface)	8.4 billion acres (3.4 billion ha, or 26% of land surface)

half the cultivated acreage. The care, feeding, and "harvesting" of the estimated 15 billion domestic animals constitute one of the most important economic activities on the planet. The primary force driving this livestock economy is the large number of the world's people who enjoy eating meat and dairy products—primarily, most of the developed world and growing numbers of people in less developed nations. As with crop farming, however, there are two patterns. In the developed world and on ranches in the developing world, livestock are raised in large herds and often under factory-like conditions. In rural societies in the developing world, however, livestock are raised on family farms or by pastoralists who are subsistence farmers.

Industrial-style animal farming can affect the environment in a host of nonsustainable ways. Since so much of the plant crop is fed to animals, all of the problems of industrialized agriculture apply to animal farming. In addition, rangelands are susceptible to overgrazing, either because of mismanagement of prime grazing land or because the land used for grazing is marginal dry grasslands, used in that manner because the better lands have been converted to producing crops. For example, overstocking on the rangelands of the western United States has reduced the carrying capacity by an estimated 50%. Much of the western rangeland is public land leased at subsidized fees that easily lead to overgrazing (Chapter 8). Another serious problem is the management of animal manure. In developing countries, manure is a precious resource that is used to renew soil fertility, build shelters, and provide fuel. In the developed countries, it is a wasted resource. Close to 1.3 billion tons of animal waste is produced each year in the United States, some of which leaks into surface waters and contributes to fish kills, pathogen contamination, and algal proliferation. With factory farms on the sharp increase, wastes from manure and meat processing are either bypassing or overwhelming the often inadequate treatment systems available and polluting the nation's waterways. The EPA asserts that animal-based agriculture is the most widespread source of pollution in the nation's rivers.

In Latin America, more than 49 million acres (20 million ha) of tropical rain forests have been converted to cattle pasture. Even though most of this land is best suited for growing rain forest trees, some of it could support a rural population of subsistence farmers producing a diversity of crops. Instead, it is held by relatively few ranchers who own huge spreads. According to the Intergovernmental Panel on Climate Change, deforestation and other land-use changes in the tropics releases an estimated 1.6 billion tons of carbon to the atmosphere annually, contributing a significant amount of carbon dioxide to the greenhouse effect.

Also, because their digestive process is anaerobic, cows and other ruminant animals annually belch and eliminate some 100 million tons of methane, another greenhouse gas (see the "Environment on the Web" essay, Chapter 21, p.522). Anaerobic decomposition of manure leads to an additional 30 million tons of methane per year. All this methane released by livestock makes up about 3% of the gases causing global warming (Chapter 21).

Even though their animals also contribute to the methane problem, it is callous to fault the subsistence farmers whose domestic animals enhance their diet and improve their quality of life. In fact, one of the most important kinds of sustainable development aid brought to rural families is the gift of a cow or a few goats or rabbits, as carried out by Heifer Project International (HPI). Working in 115 countries, HPI has distributed large animals, beehives, fowl, and fish fingerlings to families. HPI works with grassroots groups of local people who oversee a given project. The project must have the following characteristics: genuine need, environmental improvement, effective training, and a commitment to passing on the gift in the form of livestock offspring to other needy people.

The lives of millions in the developing world are tied very directly to the animals they raise, and their impact on the environment is often sustainable. Animals that are well managed can enhance the soil and enable rural farmers to maintain a balanced farm ecosystem. In summary, animal farming is far more likely to be sustainable in the context of rural farms and pastoral herding, whereas in the beef ranches and hog pens of the developed world, there continues to be the pressing need to address problems of pollution, overgrazing, and deforestation.

Prospects for Increasing Food Production

One way to evaluate the ability of regions to meet their food needs is to examine per capita food production—that is, food production divided by population numbers (Fig. 10–7). The data show that per capita food production is rising in some regions but not in others (Fig. 10–7a). Two regions stand out as being in serious trouble, for entirely different reasons: Africa, and the countries of the former Soviet Union. On balance, food production per capita in Africa has been on a downward course since 1970, a consequence of rapid population growth and an agriculture that has not undergone modernization. The breakup of the Soviet Union led to the loss of government subsidization of agriculture as well as a chaotic economy in most of the regions. Most observers believe that these countries will get back on track soon.

On a world basis, per capita grain production (grains are the staple food for most people and are the only foodstuffs for which global data are available) has been slowly declining since 1984 (Fig.10–7b). In spite of a record harvest in 1997 of 1.88 billion tons, per capita production registered a slight decline due to population growth. Nonetheless, there is no shortage of food in the world as a whole: Food production appears to be keeping up with demand. Indeed, production of meat has risen dramatically in recent years, indicative of the dietary shifts to higher protein consumption as living standards continue their rise. The greatest concern continues to be about the future: How will we manage to produce enough food for twice as many people in a few decades?

Because the end of the increases in yield brought about by the Green Revolution is in sight, and essentially all arable land in the world either is now or has recently been in cultivation, we have only two prospects for increasing food production:

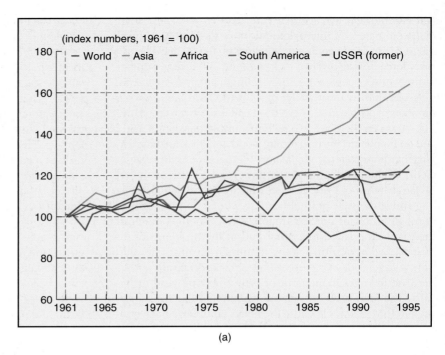

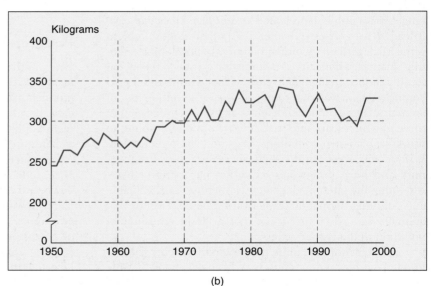

▶ **FIGURE 10–7** *Per capita food production.* (a) Changes in per capita food production by region (1961 = 100). (b) World per capita grain production, 1950–1998.

(1) continue to increase crop yields, or (2) begin growing food crops on land that is now used for feedstock crops or cash crops.

As we have seen, a dramatic rise in grain yields in the developing countries was the major accomplishment of the Green Revolution. Surprisingly, grain yields have continued to rise in some of the developed countries; in France, wheat yields have quadrupled since 1950, reaching 7 tons per hectare. Rice yields in Japan have risen 67% in the same period. The genetic potential exists for yields as high as 14 tons per hectare or higher for wheat. Can we expect yields to continue to increase up to their genetic potential?

Actually, the great differences in yields of grain between regions (Table 10-2) have little to do with the genetic strains used. Egypt and Mexico, for example, achieve higher yields

than the United States because most of their wheat is irrigated, while U.S. wheat is rain-fed and grown on former grasslands. Australian wheat yields are even lower as a consequence of their sparse rainfall. Once the agricultural land is planted with high-yielding strains and fertilized to the maximum, other factors place limits on productivity—soil, rainfall, and available sunlight. These environmental limits are a reminder that agricultural sustainability is highly dependent on soil and water conservation (Chapter 8) and the weather. For example, the climate between 1940 and 1980 was exceptionally stable and ideal for agriculture in most parts of the world. Then three severe droughts occurred, drastically affecting harvests in North America in 1980, 1983, and 1988. Just the opposite—cool, rainy weather—in the

Table 10-2 Annual Wheat Yield per Hectare in Various Countries	
United Kingdom	7.7
France	6.8
Egypt	5.6
Mexico	4.1
China	3.6
Poland	3.4
Ukraine	2.7
India	2.5
United States	2.5
Canada	2.3
Argentina	2.1
Pakistan	2.0
Australia	1.6
Russia	1.4
Kazakhstan	0.6

United States and northeast Asia brought a decline in grain harvest in 1993. Fortunately, the world, and particularly the United States, had ample stocks from previous surpluses. A year of ideal weather in North America in 1994 produced bumper crops, driving prices down and prompting farmers to put their grains in storage. Then 1996 and 1997 saw record grain harvests worldwide. This seesaw harvest situation points to the instability in food production that makes future planning practically impossible. Another imponderable in future food production is global climate change. As the predicted warming of the twenty-first century occurs, it is impossible to predict how rainfall patterns will change.

In the developing world, sub-Saharan Africa has the greatest need for increasing crop yields. With much of its land arid and populations continuing to grow at exponential rates, this region continues to fall behind in per capita food production and is becoming increasingly dependent on food imports. Yields would have to increase at 3.3% annually just to keep up with projected population growth. Yet, the potential is there to make great progress with some very basic steps. With the aid of an extension service promoted by former President Jimmy Carter, Ethiopia harvested record crops in 1995–96, showing a 32% increase in production and a 15% increase in yield in one year. The key? Use of a simple fertilizer providing nitrogen and phosphorus to the nutrient-starved soils. Poor agricultural support and a lack of capital to purchase fertilizer and new seeds are preventing many other African countries from making similar improvements.

What of the possibility of switching from the production of feed grain and cash crops to food for people? Many observers have pointed out the inefficiency of using grain to feed livestock and then eating the livestock or their products (milk, eggs, butter). Fully 70% of domestic grain in the United States is used to feed livestock; the percentages drop over other regions of the world, in proportion to the economic level of the region. Sub-Saharan Africa and India, for example, use only 2% of their grain for livestock feed. Feed grain can be considered a buffer against world hunger; if the food supply becomes critical, it might force more of the world's people to eat lower on the food chain (less meat, more grain). As we have seen, however, the trend is in exactly the opposite direction. Converting land use from cash crops to food crops is possible but is a complex undertaking, for it involves such issues as land reform and maintaining a balance of trade.

The Promise of Biotechnology

Biotechnology has raised the prospect of making some remarkable advances in food production. The revolutionary potential of genetic engineering has made it possible to crossbreed genetically different plants and to incorporate desired traits into crop lines and animals. It has also led to the cloning of domestic animals like cows and goats. Without a doubt, this technology can help the developing world to produce more food. The first products to be marketed were: (1) the Flavr Savr, a tomato that can be vine ripened and subsequently brought to market and kept fresh much longer than the locally produced products; (2) cotton plants with built-in resistance to insects that comes from genes taken from a bacterium; (3) bacteria that produce bovine somatotropin (BST), a hormone that induces greater milk production in cows. More recently, *genomics*—as the technique has come to be known—has developed the following: sorghum (an important African crop) resistant to a parasitic plant known as witchweed, which infests many crops in Africa; numerous crop plants resistant to the herbicide Roundup, allowing farmers to employ no-till techniques; corn with resistance to the European corn borer; and rice resistant to bacterial blight disease, to name a few. In just three years since bioengineered seeds were commercially available, 1/3 of soybeans, 1/2 of the cotton, and 1/5 of corn acreage in the United States are planted new breeds.

Biotech crop research is proceeding at a rapid pace in developing countries. Products under development include incorporating virus resistance into sweet potatoes, melons, and papaya, vitamin A production into rice, protein enhancement into corn and soybeans, and drought tolerance into sorghum and corn. The potential for *transgenic* crops and animals—that is, organisms with genes from another species—seems almost unlimited.

There is a downside to biotechnology, however. It is unlikely that the majority of the developing world's farmers can afford the new products and, therefore, will have to depend on aid from the developed world. Another concern is the fear that some of the bioengineered products are unsafe. This concern is high in European countries, which have been much more skeptical of the techniques than Americans. A recent poll in the United Kingdom indicated that 77% want a total ban on bioengineered crops. Giving some substance to these concerns, a recent report indicated that genes for resistance to herbicides put in crop plants can spread to weedy relatives, creating the possibility of weeds with resistance to the herbicides normally used to kill them.

In spite of these drawbacks, a look to the future almost certainly will include major advances in food production from biotechnology. If food production is to keep pace with population growth, such advances will be essential, in the view of most observers.

10.2   Food Distribution and Trade

For centuries, the general rule for basic foodstuffs—grains, vegetables, meat, and dairy products—was *self-sufficiency*. Whenever climate, blight (as in the nineteenth-century Irish potato famine), or war interrupted the agricultural production of a nation or region, the inevitable result was famine and death, sometimes on the scale of millions. Once colonies were established in the New World, timber, furs, tobacco, fish, and later, sugar, coffee, cotton, and other raw materials began to flow back to the Old World. In turn, the Old World exported manufactured goods, which helped to transform the colonies into societies much like the European ones that had given birth to them. With the Industrial Revolution, trade between nations intensified, and soon it became economically feasible to ship basic foodstuffs from one part of the world to another. In time, a lively and important world trade in foodstuffs arose; as it did, the need for self-sufficiency in food diminished. Like other sectors of the economy, food has become globalized.

Patterns in Food Trade

Today, agricultural production systems do much more than supply a country's internal food needs. For some nations (such as the United States and Canada), the capacity to produce more basic foodstuffs than the home population needs represents an extremely important entry into the international market. And for many countries (especially those of the developing world), special commodities such as coffee, fruit, sugar, spices, cocoa, and nuts provide their only significant export product (Fig. 10–8). This trade clearly helps the exporter, and it allows importing nations to use foods that they are not able to raise. Given the realities of a market economy, the exchange works well only as long as the importing nation can pay cash for the food. Cash is earned by exporting raw materials, fuels, manufactured goods, or special commodities. In this way, for example, Japan imports $36 billion worth of food, livestock feed, and other raw materials each year but more than makes up for it by exporting $350 billion worth of manufactured goods (cars, electronic equipment, and so on) annually.

The most important foodstuff on the world market is grain: wheat, rice, corn, barley, rye, and sorghum. To be sure, much of this is feed grain imported by some high- and middle-income countries to satisfy the rising demand for animal protein. It is instructive to examine the pattern of global trade in grain over the past half century (Table 10-3). In 1935, only Western Europe was importing grain; Asia, Africa, and Latin America were self-sufficient. By 1950, new patterns were emerging, and today the trade in grains—as well as other basic foodstuffs—represents a development with great economic and political implications. As the table shows, North America has become the major source of exportable grains—the world's "breadbasket" or, in another sense, the world's "meat market." Over the past 45 years, Asia, Latin America, and Africa have shown an increasing dependence on imported grain. These three regions also have in common 45 years of continued rapid population growth. For example, Mexico, birthplace of the Green Revolution, now imports 7 million metric tons of grain per year; population growth has eaten up all the gains of the Green Revolution. Although most of the food needs of these regions are met by internal production, the trend toward greater dependency is an ominous signal.

At no time in recent history has the world grain supply run out; on the average, carryover stocks (the amount in storage as a new harvest begins—in 1998, 293 million tons) are enough to supply more than a year's worth of international trade and aid. In other words, enough food is produced to satisfy the world market—enough to feed millions of animals and keep a healthy surplus in storage. Why, then, are there people in every nation who are hungry and malnourished? Shouldn't every nation make an all-out effort to

▼ **FIGURE 10–8** *A coffee harvest in East Java.* Coffee is one of many commodity crops that produce important income for nations of the developing world.

Table 10-3 World Grain Trade since 1935

Region	Amount Exported or Imported (million metric tons)[1]						
	1935	1950	1960	1970	1980	1990	1995
North America	5	23	39	56	131	123	108
Latin America	9	1	0	4	−10	−12	−4
Western Europe	−24	−22	−25	−30	−16	28	26
Eastern Europe and Former U.S.S.R.	5	0	0	0	−46	−38	−5
Africa	1	0	−2	−5	−15	−31	−31
Asia	2	−6	−17	−37	−63	−82	−78
Australia and New Zealand	3	3	6	12	19	15	13

[1]A minus sign in front of a figure indicates net import; no sign indicates net export.

provide food for its people? And if it can't, shouldn't the rest of the world assume some of the responsibility for providing food to a hungry nation? Where does the responsibility lie for meeting the need for this most basic resource?

Levels of Responsibility in Supplying Food

To begin to answer the question of who is responsible for meeting food needs, it is helpful to examine Fig. 10–9. The figure displays three major levels of responsibility for meeting food needs: the family, the nation, and the globe. At each level, the players are part of a market economy as well as a sociopolitical system. In the market economy, food flows in the direction of economic demand. Need is not taken into consideration. In the event that there are hungry cats and hungry children, the food will go to the cats if the owners of the cats have money and the children's parents don't. In other words, the cash economy, following the rules of the market, provides the *opportunity* to purchase food but not the food itself. Where the economic status of the player (a destitute breadwinner, a poor country) is very low, the sociopolitical system may be able to provide the needed purchasing power or the food.

In the United States, this help is described as the "safety net," and it is represented by a variety of welfare measures such as the Food Stamp Program and the Supplemental Security Income program. Recently, this safety net was radically transformed by the Personal Responsibility and Work Opportunity Reconciliation Act of 1996. The new law is intended to move welfare recipients from public assistance to independence by requiring that they join the work force or perform "workfare" and by limiting the time of public support. Eligibility for food stamps for adults also has work requirements. Critics of the new law state that the law represents an unprecedented and unjust attack on the poor, one that has resulted in pushing

many families and children into severe poverty—in effect, pulling the safety net out from under them.

The most important level of responsibility is the family. The *goal* at this level is **food security**: the ability to meet the food needs of everyone in the family at a nutritional level that grants freedom from hunger and malnutrition.

For an individual, there are three legitimate options for attaining food security: Purchase the food, raise the food or gather it from natural ecosystems, or have it provided by someone (dependency), usually in the context of the family.

In the event of economic or agricultural failure at the family level, the third option implies that there is an effective safety net—that at the national (or local) level there exist policies with the objectives of meeting the food-security needs of all individuals in the society. Thus, appropriate goals at the national level would be *self-sufficiency in food*—enough food to satisfy the nutritional needs of all of a country's people, and *an effective safety net*. The nation can either produce all the food its people needs or buy it on the world market. This goal implies policies to eliminate chronic hunger and malnutrition in the society.

Many nations are not self-sufficient in food, however, and must turn to the global community for unmet food needs. The United States, Canada, Australia, and the nations of the European Community have been the sources of most of the donated food. However, according to Worldwatch Institute, international food aid was cut in half between 1993 and 1996, resulting in only 7.6 million tons available for distribution. This reflects economic restraints in the donor countries as well as a weakening political support for such aid.

There are some less obvious factors to consider when we are talking about global food needs. One of the most serious is the debt crisis in developing countries, discussed in Chapter 7. Another factor in meeting global nutrition needs is the trade imbalance between the industrial and developing countries. The developing countries typically export

Family	Country	Globe

  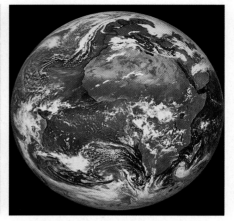

Goal: Personal and family food security
Policies:
—*Employment security*
—*Adequate land or livestock*
—*Good health and nutrition*
—*Adequate housing*
—*Effective family planning*
—*Access to food*

Goal: Self-sufficiency in food and nutrition
Policies:
—*Just land distribution*
—*Support of sustainable agriculture*
—*Effective family planning*
—*Promotion of market economy*
—*Avoidance of militarization*
—*Effective safety net*

Goal: Sustainable food and nutrition for all
    countries
Policies:
—*Food aid for famine relief*
—*Appropriate technology in development aid*
—*Aid for sustainable agricultural
    development*
—*Debt relief*
—*Fair trade*
—*Disarmament*
—*Family planning assistance*

▲ **FIGURE 10–9** *Responsibility for food security.* Goals and strategies for meeting food needs at three levels of responsibility: the family, the country, and the globe.

commodities like cash crops, mineral ores, and petroleum and import more sophisticated manufactured products like aircraft, computers, machinery, and the like. Prices for the latter have risen while those for the commodities from developing countries have declined. To some extent, the increasing use of labor in the developing countries for electronic assembly and clothing manufacture has offset this imbalance. Indeed, the globalization of markets has begun working to the advantage of the developing countries, which have a surplus of labor and low labor costs and have been the recipients of much "outsourcing" of manufacture by industrial corporations. What the poor need is jobs, and even if the pay is far below what workers in industrial countries receive, it is often enough to help workers in developing countries improve their living conditions and their diets.

Thus, being part of the global market system is a major opportunity for many developing countries, but there is often a downside to this process. Exploitation of workers remains a perennial problem in many developing countries, and this is practiced by both domestic and foreign-owned businesses. Labor laws are often ignored, workers often live in squalor, and in some countries, children are drafted into the work force. According to the International Labor Organization, some 120 million children ages 5 to 14 were fully employed in 1996 in the developing countries. To their credit, most industrial countries are trying to reduce this scourge by prohibiting the importation of goods manufactured with child labor.

Another consequence of the trade imbalance is the tendency of the developing countries to "mine" their natural resources for the external market. Thus, lumber, fish products, mineral resources, and others are exploited in order to generate a cash flow. This is often an environmentally destructive process.

Appropriate global goals for the wealthy nations are listed in Fig. 10–9. These are policies that promote self-sufficiency in food production and sustainable relationships between the poorer nations and their environments, as well as addressing the trade imbalance and the human-exploitation problems.

10.3 Hunger, Malnutrition, and Famine

At a U.N. World Food Conference in 1974, delegates from all nations subscribed to the objective "that within a decade no child will go to bed hungry, that no family will fear for its next day's bread, and that no human being's future and capacities will be stunted by malnutrition." Twenty-two years later, the United Nations held the World Food Summit to address continuing hunger, malnutrition, and famine in the world (see the "Global Perspective" essay, p. 249). In light of the evident availability of enough food to feed everyone and a global market in foodstuffs, why is this U.N. objective still unfulfilled?

GLOBAL PERSPECTIVE

WORLD FOOD SUMMIT

Under the leadership of the Food and Agricultural Organization (FAO) of the United Nations, representatives and heads of state from 100 countries met from November 13–17, 1996. Called the World Food Summit, the meeting was the result of two years of planning and negotiations. The objective of the meeting was to bring about a renewed commitment around the world to eradicate hunger and malnutrition and to promote conditions leading to food security for individuals, families, and countries everywhere. The need for this high-level attention was emphasized by new data from FAO indicating that 840 million people, or 18% of the population of the developing world, are malnourished or hungry. In light of continuing increases in population, rising costs of grain, and declining per capita grain production, the summit was a unique opportunity for world leaders to take a new look at the meaning of sustainability.

At the start of the meeting, the delegates adopted, by acclamation, two major documents: the Rome Declaration on World Food Security and the World Food Summit Plan of Action. These documents, in the words of the FAO, "set forth a seven-point plan stipulating concrete, political actions to ensure:

1. conditions conducive to food security;
2. the right of access to food by all;
3. sustainable increases in food production;
4. trade's contribution to food security;
5. emergency relief when and where needed;
6. the required investments to accomplish the plan;
7. concerted efforts to achieve results by countries and organizations."

One concrete objective of the plan is a 50% reduction in the number of hungry people by the year 2015. As a measure of success, this objective was criticized by some participants as being too timid, since it assumed that 420 million people would still be malnourished or hungry. The plan emphasizes the responsibilities of countries (especially developing countries) to enact reforms that would promote food security: policies that do not discriminate against agriculture or small farmers, open trade, greater efforts to bring down population growth, and investing in infrastructure.

Unlike the 1992 Earth Summit and the 1994 Population Summit, the 1996 Food Summit failed to draw many heads of state, particularly from the industrialized countries. However, if countries take the 20-page plan of action seriously, there is no doubt that significant progress will be made in holding back the serious threat of a rising tide of hunger and starvation.

Nutrition vs. Hunger

Hunger is the general term referring to a lack of basic food required for energy and for meeting nutritional needs such that the individual is unable to lead a normal, healthy life. **Malnutrition** is the lack of essential nutrients such as amino acids, vitamins, and minerals, and **undernutrition** is the lack of adequate food energy (usually measured in Calories). What is adequate nutrition, and how can individuals be sure that what they eat provides it? The U.S. Department of Agriculture has recently developed the **Food Guide Pyramid** to help Americans toward better nutrition (Fig. 10–10). The problem of overnutrition is widespread (22.5% of the U.S. population is clinically obese), and to address that and other nutritional problems, the USDA established six food groups and arranged them in a pyramid in order to indicate the relative proportions of each that should be consumed.

Following these guidelines facilitates weight control and cuts down on the chances of developing heart disease, high blood pressure, and some cancers and diabetes. However, these disorders are not very common in most parts of the developing world. There, the major nutritional problems are a lack of proteins and some vitamins (malnutrition), especially in children, and a lack of Calories for food energy (undernutrition) for all ages.

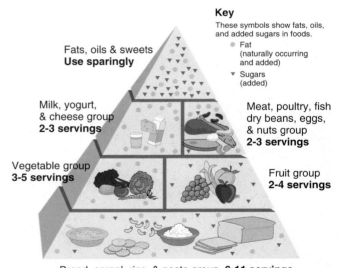

▲ **FIGURE 10–10** *The Food Guide Pyramid.* A guide produced by the USDA to help people evaluate their food needs and food intake so as to provide adequate nutrition and keep weight under control. Suggested numbers of servings are given as a range, because energy requirements vary for people depending on their size, age, and level of activity. Thus, older adults and many women should use the minimum number of servings in order to cut down on calories (they might only need 1600 calories per day); teenage boys and active men might need the upper range to get the 2800 or so calories they need daily.

Extent and Consequences of Hunger

Absolute reliable figures on the worldwide extent of hunger are unavailable, mainly because few governments make any effort to document such figures. The World Health Organization has estimated that 828 million people are underfed and under-nourished—some 18% of the population of the developing world. The regions most seriously affected are southern Asia (especially Bangladesh), Latin America, and sub-Saharan Africa. It is fair to say that hunger and malnutrition still take a terrible toll on human life and productivity. Thus, we see that 25 years later, the United Nation's 1974 objective is still unfulfilled.

The consequences of malnutrition and undernutrition vary. The effects are greatest in children and next greatest in women. Hunger can prevent normal growth in children, leaving them thin, stunted, and often mentally and physically impaired (Fig. 10–11), a condition known as protein-energy malnutrition. The U.N. Subcommittee on Nutrition has identified malnutrition and hunger in the children of southern Asia as the world's most serious nutritional problem, involving at least 108 million children who are underweight because of lack of food. Research has shown that undernutrition in early childhood can seriously limit intellectual development throughout life. Women in the developing world are especially vulnerable because of widespread patterns of low status and heavy labor for women in many countries.

Sickness and death are companions of hunger. As poor nutrition lowers resistance to disease, measles, malaria, and diarrheal diseases become common and are major causes of death in the malnourished and undernourished. According to

▲ **FIGURE 10–11** *Serious malnourishment.* A child in the Denunay Feeding Center, Somalia, operated by World Vision International. Famines and food shortages in Somalia were largely a consequence of civil conflict, not drought.

Bread for the World Institute on Hunger and Development, "Almost 40,000 children under five die each day from malnutrition and infection . . . The number of deaths is the same as if one hundred jumbo jets, each loaded with 400 infants and young children, crashed to Earth each day—one every 14 minutes." Hunger is often a seasonal phenomenon in rural areas that are supported by subsistence agriculture, as people are forced to ration their stored food in order to survive until the beginning of the next harvest. Anyone who travels in the developing world cannot help but notice the fact that few people in the rural areas look well fed; most are thin and spare from a lifetime of hard work and limited access to food.

Root Cause of Hunger

In the view of most observers, *the root cause of hunger is poverty.* Our planet produces enough food for everyone alive today. However, hungry and malnourished people lack either the money to buy food or adequate land to raise their own.

Lack of food is only one of many consequences of poverty. Alan Durning of the Worldwatch Institute defines **absolute poverty** as "the lack of sufficient income in cash or kind to meet the most basic biological needs for food, clothing and shelter"; insufficient income is measured as income of less than $1 a day. On the basis of this definition, 1.3 billion people live in absolute poverty today. Most of these people reside in rural villages, are illiterate, spend half or more of their income on food, and represent races, tribes, or religions that suffer discrimination. They are powerless to do anything about their plight, and quite often the society in which they live is content to keep them that way. Millions are trapped in a cycle of poverty that leads to the degradation of resources and perpetuates the poverty—all made worse by the high fertility of these rural populations.

Hunger and poverty do not always go from bad to worse. A number of Asian countries, including China, Indonesia, and Thailand, significantly reduced the extent of poverty and hunger during the 1980s. Deliberate public policies and social services have greatly improved the welfare of millions in China, where food security is a matter of high national priority. Indonesia has benefited from oil exports and Green Revolution technology, and it continues to put major emphasis on rural development and social infrastructure. Clearly, it is possible for societies to address the needs of the hungry poor with appropriate public policies and to make progress in reducing the extent of absolute poverty and hunger. On the other hand, the most severe kind of hunger—famine—is found in societies that are regressing into disorder and chaos, and it is here that international responsibility comes most sharply into focus.

Famine

By definition, a **famine** is a severe shortage of food accompanied by a significant increase in the death rate. Famine is a clear signal that a society is either unable or unwilling to distribute food to all segments of its population. Two factors

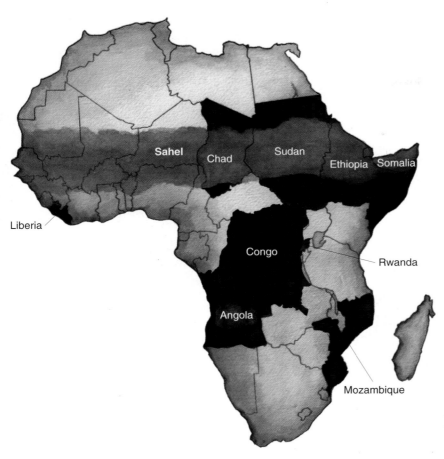

◀ **FIGURE 10–12** *The geography of famine.* Famines have occurred repeatedly in sub-Saharan Africa, especially in the Sahel (a band of dry grasslands that stretches across the continent). The map shows the countries where civil wars and droughts have recently brought on serious famines.

have been the immediate causes of famines in recent years: drought and warfare.

Drought is blamed for the famines that occurred in 1968–74 and again in 1984–85 in the Sahel region of Africa (Fig. 10–12). The Sahel is a broad belt south of the Sahara desert, occupied by 50 million people who practice subsistence agriculture or tend cattle, sheep, and goats. (Such people are called *pastoralists*.) Although the region normally has enough rainfall to support dry grasslands or savannah ecosystems, the rainfall is seasonal, undependable, and prone to failure. Making matters worse, population increases in the region have led to unsound agricultural practices and overgrazing by the expanding herds of livestock. Beginning in 1965, the region experienced 20 years of subnormal rainfall with tragic results. Crops withered, forage for livestock declined, watering places dried up, and livestock died. Both farmers and pastoralists began abandoning their land and migrating toward urban centers, where they were often herded into refugee camps (Fig. 10–13). Unsanitary conditions in the camps and the already weakened condition of the refugees led to the spread of infectious diseases such as dysentery and cholera, and many thousands died before effective aid could be organized. The latest Sahelian famine is thought to have been responsible for 100,000 deaths; the number would have been in the millions except for the aid extended by Africans and numerous international agencies. The rains have returned to the Sahel, removing the immediate threat of famine, but the region still lacks an environmentally sustainable structure for food security.

▼ **FIGURE 10–13** *A refugee camp.* Such camps represent the last resort from hunger and are often scenes of unthinkable human suffering and death. This shows Rwandans in a Ugandan camp.

Famines in which the common factor was not drought, but war, threatened several African nations in the 1990s: Ethiopia, Somalia, Rwanda, Sudan, Mozambique, Angola, and Congo (Fig. 10–12). Devastating and prolonged civil warfare has put millions of Africans at risk of famine. The civil wars disrupt the farmers' normal planting and harvesting and force the displacement of millions from their homes and food sources. Governments in power maintain control over food and relief supplies; relief agencies operate under dangerous conditions and frequently experience casualties. In some areas, the problem is made worse by persistent droughts. In Mozambique alone, 900,000 people died from direct military action or from indirect effects of the war there.

Clearly, famines from drought and war are preventable. India, Brazil, Kenya, and southern Africa have coped with droughts in recent years by mobilizing effective relief in the form of food, clothing, and medical assistance. Indeed, it was the drought of the early 1990s that accelerated the peace process in Mozambique and helped change the political landscape in southern Africa. Cooperation between South Africa and the 10 nations of the Southern Africa Development Community to prevent famine lowered barriers between South Africa and its neighbors prior to the coming of democracy to that troubled nation.

The drought and potential for famine in southern Africa were first predicted by a high-tech satellite system operated by the United States Agency for International Development: the Famine Early Warning System (FEWS). The satellite measures trends in vegetation and rainfall in Africa. In 1994, the system alerted the world to conditions in the Horn of Africa, where more than 20 million were threatened by famine. Food aid to this region was mobilized early in order to prevent the migrations of people that usually result when crops fail. The FEWS issues regular bulletins and special reports on the World Wide Web, giving governments and relief agencies accurate and timely assessments of the status of food security in African countries.

The two countries most recently impacted by famine are Sudan and North Korea. In Sudan, civil war and drought combined to put almost 1 million people in peril, triggering a massive airdrop by the U.N. Food Program to provide almost 10,000 tons a month to the southern parts of Sudan (Fig. 10–14). Many humanitarian aid organizations, such as Doctors Without Borders and World Vision, are doing what they can to help, but the situation will not get better until there is peace in the region. In North Korea, another set of circumstances has brought that country into famine conditions. The primary cause has been the failure of centrally planned agricultural policies in a country that is one of the last bastions of communism. Floods and droughts alternatively have made conditions worse, to the point where millions are judged to be close to starvation and more than a million have already perished. For years, North Korea refused to admit that there was a problem, but in 1997 and 1998 the country permitted the United States to deliver several hundred thousand tons of food and medical assistance.

These two recent cases open the question of the function and effectiveness of food aid.

▲ FIGURE 10–14 *Food Aid.* Women receive aid from the World Food Program in Sudan.

Food Aid

Food aid is being distributed to countries all over the world, not just where famines are threatened. This raises the question, What is the proper role of food aid? Clearly, aid is vital in saving lives where famines occur. But what about the people who suffer from chronic hunger and malnutrition? The basic question is, When should food be given to those in need instead of being distributed according to market economics?

Numerous humanitarian campaigns to end world hunger have been mounted in the last 50 years. The United States and Canada have been world leaders in giving away food (which is first purchased from farmers and therefore represents a subsidy). As noted, a number of serious famines have been moderated or averted by these efforts. As virtuous as such efforts seem on the surface, however, routinely supplying food aid in an attempt to alleviate chronic hunger in developing countries may be the worst thing to do.

The problem is, people will not pay more than they must for food. Therefore, free or very cheap foreign food undercuts the local market. In effect, local farmers must compete economically with free or low-cost imported food. When they cannot earn a profit, they stop producing and eventually enter the ranks of the poor. The cycle continues, as people who sell goods to the farmer also suffer when the farmer loses buying power. In the long run, the entire local economy deteriorates. Hence, the donation of food, while well intended, often aggravates the very conditions that it is meant to alleviate. Meanwhile, population pressures continue to build and the magnitude of the problem increases (see the "Ethics" essay, p. 253).

Food aid in grains averaged 10 million metric tons per year in the mid 1990s, but as noted, it is on the decline. Some aid is strictly humanitarian; Bangladesh receives 800 thousand tons per year, about half as much as it purchases on the market. The

ETHICS

THE LIFEBOAT ETHIC OF GARRET HARDIN

Biologist Garret Hardin has published several provocative essays addressing the worldwide food-vs.-population issue. Here we give you an opportunity to respond to Hardin's thinking.

We begin with the concept of carrying capacity—the number of a species that can be supported indefinitely without degrading the environment. For human societies, carrying capacity means the ability to meet food needs over the long term—that is, sustainably. If ecology had a decalogue, Hardin says, the first commandment would be "Thou shalt not transgress the carrying capacity." A look at the world scene reveals that numerous countries are pressing against the limits of, or have exceeded, their carrying capacity. This, says Hardin, is their problem, not ours, and he uses the lifeboat metaphor to show why.

Picture a number of lifeboats in the sea after a ship has sunk—some crowded with people and some where people are riding in relatively uncrowded luxury. Each lifeboat has a limited capacity. The people in the crowded boats are continually falling overboard, leaving the people on the uncrowded boats with the problem of whether to take them on board or not. Imagine an uncrowded boat with 50 on board and room for 10 more, with 100 people treading in the water and begging to be taken on board. There are several options: (1) Assume that all people have an equal right to survival and take everyone on board. This, says Hardin, would lead to catastrophe for all. (2) Admit only 10, filling all the space on the boat. Two problems with this option are that you lose your margin of safety and that you must find a basis on which to discriminate among all the people in the water. (3) Admit no more to the boat. This preserves your margin of safety and guarantees the long-term survival of the people on your boat; it is the rational solution to the lifeboat problem.

The metaphor, of course, is to be applied to the problem of food aid. Some people would argue that if less grain went to feeding animals, there would be more available to feed the hungry in poor countries. Or, since there are often agricultural surpluses in the rich nations, we could use these surpluses to feed the hungry. The real problem, then, is a problem of food distribution, not of the quantity produced.

These arguments, says Hardin, are foolish. In giving away food, we would only be encouraging the population escalator: A population growing rapidly reaches the limits of its food capacity and is supplied with food from abroad, encouraging still further growth and necessitating still further food aid, and so on. Our hearts, says Hardin, tell us to send food, but our heads should tell us not to. We only postpone the day of reckoning, and in the end, the amount of suffering will be greater. Overpopulated, food-poor countries have transgressed the first commandment of ecology. If we want to help, we should direct our aid toward limiting population growth, according to Hardin. What do you think? (Sample the opinion of those to whom you speak.)

African continent receives 3 million tons per year, which is about 10% of the total imported. All the signs indicate that this figure will increase. A significant amount of aid (1.6 million tons/yr) is flowing to Eastern Europe and the countries of the former Soviet Union—largely to bolster their fragile economies.

Food aid will undoubtedly continue to be an international responsibility. It is at best a buffer against famine, and it will probably continue to be awarded to some countries for political reasons. As part of the solution to the chronic hunger and malnutrition among the poor, however, continued food aid is clearly counterproductive. Much more good will be accomplished by a restructuring of the economic arrangements between rich and poor nations and by the extension of loans and aid directed toward fostering self-sufficiency in food and sustainable interactions with the environment.

10.4 Building Sustainability into the Food Arena

In an ecosystem context, farmers should be viewed as herbivores who manage their producers and pastoralists as predators who manage their prey. The principles of ecosystem structure and function presented earlier in the book apply perfectly well to farming and animal husbandry. The major difference between human systems and natural ecosystems is that we do not have to allow nature to take its course—indeed, we cannot do so and expect to harvest crops instead of weeds or hope to have our livestock flourish. Perhaps for this reason, we tend to forget that our manipulations of plants and animals are nevertheless subject to ecosystem laws, and if we disregard those laws, our human systems are likely to behave counter to our best interests.

Food production and distribution and hunger and famine are very much a matter of how human societies interact with their environment as well as with each other. In this last section, we examine the approaches of sustainable agriculture and look once more at the socioeconomic and political dimensions of hunger.

Sustainable Agriculture

In *A Green History of the World*, Clive Ponting shows how the downfall of several past civilizations was nonsustainable farming and animal grazing. The Sumerians of Mesopotamia raised crops under intense irrigation, and in time, salinization led to the collapse of the agricultural base of their society, followed soon by the decline of the Sumerians. Overgrazing and deforestation throughout the Mediterranean basin, beginning as far back as 650 B.C. in Greece, led to soil erosion that ruined agricultural land and greatly lowered the carrying capacity for livestock. The empires occupying the basin declined accordingly. It is Ponting's view that what happened in the past on a local scale is now occurring in global proportions.

The goal of sustainable agriculture is to maintain agricultural production while not ruining any part of the supporting system or degrading the environment. To be successful, stewardship of both human and natural resources is essential at all levels, from the individual to the national. For example, the needs of the people who work on farms and the animals have to be met. On many farms, migrant workers (human resources) provide labor at crucial planting and harvest times. Frequently they have unmet needs for housing and year-round employment. On a different note, farm policy must also address the financial needs of farmer-owners. For instance, in 1996 the FAIR Act (Chapter 8) removed many of the subsidies that farmers had depended on during lean years and allowed farmers to plant what they wanted. In 1998, severe economic recession in Asia greatly reduced the demand for U.S. farm exports. As Asians tightened their belts, prices dropped, and the effect was felt in midland America. Thus, emergency farm legislation was enacted to enable many farmers to stay in business until things improve.

One way to measure the sustainability of an agricultural system is to apply *environmental accounting* in addition to the conventional cost accounting that balances crop yields against fertilizers, pesticides, and other costs directly absorbed by the farmer (Chapter 23). This would mean measuring such natural resource costs as erosion, salinization, groundwater and surface-water pollution by fertilizers and pesticides, and health costs. When this is done, resource-conserving practices will be given the credit they deserve, and real sustainability is more likely to result. Unfortunately, there is little incentive for using such an accounting method, with the result that sustainability is quite difficult to demonstrate.

There is a growing consensus that agricultural sustainability must be patterned after that of natural ecosystems in order to be successful. In Chapters 3 and 4, we presented four principles of sustainability derived from our studies of natural ecosystems. Let us apply these principles to our analysis of agricultural sustainability, differentiating where necessary between *industrialized agriculture* and *subsistence agriculture*.

1. **For sustainability, ecosystems use sunlight as their source of energy.** Industrialized agriculture will continue to be dependent on mechanization. Subsistence agriculture, however, will do well to continue to use animal energy for working the land because the animals are fed locally and because dependence on costly fossil fuels is avoided. Wind and solar energy can be employed for many farming tasks; look for a return of the windmills once used to pump water and generate small amounts of electricity on farms all over the United States.

Sustainability is encouraged in the wealthier nations when people use small plots of land to grow their own fruits, vegetables, and small animals. This is a much more desirable use of the land than simply growing grass to mow. In the burgeoning cities of the developing nations, urban agriculture is taking hold and providing food and income for hundreds of millions. People are "farming" rooftops and small yards, raising crops that contribute substantially to their nutrition and their income. For example, in Valparaiso, Chile, a poor city housewife grows one-third of her family's food on a tenth of an acre. Organic waste is recycled, and sunlight is harvested where it normally only serves to heat buildings and streets. With the trend toward urbanization that is occurring throughout the developing world, this is one development that can make city living more sustainable.

2. **For sustainability, ecosystems dispose of wastes and replenish nutrients by recycling all elements.** This second principle is augmented by the tendency of wastes and nutrients to be held in place by plants and soil. Sustainable practices, therefore, emphasize soil health and stability. Chapter 8 discusses soil and the efforts needed to prevent erosion, salinization, and desertification.

One option that is increasing in popularity is **organic farming** (also referred to as low-input farming). This involves adding crop residues and animal manure to build up the organic matter in soil, instead of using inorganic fertilizers. When crops are harvested, vital mineral elements are removed from the soil. They are returned to the soil through the application of animal wastes and **green manure** (grasses or legumes that are plowed into the soil at the end of a growing season).

In a society that depends on subsistence agriculture, it is vitally important that the people not use animal dung for fuel; instead, the dung must be returned to the soil. Also, appropriate technologies can be adopted that address the problems of loss of moisture and erosion, such as using rock dams to hold water and protecting the forest cover around croplands. Nutrients can be cheaply added to the soil by mixing legumes with grain or root crops.

3. **For sustainability, the sizes of consumer populations are maintained so that overgrazing or other overuse does not occur.** The most obvious application of this principle is in livestock management. Mismanagement of herds and overgrazing leads to the deterioration of rangeland. Therefore, sustainable livestock management all over the world must recognize the carrying capacity of rangeland ecosystems and preserve the soils and plant cover in them. In some regions of the developing world, forests and woodlands provide fodder for livestock, and in others (tropical forests) the forests are removed to make way for cattle pasture. The sustainable approach is to protect forested areas, recognizing that they are the most stable ecosystems for the site and will benefit human populations more if they are maintained as forests.

Pests are consumers, and pest control is vital to most agriculture. Chapter 17 presents the issues surrounding the use of pesticides. Alternatives to absolute dependence on pesticides are now available, including integrated pest-management programs using natural predators. The selection of planting times, crop rotations, and plant-residue management can provide the beneficial insects with optimal habitats. The basic strategy is to maintain biological control of pests and diseases.

One obvious application of the third principle is for human populations to come to terms with the carrying capacity of the land they occupy. This chapter and previous ones give abundant evidence that some parts of the world already have violated this principle. Clearly, efforts to lower fertility rates will in themselves promote sustainable agriculture by reducing the pressure put on the land to produce food. Only if these efforts are successful will the world's subsistence farmers have any hope of meeting their family's food needs in the future.

4. **For sustainability, biodiversity is maintained.** Crop rotation is a vital part of sustainable agriculture. For example, the farmer might plant three seasons of alfalfa plowed under (green manure), followed by four successive crop seasons of first wheat, then soybeans, then wheat, and then oats. In this way, weeds and insects are more easily controlled, and plant diseases do not build up in the soil. The use of cover crops (e.g., oat and rye overseeded into corn or soybeans) increases soil organic matter and reduces erosion. Combining crops and mixing crops, trees (agroforestry), and livestock provide a diversity of marketable products that can be an effective buffer against economic and biological risks (Fig. 10–15). In more arid climates, *alley cropping*, in which rows of shade-producing trees are alternated with rows of food crops, can be adopted.

Final Thoughts on Hunger

In 1980, a Presidential Commission on World Hunger delivered its report to President Jimmy Carter, with the major recommendation "that the United States make the elimination of hunger the primary focus of its relations with the developing world." The thrust of the report was that if we respect human dignity and have a sense of social justice, we must agree that hunger is an affront to both. The right to food must be considered a basic human right. It follows, then, that we as a nation have a moral obligation to respond to world hunger, the report concluded. Today the question is, Has that moral obligation made its way into public policy in the ensuing years?

By now it should be clear that although food is our most vital resource, we do not treat it as a commons (free to all that need it). Indeed, the production and distribution of food is one of the most important economic enterprises on Earth. Alleviating hunger, as we have seen, is primarily a matter of addressing the absolute poverty that afflicts one of every five people on Earth. To treat food as a commons would be to treat wealth as a commons. Even though this is one of the tenets of socialist systems, it has proved to be a completely unworkable one. We are part of the world economy now, and it is a market economy—for the perfectly good reason that nothing else seems to work, given the realities of human nature.

What has not been done, however, is to bring this market economy under the discipline of sustainability. Short-term profit crowds out long-term sustainable restraint. Self-interest at every level—from the individual to the global community—generates decisions that prevent the sharing of political power, economic goods, knowledge, and technology, and in the process guarantees that the environment

◀ **FIGURE 10–15**
Sustainable agriculture.
A modern farm in Schuylkill County, Pennsylvania, shows a healthy diversity of crops, woodlands, and hedgerows, important in maintaining the ecological stability of the farm countryside.

will continue to be degraded. Why isn't land reform carried out in developing countries? Why are the rich nations content to maintain the current imbalance between prices for commodities and for manufactured goods?

Our understanding of the situation is quite well developed. No new science or technology is needed in order to alleviate hunger and at the same time promote sustainability as we grow our food. The solutions lie in the realm of political and social action at all levels of responsibility. Given the current groundswell of concern about the environment and the disappearance of the cold war between capitalism and communism, there may never be a better time to turn things around and take more seriously our responsibilities as stewards of the planet and as our brothers' keepers.

ENVIRONMENT ON THE WEB

GETTING THE FOOD WHERE IT'S NEEDED

It is sometimes suggested that there is enough food produced in the world to feed all of Earth's population, yet the real difficulty is to get the food where it is most needed. Indeed, close examination of recent famine situations reveals that food aid is often hindered by one of three factors: 1) poor infrastructure, such as roads and railways, for food distribution; 2) political instability or war, causing disruptions in the food distribution system; or 3) corruption within the system, so that the most powerful, rather than the poorest, individuals receive the food. In warring nations, food aid is sometimes misdirected to feed soldiers of the regime in power.

The recent famine in the Democratic People's Republic of Korea (North Korea) illustrates these points. Devastating floods in 1995 displaced more than half a million North Koreans and left more than 2.5 million critically short of food. Subsequent droughts and the Asian economic crisis have deepened the problem. Now, more than 6 million North Koreans rely on food aid. Though North Korea has experienced chronic food shortages for many years, Catherine Bertini, head of the U.N. World Food Programme, says that now "there is almost no food in the country." Western governments and relief organizations have been quick to respond, shipping more than 800,000 tons of food to the country since January 1995. Analysts believe, however, that the true need may be more than a million tons a year, just to meet minimum survival needs.

The roots of this famine lie in North Korea's unstable agricultural economy; the weak state of its agricultural policies seems to indicate that more than food aid will be needed to provide a permanent solution to the problem. A stable agricultural economy will also require political stability, particularly with neighboring South Korea. Currently, South Koreans fear that their neighbors to the north will attack out of desperation for food. The "cold war" context of the conflict between North and South Korea further complicates the distribution of food aid—and increases the risk that food will be misdirected to serve political ends.

In recent years, North Korea has strictly limited the number of international aid monitors and restricted the movement of those currently in the country. Some aid agencies are therefore calling for food contributions to be contingent on adequate needs assessment and monitoring of distribution systems to avoid the problems of persistent famine experienced in Congo, Zaire, and Bosnia in recent years.

Web Explorations

The Environment on the Web activity for this essay discusses the ethics of food aid and the need for systems that prevent famine. Go to the Environment on the Web activity (select Chapter 10 at http://www.prenhall.com/nebel) and learn for yourself:

1. about the ethics of food aid;
2. about international rights and responsibilities in the management of famine; and
3. about the underlying causes of famine.

A suggested time frame for each of these exercises is 10–30 minutes.

REVIEW QUESTIONS

1. Describe the five components of the agricultural revolution. What are the environmental costs of each?
2. What is the Green Revolution? What have been its limitations?
3. Describe subsistence agriculture, and discuss its relationship to sustainability.
4. How do sustainable animal farming and industrial-style animal farming affect the environment on different scales?
5. What are our major prospects for increasing food production in the future?
6. Consider the benefits of eating lower on the food chain.
7. What are the positive and negative aspects of using biotechnology in food production?
8. Trace the patterns in grain trade between different world regions over the last 60 years (see Table 10-3, p. 247).
9. Describe the three levels of responsibility for meeting food needs. At each level, list several ways food security can be improved.
10. Define hunger, malnutrition, and undernutrition. What are their consequences?

11. How are hunger and poverty related?

12. Discuss the causes of famine, and name the geographical areas most threatened by it.

13. Why does food aid often aggravate poverty and hunger?

14. Define sustainable agriculture. How do each of the four principles of sustainability apply to this type of agriculture?

THINKING ENVIRONMENTALLY

1. Imagine that you have been sent as a Peace Corps volunteer to a poor African nation experiencing widespread hunger. How would you begin to address the needs of the people of that nation?

2. Record your food intake over a two- to three-day period. Analyze the nutritional value of your diet. Which nutrients are lacking? Which are in excess? What changes in your diet would reconcile these differences?

3. Of the following methods for increasing food production, which do you feel are viable options? Why?

   clearing forests
   increasing yield on farmland already in production
   irrigating arid lands
   developing transgenic crops with biotechnology
   aquaculture

WEB REFERENCES

On-line resources for this chapter can be found on the World Wide Web at: **http://www.prenhall.com/nebel**. Click on Chapter 10 on the chapter selector.

Wild Species: Biodiversity and Protection

Key Issues and Questions

1. Preserving wild species may require that we find a way to show that they have value. How can we establish the value of natural species?

2. Wild species and their habitats are threatened and endangered by human activities, a matter of concern for many people. How does public policy in the United States protect endangered species?

3. Many naturalists claim that we are losing much of the biodiversity that has enriched Earth for millions of years. What is biodiversity, and what is the current extent of it?

4. It seems certain that humans are the cause of the decline in biodiversity. What human enterprises in particular are responsible for this decline?

5. Much of the loss in wild species and biodiversity is occurring outside the United States. How has the international community acted to protect biodiversity?

Aldo Leopold was a pioneer in the field of conservation and environmental ethics. In his 1949 classic, *A Sand County Almanac*, Leopold tells the story of a wilderness trip he took with some friends in the southwestern United States. They were eating lunch on a hillside when they noticed an animal crossing a small river below them. They watched as the animal, evidently a female wolf, was met on the other side of the river by a half-dozen grown pups, which tumbled all over her with their greetings.

In those days we had never heard of passing up a chance to kill a wolf. In a second we were pumping lead into the pack, but with more excitement than accuracy: how to aim a steep downhill shot is always confusing. When our rifles were empty, the old wolf was down, and a pup was dragging a leg into impassible slide-rocks.

We reached the old wolf in time to watch a fierce green fire dying in her eyes. I realized then, and have known ever since, that there was something new to me in those eyes—something known only to her and to the mountain. I was young then, and full of trigger itch; I thought that because fewer wolves meant more deer, that no wolves would mean hunter's paradise. But after seeing the green fire, I sensed that neither the wolf nor the mountain agreed with such a view.

This incident not only changed Leopold's *attitude* toward shooting wolves, it also changed his *valuation* of wolves and other wild species. He began to understand the importance of wolves in keeping deer herds from overeating their food supply, and in time, Leopold articulated an ethic that stressed the *value* of natural species and the land. By value, he meant more than economic worth; he meant that wild species and their habitats have a right to exist, and protecting that right is a matter of morality. The fierce green fire in the old wolf's eyes was symbolic of wild nature, and when it went out, Leopold knew that something very precious was gone.

In this chapter, we give attention to wild species. We will consider their importance to humans, examine the perspective of biodiversity, and look at the public policies that seek to protect wild species.

◀ *A small pack of gray wolves rests on the snow.* Wolves are being reintroduced to areas of the U.S. and are thriving in many of the new sites.

11.1   Value of Wild Species

We saw in Chapter 3 that ecosystems and the wild species living in them are of enormous value to humankind, providing goods and services that are conservatively estimated to be worth $33 trillion a year. We also saw that if we wish to maintain the sustainability of these ecosystems, we must engage in their active management; in particular, we must preserve their productivity and maintain their biodiversity. Thus, it seems only sensible to put a high priority on protecting wild species and the ecosystems in which they live. Yet, there are widely divergent views on the kind of protection that we might afford to wild species. Some want them protected so as to provide recreational hunting. Others feel strongly that we should not engage at all in hunting for sport. Many have a deeper concern for the ecological importance of wild species, and they view the ongoing loss of biodiversity as vital ecosystem services erode away as an impending tragedy. On the other hand, many in the developing world depend on collecting or killing wild species in order to eat or make money; their personal survival is at stake. These are all attitudes involving values we place on wild species. Can these differing values be reconciled in a way that still leads to sustainable management of these important natural resources?

Biological Wealth

About 1.75 million species of plants, animals, and microbes have been examined, named, and classified, but scientists estimate that between 4 million and 112 million additional species have not yet been systematically explored (a working number of 13.6 million is used). The latest estimates are from the 1995 U.N. Environmental Program (UNEP) survey, "Global Biodiversity Assessment."

These natural species of living things, collectively referred to as **biota**, are responsible for the structure and maintenance of all ecosystems. They and the ecosystems they form represent a standard of wealth—the **biological wealth**—that sustains human life and economic activity. It is as if the natural world were an enormous bank account, with biological wealth capable of paying vital, life-sustaining dividends indefinitely, but only as long as the capital is maintained. The biota, as it is found in each country, represents a major component of the country's wealth, and we have referred (Chapter 3) to the stocks of biota in a country as its **natural capital** (see Chapter 23 for an economic view of this concept). More broadly, this richness of living species is referred to as Earth's **biodiversity**.

Humankind began spending this biological wealth many centuries past. Some 10,000 years ago, humans began learning to select certain plant and animal species from the natural biota and to propagate them, and the natural world has never been the same. Over time, vast areas of forests, savannahs, and plains were converted to fields and pastures as the human population grew and human culture flourished. In the process, many living species were exploited to extinction, and others disappeared as their habitats underwent development. At least 500 plant and animal species have become extinct in the United States alone, and thousands more are at risk. We have been drawing down our biological capital, with unknown consequences.

Now, living in cities and suburbs in the industrialized world and getting all our food from supermarkets, our connections to nature seem remote. That is an illusion: Our interactions with the natural world have changed, but we are still dependent on biological wealth. Millions of our neighbors in the developing world are not so insulated from the natural world; their dependence is much more immediate, as they draw sustenance and income directly from forests, grasslands, and fisheries. However, because of overwhelming economic pressures, these people, too, are often engaged in unsustainable practices by drawing down their biological capital, with consequences that are obvious and grave. A root source of our problem is the way we regard and value wild nature.

Two Kinds of Value

It was not so long ago that hunters on horseback would ride out to the vast herds of bison roaming the North American prairies and shoot them by the thousands, often taking only the tongues for markets back in the East. The passenger pigeons that darkened the skies in huge flocks were ruthlessly killed at their roosts to fill a lively demand for their meat, until the species was gone (Fig. 11–1). Plume hunters decimated egrets and other shorebirds to satisfy the demands of fashion in the late 1800s. Appalled at this wanton destruction, nineteenth-century naturalists called for an end to the slaughter, and the U.S. public began to be sensitized to the

▲ **FIGURE 11–1** *Extinct passenger pigeon.* Clouds of passenger pigeons darkened American skies during the eighteenth and nineteenth centuries, but relentless hunting extinguished the species in the early twentieth century.

EARTH WATCH

RETURN OF THE GRAY WOLF

The state of Alaska has between 5000 and 7000 gray wolves and not a great deal of sympathy for conservationists who want to protect Alaskan wolves. In 1992, the state's Board of Game announced a plan for an aerial wolf hunt in order to boost hunters' chances of bagging more moose and caribou. The hunt never came off, however, due to an intensive protest and a short tourism boycott in Alaska. The state then proceeded with a less publicized program of wolf control using snares. This program, too, came to a halt: In late 1994, a biologist produced a videotape of four snared wolves, one of which had chewed its leg off to a stump trying to escape the snare. Also on the tape was a scene of a state biologist firing four shots into a snared wolf with a small handgun in an attempt to kill it. After seeing the videotape played on KTUU-TV in Anchorage, Alaska's new governor, Tony Knowles, stopped the control program. The net result was the removal of 700 snare traps previously set out in a 1000-square-mile area of the state. Alaska is now considering a more humane way to control unwanted wolf increases—sterilization.

Alaska is the only state in which the wolf is not endangered and therefore not protected by the ESA. Prior to the act, wolves had been exterminated from all but two of the lower 48 states. Currently, gray wolf populations are on the increase; more than 2500 are now found in five states. Symptomatic of a profound shift in public attitude toward wolves, instead of states poisoning and trapping wolves and paying bounties for them, Defenders of Wildlife pays ranchers for any losses they suffer from wolf predation (over three years, Defenders has compensated nine ranchers $12,700 for such losses).

In a highly publicized program, the U.S. Fish and Wildlife Service began releasing captured Canadian wolves in Wyoming's Yellowstone National Park and in central Idaho. The 14 wolves released in Yellowstone in 1995 demonstrated their approval of their new home by producing nine pups, successfully preying on the abundant elk and bison in the park, aggressively harassing the numerous coyotes in the park, and rewarding some 6000 visitors with views of their activities. Encouraged by their success, the wildlife biologists added 17 more wolves to Yellowstone in 1996 and believe that this may end the need for reintroductions there.

As of mid-1998, the Yellowstone wolf population numbered 114, signaling their successful adaptation to the national park. This is bad news for the coyote population; wolves have reduced coyote populations by 50% in some areas as they regain their status as top dogs in the park. The USFWS has recently announced plans to delist the Great Lakes wolf populations, based on their successful recovery during ESA protection.

Research biologist David Mech, probably the leading authority on wolves, believes that the key to continuing the wolf's comeback is to stop short of complete control. The conflicting interests of protectionists on the one hand and farmers and ranchers on the other will have to be worked out with a careful program of regulation of wolf populations—just as many other large animals, such as bears, cougars, and coyotes, are kept under control by regulations and hunting quotas. When the wolf is removed from the endangered list, in the words of David Mech, it "can be accepted as a regular member of our environment, rather than as a special saint or sinner, [and] this will go a long way toward ensuring that the howl of the wolf will always be heard throughout the wild areas of the northern world."

---

losses that were occurring. People began to see natural species as worthy of preservation, and naturalists began to look for ways to justify their calls to conserve nature.

Thus, there was an emerging sense that species should not be hunted to extinction. But why? Were these early conservationists just concerned that there might not be any animals left to hunt or trees left to chop down? Their problem then, and ours now, is to establish that wild species have some *value* that makes it essential that they be preserved. If we can identify that value, then we will be able to justify the action we must take to preserve them.

Philosophers who have addressed this problem inform us that two kinds of value should be considered. The first is **instrumental value**. A species or individual organism has instrumental value if its existence or use benefits some other entity. This kind of value is usually *anthropocentric*; that is, the beneficiaries are human beings. Clearly, many species of plants and animals have instrumental value to humans and will tend to be preserved (or conserved, as we would say) so that we can continue to enjoy the value we derive from them.

The second kind of value we must consider is **intrinsic value**. We assign intrinsic value to something when we agree that it has value for its own sake; that is, it does not have to be useful to us to possess value. How do we know that something has intrinsic value? That is a philosophical question, and it comes down to a matter of moral reasoning. People often disagree about intrinsic value, as illustrated by the animal-rights controversy.

As we study the problem of loss of species, we will find that some claim that no species on Earth except *Homo sapiens* has any intrinsic value. However, if there is no recognition of the instrinsic value of species, it is difficult to justify preserving many that are apparently insignificant or very local in distribution. In spite of the problems inherent in establishing intrinsic value for species, there is growing support in favor of preserving species that not only may be useless to humans but also may never be seen by anyone except a few naturalists or **systematists**—biologists who are experts on classifying organisms.

The value of natural species can be categorized into four areas, which we examine in this chapter:

* sources for agriculture, forestry, aquaculture, and animal husbandry.

- sources for medicines, pharmaceuticals.
- recreational, aesthetic, and scientific value.
- intrinsic value.

The first three categories mostly reflect instrumental value. In the case of aesthetic and scientific value, it could be argued that these sometimes represent intrinsic value.

Sources for Agriculture, Forestry, Aquaculture, and Animal Husbandry

Since most of our food comes from agriculture, we tend to believe that it is independent of natural biota. This is not true. Recall that in nature, both plants and animals are continuously subjected to the rigors of natural selection. Only the fittest survive. Consequently, wild populations have numerous traits for resistance to parasites, competitiveness, tolerance to adverse conditions, and other aspects of **vigor**.

Conversely, populations grown for many generations under the "pampered" conditions of agriculture tend to lose these traits because they are selected for production, not vigor. For example, a high-producing plant that lacks resistance to drought is irrigated, and the drought resistance is ignored. Also, in the process of breeding plants for maximum production, virtually all genetic variation is eliminated. Indeed, the cultivated population is commonly called a **cultivar** (for *culti*vated *vari*ety), indicating that it is a highly selected strain of the original species, with a *minimum* of genetic variation. When provided with optimal water and fertilizer, cultivars do give outstanding production under the specific climatic conditions to which they are adapted. With their minimum genetic variation, however, they have virtually no capacity to adapt to any other conditions.

To maintain vigor in cultivars and to adapt them to various climatic conditions, plant breeders comb wild populations of related species for the desired traits. When found, these traits are introduced into the cultivar through crossbreeding or genetic engineering. For example, in the 1970s the U.S. corn crop was saved from blight by genes from a wild strain of maize. Keep in mind, however, that such a trait comes from a related wild population—that is, from natural biota. If natural biota with wild populations are lost, the options for continued improvements in food plants will be greatly reduced.

Also, the potential for developing *new* agricultural cultivars will be lost. From the hundreds of thousands of plant species existing in nature, humans have used perhaps 7000 in all, and modern agriculture has tended to focus on only about 30. Of these, three species—wheat, maize (corn), and rice—fulfill about 50% of global food demands. This limited diversity in agriculture makes it ill suited to production under many environmental conditions. For example, we tend to think of arid regions as being unproductive without irrigation. However, many wild species belonging to the bean family produce abundantly under dry conditions. Scientists estimate that 30,000 plant species with edible parts might be brought into cultivation. Many of these could increase production in environments that are less than ideal. For example, consider the winged bean, native to New Guinea (Fig. 11–2). This plant is a veritable supermarket, with every part edible: pods, flowers, stems, roots, and leaves. Recently introduced to many developing countries, it has already made a significant contribution to improving nutrition. Loss of biological diversity undercuts similar future opportunities.

Another area in which biodiversity has instrumental value to humans is pest control. In Chapter 17, we will discuss the tremendous and invaluable opportunities to control pests by introducing natural enemies and increasing genetic resistance. Natural enemies and genes for increasing resistance can come only from natural biota. Destroying natural biota will destroy such opportunities.

Since we select species from nature for animal husbandry, forestry, and aquaculture, essentially all the same arguments can be made in connection with those important enterprises.

To use our concept of biological wealth, we can look at natural biota as a bank in which the gene pools of all the species involved are deposited. As long as natural biota are preserved, we have a rich endowment of genes in the bank that we can draw upon as needed. Thus, natural biota are frequently referred to as a **genetic bank**. Depleting this bank cannot help but deplete our future.

▶ **FIGURE 11–2** *The winged bean.* A climbing legume with edible pods, seeds, leaves, and roots, this tropical species is an example of the great potential of wild species for human use.

▲ **FIGURE 11–3** *The rosy periwinkle.* This native to Madagascar is a source of two anticancer agents that are highly successful in treating childhood leukemia and Hodgkin's disease.

Sources for Medicine

Earth's genetic bank also serves medicine, as the following example illustrates. For thousands of years, the indigenous people of the island of Madagascar used an obscure plant, the rosy periwinkle, in their folk medicine (Fig. 11–3). If this plant, which grows only on Madagascar, had become extinct before 1960, hardly anyone outside Madagascar would have cared. In the 1960s, however, scientists extracted two chemicals called vincristine and vinblastine, with medicinal properties, from the plant. These chemicals have revolutionized the treatment of childhood leukemia and Hodgkin's disease. Before their discovery, leukemia was almost always fatal in children; today, with vincristine treatment, there is a 95% chance of remission. These two drugs now represent a $100-million-a-year industry.

The story of the rosy periwinkle is just one of hundreds. The venom from a Brazilian pit viper (a poisonous snake) led to the development of the drug Capoten, used to control high blood pressure. Paclitaxel (trade name Taxol), an extract from the bark of the Pacific yew, has proved to be valuable for treating ovarian, breast, and small-cell cancers. For a time, this use threatened to decimate the Pacific yew; six trees were required to treat one patient for a year. The substance is now extracted from the leaves of the English yew tree, a horticultural plant easy to maintain. Stories like these have created a new appreciation for the field of *ethnobotany*, the study of the relationships between plants and people. To date, some 3000 plants have been identified as having anticancer properties. Drug companies are now financing field studies of the medicinal use of plants by indigenous peoples and are even funding the creation of parks and reserves to promote the preservation of natural ecosystems that are home to both the people and the plants.

It is a fact that 25% of pharmaceuticals in the United States contain ingredients originally derived from native plants, representing $8 billion of annual revenue for drug companies and better health and longevity for countless people. Table 11-1 shows a number of well-established drugs that were discovered as a result of analyzing the chemical properties of plants used by traditional healers. It is likely that the search for such chemicals has barely scratched the surface.

Table 11-1 Modern Drugs from Traditional Medicines		
Drug	**Medical Use**	**Source**
Aspirin	Reduces pain and inflammation	*Filipendula ulmaria*
Codeine	Eases pain; suppresses coughing	*Papaver somniferum*
Ipecac	Induces vomiting	*Psychotria ipecacuanha*
Pilocarpine	Reduces pressure in the eye	*Pilocarpus jaborandi*
Pseudoephedrine	Reduces nasal congestion	*Ephedra sinica*
Quinine	Combats malaria	*Cinchona pubescens*
Reserpine	Lowers blood pressure	*Rauwolfia serpentina*
Scopolamine	Eases motion sickness	*Datura stramonium*
Theophylline	Opens bronchial passages	*Camellia sinensis*
Vinblastine	Combats Hodgkin's disease	*Catharanthus roseus*

Recreational, Aesthetic, and Scientific Value

The species in natural ecosystems also provide the foundation for numerous recreational and aesthetic interests, ranging from sportfishing and hunting to hiking, camping, bird watching, photography, and so on (Fig. 11–4). Interests may range from casual aesthetic enjoyment to serious scientific study.

Virtually all our knowledge and understanding of evolution and ecology have come from studying wild species and the ecosystems in which they live. Pleasure and satisfaction may even be indirect. For instance, one may never see a whale, but knowing that whales and similar exciting animals exist provides a certain aesthetic pleasure. The great popularity of nature films attests to this. Further, knowing that the Earth and its biosphere continue to support and maintain such wildlife provides a sense of well-being.

Recreational and aesthetic values constitute a very important source of support for maintaining wild species. Recreational or aesthetic activities support commercial interests. **Ecotourism**—where tourists visit a place in order to observe wild species or unique ecological sites—represents the largest foreign exchange-generating enterprise in many developing countries. As the amount of leisure time available to people increases, more and more money is spent on recreation. Since some percentage of these recreational dollars will be spent on activities related to the natural environment, any degradation of that environment affects commercial interests. These activities involve a great number of people and represent a huge economic enterprise. To cite one example, 84% of the Canadian people take part in some form of wildlife-related recreation, spending an estimated $9.4 billion yearly. In a recent national survey, it was found that Americans spent $104 billion on wildlife-related recreation during 1996; 63 million Americans spent $31 billion observing, photographing, or feeding wildlife. Very likely, the broadest public support for preserving wild species and habitats is traceable to the aesthetic and recreational enjoyment people derive from them.

Intrinsic Value

The usefulness (instrumental value) of many wild species is apparent. But what about those other species that have no obvious value to anyone—probably the majority of plant and animal species, many of which are rare or inconspicuous in the environment? Some observers believe that the most important strategy for preserving all wild species is to emphasize the *intrinsic* value of species rather than the unknown or uncertain ecological and economic instrumental values. Thus, we should recognize that the extinction of a species per se is an irretrievable loss of something of value.

▲ **FIGURE 11–4** *Recreational, aesthetic, and scientific uses.* Natural biota provide numerous values, a few of which are depicted here.

Philosophers who view wild species as having a basic right to exist claim that humans have no right to terminate a species that has existed for thousands or millions of years, and that represents a unique set of biological characteristics. They argue that long-established existence carries with it a right to continued existence. Some support this view by arguing that there is value in every living thing and that one kind of living thing (e.g., human) has no greater value than any other. This argument, however, can lead to some difficulties, such as having to defend the rights of pathogens and parasites. A more common viewpoint held by ethicists is that because humans have the ability to make moral judgments, they also have a special responsibility toward the natural world, and that responsibility includes concern for other species. It should be pointed out that until recently, Western philosophers argued that only humans were worthy of ethical consideration; the Western philosophical tradition has been strongly anthropocentric.

Many ethicists find their basis for intrinsic value in religion. For example, Old Testament writings express God's concern for wild species when He created them. Jewish and Christian scholars alike maintain that by declaring His Creation good (see *Genesis* 1) and giving it His blessing, God was saying that all wild things have intrinsic value and therefore deserve moral consideration. They are created for Him, not for humans alone, and we ought to regard the Creation with respect and stewardly care. The Islamic Quran (Koran) proclaims that the environment is the creation of Allah and should be protected because it praises the Creator. This ethical concern for wild species underlies many religious traditions and represents a potentially powerful force for preserving biodiversity.

Thus, we see that even if species have no demonstrable use to humans, it can still be argued that they have a right to continue to exist. Only rarely (as in the case of parasites and pathogens) can we claim that there is any moral justification for driving other species to extinction.

11.2 Saving Wild Species

When we place a value on wild species, we are saying that we want to have them around for some reason. One compelling reason is that they provide recreation and food for people who hunt them.

Game Animals in the United States

Game animals are those traditionally hunted for sport, meat, or pelts. In the early days of the United States, there were no restrictions on hunting, and a number of species were hunted to extinction (great auk, heath hen, passenger pigeon) or near extinction (bison, wild turkey). As game animals became scarce in the face of increasing pressure from hunting, regulations were enacted. State governments, backed up by the federal government, enacted laws establishing hunting seasons and bag limits and hired wardens to enforce them. Some species were given complete protection in order to allow populations to build up to numbers that would once again allow hunting.

One success story is the wild turkey. A favorite game species, this bird was hunted to the brink of extinction but was making a slow comeback by the 1930s as a result of hunting restrictions. At that time, there was a total population of about 30,000 individuals in a few scattered states. After World War II, state and federal programs addressed the need for protecting turkey habitats. The birds were reintroduced into areas they had once inhabited, and hunting quotas were strictly limited. The turkey is now found in 49 states, and its population has risen to a total of 4.5 million as a result of these measures.

Using hunting and trapping fees as a source of revenue, state wildlife managers enhance the habitats supporting important game species and provide special areas for hunting. They monitor game populations and adjust seasons and bag limits accordingly. Game preserves, parks, and other areas where hunting is prohibited are maintained to protect habitats as well as certain breeding populations.

In spite of what seems on the surface to be a destruction of wildlife, hunting has many positive aspects. Besides the fees hunters pay, many hunters belong to organizations dedicated to the game they are interested in hunting. Organizations like Ducks Unlimited, The National Wild Turkey Federation, and Pheasants Forever raise funds that are used for the restoration and maintenance of natural ecosystems vital for the game they are interested in hunting. For example, Ducks Unlimited recently announced that 8 million acres of vital habitat for wildfowl have been conserved in North America. Since these are wetlands, their activities have resulted in stemming the tide of wetland loss that has cut North America's wetlands by more than 50% over the years. Their activities are in great measure responsible for the remarkable fall flights of ducks migrating from northern breeding grounds to wintering areas in the southern U.S., Mexico, and points south—a record 92 million ducks in 1997.

Defenders of hunting and trapping point out that their prey are often animals that lack natural predators and would increase to the point of destroying their own habitat. This is often the case with larger animals, like deer and elk. However, many members of the nonhunting public object to the killing of wildlife, and some groups such as the Humane Society of the United States and People for the Ethical Treatment of Animals (PETA) actively campaign to limit or end hunting and trapping. Some practices that are regarded as especially cruel, such as the use of leg-hold steel traps, have been banned in several states as a result of ballot initiatives.

Common game animals, such as deer, rabbits, doves, and squirrels, are well adapted to the rural field and woods environment. Thus, adapted and protected from overhunting, viable populations of these animals are being maintained. Some predatory animals are also on the increase. However, some serious problems have emerged:

1. The number of animals killed on roadways now far exceeds the number killed by hunters. As rural areas are developed, increasing numbers of animals found on roadways are a serious hazard to motorists. For example, in Ohio, motorists hit 24,000 deer each year. In Maine and New Hampshire, over 800 moose are killed annually

in collisions that often also kill motorists because of the size of the animals.

2. Many nuisance animals are thriving in highly urbanized areas, creating various health hazards. For instance, in 1992, a rabies epidemic among "urban" raccoons was a public health risk in several states in the eastern United States. Opossums, skunks, raccoons, and deer are attracted to urban areas by opportunities for food; unsecured garbage cans and pet food left outside will ensure visits from raccoons. Gardens are wonderful sources of food for groundhogs.

3. Some game animals have no predators except hunters and tend to reach population densities that push them into suburban habitats, where they cannot be effectively hunted. The white-tailed deer, for example, has become a pest to gardeners and fruit nurseries; it also poses a public health risk because it is often infested with Lyme disease ticks (Fig. 11–5). Many public parks, where hunting is not feasible, are home to high densities of deer. For example, Sharon Woods, a one-square-mile park near Columbus, Ohio, had a deer population of 500 and became an ecological disaster. Deer browsed virtually every tree from six feet down, stripped bark from many, and removed the understory. In wilder areas, rising deer populations have reduced populations of many endangered or threatened plant species by their grazing.

4. In recent years, there have been an increasing number of attacks on suburbanites by cougars or mountain lions as urbanization encroaches on the wild (Fig. 11–6). California and Oregon have cougar populations numbering in the thousands as a result of diminishing hunting, with resulting increases in contact between people and these large predators.

5. Coyotes, which roamed only in the Midwest and Western states, are now found in every state and are increasing in numbers. A highly adaptable predator, coyotes will eat almost anything. In the Adirondacks, they eat mostly deer; in suburban neighborhoods, they often dine on cats, small dogs, and human garbage. (One particularly adaptive pair had a den on the median strip of Boston's beltway, Route 128, where they lived off roadkill.) Very difficult to control, the animals occasionally attack small children.

6. Suburban parks, lawns, college campuses, and golf courses have become home to exploding flocks of Canada geese. These large birds are, ecologically, grazing herbivores, able to consume large quantities of grass and defecating as often as every eight minutes. Protected by wildlife laws, the geese present a challenge to control efforts. Few predators are able to tackle these large birds, which originated from geese captured and bred to act as decoys to wild migrating flocks. One imaginative solution used by golf courses and colleges is to "hire" border collies (and their trainers), which are trained to harass the geese and keep them from comfortably grazing the lush grass they relish.

One highly controversial effort for controlling unwanted animals is carried out by the Animal Damage Control (ADC) program, an agency of the U.S. Department of Agriculture (the agency changed its name in 1998 to Wildlife Services in an effort to soften its image). This agency responds to requests from livestock owners, farmers, and others concerned with human health and safety to remove nuisance animals and birds. "Removal" virtually always means killing, and the ADC routinely uses poisons, traps, and other devices to kill up to 1 million animals and birds yearly. In 1996, for example, the ADC killed 90,000 coyotes, 25,000 beavers, 2000 bobcats, 1000 domestic dogs, and 5000 doves, among other animals. In spite of attempts by conservation organizations to limit the budgets and activities of the ADC, it is still drawing $30 million to $40 million of taxpayers' money and enjoys strong support from Congress. Interestingly, this is the same Congress that for several years has been unable to agree to renew the major piece of legislation that provides protection to wild animals—the Endangered Species Act.

The Endangered Species Act

In colonial days, huge flocks of snowy egrets inhabited coastal wetlands and marshes of the southeastern United States. In the 1800s, when fashion dictated fancy hats adorned with feathers, egrets and other birds were hunted for their plumage. By the late 1800s, egrets were almost extinct. In 1886, the newly formed Audubon Society began a press campaign to shame "feather wearers" and end the "terrible folly." The campaign caught on, and gradually attitudes changed and new laws followed.

Florida and Texas were first to pass laws protecting plumed birds. Then, in 1900, Congress passed the **Lacey Act**, forbidding interstate commerce to deal in illegally killed wildlife, making it more difficult for hunters to sell their kill. Since then, numerous wildlife refuges have been established to protect the birds' breeding habitats. With millions of people visiting these refuges and seeing the birds in their natural locales, attitudes

▼ **FIGURE 11–5** *Suburban deer.* A white-tailed deer buck grazing in a populated area of Fire Island, New York. Deer have adapted well to suburban habitats and can often become pests there.

◀ **FIGURE 11–6** *The cougar.*
As suburbs encroach on cougar habitats in the West, attacks on humans have become more numerous. This particular cougar has been shot with a tranquilizer dart while on the roof of a home in Colorado.

have changed significantly. Today the thought of hunting these birds would be abhorrent to most of us, even if official protection were removed. Thus protected, egret populations were able to recover substantially. In the meantime, the Lacey Act has become the most important piece of legislation protecting wildlife from illegal killing or smuggling. In 1994, for example, prosecution of over 700 Lacey Act wildlife crimes resulted in 26 years of jail time and over $1 million in fines.

Congress took another major step when it passed a series of acts to protect endangered species. The most comprehensive and recent of these acts is the **Endangered Species Act (ESA)** of 1973 (reauthorized in 1988). As we learned in Chapter 4, an **endangered species** is a species that has been reduced to the point where it is in imminent danger of becoming extinct if protection is not provided (Fig. 11–7). The act also provides for the protection of **threatened species**, which are judged to be in jeopardy but not on the brink of extinction. When a species is officially recognized as being either endangered or threatened, the law specifies substantial fines for killing, trapping, uprooting (plants), or engaging in commerce in the species or its parts. The legislation forbidding commerce includes wildlife threatened with extinction anywhere in the world.

The ESA requires the U.S. Fish and Wildlife Service (USFWS), under the Department of the Interior, to draft recovery plans for protected species. Habitats must be mapped and a program for the preservation and management of critical habitats must be designed such that the species can rebuild its population. A 1995 Supreme Court decision made it clear that federal authority to conserve critical habitats extended to privately held hands. As Table 11-2 indicates, 1143 U.S. species are currently listed for protection under the act, and recovery plans are in place for 780 of those.

Some critics of the ESA believe that the act does not go far enough. A major shortcoming is that protection is not provided until a species is officially listed as endangered or threat-

ened by the USFWS and a recovery plan is established. Species usually will not make the list until their populations have become dangerously low. Now that a congressional moratorium on listing, imposed in 1995, has been lifted, the USFWS has been working intensely on both listing and delisting species and developing recovery plans. Recently, Interior Secretary Bruce Babbitt announced that 17 animals and 12 plants would be removed or downgraded from endangered to threatened over the next two years. One of the species to be removed from the list, and an amazing recovery story, is the American peregrine falcon (Fig. 11–8).

Both the peregrine falcon and the bald eagle were driven to extremely low numbers because of the use of DDT as a pesticide from the 1940s through the 1960s. DDT, carried up to these predators through the food chain, caused a serious eggshell thinning that led to nesting failures in these two species and numerous other predatory birds (Chapter 17). By 1975, a survey indicated an existence of only 324 nesting pairs of peregrines in North America. DDT use was banned in both the United States and Canada in the early 1970s, and the stage was set for recovery of the bird. Working with several nonprofit captive-breeding institutions such as the Peregrine Fund, the USFWS sponsored efforts that resulted in the release of some 6000 captive-bred young falcons in 34 states over a period of 23 years. There are now about 1600 known breeding pairs in the United States and Canada, well above the targeted recovery population of 631 pairs.

A few species have gained exceptional public attention, and heroic efforts have been mounted to save them. Efforts to save the whooping crane, for example, include virtually full-time monitoring and protection of the single remaining flock, which for years numbered only in the teens. To obtain a higher reproduction and recruitment rate, eggs were collected and artificially incubated. (When eggs are continually removed from a nest, the female can produce up to 14!) The chicks hatched in incubation and were then placed in

(a) Male Pine Barrens Tree Frog

(b) Silversword

(c) Red-cockaded Woodpecker

(d) Karner Blue Butterfly

(e) Swamp Pink

(f) Whooping Cranes

(g) Devil's Hole Pupfish

(h) White Oryx

(i) Manatee

▲ **FIGURE 11–7** *Endangered species.* Some examples of endangered species—species whose populations in nature have dropped so low that they are in imminent danger of becoming extinct unless protection is provided.

nests of related sandhill cranes to be raised by them as foster parents. However, these efforts have been unsuccessful, since the chicks raised by the sandhill crane foster parents do not mate with other whooping cranes. In addition, several captive-breeding flocks are being maintained. The migratory wild flock, however, is increasing steadily and has reached a population of 190 individuals from a low of 14 cranes in 1939. This flock migrates from wintering grounds on the Texas Gulf coast to Wood Buffalo National Park in Canada; the migration is a hazardous affair, and accidents along the way take a toll on the migrating birds. Nevertheless, the whooping crane recovery program has been so successful that other countries

Table 11-2 1998 Federal Listings of Threatened and Endangered U.S. Plant and Animal Species

Category	Endangered	Threatened	Total Listings*	Species with Recovery Plans†
Mammals	59	8	334	44
Birds	75	15	274	75
Reptiles	14	21	114	29
Amphibians	9	7	25	11
Fishes	68	40	118	86
Snails	15	7	23	19
Clams	61	8	71	45
Crustaceans	16	3	23	7
Insects	28	9	41	22
Arachnids	5	0	5	4
Plants (flowering, conifers, ferns, others)	557	118	678	438
Total	907	236	1702	780

Total endangered U.S. species	907	(350 animals, 557 plants)
Total threatened U.S. species	236	(118 animals, 118 plants)
Total listed U.S. species	1143	(468 animals, 675 plants)
Total listed non-U.S. species	559	(556 animals, 3 plants)

Source: Department of the Interior, U.S. Fish and Wildlife Service, Division of Endangered Species, Aug. 31, 1998.
*Total listings refer to both the U.S. and non U.S. species.
†Recovery plans only pertain to U.S.-listed species.

with endangered crane species have adopted many of the strategies developed by the USFWS.

Some critics of the ESA claim that it goes too far. The most famous case in point is the northern spotted owl. This

▲ FIGURE 11–8 *The peregrine falcon.* The Endangered Species Act has worked well in bringing this magnificent falcon to the point where it can be dropped from the list.

species became a focal point in the battle to save some of the remaining old-growth forests of the Pacific Northwest. The owl is found only in these forests and has dwindled to a population between 6000 and 8000. In June 1990, the USFWS listed the owl as a threatened species. The listing prompted the service to set aside 6.9 million acres (2.8 million ha) of the federally owned old-growth forests, enough to guarantee the bird protection into the future. Years of bitter controversy followed, pitting environmentalists against the timber industry, until a compromise was worked out (the Northwest Forest Plan), which both factions criticized.

Critics of the ESA also object to the powers the act gives to federal agencies to limit development and prohibit some activities on private lands where protected species are found. Although these powers exist, they have been seldom used; of the 98,000 projects reviewed by the USFWS over a five-year period, only 54 were curtailed or prohibited because of endangered-species considerations. Critics question the taxonomic status of some of the animals protected by the act. For example, the northern spotted owl is a subspecies of the single species designated by taxonomists as the spotted owl; the California spotted owl and the Mexican spotted owl are also subspecies of this species, and neither of them is listed as endangered or threatened. However, the USFWS is on solid ground when it acts to protect subspecies and even distinct populations, as the ESA defines a species to include "any subspecies of fish or wildlife or plants, and any distinct population segment of any species of vertebrate fish or wildlife which interbreeds when mature."

The ESA was scheduled for reauthorization in 1992, but Congress decided to defer consideration of it until after the coming presidential election, fearing a veto from President Bush. With the election of Bill Clinton, environmentalists were optimistic that the ESA would get swift consideration—and were disappointed when it did not. With the shift of Congress to Republican control in 1994, the ESA has been in continuous trouble and has been operating on year-to-year extensions. Opponents such as development, lumber, and mineral interests claim that the act is costly, inefficient, and favors the rights of owls and flowers over private-property owners and legitimate business interests. They have strong support in the Congress, which at the time of writing was considering a bill that would weaken the act (the Kempthorne Bill—S. 1180) and one that would strengthen it (the Miller Bill—H.R. 2351).

In recent years the USFWS has made progress in bringing together landowners and regulators by making increasing use of a procedure authorized in a 1982 amendment to the ESA—the **Habitat Conservation Plan (HCP)**. The USFWS can issue a permit for activities that potentially harm a protected species, where the landowners take special steps to protect the species over the critical part of the land to minimize any harm done. There are now more than 200 such plans in effect and more pending. As an example, the Plum Creek Timber Company and the USFWS have agreed on a plan covering 170,000 acres of land in the northern Cascade Mountains of Washington, home to spotted owls and several other endangered or threatened species. A small amount of this is old growth forest, and the company has agreed to leave this alone and, in addition, allow extra-wide forested buffers to remain around streams and employ selective logging to reserve the more mature trees. In return, the company will gradually log its mature forests over a 30-year period. The one aspect of the HCPs that is objectionable to many is that the plan is in effect for 50 years, guaranteeing the timber company that no new restrictions will be imposed even if new evidence of ESA concern surfaces. A recent study of the HCPs indicates that except for a few very bad exceptions, the HCPs are making good use of sound science and are moving in the right direction, although monitoring of results is still inadequate.

In the final analysis, the ESA is a formal recognition of the importance of preserving wild species, regardless of any economic importance. Species listed as endangered or threatened have legal rights to protection under the law. The act is something of a last resort for wild species, but it embodies an encouraging attitude toward nature that has now become public policy. Unexpected opposition to weakening the ESA came from the religious community, calling the ESA analogous to a "modern Noah's Ark." Actions taken under auspices of the act have demonstrated a unique commitment to preserve wild species. Reauthorization of a strong ESA will be a major signal of our continuing commitment to consider the value of wild species on grounds other than economic or political.

We turn now to the broader topic of biodiversity, or the sum of richness of species and habitats on Earth, and examine reasons for its decline and strategies for its preservation.

11.3 Biodiversity

We saw in Chapter 5 that over time, natural selection leads to **speciation**—the creation of new species—as well as **extinction**—the disappearance of species. Over geological time, the net balance of these processes has favored the gradual accumulation of more and more species—in other words, biodiversity. The concept of biodiversity is often extended to the genetic diversity within species, as well as the diversity of ecosystems and habitats. Its main focus, however, is the species.

How much biodiversity is there? Estimates range widely, and the fact is, no one knows. Today, the only two certainties are that 1.75 million species have been described and many more than that exist. Most people are completely unaware of the great diversity of species within any given taxonomic category. Groups especially rich in species are the flowering plants (270,000 species) and the insects (950,000 species), but even less diverse groups, such as birds or ferns, are rich with species that are unknown to most people (Table 11-3). Taxonomists are aware that their work in finding and describing new species is incomplete. Groups that are conspicuous or commercially important, such as birds, mammals, fish, and trees, are much more fully explored and described than, say, insects, very small invertebrates like mites and soil nematodes, fungi, and bacteria. Trained researchers who can identify major groups of organisms are few, and the task of exploring the diversity of life will require a major sustained effort in systematic biology.

Estimates of the number of species on Earth today are based on recent work in the tropical rain forests, which hold more living species than all other habitats combined. Costa Rica, less than half the size of New York State, is estimated to contain at least 5% of all living species. The upper boundary of the estimate keeps rising as taxonomists explore the rain forests more and more. Recent estimates place the high end at 112 million species. Whatever the number, the planet's biodiversity represents an amazing and diverse storehouse of biological wealth.

The Decline of Biodiversity

Known and Unknown Losses. The biota of the United States is as well known as any. At least 500 species native to the United States are known to have become extinct since the early days of colonization. Over 100 of these are vertebrates. A 1997 inventory of 20,439 wild plant and animal species in the 50 states, carried out by a team of biologists from The Nature Conservancy and the National Heritage Network, concluded that almost 1/3 (31.9%) are vulnerable or imperiled or already extinct (Fig. 11–9). Mussels, crayfish, fishes, and amphibians—all species dependent on freshwater habitats—are most at risk. Flowering plants are also of great concern, with one third of their numbers in trouble.

Across North America, general population declines of wild species are also occurring. Commercial landings of many species of fish are down; puzzling reductions in amphibian populations are occurring in North America and other parts of the world; and many North American songbird species, such

Table 11-3 Known and Estimated Species on Earth

| Species | Number of Known Species | Estimated numbers, in thousands | | | Accuracy |
		High (000)	Low (000)	Working Number (000)	
Viruses	4000	1000	50	400	Very poor
Bacteria	4000	3000	50	1000	Very poor
Fungi	72,000	2700	200	1500	Moderate
Protozoa	40,000	200	60	200	Very poor
Algae	40,000	1000	150	400	Very poor
Plants	270,000	500	300	320	Good
Nematodes	25,000	1000	100	400	Poor
Arthropods					
Crustaceans	40,000	200	75	150	Moderate
Arachnids	75,000	1000	300	750	Moderate
Insects	950,000	100,000	2000	8000	Moderate
Mollusks	70,000	200	100	200	Moderate
Chordates	45,000	55	50	50	Good
Others	115,000	800	200	250	Moderate
Total	**1,750,000**	**111,655,000**	**3,635,000**	**13,620,000**	**Very poor**

Source: United Nations Environment Programme, "Global Biodiversity Assessment" (Cambridge: Cambridge University Press, 1995): 118, table 3.1-1.

as the Kentucky warbler, wood thrush, and scarlet tanager, have been declining and have disappeared entirely from some local regions. These birds are neotropical migrants; that is, they winter in the Neotropics of Central and South America and breed in temperate North America. Almost 100 species of these migrants have shown a drop in numbers over the past 30 years.

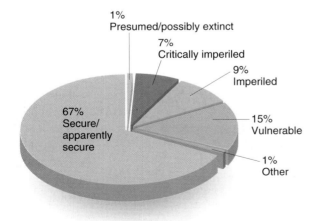

▲ **FIGURE 11–9** *The state of U.S. species.* Fully one-third of over 20,000 species of plants and animals surveyed were found by biologists to be at risk of extinction. (Source: Data from The Nature Conservancy.)

Worldwide, the loss of biodiversity is even more disturbing. At least 484 animal species and 654 plant species have become extinct since 1600. Most of the known extinctions of the past several hundred years have occurred on oceanic islands, where small landmasses limit the size of populations, and human intrusions are most severe. The known causes of animal extinctions, recently assessed by the World Conservation Union, are shown in Fig. 11–10. The "Global Biodiversity Assessment," commissioned by UNEP to provide information for the Convention on Biological Diversity, estimates that some 5400 known species of animals and 26,000 plant species are in danger of becoming extinct.

Biodiversity is richest in the tropics. This richness is almost unimaginable. Biologist E. O. Wilson identified 43 species of ants on a single tree in a Peruvian rain forest, a level of diversity equal to the entire ant fauna of the British Isles. Other scientists found 300 species of trees in a single 2.5-acre (1-ha) plot and as many as 10,000 species of insects on a single tree in Peru. Assuming the existence of 2 million species in the tropical forests (a conservative estimate) and a clearance rate of 1.8% per year for those forests, Wilson calculated that tropical deforestation is responsible for the loss of 4000 species a year. Other scientists have projected that as many as 17,000 species are lost per year, or between 500,000 and 2 million species over the last half of the twentieth century! Not all scientists agree with these calculations, however, as the estimates are based on many

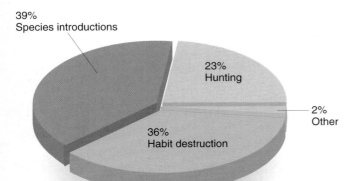

▲ **FIGURE 11–10** *Known causes of animal extinctions since 1600.* Species introductions, habitat destruction, and hunting have brought about the majority of known animal extinctions since 1600. (Source: Data from World Conservation Monitoring Centre.)

assumptions (see the "Global Perspective" essay, p. 273). The most important conclusion we can draw is that many species are in decline and some are becoming extinct. No one knows how many, but the loss is real, and it represents a continuing depletion of the biological wealth of our planet.

Reasons for the Decline

Physical Alteration of Habitats. Although multiple causes are usually the rule in losses of biodiversity, the greatest loss is caused by the physical alteration of habitats through the processes of conversion, fragmentation, and simplification. Habitat destruction has already been responsible for 36% of known extinctions (Fig. 11–10).

Conversion. Natural areas are converted to farms, housing subdivisions, shopping malls, marinas, and industrial centers. When, for example, a forest is cleared, it is not just the trees that are destroyed; every other plant and animal that occupies the destroyed ecosystem, either permanently or temporarily (e.g., migrating birds), also suffers. The idea that this

wildlife simply will move "next door" and continue to live in an undisturbed section is erroneous. Any loss of natural habitat can result in only one thing: a proportional reduction in all populations that require that habitat. Thus, the decline in songbird populations cited earlier has been traced to a combination of the loss of winter forest habitat in Central and South America and the increasing fragmentation of summer forest habitat in North America.

Fragmentation. Natural landscapes generally have large patches of habitat that are well connected to other similar patches. Human-dominated landscapes, however, consist of a mosaic of different land uses, resulting in small, often geometrically configured patches that often highly contrast with neighboring patches (Fig. 11–11). For the continued survival of any natural population, the number of individuals must never fall below a *critical number*, and that requires a certain minimum area. This minimum area must be large enough to compensate for years of adverse weather. That is, more area will be required during a dry year than during a normal year. If development reduces the habitat to a point where it cannot support the critical number during an adverse year, the entire population will perish. Similarly, development (such as a highway) that fragments a territory and prevents migration between the two fragments will cause a population to perish if neither area can adequately support the critical number. Also, reducing the size of a habitat creates a greater proportion of edges, a situation that favors some species but may be detrimental to others. For example, the Kirtland's warbler is an endangered species that is highly dependent on large patches of second-growth jack pines. It is endangered because its habitat has been greatly fragmented, creating edges that favor the brown-headed cowbird, a nest parasite that can invade the forest and lay its eggs in the nest of the rare warbler.

Simplification. Human use of habitats often simplifies them. We might, for example, remove fallen logs and dead trees from woodlands for firewood, thus diminishing an important microhabitat on which several species depend. When a forest is managed for the production of a few or one species of tree, tree diversity obviously declines, and with it so does the diversity of a cluster of plant and animal species dependent on the

▶ **FIGURE 11–11**
Fragmentation. Development usually results in the breaking up of natural areas, resulting in a mosaic of habitats that may not support local populations of species.

GLOBAL PERSPECTIVE

THE MEGA-EXTINCTION SCENARIO

Biodiversity is under assault worldwide, and to counter this assault, a number of prominent biologists have gone on record claiming that an unprecedented rate of extinction is under way. In support of this assertion, paleontologists examined the fossil record and determined that the natural rate of extinction over the past millions of years is about one to 10 species per year. Current losses are estimated to be in the range of 1000 species a year, which is far above the expected rate; indeed, some put the number as high as 17,000 species a year lost in the tropical forests alone. At the present rates of clearing land, some scientists estimate that we will lose one-fourth of all species within 50 years. If there are 13.6 million species (a working number), as some taxonomists suspect, this means the loss of 3.4 million of them! In light of these possibilities, calls are going out for such responses as stopping the development of all remaining natural lands in the developed world. Many advocate giving huge sums of money to the poor nations so that they can quickly halt population growth and convert their agriculture to sustainable and high-yielding.

This sounding of the alarm, according to some critics, has become a dogma among many biologists and is based on a number of highly questionable assumptions. Three key assumptions lie behind the mega-extinction scenario. One is that losses of habitat always mean losses of species, and often in direct proportion. However, forests cleared are not necessarily forests gone forever; many are converted to second-growth forests, which support many of the original species.

A second assumption is that the larger a geographical area, the larger the number of species it holds. In many areas, however, the number of species levels off, even though the geographical area is large. So there may be no net loss of species, even though some of an area is lost.

The third questionable assumption is that the number of species in existence is far larger than the number of species we know about; hence, estimates of the number of extinctions are similarly high. Thus, the catastrophic extinction rates predicted by biologists are based largely on species that no one has ever seen. Also, there is a disturbing lack of agreement among observers who calculate the number of existing species; one calculation may be 10 or more times another. Literally, there is an unknown genetic richness out there!

Support for the mega-extinction scenario is so strong that those who challenge it from within the biological research community risk their reputations and the normally cordial relationships expected among colleagues. Challengers do exist despite the large stakes, however, and these critics are calling for their colleagues to examine their data and base their predictions only on supportable evidence. They also ask that only realistic public-policy alternatives be advocated.

The problem is, if the dire predictions turn out to be false, some of the real problems—tropical deforestation and loss of wetlands, for instance—will also get less than the attention they deserve. And the public will be less prone to believe other, more well-founded scientific predictions of environmental disasters.

---

less favored trees. Streams are sometimes "channelized"—their beds are cleared of fallen trees and riffles, and sometimes the stream is straightened out by dredging. Such alterations inevitably bring on a loss of diversity of fish and invertebrates that live in the stream.

The Population Factor. Past losses in biodiversity can be attributed to the expansion of the human population over the globe. Continuing human population growth will bring on continued alteration of natural ecosystems and the inevitable loss of more wild species. The losses will be greatest where biodiversity is greatest and human population growth is highest—in the developing world. Africa and Asia have lost almost two-thirds of their original natural habitat. People's desire for a better way of life, the desperate poverty of rural populations, and the global market for timber and other natural resources are powerful forces that will continue to draw down biological wealth.

In Kenya, where human population growth has been explosive for several decades, the conversion of savannah and woodlands to cultivation or intensive grazing by goats and cattle has driven most of the African elephant population into the existing wildlife reserves, causing a great reduction in their numbers. The other large African mammals have experienced similar reductions, as the needs of the rural population inevitably conflict with those of the large wild animals of eastern Africa. Similarly, as North America was gradually transformed into a great agricultural and industrial continent, similar losses in large mammal populations occurred (wolves, bison, elk, black bear, cougar, etc.).

One key to holding down the loss in biodiversity lies in bringing human population growth down. There is an inverse relationship between the size of the human population and the survival of species worldwide, illustrated in Fig. 11–12. If the human population increases to 11 billion, as some demographers believe that it will, the consequences for the natural world are frightening.

Pollution. Another major factor causing loss of biodiversity is pollution. Pollution can directly kill many kinds of plants and animals. For example, the *Exxon Valdez* oil spill of 1989 killed at least 300,000 birds and 3000 sea otters (Fig. 11–13). High levels of toxic metals traced to industry and agriculture are believed to have killed 150,000 grebes (waterfowl) in California's Salton Sea in 1992. Recent declines in amphibian populations and reproductive disorders in alligators have been tentatively traced to the widespread presence of chemicals known as **endocrine disrupters**. These are commonly used chlorine compounds and plastics, as well as some pesticides. Because of their molecular structure, they may mimic the effects of

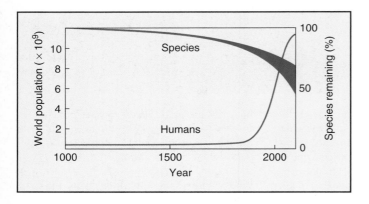

▲ **FIGURE 11–12** *Human population and species survival.* There is an inverse relationship between human population size and the survival of species worldwide. Uncertainty about the extent of species becoming extinct is reflected in the width of the species curve. (Source: M.E. Soule, "Conservation: Tactics for a Constant Crisis," *Science*, 253 [1991]: 744. Copyright by the AAAS.)

some hormones and thus disrupt the normal course of embryological development or sexual activity of animals.

Pollution destroys or alters habitats, with consequences just as severe as those caused by deliberate habitat conversions. Acid deposition and air pollution cause forests to die; sediments and nutrients kill species in lakes, rivers, and bays; DDT devastates wild-bird populations; and depletion of the ozone layer increases the impact of ultraviolet light on wild species. The list is endless. Some scientists project that global warming from the greenhouse effect may be the greatest catastrophe to hit natural biota in 65 million years. (The fossil record reveals that a massive extinction of plants and animals occurred at that time, probably caused by a major asteroid hitting Earth.)

▼ **FIGURE 11–13** *Oil spill kills.* Oil-soaked seabirds killed by the *Exxon Valdez* oil spill that left 11 million gallons of crude oil spread over Prince William Sound in Alaska.

Scientists speculate that the rate of climatic warming will far outpace the ability of most species of trees to migrate northward, thus trapping them in inhospitable climates. Every species that dies out will doubtlessly take others with it. If the forest trees are unable to migrate, neither can the rest of the wildlife that depend on them for food and habitat.

Most of the global pollution problems can be traced to the industrialized world, where energy-generating and other technologies continue to pour pollutants into the air and water at increasing rates. For this reason, it is important not to point the finger of blame primarily at the developing world, where population growth is such a problem. Global climate change and toxic-substance pollution are the legacy of the already developed nations.

Exotic Species. An **exotic species** is a species introduced into an area from somewhere else, often a different continent. Because the species is not native to the new area, it is often unsuccessful in establishing a viable population and quietly disappears. This is the fate of many pet birds, reptiles, and fish that escape or are deliberately released away from their native habitats. Occasionally, however, an introduced species finds the new environment very much to its liking. As we will see in Chapter 17, most of the insect pests and plant parasites that plague agricultural production were accidentally introduced. Exotic species are major agents in driving native species to extinction and are responsible for an estimated 39% of all extinctions of animals since 1600. A 1993 study estimated that 4500 exotic species are now in the United States; approximately 15% of these are damaging to ecosystems.

The transplantation of species by humans has occurred throughout history, to the point where most people are not aware of the distinction between native and exotic species living in their lands. The European colonists brought literally hundreds of weeds and forage plants to the Americas, so that now most of the common field, lawn, and roadside plant species in eastern North America are exotics. Fully one-third of the plants in Massachusetts are alien or introduced—about 900 species! Among the animals, the most notorious have been the house mouse and the Norway rat; others include the wild boar, donkey, horse, nutria, and even the red fox, which was brought to provide better fox hunting for the early colonial equestrians. One of the most destructive exotics is the house cat. The 60 million domestic and feral cats in the United States are very efficient at catching small mammals and birds. One recent study of cat predation in Wisconsin indicated that cats kill 19 million songbirds and 140 million gamebirds each year.

Deliberate introductions sanctioned by the U.S. Natural Resources Conservation Service (in the interest of "reclaiming" eroded or degraded lands) have brought us kudzu and other runaway plants such as autumn olive, multiflora rose, and amur honeysuckle. Many other exotic plants are introduced as horticultural desirables; one notorious example is purple loosestrife, which infests wetland habitats and replaces many plants valuable to wildlife (Fig. 4–8 illustrates purple loosestrife, kudzu, and other exotic plants). Another is Oriental bittersweet, an aggressive vine that chokes and shades small shrubs and trees. Often, it smothers young trees and prevents the forest from regrowing. In Oregon and California, the introduced star thistle invades and ruins pasture land, grows wildly, and is a fire hazard.

We might think that introduced species would add to the biodiversity of a region, but many introductions have exactly the opposite effect. Foreign plants crowd out native plants. New animal species are often very successful predators that eliminate native species not adapted to their presence. For example, the brown tree snake was introduced to the island of Guam as a stowaway on cargo ships during World War II (Fig. 11–14). This snake is mildly poisonous, grows to a length of 13 feet, and eats almost anything smaller than itself. In a little over 40 years, the snakes have eliminated most of the island's birds; nine of Guam's twelve native species of birds have become extinct, and populations of the remaining birds are so decimated that survivors are rarely seen or heard. Efforts at controlling the snakes have proved futile. There are no natural predators on the island, and the snake population has reached such a high density that snakes have invaded houses in search of prey (such as puppies and cats). At present, wildlife officials are concentrating their efforts on preventing the snakes from spreading to other islands in the Marianas chain.

Introduced exotic species may also drive out native species by competing with them for resources, as documented in Chapter 4. An Asian weed, hydrilla, was brought to Florida from India by a man selling plants for home aquariums. Since 1950, it has spread throughout lakes and reservoirs in the South, and federal agencies are spending $5 million a year in attempts to control it. The weed chokes intakes of power plants and waterworks and, because of its tendency to spread over the water surface, causes fish and invertebrates to die by preventing oxygen from reaching the deep water.

▼ **FIGURE 11–14** *The brown tree snake.* This snake, accidentally introduced to Guam, has decimated bird life on the island. Often lacking in natural predators, islands are especially vulnerable to the harmful effects of exotic species.

One of the most recent exotic invaders is the zebra mussel, which came to the United States in 1986 in the ballast water of oceangoing vessels (Fig. 4–7a). The zebra mussel is expected to cause $4 billion in damage in the Great Lakes alone over the next decade; its impact on native species is likely to be devastating.

Overuse. It should be obvious that killing whales, fish, or trees faster than they can reproduce will lead to the ultimate extinction of the species. In spite of that, overuse is another major assault against wild species, responsible for 23% of recent extinctions. Overuse is driven by a combination of economic greed, ignorance, and desperation. The plight of birds in Europe is a good example. It is said, with some justification, that a line can be drawn across the continent north of which people watch birds and south of which they eat them. Some 700,000 "protected" birds are shot in Greece each year, Malta accounts for about 3 million a year, and Italy holds a shameful record of an astronomical 50 million birds killed and eaten each year! Most of these are small songbirds.

Another prominent form of overuse is the trafficking in wildlife and products derived from wild species. Much of this trade is illegal—worldwide, this "trade" is estimated to be at least $10 billion a year. It flourishes because some consumers are willing to pay exorbitant prices for such things as furniture made from tropical hardwoods like teak, exotic pets, furs from wild animals, traditional medicines from animal parts, and innumerable other "luxuries," including polar bear rugs, baskets made from elephant and rhinoceros feet, ivory-handled knives, and reptile-skin shoes and handbags. For example, some Indonesian and South American parrots sell for up to $10,000 in the United States, a panda-skin rug can bring in $25,000, and the gall bladder of the North American black bear, which is valued as folk medicine in Korea, can fetch as much as $2000. Such prices create a powerful economic incentive to exploit the species involved.

The long-term prospect of extinction does not curtail the activities of exploiters, because the prospect of a huge immediate profit outweighs it. Even when the species is protected, the economic incentive is such that poaching and black-market trade continue. Tiger and rhinoceros populations in the wild have declined drastically in the last two decades (a 90% decline for rhinos) and are on the brink of extinction. Driving this decline is the widespread but unfounded belief in Far Eastern countries that parts from these animals have medicinal or aphrodisiac properties. Customs records from South Korea revealed imports from China of 1.6 tons of tiger bones—the remains of at least 340 tigers.

The USFWS is the agency that has been given jurisdiction over the illegal trade in wildlife in the United States. Its Special Operations force of 235 agents actively investigates cases of wildlife smuggling and poaching. A major recent investigation into parrot smuggling, Operation Renegade, revealed an international ring extending to Australia, Africa, and South America involved in the capture and transport of many rare parrot species. The operation resulted in scores of arrests and convictions; the kingpin of the smuggling ring received 82 months of jail time and a $100,000 fine.

It is easy to place the blame on the economic greed of the persons doing the killing. However, equal blame must be shared by the consumers who offer the "reward." It is their ignorance or insensitivity to the fact that their money is fostering the extinction of invaluable wildlife that is fueling the situation.

Particularly severe is the growing fad for exotic "pets": fish, reptiles, birds, and house plants. In many cases, these plants and animals are taken from the wild; often only a small fraction even survives the transition. When they are removed from the natural breeding populations, they are no better than dead, in terms of maintaining the species. At least 40 of the world's 330 species of parrots face extinction because of this fad. Numerous species of tropical fish, birds, reptiles, and plants are thus headed toward extinction because of exploitation by the pet trade.

Poor management represents another form of overuse leading to a loss of biodiversity. Forests and woodlands are overcut for firewood, grasslands are overgrazed, game species are overhunted, fisheries are overexploited, and croplands are overcultivated. These practices not only deplete the resource in question but often set into motion a cycle of erosion and desertification, with effects far beyond the exploited area. The problem is explored in depth in Chapter 12.

Consequences of Losing Biodiversity

What will happen as more and more rare or unknown species pass from the Earth because of our activities? A few people will mourn their passing, but as one observer put it, the Sun will continue to come up every morning, and the newspaper will appear on our doorstep. Currently, we seem to be getting away with it, as evidenced by the fact that many species have already become extinct as a consequence of human activities, and life goes on. However, we really do not know what we are losing when we lose species. Some ecologists have likened the loss of biodiversity to an airplane flight, where we continually pull out rivets as the plane cruises along. How many rivets can we pull out before disaster occurs? Is this a wise activity? So far, we have gotten away with driving species to extinction, but the natural world is certainly less beautiful and more monotonous as a result.

Can ecosystems lose some species and still remain productive? Research evidence indicates that it is the dominant plants and animals that determine major ecosystem processes like energy flow and nutrient cycling. Thus, simplifying ecosystems by driving rarer species to extinction is not expected to lead to certain ecosystem decline. We must be careful not to justify preserving wild species and biodiversity with predictions of disasters. However, it is possible to lose what ecologists call **keystone species**—species whose role is absolutely vital for the survival of many other species in an ecosystem. For example, a group of insects called orchid bees play a vital role in tropical forests by pollinating trees. These bees travel great distances and are able to pollinate trees and other plants that are often widely separated from one another. If they disappear, many other plants, including major forest trees, will eventually follow. Or perhaps the keystone species in an ecosystem is a predator that keeps herbivore populations under control. Often,

keystone species are the largest animals in the ecosystem, and because of their size, they have the greatest demands for unspoiled habitat. Thus, wolves, elephants, tigers, and other large animals are considered "umbrella" species for whole ecosystems. If they can survive, there will likely still be enough undisturbed habitat for the other living species.

As we have seen, it is also possible to introduce species (e.g., purple loosestrife) to ecosystems that can become new dominants; they will impact both biodiversity and functionality as they crowd out existing species, and the results are very likely to be undesirable.

Biodiversity will continue to decline as long as we continue to remove and constrict the natural habitats in which wild species live. This loss is bound to be costly, because natural ecosystems provide vital services to human societies. Recreational, aesthetic, and commercial losses will be inevitable. The loss of wild species, therefore, will bring certain and unwelcome consequences, because it is linked directly to the degradation or disappearance of ecosystems (our focus in Chapter 12).

International Steps to Protect Biodiversity

Convention on Trade in Endangered Species. Endangered species outside the United States are only tangentially addressed by the ESA. However, under the leadership of the United States, the Convention on Trade in Endangered Species of Wild Fauna and Flora, or CITES, was established in the early 1970s. CITES is not specifically a device to protect rare species; instead, it is an international agreement (by 118 nations) that focuses on trade in wildlife and wildlife parts. The treaty recognizes three levels of vulnerability of species, the highest being species threatened with extinction. Covering some 30,000 species, restrictive trade permits and agreements between exporting and importing countries are applied to those species, sometimes with the result of a complete ban on trade if the CITES nations agree. Every two or three years the signatory countries meet at a Conference of Parties (COP). The latest (10th COP) was held in Zimbabwe in June 1997.

Perhaps the best known act of CITES was to ban the international trade in ivory in 1990 in order to stop the rapid decline of the African elephant (from 2.5 million animals in 1950 to about 350,000 today). In response to requests from the African nations, the 10th COP voted to lift the ban on ivory trade for three African countries (Namibia, Botswana, and Zimbabwe) that demonstrated the ability to maintain sustainable populations of their elephants. Opponents to this move fear that the trade will trigger a new wave of poaching and smuggling that will further decimate elephants in other African countries. Others argue that it is better to encourage the limited harvesting of animals than to risk having many African countries withdraw from CITES and initiate ivory trade on their own.

Convention on Biological Diversity. Although CITES provides for some protection of species that might be involved in international trade, it is inadequate to address broader issues pertaining to the loss of biodiversity. People of the industrialized nations have expressed a growing concern about the decline of well-loved species such as pandas, elephants, and rhinos, the rapid rate of deforestation in tropical rain forests and the feared loss of many species as a consequence, and inroads into lands occupied by indigenous peoples. It is clear that the main target of their concern is the plight of biological diversity in the developing countries, where most biodiversity is found. On the other hand, people in the developing countries regard their forests and animals as part of their natural resources; millions of rural individuals depend on access to these resources, and many governments look to their development as the key to more rapid economic growth. Thus, a problem arises wherein some of the world wants to see the protection of resources that are not in its territory but which it values nevertheless.

With the support of the UNEP, an ad hoc working group proposed an international treaty that would resolve these differences and form the basis of action to conserve biological diversity worldwide. After several years of negotiation, the **Convention on Biological Diversity** was drafted and became one of the pillars of the 1992 Earth Summit in Rio de Janeiro (Fig. 11–15). The Biodiversity Treaty, as the convention is called, was ratified in December 1993 and is now in force. The treaty establishes a Conference of the Parties as the agency that will provide oversight and report on its task during periodic meetings. The preamble sets forth basic guidelines for the treaty: a concern for the intrinsic value of biodiversity, its significance for human welfare, the sovereignty of a nation over its biodiversity, and the nation's obligations to protect and conserve biodiversity.

The treaty requires the following from the signatory countries:

- Adopt specific national biodiversity action plans and strategies.
- Establish a system of protected areas and ecosystems within the country.
- Establish policies that provide incentives to promote sustainable use of biological resources.
- Restore habitats that have been degraded.

▲ **FIGURE 11–15** *1992 Earth Summit.* The Earth Summit in Rio de Janeiro passed the Convention on Biological Diversity over the "no" vote of President Bush.

- Protect threatened species.
- Respect and preserve knowledge and practices of indigenous peoples.
- Respect the ownership of genetic resources by countries, and share the technologies developed from those resources.

Concern about access to genetic resources by the U.S. biotechnology industry led President Bush to refuse to sign the treaty in Rio in 1992. President Clinton signed the treaty in June 1993, but in order to become binding, the treaty must be ratified by a two-thirds majority vote of the U.S. Senate. Continued objections raised by the agricultural and biotechnology lobbies have led the Senate to shelve ratification, leaving both the treaty and U.S. involvement in it in a state of limbo. Proponents of the treaty have claimed that these concerns are not well founded; indeed, many leaders of the supposedly endangered industries support the treaty.

Stewardship Concerns

What we do about wild species and biodiversity reflects on our *wisdom* and our *values*. *Wisdom* dictates that we take steps to protect the biological wealth that sustains so many of our needs and economic activities. The scientific team that put together the "Global Biodiversity Assessment" focused on four themes in its prescriptive recommendations:

1. Reforming policies that often lead to declines in biodiversity. For example, many governments subsidize exploitation of natural resources and agricultural activities that consume natural habitats. These policies are often counterproductive, leading to the decline of important natural goods and services, which tend to be undervalued economically because they do not enter the market.
2. Addressing the needs of people who live adjacent to or in high-biodiversity areas or whose livelihood is derived from exploiting wild species. Although protecting biodiversity benefits entire societies, it is the people who are closest to the protected areas who are key to the success of efforts to maintain sustainability of the natural resources. They must be given ownership rights or communal-use rights to natural resources; they must be involved in the protection and management of wildlife resources. Experience has shown that when local people benefit from protection of wildlife, they will more readily take up stewardship responsibility for the wildlife.
3. Practicing conservation at the landscape level. Too often we tend to believe that a given refuge or park will adequately protect the biodiversity of a region, when in reality many wild species require larger contiguous habitats or a variety of habitats. For example, to address the needs of migratory birds, summer and winter ranges and flyways between all must be considered. Buffer zones should be created between major population centers and wildlife preserves. This requires cooperation between governments, private property owners, and corporations.

4. Promoting more research on biodiversity. We know far too little about how to manage diverse ecosystems and landscapes; much remains to be known about the species within the boundaries of different countries. For example, Costa Rica is engaged in a scientific inventory of its biodiversity, with an eye especially to the development of new pharmaceutical or horticultural products. Toward that end in the United States, the National Biological Service of the U.S. Department of Interior coordinates research like taxonomic studies designed to inventory the nation's biological resources.

Our *values* are also demonstrated in our approach to wild species. Do we consider the natural world simply as grist for our increasing consumptive demands? Or do we treat nature as something we have received as a stewardship responsibility and are challenged to manage sustainably and pass on to our children and their children? Because it varies how different people answer these questions, and why, some people feel a need for alternative sets of values, other than or in addition to simply survival or utilitarianism to guide our course into the future. What each individual turns to will depend on their beliefs, needs, education, religious, or ethnic background. For instance, some environmentalists have embraced Deep Ecology, a movement that reminds humans that we are a part of the Earth and not separate from it, and thus places nature at the forefront of human awareness. For others, religious faith forms the basis for their interaction with nature. Significantly, Harvard University's Center for the Study of World Religions has held a series of conferences exploring each of the major world faiths and its ecological views. The American Academy of Arts and Sciences, the U.N., and the American Museum of Natural History have held interdisciplinary gatherings on faith and the environment.

Further strategies to address the loss of biodiversity must focus on preserving the natural ecosystems that sustain wild species. In the next chapter, we turn our attention to those ecosystems—how we depend on them, what we are doing to them, and what we must do to resist the forces that seem to be leading to a future in which all that is left of wild nature is what we have managed to protect in parks, preserves, and zoos.

ENVIRONMENT ON THE WEB

MEASURING BIODIVERSITY

Ecologists have long argued about the best ways to track the diversity and the integrity—soundness—of ecosystems. The debate centers on whether we should measure ecosystem structure or function, or both. This is like trying to decide whether you will measure the adequacy of your house by counting the number of rooms, or by assessing the amount of energy the house uses annually—both measurements provide useful insight.

The traditional way of measuring structural diversity is simply to count the number of species present and the number of individuals in each. A system with many species and few individuals in each is considered diverse relative to one with few species, each with many individuals. Such an approach, however, ignores the complexity of the abiotic environment and overemphasizes the importance of common species.

More recent approaches combine measurements of several factors that together are thought to reflect ecosystem integrity. For example, in recent years, Canada and the United States have been grappling with the challenge of measuring biodiversity in the Great Lakes basin, as part of their joint responsibility for managing the health of that shared ecosystem. Several key points have emerged in their discussions. First, because ecosystems are organized in a hierarchical—or layered—fashion (for example, a food chain), it may be necessary to measure biodiversity at several levels. This layering makes it possible to observe good species richness and primary productivity at a soil microorganism level, but poorer structure and function would be found at a scale of hundreds of hectares. Since the behavior of the ecosystem depends on the integrity of its component parts, overlooking any one layer can skew any measurements of diversity.

Second, the "right" measurement depends on the question being asked. If, on the one hand, the question is "What species are present?" then traditional species counts are a useful tool. Conversely, if the question is "How genetically diverse are these species?" it may be necessary to map the genetic "fingerprint" of each population.

However, if you ask "How well is the ecosystem functioning?" or "How well is the ecosystem able to withstand stress?" then more integrative measurements, such as net primary productivity (the ecosystem's ability to produce new biomass), community composition (e.g. relative proportions of trees, shrubs, and herbaceous species), or even gross soil erosion provide insight into how the ecosystem works. Hence, several structural and functional measurements may be needed to develop an adequate assessment of a system's biodiversity.

The fourth principle of ecosystem sustainability—*For sustainability, biodiversity is maintained*—implicitly denotes that more biodiversity is better than less biodiversity and that biodiversity can be measured. However, sound science doesn't always give us clear answers about what level of biodiversity exists in a given area. Regardless, measuring the biodiversity of a system is a form of developing a baseline condition against which to evaluate future trends. It is a form of understanding the cause-and-effect relationships that are becoming more and more essential to us in our stewardship of Earth's systems.

Web Explorations

The Environment on the Web activity for this essay describes the challenges involved in transboundary management of ecosystem diversity. Go to the Environment on the Web activity (select Chapter 11 at http://www.prenhall.com/nebel) and learn for yourself:

1. how the World Wildlife Fund has measured North American biodiversity;
2. about global patterns of biodiversity; and
3. how biodiversity protection can sometimes conflict with other transboundary issues.

A suggested time frame for each of these exercises is 10–30 minutes.

REVIEW QUESTIONS

1. Define biological wealth and apply the concept to human use of that wealth.
2. Define instrumental value and intrinsic value as they relate to determining the worth of natural species.
3. What are the four categories into which human dependence on natural species can be divided?
4. What means are used to preserve game species, and what are some problems emerging from the adaptations of many game species to the humanized environment?
5. How does the Endangered Species Act (ESA) preserve threatened and endangered species in the United States? What are the shortcomings and controversies of this act? Illustrate with examples.
6. What is biodiversity? What are the best current estimates of its extent?
7. Describe both the known and estimated losses in biodiversity.
8. How do habitat conversion, fragmentation, and simplification affect biodiversity?
9. How do population, pollution, exotic species, and overuse affect biodiversity?
10. What are the possible consequences of species extinctions?
11. What is CITES, and what are its limitations? Give three requirements of the Convention on Biological Diversity.
12. Describe ways in which our wisdom and values are reflected through our actions as stewards of biodiversity.

THINKING ENVIRONMENTALLY

1. Choose an endangered or threatened animal or plant species and research what is currently being done to preserve it. What dangers is it subject to? Write a protection plan for it.
2. You have been given the task of maintaining or increasing biodiversity on a small island. What steps will you take?
3. Some people argue that each individual animal has an intrinsic right to survival. Should this right extend to plants and microorganisms? What about the *Anopheles* mosquito, which transmits malaria? Tigers that kill people in India? Bacteria that cause typhoid fever? Defend your position.
4. Monies for enforcing the Endangered Species Act are limited. How should we decide which species are the most important to save?
5. Is it right to use animals for teaching and research? Justify your answer.

WEB REFERENCES

On-line resources for this chapter can be found on the World Wide Web at: **http://www.prenhall.com/nebel**. Click on Chapter 11 on the chapter selector.

PART FOUR

Energy

The Alaska oil pipeline snakes its way down through tundra and forest to the port of Valdez, an intrusion into a beautiful and fragile wilderness that is symbolic of the dilemma facing our society. The industrial countries are so deeply committed to fossil fuel and nuclear energy that we cannot imagine an economy apart from our rising use of these energy sources. Developing countries are following the same example, hoping to raise the wellbeing of their people. Fuel oil prices are at their lowest in decades, leading many to assume that the supply of oil is unlimited. However, a recent analysis of the oil reserves suggests that oil production will peak within ten years and then begin a decline that will eventually drive prices up to new heights. Looming in the background is the specter of global climate change and the struggle of dealing with this threat that is tied so closely to the use of fossil fuels.

Many deny that this threat exists-the power industry, the fossil fuel industry, automobile manufacturers, and some highly placed politicians. Recently though there have been significant changes to this attitude. For example, the new chairman of Ford Motor Company warns his fellow automakers not to dismiss global warming; Royal Dutch/Shell will invest more than half a billion dollars on renewable energy technology; and British Petroleum will cut their greenhouse gas emissions well below the amount called for in the Kyoto Protocol. But what about other forms of energy? Nuclear energy is an option that does not generate greenhouse gases, and is already well entrenched (but slipping) worldwide. Or is renewable energy the energy of the future? How can we develop an alternate energy source to fuel our vehicles? These are some of the vital questions we address in this section.

The Alaskan Pipeline.

13

Energy from Fossil Fuels

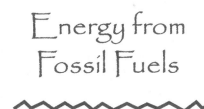

Key Issues and Questions

1. The development of modern civilization has depended greatly on the development of energy sources. How have the three primary fossil fuels been harnessed in recent history?

2. Much of total energy use is devoted to generating electrical power. How are fossil fuels coupled with electrical power, and what are the major environmental impacts of these processes?

3. What are the major categories of primary energy use and the energy sources that match those uses? How does an analysis of end-use energy demand help us to design energy supplies?

4. Crude oil is an essential energy source for transportation. What are the reserves of this resource, and how are they estimated?

5. The United States now imports more than 50% of the crude oil it uses. What led to this dependency on foreign oil? What are the problems of acquiring and shipping this quantity of oil?

6. Oil shortages and oil gluts have rocked the world economy in recent years. What are the reasons for these violent shifts, and what are their long-term consequences?

7. Coal and natural gas are also important fossil fuels in the United States. What are the reserves of these resources, and how are they presently being put to use?

8. Sustainable energy options are clearly desirable for the development of a sustainable future. What are some options that will meet our energy needs with the least economic and environmental consequences?

The *Sea Empress*, a 147,000-ton tanker, left the Firth of Forth in Scotland in early February, 1996, laden with 131,000 tons of crude oil bound for the Texaco Refinery in Milford Haven, Wales. The ship approached the entrance to the harbor shortly after 8 P.M. on February 15, when it ran aground and immediately began leaking oil (Fig. 13–1). After pulling off of the rocks, the ship lost steerage and ran aground again, still spilling oil. Tugs managed to pull it away from these rocks, but by that time the ship was listing and severely damaged. Subsequently, heavy weather set in and battered the salvage vessels attempting to recover oil and stabilize the ship; two days later the *Sea Empress* was swept back on to the rocks, damaging her again and releasing even more oil. By the time tugs were finally able to tow the ship to

harbor and the remaining oil was offloaded, some 72,000 tons of crude oil had been spilled into St. George's Channel. The oil impacted 200 kilometers of the Welsh shoreline and hundreds of km^2 of coastal ocean waters. Thousands of seabirds died in the days following the spill and for several months afterward. Some 7000 dead and dying birds were recovered, and wildlife personnel estimated that an additional 9000 had perished. Unknown numbers of seals and porpoises died. A $20 million fishing industry was devastated, putting some 700 fishers out of work indefinitely. An enormous and costly cleanup was initiated, and the coastline was freed of large oil accumulations within a few months.

The cause of the accident is still under criminal investigation, but human error was clearly indicated. Two years later, the ban on fishing off of southwest Wales was lifted, and the

Off loading of crude oil tanker in Louisiana. Over 50% of U.S. crude oil consumed must now be imported.

309

▶ **FIGURE 13–1** *The Sea Empress oil spill.* On February 15, 1996, the tanker *Sea Empress* ran aground on the coast of Wales, spilling 72,000 tons of crude oil and killing thousands of seabirds, fish, and porpoises.

seabird populations appear to be reestablished and increasing on their nesting islands. The spill could have been much worse if the weather had been calmer; the heavy seas and offshore winds dispersed the majority of the oil, keeping it from being deposited on the fragile intertidal shoreline. On the other hand, if the *Sea Empress* had been a double-hulled vessel, it might not have ruptured at all.

This spill is just one of many that have occurred in recent years, and it certainly will not be the last. Oil for the global economy is moved from the oil fields and ports and transported by oil tankers to refineries around the world. Millions of gallons are in transit at any given time, and it is inevitable that

oil spills will occur. The environmental consequences of oil spills are chalked up as one of the many costs of doing business as we mine the earth for fossil fuels and employ them in generating energy to fuel our economic enterprises. In future chapters we will encounter some of the other costs, such as air pollution and its impact on human and environmental health, and rising greenhouse gases leading to global climate change. Our objective in this chapter is to gain an overall understanding of the sources of energy that support our transportation, homes, and industry and of their relative sustainability. We will consider the consequences of our dependence on fossil fuels and begin to look at options for the future.

13.1 Energy Sources and Uses

Try to imagine what life would be like without all the energy we use. There would be no transportation beyond walking, horseback, or boats powered by paddles or wind. There would be no home appliances, air conditioners, lighting, or communications equipment. There would be little manufacturing beyond what could be done by hand, on looms, or on potters' wheels. Farming would depend on hand labor and on

draft animals such as horses and oxen. In short, anything beyond a primitive existence is irrevocably tied to harnessing energy sources. Let's look at this matter in a little more detail.

Harnessing Energy Sources: An Overview

Throughout human history, the advance of technological civilization has been tied to the development of energy sources (Fig. 13–2). In early times (and even today in less-

▼ **FIGURE 13–2** *Energy and time.* The development of energy sources has in large part supported the development of civilization.

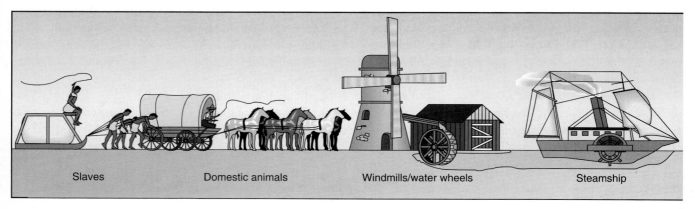

| Slaves | Domestic animals | Windmills/water wheels | Steamship |

developed regions), the major energy source was muscle power. Some people lived in relative luxury by exploiting the labor of others—slaves, indentured servants, and minimally paid workers. Human labor was supplemented to some extent by the use of domestic animals for agriculture and transportation, by water or wind power for milling grain, and by the Sun.

By the early 1700s, inventors had already designed many kinds of machinery. The limiting factor was a continual power source to run them. The breakthrough that launched the Industrial Revolution was the development of the steam engine in the late 1700s. In a steam engine, water is boiled in a closed vessel to produce high-pressure steam, which pushes a piston back and forth in a cylinder. Through its connection to a crankshaft, the piston turns the drivewheel of the machinery. Steam engines rapidly became the power source for steamships, steam shovels, steam tractors, steam locomotives, and stationary engines to run sawmills, textile mills, and virtually all other industrial plants.

At first, the major fuel for steam engines was firewood. Then, as demands for energy increased and firewood around industrial centers became scarce, coal was substituted. By the end of the 1800s, coal had become the dominant fuel and remained so into the 1940s. In addition to being used as fuel for steam engines, coal was widely used for heating, cooking, and industrial processes. In the 1920s, coal provided 80% of all energy used in the United States, with similar percentages in other industrializing countries.

Though coal and steam engines powered the Industrial Revolution that greatly improved life for most people, they had many drawbacks. The smoke and fumes from the numerous coal fires made air pollution in cities far worse than anything seen today. Writers have recorded that often they could not see as far as a city block away because of the smoke. Coal is also notoriously hazardous to mine and dirty to handle, and burning coal results in large quantities of ashes that must be removed. As for steam engines, because of the size and bulk of the boiler, the engines were heavy and awkward to operate (Fig. 13–3). Often, the fire had to be started several hours before the engine was put into operation in order to heat the boiler sufficiently.

In the late 1800s, simultaneous development of three technologies—the internal combustion engine, oil-well drilling, and the refinement of crude oil into gasoline and other liquid fuels—combined to provide an alternative to steam power. The replacement of coal-fired steam engines and furnaces with petroleum-fueled engines and oil furnaces was an immense step forward in convenience. Also, air quality greatly improved as cities were gradually rid of the smoke and soot from burning coal. (It was only in the 1960s, with the tremendous proliferation of cars, that pollution from gasoline engines became a problem.) Further, the gasoline internal-combustion engine provides a valuable power-to-weight advantage that allowed rapid advances in technology. A 100-horsepower gasoline engine weighs only a tiny fraction of what a 100-horsepower steam engine and its boiler weigh, and jet engines have an even greater power-to-weight ratio. Automobiles and other forms of transportation would be cumbersome, to say the least, without this power-to-weight advantage, and airplanes would be impossible.

Development throughout the world, so far as it has occurred since the 1940s, has been largely predicated on technologies that consume gasoline and other fuels refined from crude oil. Replacement of one energy technology with another is a very gradual process, however, because it is more cost effective to use existing machinery until it wears out before replacing it. It was not until the late 1940s that crude oil became the dominant energy source for the United States. Since then, however, its use has become the mainstay not only for the United States but also for most other countries, both developed and developing. Crude oil currently provides about 39% of the total U.S. energy demand. The story is similar throughout the rest of the world, although the timing of events differs. Earlier we noted that many poor regions of the world remain dependent on firewood. In the countries of Eastern Europe and in China, coal is still the dominant fuel. In the United States, coal is still a major source of energy, providing about 23% of total energy fuel input.

Natural gas, the third primary fossil fuel (the other two being oil and coal), is found in association with oil or is found during the search for oil. As it is largely methane, which produces only carbon dioxide and water as it burns, natural gas burns more cleanly than oil; thus, in terms of pollution, it is a more desirable fuel. Despite the obvious fuel potential of natural gas, at first there was no practical way to transport it from wells to consumers. Any gas released from oil fields was

▶ FIGURE 13–3 *Steam power.*
Steam-driven tractors marked the beginning of the industrialization of agriculture. Today, tractors half the size do 10 times the work and are much easier to operate.

▶ FIGURE 13–4 *Energy consumption in the United States.* Total consumption and major primary sources are shown here, from 1860 to 1997. Note how the mix of primary sources has changed over the years and how the total amount of energy consumed has continued to grow. Note also the skyrocketing increase in use of oil after World War II (1945) as private cars and car-dependent commuting became common. (Source: Data from Energy Information Administration, U.S. Department of Energy, *Annual Energy Review 1997,* July 1998.)

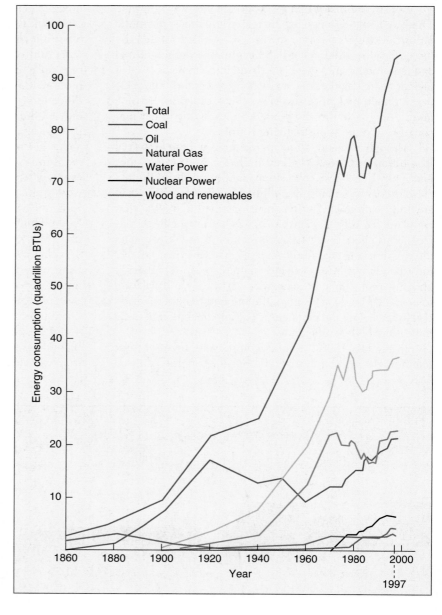

(and in many parts of the world still is) flared—that is, vented and burned in the atmosphere, a tremendous waste of valuable fuel. Gradually the United States constructed a network of pipelines connecting wells and consumers. With the completion of these pipelines, the use of natural gas for heating, cooking, and industrial processes escalated rapidly because of its cleanliness, convenience (no storage bins or tanks are required on the premises), and relatively low cost. Currently, natural gas—provides about 24% of U.S. energy demand.

Thus, three fossil fuels—crude oil, coal, and natural gas—provide 86% of U.S. energy fuel input. The remaining 14% is provided by nuclear power, water power, and renewables, which, along with coal, are used for the generation of electrical power. This changing pattern of energy sources in the United States is shown in Fig. 13–4. The picture is similar for other developed countries of the world, although the percentages differ somewhat depending on the energy resources of any given country.

Electrical Power Production

A considerable portion of total energy used is electrical power. Electrical power is called a **secondary energy source** because it depends on a **primary energy source** (coal or water power, for instance) to turn the generator.

Generators were invented in the nineteenth century. In 1831, the English scientist Michael Faraday discovered that passing a coil of wire through a magnetic field causes a flow of electrons—an electrical current—in the wire. An electric generator is basically a coil of wire that rotates in a magnetic field or that remains stationary while a magnetic field is rotated around it (Fig. 13–5). This seems like cost-free energy until we consider the first and second laws of thermodynamics (Chapter 3). The phenomena described by these important

energy laws prevent us from getting something for nothing, or even from breaking even. As the current flows in the wire, it creates a new magnetic field that is in opposition to the first and thus resists the movement. Also, some of the energy created is converted to thermal heat energy and lost. For every 100 calories of electrical energy produced, more than 100 calories of primary energy must be expended to turn the generator. This energy-losing proposition is justified only because electrical power is far more useful than coal, for example. Therefore, it is worth burning 300 calories of coal to obtain 100 calories of electrical power, a typical conversion ratio for electric power-generating plants.

In the most widely used technique for generating electrical power, a primary energy source is employed to boil water, creating high-pressure steam that drives a **turbine**—a sophisticated "paddle wheel"—coupled to the generator (Fig. 13–6a). The combined turbine and generator are called a **turbogenerator**. Any primary energy source can be harnessed for boiling the water; coal, oil, and nuclear energy are most commonly used at present, but burning refuse, solar energy, and geothermal energy (heat from Earth's interior) may be more widely used in the future.

In addition to steam-driven turbines, gas and water (hydro-) turbines are also used. In a gas turbogenerator, the high-pressure gases produced by the combustion of a fuel (usually natural gas) drive the turbine directly (Fig. 13–6b). For hydroelectric power, water under high pressure—at the base of a dam or at the bottom of a pipe from the top of a waterfall—is used to drive a hydroturbogenerator (Fig. 13–6c). Wind turbines are also coming into use (see Fig. 15–20). A utility company supplying electricity to its customers will employ a mix of power sources depending on availability and the demand cycle.

Fluctuations in Demand. Where does your electricity come from? The answer to this question is: It depends. Most utility companies are linked together in what are called *pools*. In New England, for example, The New England Power Pool (NEPOOL) ties together, through 8100 miles of transmission lines, some 400 electric power-generating plants. NEPOOL is responsible for balancing electricity supply and demand regardless of daily or seasonal fluctuations. Generating capacity of NEPOOL utilities is 24,300 megawatts (MW). A variety of energy sources are responsible for this capacity, as shown in Table 13-1. The greatest demands on electrical power are in the summer months, when air-conditioning is at its highest use. The peak load (highest demand) for 1998 came during a hot day in July, at 21,400 MW. A power pool must accommodate the daily and weekly electricity demand as well as seasonal fluctuations and will employ different mixes of energy sources depending on the demand cycle.

The typical pattern of daily and weekly demand is shown in Fig. 13–7. Baseload plants that provide the constant supply of power are typically the large coal-burning and nuclear plants, often with capacities as high as 1000 MW. As the demand rises during the day, the utility draws on additional plants that can be turned on and off—the intermediate and peakload power sources. These are the gas turbines and diesel

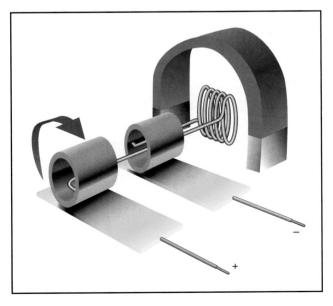

▲ **FIGURE 13–5** *Principle of an electric generator.* Rotating a coil of wire in a magnetic field induces a flow of electricity in the wire.

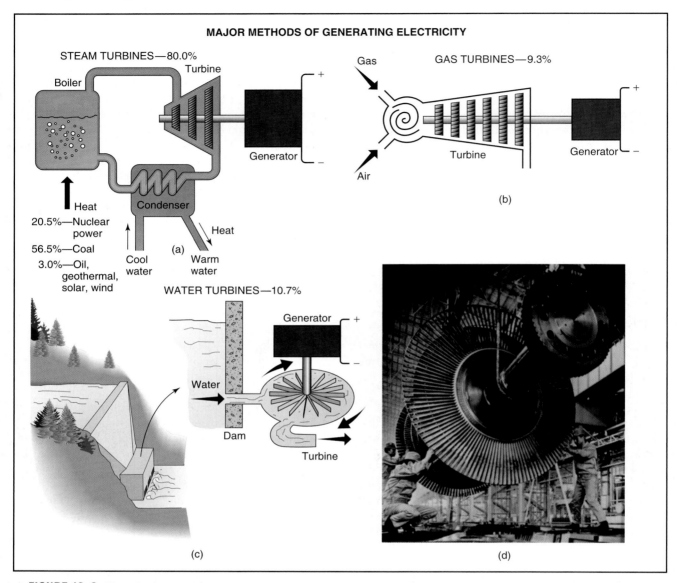

▲ FIGURE 13–6 _Electrical power generation._ Electricity is produced commercially by driving generators with (a) steam turbines, (b) gas turbines, and (c) water turbines. The percentage of the U.S. electricity supply derived from each source (1997) is indicated. Small amounts of power are now also coming from solar wind energy, solar thermal energy, and photovoltaics (solar cells). (d) Components of a steam turbine being assembled.

Table 13-1 Energy Sources for Generating Electrical Power (% of total)			
Source	New England	United States	OECD Countries*
Nuclear power	30.7	20	24
Natural gas	20.1	9	7
Coal	19.3	57	31
Oil	14.2	2	5
Hydro	9.5	11	5
Other (wood, refuse, renewables)	6.2	<1	28

*Organization for Economic Cooperation and Development (includes all the developed countries except for those of the former Soviet Union)

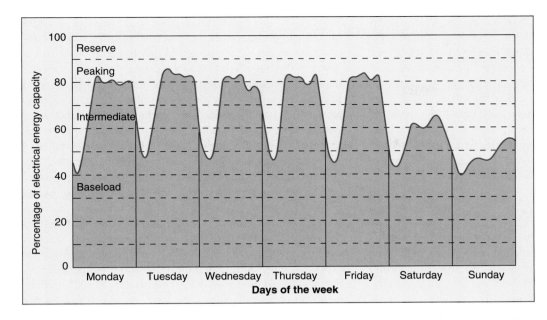

◀ **FIGURE 13–7** *Weekly electrical demand cycle.* Electrical demand fluctuates daily, weekly, and seasonally. A variety of generator types must be employed to meet baseload, intermediate, and peakload electricity needs.

plants, or the use of pumped storage hydroelectric power (water is pumped to a reservoir using off-peak power and allowed to flow through turbines when needed). Occasionally, a deficiency in available power will prompt a brownout (reduction in voltage) or blackout (total loss of power), throwing whole regions into a temporary loss of power. Such events are most likely during times of peak power demand and may be precipitated by a sudden loss of power by the shutdown of a baseload plant.

Clean Energy? Electrical power is often promoted as being the ultimate clean, nonpolluting energy source. This is true at the point of use; using electricity creates essentially no pollution. Hidden, however, is the fact that pollution is simply transferred from one part of the environment to another: Switching to "clean" electric heat implies more demand for electricity, and it must be generated from coal, hydropower, nuclear energy, or alternative energy sources. Coal-burning power plants are the major source of acid deposition and contribute substantially to other air-pollution problems (Chapter 22). If hydroelectric power is used, creating a dam and reservoir involves displacing people, farmland, and wildlife and disrupts the migration of fish, such as salmon. Finally, nuclear power, discussed in Chapter 14, is a much feared technology because of the potential for accidents and the problem of disposal of the nuclear waste products. Additional adverse effects of using electrical energy apply to other parts of the fuel cycle, such as mining coal or processing uranium ore.

This problem of transferring pollution from one place to another becomes even more pronounced when we consider efficiency: Nearly 300 Calories of coal, for example, must be burned to produce 100 Calories of electricity. In other words, the thermal production of electricity has an *efficiency of only 30%–40%*. The 60%–70% loss is accounted for partly by some heat energy from the firebox going up the chimney and partly by a large amount of heat energy remaining in the

spent steam as it comes out the end of the turbine. (This is unavoidable, a consequence of the need to maintain a high heat differential between the incoming steam and the receiving turbine so as to maximize efficiency.) Heat energy will go only toward a cooler place, so this heat energy cannot be recycled into the turbine. The most common practice is simply to dissipate it into the environment by means of a condenser. In fact, the most conspicuous features of coal-burning or nuclear power plants are the huge cooling towers used for this purpose (Fig. 13–8).

As an alternative to using cooling towers, water from a river, lake, or ocean can be passed over the condensing system (Fig. 13–6a). Thus, the waste heat energy is transferred

▲ **FIGURE 13–8** *Cooling towers.* These structures that are the most conspicuous feature of both coal-fired and nuclear power plants are necessary to cool and recondense the steam from the turbines before returning it to the boiler. Air is drawn in at the bottom of the tower, over the condensing apparatus, and out the top by natural convection. In some systems, cooling may be aided by the evaporation of water that condenses in a plume of vapor as the moist air exits the top of the tower, as seen in this photo.

to the body of water. All the small planktonic organisms, both plant and animal, drawn through the condensing system with the water are cooked, and the warm water added back to the waterway may have deleterious effects on the aquatic ecosystem. Therefore, waste heat energy discharged into natural waterways is referred to as **thermal pollution**.

Electricity is a form of energy that is indispensable to modern societies as they are currently structured. Worldwide, the electrical power industry generates revenues valued at more than $800 billion a year and at the same time generates a disturbing array of environmental problems.

Matching Sources to Uses

In addressing questions of whether energy resources are sufficient and where we can get additional energy, we need to consider more than just the source of energy, because some forms of energy lend themselves well to one use but not to others. For example, the U.S. transportation system of highways, cars, and trucks is virtually 100% dependent on liquid fuels. Virtually all other machines for moving, such as tractors, airplanes,

locomotives, and bulldozers, are likewise dependent on liquid fuels. Reserves of crude oil are needed to support this transportation system. The nuclear industry has promoted nuclear power as the major solution to the energy shortage. However, nuclear power will do little to mitigate our demand for crude oil because it is suitable only for boiling water to drive turbogenerators, which will do essentially nothing for our transportation system. Although this situation might change with more widespread use of electric vehicles, their practicality for long distances still requires a technological breakthrough in the form of a low-cost, lightweight battery that will store large amounts of power.

Primary energy use is commonly divided into four categories: (1) transportation, (2) industrial processes, (3) commercial and residential uses (heating, cooling, lighting, and appliances), and (4) generation of electrical power, which is used in categories (2) and (3). The major pathways from the primary energy sources to the various end uses are shown in Fig. 13–9. Observe again that transportation is 100% dependent on petroleum, whereas nuclear power, coal, and water power are largely limited to the production of electrical power. Natural gas and oil are more versatile energy sources.

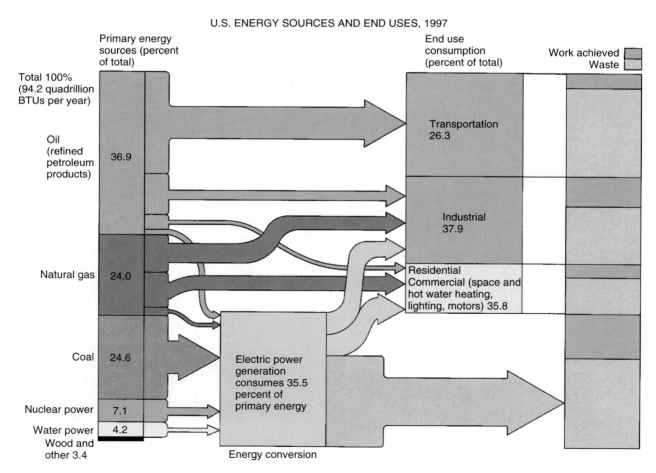

U.S. ENERGY SOURCES AND END USES, 1997

▲ **FIGURE 13–9** *Pathways from primary energy sources to end uses.* Only major pathways are shown. Note that end uses are connected to primary sources in specific ways. Also note the large percentage of energy that is wasted as a large portion of the energy consumed is converted to heat and lost. (Source: Data from Energy Information Administration, U.S. Department of Energy, *Annual Energy Review 1997*, July 1998.)

Figure 13–9 also shows the proportion of consumed energy that goes directly to waste heat energy rather than for its intended purpose. Some waste is inevitable, as dictated by the second law of thermodynamics, but the current losses are much greater than necessary. The efficiency of energy use can be at least doubled for some uses—we can make cars that get twice as many miles per gallon and appliances that consume half as much power, for example. For uses like electric power generation, however, somewhat lower increases in efficiency can be expected. Increasing the efficiency of energy use would have the same effect as increasing the energy supply, at far less expense. Such increases in efficiency are the essence of energy conservation and are discussed in more detail later in this chapter.

In addressing our energy future, we need to consider more than gross supplies and demands. We must examine how particular primary sources can be matched to particular end uses. Further, we need to consider the most cost-effective ways of balancing supplies and demands—by increasing supplies or by decreasing consumption through *conservation*, *efficiency*, and *demand management*.

13.2 The Exploitation of Crude Oil

The United States has abundant reserves of coal, adequate reserves for the time being of natural gas, and the potential for expansion of nuclear power. Therefore, there are no shortage crises looming in these areas now. (Note that we are setting aside the environmental concerns of coal and nuclear power for the moment.) Petroleum to fuel our transportation system is another matter. U.S. crude oil supplies fall far short of meeting demand; we are now dependent on foreign sources for over 50% of our crude oil, and this dependency is steadily growing as our reserves are diminished. These foreign sources come at increasing costs: trade imbalances, necessary military actions, pollution of the oceans, and coastal oil spills. Let us examine this situation, starting with where crude oil comes from and how it is exploited.

How Fossil Fuels Are Formed

The reason crude oil, coal, and natural gas are called *fossil fuels* is that all three are derived from the remains of living organisms (Fig. 13–10). In the geological era, 200 to 500 million years ago, freshwater swamps and shallow seas, which supported abundant vegetation and phytoplankton, covered vast areas of the globe. Anaerobic conditions in the lowest layers of such bodies of water impeded the respiration of decomposers and hence the breakdown of detritus. As a result, massive quantities of dead organic matter accumulated. Over millions of years, this organic matter was gradually buried under layers of sediment and converted by pressure and heat to coal, crude oil, and natural gas. Which fuel was formed depended on the nature of the organic material buried, the specific conditions, and time involved. Just as anaerobic treatment (digestion) of sewage sludge yields methane gas and a residual organic sludge, a similar anaerobic process occurred

with the buried vegetation. Natural gas is the trapped methane, and crude oil represents the residual "sludge." Coal is highly compressed organic matter—mostly leafy material from swamp vegetation—that decomposed relatively little.

Formation of additional fossil fuels by natural processes may be continuing to this day. However, they cannot possibly be considered renewable resources: We are using fossil fuels far faster than they ever formed. To accumulate the amount of organic matter that the world now consumes each day, 1000 years are needed. Recognizing how fossil fuels were formed, we can easily see that these resources are finite; sooner or later we will run out. The pertinent question is, When?

Crude Oil Reserves vs. Production

The total amount of crude oil remaining represents Earth's oil resources. Finding and exploiting these resources is the problem. The science of geology provides information about the probable locations and extents of ancient shallow seas. On the basis of their knowledge and their field experience, geologists make educated guesses as to where oil or natural gas may be located and how much may be found. These educated guesses are the world's **estimated reserves**. Of course, such estimates may be far off the mark. There is no way to determine whether estimated reserves actually exist except by the next step, exploratory drilling.

If exploratory drilling strikes oil, further drilling is conducted to determine the extent and depth of the **oil field**. From this information, a fairly accurate estimate can be made of how much oil can be economically obtained from the field. This amount then becomes **proven reserves**. Amounts are given in barrels of oil (1 barrel = 42 gallons), and the content of each reserve field is expressed in probabilities. Thus, a 50% probability or P50 of 700 million barrels for a field means that 700 million barrels is just as likely as not to come from the field. Proven reserve estimates can range from P10 to P90; obviously, a P90 estimate is the more reliable figure, but oil producers are often vague about probabilities and might prefer to use a P10 estimate in order to leave the impression of a large reserve for political or economic reasons. Also, proven reserves hinge on the economics of extraction. Thus, such reserves may actually increase or decrease with the price of oil, because higher prices justify exploiting resources that would not be worth extracting for lower prices. The final step, the withdrawal of oil or gas from the field, is called **production** in the oil business. Of course, *production* as used here is a euphemism. "It is production," says ecologist Barry Commoner, "only in the sense that a boy robbing his mother's pantry can be said to be producing the jam supply." In reality, this step is *extraction* from Earth.

Production from a given field cannot proceed at a constant rate, because crude oil is a viscous fluid held in spaces in sedimentary rock as water is held in a sponge. Initially, the field may be under so much pressure that the first penetration produces a gusher. Gushers, however, are generally short-lived, and the oil then seeps slowly from the sedimentary rock

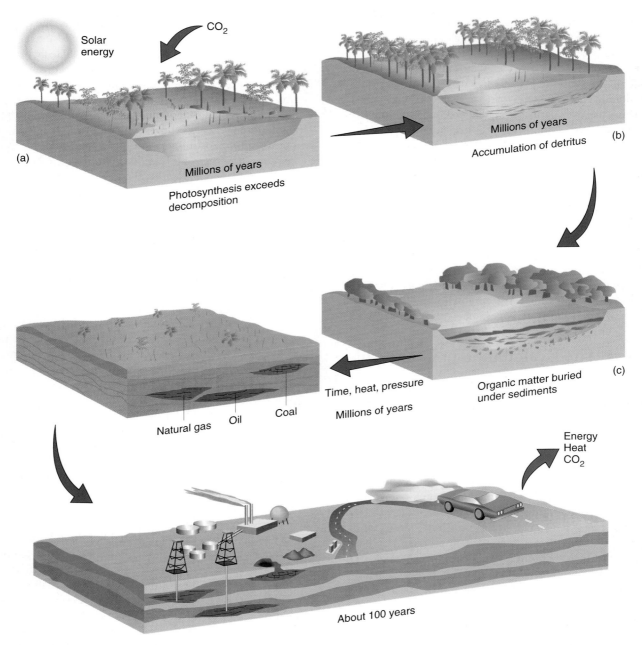

▲ **FIGURE 13–10** *Energy flow through fossil fuels.* Coal, oil, and natural gas are derived from biomass that was produced many millions of years ago. Deposits are finite and, since formative processes require millions of years, they are nonrenewable.

into a well from which it is withdrawn. In general, *maximum annual production is limited to about 10% of the remaining reserves.* Consider this rule in terms of maximum production from a field having 100 million barrels of proven reserves. As shown in Fig. 13–11, the first year's maximum production is 10 million barrels (10% of 100). Then, production in subsequent years decreases as the reserve is drawn down.

Observe that there is no sudden "running out" in the sense that a car runs out of gas. Instead, the production from given proven reserves gradually diminishes according to the fact that annual production can be no more than 10% of

remaining reserves. Production can be increased or even kept at the first year's level, only if additional reserves are found, so proven reserves are always at least 10 times the desired annual rate of production.

Then, economics—the price of a barrel of oil—comes into play. No oil company will spend more money for extracting oil than they expect to make selling it. At the current price of crude oil, around $14 per barrel, it is economical to extract no more than 35% of the oil resource in a given field. Further extraction involves more costly techniques that current prices cannot justify. Thus, an increase in price does indeed

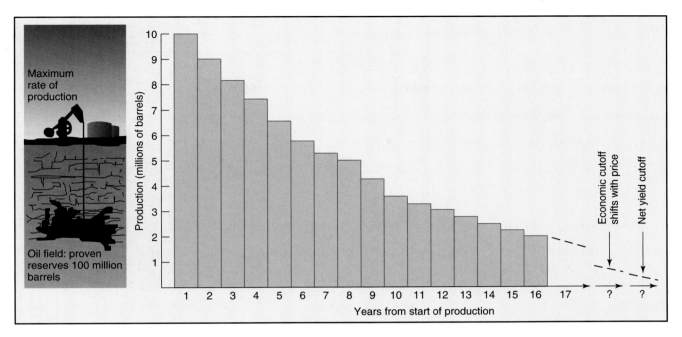

▲ **FIGURE 13–11** *The 10% rule.* No more than about 10% of the oil remaining in a field can be extracted in any given year. Consequently, maximum production from proven reserves invariably declines, but since there is always some oil remaining, there is no "running out" as such.

make more reserves available. The reverse may also occur. A practical example was experienced in the 1970s and early 1980s, when increasing oil prices justified reopening old oil fields and created an economic boom in Texas and Louisiana. Then an economic bust occurred in the late 1980s, as a drop in oil prices caused the fields to be shut down again.

Declining U.S. Reserves and Increasing Importation

Consideration of the relationships among production, price, and exploitation of reserves will help us understand the U.S. energy dilemma. A brief history is in order. Up to 1970, the United States was largely oil-independent. That is to say, exploration was leading to increasing proven reserves, such that production could basically keep pace with growing consumption (Fig. 13–12). But 1970 marked a significant turning point; new discoveries and hence reserve additions fell short, and production turned down while consumption continued its rapid growth. This downturn was a complete vindication of the predictions of late geologist M. King Hubbart, who proposed that oil exploitation in a region would follow a bell-shaped curve. Hubbart observed the pattern of U.S. exploitation and predicted that U.S. production would peak between 1965 and 1970. At that point, about half of the available oil will have been withdrawn and production would decline gradually as reserves were exploited. Fig. 13–12 demonstrates a "Hubbart" curve in progress for U.S. oil production.

To fill the energy gap between increasing consumption and falling production, the United States became increasingly dependent on imported oil, primarily from the Arab countries of the Middle East. European countries and Japan did likewise. Since imported oil cost only about $2.30 per barrel in the early 1970s and Middle Eastern reserves were more than adequate to meet the demand, this course seemed to present few problems.

The Oil Crisis of the 1970s

Thus, in the early 1970s, the United States and most other highly industrialized nations were becoming increasingly dependent on oil imports. A group of predominantly Arab countries known as **OPEC**, the Organization of Petroleum Exporting Countries, formed a cartel and agreed to restrain production and raise prices. Imported oil began to cost more in the early 1970s, and then, in conjunction with an Arab-Israeli war in 1973, OPEC initiated a boycott of oil sales to countries like the United States that gave military and economic support to Israel. The effect was almost instantaneous, because we depend on a fairly continuous flow from wells to points of consumption. (Compared with total use, little oil was held in storage.) Spot shortages occurred, and because we are so dependent on our autos, the spot shortages quickly escalated into widespread panic. We were willing to pay almost anything to have oil shipments resumed—and pay we did. OPEC resumed shipments at a price of $10.50 a barrel, more than four times the previous price.

Then, by continuing to limit oil production all through the 1970s, OPEC was able to keep supplies tight enough to force prices higher and higher. In the early 1980s, a barrel of oil delivered to the United States cost $35, and the United States was paying accordingly. Purchase of foreign oil became

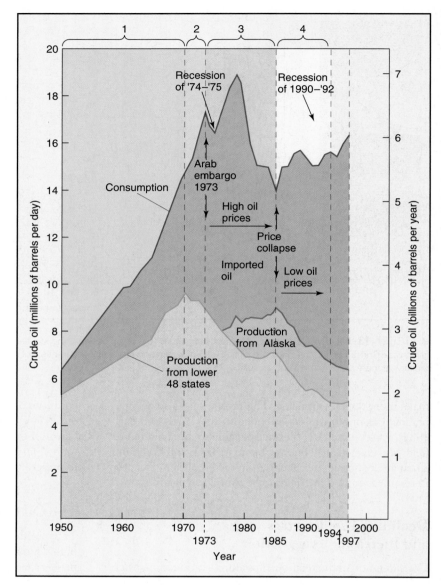

▶ **FIGURE 13–12** *Oil production and consumption in the United States.* Four stages can be seen: (1) Up to 1970, discovery of new reserves allowed production to parallel increasing consumption. (2) From 1970 to 1973, lack of new oil discoveries caused production to turn downward, while consumption continued to climb, causing a vast increase in oil imports. (3) In the late 1970s to early 1980s, high oil prices promoted both lowered consumption and increased production—which included bringing the Alaskan oil field into production. (4) From 1986, when oil prices declined sharply, until the present, consumption has again increased while production has resumed its decline, making us increasingly dependent on foreign oil. (Source: Data from Energy Information Administration, U.S. Department of Energy, *Annual Energy Review 1997*, July 1998.)

and remains the single largest factor in our balance-of-trade deficit (Fig. 13–13).

Adjusting to Higher Prices

In response to the higher prices and the recognition that oil is not forever, the United States and other industrialized countries quickly made a number of responses and adjustments during the 1970s. The most significant changes made by the United States are listed below.

To increase domestic production of crude oil:

- Exploratory drilling was stepped up.
- The Alaska pipeline was constructed. (The Alaskan oil field had actually been discovered in 1968, but it remained out of production because of its remoteness.)

- Fields that had been closed down as uneconomical when oil was only $2.30 a barrel were reopened (e.g., in Texas).

To decrease consumption of crude oil:

- Standards were set for automobile fuel efficiency. In 1973, cars averaged 13 mpg. The government mandated stepped increases for new cars to average 27.5 mpg by 1985. Speed limits were set at 55 mph for all U.S. highways.
- Other conservation goals were promoted for such things as insulation in buildings and efficiency of appliances.
- Development of alternative energy sources was begun. The government supported research and development efforts and gave tax breaks to people for installing alternative energy systems.

To protect against another OPEC boycott:

- A strategic oil reserve was created. The United States has "stockpiled," as of 1998, 574 million barrels of oil (equivalent to 58 days' of imports) in underground caverns in Louisiana.

It is noteworthy that, with the exception of the stockpile, all these steps are economically appealing only when the price of oil is high. If the price is low, the economic returns or savings may not justify the necessary investments.

Results of these actions were not immediate, nor could quick returns have been expected. For example, it takes at least three to five years to design a new car model and start production. Then another six to eight years pass before enough new models are sold and old ones retired to affect the overall highway "fleet."

As we entered the 1980s, however, the efforts were definitely having an impact. The United States remained substantially dependent on foreign oil, but oil consumption was headed down and production, up after completion of the Alaska pipeline, was holding its own (see Fig. 13–12). Elsewhere in the world significant discoveries in Mexico and the North Sea (between England and Scandinavia) made the world less dependent on OPEC oil. As a result of all these factors, plus OPEC's inability to restrain its own production, world oil production exceeded consumption in the mid-1980s, and there was an oil "glut."

What happens when supply exceeds demand? In 1986, world oil prices crashed from the high $20s to close to $10 per barrel. Since that time, oil prices have fluctuated in the range of $10 to $20 per barrel, and due to inflation, are lower now than they have been since the early 1970s (Fig. 13–13). The effect of the lower oil prices was to the apparent benefit of consumers, industry, and oil-dependent developing nations, but it hardly solved the underlying problem. Indeed, it has set the stage for another crisis.

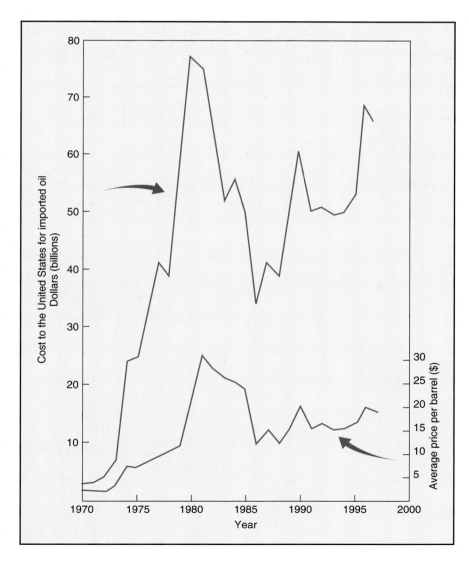

◀ **FIGURE 13–13** *The cost of oil imports.* The cost (upper curve) fluctuates sharply with the delivered price of foreign oil (lower curve) and the amount imported. (Source: Data from Energy Information Administration, U.S. Department of Energy, *Annual Energy Review 1997*, July 1998.)

Victims of Our Success

Whereas high oil prices stimulated constructive responses, the collapse in oil prices undercut those responses:

- Exploration, which had become more costly as wells were drilled deeper and in more remote locations, was sharply curtailed. In 1982, 3105 drilling rigs were operating in the United States; in 1992 there were 660.

- Production from older oil fields, which was costing around $10–$15 per barrel to pump, was terminated. The resulting drop in production caused economic devastation in Texas and Louisiana.

- Conservation efforts and incentives were abandoned. Government standards for automobile fuel efficiency were frozen at 27.5 with no intention to increase them further. Speed limits were set to 65 and even 75 mph.

- Tax incentives and other subsidies for development or installation of alternative energy sources were terminated, destroying many new businesses engaged in solar and wind energy. Grants for research and development of alternative energy sources (except for nuclear power) were sharply cut.

- The need for conservation and for development of alternative ways to provide transportation seems to have largely passed from the mind of the public.

U.S. production of crude oil continues to drop as proven reserves are drawn down. The declining production now includes Alaska, which reached its peak production in 1988. At the same time, consumption of crude oil is rising again, because both the total number of cars on the highways and the average miles per year that each car is driven are increasing. Adding to that consumption, consumers are buying vehicles that are less fuel-efficient. Minivans, four-wheel-drive sport-utility vehicles, light trucks, and other less fuel-efficient models now constitute 40% of the new-car market.

Altogether, following considerable improvement in our energy situation in the late 1970s and early 1980s, the United States has reverted to old patterns. Dependence on foreign oil has grown steadily since the mid-1980s. At the time of the first oil crisis, in 1973, the United States was dependent on foreign sources for about 30% of its crude oil. Now we are even worse off: In 1989, U.S. dependency on foreign sources passed the 50% mark, and it is continuing upward at about 1 1/2 percentage points per year (Fig. 13–14). The United States is not alone in this situation. Much of the rest of the world, both developed and developing countries, have a growing dependency on oil from those relatively few nations with surplus production. What problems does this growing dependency on foreign oil present?

Problems of Growing U.S. Dependency on Foreign Oil

As the U.S. dependency on foreign oil grows again, we are faced with problems at three levels: (1) the costs of purchase, (2) the risk of supply disruptions due to political instability in the Middle East, and (3) ultimate resource limitations in any case. Let's consider each in more detail.

Costs of Purchase. The yearly cost to the United States of purchasing foreign oil is shown in Fig. 13–13. The cost has dropped considerably from its high of over $70 billion in the early 1980s, the decline being brought about both by falling oil prices and by conservation. However, the current $62 billion per year is no small number, and it is now climbing again with no downturn in sight as we become increasingly dependent on foreign oil.

▶ **FIGURE 13–14** *Imported oil as a percentage of total consumption.* After the oil crisis of the 1970s, the United States became less reliant on foreign oil. But since the mid-1980s, consumption of foreign oil has for the most part been rising continuously. (Source: Data from Energy Information Administration, U.S. Department of Energy, *Annual Energy Review 1997*, July 1998.)

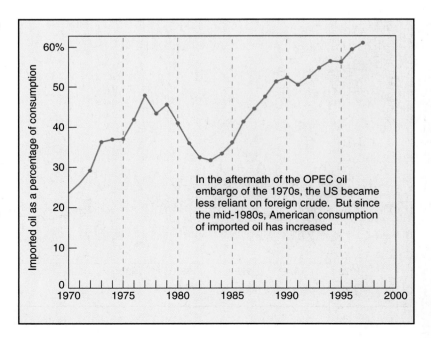

In the aftermath of the OPEC oil embargo of the 1970s, the US became less reliant on foreign crude. But since the mid-1980s, American consumption of imported oil has increased

This cost for crude oil represents about 59% of our current balance-of-trade deficit ($108 billion in 1996), which has manifold consequences for our economy. The price we pay at the pump is basically the same whether the oil is produced here or abroad. However, if the oil is produced at home, the money we pay for it circulates within our own economy, providing jobs and producing other goods and services. When we buy oil from a foreign country, we are effectively providing "foreign aid" to that country and that much of a drain on our own economy.

Of course, the ecological costs of oil spills such as the *Sea Empress* and the estimated $15 billion *Exxon Valdez* incident and of other forms of pollution from the refining and consumption of fuels must not go unnoticed—although estimating total damage in dollars is sometimes difficult. There are also military costs, as we will describe in the next section.

Risk of Supply Disruptions. The Middle East is a politically unstable, unpredictable region of the world. As noted, it was the unexpected Arab boycott that plunged us into the first oil crisis in 1973. Recognizing this political instability, the United States maintains a military capacity to ensure our access to Middle Eastern oil. This military capacity was tested in the fall of 1990 when Saddam Hussein of Iraq invaded Kuwait, then producing 6 million barrels per day. The Persian Gulf War of early 1991 prevented further advances. Nevertheless, Kuwait's oil production was knocked out for more than a year after the war, because the Iraqis dynamited and set fire to nearly all of Kuwait's almost 700 wells before leaving (Fig. 13–15).

Military costs associated with maintaining access to Middle Eastern oil can be looked at as the U.S. government's subsidizing our oil consumption. Our military presence there costs the United States $50 billion per year. On the basis of importing about 635 million barrels a year from Persian Gulf sources (in 1997), the dollar amount of this "subsidy" comes to about $80 per barrel. In other words, when military "support services" are added to the $14 market price, we are actually paying about $94 per barrel!

Maintaining access to Middle Eastern oil is also a major reason for our efforts toward negotiating peace agreements between various parties in that region. Even with all this effort, however, continuing risks are present. The rise of the Islamic radicals, who threaten to take over governments of more Middle Eastern countries, may pose a great danger.

It is significant that the United States has recently turned to oil suppliers in the Western Hemisphere; Venezuela has recently replaced Saudi Arabia as the number one exporter of oil to the United States, with about 20% of the market. Canada and Mexico are also major oil sources. The Western Hemisphere sources now provide about 50% of our imports. The political stability and proximity of these sources are clear reasons for the United States to turn to them and away from the Middle East.

Resource Limitations. No responsible geologist questions the fact that U.S. crude oil production is decreasing because of diminishing domestic reserves. Nor do geologists hold out hope for major new finds. The dim hope for new U.S. finds derives

▲ **FIGURE 13–15** *The Gulf War of 1991.* Even military might may not guarantee uninterrupted access to world oil supplies. Although we were the winner of the Persian Gulf War of 1991, Iraq's forces nevertheless succeeded in setting fire to nearly 700 of Kuwait's oil wells before their retreat. Six million barrels per day were thus taken out of production for nearly a year.

from what is called the "Easter-egg hypothesis." As your searching turns up fewer and fewer eggs, you draw the logical conclusion that nearly all the eggs have been found. In terms of oil, North America is already the most intensively explored of any continental landmass. The last major find was the Alaskan oil field in 1968 (see "Ethics" essay, p. 324). Discoveries since then have been small in comparison, with increasing numbers of dry holes in between. Indeed, most of the "new" oil in the United States is coming from innovative computer mapping of geological structures and horizontal drilling technology that enables the identification and withdrawal of oil from small isolated pockets within old fields that were previously missed. Yet an increasing percentage of our needs must continue to be met by imported oil.

How much oil is still available for future generations? To put this question in perspective, it is instructive to explore what has already been used and the current world rate of use. About 800 billion barrels of oil has already been used, almost all in the twentieth century. Currently, the nations of the world are using about 67 million barrels per day, or some 24.5 billion barrels a year. Rising demand by the developing countries, and the continued rise in oil demand by the developed countries, point to an increase of about 2% per year, reaching 40 billion barrels a year by 2020. How long can the existing oil reserves provide this much oil?

Two oil geologists, Colin Campbell and Jean Laherrère, have recently provided an analysis that puts the world's

ETHICS

TRADING WILDERNESS FOR ENERGY IN THE FAR NORTH

The search for new energy sources to satisfy the enormous energy appetite of the U.S. economy has turned northward in recent years. Discovery and development of the huge Prudhoe Bay oil field in Alaska and the building of the 800-mile (1230 km) Alaska pipeline have dramatized the stark contrasts between wilderness and modern technology. Scenes of caribou and grizzly bears are shown with a background of oil rigs and the pipeline carrying oil south to Valdez. East of Prudhoe Bay is the Arctic National Wildlife Refuge (ANWR), the largest single wilderness area in the United States. ANWR is home to 180,000 caribou and several hundred native people whose subsistence lifestyle is tied to the caribou. ANWR is also the summer breeding ground for innumerable birds and other wildlife, but it is also believed to be on top of another large oil field that industrial interests are anxious to exploit.

ANWR became a major environmental battleground during the 1990s. On one side, the energy and political interests see values in terms of energy supplies for a thriving economy and national security. On the other side, native people and environmentalists see values in terms of the beauty and biodiversity of the northern wilderness, as well as the human rights of native peoples. Those favoring development claim that it will be done in an environmentally sound way and the native people will be compensated. Those favoring preservation are not satisfied with good intentions and poor track records. They point to the *Exxon Valdez* disaster, in which 11 million gallons of crude oil were spilled in the pristine waters of Prince Edward Sound, despite all safeguards. They point out that present oil development in Alaska has introduced much of the worst of American culture to native peoples; a sense of lost identity is apparent as their traditional culture is swamped by consumer goods and pop culture. Environmentalists also question development on the basis of its sustainability. The estimated oil reserve underlying ANWR is 3.5 billion barrels; this is about six months' supply at the current rate of use in the United States. Are there ways to have adequate power without purchasing it at the expense of wilderness?

For the time being, ANWR has been declared off limits to oil exploration by the Clinton administration. Some of the pressure was taken off when part of the National Petroleum Reserve, an area of almost 4 million acres just west of Prudhoe Bay, was opened for exploration in the summer of 1998. This area was set aside 75 years ago for oil exploration by President Warren Harding and is thought to contain smaller amounts of oil than ANWR but nevertheless is worth exploiting. The area is also environmentally sensitive, a famous breeding ground for many migratory waterfowl and home to caribou, moose, musk oxen, and polar bears.

How do we balance our profligate need for energy, jobs, and economic growth against caribou, musk oxen, and polar bears? How do we weigh the traditions and cultures of native peoples against the impacts of our own?

Sources: H. Thurston, "Power in a Land of Remembrance," *Audubon*, Nov.–Dec. 1991; "Arctic Wars: The Fight for the Arctic National Wildlife Refuge," *National Audubon Society* film, 1990; "West's Balancing Act: Economy, Nature," *Christian Science Monitor*, Aug. 10, 1998.

---

proven reserves at about 850 billion barrels. This is less than the 1020–1160 billion barrels claimed by some oil industry reports (Table 13-2); Campbell and Laherrère made their estimate based on the P50 values of proven reserves and have discounted some reports of reserves that were inflated for political reasons. Based on their estimates, and on the known amount already used, the two authors present a Hubbart curve indicating that peak oil production will occur sometime in the next decade (Fig. 13–16). When that happens, production will begin to taper off in spite of rising demand, and the world will enter a new era of rising oil prices. Even if Campbell and Laherrère are wrong and the oil industry estimates are right, it would only put off the peak by a decade at most. To quote the authors, "The world is not running out of oil—at least not yet. What our society does face, and soon, is the end of abundant and cheap oil on which all industrial nations depend." Clearly, the twenty-first century will become known as an era of declining oil use, just as the twentieth century has seen continuously rising oil use.

Table 13-2 Proven Reserves of Petroleum (in billions of barrels)

Region	Reserves
North America	75.7
South & Central America	79.1
Europe	20.2
Former Soviet countries	57.0
Middle East	676.4
Africa	67.6
Far East & Oceania	42.3
Total	**1018.3**

Source: Data from *Oil and Gas Journal*, Energy Information Administration, U.S. Department of Energy.

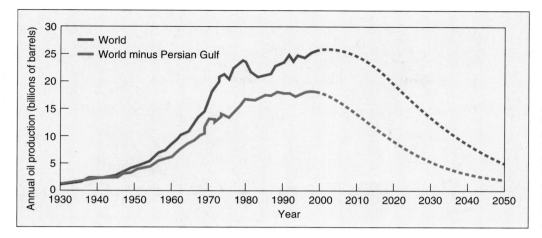

◀ **FIGURE 13–16** *Hubbart curves of oil production.* Oil production in a region follows a bell-shaped curve. The evidence indicates that world oil production will peak sometime in the next decade. (Source: Data from Richard A. Kerr, "The Next Oil Crisis Looms Large—and Perhaps Close," *Science* 281 [1998]:1128–1131.)

As oil becomes more scarce, the OPEC nations will once again achieve dominance of the world oil market. The Middle East possesses 64% of the world's proven reserves, and eventually the developed world—and in particular the United States—will have to turn to them for its oil imports (Table 13-2). It would seem the path of wisdom to begin to prepare for that day by reducing our dependency on foreign oil now. We can become more independent in three ways: (1) We can use the other fossil-fuel resources we have to make fuel for vehicles; (2) we can reduce our energy demand through conservation and energy efficiency; and (3) we can develop alternatives to fossil fuels, many of which are ready now. The best answer probably lies in a suitable mix of all three choices.

13.3 Alternative Fossil Fuels

Unlike crude oil, considerable reserves of natural gas and comparatively vast reserves of coal and oil sands are still available in the United States and elsewhere in the world. Natural gas is gradually coming into use as a vehicle fuel, and there is some potential in this respect for coal and oil sands.

Natural Gas

In the United States, natural gas has been rising as a fossil fuel of choice, so much so that we are now importing 13% of annual use. Although most of its use is for space heating and cooking, natural gas is increasingly being employed for electrical power generation. At the current rate of use, proven reserves in the United States hold only a nine-year supply of gas. However, this is somewhat deceiving, because natural gas is continually emitted from oil and gas-bearing geological deposits. The expected long-term recovery of natural gas in the United States is about six times greater than the proven reserves, which puts the supply closer to 50 years.

Worldwide, the estimated natural gas resource base is even more plentiful. Almost four times as much equivalent gas (energy equivalency) is likely to be available as oil.

Much of it, however, is relatively inaccessible. Gas must either be pumped through a pipeline (a process that is limited by pipeline length) or highly pressurized so as to remain a liquid at room temperature. Nevertheless, liquefied natural gas is increasing in production and use.

Natural gas can also meet some of the fuel needs for transportation. With the installation of a tank for compressed gas in the trunk and some modifications of the engine fuel-intake system—costing about a thousand dollars—a car will run perfectly well on natural gas, although trunk space is a problem. Such cars are in widespread use in Buenos Aires, where service stations are equipped with compressed gas to refill the tank. Natural gas is a clean-burning fuel; hydrocarbon emissions are nil. Indeed, natural gas is coming into use in the United States as a vehicle fuel, and it is already being used by many buses and a number of both private and government car fleets.

With the aid of a chemical process, natural gas can be converted to a hydrocarbon that is liquid at room temperature and pressure—basically, a synthetic oil. Researchers are hotly pursuing ways to more efficiently convert natural gas to synthetic oil. Presently, this process produces a fuel that is only about 10% more costly than oil. Already, some diesel fuel in California is partly derived from this process. Estimates indicate that enough natural gas is available to produce 500 billion barrels of synthetic oil.

These two applications of natural gas to the transportation sector promise to extend the oil phase of our economy, but they will not be a sustainable solution, because the natural gas reserves are also limited. However, what about coal, the most abundant fossil fuel?

Coal

In the United States 57% of our electricity comes from coal-fired power plants, and current reserves are calculated to be 462 billion metric tons (Fig. 13–17). At the current rate of use (991 million tons/yr), we have over 400 years of supply.

Coal can be obtained by two methods—strip mining and underground mining. Both these methods have substantial

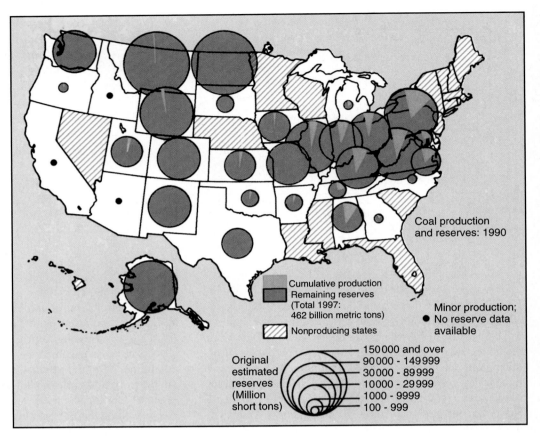

Coal production
and reserves: 1990

Cumulative production
Remaining reserves
(Total 1997:
462 billion metric tons)

Nonproducing states

Minor production;
● No reserve data
available

Original
estimated
reserves
(Million
short tons)

150 000 and over
90 000 - 149 999
30 000 - 89 999
10 000 - 29 999
1000 - 9999
100 - 999

▶ **FIGURE 13–17** *Major coal deposits in the United States.* With reserves of at least 462 billion metric tons, and a 400-year supply, coal is our most abundant fossil fuel. (U.S. Geological Survey)

environmental impacts. In underground mining, at least 50% of the coal must be left in place to support the mine roof. In the more common strip mining, gigantic power shovels turn aside the rock and soil above the coal seam and then remove the coal (Fig. 13–18). Obviously, this procedure results in total ecological destruction. Although such areas may be reclaimed—that is, graded and replanted—it takes many decades before an ecosystem resembling the original can develop. In arid areas of the West, the water limits may prevent the ecosystem from ever being reestablished. Consequently, strip-mined areas may be turned into permanent deserts. Furthermore, erosion and acid leaching from the disturbed ground have numerous adverse effects on waterways and groundwater in the surrounding area.

Once it is mined, coal is transported to large, baseload power plants. A typical 1000 megawatt power plant burns 8000 tons of coal a day; if it is delivered by train, this means a mile-long train's worth of coal every day. Combustion of this much coal releases 20,000 tons of carbon dioxide and 800 tons of sulfur dioxide (most of this is removed if the power plant is well provided with scrubbers—see Chapter 22). Then there is the waste—800 tons of fly ash and 800 tons of boiler residue ash daily, requiring a special landfill.

There are chemical processes that will convert coal to a liquid or gas fuel. Such coal-derived fuels are referred to as **synthetic fuels** or **synfuels**. With both government and

corporate support, a great deal of research went into upscaling these chemical processes to commercial production in the late 1970s and early 1980s. The projects were all abandoned, however, because they proved too expensive, at least for the present. Profitable production of coal-based synthetic fuels could occur only if oil prices rose to $60 to $70 per barrel.

Thus, there are substantial hazards and pollutants involved in mining and burning coal. The thought of doubling or tripling this impact to run vehicles on coal-derived fuel makes a mockery of the country's progress in environmental protection. In addition, converting coal to liquid fuel creates polluting byproducts that would be difficult and expensive to control.

Oil Shales and Oil Sands

The United States has extensive deposits of oil shales in Colorado, Utah, and Wyoming. **Oil shale** is a fine sedimentary rock containing a mixture of solid, waxlike hydrocarbons called *kerogen*. When shale is heated to about 1100°F (600°C), the kerogen releases hydrocarbon vapors that can be recondensed to form a black, viscous substance similar to crude oil, which can be refined into gasoline and other petroleum products. However, it requires about a ton of oil shale to yield little more than half a barrel of oil. The mining, transportation, and disposal of wastes necessitated by an operation producing, for instance, a million barrels a day (5% of U.S. demand)

▲ FIGURE 13–18 *Strip mining of coal.* It is only economical to exploit most coal deposits by strip-mining, where gigantic power shovels remove as much as 100 feet (32 m) of rock and soil from the surface to get at the coal seam.

would be a Herculean task, to say nothing of its environmental impact. Perhaps to our environmental advantage, oil shale, like oil from coal, has proved economically impractical for the present. Also, it requires large amounts of water, something in short supply in the West.

Oil sands are a sedimentary material containing bitumen, an extremely viscous, tarlike hydrocarbon. When oil sands are heated, the bitumen can be "melted out" and refined in the same way as crude oil. Northern Alberta, Canada, has the world's largest oil-sand deposits (300 billion barrels), and Canada is commercially exploiting them, because the cost is competitive with current oil prices. Suncor Energy of Canada sells 28 million barrels of this unconventional oil a year. The United States has smaller, poorer oil-sand deposits, mostly in Utah, that might produce oil for $50–$60 per barrel.

Worldwide, the combination of oil shales and oil sands has been estimated to hold several trillion barrels of oil. Large-scale commercial development of these resources will not occur until oil prices rise dramatically. Both sources involve substantial environmental impacts.

13.4 Sustainable Energy Options

Fossil fuel use has been rising at rates of 2–3% per year, with no end in sight. Yet, at the same time we are rethinking our energy future on the basis of resource availability and managing our demands, we need to consider environmental factors looming on the horizon, primarily global climate change. Note again that all fossil fuels are carbon-based compounds and none can be burned without carbon dioxide (CO_2) resulting as a byproduct. Of the top three—coal, oil, and natural gas—coal produces the most, and natural gas produces the least CO_2 (Table 13-3). Because CO_2 is a greenhouse gas, the increasing atmospheric concentration of CO_2 is likely to bring about major changes in the global climate (see Chapter 21). This threat of climatic change—mounting evidence indicates that it is already occurring—is likely to force a curtailment of fossil-fuel consumption long before limited resources do. Therefore fossil-fuel contributions to global warming (Fig. 13–19) are probably the most compelling reason to reduce energy demand through conservation and to examine non-fossil-fuel alternatives.

Also, in rethinking energy strategies, we need to start with the concept that energy has no real meaning or value by itself. Its only worth is in the work it can do. Thus, what we really want is not energy but comfortable homes, transportation to where we need to go, manufactured goods, and so forth. Therefore, many energy analysts are saying that we should stop thinking in terms of where we can get additional supplies of the old fuels to keep things running as they are. Instead, we should be thinking in terms of how we can satisfy these needs with minimum expenditure of energy and environmental impact.

Conservation

Imagine the discovery of a new oil field having a production potential of at least 6 million barrels a day—three times the capacity of the Alaskan field even at its height. Furthermore, assume that this new-found field is inexhaustible and that its exploitation will not adversely affect the environment. Of course, such an oil field sounds like a dream, but this is effectively what can be achieved through fossil-fuel conservation.

Table 13-3 CO_2 Emissions Per Unit Energy for Fossil Fuels (Natural Gas = 100%)	
Natural Gas	100%
Gasoline	134%
Crude Oil	138%
Coal	178%

Source: Data from Energy Information Administration, U.S. Department of Energy.

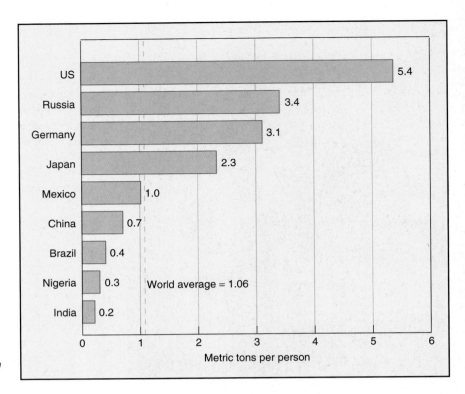

▶ **FIGURE 13–19** *Carbon emissions per capita from fossil-fuel burning for selected countries.* (Source: Data from C. Flavin and D. Tunali, "Getting Warmer," *Worldwatch* [March–April 1995]:17.)

As noted earlier in this chapter, this "conservation reserve" has already been tapped to some extent, especially as the average fuel efficiency of cars was doubled from 13 to 27.5 mpg. Thus, drivers of those autos are satisfying their transportation needs with only half the expenditure of energy and dollars. This and other conservation measures are already saving about $100 billion per year in oil imports. In short, advances in conservation have been extremely cost-effective as well as environmentally benign.

A number of energy experts estimate that at least 80% of the "conservation reserve" remains to be tapped. Cars are currently available that get at least double the current average of 27.5 mpg, and Peugeot, Toyota, Volkswagen, and Volvo all have prototypes that get from 70 to 100 mpg. The Toyota Prius, a gasoline-electric hybrid that gets 60 mpg, was introduced to the United States in 1999. To guarantee high sales, however, most automakers are waiting for government mandates, consumer demand, or fuel shortages before putting these models into production. Of course, carpooling with just one person in addition to the driver effectively doubles the mpg.

A technique known as **cogeneration** (see the "Earth Watch" essay, p. 329) is another way of gaining large amounts of energy from the conservation reserve. In recent years, a new technology has been increasingly employed to generate electricity—the **combined-cycle natural gas unit**. Two turbines are employed, the first burning natural gas in a conventional gas turbine (Fig. 13–6b), and the second a steam turbine (Fig. 13–6a) that runs on the excess heat from the gas turbine. Use of this technology boosts the efficiency of conversion of fuel energy to electrical energy from around 30%

to as high as 50%. The cost of the combined-cycle systems is half that of conventional coal-burning plants, and the pollution from such systems is much lower. A large proportion of new units being built and planned for the future is combined-cycle systems.

Other major changes are under way for the entire electric power industry, as an increasing number of states adopt **deregulation** of the industry. Deregulation requires the utilities to maintain distribution services (transmission lines, telephone poles, etc.) but to divest themselves of the power-generating facilities. Thus, consumers will be able to choose energy suppliers as the utilities lose their monopolies on electrical power supply. In Massachusetts, deregulation has spawned a number of new power plants, all natural gas-fired. Consumers can, for example, opt to purchase energy from AllEnergy, a company developing power from landfill methane (see Chapter 19) and other renewable sources. One interesting outcome of deregulation is the attempt on the part of the Conservation Law Foundation to join with a power company on the bidding for a group of power plants, and if they are successful, they plan to close down a group of the most polluting ones. The final outcome of deregulation is still uncertain, but energy conservation and renewable energy appear to be much more achievable as a consequence.

Other readily available conservation measures are underutilized: Improved insulation and double-pane windows for homes and buildings would save on heating fuel. Substitution of fluorescent lights, which are between 20% and 25% efficient (Fig. 13–20), for conventional incandescent light bulbs, which are only 5% efficient, would cut back on our

EARTH WATCH

COGENERATION: INDUSTRIAL COMMON SENSE

In many situations a single energy source can be put to use to produce both electrical and heat energy. This process is called cogeneration. Cogeneration can be employed to conserve energy in two fundamental ways: (1) A power plant whose primary function is to generate electricity also can be used to heat surrounding homes and buildings. As one example, the Consolidated Edison electric utility has been using steam from waste heat in New York City for years. (2) A facility whose primary function is to produce heat or mechanical energy can also generate electricity, which is used by the facility and can also be sold to the local utility. The system can work for all sorts of fuels and heat engines. This is by far the most common application of cogeneration.

The arrangement makes perfect sense. Every building and factory requires both electricity and heat. Traditionally, electricity and heat come from separate sources—the electricity from power lines brought in from some central power station and the heat from an on-site heating system, usually fueled by natural gas or oil. Ordinary power plants are at best 30% efficient in their use of energy. Combining the two in cogeneration can provide an overall efficiency of fuel use as high as 80%. Besides providing savings on fuel, the cogenerating systems have at least 50% lower capital costs per kilowatt generated than do conventional utilities. This cost savings makes cogeneration quite competitive with utility-produced power.

The Santa Barbara (California) County Hospital recently installed a cogeneration system based on natural gas as the fuel source. Natural gas drives a turbine with the capacity to generate 7 MW of electricity, much of which is used by the hospital. Boilers use the waste heat from the turbine to produce high-pressure steam, which provides all of the hospital's heating and cooling needs. Any steam not required by the hospital is directed back to the turbine to increase its power. Electrical power not used by the hospital is sold to the local utility. The system has the added advantage of protecting the hospital against widescale blackouts or brownouts that may occur in the centralized utility system.

Enabling legislation known as PURPA (Public Utility Regulatory Policies Act of 1978) was essential for stimulating use of cogeneration. Under this act, a facility with cogenerative capacity is entitled to use fuels not usually available to electricity generating plants (for example, natural gas), must produce above-standard efficiency of fuel use, and is entitled to sell cogenerated electricity to the utility at a fair price. After many challenges by the utilities (which considered the procedures an interference with their business), the Supreme Court decided in favor of PURPA, and the legislation is now firmly established.

Cogeneration is becoming increasingly popular in U.S. industries as a major strategy for cutting energy costs. Given the future possibilities of major price increases for petroleum and CO_2 emissions reductions, pressures favoring energy conservation will increase, and use of cogeneration will continue to grow. The Federal Energy Regulatory Commission believes that cogeneration may provide as much as 25% of our electricity by the year 2000.

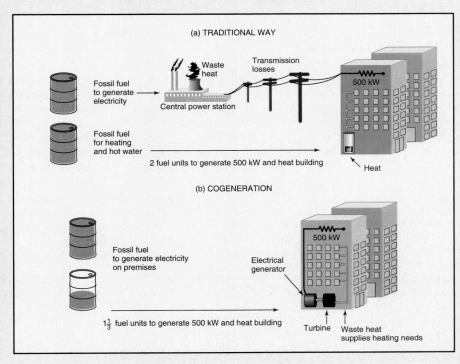

Cogeneration. (a) The traditional method of providing the energy needed for buildings. (b) In cogeneration, the heat lost from an on-site power-generating system provides heating needs.

(a) TRADITIONAL WAY

Fossil fuel to generate electricity

Waste heat

Central power station

Transmission losses

500 kW

Fossil fuel for heating and hot water

2 fuel units to generate 500 kW and heat building

Heat

(b) COGENERATION

Fossil fuel to generate electricity on premises

500 kW

Electrical generator

$1\frac{1}{3}$ fuel units to generate 500 kW and heat building

Turbine

Waste heat supplies heating needs

▲ **FIGURE 13–20** *Energy-efficient light bulbs.* Replacing standard incandescent light bulbs with screw-in fluorescent bulbs shown here can cut the energy demand for lighting by 75–80%. The higher cost of the fluorescent bulbs is more than offset by greater efficiency and much longer lifetime of the bulb.

electricity demands for lighting by 75%–80%. Similarly, more efficient refrigerators and other appliances can reduce energy demands for those purposes by 50%–80%. Energy-conserving light bulbs and appliances are more expensive than conventional models, but calculations show that they more than pay for themselves in energy savings. Also, some utilities are subsidizing the installation of fluorescent light bulbs and energy-efficient appliances as a cost-effective alternative to building additional power plants.

Many observers have pointed out that demanding more energy while failing to conserve is like demanding more water to fill a bathtub while leaving the drain open. To be sure, conservation and efficiency strategies by themselves will not eliminate demands for energy, but it can make the demands much easier to meet regardless of what options are chosen to provide the primary energy.

Development of Non-Fossil-Fuel Energy Sources

Lying before us are two major pathways for developing non-fossil-fuel energy alternatives. One is to pursue nuclear power; the other is to promote solar-energy applications. Of course, nuclear power has been in use for 35 years, and various solar-energy systems have also been developed. In other words, both nuclear and solar alternatives are at a stage where they could be rapidly expanded to provide further energy needs if suitable technological solutions and public acceptance (nuclear power) or pressure and government support (of solar energy) were provided. The following chapter focuses on the nuclear option, and Chapter 15 focuses on solar options. We will address public policy needs at the end of Chapter 15.

ENVIRONMENT ON THE WEB

THE HIDDEN COSTS OF FOSSIL FUEL UTILIZATION

Fossil fuel use incurs real costs for society. Some of these costs, such as those related to exploration and the construction of generating plants, are "direct" and have a monetary value passed on to the consumer through electricity bills. However, there are also many hidden or "indirect" costs of fossil fuel use, often related to the environmental or human health impacts of fossil fuel use. These include the tangible or intangible costs of additional health care, diminished quality of life (including shorter lives), loss of wildlife habitat, and loss of aesthetic value (such as loss of a scenic vista).

Because these costs are externalities—not paid directly by consumers or producers—and often difficult to quantify, they are usually excluded from energy pricing. Yet they are nevertheless real costs, with real implications for society. The National Academy of Sciences estimates, for instance, that air pollution causes billions of dollars a year in environmental and structural damage. The health care burden of air pollution related to fossil fuel combustion adds another $20 billion a year. Although the potential impacts of global warming are still poorly understood, they are thought to include effects on crop yields, water supply, redistribution of populations, and the impacts of altered weather patterns on hurricanes, typhoons, and similar phenomena. All of these are likely to carry a cost to society.

Although some jurisdictions are now moving toward "full-cost pricing" of electricity and other utilities to include a wider range of costs, such as anticipated maintenance and replacement of costly generating equipment, the full hidden costs of fossil fuel use still may not be apparent to consumers. This is because few, if any, utilities include externalities such as the societal costs of air pollution and increased health care in their pricing, even though these are real costs borne by real people and directly related to fossil fuel use.

Web Explorations

The Environment on the Web activity for this essay explores some of the methods used to put a price on some externalities. Go to the Environment on the Web activity (select Chapter 13 at **http://www.prenhall.com/nebel**) and learn for yourself:

1. about other externalities associated with fossil fuel use;
2. how to compare the external costs of different sources of energy; and
3. about the advantages and disadvantages of retaining and upgrading existing fossil fuel generating facilities.

🕐 **A suggested time frame for each of these exercises is 10–30 minutes.**

REVIEW QUESTIONS

1. How have the fuels to power homes, industry, and transportation changed from the beginning of the Industrial Revolution to the present?

2. What are the three primary fossil fuels, and what percentage does each contribute to the United States energy supply?

3. Electricity is a secondary energy source. How is it generated, and how efficient is its generation?

4. Name the four major categories of primary energy use in an industrial society, and match the energy sources that correspond to each.

5. What is the distinction between estimated reserves and proven reserves of crude oil, and what factors cause the amounts of each to change?

6. How does the price of oil influence the amount produced?

7. What were the trends in oil consumption, in discovery of new reserves, and in United States production before 1970? After 1970?

8. What did the United States do in the early 1970s to resolve the disparity between oil production and consumption? What events then caused the sudden oil shortages of the mid-1970s and then the return to abundant but more expensive supplies?

9. What responses were made by the United States and

other industrialized nations to the "oil crisis" of the 1970s, and what were the impacts of those responses on production and consumption?

10. Why did an oil glut occur in the mid-1980s, and what were its consequences in terms of oil prices and the continuing pursuit of the responses listed in Question 9?

11. What have been the directions of United States production, consumption, and importation of crude oil since the mid-1980s?

12. What are the liabilities and risks of our growing dependence on foreign oil? How do these relate to the Persian Gulf War of 1991?

13. What alternative fossil fuels might the United States exploit to supplement declining reserves of crude oil? What are the economic and environmental advantages and disadvantages of each?

14. What phenomenon may restrict consumption of fossil fuels before resources are depleted?

15. To what degree has energy conservation served to mitigate energy dependency? What are some prime examples of energy conservation? To what degree may conservation alleviate future energy shortages and the cost of developing alternatives?

THINKING ENVIRONMENTALLY

1. Statistics show that less developed countries use far less energy per capita than developed countries. Explain why this is so.

2. Between 1980 and 1999, gasoline prices in the United States declined considerably relative to inflation. Predict what they will do in the next 10 years, and give a rationale for your prediction.

3. Suppose your region is facing power shortages (brownouts). How would you propose solving the problem? Defend your proposed solution on both economic and environmental grounds.

4. List all the environmental impacts that occur during the "life span" of gasoline—that is, from exploratory drilling to consumption of gasoline in your car.

WEB REFERENCES

On-line resources for this chapter can be found on the World Wide Web at: **http://www.prenhall.com/nebel**. Click on Chapter 13 on the chapter selector.

Renewable Energy

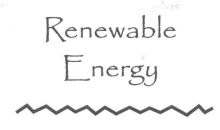

Key Issues and Questions

1. The total amount of solar energy reaching Earth is enormous. What is the potential for harnessing this energy?

2. Solar water-heating and space-heating systems for buildings represent well-developed technologies. What is preventing the more widespread adoption of these forms of solar heating?

3. Solar energy via photovoltaic cells and solar-tough collectors is used to produce electrical power. What are the current applications of these technologies, and what is their promise for the future?

4. There is a great need to develop solar energy sources that can be coupled to fuel for transportation. What is the potential for fuel from solar hydrogen production?

5. Water, fire, and wind have provided energy for centuries. What are sustainable ways of expanding these options in the near future?

6. The options for energy into the twenty-first century are many. In light of global climate change, moving in the direction of renewable energy sources seems essential. What is in the way of such a move, and what is being done now to move in this direction?

The Hummingbird Highway is one of the main roads in Belize, Central America, running southeasterly from the capital of Belmopan through second-growth rain forests and orange orchards to Dandriga on the coast. About 15 miles down the highway from Belmopan, a small dirt road leads in to Jaguar Creek, an environmental research and education center run by TargetEarth. With 14 buildings totaling nearly 10,000 square feet, Jaguar Creek accommodates up to 40 people who might be there for short-term conferences or semester-long college courses. Belize is a developing country, and rural electrification is one facet that is still developing. Jaguar Creek is many miles from any electrical power, yet carries on with refrigeration, lighting, washing machines, and fans in every living space. This "off the grid" facility has its own solar electrical system, which covers the roof of the operations building (Fig. 15–1), capable of generating 50,000 watts (50 KW) of power per day. Backup for rainy periods is provided by a propane-powered generator. The

total installed cost for the system was $100,000; this breaks down to about $5.00 per watt, and the system has a life expectancy of at least 25 years. Compared to the cost of a utility system with poles, wires, transformers, and the power supply, this is a bargain. Local utility service is estimated to cost at least $25,000 per mile, and Jaguar Creek is 10 miles from the nearest power line. This sustainable energy system is consistent with Jaguar Creek's Earth-friendly approach, which also includes composting toilets and graywater leaching fields.

Similar panels of solar cells are providing electricity in both developed and developing countries around the world. Throughout Israel and other countries in warm climates, it is now commonplace to have hot water heated by the Sun (Fig. 15–2). Even in temperate climates, many people have discovered that standard furnaces can be made obsolete with suitable designs to utilize the Sun's warmth. In the desert northeast of Los Angeles are "farms" with rows of trough-shaped mirrors tipped toward the Sun (Fig. 15–3).

◀ **Wind Power in Vermont.** Green Mountain Power Corporation installed 11 wind turbines in Searsburg, Vermont, in 1997, making it the largest wind power generating station in the eastern United States. On moderately windy days, the turbines provide enough power to satisfy the needs of 2000 homes.

▶ **FIGURE 15–1** *Jaguar Creek.* Panels of photovoltaic cells power the Jaguar Creek facility operated by TargetEarth in Belize. The system generates 50 KW of power a day, enough to power this facility which houses up to 40 people.

These reflectors are focusing the Sun's rays to boil water or synthetic oil and drive turbogenerators. In the hills east of San Francisco, regiments of "windmills" (more properly called *wind turbines*) standing in rows up the slopes and over the crests of the hills are producing electrical power equivalent to that produced by a large coal-fired power plant. Wind turbines are also becoming commonplace across northern Europe and sprouting up in India, Mexico, Argentina, New Zealand, and other countries around the world.

The inertia of "business as usual" has kept the world on a track of growing dependence on fossil fuels, with nuclear power retaining its share of the power for the present. In the meantime, however, these examples demonstrate how the use of energy from the Sun and wind has been making quiet but steady progress, to the point where these renewable sources of energy are becoming cost-competitive with traditional energy sources—and in many situations are far more practical. Thus, the world has at its disposal the potential to move toward a nonpolluting, inexhaustible energy economy

▼ **FIGURE 15–2** *Rooftop hot water.* In warm climates, solar hot water heaters are becoming commonplace. Note the solar water-heating panels on each of the homes in this development in southern California.

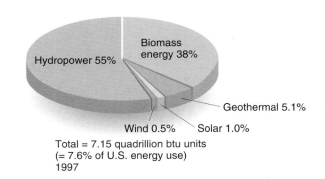

◀ **FIGURE 15–3** *Solar-thermal power in southern California.* Sunlight striking the parabolic shaped mirrors is reflected onto the central pipe, where it heats a fluid that is used, in turn, to boil water and drive turbogenerators.

based on using the Sun's current energy output. In short, we have the capability of abiding by the first principle of sustainability—running on solar energy.

Our objective in this chapter is to give you a greater understanding of the potential for sustainable energy from sunlight, wind, biomass, and other sources. This is not only the energy of the future, it is also a viable option for the present, given equitable policy considerations that level the playing field between sustainable sources and fossil fuels. Currently, renewable energy use in the United States provides 7.6% of the nation's energy (Fig. 15–4); this percentage has changed little in recent years. We will address energy policy at the end of the chapter as we compare the pros and cons of fossil fuels, nuclear power, and sustainable energy sources.

Total = 7.15 quadrillion btu units
(= 7.6% of U.S. energy use)
1997

▲ **FIGURE 15–4** *Renewable energy use in the United States.* A mix of sources of renewable energy provided 7.6% of the nation's energy use in 1997. (Source: Data from Department of Energy, Energy Information Agency, *Renewable Energy Annual 1997*, July 1998.)

15.1 Principles of Solar Energy

Before turning our attention to the practical ways we capture and use solar energy, let's consider some general parameters of solar energy.

Solar energy originates with thermonuclear fusion reactions occurring in the Sun. (Importantly, all the chemical and radioactive polluting byproducts of the reactions remain behind on the Sun.) The solar energy reaching Earth is radiant energy, ranging from ultraviolet light, which is largely screened out by ozone in the stratosphere, to visible light and to infrared (heat energy)(Fig. 15–5).

The total amount of solar energy reaching Earth is vast almost beyond belief. Just 40 minutes of sunlight striking the land surface of the United States yields the equivalent energy of a year's expenditure of fossil fuel. If less than one-tenth of 1% of the Earth's surface were dedicated to solar energy collection, it could supply the electricity needs of the world.

Moreover, utilizing some of this solar energy will not change the basic energy balance of the biosphere. Solar energy absorbed by water or land surfaces is converted to heat energy and eventually lost to outer space. Even what is absorbed by vegetation and used in photosynthesis is ultimately given off again in the form of heat energy as various consumers break down food (Chapter 3). Similarly, if humans were to capture and obtain useful work from solar energy, it would still ultimately be converted to and lost as heat in accordance with the second law of thermodynamics (see page 60). The overall energy balance would not change.

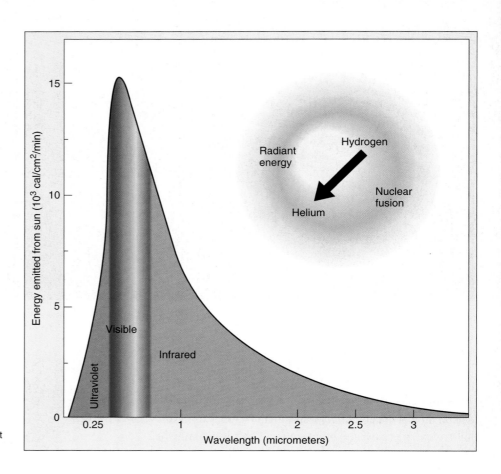

▶ **FIGURE 15–5** *The solar energy spectrum.* Equal amounts of energy are found in the visible-light and the infrared-light regions of the spectrum.

15.2 Putting Solar Energy to Work

Solar energy is a diffuse (widely scattered) source; solar radiation has the energy intensity of about 1 KW per square meter. The problem of using solar energy is one of taking a diffuse source and concentrating it into an amount and form, such as fuel or electricity, that we can use as heat and to run vehicles, appliances, and other machinery. Then there is the obvious problem: What does one do when the Sun is not shining? These problem areas may be categorized as collection, conversion, and storage. Also, in the final analysis, overcoming these hurdles must be cost-effective.

Consider how natural ecosystems handle these three aspects: Leaves collect light over a wide area. Photosynthesis converts and stores light energy as chemical energy—namely, glucose and other forms of organic matter—in the plant. The plant biomass then "fuels" the rest of the ecosystem. In the following sections, you will see how we can overcome or, even better, sidestep these hurdles in various ways to meet our needs from solar energy in cost-effective ways.

Solar Heating of Water

As noted, solar hot-water heating is already popular in warm, sunny climates. A solar collector for heating water consists of a thin, broad "box" with a glass or clear plastic top and a

black bottom with water tubes embedded within (Fig. 15–6). Such collectors are called **flat-plate collectors**. Faced to the

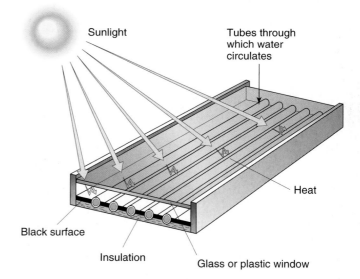

▲ **FIGURE 15–6** *The principle of a flat-plate solar collector.* As it is absorbed by a black surface, sunlight is converted to heat. A clear glass or plastic window over the surface allows the sunlight to enter but traps the heat. Air or water is heated as it passes over and through tubes embedded in the black surface.

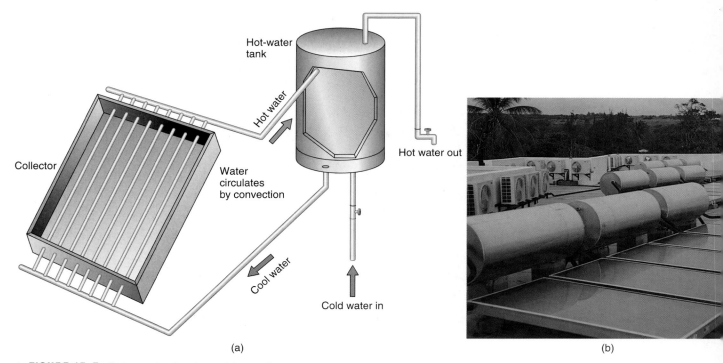

▲ **FIGURE 15–7** *Solar water heaters.* (a) In nonfreezing climates, simple water-convection systems may suffice. Where freezing occurs, an antifreeze fluid is circulated. (b) Solar panels (flat black collectors) and hot water tanks on the roof of a hotel in Barbados.

Sun, the black bottom gets hot as it absorbs sunlight—similar to how black pavement heats up in the Sun—and the clear cover prevents the heat from escaping. Water circulating through the tubes is thus heated and goes to the tank where it is stored.

The heated water may be actively moved by means of a pump—*active systems.* Or natural convection currents may be used—*passive systems.* Passive systems, as you can guess, are more economical.

For a passive solar water-heating system, the system must be mounted so that the collector is lower than the tank. Thus, heated water from the collector rises by natural convection into the tank while cooler water from the tank descends into the collector (Fig. 15–7). There are no pumps to buy, maintain, or remember to turn on and off. The size and number of collectors, of course, is adjusted to the need. The greatest economy is gained by careful management of requirements for hot water, so that the size of the tank and collectors can be minimized.

In temperate climates, where water in the system might freeze, the system may be adapted to include a heat-exchange coil within the hot-water tank. Then, antifreeze fluid is circulated between the collector and the tank. In the United States, there are an estimated 800,000 solar hot-water systems operating, but this number is still only about 0.5% of the total of all hot-water heaters.

In recent years, the most common end use of solar thermal systems has been for heating swimming pools; these are very inexpensive systems consisting of black plastic or rubber-like sheets with plastic tubing that circulates water through the systems.

Solar Space Heating

The same concept as in solar water heating can be applied to heat spaces. Flat-plate collectors like those used in water heating can be used for space heating. Indeed, the collectors for space heating may even be less expensive, home-made devices because it is necessary only to have air circulate through the collector box. Again, efficiency is gained if the collectors are mounted to allow natural convection to circulate the heated air into the space to be heated (Fig. 15–8).

The greatest efficiency in solar space heating, however, is gained by designing the building to act as its own collector. The basic principle is to have Sun-facing windows. In the winter, because of the Sun's angle of incidence, sunlight can come in and heat the interior (Fig. 15–9a). At night, insulated drapes or shades can be pulled to trap the heat inside. The well-insulated building, with appropriately made doors and windows, would act as its own best heat-storage unit. Beyond good insulation, other systems of storing heat, such as in tanks of water or masses of rocks, have not proved cost-effective. Excessive heat load in the summer can be avoided by using an awning or overhang to shield the window from the high summer Sun (Fig. 15–9b).

Along with design, positioning, and improved insulation, appropriate landscaping can also contribute to the heating and cooling efficiency of both solar and nonsolar space-heating designs. In particular, deciduous trees or vines on the sunny side of a building can block much of the excessive summer heat while letting the desired winter heat pass through.

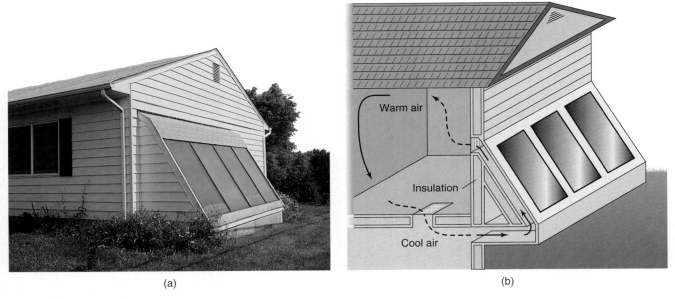

▲ FIGURE 15–8 *Passive hot air solar heat.* (a) Many homeowners could save on fuel bills by adding homemade solar collectors as shown here. (b) Air heated in the collector moves into the room by passive convection.

An evergreen hedge on the shady side can provide protection from the cold (Fig. 15–10).

One approach that combines insulation and solar heating is **earth-sheltered housing** (Fig. 15–11). The principle is to use earth as a form of insulation and orient the building for passive solar energy. Two strategies are employed: one is to build up earth against walls (earth berms) in a more conventional house design; the other is to cover the building to varying degrees with earth, exposing the interior to the outside with south-facing windows. Earth has a high capacity for heat storage, and since the walls and floors of an earth-sheltered home are usually built with poured concrete or masonry, at night the building will radiate the heat stored

during the day and warm the house. In the summer, the building is kept cool by its contact with earth, although moisture often will have to be controlled with dehumidifiers.

The Center for Renewable Resources estimates that in almost any climate, a well-designed passive solar home can reduce energy bills by 75% with an added construction cost of only 5% to 10%. About 25% of our primary energy is used for space and water heating. If solar heating were optimally used in this sector, it would save considerably on the oil, natural gas, and electrical power currently used.

A common criticism of solar heating is that a back-up heating system is still required for periods of inclement weather. Good insulation is a major part of the answer to this criticism.

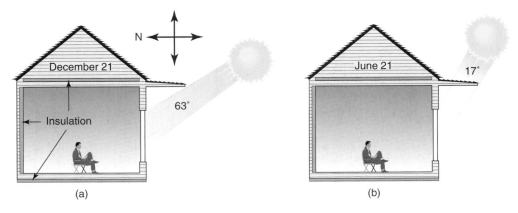

▲ FIGURE 15–9 *Solar building siting.* In contrast to expensive and complex active solar systems, solar heating *can* be achieved by suitable architecture and orientation of the home at little or no additional cost. (a) The fundamental feature is large, Sun-facing windows that permit sunlight to enter during winter months. Insulating drapes or shades are drawn to hold in the heat when the Sun is not shining. (b) Suitable overhangs, awnings, and deciduous plantings will prevent excessive heating in the summer.

Summer

Evergreens

Deciduous trees in full leaf shade home

Vines on trellis largely covering front of house

(a)

Winter

Trees have lost all their leaves

Northwind

(b)

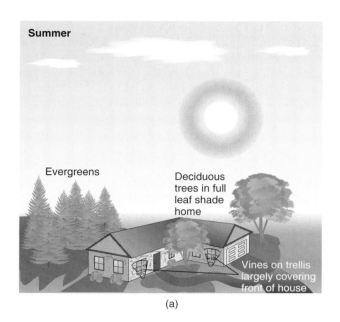
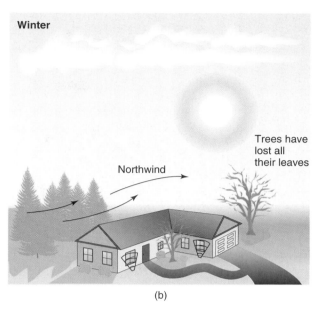

▲ **FIGURE 15–10** *Landscaping in solar heating and cooling.* (a) In summer, the house may be shaded with deciduous trees or vines. (b) In winter, leaves drop, and the bare trees allow the house to benefit from sunlight. Evergreen trees on the opposite side protect and provide insulation from cold winds.

People with well-insulated solar homes find that they have minimal need for back-up heating. When back-up is needed, a small wood stove or gas heater suffices. In any case, the criticism concerning the need for back-up heating misses the point. Remember that the objective of solar heating is to reduce our dependency on traditional fuels. Even if solar heating and improved insulation reduced the demand for conventional fuels by a mere 20%, this would still represent sustainable savings of 20% of the traditional fuel and its economic and environmental costs. With any systematic effort, the savings could be much greater.

However, this effort to reduce the use of fossil fuels has not pleased some sectors of our society. As a result, in the 1980s, intensive advertising campaigns by utility and oil companies purporting that solar energy is impractical and not cost-effective did much to curtail the growth of the solar energy industry. These fossil-fuel interests successfully lobbied the government to cancel research and end incentive programs for renewable energy. Now, in the 1990s, the major factor in discouraging the use of solar heating is relatively low energy prices, which are a disincentive to making any change. But the opportunity is open.

Solar Production of Electricity

Solar energy can also be used to produce electrical power, thus providing an alternative to coal and nuclear power. Currently, there are a few methods that are feasible—two methods in particular, *photovoltaic cells* and *solar-trough collectors*, are proving to be economically viable.

Photovoltaic Cells. A solar cell, more properly called a **photovoltaic**, or **PV cell**, looks like a simple wafer of material

with one wire attached to the top and one to the bottom (Fig. 15–12). As sunlight shines on this "wafer," it puts out an amount of electrical current roughly equivalent to a flashlight battery. Thus, PV cells accomplish the collection of light and its conversion to electrical power in one step. One PV cell is no more than about 2 inches (5 cm) in diameter. However, almost any

▲ **FIGURE 15–11** *Earth-sheltered house.* With the strategic use of earth berms and siting, such houses can be heated and cooled with a fraction of the energy normally required for home energy.

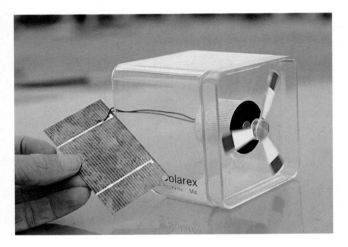

▲ **FIGURE 15–12** *Photovoltaic cell.* Converting light to electrical energy, this cell provides enough energy to run the small electric motor needed to turn these fan blades.

amount of power can be produced by wiring cells together in panels (recall Fig. 15–1).

The deceptively simple appearance of PV cells belies a very sophisticated materials science and technology. Each cell consists of two very thin layers of material. The lower layer has atoms with single electrons in the outer orbital that are easily lost. The upper layer has atoms lacking electrons from their outer orbital; such atoms readily gain electrons. (See Appendix C for some basic chemical principles.) The kinetic energy of light photons striking this "sandwich" dislodges electrons from the lower layer, creating an electrical potential between the two layers. This potential provides the energy for an electrical current through the rest of the circuit. Electrons from the lower side flow through a motor or other electrical device back to the upper side. Thus, with no moving parts, solar cells convert light energy directly to electrical power.

Because there are no moving parts, solar cells do not wear out; however, deterioration due to exposure to the weather limits their present life span to about 20 years. The major material used in PV cells is the element silicon, one of the most abundant elements on Earth, so there is little danger that production of PV cells will ever suffer because of limited resources. The cost of these cells lies mainly in their sophisticated design and construction.

Photovoltaic cells are already in common use in pocket calculators, watches, and numerous toys. Panels of PV cells are providing power for rural homes, irrigation pumps, traffic signals, radio transmitters, lighthouses, offshore oil-drilling platforms, and other installations that are distant from power lines. It is not hard to imagine a future in which every home and building has its own source of pollution-free, sustainable electrical power from an array of PV panels on the roof. Current installations in many countries and in over half of the states in the United States involve "net metering," where the rooftop electrical output is subtracted from the customer's use of power from the power grid.

The cost of PV power (cents per kilowatt-hour) will be the cost of the PV cells divided by the total amount of power they may be expected to produce over their lifetime. Then the cost of PV power must be compared with that of other power alternatives. The first PV cells had a cost factor several hundred times that of electricity from conventional power stations transmitted through the power grid, so they have been used mainly in areas far from the grid. It is easy to see why PV power had its first significant application in the 1950s, in the solar panels of space satellites. This application in satellites, in fact, started the development cycle rolling. As more efficient cells and less expensive production techniques evolved, costs came down. As costs came down, applications, sales, and potential markets expanded, creating the incentive for further development (Fig. 15–13). In 1998, PV power was still about three times more costly than electricity from the power grid; however, the technology does not share in the same subsidies and incentives that are given to fossil fuels and nuclear power. The price is still hovering around $4000 per kilowatt. In spite of this, the industry is growing rapidly. Sales of solar panels were $395 million in the United States in 1996, with the bulk of sales going to overseas customers, mostly in the developing world. Globally, the market has surpassed $1 billion and is expected to continue to grow 15% annually.

What will utility companies do as PV power becomes competitive? A demonstration PV power plant has been built in southern California (Fig. 15–14). However, the more promising future for PV power would appear to be installation of home-sized systems on rooftops, thus avoiding additional land and transmission costs. (A new technique for PV cell design—the "thin film" collectors—has the potential for constructing roof shingles that generate electrical power at a fraction of the cost of conventional PV cells.) Toward this end, a local utility in Sacramento, California, has launched its PV Pioneers program. The utility installs the systems and pays homeowners $4 per month to maintain a 3 to 4 KW PV system on their roofs. These systems feed into the power grid; the installed costs of the systems are expected to continue to drop until, by 2002, the power they generate will be similar to the cost of peakload power. A similar program in Japan has led to the installation of more than 20,000 systems in homes. Before such systems are widespread in the United States, policy changes will be needed to provide incentives through tax credits and other means.

Solar-Trough Collectors. The second method that is proving cost-effective is the **solar-trough** collector system, which is so named because the collectors are long, trough-shaped reflectors tilted toward the Sun (Fig. 15–15). The curvature of the "trough" is such that sunlight hitting the collector is all reflected onto a pipe running down the center. Oil or other heat-absorbing fluid circulating through the pipe is thus heated to very high temperatures. The heated fluid is passed through a heat exchanger to boil water and produce steam for driving a turbogenerator.

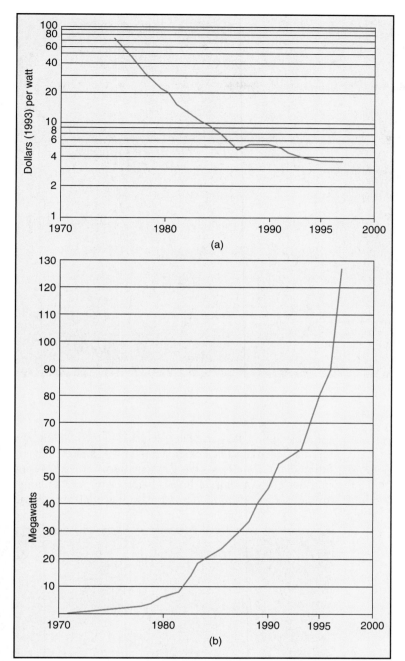

◀ **FIGURE 15–13** *The market for PV cells.*
(a) Research and development has brought the price of photovoltaic cells down about 50-fold in the past 25 years. (b) As prices have come down, sales have increased dramatically. Note the surge in sales for 1997—to 127 MW. (Source: Data from C. Flavin, N. Lenssen, *Power Surge: Guide to the Coming Energy Revolution* [New York: W. W. Norton, 1994] fig. 8–1 and 8–2; and Lester Brown, et. al., *Vital Signs 1998* [Washington, D.C.: Worldwatch Institute, 1998] 61.)

Nine solar-trough facilities in the Mojave Desert of California are now connected to the Southern California Edison utility grid, with a combined capacity of 350 MW, about one-third the capacity of a large nuclear power plant. The facility, which converts a remarkable 22% of incoming sunlight to electrical power, is producing power at a cost of 10 cents per kilowatt-hour, barely more than the cost at coal-fired facilities.

Experimental Technologies. Two additional methods of harnessing thermal energy from the Sun have been developed with government funding. The first is what is commonly called a "power tower," where an array of Sun-tracking mirrors focus the sunlight falling on several acres of land onto a receiver mounted on a tower in the center (Fig. 15–16). The receiver transfers the heat energy to a molten salt liquid, which then flows either to a heat exchanger to drive a conventional turbogenerator, or to a tank at the bottom of the tower to store the heat for later use. The latest pilot plant model of this technology, Solar Two, is now connected to the utility grid and is generating 10 MW of electricity, enough for 10,000 homes.

The second is the dish/engine system (Fig. 15–17). This is a smaller system, consisting of a set of parabolic concentrator dishes that focus sunlight on a receiver. Fluid (often molten sodium) in the receiver is transferred to an engine

▶ **FIGURE 15–14** *PV power plant.* The world's first photovoltaic (solar cell) power plant located near Bakersfield, California. The array of 220 34-foot (11-m) panels produces 6.5 MW at peak, enough for 2400 homes.

(Stirling engine) that generates electricity directly. These systems have demonstrated efficiencies of 30%, higher than any other solar technology. They generate from 5 to 40 KW; any number of them can be linked together to provide larger amounts of electricity. They are well suited for off-grid

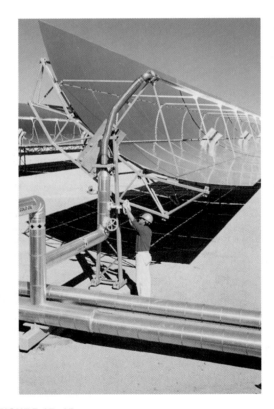

▲ **FIGURE 15–15** *Solar-trough power plant.* Several solar-trough facilities are now in operation in southern California. The curved reflector focuses sunlight on and heats oil in the pipe. The heated oil is then used to boil water and generate steam for driving a conventional turbogenerator.

▲ **FIGURE 15–16** *Power tower Solar Two.* Sun-tracking mirrors are used to focus a broad area of sunlight onto a molten salt receiver mounted on the tower in the center. The hot salt is stored or pumped through a steam generator that drives a conventional turbogenerator.

▲ **FIGURE 15–17** *Solar dish-engine system.* A circular mirror array focuses solar energy on a receiver, which transmits the energy to an engine that generates electric power.

applications to provide power to remote sunny areas. Both of these new technologies have the advantage of requiring little maintenance and, of course, generate no polluting gases.

The Promise of Solar Energy

Despite the promise of solar energy, there do exist disadvantages. First, the available technologies are still more expensive than conventional energy sources, although the costs continue to come down. Second, solar energy only works during the day, requiring a back-up energy source or a storage battery, and the climate is not sunny enough in the winter in many parts of the world. However, it should be considered that there are hidden costs of such things as air pollution, strip-mining, and nuclear waste disposal that are not included in the cost of power from traditional sources. Likewise, criticism that solar power requires an exorbitant amount of land to "harvest" the sunlight rings hollow when one considers the thousands of acres that are strip-mined for coal each year or that might be made uninhabitable by a nuclear accident.

Again, the fact that the Sun provides power only during the day can be countered with the argument that 70% of the electrical demand occurs in daytime hours when industries, offices, and stores are in operation. Thus, proponents of solar energy argue that potential savings may be achieved by using solar panels just for daytime needs and continuing to rely on conventional sources at night. All the solar power that can conceivably be built in the next 20 years or more could be used to supply the daytime demand, saving traditional fuels to provide nighttime demand. In particular, the demand for air conditioning, which, after refrigeration, is the second largest power consumer, is well matched to energy from PV cells. In addition, solar-powered air conditioners could operate independently from the rest of the electrical system, thus avoiding the costs of interconnecting systems. (Such air conditioners may be expected to come on the market within the next few years.) In the long run, we might envision the nighttime load being carried by forms of indirect solar, energy such as wind power or water power (discussed later in the chapter).

You observed in Table 13-1 that about 77% of our electrical power is currently generated by coal-burning and nuclear power plants. Therefore, development of solar electrical power can be seen as gradually reducing the need for coal and nuclear power (see "Earth Watch" essay, p. 368). Many argue that solar electrical power development is our best strategy for mitigating the acid rain, carbon dioxide, and other environmental impacts of burning coal and for countering the risks of nuclear power. It has also proved suitable for meeting the needs of electrifying villages and towns in the developing world. Already, rural electrification projects based on photovoltaic cells are beginning to spread throughout the developing world where centralized power is not available. The cost of the alternative—putting roughly 2 billion people that are still without electricity on the **power grid** (central power plants, high-voltage transmission lines, poles, wires, and transformers)—is unimaginable.

Solar Production of Hydrogen— The Fuel of the Future

So far the current methods of harnessing solar energy that have been discussed have not addressed one of our greatest resource concerns—the fact that our oil reserves needed to fuel transportation are dwindling. However, there may be an answer here as well. One obvious possibility is hydrogen gas. Conventional cars can be run on hydrogen gas (H_2) as a fuel in the same manner as they are now beginning to be run on natural gas (methane, CH_4) (see Chapter 13). This fact has been amply demonstrated. Furthermore, there is no carbon dioxide or hydrocarbon pollution produced in burning hydrogen. The only byproduct is water vapor:

$$2H_2 + O_2 \rightarrow 2H_2O + energy$$

(Some nitrogen oxides will be produced, however, because the burning still uses air, which is nearly four-fifths nitrogen.)

If hydrogen gas is such a great fuel, why are we not using it? The answer is that there is virtually no hydrogen gas on Earth. Any hydrogen gas in the atmosphere has long since been ignited by lightning and burned to form water. And, although there are many soil bacteria that produce hydrogen in fermentation reactions, other bacteria are quick to use the hydrogen because it is an excellent energy source. Thus, there are abundant amounts of the *element* hydrogen, but it is all combined with oxygen in the form of water (H_2O) or other low-energy compounds.

Although we can re-isolate hydrogen from water, this process *uses* energy. Recall that the first law of thermodynamics dictates that energy cannot be created—"You can't get something for nothing." In fact, it requires *more* energy to

EARTH WATCH

ECONOMIC PAYOFF OF SOLAR ENERGY

In addition to being a nonpolluting renewable energy source, photovoltaic, solar-trough, and wind-powered energy-generating facilities have another advantage: They can be installed quickly and added to the utility system in relatively small increments. To understand this advantage, let's draw a comparison with a nuclear power plant.

A nuclear power plant can be cost-effective only if it is large—about 1000 MW capacity, enough to power a million average homes. Such a plant, assuming public acceptance, would require 10 to 15 years from the time of the decision to build until it is in operation, and would cost around $5 to $10 billion. Therefore, a utility taking this route must borrow billions of dollars and keep that money tied up for 10 to 15 years before a return on the investment is realized by selling power from the new plant. Even then, if the 1000 MW of additional power is not needed when the new plant comes on-line, the return on the investment may be meager indeed. In the late 1970s and early 1980s a number of utility companies went bankrupt (or nearly so) because they had huge amounts of money tied up in nuclear power plants but a slack demand for the power produced. Clearly, building large power plants (nuclear or coal) involves considerable financial risk on the part of both utility and consumers, as consumers are the ones who ultimately pay.

On the other hand, solar or wind facilities can be operational within a few months of the decision to build; they can be installed in small increments, and additional modules can be added as demand requires. The utility does not have to guess what power demands will be in 10 or 15 years. Through solar or wind power, a utility can add capacity as it is needed, make relatively small investments at any given time, and have those investments start paying back almost immediately. You can see that this approach involves much less financial risk both for the utility and for consumers.

---

convert water into hydrogen and oxygen because of inevitable losses (the second law of thermodynamics). Therefore, the use of hydrogen as a fuel must await an energy source that is itself suitably cheap, abundant, and nonpolluting. Do these specifications sound like solar energy?

From eons past, the natural world has had a method of splitting water into hydrogen and oxygen using light energy. This is what happens in photosynthesis. (You will recall that the oxygen from photosynthesis is released into the atmosphere, and the hydrogen is attached to carbon dioxide to form sugars. See page 61.) Scientists, however, have not yet developed a means of mimicking the photosynthetic reaction on a scale sufficient to produce commercial amounts of hydrogen.

However, hydrogen can be produced with very simple equipment by means of *electrolysis of water*, in which a direct current passed through water causes water molecules to dissociate. Hydrogen bubbles come off at the negative electrode, while oxygen bubbles come off at the positive electrode. (You can demonstrate this process for yourself with a battery, some wire, and a dish of water.) In addition, a small amount of battery acid in the water facilitates the conduction of electricity and the release of hydrogen.

An economic analysis projects that the cost of solar-produced hydrogen gas would be equivalent to gasoline costing $1.65 to $2.35 per gallon. This may not sound cheap, but it is less than what gasoline will cost as inevitable constraints come into play (see Chapter 13). Consumers in European nations and Japan are already paying much more. One important drawback to this technology is the fact that it is difficult to store enough hydrogen in a vehicle to allow for a practical operating range. Compressing it into a liquid would require energy, thus reducing the energy efficiency involved.

A rather elaborate plan has been suggested were we to take the concept of using solar energy-produced hydrogen for vehicular fuel to the ultimate. We would start by building arrays of solar-trough or photovoltaic generating facilities in the deserts of the southwestern United States, where land is cheap and sunlight plentiful. The electrical power produced by these methods would then be used to produce hydrogen gas by electrolysis. Conveniently, Texas is the hub of a network of natural-gas pipelines that were originally constructed to transport natural gas from the Texas oil fields. Many of these pipelines throughout the nation are now underutilized because those fields have been largely depleted. Therefore, the hydrogen gas produced in the Southwest could be transported throughout the nation by way of these pipelines. As vehicles are already being adapted to run on natural gas, hydrogen could slowly be phased in by mixing gradually increasing proportions of hydrogen gas with natural gas. Thus, there could be a smooth transition from present fuel through natural gas to hydrogen.

An alternative to burning hydrogen in conventional internal combustion engines uses hydrogen in *fuel cells* to produce electricity and power the vehicle with motors—just the opposite of the previous technology! Fuel cells are devices in which hydrogen or other fuels are recombined with oxygen chemically in a manner that produces an electrical potential rather than burning (Fig. 15–18). Because fuel cells create much less waste heat than conventional engines do, there is a more efficient transfer of energy from the hydrogen to the vehicle. Both of these technologies are in the development stage for transportation and are at least a decade or more away from any commercial application to transportation.

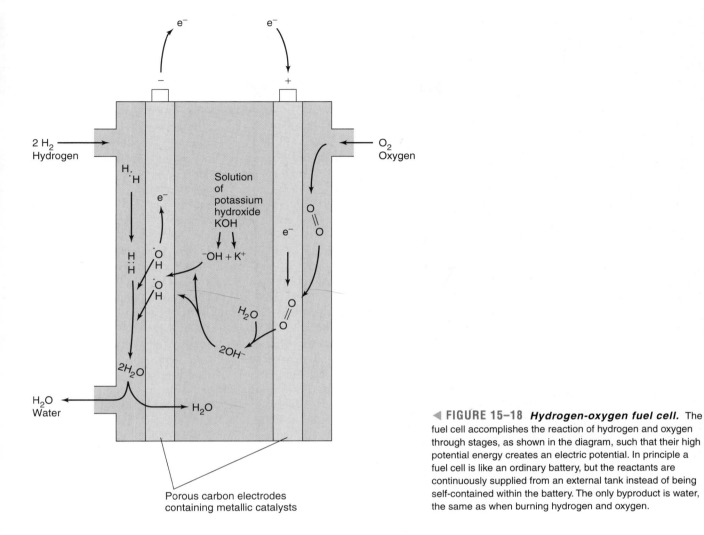

◀ **FIGURE 15–18** *Hydrogen-oxygen fuel cell.* The fuel cell accomplishes the reaction of hydrogen and oxygen through stages, as shown in the diagram, such that their high potential energy creates an electric potential. In principle a fuel cell is like an ordinary battery, but the reactants are continuously supplied from an external tank instead of being self-contained within the battery. The only byproduct is water, the same as when burning hydrogen and oxygen.

15.3 Indirect Solar Energy

Water, fire, and wind have provided energy for humans throughout history. We are all familiar with dams, firewood, and windmills. We group these age-old sources of energy together under "indirect solar energy" because a moment's reflection shows us that energy from the Sun is the driving force behind each. The question is, What is the potential to expand these options from the past into major sources of sustainable energy for the future?

Hydropower

It was discovered early in technological history that the force of falling water could be used to turn paddle wheels, which in turn would drive machinery to grind grain, saw logs into lumber, and do other laborious tasks. The modern culmination of this use of water power, or hydropower, is huge hydroelectric dams, where water under high pressure flows through the bottom of the dam, driving turbogenerators as it passes through (Fig. 15–19). The amount of power generated is proportional to both the height of water behind the

dam—that is what provides the pressure—and the volume of water that flows through.

About 11% of the electrical power generated in the United States currently comes from hydroelectric dams, most of it from about 300 large dams concentrated in the Northwest and Southeast. Worldwide, hydroelectric dams have a generating capacity of 714,600 MW; an additional 50,000 MW is in the construction or planning stage. Hydropower generates 19% of electrical power throughout the world and is by far the most common form of renewable energy in use.

Whereas water power is basically a nonpolluting, renewable energy source, harnessing it by means of hydroelectric dams still involves high ecological, social, and cultural trade-offs:

1. The reservoir created behind the dam inevitably drowns farmland or wildlife habitat and perhaps towns or land of historical, archaeological, or cultural value. Glen Canyon Dam (on the border between Arizona and Utah) drowned one of this world's most spectacular canyons. A number of dams have obliterated sacred lands of Native Americans.

▲ **FIGURE 15–19** *Hoover Dam.* About 11% of the electrical power used in the United States comes from large hydroelectric dams such as this.

2. Migration of fish, even when fish ladders are provided, is impeded or prevented. Federal surveys show that fish habitat is suffering in the majority of the nation's rivers because of damming. Salmon fishing, one of the great industries of the Northwest, has been heavily impacted there because of this factor.

3. Changing from a cold-flowing river to a warm-water reservoir can have unforeseen ecological consequences. The reservoir behind the Aswan High Dam in Egypt has fostered the spread of a parasitic worm that causes a debilitating disease. The reservoir has also increased humidity over a widespread area, which is now accelerating the deterioration of ancient monuments and artifacts that have stood virtually unchanged for many centuries.

4. Ecological consequences occur below the dam as well. Because water flow is regulated according to the need for power, dams play havoc downstream; water levels may go from near flood levels to virtual dryness and back to flood levels in a single day. Other ecological factors are also affected, because sediments with nutrients settle in the reservoir, and smaller amounts reach the river's mouth.

As if such trade-offs were not enough, thoughts of greatly expanding water power in the United States are nullified by the fact that few sites conducive to large dams remain. There are already 65,000 U.S. dams; only 2% of the nation's rivers remain free-flowing, and many of these are now protected by the Wild and Scenic Rivers Act of 1968, a law that effectively gives certain scenic rivers the status of national parks.

In short, proposals for new dams in the United States and elsewhere are embroiled in controversy over whether the projected benefits justify the ecological and sociological trade-offs. In 1997, the World Bank gave preliminary approval to a loan to the Lao People's Democratic Republic for construction of the Nam Theun Two Dam, a 680 MW project that is expected to help Laos begin a process of economic development. The World Bank's decision has come under fire because the project will displace 4500 people and destroy 2485 square miles of sensitive environments. The loan is contingent on reception of a thorough study of the dam's impacts. Another huge dam under construction is the Three Gorges Dam on the Yangtze River in China. When completed, it will be the largest dam ever built, displacing some 1.9 million people in order to generate 18,200 MW of electricity. Offsetting the human and economic costs is the fact that it would take more than a dozen large coal-fired power plants to produce the same amount of electricity.

The enormous costs of the large dam projects suggest another (perhaps more feasible) alternative—the "microhydro" project. Such projects develop rivers in a more benign way, by way of construction of much smaller dams and strategic placement of turbines in turbulent stretches of rivers. Some have estimated that the potential for the microhydro projects could be as high as 20,000 MW.

Wind Power

Throughout much of history, wind power—in addition to propelling sailing ships—was widely used for grinding grain, hence the term *windmill*. Then, wind-driven propellers went on to perform other tasks. Until the 1940s, most farms in the United States used windmills for pumping water and generating small amounts of electricity. In the '30s and '40s, windmills fell into disuse, however, as transmission lines brought abundant lower-cost power from central generating plants. Not until the energy crisis and rising energy costs of the '70s did wind begin to be seriously considered again as a potential source of sustainable energy.

Many different designs of wind machines have been proposed and tested, but what has proved most practical is the age-old concept of wind-driven propeller blades. The propeller shaft is geared directly to a generator. (Wind driving a generator is more properly called a **wind turbine** than a windmill.)

As reliability and efficiency of wind turbines have improved, the cost of wind-generated electricity has come down. **Wind farms**, arrays of up to several thousand such machines, are now producing pollution-free, sustainable power for as little as 4 cents per kilowatt-hour, competitive with traditional sources. Moreover, the amount of wind that potentially can be tapped is immense. The American Wind Energy Association calculates that wind farms located throughout the Midwest could meet the electrical needs for the entire country, while the land beneath the turbines could still be used for farming. This situation is similar throughout the world.

For example, Searsburg, in southern Vermont, is the site of Green Mountain Power Corporation's new "wind farm." In 1997, the wind farm of 11 turbines was opened and began generating enough electricity to provide power to 2000 homes. This new facility, built with the help of a $3.5 million subsidy from the DOE, joins a rapidly rising number of wind turbines whose capacity has now surpassed 7600 MW worldwide (Fig. 15–20). In Denmark, hard-pressed by the lack of fossil-fuel energy, wind now supplies 7% of electricity. California leads the United States in wind energy; the Tehachapi-Mojave area alone contains more than 5000 turbines generating enough electricity to power 500,000 homes.

There are still some drawbacks with wind power, however. First, as it is an intermittent source, problems of back-up or storage must be considered. Second is the aesthetic consideration. One or two windmills can be charming, but a landscape covered with them can be visually tiresome. Third, they are a hazard to birds; location of wind farms on migratory routes could be a problem for some avian populations.

Biomass Energy

Burning firewood for heat is perhaps one of the oldest forms of energy that humans have used throughout history. However, there is nothing like a new name to put life into an old concept. Thus, "burning firewood" has become "utilizing biomass energy." Biomass energy or bioconversion means deriving energy from present-day photosynthesis. As Fig. 15-4 indicates, biomass energy is second only to hydropower in renewable energy production in the United States, most of which is for heat.

In addition to burning wood in a stove, major forms of biomass energy include: obtaining energy from the burning of municipal waste paper and other organic waste; methane production from anaerobic digestion of manure and sewage sludges; and alcohol production from fermenting grains and other starchy materials. Let's look further at the potential of each of these techniques.

Burning Firewood. Where forests are ample relative to population, firewood or fuelwood can be a sustainable energy resource, and indeed wood has been a main energy resource over much of human history. In the United States, wood stoves have enjoyed a tremendous resurgence in recent years—about 5 million homes rely entirely on wood for heating and another 20 million use it for partial heating. Sales of wood stoves continue at a level of 400,000 units per year.

The most recent development in wood stoves is the *pellet stove*, a device that burns compressed wood pellets made from wood wastes (Fig. 15–21). The pellets are loaded into

◀ **FIGURE 15–20** *Wind farm at Altamont Pass, California.* Such arrays of wind turbines are proving to be an economically competitive way of generating electrical power. California now leads the United States in use of wind energy.

▲ **FIGURE 15–21** *Pellet stove.* Pellet stoves use pelletized wood waste to achieve an efficient burn that requires much less care than conventional wood stoves.

a hopper (controlled by computer chips) that feeds fuel when needed. These appliances require very little attention and burn very efficiently, relying on a fan to draw air into the stove and another fan to distribute heated air to the room.

The Fuelwood Crisis. For over a billion people in developing countries, firewood is their only source of fuel for cooking. In fact, 90% of the world's fuelwood is produced and used in the developing countries. Recall that two patterns of use determine how the forest resources will be exploited: consumptive use and productive use (Chapter 12). Most fuelwood use in developing countries fits the consumptive pattern; people simply forage for their daily needs in local woodlands and forests. Unfortunately, this kind of use often results in a deforestation that spreads outward from population centers. One example is the Central American country of Honduras. PRO-LENA, a citizen's group, estimates that Hondurans cut 7 million cubic meters of firewood every year as they obtain wood to cook their meals, resulting in a high rate of deforestation. The country is in desperate need of a fuelwood policy that would encourage replanting and sustainable use of wood in a land that is ecologically ideal for growing trees.

The story is similar in many other parts of the developing world: Growing numbers of people move out from villages and towns daily to gather wood. The solution? Develop a market in wood, which would encourage people with land to plant trees as a cash crop. Where people have to pay for fuelwood, trees become an important resource that can be put under the stewardship of local communities or private landowners, and sustainable forest use can result. Then all of the other benefits of goods and services provided by forests are preserved or restored. This is the trend in many parts of the developing world, but often it requires encouragement from

governments or NGOs (see Chapter 12). Toward that end, the U.N.-sponsored Forest Trees and People Program (FTPP) works with local institutions and communities to encourage the management of forest resources for both conservation and sustainable development.

Burning Wastes. Facilities to generate electrical power from the burning of municipal wastes (waste-to-energy conversion) are discussed in Chapter 19. Many sawmills and woodworking companies are now burning wood wastes, and a number of sugar refineries are burning the cane wastes to supply all or most of their power. The primary virtue in this method is in finding a productive or less costly way to dispose of wastes. Power from these sources could never meet more than a few percent of total electrical needs. Also, such schemes must be weighed against the potential for recycling or for other uses of such wastes. Wood wastes, for example, are made into pressed plywood and pellets for wood stoves.

Producing Methane. You will see in Chapter 18 that anaerobic digestion of sewage sludge yields *biogas*, which is two-thirds methane, plus a nutrient-rich treated sludge that is a good organic fertilizer. Animal manure can be digested likewise. Where these three aspects—manure disposal, energy production, and fertilizer creation—can be combined in an efficient cycle, great economic benefit can be achieved.

This concept is demonstrated on the Mason-Dixon Dairy farm located near Gettysburg, Pennsylvania (Fig. 15–22). Manure from 2000 cows that are fed in a barn drops through a grated floor and slowly flows by gravity into anaerobic digesters. The biogas produced (purification of the methane has proved unnecessary) fuels engines that drive generators supplying not only all the electrical power for the dairy but also considerable excess that is sold to the utility company. Waste heat from the engines is used to heat the digesters and the buildings. The nutrient-rich sludge remaining after digestion is recycled back to the land in order to maintain the fertility of the fields growing feed for the cows.

With the savings and proceeds from selling energy, as well as sale of dairy products and additional savings on fertilizer, the Mason-Dixon Dairy is probably the most profitable dairy in the country. In China, millions of small farmers maintain a simple digester in the form of a sealed pit into which they put agricultural wastes. The biogas produced is used as a source of cooking fuel. This concept could be expanded throughout the developing world to provide an alternative to fuelwood.

Producing Alcohol. Alcohol is produced by the fermentation of starches or sugars. The usual starting material is grain, sugar cane, or various sugary fruits such as grapes. Obviously, this is the process used in the production of alcoholic beverages. The only new part of the concept is that instead of drinking the brew, we distill it and put it in our cars. *Gasohol,* a mixture of gasoline and alcohol, has been promoted and marketed in the Midwest since the late 1970s.

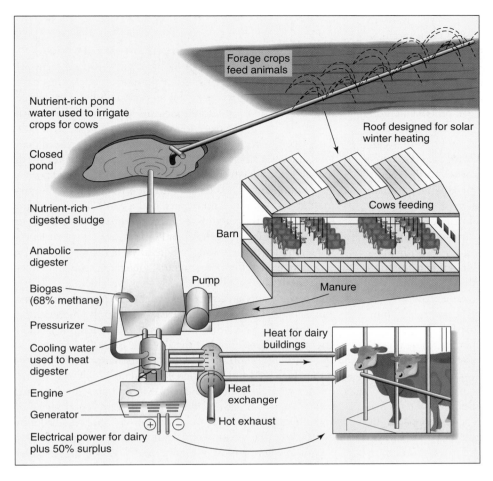

◄ FIGURE 15–22 *Biogas power.*
The total power needs for the Mason-Dixon
Dairy, located near Gettysburg,
Pennsylvania, are obtained as a byproduct
of cow manure, and nutrients are recycled
in the process. Excess power, nearly half of
what is produced, is sold to the local utility.

However, the enthusiastic support of gasohol in the Midwest is not motivated purely by the desire to save fossil fuel. This product provides another market for corn surpluses and thereby improves the economy of the region. If farmers get higher prices for their corn as feedstock (for animals), alcohol production declines because its production is not cost effective.

Thus, where grain surpluses do not exist—and this includes most of the world—alcohol as a fuel is a dubious goal at best. For example, Brazil promoted the large-scale production of alcohol from sugarcane in the 1980s. Sugarcane production rose to record levels, while food crops declined by 10%, despite widespread malnutrition and a rapidly growing population. Brazil has now abandoned its promotion of alcohol fuel.

Another problem with alcohol as fuel is pollution. Alcohol is promoted as clean-burning, and it is. However, producing the alcohol generates much pollution because inexpensive, dirty-burning fuels such as soft coal are used to distill the alcohol.

Finally, proposals that crops be grown specifically for biomass energy—either for direct burning or for conversion to alcohol or methane—must be weighed against the impacts of soil erosion and the loss of other important ecosystem services that such use generally entails. Also, energy crops will invariably compete with food crops because either may be grown on the same land.

15.4 Additional Renewable Energy Options

Several additional, theoretically sustainable energy options have strong support from certain sectors and deserve consideration: geothermal power, tidal power, and ocean thermal energy conversion.

Geothermal Energy

In various locations in the world, such as the northwestern corner of Wyoming (now Yellowstone National Park), one finds springs that yield hot, almost boiling water. Natural steam vents and other thermal features are also found in such areas. They occur where the hot molten rock of Earth's interior is close enough to the surface to heat groundwater, as may occur in volcanic regions. Using such naturally heated water or steam to heat buildings or drive turbogenerators is the basis of **geothermal energy**.

In 1998, geothermal energy was being used to generate electrical power (8029 MW capacity) equivalent to eight large nuclear or coal-fired power stations in countries as widely diverse as Nicaragua, the Philippines, Kenya, Iceland, and New Zealand. Today, the largest single facility is in the United States at a location known as The Geysers, 70 miles (110 km) north of San Francisco (Fig. 15–23). As impressive as this application for power generation is, nearly double the amount of geothermal energy used for power generation is being used to directly heat homes and buildings largely in Japan and China.

Many experts feel that the potential for geothermal energy has been barely tapped, although its development will probably remain restricted to regions such as Japan, the Philippines, Iceland, and New Zealand, which have suitable underlying volcanic geology. Another concern is that recent experience has shown that geothermal energy may not always be sustainable. The Geysers in California was steadily expanded until in 1988 it reached about 2000 MW (equivalent to two large nuclear power plants); it was projected that by the year 2000 the facility would be producing 3000 MW. In fact, output from The Geysers has been steadily declining since 1988; in 1991 it was producing just 1500 MW. The problem is that although the source of geothermal heat may be unlimited, the amount of groundwater isn't. Tapping the hot steam is depleting the groundwater. To restore some of the groundwater, a 29-mile pipeline has been constructed to conduct treated wastewater from Middletown to one of the Geysers plants, where it is injected into the ground to raise the level of groundwater needed to produce steam. If this is successful, other communities may also be tapped for their wastewater.

Geothermal power presents some pollution problems that must be mitigated if the process is not to be environmentally ruinous. The hot steam and water brought to the surface are frequently heavily laced with salts and other contaminants, particularly sulfur compounds leached from minerals in the bedrock. These contaminants are highly corrosive to turbines and other equipment, and they cause air pollution if the steam is released into the atmosphere. Sulfur dioxide pollution from a geothermal plant can be equivalent to that from a plant burning high-sulfur coal. Hot brines from geothermal sources released into streams or rivers would be ecologically disastrous.

Tidal Power

A phenomenal amount of energy is inherent in the twice-daily rise and fall of ocean tide, brought about by the gravitational pull of the moon and Sun. Many imaginative schemes have been proposed for capturing this eternal, pollution-free source of energy. The most straightforward idea is to build a dam across the mouth of a bay and mount turbines in the structure. The incoming tide flowing through the turbines would generate power. As the tide shifted, the blades would be reversed, so that the outflowing water would continue to generate power.

Tantalizing as this idea sounds, it is fraught with a fundamental problem: In most regions of the world, the maximum difference between high and low tide is a matter of only 40 to 50 cm (16 to 20 inches). A head of pressure 50 cm or less is not enough to drive turbines efficiently. There are about 30 locations in the world where shoreline topography generates tides high enough—6 m (20 ft) or more—for this kind of use. Large tidal power plants already exist at two of those places—in France and Canada. The only suitable location in North America is the Bay of Fundy, where the Annapolis Tidal Generating Station has operated since 1984, with 20 MW capacity.

Thus, this application of tidal power has potential only in certain localities, and even there it would not be without adverse environmental impacts. The dams would trap sediments, impede migration of fish and other organisms, prevent navigation, and alter the circulation and mixing of salt and fresh water in estuaries, perhaps having other unforeseen ecological effects.

A new application of tidal power has been proposed by Tidal Electric, Inc., which promises to avoid the problems of conventional tidal power. In this technology, the tidal generator is located offshore and consists of a concrete impoundment enclosure that sits on the ocean bottom, and an adjacent turbine housing. Power generated is highly related to the vertical tidal range, as the output varies with the square of the range. The impoundment is filled with water as the tide rises, creating a "head"

▶ **FIGURE 15–23** *Geothermal energy.* One of the 11 geothermal units operated by the Pacific Gas & Electric Company at The Geysers in Sonoma and Lake counties, California.

that is gradually released when the tide is low, the water running through turbines that generate electric power. Feasibility studies are under way for two locations in Alaska and India.

Ocean Thermal Energy Conversion

Over much of the world's oceans, a thermal gradient of about 20°C (36°F) exists between surface water heated by the Sun and colder deep water. **Ocean thermal energy conversion (OTEC)** is the name of an experimental technology for using this temperature difference to produce power. It involves using the warm surface water to heat and vaporize a low-boiling-point liquid such as ammonia. The increased pressure of the vaporized liquid would drive turbogenerators. The ammonia vapor leaving the turbines would then be recondensed by cold water pumped up from as much as 300 feet (100 m) deep and returned to the start of the cycle.

Various studies show that OTEC power plants show little economic promise—unless, perhaps, they can be coupled with other, cost-effective operations. For example, in Hawaii a shore-based OTEC plant uses the cold, nutrient-rich water pumped from the ocean bottom to cool buildings and supply nutrients for vegetables in an aquaculture operation in addition to cooling the condensers in the power cycle. Even so, interest in duplicating such operations is minimal at present.

15.5 Policy for a Sustainable Energy Future

In the last three chapters, we have studied our planet's energy resources, requirements, and management options. The solar and other renewable energy alternatives we have discussed in this chapter are diagrammatically shown in Fig. 15–24. A review of the energy situation as a whole suggests that there is no reason for an immediate fear of "running out" of energy. World reserves of fossil fuels, even crude oil, are adequate for at least the next 20 to 50 years. Then, there are combinations of solar and wind options that can be developed and phased in over that time frame. And there is nuclear power if we should choose to go that route.

The major question confronting us is, Should we be taking any specific steps to promote or even force the transition to an economy based on solar energy? Or should we simply eliminate all subsidies for energy to level the playing field and let free-market forces sort out the situation?

Some policies of promoting the transition away from dependence on crude oil and toward solar energy were implemented in the 1970s. They included:

- subsidies to producers of solar energy or solar energy products
- subsidies to people for solar installations in homes and other structures
- additional taxes on traditional fuels, making them more expensive relative to solar energy

- mandates to increase fuel efficiency of cars

Our final incentives are the moral and legal agreements with other nations that we are under to stabilize emissions of greenhouse gases—the major one of which is carbon dioxide from fossil-fuel burning (see Chapter 21). However, of all of these policies implemented to promote alternate sources of energy, all were phased out in the early 1980s, except for a modest tax on gasoline, effectively returning the nation to the traditional policy—"business as usual"—with an economic bias toward traditional energy sources. Meanwhile, subsidies and incentives for fossil fuels and nuclear energy have remained in force. Although we see photovoltaic cells and wind energy making headway in today's marketplace, the business-as-usual view overlooks the fact that the field of competition between solar and traditional sources of energy is steeply tilted in favor of traditional fuels. The traditional fuel industry is heavily subsidized by:

- depletion allowances (tax write-offs as a resource is depleted)
- leasing of public lands at bargain-basement prices, encouraging exploitation of fossil fuel reserves
- military support to assure access to oil in the Middle East

Moreover, substantial costs are discounted or overlooked. These costs include:

- damage to the environment and human health from the pollution caused by burning fossil fuels
- environmental destruction from strip-mining, oil spills, and other accidents
- the continuing and increasing economic drain as we import growing amounts of crude oil

On a positive note, the winds of change are sweeping through U.S. energy policy and elsewhere. Consider the following developments:

- The EPA has successfully initiated two energy-efficiency programs: the **Energy Star** program and the **Green Lights** program. These are voluntary programs where EPA provides companies and institutions with technical and informational support, enabling them to achieve significant savings in energy. Thus, public interest and self-interest are wedded in a way that has been achieving better results than could ever be obtained through the heavy hand of regulation. Over 2300 entities signed up to participate in these programs by mid-1997, saving more than $600 million on their electric bills, and eliminating 1.3 million metric tons of carbon and other pollutants.
- The **Climate Change Technology Initiative** is the administration's package of research and development investments directed toward reducing greenhouse gas emissions that has been funded for several years. Congress approved over $1 billion for this important program in the

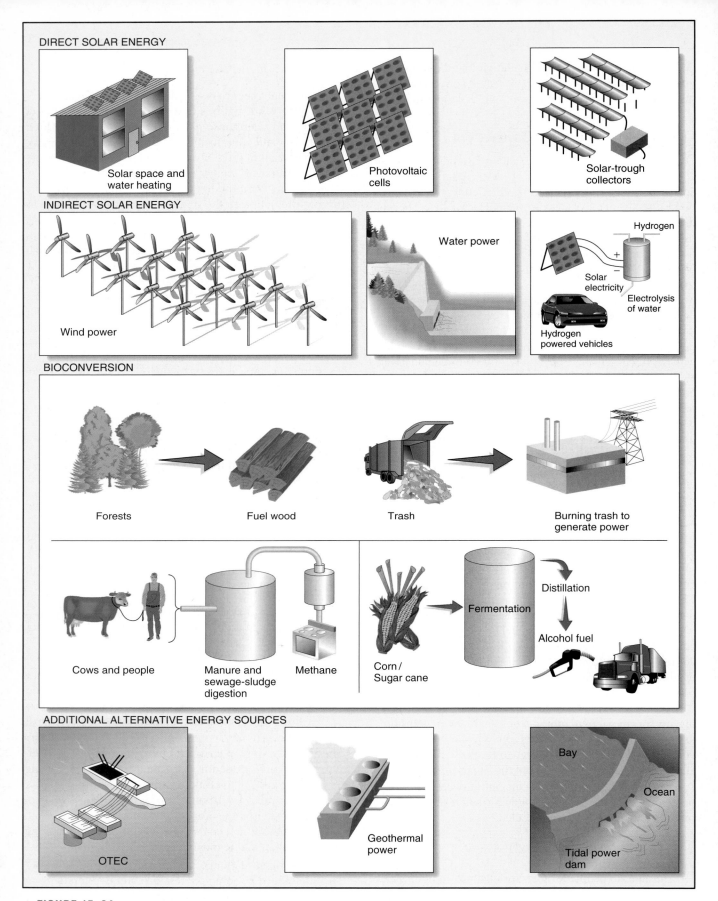

▲ FIGURE 15–24 *Renewable energy resource alternatives.* At the present time, direct solar heating of space and hot water, photovoltaic cells, solar-trough collectors, wind power, production of hydrogen from solar or wind power, and production of methane from animal manure and sewage sludges seem to offer the greatest potential for supplying sustainable energy with a minimum of environmental impact.

ETHICS

TRANSFER OF ENERGY TECHNOLOGY TO THE DEVELOPING WORLD

The nations of the industrialized Northern Hemisphere have achieved their level of development using energy technologies based largely on fossil fuels (and, to a lesser extent, nuclear energy). Their development took place during a time when fossil fuels were inexpensive; only as the expense of those fuels has risen have the nations of the North begun to get serious about technologies to make energy use more efficient. The specter of global climate change has brought a new perspective to the need to wean the industrialized North away from fossil fuel energy in the next century.

The same traditional fossil-fuel-based technologies have been adopted by the developing nations of the Southern Hemisphere, except that they are lagging behind in their ability to implement the technology on a large scale. As we have seen in Chapter 13, however, the fossil-fuel-

based energy pattern of the twentieth century may have to largely be replaced with renewable energy systems in the next century. If the United States and other developed nations are willing to play a significant role in helping the developing nations develop, it is questionable whether the same developmental path the North has taken should be promoted. Instead, there is the opportunity to engage in some "leapfrogging" technology transfer. The North can put its development dollars into such technologies as electrification of rural areas via photovoltaics and efficient public transport systems for the cities.

The solar route is especially attractive for many of the climates in the developing world, where an estimated 2 billion people lack electricity. A model program is the work of Richard Hansen, whose nonprofit development group Enersol helped to establish the Asociación de Desarrollo de Energía Solar (ADES) in the Dominican Republic.

ADES has established a revolving low-interest loan fund and the necessary equipment to help rural people electrify their homes. Rooftop photovoltaic units the size of two pizza boxes produce enough power for five electric lights, a radio, and a television set. Similar programs have been started in Sri Lanka and Zimbabwe.

It is entirely possible that if this strategy of development aid is followed, the nations of the South may end up pointing the way to a sustainable energy future for the rest of the world. The Rio Declaration, signed at the Earth Summit in June 1992, commits the industrial countries to providing greater levels of development aid to the poor countries. It also stipulates that the development should be sustainable. As the details of this commitment get fleshed out, it will be interesting to see if solar energy receives the attention it deserves. If it does, there is much the donor countries can learn from the results. What do you think?

---

fiscal year 1999 budget. Much of this ($336 million) is for the DOE's work on renewable technologies, including some significant public-private partnerships. Some $526 million is for DOE's energy-efficiency programs. For example, one is the **Partnership for a New Generation of Vehicles** a collaborative effort with U.S. auto manufacturers to develop a family car that would get 80 miles per gallon by 2004. (The Energy Star and Green Lights programs are also funded at $109 million.)

- Basic **climate change research** has been federally funded for several years at around $1.8 billion per year. This research by academic and independent research institutions is providing much of the foundational information employed by the Intergovernmental Panel on Climate Change in its reporting.

- The ongoing **deregulation of the electrical industry** is not only saving consumers on their electric bills, it is offering them the option of choosing electricity suppliers. "Green power marketing" is now feasible, where customers will be able to buy their electricity from renewable energy sources or less polluting fossil fuel sources such as gas-powered plants. Many consumers prefer renewable sources, even if they have to pay a premium for them. Of course, removing the fossil fuel and nuclear energy subsidies would quickly make renewable power quite competitive in pricing.

- Some major energy producers are looking ahead to the day when greenhouse gases will have to be cut because of very clear evidence of global climate change. Toward that end, British Petroleum is planning to cut its greenhouse gas emissions 10% below 1990 levels by 2010. Royal Dutch/Shell will spend more than $500 million on renewable energy technology over the next five years. Both of these firms have withdrawn from an industry-lobbying group opposing the Kyoto Protocol on climate change.

- As we have seen, renewable energy technologies are being adopted widely in the developing world. To a great extent, this development represents a "leapfrogging" over the conventional fossil fuel-based energy technology of the developed world (see the "Ethics" essay, above).

- "Net metering" is the practice of allowing people on the power grid to subtract the energy they contribute via wind turbines or photovoltaics from their electric bills. This is now possible in many states and is an encouragement to concerned people to take action on the energy problem by installing their own renewable energy systems.

It should not escape our notice that the United States stands out as the only industrialized country that has seen fit to keep gasoline prices relatively low. Fuel in all other highly developed countries is so heavily taxed that it costs consumers $3 to $5 per gallon—U.S. prices, adjusted for inflation, remain at the 1952 price level. So one future development that a number of environmental groups promote is the implementation of a **carbon tax** a tax levied on all fuels according to the amount of carbon dioxide that is produced in their consumption. Such a tax, proponents believe, would provide both incentives for using solar sources, which would not be taxed, and disincentives for consumption of fossil fuels. It is hard to imagine any step that would be more effective in reducing greenhouse gas emissions in the United States. A number of European countries have already adopted such a tax.

Are these developments enough to enable us to achieve sustainable energy and mitigation of global climate change? Quite clearly, no. Yet, they are moving us in the stewardly direction that is vital to the future of the global environment.

ENVIRONMENT ON THE WEB

COPING WITH "SOMETIMES" ENERGY

Renewable energy sources offer a way to supplement, or replace, much of our current reliance on traditional thermal generating technologies. But these sources also differ in several important ways from conventional energy sources. For example, wind, solar, and tidal power provide only intermittent, not continuous, generation of electricity. We cannot simply turn on the wind, the Sun, or the tides to meet a surge in power demand nor can we turn off those energy sources when demand is low. Because of this inherent uncertainty, it is often suggested that renewable energy sources are most useful as an adjunct to, rather than a replacement for, conventional thermal power generation.

Another difference is that wind, solar, and tidal generating facilities must be sited or built in places where they can best take advantage of natural conditions rather than at the convenience of the consumer. Sometimes optimal sites are far from human habitation. This means that transmission lines may be longer than for most conventional generating facilities. And when lines are longer, there is more risk of voltage loss in transmission, rather like trying to send water through a very long and leaky garden hose.

Older wind systems in particular have been plagued with problems of long transmission lines, inefficient or unreliable generating equipment, and variable voltage. If unresolved, these problems have the potential to create serious problems for the consumer. Imagine how your computer—and the work you have saved on its hard disk—could be affected by wildly fluctuating electrical current or unexpected power failure. A recent study of Hawaii Electric Light Company (HELCO), which uses some older wind-energy generating equipment, showed that modest wind power output changes, and/or unscheduled changes in consumer demand, would cause the system to lose its ability to regulate current within acceptable limits. New generating and voltage-control devices have greatly reduced the problems inherent in intermittent energy sources and will allow HELCO to expand its consumer base while maintaining its reliance on wind energy.

The future of renewable energy sources will depend partly on the ability of utilities to forecast and control intermittent sources, to store energy at peak generating periods and retrieve it when it is needed, and to maintain a consistent high-quality flow of electricity to consumers—with or without the aid of conventional generating systems.

Web Explorations

The Environment on the Web activity for this essay describes the recent trend to move to "full-cost pricing" of energy in order to pass on to the consumer the hidden costs of energy use. Go to the Environment on the Web activity (**Select chapter 15 at http://www.prenhall.com/nebel**) and learn for yourself:

1. about the viability of renewable energy technologies;
2. about trends in the growth of wind and solar energy; and
3. what people are willing to pay for the choice of "green energy."

A suggested time frame for each of these exercises is 10–30 minutes.

REVIEW QUESTIONS

1. How does solar energy compare with fossil-fuel energy and basic energy needs? What are the three basic problems of harnessing solar energy?

2. How do active and passive solar hot-water heaters work?

3. How can a building best be designed to become a passive solar collector for heat? What are the barriers to more widespread adoption of this alternative?

4. How does a photovoltaic (PV) cell work, and what are some present applications of them? What is the potential for providing more energy from PV cells in the near future?

5. Describe what the solar-trough system is, how it works, and its potential for providing power.

6. How can hydrogen gas be produced using solar energy?

How might hydrogen meet the needs for fuel for transportation in the future?

7. What is the potential for developing more hydroelectric power in North America, and what would be the environmental impacts of such development?

8. How is wind power being harvested, and what is the future potential for wind farms?

9. What are four ways of converting biomass to useful energy, and what is the potential for and environmental impact of each?

10. What is meant by geothermal energy, and how may it be harnessed? What has been the recent experience with geothermal power in California, and how has it changed the outlook for geothermal energy?

11. What is the potential for developing tidal power in the United States?

12. What policies were moving the United States toward solar energy in the 1970s?

13. How is the traditional fossil-fuel-based system subsidized? What are the "hidden costs" of continuing to rely on fossil-fuel energy systems?

14. What are some developments in policy that might move us in a direction of a sustainable energy future?

THINKING ENVIRONMENTALLY

1. A sustainable society will ultimately be one that gains all its energy needs from what source(s)? Consider all energy sources, both traditional and alternative, and discuss each in terms of long-term sustainability. What conclusions do you reach? Defend your conclusions.

2. Design what you feel should be the energy policy for the United States based on the total range of energy options, including fossil fuels and nuclear power as well as various solar alternatives. Which energy sources should be promoted, and which should be discouraged? Suggest laws, taxes, subsidies, and so forth that might be used to bring to fruition the policy you've designed. Give a rationale for each of your recommendations.

3. Various solar energy alternatives seem to promise a sustainable energy future. Yet, the U.S. government seems locked into a policy of maintaining dependence on fossil fuels and nuclear power. Does this mean that solar advocates are wrong? Discuss the economic and political forces that might be at work in maintaining the status quo.

4. Consider all the energy needs in your daily life. Describe how each might be provided by one or another solar option.

MAKING A DIFFERENCE

PART FOUR: Chapters 13, 14, 15

1. Examine your transportation uses: Make a high MPG car a top priority; get behind car pools; use alternative transportation; keep your car tuned up and tires inflated and balanced properly.

2. Take advantage of services from your electric company on energy conservation, home-energy audits, and subsidized compact fluorescents.

3. Keep informed about any nuclear power plants in your vicinity, and lobby for their strict adherence to NRC guidelines; if you do not support nuclear power, join an organization that lobbies against it.

4. Study your own home (or plans for a future one) in terms of thermal efficiency (insulation, appliances, etc.) and the potential for using solar water or space heating. Improve your home's thermal efficiency and add solar in whatever ways you can.

5. Adopt ways to use less fuel and electricity both at home and where you work (turn thermostats back, turn off lights when leaving a room, wear sweaters, use energy-efficient lightbulbs, etc.).

6. Using the Web, research EPA's Green Lights and Energy Star programs and promote their use for saving energy in an institution you are familiar with: your school, college, church, business.

WEB REFERENCES

On-line resources for this chapter can be found on the World Wide Web at: **http://www.prenhall.com/nebel**. Click on Chapter 15 on the chapter selector.

PART FIVE

Pollution and Prevention

An industrial plant sends a steady stream of exhaust gases into the air, and the wind carries the pollutants away. If it didn't, there would be no flowers in the surrounding fields. But where is "away"? The next county, or state, or country? The upper atmosphere? In reality there is no "away" for the pollution of land, water and air resulting from the activities of the six billion people inhabiting Earth. As the human population continues to grow and develop economically, we are increasingly constrained by the wastes we produce.

In this section, a new chapter (Chapter 16) starts us off with a look at the links between environmental hazards and human health. In the chapters that follow we will look at the major forms of pollution and we will see that whether we are talking about water pollution, hazardous wastes and household wastes, or air pollution, there are sustainable ways of dealing with the problems. We will explore the possibility of a sustainable system where people and economic enterprises do not overburden land, air and water with wastes. We will also see that dealing responsibly with our wastes may sometimes call for international action, as in the case of global climate change and ozone depletion. In every chapter we consider the public policies that have emerged to deal with each of the problems raised by pollution, and evaluate them in the light of our three principles of stewardship, sustainability and sound science. Without exaggeration, the problems of pollution pose some of the most difficult and important issues facing human society as we move into a new century.

Paper mill at sunset, Georgetown, South Carolina.

16

Environmental Hazards and Human Health

~~~~~~~~~

Key Issues and Questions

1. Life expectancy is rising worldwide, yet 10 million deaths occur yearly in children under five in the developing world. What are the major causes of death in developing and developed countries?
2. Exposure to hazards in the human environment brings about the risk of injury, disease, and death. What kinds of cultural factors, infectious diseases, physical factors, and toxic chemicals are most important in this regard?
3. Hazards take many pathways in mediating environmental risks to humans. In what ways do the following bring harm: poverty, smoking, malarial disease, and indoor air?
4. Risk analysis is a scientific tool that the EPA is applying to its regulatory work. How is risk analysis practiced by scientists and employed in policy development?
5. The public often perceives risks differently from the experts. What is the significance of risk perception in policy development?

No larger than a pinhead, with a light brown abdomen and darker brown head and thorax, the black-legged tick, *Ixodes scapularis*, is easy to overlook (Fig. 16–1a). These eight-legged agile creatures can be picked up by way of a casual walk through the brush in many parts of the United States where deer and white-footed mice are common (Fig. 16–1b). If a tick has fed on a white-footed mouse infected with the spirochaete bacterium, *Borrelia burgdorferi*, it can become the **vector**, or carrying agent, for Lyme disease in humans, just as it is for deer. A new bite from an infected tick usually leaves a circular red rash, which may be easy to miss. In the next few weeks, the victim develops fever, chills, fatigue, and headache, very general symptoms that may be mild at onset and hard to diagnose. If treatment is delayed, Lyme disease can become chronic and lead to long-term arthritis, memory impairment, numbness, and pain. Discovered first in Lyme, Connecticut, Lyme disease is on the increase in many parts of the United States; 12,801 cases were reported in 1997.

A recent study by a team from the Institute of Ecosystem Studies in Connecticut, led by Clive Jones, has uncovered an intriguing relationship between oak trees, gypsy moths, and Lyme disease. The gypsy moth is an introduced pest that can so ravage a forest that it looks like winter in the middle of summer. The study indicated that when the forest thrives because of natural control of the gypsy moths by mouse predation, it also presents a much greater risk of Lyme disease to people in the vicinity. If an oak forest produces a bumper crop of acorns, which happens every few years, the white-footed mouse population will thrive on the acorns. Unfortunately, the acorns also attract deer, and as a result, all the ingredients of a Lyme disease cycle are present. However, the mice also eat the over-wintering pupae left behind by the gypsy moths and thereby protect the forest from being attacked the following summer and allowing it to thrive. Paradoxically, gypsy moth defoliation, not a good idea for the forest, leads to poor acorn crops and declines in mouse populations, therefore greatly lowering the risk of Lyme disease, which is good for the human population in the vicinity. The dilemma seems to be a choice between a healthy forest and healthy people, although as yet no attempt has been made to intervene in this cycle.

Lyme disease is only one of a number of emerging new diseases that continually remind us that we are not always

Rural health clinic. Health workers administer a polio vaccine to a child in Emelo, Ethiopia. In November 1997, 8.5 million children across Ethiopia received an oral polio vaccine as part of the country's program to eliminate polio.

383

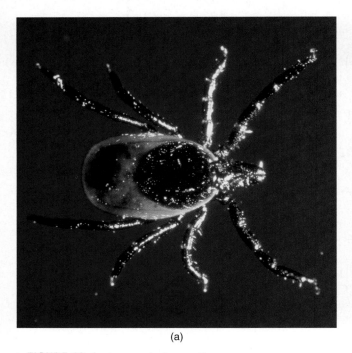

(a)

(b)

▲ **FIGURE 16–1** *Agents in Lyme disease.* (a) The deer tick, *Ixodes scapularis,* carries the spirochaete bacteria that cause Lyme disease. (b) The white-footed mouse is host to the nymph stage of the tick. The existence of mice and deer in woodlands in the eastern United States provides suitable conditions for the continued presence of Lyme disease.

able to control the hazards that are posed by our inevitable contacts with the natural environment. Lassa fever, the Ebola virus, Legionnaires' disease, and the Hanta virus all have taken human lives in recent years, and there is every reason to expect more new diseases to appear as we continue to manipulate the natural environment and change ecological relationships. However, to put these diseases in perspective, it is not the new, emerging diseases that represent our greatest threat. It is the common, familiar ones that take the greatest toll of human life and health—diseases like malaria, diarrhea, respiratory viruses, and worm infestations, which have been kept at bay in the developed countries but continue to ravage the developing countries. In the developed countries, cancer is the killer that is most closely linked

to the environment, leaving us with questions about the chemicals we are exposed to on a daily basis.

The study of the connections between hazards in the environment and human disease and death is often called **environmental health**, but it should be realized that it is *human* health that is in focus here, not the health of the natural environment. In this chapter, we will examine the nature of environmental hazards and the consequences of exposure to those hazards. Then we will consider the pathways whereby humans encounter the hazards, some interventions that can alleviate the risks to health, and how public policy in the form of risk analysis can address the need for continued surveillance and improvement in our management of environmental hazards.

16.1 Links Between Human Health and the Environment

Given our focus on human health, we can more precisely define how we will be using the common term "**environment**" in this chapter. By environment, we mean the whole context of human life—the physical, chemical, and biological setting of where and how people live. Thus, home, air, water, food, neighborhood, workplace, and even climate constitute elements of the human environment. Within human environments there are *hazards* that can make us sick, cut our lives short, or contribute in other ways to human misfortune. In the

context of environmental health, a **hazard** is defined as anything that can cause (1) injury, disease, or death to humans, (2) damage to personal or public property, or (3) deterioration or destruction of environmental components. The existence of a hazard does not mean that undesirable consequences inevitably follow. Instead, we speak of the connection between a hazard and something happening because of that hazard as a **risk**. Here, a risk is defined as the *probability* of suffering injury, disease, death, or other loss as a result of exposure to a hazard. Since risks are expressed as probabilities, the analysis of the probability of suffering some harm from a hazard becomes an exercise that must be carried out by scientists or other experts, and we will eventually deal with the analysis of risk as it becomes the basis of policy decisions. For now, however, our focus is on the nature of environmental hazards.

The Picture of Health

Health is difficult to define, because of its many dimensions—physical, mental, spiritual, and emotional. For our focus on environmental health, we will consider health to be simply the absence of *disease.* Two terms are used to indicate the absence of *health* in studies of disease: morbidity and mortality. **Morbidity** is the incidence of disease in a population and is commonly used to trace the presence of a particular type of illness, like influenza or diarrheal disease. **Mortality** refers to the incidence of death in a population; records are usually kept of the cause of death, making possible an analysis of the relative roles played by infectious diseases and factors like cancer and heart disease. **Epidemiology** is the study of the presence, distribution, and control of disease in populations.

One universal indicator of health is human life expectancy. In 1955, average life expectancy globally was 48 years. It is now 68 years and rising and is expected to reach 73 years by 2025. This progress is the result of social, medical, and economic advances in the latter half of the twentieth century that have substantially increased the well-being of large segments of the human population. In Chapter 6 we discussed the *epidemiologic transition* as the trend of decreasing death rates that is seen in countries, as they modernize. We saw that as this transition occurred in the developed countries it involved a shift from high mortality (due to infectious diseases) toward the present low mortality rates (due primarily to diseases of aging like cancer and cardiovascular diseases).

However, countries of the world have undergone this transition to different degrees, with very different consequences. In spite of the progress indicated by the general rise in life expectancy, one-fifth (10 million) of the annual deaths in the world occur to children under the age of five in the developing

world. The discrepancy between the developed and developing countries can be seen quite clearly in the distribution of deaths by main causes, as seen for 1997 in Fig. 16–2. Infectious diseases were responsible for 43% of mortality in the developing countries versus only 2% of mortality in the developed countries.

An entirely different perspective is brought into play as we consider some of the environmental hazards that accompany industrial growth and intensive agriculture. In addition, some of the most lethal hazards in the developed world are the outcome of purely voluntary behavior—in particular, smoking and obesity. Let us look into these different categories of environmental hazards.

Environmental Hazards

There are two fundamental ways to consider hazards to human health. One way is to regard the *lack of access* to necessary resources as a hazard. No one would argue with the fact that lack of access to clean water and nourishing food is harmful to a person. Investigating hazards from this perspective means considering the social, economic, and political factors that prevent a person from having access to such basic needs. Although this is a fundamentally important perspective, much of it is outside the scope of this chapter. Instead, our focus will be primarily on the second way of considering hazards, namely, *exposure to hazards* in the environment. What is it in the environment that brings the risk of injury, disease, or death to people? Here we will consider four classes of hazards: cultural, biological, physical, and chemical.

Cultural Hazards. Many of the factors that contribute to mortality and disability are a matter of choice, or at least can be influenced by choice. People engage in risky behavior and subject themselves to hazards. Thus, we may eat too much, drive too fast, use addictive and harmful drugs, consume alcoholic beverages, smoke, sunbathe, hang glide, engage in risky sexual practices, get too little exercise, or choose hazardous occupations (Fig. 16–3). Why do we subject ourselves to these hazards? Basically,

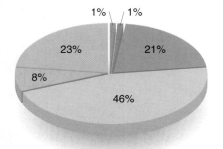

Total number of mortalities = 12 million
Developed countries

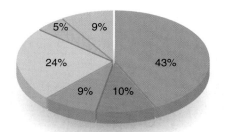

Total number of mortalities = 40 million
Developing countries

◀ **FIGURE 16–2** *Leading causes of mortality in developing and developed countries for 1997.* Infectious and parasitic diseases are prominent in the developing countries, while cancer and diseases of the circulatory system predominate in the developed countries. (Source: Data from World Health Organization, *1998 World Health Report.*)

- Infectious and parasitic diseases
- Perinatal and maternal causes
- Cancers
- Diseases of the circulatory system
- Diseases of the respiratory system
- Other and unknown causes

because we derive some pleasure or other benefit from them. Wanting the benefits, we are willing to take the risk that the hazard will not harm us. Factors like living in inner cities, engaging in criminal activities, and so on can be added to the list of cultural sources of mortality. As Fig. 16–4 indicates, a large fraction of the mortality in the United States can be traced to such cultural factors. It goes without saying that most of these deaths are preventable. However, trying to convince people that they should be more careful is often a losing battle.

The connection between behavior and risk is especially lethal in the case of AIDS (acquired immune deficiency syndrome). AIDS is caused by the human immunodeficiency virus (HIV). It is one of a group of sexually transmitted diseases (STDs), which includes gonorrhea, syphilis, and genital herpes. As we saw in Chapter 7, AIDS is taking a terrible toll in the developing world, where over 90% of the HIV-infected people live. At present, there are no cures or vaccines for the HIV virus.

Biological Hazards. Human history can be told from the perspective of the battle with pathogenic bacteria and viruses. It is a story of epidemics like the black plague and typhus that ravaged Europe in the Middle Ages, killing millions in every city; of smallpox that swept through the New World as the explorers brought not only their weapons of warfare but

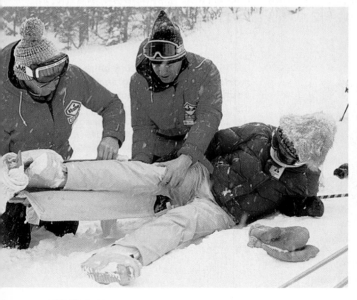

▲ FIGURE 16–3 *Environmental hazards.* Hazards are an unavoidable part of life. Some are a matter of choice of lifestyle, and some are a consequence of where we live or work. All carry a risk that something unpleasant will happen.

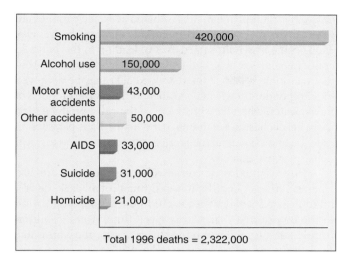

▲ **FIGURE 16-4** *Deaths from various cultural hazards in the United States, 1996.* Many hazards in the environment are a matter of personal choice and lifestyle. Others occur because of accidents and the lethal behavior of others. (Source: Data from Centers for Disease Control, National Center for Health Statistics.)

Table 16-1 **Leading Infectious Diseases, 1997**

Cause of Death	Estimated Yearly Deaths*
Acute respiratory infections	3,700,000
Tuberculosis	2,900,000
Diarrheal diseases	2,500,000
HIV/AIDS	2,300,000
Malaria	1,500,000–2,700,000
Hepatitis	1,000,000–2,000,000
Measles	1,000,000
Pertussis (whooping cough)	200,000–300,000
Meningitis, bacterial	135,000
Amebiasis	70,000
Hookworm	65,000
Rabies	40,000–70,000
Yellow fever	30,000
Trypanosomiasis (sleeping sickness)	20,000
Schistosomiasis	20,000

*Total deaths from infectious diseases in 1997 is estimated at 17.3 million by the World Health Organization. Source: Data from World Health Organization.

also their much more lethal diseases. The story continues in the nineteenth century with the first vaccinations and the "golden age" of bacteriology (a span of only 30 years), when bacteriologists discovered most of the major bacterial diseases and brought the bacteria into laboratory culture. The twentieth century saw the advent of virology (the discovery and research of very small viruses); the great discoveries of antibiotics; immunizations that led to global eradication of smallpox, the defeat of polio, and the victory over many childhood diseases; and the growing influence of molecular biology as a powerful tool in the battle against diseases.

The battle is not over, however, and never will be. Pathogenic bacteria, fungi, viruses, protozoans, and worms continue to plague every society and, indeed, every person. They are inevitable components of our environment; many are there regardless of human presence, others are uniquely human pathogens whose access to new, susceptible hosts is mediated by the environment. Table 16-1 shows the leading killers and the annual number of deaths they cause.

Approximately one-third of the 52 million deaths in 1997 were due to infectious and parasitic diseases. The leading causes of death in this category are the *acute respiratory infections* (e.g., pneumonia, diphtheria, tuberculosis, whooping cough, influenza, strep throat). These are both bacterial and viral; of this group of diseases, pneumonia is by far the most deadly form (Fig. 16–5). Other bacterial and viral respiratory infections represent the most common reasons for visits to the pediatrician in the developed countries, but it is primarily in the developing countries that the respiratory infections lead to death, mostly in children who are already weakened by malnourishment or other diseases.

Tuberculosis is the largest cause of adult deaths from a single disease. Even though the bacterial pathogen that causes it has been known for over 100 years, complacency about

▲ **FIGURE 16-5** *Jim Henson.* The famous creator of the Muppets succumbed in 1990 to a virulent pneumonia caused by the common bacterium *Streptococcus pneumoniae*.

treatment and prevention in recent years has led to a resurgence of this infectious disease. Some of that resurgence can be traced to the AIDS epidemic; HIV infection eventually leads to a compromised immune system, which allows diseases like tuberculosis to flourish.

Diarrheal diseases (e.g., cholera, dysentery, salmonellosis, giardiasis, and many viral and bacterial infections), responsible for 2.5 million deaths in 1997, have the most obvious link to the environment. Most serious cases are the consequence of ingesting food or water contaminated with pathogens, such as *Salmonella, Campylobacter,* and *E. coli,* from human wastes. Infection in small children often leads to dehydration, which can quickly be life threatening; most deaths from diarrhea occur to children under five.

Of the infectious diseases present in the tropics, malaria is by far the most serious, accounting for an estimated 300 to 500 million cases each year and over 1.5 million deaths. Caused by protozoan parasites of the genus *Plasmodium,* malaria is propagated by mosquitos of the genus Anopheles. The life cycle begins with a mosquito biting an infected person and then incubating the parasite within itself (Fig. 16–6). The infected mosquito can then transmit the disease to human hosts with every bite. Within humans, the parasite then invades the liver, where it multiplies and eventually breaks out to invade red blood

cells. After multiplying in the red blood cells, the parasites are released, destroying the cell, and invade other red blood cells. This occurs in synchronized cycles, leading to the periodic episodes of fever, chills, and malaise typical of malaria. Eventually, as high as 10% of the red blood cells may be destroyed in one episode, often leading to anemia. The waste products of the parasites, when they break out of the cells, produce high fever, headaches, and vomiting in the victim. If the parasites invade the brain (cerebral malaria), death occurs unless medical help is reached quickly.

Most of the diseases responsible for mortality are also leading causes of debilitation in humans of all ages. Over 4 billion episodes of diarrhea occur to the human population in the course of a year; at any given time, over 3.5 billion people suffer from the chronic effects of parasitic worms like hookworm and schistosomiasis (Fig. 16–7).

Physical Hazards. Natural disasters take a toll of human life and property every year, from hurricanes, tornadoes, floods, earthquakes, landslides, and volcanic eruptions (Fig. 16–8). It can be argued that some of the loss brought on by natural disasters is a consequence of knowingly being in harm's way. People build homes and towns on flood plains; villages are nestled up to volcanic mountains; cities are constructed on

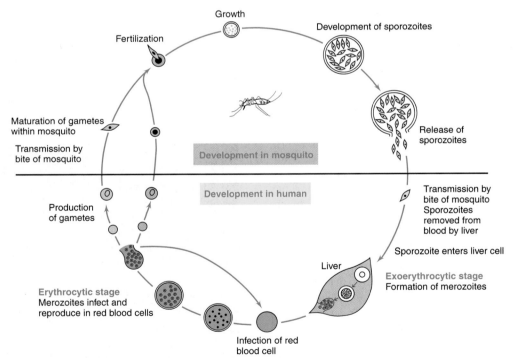

▲ **FIGURE 16–6** *Life cycle of malarial parasite.* Part of the life cycle of the protozoan parasite *Plasmodium,* which causes malaria in humans, is in the *Anopheles* mosquito. In the mosquito, the protozoan develops from gametes (sex cells) into small, elongated cells called sporozoites, which multiply and migrate to the mosquito's salivary glands. The sporozoite stage can then be transmitted to humans with a mosquito bite. Sporozoites migrate to the liver and develop into small cells called merozoites, which multiply and then are released into the blood. There they enter red blood cells (erythrocytes) and further multiply. The parasite harms the human host by destroying red blood cells and by releasing fever-inducing substances into the bloodstream. Protozoal cells in the blood can then infect another *Anopheles* mosquito when it bites the infected person, thus beginning the whole cycle again. (After Madigan, Michael T., Martinko, John M., Parker, Jack, *Brock Biology of Microorganisms,* 8th ed., 1997. Reprinted by permission of Prentice-Hall, Inc., Upper Saddle River, NJ.)

(a)

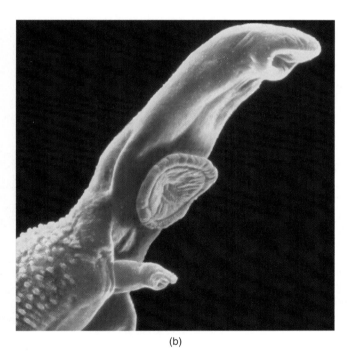

(b)

▲ **FIGURE 16–7** *Chronic biological hazards.* Parasitic worms like (a) hookworms and (b) schistosomes infest millions and cause serious disease and debilitation.

known geological fault lines. A great human tendency is to assume that disasters only happen to other people in other places, and even if there is a risk, say, of a hurricane striking a coastal island, many people are willing to take that calculated risk in order to enjoy life on the water's edge.

Some disasters, however, are impossible to anticipate. Tornadoes are a good example. Each year in the United

States an average of 780 tornadoes strike. These are spawned from the severe weather accompanying thunderstorms. Because of the tendency for cold, dry air masses from the North to mix with warm, humid air from the Gulf of Mexico, the central United States generates more tornadoes than anywhere else on Earth. Of short duration, they develop into some of the most destructive forces known in nature as winds reach as high as 300 miles per hour. The most intense tornadoes have killed hundreds of people; on April 3, 1974, a line of tornadoes from Canada to Georgia took more than 300 lives and caused untold property damage.

The most devastating consequences of the hazards are disproportionately felt by those who are least capable of anticipating them and dealing with their effects. In October 1998, Hurricane Mitch developed in the Caribbean Sea and slowly made its way across Central America. The combination of storm surge, winds, and rain devastated the developing countries of Honduras, El Salvador, Nicaragua, and Guatemala, killing over 10,000 people. Some locations received a year's worth of rain in one day, and the result was mudslides that obliterated whole villages. It was judged that the impact of the hurricane on the infrastructure of these struggling countries set them back 40 years. Three hundred and fifteen schools in Honduras alone were wiped out. Leaders of these Central American countries called for massive international relief on the order of the Marshall Plan that rescued Western Europe after World War II.

Chemical Hazards. Industrialization has brought with it a host of technologies that employ chemicals: cleaning agents, pesticides, fuels, paints, medicines, and many directly used in industrial processes (Fig. 16–9). The manufacture, use, and disposal of these chemicals often bring humans into contact with them. Exposure is either through ingestion of contaminated food and drink, breathing of contaminated air, absorption through the skin, direct use, or by accident (discussed at length in Chapter 20). Toxicity (the capability of being harmful, deadly, or poisonous) depends not only on exposure but on the actual *dose* of a toxic substance—the amount actually absorbed by a sensitive organ. Further, it is known that for most substances there is a threshold level below which no toxicity can be detected (see Fig. 22–3). Making things even more difficult, different people have different thresholds of toxicity for given substances. For example, children are at greater risk than adults because they are growing rapidly and are incorporating more of their food (and any contaminants) into new tissue. Even greater sensitivity occurs during embryonic development. Substances transmitted across the mother's placenta can have grave impacts on development, especially in the early stages.

Many of the chemicals, like heavy metals, organic solvents, and pesticides, are hazardous to human health even at very low concentrations (Table 16-2, p. 392). Acute poisoning episodes are easy to understand and are clearly preventable. It is the long-term exposure to low levels of many of these substances that presents the greatest challenge to our understanding. Some of these chemical hazards are

▲ **FIGURE 16–8** *Dangerous physical hazards.* Natural disasters bring death and destruction in their train. (a) Effects of Hurricane Mitch in Tegucigalpa, Honduras; (b) aftermath of an earthquake in Kobe, Japan; (c) a tornado as it swept through Cantrall, Illinois, in early May, 1995; and (d) Popocatépetl, volcano located 50 miles southeast of Mexico City, erupted again on November 27, 1998.

known carcinogens (cancer-causing). As Fig. 16–2 indicates, cancer takes a very large toll in the developed countries, accounting for one-fourth of mortality. Because cancer often develops over a period of 10 to 40 years, it is often hard to connect the cause with the effect. Yet 74 chemicals are now known to be human carcinogens, and many of these are substances used in both developed and developing countries. Other potential effects of toxic chemicals include impairment of the immune system, brain impairment, infertility, and birth defects.

Many developing countries are in double jeopardy because of both the high level of infectious disease and the rising exposure to toxic chemicals due to industrial

processes and home uses. In both cases, the risks are largely preventable. Where industrial growth is especially rapid, it is often the case that insufficient attention is given to the pollutants that accompany that growth. Thirteen of the 15 cities with the worst air pollution are found in Asia, where economic growth has risen rapidly.

Cancer. The word carries a burden of fear and uncertainty in its train. In the United States, 23% of all deaths—over 540,000 in 1997— are traced to cancer, and as everyone knows, even though it is largely a disease of older adults, it can strike people of all ages. How is it that a simple chemical like benzene or arsenic can cause cancer?

(a)

(b)

(c)

(d)

▲ FIGURE 16–9 *Industrial processes and hazards.* Many industries produce chemicals and pollutants that contribute to the impact of toxic substances on human health. Some examples are: (a) A paper mill in Port Angeles, Washington; (b) a tannery in Marrakech, Morocco; (c) an oil refinery in Philadelphia, Pennsylvania; and (d) a plastics factory.

The answer involves an understanding of cellular growth and metabolism. In Chapter 5 we saw that a mutation is a change in the informational content of the DNA (the genes) that controls a living cell. During normal growth, cells develop into a variety of specific cell types that perform unique functions in the body: blood cells, heart muscle cells, liver cells, absorptive cells that line the small intestine, and so forth. Some cells, like the stem cells that form the cells of the blood, keep on dividing until death. Other cells, however, are programmed to stop dividing and simply perform their functions, often throughout life. A cancer is a cell line that has lost its normal control over growth. Recent research has revealed the presence of as many as three dozen genes that can bring about a malignant (out-of-control) cancer; they are like ticking time bombs that may or may not go off.

Carcinogenesis (the development of a cancer) is now known to be a process with several steps, often with long periods of time in between those steps. In most cases, a sequence of five or more mutations must occur in order to initiate a cancer. Environmental carcinogens are either chemicals that are able to bind to DNA and prevent it from functioning properly or agents like radiation that can strike the DNA and disrupt it. When such an agent initiates a mutation, the cell involved may go for years before the next step—another mutation—occurs. Thus, it is easy to see how it could take upward of 40 years for some cells to accumulate the sequence of mutations that leads to a malignancy. When one occurs, the cells grow out of control and form tumors that may spread (metastasize) to other parts of the body. Early detection of cancers has improved greatly over the years, and many people survive cancer who would have succumbed to it a decade ago. The best strategy, however, is prevention, and so great attention is now given to the environmental carcinogens and habits like smoking that are clearly related to cancer development. The World Health Organization estimates that 25% of cancers can be traced to environmental causes.

Table 16-2   Hazardous Chemicals in the Environment

Chemicals found in the air in urban environments

- lead (vehicle exhaust and industrial processes)
- hydrocarbons and other volatile organic compounds (vehicle exhaust, industrial processes)
- nitrogen oxides and sulfur oxides (vehicle exhaust and fixed combustion processes like power plants)
- particulate matter (dust, soot, metal particles from vehicle exhaust, industrial processes, and construction)
- ozone (secondary pollutant from the interaction of hydrocarbons and nitrogen oxides in sunlight)
- carbon monoxide (vehicle exhaust and other combustion processes)

Chemicals found in food and water

- pesticides and herbicides (agricultural applications to food crops)
- heavy metals such as arsenic, aluminum, and mercury (soil and water contaminated with industrial chemicals)
- lead (water pipes)
- aflatoxins and other natural toxins (foods contaminated by fungal growth)
- nitrates (drinking water with high nitrogen content)

Chemicals found in the home and workplace (indoors)

- lead (household paints)
- smoke and particles (combustion of biofuels, coal, and kerosene)
- carbon monoxide (incomplete combustion of fuels)
- asbestos (insulation)
- tobacco smoke
- household products (poisonings and exposure from improper use)

Chemicals contaminating some land sites

- cadmium, chromium, other heavy metals (industrial wastes)
- dioxins, PCBs, other persistent organic chemicals (industrial wastes)

16.2   Pathways of Risk

The environmental hazards previously identified are clearly responsible for untold human misery and death. If many of the consequences of exposure to the different hazards are preventable, how is it that the hazards turn into mortality statistics? In other words, what are the pathways whereby risks of infection, exposure to chemicals, and vulnerability to physical hazards are mediated? If prevention is our goal, and it should be, this knowledge is crucial. We will consider a few examples of these pathways, realizing that a comprehensive examination of this question would take volumes.

The Risks of Being Poor

One major pathway for hazards is poverty. Indeed, the World Health Organization (WHO) has stated that poverty is the world's biggest killer. Environmental risks are borne more often by the poor, not only in the developing countries but also in the advanced countries of the world. The people who are dying from infectious diseases are people who lack access to adequate health care, clean water, nutritious food, healthy air, sanitation, and shelter. Because of this, they are exposed to more environmental hazards and thus encounter greater risks.

Once again we are reminded of the role of wealth in determining the well-being of people, and of the gap between wealthy and poor countries or, as is the case, between developed and developing countries. Overall, the wealthier a country becomes, the healthier is its population. However, those with wealth in virtually all countries are able to protect themselves from many of the environmental hazards. People in developed countries tend to live longer, and when they die, the chief causes of death are diseases of old age. Sixty-seven percent of the death toll in the developed countries can be traced to the so-called degenerative diseases such as cancers and diseases of the circulatory system (heart disease, strokes, etc.). Only 33% of the deaths in the developing countries are indicative of people who are living out their life span. Another indicator is life expectancy; two-fifths of the annual deaths in the world (more than 20 million) occur to people under the age of 50. Most of this mortality is due to infectious disease and is preventable.

There are many reasons for this discrepancy. One is education; advances in socioeconomic development and advances in education are strongly correlated. As people—and especially women—are educated, they act on their knowledge to secure relief from hazards to health. They may improve their hygiene, for example, immunize their children, or recognize dangerous symptoms such as dehydration and seek oral rehydration therapy (replacement of lost fluids with balanced saline solutions) (Fig. 16–10). Another reason for the connection between health and wealth is nutrition. The

◀ **FIGURE 16–10** *Public health clinic.* This outreach immunization center in Bangladesh provides basic health information and services to people in the surrounding area.

poor in the poorest of the developing countries are most susceptible to malnutrition and hunger, as we saw in Chapter 10. Malnutrition acts in tandem with disease to make the impacts of ordinary diseases like diarrhea and respiratory infections life-threatening. Indeed, malnutrition is thought to contribute to at least one-half of deaths of children in the developing countries.

Education and nutrition and, indeed, the general level of wealth in a society do not tell the whole story, however. A nation may make a deliberate choice to put its resources into improving the health of its population above, say, militarization or the development of modern power sources. Countries like Costa Rica, China, and Sri Lanka have a much longer life expectancy and lower infant mortality than their gross domestic product data would predict. They have done this by focusing public resources on such concerns as immunizations, upgrading sewer and water systems, and land reform.

Another factor in the equation of environmental health is the relative distribution of wealth in a society. Where the gap between the rich and the poor is lower and the general well-being of the society is higher, the health status of the society is also higher. For example, both Japan and the United States have similarly high economic status. Japan, however, now has the highest life expectancy in the world (at 80 years) and the most equitable income distribution. The United States, on the other hand, ranks twenty-first out of 157 in life expectancy and has perhaps the greatest inequities anywhere between the wealthy and the poor.

In spite of this picture of high environmental hazards in the developing countries, there is no reason to conclude that hazards are nonexistent in the developed world. For one thing, it is vitally important to maintain surveillance over food, air, and water to keep pathogens under control.

It is also a fact that in the developed world we are able to contain potentially lethal pathogens like respiratory and diarrheal diseases with the aid of appropriate prevention and treatment.

The Cultural Risk of Smoking

Lifestyle choices like lack of exercise, overeating, driving fast, alcohol consumption, mountain climbing, and so forth carry with them a significant risk of accidents and death. The cultural hazard that carries the highest risk of harm, however, is **smoking**. Smoking, like alcohol and other drugs, is considered to be a *personal pollutant* because it is a cultural hazard people expose themselves to on a voluntary basis. Unfortunately, recent trends show that people are likely to start smoking at a younger age. Lured by the image of smoking as "cool" or by peer pressure, youngsters are soon addicted to the nicotine in cigarettes, an addiction that is difficult to break. Recent research has indicated that nicotine and cocaine share the same neural pathways, putting nicotine in the same class as the dangerous and illegal drug cocaine.

Smoking is, essentially, portable air pollution. The smoker is the primary recipient of the pollution, but the effects spread to all who breathe the smoke-clouded air and, in the case of pregnant women, to the unborn fetus. Worldwatch Institute's William Chandler maintains that "Tobacco causes more death and suffering among adults than any other toxic material in the environment." Cigarette smoking has been clearly and indisputably correlated with cancer and other lung diseases. Smoking has been shown to be the major single cause of cancer deaths in the United States, responsible for 30% of them. It is the leading cause of preventable deaths in the United States; at least 420,000 people die each year from smoking-related causes (Fig. 16–11). The WHO puts

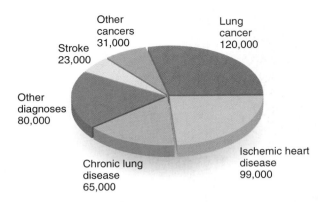

▲ **FIGURE 16–11** *Deaths caused by smoking.*
Approximately 420,000 deaths in the United States in 1990 are
attributed to smoking. (Source: Data from Centers for Disease Control,
Tobacco Information and Prevention Source.)

the number of smokers worldwide at 1.1 billion, with 3.5 million deaths per year due to smoking. Smoking is increasing in the developing world at a rate of 2% per year, and in China, where smoking is practically epidemic, 63 percent of men are smokers. Studies have shown that smokers living in polluted air experience a much higher incidence of lung disease than smokers living in clean air—demonstrating a synergistic effect. Certain diseases typically associated with occupational air pollution show the same synergistic relationship with smoking. For example, black lung disease is seen almost exclusively among coal miners who are also smokers, and lung disease predominates in smokers who are also exposed to asbestos. Smoking also kills by impairing respiratory function and by impacting cardiovascular function. Tobacco-related illness is estimated to cost almost $200 billion a year in lost work capacity and health-care expenses. Huge increases in smoking-related deaths will occur in the next two decades, owing to the rapid rise in smoking.

Several things have been attempted in order to regulate this cultural hazard. Surgeons General of the United States have issued repeated warnings against smoking since 1964, and public policy has taken them seriously by requiring warning labels on smoking materials, banning cigarette advertising on television, promoting smoke-free workplaces, requiring nonsmokers' areas in restaurants, and banning smoking on all domestic airline flights. Since the warnings began, the U.S. smoking population has gradually dropped from 40% to 26%.

Several recent developments in public policy in the United States are bringing major changes in the smoking arena. In January 1993, the EPA classified ETS (environmental tobacco smoke, meaning secondhand or sidestream smoke) as a Class A (known human) carcinogen. In the words of EPA administrator Carol Browner, "Widespread exposure to secondhand smoke in the United States presents a serious and substantial public health risk." The EPA's action was unanimously supported by its science advisory board. Fallout from this action has been substantial: Smoke-free zones in public areas such as shopping malls, bars, restaurants, and workplaces

are now common. In particular, the EPA has taken specific steps to protect children from ETS in all public places, and OSHA (Occupational Safety and Health Administration) is promoting policies to protect workers from involuntary exposure to ETS in the workplace. The impact of these actions on public health should be substantial: One assessment indicated that just a 2% decrease in ETS would be the equivalent, in terms of human exposure to harmful particulates, of eliminating all the coal-fired power plants in the country.

Another significant development came from the FDA (Food and Drug Administration). Citing the addictive properties of nicotine, then FDA administrator David Kessler proposed in February 1994 that the agency have the authority to regulate cigarettes as a drug. At congressional hearings following the FDA's initiative, tobacco executives denied that cigarette smoking is addictive. The hearings included allegations of a cover-up of the tobacco industry's own research, which revealed the addictive properties of nicotine and the further allegation that tobacco manufacturers spike their cigarettes with additional nicotine in order to "hook" smokers. A year later, President Clinton declared nicotine an addictive drug, which automatically gave the FDA regulatory authority. The FDA's initial efforts are concentrating on curbing tobacco sales to the underaged. One early casualty was Joe Camel, a cartoon character that seemed to be targeted to the young (Fig. 16–12).

The Attorneys General of a number of states propelled the most recent development. The states are trying to recover some of their Medicaid expenses brought about by smoking-related illnesses. In March 1997, a lawsuit brought by 22 states was settled with the Liggett Group, one of the "big five" U.S. tobacco companies. Liggett agreed to turn over documents expected to incriminate the tobacco industry and publicly admitted that nicotine is addictive and cigarettes are carcinogenic. Liggett also admitted that industry advertising targets teenagers, recognizing that they are the future of the industry. In late 1998, eight states reached a $206 billion settlement with the top four tobacco companies and opened the door for other states already suing the companies to join the settlement. The money will be used not only to reimburse the states but also to help finance programs to discourage people—especially young people—from smoking.

The "resolution" of this controversy, if indeed it is a resolution, is symptomatic of our ambivalence toward tobacco in the United States. On the one hand, we try to protect people from the serious health effects of smoking; on the other hand, we continue to subsidize the tobacco industry and tobacco exports. Overseas sales of American cigarettes total $5.3 billion and have been vigorously protected by U.S. trade representatives. Although the links between cigarette smoking and illness are now firmly established, the tobacco industry has largely managed to avoid liability for the deaths caused by smoking. As public policy moves toward banning smoking in more and more locations, the industry is fighting back with propaganda emphasizing smoking as an important freedom (see the "Ethics" essay, p. 396).

◀ **FIGURE 16–12** *Joe Camel.* This cartoon character has been featured for years in ads. It has a very high recognition rate among children, an indication that advertising by the manufacturer, R. J. Reynolds, has been successful in reaching them.

Risk and Infectious Diseases

Epidemiology is the branch of environmental health science that is directed toward the study of diseases in human populations. It has been described as "medical ecology," because the epidemiologist traces the disease as it occurs in geographic locations, as well as the mode of transmission and consequences of the disease. Epidemiologists point to many reasons why infectious diseases and parasites are more prevalent in the developing countries. One major pathway of risk is contamination of food and water due to the lack of adequate sewage treatment and hygiene. If drinking water is not thoroughly treated, it becomes the vehicle for transmission of human pathogens from feces. Some of this can be traced to the lack of adequate practical education to inform people about diseases and how they may be prevented. Much of it, however, is simply a consequence of the general lack of resources needed to build effective public health-related infrastructure in cities, towns, and villages. Globally, an estimated 2.9 billion people are without adequate sanitation, and 1.4 billion lack access to safe drinking water. Many of the diarrheal diseases are transmitted in this way; some of these are deadly and are capable of spreading to epidemic proportions, like cholera and typhoid fever. Any diarrheal infection in small children, however, can be fatal if it is accompanied by malnutrition and/or dehydration.

These problems are not unique to the developing world. Failure of adequate surveillance in developed countries leads to major outbreaks of diarrheal diseases, with sometimes fatal outcomes. In 1993, the water supply of Milwaukee, Wisconsin, became contaminated with *Cryptosporidium parvum*, a protozoal parasite originating in farm animal wastes. Some 370,000 people developed diarrheal illnesses from the outbreak, and over 4000 were hospitalized. The factory-like conditions involved in food processing in the developed countries often give pathogens opportunity for widespread contamination, resulting in diarrheal outbreaks and major recalls of food products. The occurrence of such outbreaks, the resurgence of known diseases, and the emergence of new ones like AIDS and the Ebola virus are the concern of public health agents and epidemiologists in organizations like the World Health Organization and the Centers for Disease Control. (For a look into their work, see the "Global Perspective" essay, p. 398.)

The tropics, where most of the developing countries are found, have climates ideally suited for the year-round propagation of disease by **insects**. Mosquitoes are vectors for several deadly and debilitating diseases: yellow fever, dengue fever, elephantiasis, Japanese encephalitis, and malaria. Of these, malaria is by far the most serious. Public health attempts to control malaria have been aimed at eradicating the Anopheles mosquito with the use of insecticides. Malaria was once prevalent in the southern United States, but an aggressive campaign to eliminate Anopheles mosquitoes and identify and treat all human cases of malaria in the 1950s led to the complete eradication of the disease. Unfortunately, the mosquitoes have developed resistance to all of the pesticides employed, and eradication has remained an elusive goal in the tropics. Further complicating control of the disease, the *Plasmodium* protozoans have also developed resistance to one treatment drug after another. Until recently, chloroquine was quite effective against malaria in Africa. Now it is ineffective, and mefloquine is the drug of choice for treating cases of malaria and for prophylactic protection.

One very promising effort funded by WHO's Tropical Disease Research found that children provided with insecticide-treated bed nets experienced a substantial reduction in mortality from all causes. This thrust is being recommended for large-scale intervention throughout Africa as a

ETHICS

THE RIGHTS OF SMOKERS?

Smoking in public is becoming socially unacceptable in many parts of the country, forcing many smokers to retreat to rest rooms, specially designated smoking areas, or sidewalks. Almost all states have laws restricting smoking in public places, and federal law now prohibits smoking on all domestic airline flights.

Although it was once considered simply a nuisance to have to breathe secondhand smoke, evidence has shown that consistent breathing of sidestream smoke can lead to some of the same consequences experienced by smokers. One study showed that nonsmoking wives of men who smoke are twice as likely to die of lung cancer as are nonsmoking wives of men who do not smoke. Some workers (bartenders, food servers, and musicians, for instance) involuntarily inhale the equivalent of 10 or more cigarettes per day. All of this new knowledge has put smokers on the defensive, as nonsmokers adopt slogans such as "Your right to smoke stops where my nose begins" and "If you smoke, don't exhale!"

Smoking, of course, has its defenders. Smokers feel that they have a right to indulge in their habit and would like nonsmokers to understand that smokers have a physical addiction that is quite difficult to break. Some smokers have become belligerent about the restrictions and are willing to take their rights into court. The newest strategy from the smoking lobby emphasizes smoking as an issue of personal freedom, like the right to use alcohol or caffeine. However, court challenges regarding the restrictions on smoking have not gone well for smokers. For example, a federal court upheld a ruling that prohibited a firefighter from smoking, whether on or off the job, because it was determined that firefighters must be in top physical condition and are subjected to hazards that smoking could aggravate. The judges ruled that smoking was not protected by the constitutional right to privacy. Civil libertarians object to such restrictions on the basis that a person has a right to smoke; others counter with arguments that individual liberty can be limited if there is a rational purpose to doing so.

In the United States, nonsmokers are in the majority, and they seem to be gaining the upper hand in the controversy over public smoking. Have we gone too far, or should we do even more to curb sidestream smoke? What do you think?

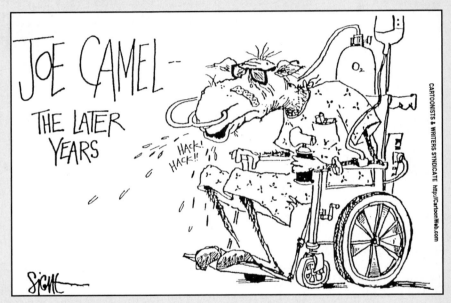

Confined to a wheelchair and breathing oxygen, Old Joe Camel displays the growing public awareness of the health hazards of many lifetime smokers as well as of those who are exposed to sidestream smoke. The medical costs associated with this level of health care have prompted the many lawsuits mounted against the tobacco companies.

cost-effective way of reducing the high malaria mortality in African children. Bed nets are especially effective against the Anopheles mosquitoes, which emerge from hiding and feed primarily at night. Elsewhere, WHO has documented increases in the incidence of malaria in association with land-use changes like deforestation, irrigation, and the creation of dams (Fig. 16–13). In semi-arid areas of Africa, there was a 17-fold increase in malaria resulting from development projects involving irrigation and other high-intensity agriculture. The basic problem cited by WHO is that the agencies promoting the land-use changes did not take the possible health impacts into consideration.

Research continues for the development of new, more effective anti-malarial drugs and for the development of an effective vaccine. The parasite's complicated life cycle and the fact that malaria does not readily induce long-term immunity have made vaccine development an elusive goal, and so the current public health strategy is to inhibit access of mosquitoes to hosts (with bed nets and spraying) and provide prompt treatment and prophylaxis to a greater proportion of affected populations. The WHO has recently announced a "Roll Back Malaria" campaign, designed not to eradicate the disease but to reduce by half the malaria-caused mortality by 2010 and then to halve mortality again by 2015.

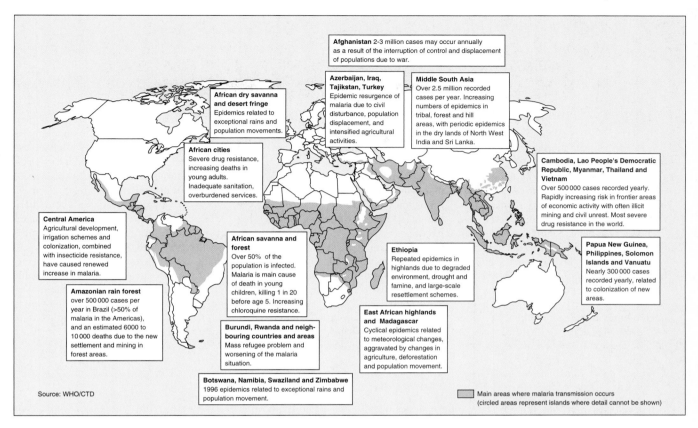

▲ **FIGURE 16–13** *Malaria distribution and problem areas.* Malaria is found in the tropics and subtropics around the world. Drug resistance and environmental disruptions are contributing to the strong resurgence of malaria in the last two decades. (Source: World Health Organization, *Tropical Disease Research: Progress 1995–96: Thirteenth Programme Report* [Berne, Switzerland: Berteli Printers, 1997] 42.)

Toxic Risk Pathways

How is it that people are exposed to chemical substances that can bring them harm? Airborne pollutants represent a particularly difficult set of chemical hazards to control, as they are chemicals that are more difficult to measure and also more difficult to avoid. Much attention is given to air pollutants released to the atmosphere, and rightly so (Chapter 22 deals with these). However, **indoor air pollution** can pose an even greater risk to health. It turns out that air inside the home and workplace often contains much higher levels of hazardous pollutants than outdoor air does. Environmental health science has recently turned to the concept of **total exposure assessment (TEA)** in order to evaluate more accurately the impact of air pollutants on human health. To accomplish TEA, scientists measured the pollutants in the air spaces occupied by people and then calculated exposure on the basis of the time spent in those spaces. This approach of direct measurement of exposure has placed increased emphasis on indoor air pollution.

However, in the developed countries, the overall indoor air pollution problem is threefold. First, increasing numbers and types of products and equipment used in homes and offices give off potentially hazardous fumes. Second, buildings have become increasingly well insulated and sealed; hence, pollutants are trapped inside, where they accumulate to potentially dangerous levels. Third, people are exposed more to indoor pollution than to outdoor pollution. The average person spends 90% of his or her time indoors, and the people who spend the most time inside are those most vulnerable to the harmful effects of pollution: small children, pregnant women, the elderly, and the chronically ill.

The most serious indoor air pollution threat is found in the developing world, where at least 3.5 billion people continue to rely on biofuels like wood and animal dung for cooking and heat. Fireplaces and stoves are often improperly ventilated (if at all), exposing the inhabitants to very high levels of particulates and fumes. Four major problems have been associated with indoor air pollutants in the developing world: acute respiratory infections in children, chronic lung diseases such as asthma and bronchitis, lung cancer, and birth-related problems. One study in Mexico showed that women who are continually exposed to indoor smoke had 75 times the risk of developing chronic lung disease than women who were free of this kind of exposure. Public policy solutions to this problem include the design of ventilated stoves that burn more efficiently and the conversion to bottled gas or liquid fuels like kerosene.

GLOBAL PERSPECTIVE

AN UNWELCOME GLOBALIZATION

We live in a time when globalization is rapidly encompassing travel, information, trade, and investment. The World Wide Web ties people together in ways unimagined a few years ago. Globalization of health, however, remains an elusive goal, similar to globalization of economic well-being. Laurie Garrett, in her book *The Coming Plague*, describes an unwelcome form of globalization: the globalization of disease. Garrett examines the recent history of emerging diseases like AIDS, Ebola, Hantavirus, Rift Valley Fever, Legionnaires' disease, and others. She also provides explanations for the resurgence of familiar diseases like tuberculosis, cholera,

and pneumonia as a consequence of the widespread and unwise use of antibiotics. Many of the new diseases are clearly linked with changes in land use, which bring humans into close contact with rodents or other animals that harbor viruses previously unknown to medicine and often deadly to humans. Resurgent diseases, on the other hand, are a creation of our medical practice. By treating people with antibiotics without restraint, we unknowingly selected strains that are immune to the antibiotics and which passed on their resistant genes to unrelated bacteria by way of plasmid transfer (plasmids are small circular pieces of DNA that can move from one bacterium to another by a semi-sexual process).

The heroes of the book are the women and men on the front lines of epidemiology, people like Joe McCormick of the CDC, who doggedly pursued lethal viruses in Africa, or Patricia Webb, who worked in maximum containment laboratories on the deadly viruses back in the United States. Garrett makes a plea for a greater commitment from our universities, medical schools, and government agencies to train workers who will be capable of recognizing new diseases and who will be able to move about equally well in the laboratory, the hospital, and the field in pursuit of knowledge and public health intervention around the world.

In developed countries, the sources of indoor air pollution are more varied, as demonstrated in Fig. 16–14. They include:

- Formaldehyde and other synthetic organic compounds emanating from plywood, particle board, foam rubber, "plastic" upholstery, and no-iron sheets and pillowcases.
- A wide range of compounds from foods burned on the stove or in the oven.
- Incomplete combustion and impurities from fuel-fired heating systems such as gas or oil furnaces, kerosene heaters, and wood stoves.
- Fumes from household cleaners and other cleansing agents.
- Fumes from glues and hobby materials.
- Pesticides. (These will be addressed in Chapter 17.)
- Air fresheners and disinfectants. Most air fresheners work by either dulling the sense of smell so that you don't notice noxious odors or by introducing "high-intensity" smells that cover up odors.
- Aerosol sprays of all sorts (e.g., oven cleaners, pesticides, hair sprays, cooking oils).
- Radon. This radioactive gas, a byproduct of the disintegration of uranium in rock, soil, and water, can accumulate in the basement level of homes and other buildings. As warm air escapes from the top of a house, creating a partial vacuum, radon may be drawn through the basement floor. Trapped in the house, the gas may then reach hazardous levels (see the "Earth Watch" essay, p. 400).
- Asbestos. This natural mineral has fiberlike crystals. It is mined from rock and was once used for heat insulation

and fire retardation material. Pipes in steam-heating systems were wrapped in asbestos, ceilings in many public buildings were covered with it, and it was used in ironing-board covers, paints, and roofing materials. In the 1960s, researchers determined that the inhalation of asbestos fibers is associated with a unique form of lung cancer that develops as long as 20 to 30 years after exposure to the material. The EPA began regulating asbestos in the mid-1970s and initiated intensive campaigns to remove it from schools. Because of the obvious health risks, the EPA has banned practically all uses of asbestos. In the meantime, the agency has published guidelines for controlling old asbestos in public buildings, especially schools. The guidelines caution school administrators against assuming that removal is the only option; in most cases, asbestos can be sprayed with a sealant (encapsulation) or enclosed for much lower expense than the cost of removal and with far less risk of exposure for the public.

- Smoking. Smoking carries a much higher health risk than the average exposure to any of the previously mentioned materials, and it may act synergistically to increase the risk of exposure to other pollutants. Smoking has also been shown to increase the health risks of nonsmokers subjected to ETS. ETS contains over 4000 substances, at least 40 of which are known to be carcinogenic and many of which are potent respiratory irritants.

Even though the materials in this list are clearly hazardous, the link between their presence in the air and development of a health problem is harder to establish than

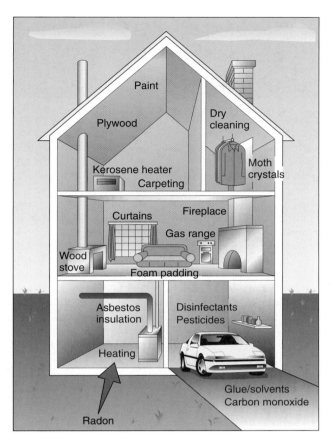

▲ **FIGURE 16–14** *Indoor air pollution.* Pollutants originate from many sources and can accumulate to unhealthy levels, leading to asthma and "sick building syndrome" (characterized by eye, nose, throat irritations, nausea, irritability, and fatigue) as well as other serious problems.

with the infectious diseases. As with cancer, often the best that can be done is to establish a statistical correlation between exposure levels and the development of an adverse effect. The science of **toxicology** addresses the need to study the impacts of toxic substances on human health and investigates the relationships between the presence of such substances in the environment and health problems like cancer. Chapter 20 will give us a more detailed look at toxic substances and the work of toxicology. However, one of the most important tools of the toxicologist in the developed countries is *risk analysis*, an approach to the problems of environmental health that is now a major element of public policy but also the subject of some controversy.

16.3 Risk Analysis

All of us would like to live long and healthy lives. When confronted with the news that our days may be cut short by the effects of radiation, pesticides, secondhand smoke, or chemicals in our drinking water, we become indignant. We expect someone in authority to keep us from dying young or from spending our later years in a hospital. The high priority we

place on environmental health is largely a reflection of this concern for hazards coming at us from the air we breathe, the food we eat, and the water we drink. At the grassroots level, it is this concern that is largely responsible for the billions of dollars spent on environmental protection.

In fact, our lives in this late twentieth-century technological society are honeycombed with hazards and the risks that these hazards present to our health, as we have seen. For our own self-interest, we would like to have some knowledge and evaluation of the risks to which we are subjected. This knowledge would enable us to make informed choices as we consider the benefits and risks of the hazards around us. We might learn, for example, that 82 million people go swimming each year and 2600 of them drown while swimming. The risk of drowning, then, is 32 in a million. **Risk analysis**—which is what we just did—*is the process of evaluating the risks associated with a particular hazard before taking some action in which the hazard is present.* To reduce the risk of drowning, for example, we might decide to go swimming at a beach where a lifeguard is on duty, or we might choose to swim in a calm lagoon rather than in the ocean.

Table 16-3 lists some everyday risks, expressed as the probability of dying from a given hazard. For example, the table indicates that the annual risk of dying from smoking

Table 16-3 Some Commonplace Hazards, Ranked According to the Degree of Risk

Hazardous Action	Annual Risk*		
Cigarette smoking, 1 pk/day	3.6	per	1,000
All cancers	2.8	per	1,000
Mountaineering (mountaineers)	6	per	10,000
Motor vehicle accident (total)	2.4	per	10,000
Police killed in line of duty	2.2	per	10,000
Air pollution, eastern U.S.	2	per	10,000
Home accidents	1.1	per	10,000
Frequent-flying professor	5	per	100,000
Alcohol, light drinker	2	per	100,000
Sea-level background radiation	2	per	100,000
Four tablespoons peanut butter/day	8	per	1,000,000
Electrocution	5.3	per	1,000,000
Drinking water containing EPA limit of chloroform	6	per	10,000,000

*Probability of dying
Source: R. Wilson and E.A.C. Crouch, "Risk Assessment and Comparisons: An Introduction," *Science*, 236 (1987). Copyright © 1987 by the AAAS.

EARTH WATCH

RADON: THE KILLER IN YOUR HOME?

A radioactive gaseous element—invisible, tasteless, and odorless—radon seeps into living spaces in homes, schools, and other buildings. As it is breathed into the lungs, some of the radon undergoes radioactive decay into solid chemicals that can become lodged in the lungs and, in time, initiate cancer. If this sounds like science fiction to you, you are not alone. Not everyone believes that radon is the hazard the EPA claims it is.

Radon comes from the natural breakdown of uranium in rock, soil, and water. It drew attention as a health risk when underground miners developed lung cancer at alarmingly high rates. Many studies have shown a clear link between the exposure of miners to radon and the incidence of lung cancer. Although most of the miners were exposed to much higher levels of radon than those found in homes, many scientists

believe that there is no threshold level for radioactive substances. Thus, the EPA has judged that a radon concentration of 4 picocuries per liter (EPA's "action level") inhaled over a lifetime is capable of causing lung cancer in more than one person per thousand. Using risk-analysis techniques, the National Academy of Sciences estimates that radon causes between 15,000 and 23,000 lung cancer deaths each year in the United States; smokers are at a higher risk of radon-induced lung cancer.

As a consequence of the estimates of radon's risk to health, the EPA considers it a significant indoor-air health problem. The agency's strategy has been directed toward informing the public of the risk and encouraging people to test their homes and other buildings for the presence of radon (a simple procedure involving a $20 test kit). Where levels are shown to be above 4 picocuries per liter, the EPA recommends a

remediation process. This usually involves sealing cracks in foundations or, in some cases, installing a ventilation system that removes radon from beneath a foundation. So far, the EPA has adopted a nonregulatory approach to radon.

Critics of the EPA's attention to radon abound. While admitting that radon does pose a health hazard, they cite the great uncertainties involved in the risk-analysis procedures employed. The critics also question the "no threshold" assumption, pointing out that humans have been exposed to low levels of radon since the beginning of the species and have likely developed resistance to such low levels of radioactivity. With an estimated $45 billion price tag for testing and remediation, the response to radon suggested by the EPA is costly and, in the judgment of the critics, largely unnecessary. Congress presently still has not taken any specific action to address radon.

---

one pack of cigarettes a day is 3.6 per 1000, meaning that in the course of a year, 3.6 out of every 1000 people who smoke a pack a day will die from smoking-related diseases.

Not very many people actually make informed choices about the hazards in their lives on the basis of such risk analysis. However, risk analysis has become an important process in the development of public policy and is being regarded as a major way of applying science to the hard problems of environmental regulation.

Risk Analysis by the EPA

Risk analysis began at the EPA in the mid-1970s as a way of addressing the cancer risks associated with pesticides and toxic chemicals. Since then, the process has continued to focus largely on risks to human health. As currently performed at the EPA, there are four steps in risk analysis: *hazard assessment, dose-response assessment, exposure assessment,* and *risk characterization.* One important limitation to the process is the fact that the analyst must work with the available information, which never seems to be sufficient. The final estimate of risk therefore includes a measure of the imprecision that always accompanies the use of imperfect information.

Hazard Assessment: Which Chemicals Cause Cancer? Hazard assessment is the process of examining evidence linking

a potential hazard to its harmful effects. In the case of accidents, the link is obvious. The use of cars, for example, involves a certain number of crashes and deaths. In these cases, *historical data,* such as the annual highway death toll, are very useful for calculating risks.

In other cases the link is not so clear, because there is a time delay between first exposure and final outcome. For example, establishing a link between exposure to certain chemicals and the development of cancer some years later is often difficult. In cases where linkage is not obvious, the data may come from two sources: epidemiological studies and animal tests. An **epidemiological study** is a study that tracks how a sickness spreads through a community. Thus, in our example of finding a link between cancer and exposure to some chemical, an epidemiological study would examine all the people exposed to the chemical and determine whether this population has more cancer than the general population. Data from such studies are considered to be the best data for risk analysis and have resulted in scores of chemicals being labeled as *known* human carcinogens.

The second data source, animal testing, is used when we want to find out *now* what might happen many years in the future. For example, we do not want to wait 20 years to find out that a new food additive causes cancer, so we accept evidence from **animal testing** (Fig. 16–15). A test involving several hundred animals (usually mice) takes about three years and costs more than $250,000. If a significant number

▲ **FIGURE 16–15** *Animal Testing.* Laboratory mice are routinely used to test the potential of a chemical to cause cancer. They are an important source of information used to assess the presence of a hazard in food, cosmetics, or the workplace.

of the animals develop tumors after being fed the substance being tested, the results indicate that the substance is either a *possible* or *probable* human carcinogen, depending on the strength of the results.

Three objections have been raised about animal testing: (1) Rodents and humans may have very different responses to a given chemical; (2) the doses used on the animals are often unrealistically high; and (3) some people are opposed on ethical grounds to the use of animals for such purposes. A point that supports the value of animal testing is that all chemicals shown by epidemiological studies to be human carcinogens are also carcinogenic to test animals. Besides, since it is ethically wrong to perform such tests on humans, there must be some way of testing chemicals that might be used as a food preservative or a cosmetic, in order diminish the risks to humans later on.

Hazard assessment is the important first step in risk analysis. Whether it involves an analysis of accident data, epidemiological studies, or animal testing, hazard assessment tells us that we *may* have a problem that requires regulation.

Dose-Response and Exposure Assessment: How Much for How Long? When animal tests show a link between exposure to a chemical and an ill effect, the next step is to analyze the relationship between the concentrations of chemicals in the test (the **dose**) and both the incidence and the severity of the response in the test animals. From this information, projections are made about the number of cancers that may develop in humans who are exposed to different doses of the chemical. This process is **dose-response assessment**.

Following dose-response assessment is **exposure assessment**. This procedure involves identifying human groups already exposed to the chemical, learning how that exposure came about, and calculating the doses and length of time of the exposure.

Risk Characterization: How Many Will Die? The final step, **risk characterization**, is to pull together all the information gathered in the first three steps in order to determine the risk and its accompanying uncertainties. More commonly, risk is expressed as the probability of a fatal outcome due to the hazard, as in Table 16-3 for common, everyday risks.

For example, the EPA expresses cancer risk as an "upper-bound, lifetime risk," meaning the highest value in the range of probabilities of the risk calculated over a lifetime. The Clean Air Act of 1990 directs the EPA to regulate chemicals that have a cancer risk of greater than one in a million $(>1 \times 10^6)$ for the people who are subject to the highest doses. That is the same standard employed by the FDA for regulating chemicals in food, drugs, and cosmetics.

Because people often have difficulty conceptualizing such data, risks may be expressed in different ways, such as reduction in life expectancy caused by engaging in a risky activity. To illustrate, such calculations show that smoking one cigarette reduces life expectancy by five minutes, which is just about the amount of time it takes to smoke the cigarette!

Risk Management

Regulatory Decisions. The analysis of hazards and risks is a task that falls primarily to the scientific community. The task is incomplete, however, if no further consideration is given to the information gathered. Risk management naturally follows, and this is the responsibility of lawmakers and administrators. It is EPA policy to separate risk analysis and risk management within the agency.

Risk management involves (1) *a thorough review of the information available pertaining to the hazard in question and the risk characterization of that hazard* and (2) *a decision on whether the weight of evidence justifies a regulatory action.* Without doubt, public opinion can play a powerful role in determining these decisions. In general, however, a regulatory decision will hinge on one or more of the following considerations:

1. **Cost-benefit analysis.** A comparison of costs and benefits relative to a technology or proposed chemical product; if done fairly, this analysis can in many cases make a regulatory decision very clear-cut.
2. **Risk-benefit analysis.** A decision can be made on the benefits versus the risks of being subjected to a particular

hazard, especially where those benefits cannot be easily expressed in monetary values. The use of medical X-rays is a good example. X-rays carry a calculable risk of cancer, but the benefit derived when you x-ray a broken bone is much greater than the small cancer risk involved.

3. **Public preferences.** As we will see, people have a much greater tolerance for risks that they feel are under their own control or are voluntarily accepted.

Risk management has been thoroughly incorporated into the EPA's policymaking process for at least 15 years. Its most common use has been in the design of regulations. The process has enabled the agency to target appropriate hazards for regulation and to determine where to aim the regulations—that is, at the source, at the point of use, or at the disposal of the hazardous agent. As was done for toxic chemicals and cancer, a given risk level may be adopted as a standard against which policymakers can measure new risks.

Risk Perception

In 1990, the EPA received a report from a science advisory board convened to evaluate different environmental risks in light of the most recent scientific data. One of the most important problems raised by the scientists was the significant difference between their evaluation of risks and that of the public. This discrepancy deserves a further look.

The U.S. public's increasing concern about environmental problems can be traced to the fear of hazards that pose a risk to human life and health. People *perceive* that their lives are more hazardous than ever before, but this is not true. In fact, our society is freer from hazards than it has ever been, as evidenced by increased longevity. Why, then, do people protest against nuclear power plants, waste sites, and pesticide residues in food when, according to experts, these hazards pose extremely small risks? The answer lies in people's **risk perceptions**—their intuitive judgments about risks. In short, people's perceptions are not consistent with the reality of the situation.

Table 16-4 indicates the top 11 environmental problems from the study presented to the EPA, compared with the top 28 concerns of the public, as indicated in a Roper poll. Clearly, the public perception of risks is quite different from that of the scientists.

Hazard vs. Outrage. The reason for the inconsistency between public perception and actual risk calculations, according to Peter Sandman of Rutgers University, is that the public perception of risks is more a matter of outrage than hazard. Sandman holds that while the term *hazard* primarily expresses concern for fatalities only, the term *outrage* expresses a number of additional concerns:

1. Lack of familiarity with a technology: how nuclear power is produced and how toxic chemicals are handled, for example.

2. Extent to which the risk is voluntary: Research has shown that people who have a choice in the matter will accept risks roughly 1000 times as great as when they have no choice.

3. Public impression of hazards: Accidents involving many deaths (Bhopal) or a failure of technology (Three Mile Island) are thoroughly imprinted into public awareness by media coverage and are not quickly forgotten.

4. Overselling of safety: The public becomes suspicious when scientists or public relations people play up the benefits of a technology and play down the hazards.

5. Morality: Some risks have the appearance of being morally wrong. If it is wrong to foul a river, the notion that the benefits of cleaning it are not worth the costs is unacceptable. You should obey a moral imperative regardless of the costs.

6. Control: People are much more accepting of a risk if they are in control of the elements of that risk, as in automobile driving, indoor air pollution, and radon.

7. Fairness: The benefits and the risks should be connected. If the benefits go to someone else, why should you accept any risk?

The public perception of risk is strongly influenced by the media, which are far better at communicating the outrage elements of a risk than the hazard. Public concern over oil spills rose precipitously following the *Exxon Valdez* oil spill in Alaska, an accident that received extensive media coverage and had a high "outrage quotient." Cigarette smoking, however, which causes 420,000 deaths a year in the United States alone, receives minimal media attention because it is not "news" and there is no outrage factor. Indeed, it has been suggested that if all the year's smoking fatalities occurred on one day, the media would have a field day, and smoking would be banned the next day!

Risk Analysis and Public Policy. The most serious problem with the discrepancy between experts and the public is that, generally speaking, *public concern* rather than cost-benefit analysis or risk analysis conducted by scientists drives public policy. The EPA's funding priorities are set largely by Congress, which reflects public concern. How important is this problem?

If public outrage is the primary impetus for public policy, some serious risks will get less attention than they deserve. In particular, risks to the environment are perceived as much less important than they really are, because of the public's preoccupation with risks to human health. As Table 16-4 indicates, two major ecological risks on the EPA's list—alteration of habitat, and extinction of species and loss of biodiversity—do not even show up on the list of public concerns. Only three of the public's top 10 concerns are on the EPA's list.

This difference of opinion points to the importance of *risk communication*, a task that should not be left to the media. The value of ecosystems and their connections to human health and welfare need far greater emphasis in the public consciousness, and this responsibility falls to the scientific and educational communities, as well as to governmental

Table 16-4 Public Concerns vs. the EPA's Top 11 Risks

The EPA's Top 11 Risks (not ranked)	Public Concerns (in rank order)*
Ecological Risks	1. Active hazardous waste sites
	2. Abandoned hazardous waste sites
Global climate change	3. Water pollution from industrial wastes
Stratospheric ozone depletion	4. *Occupational exposure to toxic chemicals*
Alteration of habitat	5. Oil spills
Extinction of species and loss of biodiversity	6. *Destruction of the ozone layer*
Health Risks	7. Nuclear power plant accidents
	8. Industrial accidents releasing pollutants
Criteria air pollutants (e.g., smog)	9. Radiation from radioactive wastes
Toxic air pollutants (e.g., benzene)	10. *Air pollution from factories*
Radon	11. Leaking underground storage tanks
Indoor air pollution	12. Contamination of coastal waters
Contamination of drinking water	13. Solid waste and litter
Occupational exposure to chemicals	14. *Pesticide risks to farm workers*
Application of pesticides	15. Water pollution from agricultural runoff
	16. Water pollution from sewage plants
	17. *Air pollution from vehicles*
	18. Pesticide residues in foods
	19. *Global warming*
	20. *Contamination of drinking water*
	21. Destruction of wetlands
	22. Acid rain
	23. Water pollution from city runoff
	24. Nonhazardous waste sites
	25. Biotechnology
	26. *Indoor air pollution*
	27. Radiation from X rays
	28. *Radon in homes*

*Items in italics also appear on the EPA's list.
Source: L. Roberts, "Counting on Science at EPA," *Science* 249 [1990], 616. Copyright © 1990 by the AAAS.

agencies. Studies have shown that the most effective risk communication occurs by starting with what people already know and what they need to know and tailoring the message so that it effectively provides people with the knowledge that helps them to make an informed decision about risks.

Nonetheless, the public's concern for more than the probabilities of fatalities has merit. Public outrage must be heard, understood, and given a reasonable response. It may not be the best source of public policy, but it reflects certain values and concerns that could easily be omitted by an "objective" risk analysis. The fact is, subjective judgments are going to play a role at every step in the risk-analysis process, from hazard assessment to risk perception to risk management.

Currently, there is legislative pressure to apply objective risk analysis to set national environmental policy goals. The concern behind this pressure is the rising cost of fulfilling environmental mandates, and the implication of the concern is that some risks simply do not merit the regulatory attention (and costs of compliance) they have received. Thus,

according to proponents of this legislation, scarce public and industrial resources are being wasted.

The environmental community views this new pressure with alarm, arguing that risk analysis is too imperfect a device on which to hang major policy decisions involving environmental health. Decisions about risks involve people's lives and health and therefore take on a moral dimension. For example, much risk analysis focuses on risk as applied to an entire society. A better approach would be to recognize the importance of **distributive justice** and to incorporate its insights in decision-making. Distributive justice is basically the ethical process of making certain that all parties affected by a process or decision receive their proper share or consideration. Thus, the interests of the poor or powerless in, say, a depressed region must be given higher priority when considering siting a landfill or hazardous industry. Or, concern for the real interests of developing countries must be foremost when considering whether to allow the export of toxic wastes from developed countries.

The uncertainties involved in the processes of risk analysis should remind us that the process is only a tool and an imperfect one at best. Therefore, as a final recommendation for the use of risk analysis to prescribe public policy, we must consciously employ the "**precautionary principle**" and act to prevent some potential environmental health problem even at the risk of being wrong or at the risk of spending more than the problem technically deserves. As a result, the burden of proof about a potential hazard should rest on those who wish to promote the perceived benefits of the hazard (a new chemical, for example), not on the public agencies enjoined to protect human welfare.

ENVIRONMENT ON THE WEB

TO CHLORINATE OR NOT TO CHLORINATE?

With the wealth of media coverage on environmental hazards and human health issues, we are often overwhelmed with fears about environmental quality, and it can be difficult to know what risks to take most seriously. In some cases, we can make lifestyle choices that reduce our risk of accident, cancer, or infectious disease. Yet, in other cases, we are involuntarily exposed to hazards beyond our control. Examples of such involuntary risks include ingestion of trace organic compounds in food and in water, exposure to low-level radiation—for instance, from the Sun—and infection by drug-resistant pathogens.

Over the last decade, human exposure to chlorine has been a particular focus of concern in the media and in public debate. First discovered in 1774, chlorine is now widely used in manufacturing, bleaching, and disinfection. Its most common uses are in the production of safe drinking water (where it is used to kill microorganisms), in the bleaching of paper and textiles, and in the manufacture of solvents, plastics, and insecticides (DDT, for instance, is a chlorine-based insecticide). In gaseous form, chlorine is a respiratory irritant and was used as a war gas in World War I. In the modern environment, chlorinated compounds are found in air, water, soil, and biological tissue.

Most important, the use of chlorine to disinfect water, or chlorination, has been the single largest factor in controlling outbreaks of waterborne disease and has undoubtedly saved millions of lives that would otherwise have been lost to cholera, typhoid, and similar diseases. However, the debate over the use of chlorine to disinfect water is that chlorination of drinking water can produce a range of chlorine byproducts, including chloroform and bromoform. These chemicals, called trihalomethanes (THMs), are produced when chlorine reacts with organic material present in the raw water and can have serious health implications.

Does this mean that we should stop chlorinating drinking water supplies? The debate about water chlorination is a good example of the factors that must be considered when we weigh the risks we take in accepting—or avoiding—biological or chemical hazards.

Web Explorations

The Environment on the Web activity for this essay expands on the debate around drinking-water chlorination and explores both the chemical and biological hazards of both sides of the argument. Go to the Environment on the Web activity (select Chapter 16 at http://www.prenhall.com/nebel) and learn for yourself:

1. about the arguments for and against chlorination of drinking water;
2. about a recent cholera epidemic in Latin America; and
3. about tradeoffs in selecting drinking water disinfection systems.

A suggested time frame for each of these exercises is 10–30 minutes.

REVIEW QUESTIONS

1. Differentiate between a hazard and a risk.
2. Define morbidity, mortality, and epidemiology.
3. Name the two fundamental ways of considering human health hazards.
4. What are the four categories of human environmental hazards? Give examples of each.
5. What are the pathways of risk described in this chapter? What are some components of each category?
6. Discuss the four steps of risk analysis used by the EPA.
7. What two steps are involved in risk management? What are the three considerations used to make a regulatory decision?
8. How does people's risk perception relate to the reality of a situation?
9. What factors have been known to generate public outrage?
10. Discuss the relationship between public risk perception and analysis and public policy.
11. How would employing the terms "distributive justice" and the "precautionary principle" better our approach to environmental public policy?

THINKING ENVIRONMENTALLY

1. Consider Table 16-4. Which would you rank as the top three ecological and health risks? Which of the seven public-outrage factors might be playing a role in your rankings?

2. Imagine that you have been appointed to a risk-benefit analysis board. Explain why you approve or disapprove the widespread use of:

   genetic engineering
   drugs to slow the aging process
   nuclear power plants
   the abortion pill

3. Suppose your town wanted to spray trees to rid them of a deadly pest. How would you use cost-benefit analysis to determine whether this is a good idea?

4. Design an advertisement, a song, or some other form of communication that can be used to inform the public about the risks of a hazardous activity (e.g. smoking) and what they might do to reduce their exposure to the hazard.

WEB REFERENCES

On-line resources for this chapter can be found on the World Wide Web at: **http://www.prenhall.com/nebel**. Click on Chapter 16 on the chapter selector.

Pests and
Pest Control

〜〜〜〜〜〜〜

Key Issues and Questions

1. For thousands of years, human enterprises have been frustrated by what we call "pests." What are pests, and why do we want to control them?

2. Different methods are used to control agricultural pests. What are the basic philosophies behind these methods?

3. As new and more effective pesticides like DDT were employed, some serious problems began to appear. What are the problems resulting from the use of chemical pesticides?

4. There are alternatives to using pesticides to control pests. How well do alternative control methods work?

5. Biotechnology has revolutionized many biological applications. How does biotechnology add to the battery of genetic methods to control pests?

6. As illustrated in the chapter opening, integrated pest management is an important way to reduce pesticide use. What is integrated pest management, and how has it worked?

7. Public policy for controlling pesticides has been revised. What are the objectives of current policy, and how do they correct past problems?

8. Large quantities of pesticides are exported to developing countries. How is pesticide export and use in developing countries regulated?

To an Indonesian rice grower in 1986, life must have seemed strange. Having been supplied in the late 1970s with the new high-yielding rice developed by the International Rice Research Institute (IRRI), the farmer had become accustomed to greatly improved yields. The government supplied pesticides at 15% of the market price, the farmer sprayed them faithfully, and rice crops flourished—for a while. Then a formerly uncommon insect called the brown planthopper began to appear in the fields, and it could not be stopped (Fig. 17–1). The more the farmer sprayed, the heavier the infestation by the hoppers, and the greater the loss of rice became.

Suddenly in 1986, rice growers were told by the government that they should no longer use the pesticides. Of 63 pesticides used on rice, 57 were banned. The IRRI scientists had found that the pesticides were killing off the predatory spiders and water striders that normally kept the planthoppers in check. They advised the government to stop trying to control the outbreaks with pesticides. In the words of one worker, spraying pesticides to control the planthopper was like pouring kerosene on a house fire! Crop advisors fanned out into the countryside and taught the farmers to protect the natural predators and spray more selectively. Once the rice farmers cut back on their use of pesticides (which now has dropped by 65% in Indonesia), the natural enemies of the planthopper returned, rice yields were 12% higher, and the government eliminated $120 million in pesticide subsidies in the first two years of use. Integrated pest management had come to Indonesia.

◄ **Locusts in Africa.** The desert locust is normally distributed across a broad band of desert and scrub regions of northern and central Africa, the Arabian peninsula, and parts of Asia to western India. Outbreaks can lead to migrations of up to 3000 km and swarms of billions of locusts covering 1000 km^2.

▲ FIGURE 17–1 *Brown planthopper.* Immature brown planthoppers, shown on the stem of a rice plant.

17.1 The Need for Pest Control

Since earliest times, humans have suffered the frustration and food losses brought on by destructive pests. To this day, both farmers and pastoralists must wage a constant battle against the insects, plant pathogens, and unwanted plants that compete with them for the biological use of crops and animals.

One dictionary defines a **pest** as "any organism that is noxious, destructive, or troublesome." This term includes a broad variety of organisms that interfere with humans or with our social and economic endeavors: pathogens, nuisance wild animals, annoying insects, molds, etc. Earlier chapters considered nuisance wild animals (Chapter 11) and pathogens (Chapter 16). Our emphasis in this chapter will be on those pests that interfere with crops, grasses, and domestic animals, namely, agricultural pests and weeds.

Defining Pests

Agricultural pests are organisms that feed on ornamental plants or agricultural crops or animals. The most notorious of these organisms are various insects, but certain worms, snails, slugs, rats, mice, and birds also fit into this category. **Weeds** are plants that compete with agricultural crops, forests, and forage grasses for light and nutrients. Some unwanted plants poison cattle or have other serious effects; others simply detract from the appearance of lawns and gardens. We try to bring these pests under control for three main purposes: to protect our food, to protect our health, and for convenience.

The Importance of Pest Control

Part of the credit for modern human prosperity can be attributed to pest control. On the credit side of the pesticide ledger, pesticides are vital elements in the prevention of the diseases that kill and incapacitate humans. In addition to having an agricultural use, pesticides have become important public health tools used to combat diseases such as malaria and sleeping sickness.

Insects, plant pathogens, and weeds destroy an estimated 37% (before and after harvest) of potential agricultural production in the United States, at a yearly loss of $64 billion. Efforts to control these losses in 1995 involved the use of 973 million pounds (443 thousand metric tons) of **herbicides** (chemicals that kill plants) and **pesticides** (chemicals that kill animals and insects considered to be pests) annually, with a direct cost of $11.3 billion per year (Fig. 17–2). Worldwide, 2.6 million metric tons of pesticides were used in 1995, at a cost of $38 billion. Many of the changes in agricultural technology, such as monoculture and the widespread use of genetically identical crops, which have boosted crop yields, have also brought on an increase in the proportion of crops lost to pests—from 31% in the 1950s to 37% today. During these past 45 years, the use of herbicides and pesticides has multiplied manyfold, leading to a disturbing and unsustainable dependency.

Different Philosophies of Pest Control

Medical practice employs two basic means of treating infectious diseases. One approach is to give the exposed patient a massive dose of antibiotics, hoping to eliminate the pathogen causing the problem or to stop the pathogen before it can get established. The other approach is to stimulate the patient's immune system with a vaccine to produce long-lasting protection against any future invasion. In practice, both means are often used to keep a particular pathogen under control.

The same basic philosophies are used to control agricultural pests. The first is **chemical technology**. Like the use of antibiotics, chemical technology seeks a "magic bullet" that will eradicate or greatly lessen the numbers of the pest organism. Although it has had much success, this approach gives only short-term protection. Furthermore, the chemical often has side effects that are highly damaging to other organisms.

The second philosophy is **ecological pest management**. Like stimulating the body's immune system, this approach seeks to give long-lasting protection by developing control agents based on knowledge of the pest's life cycle and ecological relationships. Such agents, which may either be other organisms or chemicals, work in one of two ways. Either they are highly specific for the pest species being fought, or they manipulate one or more aspects of the ecosystem. Ecological pest management emphasizes the protection of people and domestic plants and animals from damage from pests, rather than eradication of the pest organism. Thus, the benefits of pest control can be obtained while maintaining ecosystem integrity.

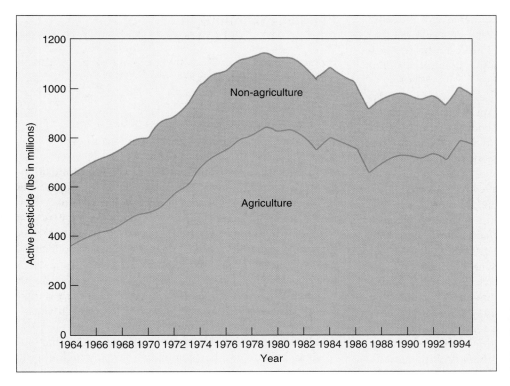

◀ **FIGURE 17-2** *Pesticide use in the United States.* Non-agricultural and agricultural use is shown for the time period of 1964–1995. (Source: Data from the Environmental Protection Agency, Office of Pesticide Programs.)

These two philosophies are combined in the approach called **integrated pest management (IPM)**. IPM is *an approach to controlling pest populations using all suitable methods—chemical and ecological—in a way that brings about long-term management of pest populations and also has minimal environmental impact.* This approach is increasing in usage, especially where pesticides are seen as undesirable because of health risks and in developing countries, where the cost of pesticides is often prohibitive.

17.2 Promises and Problems of the Chemical Approach

Pesticides are categorized according to the group of organisms they kill. There are insecticides (for insects), rodenticides (for mice and rats), fungicides (for fungi), herbicides (for plants), and so on. None of these chemicals, however, is entirely specific for the organisms it is designed to control; each poses hazards to other organisms, including humans. Therefore, pesticides are sometimes referred to as **biocides**, a name that emphasizes that they may endanger many forms of life.

Development of Chemical Pesticides and their Successes

Finding effective materials to combat pests is an ongoing endeavor. The early substances used (frequently referred to as **first-generation pesticides**) included toxic heavy metals such as lead, arsenic, and mercury. We now recognize that these

substances may accumulate in soils and inhibit plant growth. Poisoning of animals and humans is also possible. In addition, these chemicals lost their effectiveness as pests became increasingly resistant to them. For example, in the early 1900s, citrus growers were able to kill 90% of injurious **scale insects** (minute insects that suck the juices from plant cells) by placing a tent over an infested tree and piping in deadly cyanide gas for a short time. By 1930, this same technique killed as few as 3% of the pests.

The next step had its origins in the science of organic chemistry in the early 1800s. During the nineteenth century, chemists synthesized thousands of organic compounds, but for the most part, these compounds sat on shelves because uses for them had not been found. By the 1930s, however, with agriculture expanding to meet the needs of a rapidly increasing population, and with first-generation (inorganic) pesticides failing, farmers were begging for new pesticides. In time, **second-generation pesticides**, as they came to be called, were found as a result of synthetic organic chemistry.

The DDT Story. In the 1930s a Swiss chemist, Paul Muller, began systematically testing some organic chemicals for their effect on insects. In 1938 he hit upon the chemical dichloro-diphenyltrichloroethane (DDT), a chlorinated hydrocarbon that had first been synthesized some fifty years before.

DDT appeared to be nothing less than the long-sought "magic bullet," a chemical that was extremely toxic to insects and yet seemed nontoxic to humans and other mammals. It was very inexpensive to produce. At the height of its use in the early 1960s, it cost no more than about twenty cents a pound. It was **broad spectrum**, meaning that it was effective against a multitude of insect pests. It was also **persistent**, meaning that it did not break down readily in the environment and hence provided lasting protection. This last attribute provided additional economy by eliminating both the material and labor costs of repeated treatments.

DDT quickly proved successful in controlling important insect disease carriers. During World War II, for example, the military used DDT to control body lice, which spread typhus fever among the men living in dirty battlefield conditions. As a result, World War II was one of the first wars in which fewer men died of typhus than of battle wounds. The World Health Organization of the United Nations used DDT throughout the tropical world to control mosquitoes and greatly reduced the number of deaths caused by malaria. There is little question that DDT saved millions of lives. In fact, the virtues of DDT seemed so outstanding that Muller was awarded the Nobel Prize in 1948 for his discovery.

Postwar uses of DDT expanded dramatically. It was sprayed on forests to control defoliating insects such as the spruce budworm. It was routinely sprayed on salt marshes to deal with nuisance insects. It was sprayed on suburbs to control the beetles that spread Dutch elm disease. And, of course, DDT proved very effective in controlling agricultural insects (Fig. 17–3). It was so effective, at least in the short run, that many crop yields increased dramatically. Growers could ignore other, more painstaking methods of pest control such as crop rotation and destruction of old crop residues. They could grow varieties that were less resistant but more productive. They could grow certain crops in warmer or moister regions, where formerly the damage from pests had been devastating. In short, DDT gave growers more options for growing the most economically productive crop.

It is hardly surprising that DDT ushered in a great variety of synthetic organic pesticides. Examples and some characteristics of insecticides and herbicides are listed in Table 17-1.

▲ **FIGURE 17–3** *Aerial spraying.* A crop is being sprayed with pesticides to keep insects under control. Spraying is the basic technique still used in most control programs, although nowadays the pesticides in use do not persist in the environment for more than a week or two.

Problems Stemming from Chemical Pesticide Use

Problems associated with synthetic organic pesticides can be placed in three categories:

* development of resistance by pests
* resurgences and secondary pest outbreaks
* adverse environmental and human health effects

Development of Resistance by Pests. The most fundamental problem for growers is that chemical pesticides gradually lose their effectiveness. Over the years, it becomes necessary to use larger and larger quantities, new and more potent pesticides, or both to obtain the same degree of control. Synthetic organic pesticides fared no better than first-generation pesticides in this respect. For example, in 1946, 1 kg (2.2 lbs) of pesticides provided enough protection to produce about 60,000 bushels of corn. By 1971, 64 kg (141 lbs) were used for the same production, and losses due to pests actually *increased* during the intervening years.

Resistance builds up because pesticides destroy the sensitive individuals of a pest population, leaving behind only those few that already have some resistance to the pesticide. Resistant insect populations develop rapidly because insects have a phenomenal reproductive capacity. A single pair of houseflies, for example, can produce several hundred offspring that may mature and reproduce themselves only two weeks later. Consequently, repeated pesticide applications result in the unwitting selection and breeding of genetic lines that are highly, if not totally, resistant to the chemicals that were designed to eliminate them. Cases have been recorded in which the resistance of a pest population has increased as much as 25,000-fold.

Over the years of pesticide use, the number of resistant species has climbed steadily. Many major pest species are resistant to all of the principal pesticides; at least 520 insects and arachnids have shown resistance to pesticides (Fig. 17–4). Interestingly, as a pest population becomes resistant to one pesticide, it also may gain resistance to other, unrelated pesticides, even though it has not been exposed to the other chemicals (see the "Earth Watch" essay, p. 413).

Resurgences and Secondary Pest Outbreaks. The second problem with the use of synthetic organic pesticides is that after a pest has been virtually eliminated with a pesticide, the pest population not only recovers but explodes to higher and more severe levels. This phenomenon is known as a **resurgence**. To make matters worse, small populations of insects that were previously of no concern because of their low numbers suddenly start to explode, creating new problems. This phenomenon is called a **secondary pest outbreak**. For example, with the use of synthetic organic pesticides, mites have become a serious pest problem, and the number of serious pests on cotton has increased from 6 to 16.

To illustrate the seriousness of resurgences and secondary pest outbreaks, a recent study in California listed a series of 25

Table 17-1 Characteristics of Pesticides

Insecticides

Type	Examples	Toxicity to Mammals	Persistence
Organophosphates	Parathion, malathion, phorate, chloropyrifos	High	Moderate (weeks)
Carbamates	Carbaryl, methomyl, aldicarb, aminocarb	Moderate	Low (days)
Chlorinated hydrocarbons	DDT, toxaphene, dieldrin, chlordane, lindane	Relatively low	High (years)
Pyrethroids	Permethrin, bifenthrin, esfenvalerate, decamethrin	Low	Low (days)

Herbicides

Type	Examples	Effects on Plants
Triazines	Atrazine, simazine, cyanizine	Interfere with photosynthesis, especially in broadleaf plants
Phenoxy	2, 4-D, 2, 4, 5, -T, methylchlorophenoxybutyrate (MCPB)	Plant hormone-like effects on actively growing tissue
Acidamine	Alachlor, Propachlor	Inhibit germination and early seedling growth
Dinitroaniline	Trifluralin, oryzalin	Inhibit cells in roots and shoots; prevent germination
Thiocarbamate	ethylpropylthiolcarbamate (EPTC), cycloate, butylate	Inhibit germination, especially in grasses

Source: U.S. Environmental Protection Agency, *Private Pesticide Applicator Training Manual*, 1993.

major pest outbreaks, each of which caused more than $1 million worth of damage. All but one involved resurgences or secondary pest outbreaks. Of course, the species appearing in secondary outbreaks quickly became resistant to pesticides, thus compounding the problem. At first, pesticide proponents denied that resurgences and secondary pest outbreaks had anything to do with the use of pesticides. However, careful investigations have shown otherwise, as with the brown planthopper and rice in Indonesia. Resurgences and secondary pest outbreaks occur because the insect world is part of a complex food web.

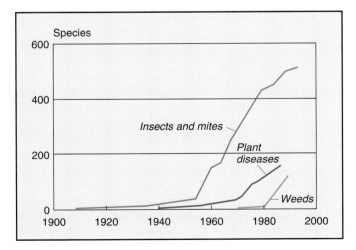

▲ **FIGURE 17–4** *Number of species resistant to pesticides, 1908–95.* Using pesticides causes selection (survival of the fittest) for those individuals that are resistant. As we continue to use pesticides, we breed insects and other pests that are increasingly resistant to the pesticides used against them. (Source: Lester Brown et al., *State of the World 1996* [Washington, DC: Worldwatch Institute, 1996].)

Thus, the chemical approach fails because it is contrary to basic ecological principles. It assumes that the ecosystem is a static entity in which one species, the pest, can simply be eliminated. In reality, the ecosystem is a dynamic system of interactions, and a chemical assault on one species will inevitably upset the system and produce other undesirable effects. Populations of plant-eating insects are frequently held in check by other insects that parasitize or prey on them (Fig. 17–5). Pesticide treatments often have a greater impact on these natural enemies than on the plant-eating insects they are meant to control. Consequently, with natural enemies suppressed, both the population of the original target pest and populations of other plant-eating insects explode. The path to sustainability demands that we understand how ecosystems work and that we adapt our interventions accordingly.

The late entomologist Robert van den Bosch coined the term **pesticide treadmill** to describe attempts to eradicate pests with synthetic organic chemicals. It is an apt term: The chemicals do not eradicate the pests; they increase resistance and secondary pest outbreaks, which lead to the use of new and larger quantities of chemicals, which in turn lead to more resistance and more secondary outbreaks, and so on. The process is an unending cycle constantly increasing the risks to human and environmental health and is clearly not sustainable (Fig. 17–6).

Human Health Effects. As with all toxic substances, pesticides can be responsible for both acute and chronic health effects. Acute poisonings involve an estimated 67,000 persons annually in the United States. Most of the victims are farm workers or employees of pesticide companies who come in direct contact with the chemicals. Some 3000 of these victims

(a)

Insect food chains

Producers	Primary consumer	Secondary consumer
Grass	Meadow grasshopper	Praying mantis
Pin oak	Caterpillar	Parasitic wasp
Alfalfa	Aphid	Lady beetle larva

(b)

▶ **FIGURE 17–5** *Insect food chains.* (a) Food chains exist among insects just as they do among higher animals. (b) Aphid lion (on right) impaling and eating a larval aphid (a small insect that sucks sap from plants).

require hospitalization, and about 20 to 30 of them die each year from pesticide toxicity. The World Health Organization conducted a survey of pesticide poisoning in humans and estimates that there are between 3.5 and 5 million cases of acute occupational pesticide poisoning each year in developing countries, of which at least 20,000 result in death. Use by untrained persons is considered to be the major cause of these poisonings, but many cases of poisoning occur where children and families come in contact with the pesticides through aerial spraying, the dumping of pesticide wastes, or the use of pesticide containers for drinking-water storage.

Chronic effects of pesticide include the potential for causing cancer, as indicated by animal testing; evidence for carcinogenicity has been recorded for eighteen pesticides. Epidemiological evidence has implicated organochlorine pesticides in various cancers, including lymphoma and breast cancer (Chapter 16). Other chronic effects include dermatitis, neurological disorders, birth defects, and infertility; male sterility has been clearly linked to the use of dibromochloropropane (DBCP), once used to control nematodes and now banned. Two disorders recently associated with pesticide use are immune system depression and endocrine system disruption. Again, the evidence for human impact comes from laboratory animal testing and epidemiological studies. Persistent exposure to pesticides among factory workers in India led to abnormally low white blood cell counts (the white blood cells are crucial elements of the immune system); similar results were found for agricultural workers in the former Soviet Union. Laboratory studies show many forms of immune suppression by different types of pesticide.

Laboratory tests have shown a number of pesticides to have estrogen-like effects: atrazine (a weed killer), DDT, endosulfan (insecticide), methoxychlor (insecticide), and others. Both the

EARTH WATCH

THE ULTIMATE PEST?

If we were to conjure up the ultimate insect pest, we might imagine the following characteristics. The ultimate insect pest would:

1. attack a broad variety of plants and fruits
2. be resistant to all of the usual pesticides
3. be highly prolific and have a rapid life cycle
4. lack natural predators and parasites
5. become a nuisance in other ways (it would interfere with breathing, cover car windshields, and so forth)

Such a pest, if unleashed on agriculture, could cause millions of dollars in crop losses and devastate many farmers.

Unfortunately, recent news from Texas and California suggests that something very close to the ultimate pest might already be with us. The silverleaf whitefly (*Bemisia argentifolii*) is a tiny white insect that emerged from Florida's poinsettia greenhouses in 1986 and has become established from coast to coast. It has all the characteristics listed and has been dubbed the "Superbug" by farmers who have encountered it. It is known to eat at least 500 species of plants—just about everything except asparagus and onions, and it thrives

on roadside weeds. Total crop losses to this insect are often greater than $500 million each year, and it continues to cause extensive damage. The insects swarm all over the plants, sucking them dry and leaving them withered and rotten. Pesticides have proved useless so far. Swarms become so dense that they interfere with breathing and vision. Entomologists are frantically searching for a natural predator or other pesticides to bring this pest under control. Such natural-enemy research continues in 25 countries. Unfortunately, no one knows where the insect originated. Unless some breakthrough occurs, we all may be eating more onions than ever before.

rise in the incidence of breast cancer and the report of abnormal sexual development in alligators and other animals in the wild have given credence to the possibility that very low levels of a number of chemicals are able to mimic the effects of estrogenic hormones (sexual hormones that are very potent at low concentrations). Concern about this has reached Congress, and the Environmental Protection Agency (EPA) is now in the process of developing procedures for testing thousands

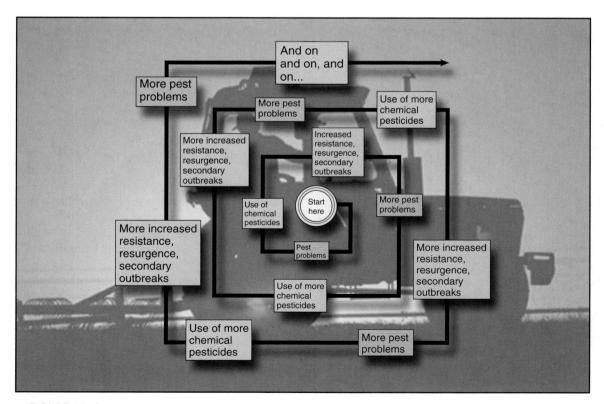

▲ **FIGURE 17–6** *The pesticide treadmill.* The use of chemical pesticides aggravates many pest problems. The continued use of these products demands ever increasing dosages of pesticides, which further aggravate pest problems and produce more contamination of foodstuffs and ecosystems.

of chemicals for "endocrine disruption" activity. Pesticides are prime suspects, and chemical manufacturers will be required to test as many as 62,000 chemicals in use in the United States, at a cost of many millions of dollars.

Environmental Effects. Aerial spraying for insect control has most clearly pointed to the adverse environmental effects of pesticides. The story of DDT, used so widely during the 1940s and 1950s (80,000 metric tons in 1962 alone), illustrates the hazards.

The Conclusion of the DDT Story. In the 1950s and 1960s, ornithologists (people who study birds) observed drastic declines in populations of many species of birds that fed at the tops of food chains. Fish-eating birds such as the bald eagle and osprey (Fig. 17–7) were so affected that their extinction was feared. Investigators at the U.S. Fish and Wildlife National Research Center near Baltimore, Maryland, showed that the problem was reproductive failure; eggs were breaking in the nest before hatching. They also showed that the fragile eggs contained high concentrations of dichloro-diphenyl-dichloroethylene (DDE), a product of the partial breakdown of DDT by the animal's body. DDE interferes with calcium metabolism, causing birds to lay thin-shelled eggs. Further study revealed that birds were acquiring high levels of DDT and DDE by bioaccumulation and biomagnification, the process of accumulating higher and higher doses through the food chain.

Because of accumulation, small, seemingly harmless amounts received over a long period of time may reach toxic levels. This phenomenon—referred to as **bioaccumulation**—can be understood as follows. Many synthetic organics are highly soluble in lipids (fats or fatty compounds) but sparingly

soluble in water. As they pass through cell membranes, which are lipid, they come out of water solution and enter into the lipids of the body. Thus, traces of synthetic organics like pesticides that are absorbed with food or water are trapped and held by the body's lipids, while the water and water-soluble wastes are passed in the urine. Since these are unnatural compounds, the body has no mechanism to excrete them or to metabolize them further. Trace levels consumed over time gradually accumulate in the body and may sooner or later produce toxic effects.

Bioaccumulation, which occurs in the individual organism, may be compounded through a food chain. Each organism accumulates the contamination in its food, so it accumulates a concentration of contaminant in its body that is many times higher than that in its food. The next organism in the food chain effectively now has a more contaminated food and accumulates the contaminant to yet a higher level. Essentially all the contaminant accumulated by the large biomass at the bottom of the food pyramid is concentrated, through food chains, into the smaller and smaller biomass of organisms at the top of the food pyramid. This multiplying effect of bioaccumulation that occurs through a food chain is called **biomagnification**. One of the most distressing aspects of bioaccumulation and biomagnification is that there are no warning symptoms until contaminant concentrations in the body are high enough to cause problems. Then, it is too late to do much about it. As is often the case, bioaccumulation and biomagnification go unrecognized until serious problems bring the phenomena to light. Fig. 17–8 shows how DDT and its metabolic products worked their way up the food chain to the top predators.

Silent Spring. In the 1950s, Rachel Carson, a U.S. Fish and Wildlife Service biologist and accomplished science writer, began reading the disturbing scientific accounts of DDT's effects on wildlife. Carson was finally galvanized into action by a letter from a friend distressed over the large number of birds killed when the friend's private bird sanctuary was sprayed for mosquito control. By 1962, she finished a book-length documentation of the effects of the almost uncontrolled use of insecticides across the United States. The book, *Silent Spring*, became an instant best-seller. Its basic message was that if insecticide use continued as usual, there might someday come a spring with no birds—and with ominous impacts on humans as well.

Silent Spring triggered a national debate that continues today. Representatives of the agricultural and chemical industries claimed *Silent Spring* was an unreasonable and unscientific account that, if taken seriously, would lead to a halt of human progress. At the same time, however, the book was hailed as an unparalleled breakthrough in environmental understanding. Eventually, concerns about environmental and long-term health effects led to the banning of DDT in the United States and most other industrialized countries in the early 1970s. Numerous other related chlorinated hydrocarbon pesticides (e.g., chlordane, dieldrin, endrin, and heptachlor) were also banned because of their propensity for bioaccumulation in the environment and suspected potential for causing cancer. In

▲ **FIGURE 17–7** *Bald eagle.* Populations of fish-eating birds like the osprey, brown pelican, and bald eagle were decimated in the 1950s and 1960s by the effects of widespread spraying with DDT. With the banning of DDT, these populations have greatly recovered. The bald eagle was taken off the endangered species list in 1994.

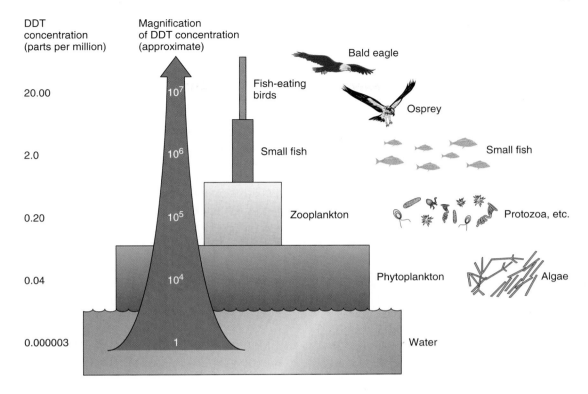

▲ **FIGURE 17–8** *Biomagnification.* Organisms on the first trophic level absorb the pesticide from the environment and accumulate it in their bodies (bioaccumulation). Then, each successive consumer in the food chain accumulates the contaminant to a yet higher level. Thus, the concentration of the pesticide is magnified manyfold throughout the food chain. Organisms at the top of the food chain are likely to accumulate toxic levels.

the years since the banning of DDT, observers have noted a marked recovery in the populations of birds that were adversely affected.

Thirty-five years later, Rachel Carson is credited with stimulating the start of the environmental movement and the creation of the EPA. *Silent Spring* has become a classic, and the regulation of insecticides and other toxic chemicals is in a sense a monument to Rachel Carson, who died of cancer only two years after her book was published.

Nonpersistent Pesticides: Are They the Answer?

A key characteristic of chlorinated hydrocarbons is their persistence—that is, their slowness to break down. Lingering in the environment for years, they can contaminate many organisms and become biomagnified. This persistence results because these chemicals have such complex structures that soil microbes are unable to metabolize them. DDT, for example, has a half-life on the order of 20 years.

Recognizing this persistence factor, the agrochemical industry has in large measure substituted *nonpersistent* pesticides for the banned compounds. For example, the synthetic organic phosphates malathion and parathion, as well as carbamates such as aldicarb and carbaryl, are now used extensively in place of chlorinated hydrocarbons (see Table 17-1). These compounds break down into simple nontoxic

products within a few weeks after application. Thus, there is no danger of their migrating long distances through the environment and affecting wildlife or humans long after being applied. When used wisely, these pesticides are an important tool for farmers and gardeners.

For several reasons, however, nonpersistent pesticides are not as environmentally sound as they might appear. First of all, total environmental impact is a function not only of persistence but also of three other important factors: toxicity, dosage applied, and location where applied. Many of the nonpersistent pesticides are far more toxic than DDT. This higher toxicity, combined with the frequent applications needed to maintain control, presents a significant hazard to agricultural workers and others exposed to these pesticides. For instance, the organophosphates are responsible for an estimated 70% of all pesticide poisonings.

Second, nonpersistent pesticides may still have far-reaching environmental impacts. For example, to control outbreaks of grasshoppers that eat sunflowers, farmers in Argentina began spraying monocrotophos, an acutely toxic organophosphate pesticide, in 1995. Shortly afterwards, thousands of dead Swainson's hawks were seen in the sunflower fields. The hawks, which spend summers in the United States, winter in great numbers in the Argentine pampas and feed voraciously on grasshoppers there. It was estimated that 20,000 hawks died in one year, about 5% of the world's population of the species. In response, the Argentine government banned the use of monocrotophos; it had already been withdrawn from use in the United States in 1988.

Third, desirable insects may be just as sensitive as pest insects to these substances. Bees, for example, which play an essential role in pollination, are highly sensitive to nonpersistent pesticides. Thus, use of these compounds creates an economic problem for beekeepers and jeopardizes pollination. Also, regular spraying of neighborhoods with malathion to control mosquitoes leads inevitably to great declines in butterflies and fireflies.

Finally, nonpersistent chemicals are just as likely to cause resurgences and secondary pest outbreaks as are persistent pesticides, and pests become resistant to nonpersistent chemicals just as quickly as they do to persistent pesticides.

17.3 Alternative Pest Control Methods

Numerous ecological and biological factors affect the relationship between a pest and its host. Ecological pest management seeks to manipulate one or more of these natural factors so that crops are protected without upsetting the rest of the ecosystem or jeopardizing environmental and human health. Since ecological pest management involves working with natural factors instead of synthetic chemicals, the techniques are referred to as **natural control** or **biological control** methods. This natural approach, unlike the chemical technology approach, depends on an understanding of the pest and its relationship with its host and with its ecosystem. The more we know about the organisms involved, the greater are our opportunities for natural control.

To illustrate, the life cycle typical of moths and butterflies is shown in Fig. 17–9. Many groups of insects have a similarly complex life cycle. The development of each stage may be influenced by numerous abiotic factors, and at each stage the insect may be vulnerable to attack by a parasite or predator. Proper completion of each stage depends on internal chemical signals provided by hormones. Locating mates, finding food, and other behaviors depend on external chemical signals. All these findings suggest ways in which pest populations may be controlled without resorting to synthetic chemical pesticides.

The four general categories of natural or biological pest control are: (1) **cultural control**; (2) **control by natural enemies**; (3) **genetic control**; and (4) **natural chemical control**. We will consider each of these in turn.

▶ **FIGURE 17–9** *Complex life cycle of insects.* Like the moth shown here, most insects have a complex life cycle that includes a larval stage and an adult stage. Biological control methods recognize the different stages and attack the insect, using knowledge of its needs and life cycle.

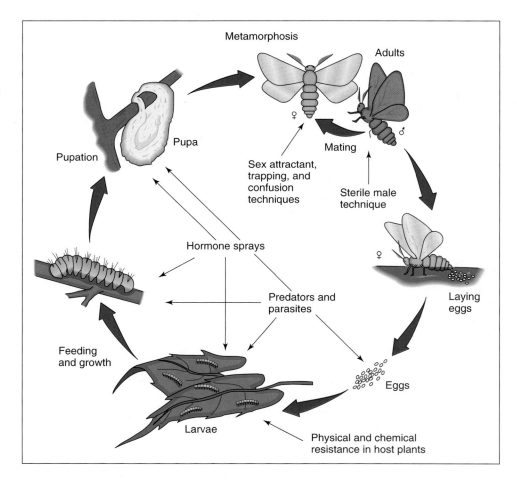

Cultural Control

A cultural control is a nonchemical alteration of one or more environmental factors in such a way that the pest finds the environment unsuitable or is unable to gain access to it.

Cultural Control of Pests Affecting Humans. We routinely practice many forms of cultural control against diseases and parasitic organisms. Some of these practices are so familiar and well entrenched in our culture that we no longer recognize them for what they are. For instance, disposing properly of sewage and avoiding drinking water from unsafe sources are cultural practices that protect against waterborne disease-causing organisms. Combing and brushing the hair, bathing, and wearing clean clothing are cultural practices that eliminate head and body lice, fleas, and other parasites. Regular changing of bed linens protects against bedbugs. Properly and systematically disposing of garbage and keeping a clean house with any cracks sealed and with good window screens are effective methods for keeping down populations of roaches, mice, flies, mosquitoes, and other pests. Sanitation requirements in handling and preparing food are cultural controls designed to prevent the spread of diseases. Refrigeration, freezing, canning, and drying of foods are cultural controls that inhibit the growth of organisms that cause rotting, spoilage, and food poisoning.

If such practices of personal hygiene and sanitation are compromised, as they usually are in any major disaster, there is very real danger of additional widespread mortality resulting from outbreaks of parasites and diseases. As we saw in Chapter 16, sickness and death are the consequences where these practices are not broadly pursued in a society, as in some less developed countries.

Cultural Control of Pests Affecting Lawns, Gardens, and Crops. Homeowners are prone to use excessive amounts of herbicides to maintain a weed-free lawn. Weed problems in lawns are frequently a result of cutting the grass too short. If grass is left at least 3 inches (8 cm) high, it will usually maintain a dense enough cover to keep out crabgrass and many other noxious weeds. Thus, monitoring the height of grass is a form of cultural weed control. Also, many homeowners allow a diversity of plants in their lawns, tolerating a variety of plants some people might consider weeds.

Some plants are particularly attractive to certain pests; others are especially repugnant. In either case, the effect may spill over to adjacent plants. A gardener may control many pests by paying careful attention to eliminating plants that act as attractants (for example, roses) and growing those that act as repellents. Marigolds and chrysanthemums are justly famous for being insect repellents.

Some parasites require an alternative host. They can be controlled by eliminating this host (Fig. 17–10). Also, hedgerows, fencerows, and shelterbelts can provide refuges where natural enemies of pests (birds, amphibians, preying mantises, and so on) can be maintained.

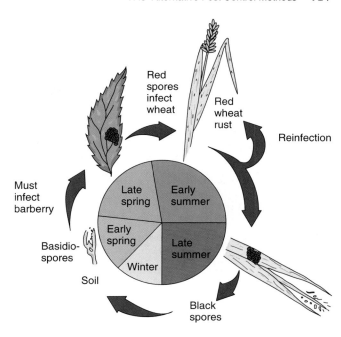

▲ **FIGURE 17–10** *Cultural control.* Part of the life cycle of wheat rust, a parasitic fungus that is a serious pest on wheat, requires that the rust infest barberry, an alternative host plant. The elimination of barberry in wheat-growing regions has been an important cultural control.

Management of any crop residues not harvested is important. Spores of plant disease organisms and insects may overwinter or complete part of their life cycle in the dead leaves, stems, or other plant residues that remain in the fields after harvesting. Plowing under or burning the material may be very effective in keeping pest populations to a minimum. In gardens, a clean mulch of material such as grass clippings or hay will keep down the growth of weeds and protect the soil from drying and erosion.

Growing the same crop on a plot of land year after year keeps the pest's food supply continuously available. Crop rotation, the practice of changing crops from one year to the next, may provide control because pests of the first crop cannot feed on the second crop and vice versa. Crop rotation is especially effective in controlling root nematodes (roundworms that live in the soil and feed on roots) and other pests that do not have the ability to migrate appreciable distances.

For economic efficiency, agriculture has moved progressively toward monoculture—the Corn Belt, Cotton Belt, and so on. When a pest outbreak occurs, monoculture is most conducive to its rapid multiplication and spread. The spread of a pest outbreak is impeded and other natural controls may be more effective if there is a mixture of crop species, some of which are not vulnerable to attack. One approach that works well in the British Isles is to intersperse cultivated with uncultivated strips that are not treated with pesticides. Natural enemies of pests are maintained in the uncultivated strips.

Most of our pests that are hardest to control were unwittingly imported from other parts of the world, and we realize that many other species in other regions would be serious pests if introduced here. Therefore, it is important to keep would-be pests out of the country. This is a major function of the U.S. Customs Bureau and of the agriculture departments of some states. Biological materials that may carry pest insects or pathogens either are prohibited from crossing the border or are subjected to quarantines, fumigation, or other treatments to ensure that they are free of pests. The cost of such procedures is small in comparison to the costs that could be incurred if these pests gained entry and became established.

Control by Natural Enemies

The following examples illustrate the range of possibilities for controlling pests with natural enemies:

- Scale insects, potentially devastating to citrus crops, have been successfully controlled by vedalia (ladybird) beetles, which feed on them.
- Various caterpillars have been controlled by parasitic wasps (Fig. 17–11).
- Japanese beetles are controlled in part with the bacterium *Bacillus thuringiensis* (Bt), which produces toxic crystals when soil containing the bacterium is ingested by the larvae.
- Prickly pear cactus and numerous other weeds have been controlled by plant-eating insects (Fig. 17–12). (More than 30 weed species worldwide are now limited by insects introduced into their habitats.)
- Rabbits in Australia are controlled by an infectious virus.
- Aquatic weeds in the American Midwest are controlled by a vegetarian fish native to South America.
- Mealybugs in Africa are controlled by a parasitic wasp (see the "Global Perspective" essay, p. 421).

The problem with using natural enemies is finding organisms that provide control of the target species without attacking other, desirable species. Entomologists estimate that of the 50,000 known species of plant-eating insects that have the potential of being serious pests, only about 1% actually are. The populations of the other 99% are held in check by one or more natural enemies, so they do not cause significant amounts of damage. Therefore, the first step in using natural enemies for control should be *conservation*—protecting the natural enemies that already exist. Conservation means avoiding the use of broad-spectrum chemical pesticides, which may affect natural enemies even more than they do the target pests. Eliminating or placing considerable restrictions on the use of broad-spectrum chemical pesticides will, in many cases, allow natural enemies to reestablish themselves and control secondary pests, those that became serious problems only after the use of pesticides.

However, effective natural enemies are not always readily available. In many cases, the lack of natural enemies is the result of accidentally importing the pest without its natural enemy. Quite often, effective natural enemies have been found by systematically combing the home region of an introduced pest and finding its various predators or parasites. To date, the U.S. Department of Agriculture (USDA) has released almost 1000 insects to control pests, with generally favorable results. Specificity of the enemy species is vital and must be carefully tested before intentional release, something that has not always been done well. Some 16% of 313 parasitoid wasp species introduced into North America have attacked native species. In one recent case, a weevil (a type of beetle) imported to control an exotic musk thistle (weed), has also attacked five native thistle species that pose no threat to agriculture. When effective natural enemies are found, however, they can provide control indefinitely, saving millions of dollars per year without further expense.

Genetic Control

Most plant-eating insects and plant pathogens attack only one species or a few closely related species. This specificity implies a genetic incompatibility between the pest and species that are not attacked. Most genetic-control strategies are designed to develop genetic traits in the host species that provide the same incompatibility—that is, resistance to attack by the pest. This technique has been utilized extensively in connection with plant diseases—fungal, viral, and bacterial parasites. For example, in the years 1845–47 the potato crop in Ireland was devastated by late blight, a fungal parasite. Nearly a million people starved, and another million people emigrated to escape the same fate. Nowadays, protection against such disasters is provided in large part by growing varieties of potato that are resistant to the blight. It is no overstatement to say that the world owes much of its production of potatoes, corn, wheat, and other cereal grains to the painstaking work of plant geneticists who selected and bred disease-resistant varieties.

The same potential exists for breeding plants that are resistant to insect pests. Traits that provide resistance may act as chemical or physical barriers.

Control with Chemical Barriers. A chemical barrier is a chemical produced by the plant we want to protect; the substance is lethal or at least repulsive to the would-be pest. Once they have identified such a barrier, plant breeders can use selection and crossbreeding to enhance this trait in the desirable plant. The relationship between wheat and the Hessian fly provides an example. This fly lays its eggs on wheat leaves, and the larvae move down the leaves and into the main stem as they feed. The weakened stem either dies or is broken in the wind. The Hessian fly was introduced into the United States in the straw bedding of Hessian soldiers during the Revolutionary War. The fly eventually spread throughout much of the Midwest, causing widespread devastation until scientists

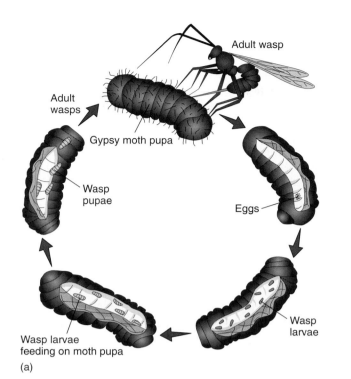

(b)

(a)

(c)

▲ **FIGURE 17–11** *Parasitic wasps.* (a) The life cycle of the parasitic wasp that uses the gypsy moth as its host. (b) A wasp depositing eggs in a gypsy moth larva. (c) Another insect parasite, a braconid wasp, lays its eggs on the pest known as the tomato hornworm, which is the larva of the sphinx moth. The wasp larvae feed on the caterpillar, and shortly before the caterpillar dies, they emerge and form the cocoons seen here.

at the University of Kansas developed a variety of wheat that produces a chemical toxic to the fly; it kills the larvae when they feed on the leaves.

Increasing resistance through breeding may not provide 100% protection, but even partial protection can make the difference between profit and loss for the grower. In addition, any degree of resistance lessens the need for chemical pesticides.

Control with Physical Barriers. Physical barriers are structural traits that impede the attack of a pest. For example, leafhoppers are significant worldwide pests of cotton, soybeans, alfalfa, clover, beans, and potatoes, but they can damage only plants that have relatively smooth leaves. Hooked hairs on the leaf surfaces of some plants tend to trap and hold immature leafhoppers until they die. Similarly, glandular

(a)

(b)

▲ **FIGURE 17–12** *Prickly pear cactus.* Biological control of the weed prickly pear cactus by a cactus-eating moth in Queensland, Australia. (a) The land shown here, once settled, had to be abandoned because of prickly pear infestation. (b) The same land was reoccupied following the destruction of the prickly pear by the moth.

hairs that exude a sticky substance fatally entrap alfalfa weevil larvae. Such traits can be enhanced in vulnerable plants through selective breeding.

Control with Sterile Males. Another genetic-control strategy involves flooding a natural population with sterile males that have been reared in laboratories. Combating the screwworm fly provides a prime illustration. This fly, which is closely related to the housefly and looks much like it, lays its eggs in open wounds of cattle and other animals. The larvae feed on blood and lymph, keeping the wound open and festering. Secondary infections frequently occur and often lead to the death of the animal.

Early in the century, this problem became so severe that cattle ranching from Texas to Florida and northward was becoming economically impossible. In studying the situation during the 1940s, Edward Knipling, an entomologist with the USDA, observed two essential features of screwworm flies: (1) Their populations are never very large, and (2) the female fly mates just once, lays her eggs, and then dies. Knipling reasoned that if the female mated with a sterile male, no offspring would be produced. His hypothesis was correct, and today sterile males are routinely used to control this pest. After huge numbers of screwworm larvae are grown on meat in laboratories, the resulting pupae are subjected to just enough high-energy radiation to render them sterile. These sterilized pupae are then airdropped into the infested area. Ideally, 100 sterile males are dropped for every normal female in the natural population, giving a 99% probability that wild females will mate with one of the sterile males.

This technique proved so successful that it eliminated the screwworm fly from Florida in 1958–59, and it continues to be used to control the problem in the Southwest to this day. The savings to the cattle industry are estimated at more than $300 million a year. The technique has recently been used to eradicate the tsetse fly from the island of Zanzibar, thus eliminating the disease trypanosomosis (sleeping sickness), which the tsetse fly carries. It has made it possible for herdsmen to raise their cattle without the devastating losses the disease causes elsewhere in sub-Saharan Africa—more than 10 million square miles are inhabited by tsetse flies. Research efforts are under way to extend this success to other parts of Africa.

Strategies Using Biotechnology. Biotechnology has recently multiplied the potential for genetic control. More complex than basic plant breeding, genetic engineering makes it possible to introduce genes into crop plants from a variety of sources: other plant species, bacteria, and viruses. The new transgenic crops are undergoing rapid development and testing; more than 40 species of food plants have been genetically engineered, and by mid-1996, 25 transgenic crops had received regulatory approval. One promising strategy is to incorporate the protein coat of a plant virus into the plant itself. When the plant "expresses" (that is, manufactures) the virus's protein coat, it becomes resistant to infection by the real virus. In this way, crop plants have been made resistant to more than a dozen plant viruses.

Resistance to insects has been engineered with the use of a potent protein produced by *Bacillus thuringiensis* (Bt). This protein kills the larvae of a number of plant-eating insects and is harmless to mammals, birds, and most other insects. Scientists have engineered the gene into a number of plants, including cotton, potatoes, and corn (Fig. 17–13). This development alone is expected to cut in half the use of pesticides on cotton—a crop that consumes more than 10% of the pesticides used throughout the world. The protection comes from a naturally occurring protein used for three decades by home gardeners, organic growers, and other farmers. Wide-scale use of the Bt cotton (Monsanto's "Bollgard" cotton) revealed that it is effective against two of three pests, but farmers had to employ pesticides to control the bollworm (a kind of moth larva) in 40% of the 2 million

GLOBAL PERSPECTIVE

WASPS 1, MEALYBUGS 0

The cassava (manioc) plant originated in South America but has been cultivated throughout the tropical world. Currently, it is the primary food for more than 200 million people in sub-Saharan Africa, one-third of the human population in that region. It is a high-yielding crop that requires no modern technology and, accordingly, is raised by subsistence farmers everywhere.

An insect not previously seen in Africa, the mealybug appeared in Congo in the early 1970s and spread across sub-Saharan Africa, leaving a trail of ruined harvests and hunger in its wake. Zaire and Congo, unable to afford pesticides, turned for help to the International Institute for Tropical

Agriculture in Nigeria. This group formed the Biological Control Program and began to look for natural enemies of the mealybug.

Returning to the land of origin of the cassava, a researcher found the mealybug in Paraguay and observed that the insect was kept under control by natural predators and parasites. After extensive testing, researchers identified a parasitic wasp, *Epidinocarsis lopezi*, as the prime candidate for controlling the bugs. No larger than a comma on this page, the female wasp seeks out mealybugs, paralyzes them with a sting, and deposits eggs that hatch inside the mealybug and eat their way through the insect.

Once this wasp-bug relationship was identified, the wasps were reared by the

thousands on captive mealybugs and then spread across the continent over a period of eight years. The Biological Control Program has trained 400 workers to monitor the progress of the program, and all indications are that the battle has been won. It is estimated that every dollar invested in the control program has yielded $149 in crops saved from destruction. The best news is that the control is permanent and does not require the repeated application of expensive and environmentally damaging pesticides. In the wake of the project, national biological control programs have been established and are now focusing on pests of other crops of sub-Saharan Africa. (Source: Jane Ellen Stevens, *The Boston Globe*, January 3, 1994.)

acres planted of the new crop. Bt corn has been developed to combat a huge problem, the European corn borer; over $1 billion in losses and control costs is traced to this one pest.

One unusual biotech strategy has been to give a crop species resistance to a broad-spectrum herbicide. Although this appears to encourage the use of herbicides, it is actually expected to make it possible to use less total amounts of herbicide and reduce the number of applications per season on a given acre and to use compounds that are more environmentally benign. One such herbicide is glyphosate (Roundup®). It inhibits an amino acid

pathway that only occurs in plants and not in animals, biodegrades rapidly, and can control both grasses and broadleaf plants. (Because it is nonspecific and will kill every plant, application and cleanup must be done carefully.) Genes resistant to glyphosate have been successfully introduced into cotton, canola, soybean, and corn. Farmers do it because they save money on herbicides; they need only one spraying of Roundup to control weeds in their soybean fields. (Over 60% of the U.S. soybean crop is now planted with "Roundup Ready" soybeans.)

There is a downside to these biotech innovations. They are not well suited to developing countries, as they foster a dependence on a costly annual supply of seeds under current corporation practices. There is also the definite possibility of developing "superweeds" that are resistant to the herbicides or plant viruses as resistance genes are shared with genetically close wild species. Insects are likely to develop resistance to Bt plants much more readily than they would to occasional spraying of the Bt insecticide directly, because they are constantly exposed to Bt toxic effects season after season. The same potential for developing resistance applies to chemical and physical control strategies. Thus, scientists have had to develop new resistant varieties of wheat seven times in the case of wheat and the Hessian fly. Nevertheless, it is likely that the applications of biotechnology will gradually reduce the burden of pesticides in the environment and their potential impact on human health.

▲ **FIGURE 17–13** *Bioengineered potatoes.* The center row shows potato plants that have the Bt gene incorporated into the plant genome. These plants are protected from the Colorado potato beetle, while surrounding rows show typical beetle devastation of nonengineered potato plants.

Natural Chemical Control

As in humans and other animals, hormones control each stage in the development of an insect. Hormones are chemicals produced in the organism that provide "signals" to control developmental processes and metabolic functions. In

▲ **FIGURE 17–14** *Trapping technique.* Adult Japanese beetles are lured into a trap by scented bait they find attractive. Once inside the trap, baffles prevent the beetles from escaping. These traps are available in hardware stores and are quite effective. (Japanese beetles are notorious lawn grubs as larvae; as adults, they are voracious consumers of garden flowers and vegetables.)

addition, insects produce many pheromones—chemicals secreted by one individual that influence the behavior of another individual of the same species.

The aim of natural chemical control is to isolate, identify, synthesize, and then use an insect's own hormones or pheromones to disrupt its life cycle. Two advantages of natural chemicals are that they are nontoxic and that they are highly specific to the pest in question (they do not affect the natural enemies of the pest to any appreciable extent). If the affected pest is eaten by another organism, it is simply digested. Following are some examples.

Scientists have discovered that caterpillar pupation is triggered by a decrease in the level of the chemical called

juvenile hormone. If this chemical is sprayed on caterpillars, pupation does not occur. The larvae simply continue to feed and grow, become grossly oversized, and eventually die. One newly developed insecticide, called Mimic, is a synthetic variation of **ecdysone**, the molting hormone of insects. Mimic begins the molting process in insect larvae but doesn't complete it. As a result, the larva is trapped in its old skin and eventually starves to death. Mimic is specific to moths and butterflies and is viewed as a potent agent in the future control of the spruce budworm, the gypsy moth, the beet army worm, and the codling moth—all highly devastating pests. Field test results with Mimic have been promising.

Adult female insects secrete pheromones that attract males for the function of mating. Once identified and synthesized, these pheromones may be used in either of two ways: the trapping technique or the confusion technique. In the trapping technique, the pheromone is used to lure males into traps or into eating poisoned bait (Fig. 17–14). In the confusion technique, the pheromone is dispersed over the field in such quantities that males become confused, cannot find the females, and thus fail to mate.

The enormous potential of natural chemicals for controlling insect pests without causing ecological damage or disruption has been recognized for at least 30 years. However, the background research and necessary testing are just now reaching fruition. Tests are extremely promising. It appears that the great expectations in this field are about to be realized, as over 800 natural chemicals have been identified and more than 250 are being produced commercially.

17.4 Socioeconomic Issues in Pest Management

The increasing availability of alternative pest control methods and the accumulating evidence of the failures of the chemical approach to pest management have led to a growing movement toward avoiding pesticide use, particularly on foods. This movement must contend with a strong tendency on the part of farmers to employ pesticides.

Pressures to Use Pesticides

With any method of pest control, it is important to keep in mind that a species becomes a pest only when its population multiplies to the point of causing significant damage. Natural controls are generally aimed at keeping pest populations below damaging levels, not at total eradication of the populations. By keeping pest populations down, natural controls avert significant damage while preserving the integrity of the ecosystem.

Therefore, the question to be asked in facing any pest species is whether it is causing significant damage. Damage should be deemed significant only when the economic losses due to the damage considerably outweigh the cost of applying a pesticide. This point is called the **economic threshold** (Fig. 17–15). If significant damage is not occurring, natural

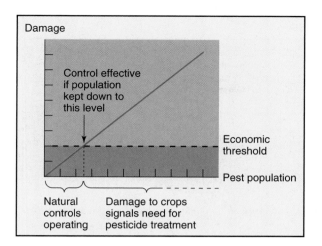

▲ **FIGURE 17–15** *The economic threshold.* The objective of pest control should not be to eradicate a pest totally. All that is needed is to keep population levels below the economic threshold.

controls are already operating, and the situation is probably best left as is. Spraying with synthetic chemicals at this stage is more than likely to upset the natural balance and make the situation worse through resurgences. On the other hand, if significant damage *is* occurring, a pesticide treatment may be in order.

It makes little difference whether the threat of loss as a result of pest infestation is real or imagined, close at hand or remote; what is important is how the grower perceives the threat. Even if there is no evidence of immediate damage from a pest, a grower who believes that his or her plantings are at risk is likely to resort to **insurance spraying**, the use of pesticides "just to be safe."

Consumers also put indirect pressure on growers to use pesticides. From customers to supermarket chains to canneries, there is a tendency to select the best looking fruits and vegetables, leaving the remainder to be sold at lower prices or trashed. Like Snow White, we all tend to reach for the apparently perfect, unblemished apple. Growers know that blemished produce means less profit, so they indulge in **cosmetic spraying**—the use of pesticides to control pests that harm only the item's outward appearance. Cosmetic spraying accounts for a significant fraction of pesticide use, does nothing to increase yield or nutritional value, and results in an increase in pesticide residues remaining on the produce.

Integrated Pest Management

Integrated pest management (IPM) aims to minimize the use of synthetic organic pesticides without jeopardizing crops. This is made possible by addressing all the interacting sociological, economic, and ecological factors involved in protecting crops. With IPM, the crop and pests are seen as part of a dynamic ecosystem; the goal is not the eradication of pests, but maintaining crop damage below the economic threshold.

Cultural and biological control practices form the core of IPM techniques. Such practices as crop rotation, polyculture (instead of monoculture), the destruction of crop residues, the maintenance of predator populations, and carefully timed planting and fertilizing are basic to IPM. "Trap crops" are often used: Early strips of a crop such as cotton are planted to lure existing pests, and then the pests are destroyed by hand or by the limited use of pesticides before they can reproduce. Spraying is not performed during the growing season, in order to encourage natural enemies of the pests to flourish. Pest populations are monitored, often by persons employed by local agricultural extension services or farm cooperatives or by persons acting as independent consultants. These **field scouts** are trained in monitoring pest populations (traps baited with pheromones are used for this purpose) and in determining whether the pest population exceeds the economic threshold. If this happens, remedial measures are taken to decrease the pest population. Here, pesticides are often used, in quantities and brands that do the least damage to natural enemies of the pest. In addition, **pest-loss insurance**, which pays the farmer in the event of loss due to pests, has enabled insured growers to refrain from unnecessary "insurance spraying."

Making the economic benefits of natural controls known to growers is an important aspect of IPM. Because pesticides increased yields and profits when they were first used, many farmers still cling to them, believing that they offer the only way to bring in a profitable crop. In some instances, however, the rising costs of pesticides, along with their tendency to aggravate pest problems, have eliminated their economic advantage.

Particularly in the developing world, government agricultural policy often determines the extent to which IPM is adopted. Governments and aid agencies usually subsidize the purchase of pesticides, thus strongly encouraging growers to step onto the pesticide treadmill. By contrast, the Indonesian experience cited earlier has provided a viable IPM model for other rice-growing countries. The success of the program can be traced to close cooperation between the Indonesian government and the FAO (Fig. 17–16). FAO workers ran training sessions for farmers (Farmers Field Schools), weekly meetings that lasted throughout the growing season. Eventually, more than 200,000 farmers were taught about the rice agro-ecosystem and IPM techniques, and these farmers formed a corps who could in turn teach others. The economic and environmental benefits of the Indonesian IPM program have been remarkable. The government has saved millions of dollars annually by not purchasing pesticides; farmers have not had to invest in pesticides and spray equipment; the environment has been spared the application of thousands of tons of pesticide; fish are once again thriving in the rice paddies; the health benefits of reduced pesticide use have been spread from applicators to consumers and wildlife. (The FAO is sponsoring similar IPM training programs in eight other rice-growing nations and has started programs for vegetable crops in six nations.)

Recently, FAO, the World Bank, UNDP, and UNEP have joined to cosponsor the IPM Facility. Its mission is to

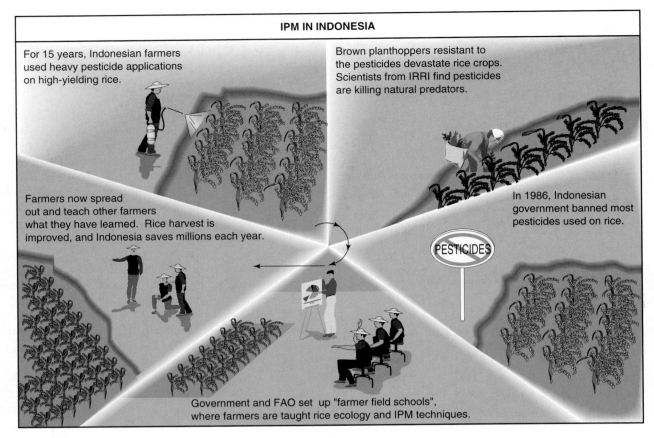

IPM IN INDONESIA

For 15 years, Indonesian farmers used heavy pesticide applications on high-yielding rice.

Brown planthoppers resistant to the pesticides devastate rice crops. Scientists from IRRI find pesticides are killing natural predators.

In 1986, Indonesian government banned most pesticides used on rice.

PESTICIDES

Farmers now spread out and teach other farmers what they have learned. Rice harvest is improved, and Indonesia saves millions each year.

Government and FAO set up "farmer field schools", where farmers are taught rice ecology and IPM techniques.

▲ **FIGURE 17–16** *Integrated pest management.* IPM has helped Indonesian rice farmers to bring the brown planthopper under control, after years of frustration on the pesticide treadmill.

carry forward the objectives of *Agenda 21* and to establish networks among farmers, researchers, and extension services. Although IPM does rely on chemical pesticides when needed, if it is broadly adopted, it is a logical pathway to a sustainable future, in which pesticides are not polluting groundwater, contaminating food, killing pollinating insects, and, in the end, creating new, more resistant pests. (For an example of IPM in practice, see the "Ethics" essay, p. 425.)

Organically Grown Food

There is often strong public feeling against the use of chemical pesticides. Cosmetic or other unnecessary spraying persists in large part because consumers are kept ignorant about the kinds or amounts of pesticides used. Experience shows that when consumers are informed, many of them abandon the Snow White attitude. Food outlets selling **organically grown** produce (produce that is grown without synthetic chemical pesticides or fertilizers) are appearing, although the produce is not as cosmetically perfect and usually costs more than chemically treated produce (Fig. 17–17). Nonetheless, an increasing number of growers are relying on natural controls and selling through these specialized markets. Sales of organics have increased 20% a year for the past decade; organically grown food is now a $4 billion enterprise.

Expanding markets for certified organic foods in the United States, Japan, and Europe have encouraged increasing numbers of growers to abandon the use of agricultural chemicals.

▲ **FIGURE 17–17** *Organic foods.* A roadside farm stand in Concord, Massachusetts, specializes in organic produce. Organically grown foods represent an expanding market in the United States.

ETHICS

THE LONG WAR AGAINST THE MEDFLY

The program to protect crops in the United States from the Mediterranean fruit fly, or medfly, exemplifies how IPM can work. It also illustrates an ethical dilemma in which state governments are caught between growers who want aerial spraying to control the medfly and citizens who strongly protest having entire metropolitan areas sprayed with pesticides.

Unlike the common fruit fly, which is attracted only to very ripe fruit, the medfly lays its eggs on unripe fruits and vegetables in the field. The feeding maggots thus cause extensive damage before harvesting and during storage and transport of the fruit. Worldwide, this insect is one of the most destructive pests known. If it became established in the United States, it would present an enormous economic threat.

Medflies (probably brought in on imported produce) have invaded Florida, Texas, and California on several occasions, and whenever they have been detected, an intensive eradication campaign has been mounted. This diligence seems to have

been successful in Florida and Texas, but there are signs that the medfly may now be established in California, leading to increasingly frequent but controversial sprayings. The 1990 appearance of medflies in the Los Angeles basin prompted an aerial spraying of the pesticide malathion, a spraying bitterly opposed by residents of the area.

Malathion spraying represents the last resort in a battery of IPM techniques. First, all imported produce is checked and then fumigated if officials detect any trace of the insect. Second, a network of traps baited with a sex attractant is maintained throughout the agricultural area. Monitoring these traps provides an early warning of the presence of medflies that have slipped through customs. If medflies are found in the traps, the sterile-male technique is called on as the third line of defense. Stocks of sterile medflies are maintained in South America, and batches of them can be delivered on short notice and dropped over the infected area. In addition, fruit is stripped from the trees in the infected area to prevent the medfly from reproducing. Officials continue to use pheromone-baited

traps to monitor the program's success. The final line of defense is aerial spraying.

In June of 1997, residents of the Tampa Bay area of Florida were subjected to intense aerial spraying of malathion to eradicate an outbreak of medfly that was centered in residential areas. Warnings for people to stay indoors during the spraying were not always enough, and doctors in the area began to treat people with symptoms of malathion exposure: skin irritations, dizziness, nausea, and sweating. Although eradication was deemed successful, the EPA pressured the USDA to withdraw its 1998 application to use malathion and urged the state and the USDA to employ a more comprehensive IPM medfly program similar to what is being used in California.

Although malathion is not very toxic to humans, Tampa Bay residents feel that aerial spraying is an unacceptable alternative for controlling the medfly. They argue that the benefits to the growers are not outweighed by the potential harm to the environment and to humans. How would you feel about it if you lived in Tampa Bay?

Certification itself has been a haphazard and complicated process, and the industry has requested help from the government. In response, Congress passed the Organic Foods Protection Act in 1990, which established the National Organic Standards Board (NOSB), under USDA auspices. After some years of work, proposed guidelines from the NOSB were aired in early 1998; possibly in response to lobbying from the food industry, the board allowed genetically engineered foods, irradiated foods (subjected to radiation to kill bacteria), and crops fertilized with sewage sludge to be included as organic. A firestorm of protest from the organic food community ensued, and the USDA backed down. Revised guidelines were made available in 1999. With this episode as introduction, let us turn to public policy as it involves the use and regulation of pesticides.

17.5 Public Policy

In regulating the use of pesticides, three concerns are obvious: (1) pesticides must be evaluated for their intended uses and for their impacts on human health and the environment; (2) those who use the pesticides, especially agricultural workers,

must be appropriately trained and protected from the risks of close contact; (3) since most of the agricultural applications involve food, the public must be protected from the risks of pesticide residues on food products.

FIFRA

The **Federal Insecticide, Fungicide, and Rodenticide Act**, commonly known as FIFRA, addresses the first two of these concerns. This law, established by Congress in 1947 and amended in 1972, is administered by the EPA; the EPA is given total jurisdiction over pesticide manufacture, sale, use, and testing. FIFRA requires manufacturers to register pesticides with the EPA before marketing them. The registration procedure includes testing for toxicity to animals (and, by extrapolation, to humans). From the test results, usage standards are set. For example, chlordane, which was used for 40 years as a pesticide on crops, lawns, and gardens, was tested and shown to be acutely toxic to animals in high doses; chronic exposure led to damaging effects on most vital systems and caused liver cancer in animals. Acute and chronic exposure of humans to chlordane showed similar results, except for cancer development. Based on these results, the EPA cancelled all uses of chlordane for food crops in 1978 but allowed continued use against termites until 1988, when all uses in the United States were cancelled.

The law also stipulates that the EPA must determine whether a product "generally causes unreasonable adverse effects on the environment," and if it so determines, the product may be restricted or cancelled. This was the basis for the cancellation of DDT.

All pesticides must have an EPA-approved label, which lists its active ingredients, instructs on its proper use, and informs users of the risks involved with its use. The large number of existing pesticides and the continuing development of new products have posed a huge challenge to the EPA's ability to provide a thorough evaluation of each substance. Since the law allows existing pesticides to remain in use unless proved hazardous, many products on the market still have not been subjected to adequate tests, and existing standards may allow hazardous levels of pesticide residues in some foods. However, there are enough complaints from the pesticide industry about the EPA's "over-aggressive regulation" to judge that the agency is doing reasonably well in carrying out its congressional mandate.

In protecting workers from risks posed by occupational exposure to pesticides, the EPA works with state environmental agencies, which have the primary enforcement responsibility. The EPA has issued regulations under its Worker Protection Standard program involving requirements for safety training, use of protective equipment, provision for emergency assistance, and permitting standards. The states must then demonstrate to the EPA that, based on the federal regulations, they have adequate regulation and enforcement mechanisms.

FQPA of 1996

Protection of consumers from exposure to pesticides on food has been a contentious and confusing scene for many years. Three agencies are involved: the EPA, the Food and Drug Administration (FDA), and the Food Safety and Inspection Service of the USDA. Basically, the EPA sets the allowable tolerances for pesticide residues based on its assessment of toxicity; the FDA enforces the tolerances on all foods except meat, poultry, and egg products, which are monitored by the USDA. In prior years, the **Federal Food, Drug, and Cosmetic Act of 1958 (FFDCA)** was the enabling legislation assigning these responsibilities. One clause of the FFDCA, the so-called **Delaney clause**, was highly controversial. The clause states that "no (food) additive shall be deemed to be safe if it is found to induce cancer when ingested by man or animal." Since pesticides represent one of the largest categories of toxic chemicals to which people are exposed, this clause was applied in prohibiting many pesticides from being used on foodstuffs when those pesticides had been found to cause cancer in laboratory tests with animals. In essence, the law states that if a given pesticide presents any risk of cancer, no detectable residue may remain on the food. As analytical chemical techniques became more and more sensitive, extremely low traces of a pesticide could be measured, and according to the Delaney clause, if that pesticide caused cancer in experimental animals, farmers could not use it.

The National Research Council examined this dilemma and recommended in 1987 that the anti-cancer clause be replaced with a "negligible risk" standard, based on risk-analysis procedures that had been developed subsequent to the Delaney clause. After years of debate, Congress passed the **Food Quality Protection Act (FQPA)** in 1996 and did away with the Delaney clause. The following are the major requirements of the act:

- The new safety standard is "a reasonable certainty of no harm" for substances applied to foods.
- Special consideration must be given to the exposure of young children to pesticides residues.
- Pesticides or other chemicals are prohibited if they can be shown to carry a risk of more than one case of cancer per million people when consumed at average levels over the course of a lifetime.
- All possible sources of exposure to a given pesticide must be evaluated, not just from food.
- A special attempt must be made to assess the potential harmful effects of the so-called "hormone disrupters."
- The same standards are to be applied to raw and processed foods (previously dealt with separately).

The focus on children is the outcome of a 1993 report by the National Research Council, *Pesticides in the Diets of Infants and Children.* The focus is on children both because they typically consume more fruits and vegetables per unit of body weight than do adults and because studies have shown that the young are frequently more susceptible than adults to carcinogens and neurotoxins.

The EPA is still working out the regulations that the act authorizes. Industries are concerned that the FQPA will give the EPA a new opportunity to raise the bar on safety levels when risks are calculated, especially in light of the special consideration that must be given to young children. Proposed regulations will be published soon.

Pesticides in Developing Countries

The United States currently exports more than 200,000 metric tons of pesticides to developing countries each year. Some 25% of this total consists of products banned in the United States itself. FIFRA requires "informed permission" prior to the shipment of any pesticides banned in the United States. This permission comes from the purchaser, which can often be a foreign subsidiary of the exporting company. The EPA must then notify the government of the importing country.

Fortunately, the international community has erected a more effective system, through the cooperative work of two

U.N. agencies: the FAO and the United Nations Environment Program (UNEP). There is now a process of prior informed consent (PIC), whereby exporting countries inform all potential importing countries of actions they have taken to ban or restrict the use of pesticides or other toxic chemicals. Governments in the importing country respond to the notifications via the U.N. agencies, which then disseminate all the information they receive to the exporting countries and follow up by monitoring the export practices of the exporting countries. The PIC process was officially approved and signed by 80 government representatives in 1998 and will be in force when 50 countries have ratified it.

The more serious problem of unsafe pesticide use in the developing countries has also been addressed by the FAO. In spite of early opposition from the pesticide industry and exporting countries, an international "Code of Conduct" has been drawn up, whereby conditions of safe pesticide use are addressed in detail. The code makes clear the responsibilities of private companies and of countries receiving pesticides in promoting their safe use. As with PIC, the code is not yet legally binding, but it has proved very useful in holding private industry and importing countries to standards of safe use. In the view of some observers, both PIC and the FAO Code of Conduct are not as strong as they should be; however, it is acknowledged that their very existence represents great progress in addressing a difficult problem in the developing countries.

New Policy Needs

A group of leading entomologists called for a 50% reduction in pesticide use in the United States, supporting its recommendation with the economic and environmental benefits of such a reduction. This goal could become public policy if it is included in legislative or regulatory law. Shortly after coming into power in 1992, heads of the Clinton administration and the EPA, FDA, and USDA issued a joint announcement of their commitment to reduce pesticide use in the United States. According to the EPA, a policy of "maximum feasible reduction" will apply to all pesticide users, from farmers to homeowners. As Fig. 17–2 indicates, we have a long way to go to achieve a 50% reduction from the 1979 high.

Pesticide reform is in the wind. As with so many of the issues we consider in this book, significant progress in pesticide control has benefited from grassroots action and pressure from public interest groups and nongovernmental organizations. Such groups are pressing for a continued movement in the direction of ecological pest management and continued progress in keeping food free of pesticide residues. Also, with three governmental agencies going on record favoring a reduction in pestidice use, there is hope that we will soon substantially move toward helping farmers jump off the pesticide treadmill and toward aiding consumers to a pesticide-free food supply.

ENVIRONMENT ON THE WEB

PESTS: KILL 'EM OR EAT 'EM

Chapter 11 discusses the value of biodiversity in the natural environment. In food production, however, diversity of plant and animal species is far from welcome. Instead, species that are not intentionally cultivated are often actively removed through physical, chemical, or biological processes. Yet a pest in one situation may be highly valued in another. For example, the common dandelion is aggressively controlled in most urban lawns but is a valued foodstuff in some cuisines. Similarly, molds are unappealing in most foods but can also be a valuable source of pharmaceuticals, such as is the case with penicillin. Thus, the situation raises the question, What is a pest?

Indeed, the definition of "pest" will depend on the social and economic context of its occurrence—there are few species that are universally regarded as objectionable. For example, in many developing countries, where sources of protein such as beef are scarce for people of limited means, insects can provide an alternative protein source. "Mopani," or the emperor moth larvae, are one of many insects that are sold and eaten in South Africa; similarly, ants, worms, locusts, and cockroaches are eaten in other developing countries across Asia and Africa.

In the United States, urban dwellers differ significantly on their interpretation of pests. This can be seen in both the fact that insects are not food found in supermarkets, and the extreme lengths people go to get rid of pests. Many would wrinkle their nose at the very thought of eating an insect. For the most part, pests are considered to be troublesome insects or weeds that are annoying to look at or destructive to property, such as a front lawn.

In urban landscapes, the desired landscape calls for lush, unblemished expanses of lawn, dotted with trees and flower beds. To maintain such a standard, however, requires intensive maintenance through fertilizer and pesticide inputs, regular mowing, aeration, and so on. It has been estimated that Americans spend $6 billion a year on lawn care. Yet in trying to solve the pest problem, Americans may be creating other environmental concerns. Lawn maintenance uses almost 70 million pounds of pesticides and 4 to 5 million tons of fertilizer every year, generates high levels of noise and air pollution (from gas-powered lawn mowers and trimmers), and contaminates urban lakes and streams with nutrient-rich runoff and high levels of residual pesticides.

Like other personal choices, how people define pests is determined by their values and the social, cultural, and economic context of their lives. However, the definition they choose can mean consequences and trade-offs for the environment.

Web Explorations

The Environment on the Web activity for this essay describes the environmental issues surrounding lawn maintenance in the urban environment. Go to the Environment on the Web activity (select Chapter 17 at http://www.prenhall.com/nebel) and learn for yourself:

1. about insects as a source of food;
2. about the differences between traditional and organic gardening methods; and
3. about integrated pest management strategies.

A suggested time frame for each of these exercises is 10–30 minutes.

REVIEW QUESTIONS

1. Define pests. Why do we control them?
2. Discuss the basic philosophies of pest control.
3. What were the apparent virtues of the synthetic organic pesticide DDT?
4. What adverse environmental and human health effects can occur as a result of pesticide use?
5. Define bioaccumulation and biomagnification.
6. Why are nonpersistent pesticides not as environmentally sound as first thought?
7. Describe the four categories of natural or biological pest control. Cite examples of each and discuss their effectiveness.
8. Define the term economic threshold as it relates to pest control.
9. How does integrated pest management work? Give examples.
10. How do FIFRA and FFDCA attempt to control pesticides? What are the shortcomings of this type of legislation?
11. Discuss recent policy regarding the export of pesticides to developing countries.
12. What recent amendments to public policy are promoting pest control that is environmentally safer than before, and what needs still have to be addressed?

THINKING ENVIRONMENTALLY

1. U.S. companies export pesticides that have been banned or restricted in this country. Should this practice be allowed to continue? Support your answer.
2. Almost one-third of the chemical pesticides bought in the United States are for use in houses and on gardens and lawns. What should manufacturers and users do to ensure their limited and prudent use?
3. Should the government give farmers economic incentives to switch from pesticide use to integrated pest management? Why or why not?
4. Investigate how bugs and weeds are controlled on your campus. Are IPM techniques being used? Organic fertilizers? Consider lobbying for their use.
5. Read or reread the "Ethics" essay entitled "The Long War Against the Medfly" (p. 425) If you were a farmer, what methods would you advocate to control this pest? Why?

WEB REFERENCES

On-line resources for this chapter can be found on the World Wide Web at: **http://www.prenhall.com/nebel.** Click on Chapter 17 on the chapter selector.

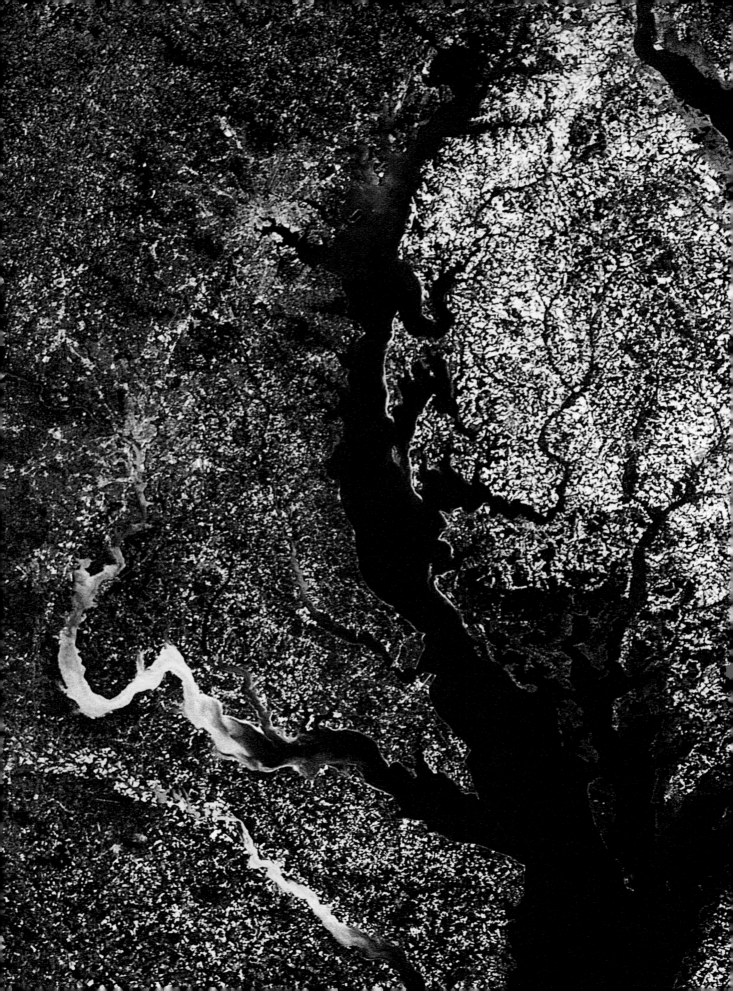

18

Water:
Pollution and Prevention

〜〜〜〜〜〜〜

Key Issues and Questions

1. Water pollution is the human-caused addition of any material in amounts that cause undesired alterations to the water. What are some point and nonpoint sources of water pollution?

2. Waterborne diseases, organic wastes, chemical pollutants, sediments, and nutrients are the most important types of water pollutants. How do these reach the water, and what are their basic impacts?

3. Eutrophication refers to the ecosystem changes that occur with nutrient enrichment. Describe these changes, including the depletion of dissolved oxygen, and why they occur.

4. Although it is possible to treat the symptoms of eutrophication, long-term control will depend on reducing inputs of nutrients and sediments. How can this best be accomplished?

5. Modern societies employ a system of sewers and wastewater treatment plants to deal with human and domestic wastes. Describe the technologies of primary and secondary treatment and biological nutrient removal.

6. Sewage sludge is a byproduct of cleaning the water. How can sludges be treated and converted into useful products?

7. The Clean Water Act of 1972 is the landmark legislation establishing public policy for water pollution. Describe both its accomplishments and problems in reauthorizing the act.

Chesapeake Bay is North America's largest estuary (opposite page). Before the 1970s, Chesapeake Bay was highly productive, yielding many millions of pounds of fish and shellfish and supporting vast flocks of waterfowl. Most of the food chains supporting this rich bounty had their origin in the seagrasses—0.5 million acres (200,000 ha) of underwater "grass" growing on the bottom, 3–6 feet (1–2 m) beneath the surface. The beds of seagrass provided food, spawning habitats, shelter for young fish and shellfish, and dissolved oxygen for them to breathe.

In the early 1970s, the seagrasses in all the tributaries leading into the bay started dying. By 1975, the die-off had spread. By 1980, the grasses were gone except in the main stem of the lower bay. Populations of the fish, shellfish, and waterfowl that had depended on the grasses soon declined also. Even more devastating, the bottom waters in deep areas of the bay became depleted of dissolved oxygen, causing huge numbers of fish

and shellfish to suffocate. What caused the die-off of seagrasses and the depletion of dissolved oxygen in the Chesapeake?

A team of scientists from the University of Maryland and the Virginia Institute of Marine Science, supported by grants from the EPA, investigated the problem. Toxic chemicals from industry were ruled out because, although they were a problem in certain locations, they could not have been responsible for a die-off throughout the bay. Herbicides used on farmlands were suspected, but tests showed that they did not reach damaging levels except in small ditches and streams receiving drainage directly from farm fields. Then investigations turned to the role of sunlight, and this proved to be the key. The waters of the Chesapeake had become increasingly **turbid** (murky or cloudy), and the turbidity was persisting over extended periods of time. It was cutting off the light required for photosynthesis, and as a result, the seagrasses were dying. What was causing the increased turbidity? It was **phytoplankton** (*phyto*, plant; *plankton*, free-floating), various forms of microscopic

Chesapeake Bay. This satellite photograph, taken of the bay by the *Earth Resources Satellite* at an altitude of 550 miles (870 km), shows the urban areas of Washington D.C. and Baltimore, MD, in the upper left gray area. The tan color of the two rivers in the lower left is a result of sediments eroding after a rain.

431

plants that grow and multiply freely suspended in the water. The growth of phytoplankton was stimulated by enrichment of the water with *nutrients*. Sediments (mainly clay particles) in suspension compounded the problem. As the phytoplankton died and sank to deeper water, they were decomposed by bacterial action. Bacterial decomposition was consuming dissolved oxygen, making it unavailable to fish and shellfish. Chesapeake Bay had fallen prey to pollution by excessive nutrients from domestic sewage and agricultural runoff in a process called **eutrophication** (literally, enrichment). We were learning that it is not possible to use vital water resources to receive pollutants and also expect them to continue to provide us with their usual bounty of goods and services.

In this chapter we will explore eutrophication and other forms of water pollution. The next four chapters will focus on pollution by solid and hazardous wastes, and air pollution.

18.1 Water Pollution

Before beginning, we need to clarify the definition of pollution. The EPA defines pollution as "the presence of a substance in the environment that because of its chemical composition or quantity prevents the functioning of natural processes and produces undesirable environmental and health effects." Any material that causes the pollution is called a **pollutant**.

Pollution Essentials

Pollution is rarely the result of wanton mistreatment of the environment; the additions that cause pollution are almost always the *byproducts* of otherwise worthy and essential activities—producing crops, creating comfortable homes, providing energy and transportation, manufacturing products—and of our basic biological functions (excreting wastes). Pollution problems have become more pressing over the years because both growing population and expanding per capita use of materials and energy increase the amounts of byproducts that go into the environment. Also, many materials now widely used, such as aluminum cans, plastic packaging, and synthetic organic chemicals, are **nonbiodegradable**. That is, they resist attack and breakdown by detritus feeders and decomposers and consequently accumulate in the environment. An overview of the categories of pollutants that result from various activities is shown in Fig. 18–1.

It is important to note the breadth and diversity of pollution. Any part of the environment may be affected, and virtually anything may be a pollutant. The only criterion is that the addition of a pollutant results in undesirable alterations. The impact of the undesirable alteration may be largely aesthetic—hazy air obscuring a distant view or the unsightliness of roadside litter, for instance. The impact may be on ecosystems as a whole—the die-off of fish or forests, for example. Or the impact may be on human health—toxic wastes contaminating water supplies. The impact also may range from very local—the contamination of an individual well, for instance—to global. We tend to think of pollution as the introduction of human-made materials into the environment. But undesirable alterations may be caused by introductions of too much of otherwise natural compounds; fertilizer nutrients introduced into waterways or carbon dioxide introduced into the atmosphere (Chapter 21) are examples.

Therefore, the slogan "Don't pollute" is a gross oversimplification. The very nature of our existence necessitates the production of wastes. Our job in remediating present and future pollution problems is parallel to the concept of sustainable development itself. It is to adapt the means of meeting our present needs so that byproducts are managed in ways that will not cause alterations that will jeopardize present and future generations. The general strategy in each case must be to:

1. identify the material or materials that are causing the pollution—the undesirable alteration;
2. identify the sources of those pollutants;
3. develop and implement pollution control strategies to prevent those pollutants from entering the environment;
4. develop and implement alternative means of meeting the need that do not produce the polluting byproduct—pollution avoidance.

Basically, the strategy for addressing pollution may be seen as finally recognizing the second principle of sustainability: *For sustainability, ecosystems dispose of wastes and replenish nutrients by recycling all elements*, thereby avoiding both pollution and resource depletion.

Water Pollution: Sources, Types

Water pollutants originate in a host of human activities; the pollutants reach surface water or groundwater through an equally diverse host of pathways. For purposes of regulation, it is customary to distinguish between **point sources** and **nonpoint sources** of pollutants (Fig. 18–2). Point sources involve the discharge of substances from factories, sewage systems, power plants, underground coal mines, and oil wells. These sources are relatively easy to identify and therefore are easier to monitor and regulate. Nonpoint sources, on the other hand, are poorly defined and scattered over broad areas. Pollution occurs as rainfall and snowmelt move over and through the ground, picking up pollutants as they go. Some of the most prominent nonpoint sources are agricultural runoff (from farm animals and croplands), stormwater drainage (from streets, parking lots, and lawns), and atmospheric deposition (from air pollutants washed to earth or deposited as dry particles). Because they are so diffuse, these sources are far more difficult to regulate.

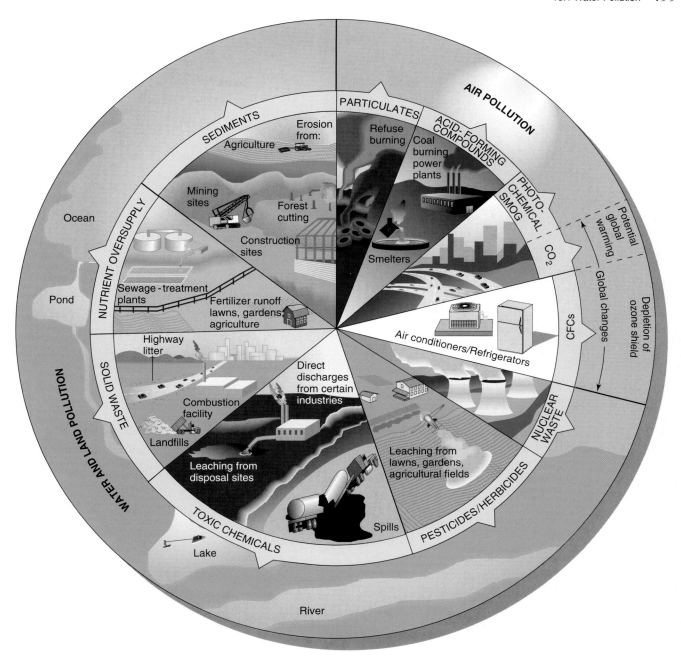

▲ **FIGURE 18–1** *Categories of pollution.* Pollution is an outcome of otherwise worthy human endeavors. Major categories of pollution and activities that cause them are shown here.

Two basic strategies are employed in attempting to bring water pollution under control: (1) reduce the sources, and (2) treat the water so as to remove pollutants or convert them to harmless forms. Water treatment is the best option for point sources; source reduction can be employed for both kinds of sources and is the best option available for the nonpoint sources.

Pathogens. The most serious water pollutants are the infectious agents that cause sickness and death (see Chapter 16). The excrement from humans and other animals infected with certain **pathogens** (disease-causing bacteria, viruses,

and other parasitic organisms) contains large numbers of these organisms or their eggs (see Table 18-1). Even after symptoms of disease disappear, an infected person or animal may still harbor low populations of the pathogen, thus continuing to act as a carrier of disease. If wastes from carriers contaminate drinking water, food, or water used for swimming or bathing, the pathogens can gain access to and infect other individuals (Fig. 18–3).

Before the connection between disease and sewage-carried pathogens was recognized in the mid-1800s, disastrous epidemics were common in cities. For example, epidemics of

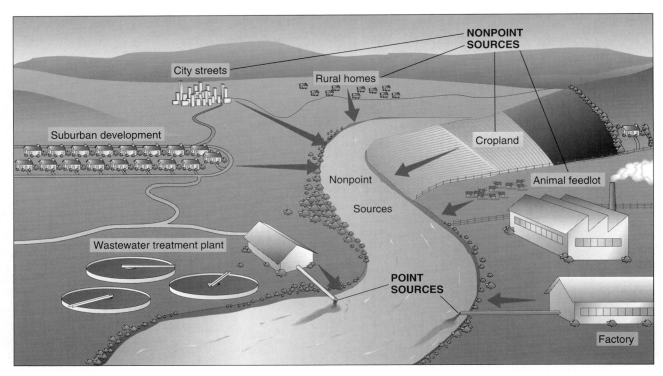

▲ FIGURE 18–2 *Point and nonpoint sources.* Point sources are far easier to identify and correct than the diffuse nonpoint sources.

typhoid fever and cholera, which killed thousands of people, were common in cities before the twentieth century. Today, public-health measures that prevent this disease cycle have been adopted throughout the developed world and to a considerable extent in the developing world. These measures involve:

- purification and disinfection of public water supplies with chlorine or other agents (Chapter 9);

- sanitary collection and treatment of sewage wastes; and
- maintenance of sanitary standards in all facilities where food is processed or prepared for public consumption.

Standards regarding the items above are set and enforced by government public health departments. A variety of other measures are enforced as well. For example, if oyster beds are contaminated with raw sewage, health departments close

▶ FIGURE 18–3 *The Ganges River in India.* In many places in the developing world the same waterways are used simultaneously for drinking, washing, and disposal of sewage. In the case of the Ganges River, religious practice and spiritual cleansing are also common. A high incidence of disease, infant and childhood mortality, and parasites is the result.

Table 18-1 Pathogens Carried by Sewage

Disease	Infectious Agent
Typhoid fever	*Salmonella typhi* (bacterium)
Cholera	*Vibrio cholerae* (bacterium)
Salmonellosis	*Salmonella* species (bacteria)
Diarrhea	*Escherichia coli,* *Campylobacter* species (bacteria)
Infectious hepatitis	Hepatitis A virus
Poliomyelitis	Poliovirus
Dysentery	*Shigella* species (bacteria) *Entamoeba histolytica* (protozoan)
Giardiasis	*Giardia intestinalis* (protozoan)
Numerous parasitic diseases	(Roundworms, flatworms)

them to harvesting. Implicit in all measures is monitoring for sewage contamination (see the "Earth Watch" essay, p. 438). Of course, our own personal hygiene, sanitation, and health precautions, such as not drinking water from untested sources and making sure foods like pork and chicken are always well cooked, remain the last and most important line of defense against disease.

Many people attribute good health in a population to modern medicine, but good health is more a result of disease prevention through public-health measures. An estimated 1.4 billion people do not have access to treated drinking water. Some 2.9 billion people live in areas having poor (or no) sewage collection or treatment. Largely because of poor sanitation regarding water and sewage, a significant portion of the world's population is chronically infected with various pathogens. More than 250 million new cases of waterborne disease are reported each year, about 10 million of them resulting in death, and about half of those deaths are among children under five. Moreover, populations in areas where there is little or no sewage treatment are extremely vulnerable to deadly epidemics of any and all diseases spread via the sewage vector. Because of unsanitary conditions, in 1990 an outbreak of cholera in Peru killed several thousand people as it spread through Latin America. Another outbreak of cholera occurred in India, in 1994.

Organic Wastes. Along with pathogens, human and animal wastes contain organic matter, which if it enters water bodies untreated, creates serious problems. Other kinds of organic matter (leaves, grass clippings, trash, etc.) can enter

water bodies as a consequence of runoff and, in the case of excessive phytoplankton, can grow within the water body. With the exception of plastics and some man-made chemicals, these wastes are biodegradable. As bacteria and detritus feeders decompose organic matter in water, they consume oxygen, which is dissolved as a gas in water. The amount of oxygen that water can hold in solution is severely limited. In cold water, dissolved oxygen (DO) can reach concentrations up to 10 ppm (parts per million); even less can be held in warm water. Compare this with the oxygen in air, which is 200,000 ppm (20%), and you can understand why even a moderate amount of organic matter decomposition in water can deplete the water of its DO. Bacteria keep the water depleted in DO as long as there is dead organic matter to support their growth.

Thus, a common water-quality test is **biochemical oxygen demand (BOD)**, a measure of the amount of organic material, in terms of how much oxygen will be required to break it down biologically or chemically or both. The higher the BOD measure, the greater the likelihood that dissolved oxygen will be depleted in the course of breaking it down. A high BOD causes so much oxygen depletion that animal life is severely limited or precluded, as in the bottom waters of Chesapeake Bay. Fish and shellfish are killed when the DO drops below 2 or 3 ppm; some are less tolerant at even higher DO levels. If the system goes anaerobic (no oxygen), only bacteria can survive, using their abilities to switch to fermentation or anaerobic respiration (metabolic pathways that do not require oxygen). A typical BOD value for raw sewage would be around 250 ppm. Even a moderate amount of sewage, added to natural waters containing at most 10 ppm DO, can deplete the water of its oxygen and cause highly undesirable consequences well beyond the introduction of pathogens.

Chemical Pollutants. Because water is such an excellent solvent, it is able to hold many chemical substances in solution that have undesirable effects on water bodies. Water-soluble **inorganic chemicals** constitute an important class of pollutants, including heavy metals (lead, mercury, cadmium, nickel, etc.), acids from mine drainage (sulfuric acid) and acid precipitation (sulfuric and nitric acids), and road salts employed to melt snow and ice in the colder climates (sodium and calcium chlorides). The **organic chemicals** are another group of substances that are found in polluted waters. Petroleum products pollute many water bodies, from the major oil spills in the ocean (Chapter 13) to small streams receiving runoff from parking lots. Other organic substances with serious impacts are the pesticides that drift down from aerial spraying or that run off from land areas (Chapter 17), and the various industrial chemicals, like polychlorinated biphenyls (PCBs), cleaning solvents, and detergents.

Many of these pollutants are toxic, even at low concentrations (Chapter 20). Some may be concentrated through passage up the food chain in the process called biomagnification (Chapter 17). Even at very low concentrations, they can render water unpalatable for humans and dangerous for

▲ **FIGURE 18–4** *Acid mine drainage.* Acidic water from mines contains high concentrations of sulfides and iron, which are oxidized when they come into the open, thus producing unsightly and sterile streams.

aquatic life. At higher concentrations, they can change the properties of the water bodies so as to prevent them from serving any useful purpose except perhaps for navigation. Acid mine drainage, for example, pollutes thousands of miles of streams in coal-bearing regions of the United States (Fig. 18–4). Stream bottoms are coated with orange deposits of iron, and the water is so acid only a few hardy bacteria and algae can tolerate it.

Sediments. As natural landforms weather, a certain amount of sediment enters streams and rivers. However, erosion from farmlands, deforested slopes, overgrazed rangelands, construction sites, mining sites, stream banks, and roads can greatly increase the load of sediment entering waterways. Sediments (sand, silt, and clay) have direct and extreme physical impacts on streams and rivers. When erosion is slight, streams and rivers of the watershed run clear. They support algae and other aquatic plants that attach to rocks or take root in the bottom. These producers, plus miscellaneous detritus from fallen leaves and so on, support a complex food web of bacteria, protozoa, worms, insect larvae, snails, fish, crayfish, and other organisms. These organisms keep themselves from being carried downstream by attaching to rocks or seeking shelter behind or under rocks. Even fish that maintain their position by active swimming occasionally need such shelter to rest (Fig. 18–5a).

Sediment entering waterways in large amounts has an array of impacts. Sand, silt, clay, and organic particles (humus) are quickly separated by the agitation of flowing water and are carried at different rates. Clay and humus are carried in suspension, making the water muddy and reducing light penetration and photosynthesis. As this material settles, it coats everything and continues to block photosynthesis. It also kills the animal organisms by clogging their gills and feeding structures. Eggs of fish and other aquatic organisms are particularly vulnerable to being smothered by sediment.

Equally destructive is the **bedload**, the sand and silt, which is not readily carried in suspension but is gradually washed along the bottom. As particles roll and tumble along, they scour organisms from the rocks. They also bury and smother the bottom life and fill in the hiding and resting places of fish and crayfish. Aquatic plants and other organisms are prevented from reestablishing themselves because the bottom is a constantly shifting bed of sand (Fig. 18–5b and c).

Sediments do not receive the attention the news media give to hazardous wastes and certain other pollution problems, but erosion is so widespread throughout the world that few streams and rivers escape the harsh impact of excessive sediment loads. The U.S. Natural Resources Conservation Agency estimates that the damage from sediments costs the United States over $6 billion each year. Finally, sediments increase eutrophication. Particles of clay and humus in suspension contribute to the turbidity that blocks the photosynthesis of aquatic plants. The same particles invariably carry nutrients that contribute to the growth of phytoplankton.

Nutrients. Some of the inorganic chemicals carried in solution in all bodies of water are classified as nutrients—essential elements required by plants. The two most important nutrients for aquatic plant growth are phosphorus and nitrogen, and they are often in such low supply in water that they are the limiting factors for phytoplankton or other aquatic plants, like the grasses of Chesapeake Bay. More nutrients mean more plant growth; therefore, nutrients become water pollutants when they are added from point or nonpoint sources and stimulate undesirable plant growth in bodies of water. The most obvious point sources of excessive nutrients are sewage outfalls. As we will see later, it takes a special effort to remove nutrients from sewage during the treatment process, and this is not always done. Of course, if the sewage is not treated at all, high levels of nutrients will enter the water; this is the situation in much of the developing world.

Agricultural runoff is the most notorious nonpoint source of nutrients. Nutrients in many forms are applied to agricultural crops: chemical fertilizers, manure, irrigation water, sludge, and crop residues. When they are applied in excess of plant needs, or when runoff from agricultural fields is heavy, the nutrients are picked up by water that eventually enters streams, rivers, and lakes. Other nonpoint sources include lawns and gardens, golf courses, and storm drains.

The net effect of excessive nutrient pollution to bodies of water is eutrophication.

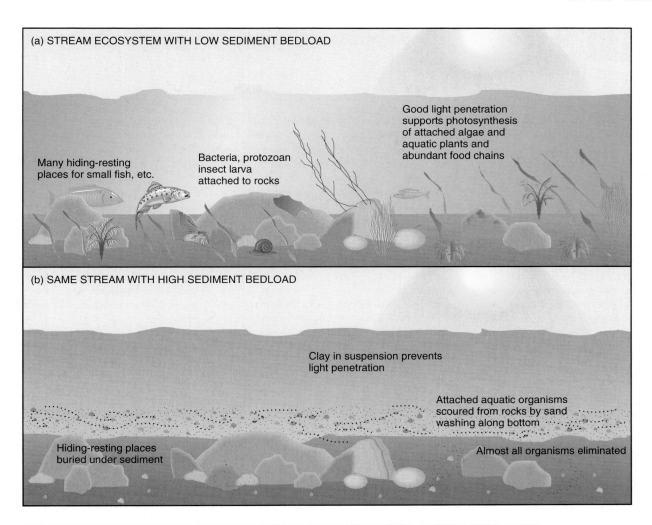

(a) STREAM ECOSYSTEM WITH LOW SEDIMENT BEDLOAD

Good light penetration supports photosynthesis of attached algae and aquatic plants and abundant food chains

Many hiding-resting places for small fish, etc.

Bacteria, protozoan insect larva attached to rocks

(b) SAME STREAM WITH HIGH SEDIMENT BEDLOAD

Clay in suspension prevents light penetration

Attached aquatic organisms scoured from rocks by sand washing along bottom

Hiding-resting places buried under sediment

Almost all organisms eliminated

(c)

◀ **FIGURE 18–5**
Sediment impact on streams and rivers. (a) The ecosystem of a stream that is not subjected to a large sediment bedload. (b) The changes that occur when there are large sediment inputs. (c) Platte River at Lexington, Nebraska. The sandbars seen here constitute the bedload; they shift and move with high water, preventing reestablishment of aquatic vegetation.

EARTH WATCH

MONITORING FOR SEWAGE CONTAMINATION

An important aspect of public health is the detection of sewage in water supplies and other bodies of water having contact with humans. It is worth understanding how this detection is done.

It is virtually impossible to test for each specific pathogen that might be present. Therefore, an indirect method called the **fecal coliform test** has been developed. This test is based on the fact that huge populations of a bacterium called *E. coli (Escherichia coli)* normally inhabit the lower intestinal tract of humans and other animals, and large numbers of the bacterium are excreted with fecal material. In temperate regions at least, *E. coli* does not last long in the outside environment. Therefore, the presumption is that when *E. coli* is found in natural waters, it indicates recent and probably persisting contamination with sewage wastes. In most situations, *E. coli* is not a pathogen itself but is referred to as an **indicator organism**. Its presence indicates that water is contaminated with fecal wastes and that sewage-borne pathogens may be present. Conversely, the absence of *E. coli* is taken to mean that water is free from such pathogens.

The fecal coliform test, one technique of which is shown in the figures here, detects and counts the number of coliform bacteria in a sample of water. Thus, the results indicate the relative degree of contamination and the relative risk of pathogens. For example, to be safe for drinking, water should have an average *E. coli* count of no more than one *E. coli* per 100 milliliters (about 0.4 cups) of water. Water with as many as 200 *E. coli* per 100 mL is still considered safe for swimming. Beyond that level, a river may be posted as polluted, and swimming and other direct contact should be avoided. By comparison, raw sewage (99.9% water, 0.1% waste) has *E. coli* counts in the millions.

   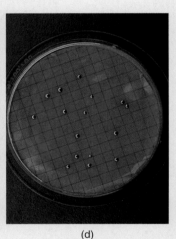

(a)   (b)   (c)   (d)

Testing water for sewage contamination—the Millipore technique. (a) A Millipore filter disk is placed in the filter apparatus. (b) A sample of the water being tested is drawn through the filter, and any bacteria present are entrapped on the filter disk. (c) The filter disk is then placed in a petri dish on a special medium that supports the growth of bacteria and will impart a particular color to fecal *E. coli* bacteria. The dish is then incubated for 24 hours at 38ºC, during which time each bacterium on the disk will multiply to form a colony visible to the naked eye. (d) *Escherichia coli* bacteria, indicating sewage contamination, are identifiable as the colonies with a metallic green sheen.

18.2   Eutrophication

Recall from Chapter 2 that trophic refers to feeding. Literally, eutrophic means well-nourished. As we will see, there are some very undesirable consequences of high levels of "nourishment" in water bodies. Although eutrophication can be an entirely natural process, the introduction of pollutants to water bodies has greatly increased the scope and speed of eutrophication. Let us investigate the causes of this problem as well as the means of preventing and eventually reversing it.

Different Kinds of Aquatic Plants

To understand eutrophication, we need to consider distinct kinds of aquatic plants—namely, *benthic* plants and *phytoplankton*.

Benthic plants (from *benthos*, deep) are aquatic plants that grow attached to or are rooted in the bottom. All common aquarium plants and seagrasses are examples (Fig. 18–6a). As shown in the figure, benthic plants may be divided into two categories: **submerged aquatic vegetation (SAV)**, which generally grows totally under water, and **emergent vegetation**, which grows with the lower parts in water but the upper parts emerging from the water.

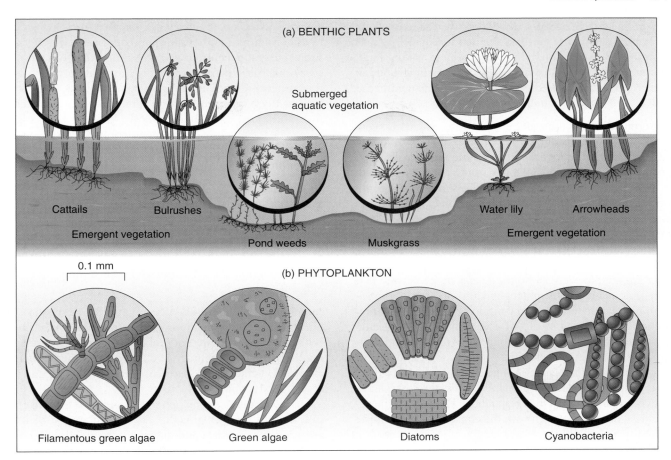

▲ FIGURE 18–6 *Aquatic photosynthesizers.* (a) Benthic, or bottom-rooted, plants. These are subdivided into submerged aquatic vegetation (SAV) and emergent vegetation. (b) Phytoplankton, various photosynthetic organisms that are either single cells, or small groups or filaments of cells, that float freely in the water.

The most important point for understanding eutrophication is this: SAVs require water that is clear enough to allow penetration of sufficient light to support their photosynthesis. As water becomes more turbid, light is reduced; in extreme situations, it may be reduced to a matter of a few centimeters. Thus, increasing turbidity decreases the depth at which SAVs can survive.

A second important feature of SAVs is that they absorb their required mineral nutrients from the bottom sediments through their roots just as land plants do. They are not handicapped by water that is low in nutrients. Indeed, enrichment of the water with nutrients is counterproductive for the SAVs because it stimulates the growth of phytoplankton.

Phytoplankton consist of numerous species of algae and chlorophyll-containing bacteria (cyanobacteria, formerly referred to as blue-green algae) that grow as microscopic single cells or small groups, or "threads," of cells. Phytoplankton live suspended in the water and are found wherever light and nutrients are available (Fig. 18–6b). In extreme situations, water may become literally pea-soup green (or tea-colored, depending on the species involved), and a scum of phytoplankton may float on the surface and absorb essentially all the light. However, phytoplankton reach such densities only in nutrient-rich water because, not being connected to the bottom, they must absorb their nutrients from the water. A low level of nutrients in the water limits the growth of phytoplankton accordingly.

Considering the requirements of phytoplankton and SAVs, you can see how the balance between them is altered when nutrient levels in the water are changed. As long as water remains low in nutrients, populations of phytoplankton are suppressed; without their presence, the water is clear, and light may penetrate to support the growth of SAVs. As nutrient levels increase, phytoplankton can grow more prolifically, making the water turbid and shading out the SAVs.

The Impacts of Nutrient Enrichment

A lake with deep light penetration, one in which the bottom is visible beyond the immediate shoreline, is **oligotrophic** (low in nutrients). Such a lake is fed by a watershed that holds its nutrients well. A forested watershed, for example, holds nitrogen and phosphorus tightly and allows very little of it to enter the water draining into the streams and rivers. If the streams feed into a lake, it will reflect their low nutrient content.

In an oligotrophic lake, the low nutrient levels limit the growth of phytoplankton and allow light penetration that supports the growth of SAVs, which draw their nutrients from the bottom sediments. In turn, the benthic plants support the rest of a diverse aquatic ecosystem by providing food, habitat, and dissolved oxygen. The oligotrophic body of water is prized for its aesthetic and recreational qualities as well as for its production of fish and shellfish.

Eutrophication. As the water of an oligotrophic body is enriched with nutrients, numerous changes are set in motion. First, the nutrient enrichment allows the rapid growth and multiplication of phytoplankton, causing increasing turbidity of the water. The increasing turbidity shades out the submerged aquatic vegetation. With the die-off of SAV, there is an obvious loss of food, habitat, and dissolved oxygen from their photosynthesis.

Phytoplankton have remarkably high growth and reproduction rates. Under optimal conditions, phytoplankton biomass can double every 24 hours, a capacity far beyond that of benthic plants. Thus, phytoplankton soon reach a maximum population density, and continuing growth and reproduction are balanced by die-off. Dead phytoplankton settle out, resulting in heavy deposits of detritus on the bottom. In turn, the abundance of detritus supports an abundance of decomposers, mainly bacteria. The explosive growth of bacteria creates an additional demand for dissolved oxygen as they consume oxygen in their respiration. The result is the depletion of dissolved oxygen with the consequent suffocation of the fish and shellfish.

In sum, **eutrophication** refers to this whole sequence of events—starting with nutrient enrichment, growth and die-off of phytoplankton, accumulation of detritus, growth of bacteria, and finally to the depletion of dissolved oxygen and suffocation of higher organisms (Fig. 18–7). From the human perspective, the eutrophic system is less than appealing for swimming, boating, and sport fishing. Also, if the lake is a source of drinking water, its value may be greatly impaired because phytoplankton rapidly clog water filters and may cause a foul taste. Also, some species of phytoplankton secrete various toxins into the water that may kill other aquatic life and be injurious to human health as well (see the "Earth Watch" essay, p. 442).

Eutrophication of Shallow Lakes and Ponds. In lakes and ponds where the water depth is 6 feet (2 m) or less, eutrophication takes a somewhat different course. Submerged aquatic vegetation may grow to a height of a meter or more, thus reaching the surface. Therefore, with nutrient enrichment, the SAV is not shaded out, but grows abundantly, sprawling over and often totally covering the water surface with dense mats of vegetation that make boating, fishing, or swimming impossible (Fig. 18–8, see p. 443). Any vegetation beneath these mats is shaded out. As the mats of vegetation die and sink to the bottom, they create a BOD that often depletes the water of dissolved oxygen, causing the death of aquatic organisms other than bacteria.

Natural vs. Cultural Eutrophication. In nature, apart from human impacts, eutrophication is part of the process of aquatic succession, discussed in Chapter 4 (p. 97). Over periods of hundreds or thousands of years, bodies of water are subject to gradual enrichment with nutrients. Thus, we can speak of *natural eutrophication* as a normal process. Since the nutrients come from sewage-treatment plants, poor farming practices, urban runoff, and certain other human activities, humans have inadvertently managed to vastly accelerate the process of nutrient enrichment. We refer to the accelerated eutrophication caused by humans as **cultural eutrophication**.

Combating Eutrophication

There are two approaches to combating the problem of eutrophication. One is to *attack the symptoms*—the growth of vegetation or the lack of dissolved oxygen or both. The other is to *get at the root cause*—excessive inputs of nutrients and sediments.

Attacking the Symptoms. Attacking the symptoms has application in certain situations where immediate remediation is the goal and costs are not prohibitive. Methods of attacking the symptoms include (1) chemical treatments (herbicides), (2) aeration, (3) harvesting aquatic weeds, and (4) drawing water down.

Herbicides are often applied to ponds and lakes to control nuisance plant growth. To control phytoplankton growth, copper sulfate and diquat are frequently used. For controlling SAVs and emergent vegetation, fluridone, glyphosate, and 2,4-D are employed. However, these compounds are toxic to fish and aquatic animals, sometimes at concentrations required to bring the vegetation under control. Fish kills often occur after herbicide applications because the rotting vegetation depletes the water of dissolved oxygen. There are required waiting periods after application of an herbicide before using the water for swimming, irrigation, or fishing. The use of herbicides in this connection is a good example of the pesticide treadmill (Chapter 17), where repeated applications are required indefinitely. Herbicides only provide cosmetic treatment, and as soon as they wear off, the vegetation grows back rapidly.

Depletion of dissolved oxygen by decomposers and consequent suffocation of other aquatic life is the final and most destructive stage of eutrophication. Therefore, it follows that *artificial aeration* of the water can avert this terminal stage. An aeration system currently gaining popularity is to lay a network of plastic tubes with microscopic pores on the bottom of the waterway to be treated. High-pressure air pumps force microbubbles from the pores that dissolve directly into the water. The system is proving effective in speeding up the breakdown of accumulated detritus, improving water quality, and enabling the return of more desirable aquatic life. Despite its high cost, the system has applicability in harbors, marinas, and in some water-supply reservoirs, where the demand for better water quality justifies the cost.

In shallow lakes or ponds where the problem is bottom-rooted vegetation reaching and sprawling over the surface

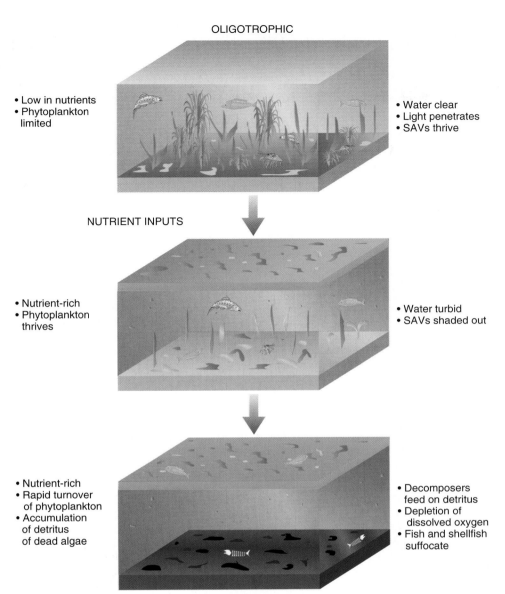

OLIGOTROPHIC

- Low in nutrients
- Phytoplankton limited

- Water clear
- Light penetrates
- SAVs thrive

NUTRIENT INPUTS

- Nutrient-rich
- Phytoplankton thrives

- Water turbid
- SAVs shaded out

- Nutrient-rich
- Rapid turnover of phytoplankton
- Accumulation of detritus of dead algae

- Decomposers feed on detritus
- Depletion of dissolved oxygen
- Fish and shellfish suffocate

EUTROPHIC

◀ **FIGURE 18–7** *Eutrophication.*
As nutrients are added from pollution sources, an oligotrophic system rapidly becomes eutrophic and undesirable.

(see Fig. 18–8), *harvesting the aquatic weeds* may be an expedient way to improve recreational potential and aesthetics. Commercial mechanical harvesters are used, and nearby residents also have gotten together to remove the vegetation by hand. The harvested vegetation makes good organic fertilizer and mulch. But even harvesting has a limited effect. The vegetation soon grows back because roots are left in the nutrient-rich sediments.

This method also does not take removing phytoplankton into consideration. To remove the microscopic cells would require filtering them from the water. Mechanically "harvesting" phytoplankton is not at all practical. The volume of water to be filtered in even a modestly sized pond is overwhelming, and the effort is further frustrated by the fact that plankton cells rapidly clog the filter, precluding the passage of water.

Because many recreational lakes are dammed, another option for shallow-water weed control is to draw the lake down for a period of time each year. This process kills most of the rooted aquatic plants along the shore; however, they will grow back in time.

All of these approaches have in common the fact that they are only temporary fixes for the problem and will have to be repeated often, at significant cost, to keep the unwanted plant growth under control.

Getting at the Root Causes. Controlling eutrophication means long-term strategies for correcting the problem, which ultimately involves reducing the inputs of nutrients and sediments—two of the types of pollution discussed at the beginning of this chapter. The first step is to identify the major point and nonpoint sources of nutrients and sediments. Then

EARTH WATCH

THE ALGAE FROM HELL

Surprises dot the environmental landscape of the last 50 years, but one of the most bizarre and unexpected of surprises emerged when in 1988 a North Carolina State University plankton biologist, Dr. Joanne Burkholder, examined some open sores that were found on fish. She identified a new organism, a microscopic dinoflagellate (a type of algae) she named *Pfiesteria piscicida*. Subsequent work in laboratory cultures of *Pfiesteria* revealed that the species has no less than 24 different life forms, unprecedented in the annals of science. Sometimes it is an amoeba, sometimes a flagellate, sometimes it is in a cyst; sometimes it secretes a toxin that stuns fish, then turns into a form that proceeds to feed on the fish, which have been disoriented by the toxin. Sometimes it behaves like a decent, ordinary alga and lives by photosynthesis. *Pfiesteria* has killed billions of fish in Albemarle and Pamlico Sounds, North Carolina, and has been identified in waters from Virginia, Maryland, Delaware, and Florida. In the summer of 1997, however,

Pfiesteria struck several tributaries of Chesapeake Bay, particularly the Pocomoke River; fish kills and catches of fish with ulcerous lesions finally led the state to close the river to all fish and shellfish taking.

Pfiesteria is not just toxic to fish and shellfish; it has now been implicated in skin rashes and neurological disorders in people who have encountered water containing the organism. The problems first appeared in Burkholder's laboratory; later, watermen and others in close contact with infested waters began reporting similar symptoms: short-term memory loss, respiratory problems, fatigue, disorientation, blurred vision. When news of the fish kills in Chesapeake Bay and the impacts on humans became public, there was a predictable panic response, and consumers all around the Chesapeake stopped eating seafood. In Maryland alone, the economic loss was estimated at between $15 and $20 million in sales during 1997.

The recent increase in *Pfiesteria* episodes joins a long list of what are called Harmful Algal Blooms (HABs). Red tides, consisting of concentrations of different

species of toxic dinoflagellates, have been occurring more and more frequently in coastal waters all around North America. Many scientists believe that nutrients—especially nitrogen and phosphorus—are the basic culprits in feeding the growth of the algae in estuaries and coastal waters. In the Delaware-Maryland-Virginia peninsula, some 600 million chickens are produced each year, and runoff from the poultry farms is believed to be a major source of nutrients to Chesapeake tributaries. North Carolina State University has just opened a new *Pfiesteria* research facility to provide Dr. Burkholder with the capability to analyze samples from suspected infestations and to continue her work with the basic biology of *Pfiesteria*. NOAA (National Oceanic and Atmospheric Administration) recently announced a $15 million research effort to study the HABs occurring all around the continent. There is little doubt that the blooms are increasing in frequency, and there is a good possibility that the explanation will be found in the pollutants that are still entering our waterways.

---

it is a matter of developing and implementing strategies for correction. Which source or factor is most significant will depend on the human population and the land uses within the particular watershed. Therefore, each watershed must be analyzed as a separate entity, and appropriate measures must be taken to reduce the nutrients and sediments exiting from that watershed. In the following sections we shall discuss major control strategies that are being implemented.

First, recall the concept of limiting factors (Chapter 2); only one nutrient need be lacking to suppress growth. In natural freshwater systems, **phosphorus** is most commonly limiting. In marine systems, the limiting nutrient is most commonly **nitrogen**. Both in the environment and in biological systems, phosphorus (P) is present as phosphate (PO_4^{-3}), and nitrogen is present in a variety of compounds, most commonly nitrate (NO_3^-) or ammonium (NH^{+4}). Therefore, we will focus primarily on phosphate and nitrogen compounds.

Controlling Strategies for Point Sources. In heavily populated areas, discharges from sewage-treatment plants are recognized as major sources of nutrients entering waterways. The levels of phosphate in effluents from sewage-treatment plants are elevated even more in areas where

laundry detergents containing phosphate are used. Phosphate contained in detergents for cleaning purposes goes through the system and out with the discharge. In regions where eutrophication has been recognized as a problem, a key step toward prevention has been to ban the sale of phosphate detergents or at least to regulate the maximum allowable level of phosphates. Total or partial phosphate bans are now in effect in 20 states and in the District of Columbia (Table 18-2). Major detergent manufacturers are shifting to general production of non-phosphate formulations; however, the consumer should read the product's active ingredients. Also, the bans do not cover dishwashing detergents, and some brands are high in phosphate. You can check the labels on these also to find a brand with little or no phosphate.

In addition, there is an ongoing program aimed at upgrading sewage-treatment plants to remove nutrients from the effluents or to handle the effluents in alternative ways to avoid having the nutrients go into waterways. These processes and alternatives will be considered shortly.

Phosphate-detergent bans and upgrading sewage treatment have brought about marked improvements in waterways that were heavily damaged by effluents from sewage-treatment plants.

▲ FIGURE 18–8 *Eutrophication in shallow lakes and ponds.* In shallow water, sufficient light for photosynthesis continues to reach the submerged aquatic vegetation. Hence, oversupply of nutrients stimulates growth so that vegetation reaches the surface and forms mats.

Table 18-2 Progress Toward Banning Phosphate from Detergents	
States that ban phosphate-containing laundry detergents	
Delaware	North Carolina
Georgia	Oregon
Indiana	Pennsylvania
Maine	Vermont
Maryland	Virginia
Michigan	Washington, D.C.
Minnesota	Wisconsin
New York	

States that restrict the level of phosphate in laundry detergents or ban phosphate detergents in some areas

Connecticut	Illinois	Ohio
Florida	Montana	

In a sense, however, these are the easy measures in that the target for correction, the sewage-treatment-plant effluent, is an obvious *point source*, and methods for correction are clear-cut.

Controlling Strategies for Nonpoint Sources. Correction becomes more difficult, but not less important, when the source is diffuse, as in the case of farm and urban runoff. Remediation of such *nonpoint sources* will involve thousands, even millions, of individual property owners' adopting new practices regarding management and use of fertilizer and other chemicals on their properties. Nevertheless, that is the challenge. When the Clean Water Act was reauthorized in 1987, a new section (section 319) was added, requiring states to develop management programs to address nonpoint sources of pollution. Thus, there is a legal mandate to address the issues of agricultural and urban runoff.

Where a significant portion of the watershed is devoted to agricultural activities, the major source of nutrients and sediments is likely to be erosion and leaching of fertilizer from croplands and runoff of animal wastes from barns and feedlots. All the practices that may be used to minimize such erosion, runoff, and leaching are lumped under a single term: **best management practices (BMP)**. BMP includes all the methods of soil conservation discussed in Chapter 8. Topping the list are keeping the ground covered with vegetation or mulch to prevent erosion, strip-cropping, using clover or other legumes for natural nitrogen addition, and using organic fertilizers (such as compost and manure) in place of inorganic fertilizers. However, other things are important, too.

Reestablishing riparian woodlands—the strip of trees, shrubs, and other vegetation along streams and drainage areas—has been found to be particularly helpful. The riparian woodland is very effective in filtering and absorbing nutrients from the water draining into the stream.

Another possibility is available where animal wastes from feedlots, dairy barns, or horse stables drain directly into waterways. There, ponds can be constructed to intercept the nutrient-rich runoff. Such water can then be recycled as irrigation water, returning the nutrients to the soil (Fig. 18–9).

The implementation of BMP on farms has three factors in its favor. First, one has to negotiate with only a limited and fairly stable number of farmers. Second, the implementation of BMP usually has long-term economic benefits for the farmer. And third, there is an existing array of various farm subsidies that can be manipulated to provide additional economic incentives. Thus, implementation of BMP on farms is moving forward.

Urban runoff is another major nonpoint nutrient source. Homeowners tend to use three to five times more fertilizers and pesticides on lawns and gardens per unit area than farmers use on crops. Therefore, the runoff and leaching of nutrients from urban and suburban areas is considerably greater than from similar areas of agricultural land. Pet excrements also contribute significantly to the nutrient content of such runoff. Golf courses are a particular issue: Managers tend to use fertilizer intensively, and soils are compacted, thus promoting runoff. In watersheds with sizable cities, the major source of nutrients causing eutrophication may be the runoff from developed areas.

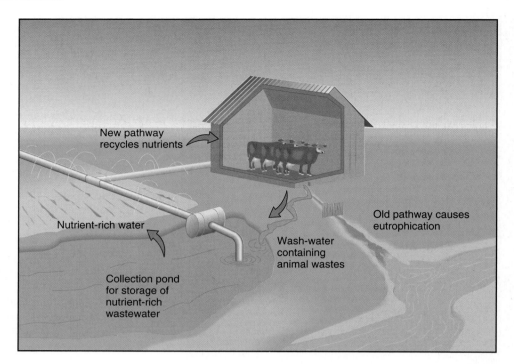

New pathway
recycles nutrients

Nutrient-rich water

Collection pond
for storage of
nutrient-rich
wastewater

Wash-water
containing
animal wastes

Old pathway causes
eutrophication

▶ **FIGURE 18–9** *A collection pond for dairy-barn washings.* When washings from animal facilities are flushed directly into natural waterways, they contribute significantly to eutrophication. This may be avoided by collecting the flushings in ponds from which both the water and the nutrients may be recycled.

For some newer developments, stormwater-retention reservoirs are constructed. These are small ponds into which runoff and erosion are channeled (Chapter 9). As nutrient-laden water enters the pond, its velocity slows, sediment settles, and the water evaporates over time or is allowed to drain through a small wetland where the nutrients are intercepted by wetland plants.

Urban runoff is growing with increasing population and development, and it promises to be the most difficult to control.

Recovery. The good news is that cultural eutrophication can be controlled and often reversed. This requires a total watershed-management approach. For a small lake, fed from a modest-sized watershed, the task may be quite straightforward. Major sources of nutrients can be identified and addressed. For large bodies of water, such as the Chesapeake Bay—the watershed of which includes some 12,000 square miles (27,000 km²) in five states and in the District of Columbia—the task is significantly greater. Nevertheless, the same approach is being used. The governmental jurisdictions involved are cooperating under a Chesapeake Bay program with the common goal of reducing nutrient inputs into the bay by 40% by the year 2000. The watershed is divided into the smaller watersheds of each of the tributaries, and each jurisdiction is taking the responsibility of reducing the nutrients emanating from its tributaries.

One remarkable success story is Lake Washington, east of Seattle. This 34-square-mile lake is the center of a large metropolitan area. During the 1940s and 1950s, 11 sewage-treatment plants were sending state-of-the-art treated water into the lake at a rate of 20 million gallons per day. At the same time, phosphate-based detergents came into wide use. The lake responded to the massive input of nutrients by developing unpleasant "blooms" of noxious algae. The water

lost its clarity, the desirable fish populations declined, and masses of dead algae accumulated on the shores of the lake.

Citizen concern led to the creation of a system that diverted the treatment-plant effluents into nearby Puget Sound, where tidal flushing would mix them with open-ocean water. The diversion was complete by 1968, and the lake responded quickly. The algal blooms diminished, the water regained its clarity, and by 1975, recovery was complete. Careful studies by a group of limnologists from the University of Washington showed that phosphate was the culprit. Before the diversion, the lake was receiving 220 tons of phosphate per year from all sources. Afterward, the total dropped to 40 tons per year, well within the lake's normal capacity to absorb phosphate by depositing it in deep bottom sediments. One clear lesson learned was that sewage-treatment effluent must never be allowed to enter a lake unless nutrient removal is part of the treatment. The major lesson, however, was that eutrophication can be reversed by paying attention to nutrient inputs and addressing the various sources. Lake management, in other words, is a matter of controlling the phosphate loading into a lake from all sources: human and animal sewage, agricultural and yard fertilizers, street runoff, and failing septic systems. Many lakes are now protected by lake associations, consisting of concerned citizens who see themselves as stewards of their lake.

In recent years, the United States and other developed nations have recognized the problem of eutrophication and have made substantial progress in developing or modifying sewage-treatment systems that will keep nutrients in a cycle on the land and not discharge them into waterways. But, at the other extreme, particularly in developing countries, much **raw sewage**—completely untreated sewage—is still discharged, creating not only eutrophication of waterways but also a significant disease hazard. We will now turn our attention to the important task of managing our organic wastes, another form of water pollution.

18.3 Sewage Management and Treatment

Before the late 1800s, the general means of disposing of human excrement was the outdoor privy. Seepage from the privy frequently contaminated drinking water and caused disease, especially in places where privies and wells were located near one another. In the late 1800s, Louis Pasteur and other scientists showed that sewage-borne bacteria were responsible for many infectious diseases. This important discovery led to intensive efforts to rid cities of human excrements as expediently as possible. Cities already had drain systems for storm water, but using these systems for human wastes had been prohibited. With the urgency of the situation, however, minds quickly changed. The flush toilet was introduced, and sewers were tapped into storm drains. Thus, Western civilization initiated the one-way flow of flushing sewage wastes into natural waterways.

The results of this practice are obvious. Receiving waters that had limited capacity for dilution became essentially open cesspools of foul odors, vermin, and filth as the overload of organic matter depleted dissolved oxygen and all aquatic life suffocated. For increasing distances around or downstream from the sewage outfall, the water became unswimmable because of the sewage contamination.

Development of Collection and Treatment Systems

In order to alleviate the problem of sewage-polluted waterways, sewage-treatment facilities were designed and constructed to treat the outflow before it entered the receiving waterway. The first treatment plants in the United States were built around 1900. However, the combined volumes of sewage and storm water soon proved impossible to handle. During heavy rains, wastewater would overflow the treatment plant and carry raw sewage into the receiving waterway. Gradually, regulations were passed requiring developers to install separate systems: **storm drains** for collecting and draining runoff from precipitation, and **sanitary sewers** to receive all the wastewater from sinks, tubs, and toilets in homes and other buildings. (Note the distinction in terms; it is incorrect to speak of storm drains as sewers.)

The ideal modern system, then, is one in which all sewage water is collected separately from storm water and fully treated to remove all pollutants before the water finally is released. But progress toward this goal has been extremely uneven. Up through the 1970s, even in the United States and other developed countries, countless communities still discharged untreated sewage directly into waterways, and for countless more the degree of treatment was minimal. Indeed, the increasing

▲ **FIGURE 18–10** *Waterways in the developing world.* Many parts of the developing world do not yet have sanitary water or sewage-collection systems. Waterways through such poor areas receive direct discharges of raw sewage yet still serve as a source of water for washing and even drinking.

sewage pollution of waterways and beaches was the major impetus behind the passage of the Clean Water Act of 1972.

In the meantime, much of the developing world still exists in the most primitive stage with regard to sewage. Innumerable poor villages and poor areas of mushrooming cities of the developing world don't even have sewer systems for collection. In such areas, it is not uncommon to find raw sewage littering the ground and overflowing gutters and streams, with children playing in the filth. Even where flush toilets and collection systems exist, the discharge of the raw sewage into waterways is still common. Many of the people living in these regions must use these badly polluted waters for bathing, laundering, and even drinking (Fig. 18–10).

Before addressing the topic of sewage treatment, let us first clarify which contaminants and pollutants we are talking about.

The Pollutants in Raw Sewage

Raw sewage is not only the flushings from toilets, it is also the collection from all other drains in homes and other buildings. A sewer system brings all tub, sink, and toilet drains from all homes and buildings together into larger and larger sewer pipes, just as twigs of a tree come together eventually into the trunk. This total mixture collected from all drains, which comes out at the end of the "trunk" of the collection system, is termed **raw sewage** or **raw wastewater**. Because we use such large amounts of water to flush away small amounts of dirt, especially as we stand under the shower or often just run the water with no waste at all, most of what goes down sewer drains is water. Raw sewage is about 1000 parts water for every 1 part of waste—99.9% water to 0.1% waste.

Given our voluminous use of water, the quantity of raw-sewage output is in the order of 150–200 gallons (600–800 liters) per person per day. That is, a community of 10,000 persons will produce on the order of 1.5 to 2.0 million gallons (6–8 million liters) of wastewater each day. With the addition of storm water, raw sewage is diluted still more. Nevertheless, the pollutants are sufficient to make the water brown and smell foul.

The pollutants generally are divided into the four following categories, which, as we shall find, correspond to techniques used for their removal.

1. Debris and grit: rags, plastic bags, coarse sand, gravel, and other objects flushed down toilets or washed through storm drains.
2. Particulate organic material: fecal matter, food wastes from garbage-disposal units, toilet paper, and other matter that will tend to settle in still water.
3. Colloidal and dissolved organic material: very fine particles of the particulate organic material, bacteria, urine, soaps, detergents, and other cleaning agents.
4. Dissolved inorganic material: nitrogen, phosphorus, and other nutrients from excretory wastes and detergents.

In addition to the four categories of pollutants in "standard" raw sewage, variable amounts of pesticides, heavy metals, and other toxic compounds may be found in sewage because people pour unused portions of products containing such materials down sink, tub, or toilet drains. Also, industries may discharge various toxic wastes into sewers.

Removing the Pollutants from Sewage

The challenge of sewage treatment is more than installing a technology that will do the job. It is finding one that will do the job at a reasonable cost. A diagram of water-treatment procedures is shown in Fig. 18–11.

Preliminary Treatment (Removal of Debris and Grit). Because debris and grit will damage or clog pumps and later treatment processes, removing them is a necessary first step and is termed **preliminary treatment**. Preliminary treatment usually involves two steps: a screening out of debris and a settling of grit. Debris is removed by letting raw sewage flow through a **bar screen**, a row of bars mounted about 1 inch (2.5 cm) apart. Debris is mechanically raked from the screen and taken to an incinerator.

After passing through the screen, the water flows through a **grit chamber**, a swimming pool-like tank, where its velocity is slowed just enough to permit the grit to settle (Fig. 18–11). The settled grit is mechanically removed from these tanks and taken to landfills.

Primary Treatment (Removal of Particulate Organic Material). After preliminary treatment, the water moves on to **primary treatment**, where it flows very slowly (about 6 feet, 2 m, per hour) through large tanks called **primary clarifiers** (Fig. 18–11). Because it flows slowly through these tanks, the water is nearly motionless for several hours. The particulate organic material, about 30% to 50% of the total organic material, settles to the bottom, from where it can be removed. At the same time, fatty or oily material floats to the top, where it is skimmed from the surface. All the material removed, both particulate organic material and fatty material, is combined into what is referred to as **raw sludge**; we shall consider its treatment shortly.

Note that primary treatment involves nothing more complicated than putting polluted water into a "bucket," letting material settle, and pouring off the water. Nevertheless, it removes much of the particulate organic matter at minimal cost.

Secondary Treatment (Removal of Colloidal and Dissolved Organic Material). **Secondary treatment** is also called **biological treatment** because it makes use of organisms—natural decomposers and detritus feeders. Basically, an environment is created to enable these organisms to feed on the colloidal and dissolved organic material and break it down to carbon dioxide and water via their cell respiration. The sewage water from primary treatment is the food- and water-rich medium. The only thing that needs to be added in addition to the organisms is oxygen to enhance their respiration and growth. Recall that the oxygen consumed in the decomposition process is called the biochemical oxygen demand (BOD). Either of two systems may be used: *trickling filters* or *activated-sludge systems*. The earliest and still widely used systems are trickling filters. In a **trickling-filter system**, the water exiting from primary treatment is sprinkled onto and allowed to percolate through a bed of fist-sized rocks 6–8 feet (2–3 m) deep (Fig. 18–12). The spaces between the rocks provide for good aeration. As in a natural stream, this environment supports a complex food web of bacteria, protozoans, rotifers (organisms that consume protozoans), various small worms, and other detritus feeders attached to the rocks. The organic material in the water, including pathogenic organisms, is absorbed and digested by these organisms as it trickles by.

The **activated-sludge system**, shown in Fig. 18–11, is the most common secondary treatment system. In this system, water from primary treatment enters a large tank that is equipped with an air-bubbling system or a rapidly churning system of paddles. A mixture of detritus-feeding organisms, referred to as **activated sludge**, is added to the water as it enters the tank, and the water is vigorously aerated as it moves through the tank. Organisms in this well-aerated environment reduce the biomass of organic material, including pathogens, as they feed. As organisms feed on each other, they tend to form into clumps, termed *floc*, that settle readily when the water is stilled. Thus, from the aeration tank the water is passed into a secondary **clarifier tank**, where the organisms settle out and the water—now with better than 90% of all the organic material removed—flows on. The settled organisms are pumped back into the entrance of the aeration tank. They are the activated sludge that is added at the beginning of the process. Surplus amounts of activated sludge, which occur as populations of organisms grow, are removed and added to the raw sludge. The organic material is oxidized

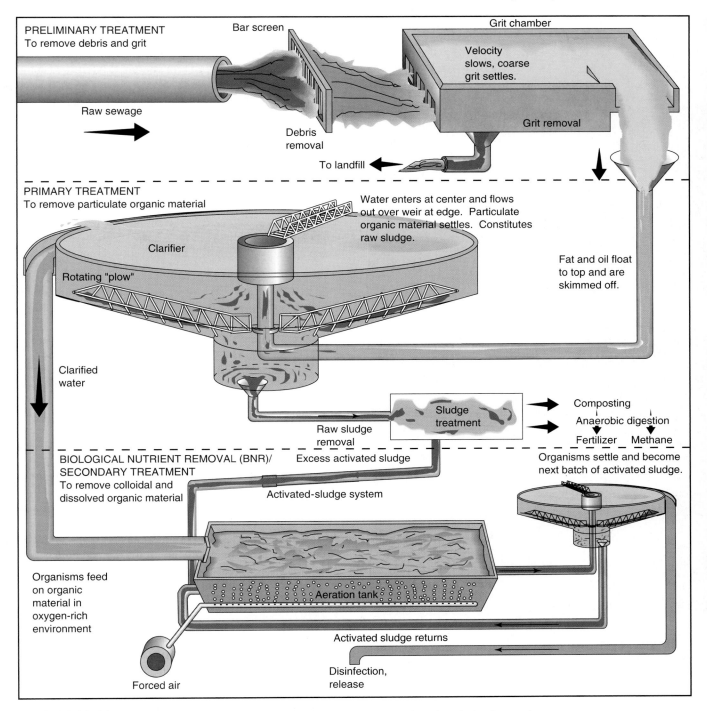

▲ **FIGURE 18–11** *A diagram of wastewater treatment.* Raw sewage moves from the grit chamber to primary treatment, where sludge is removed. The clarified water then proceeds to secondary treatment (here shown as activated sludge treatment).

by microbes to form carbon dioxide, water, and mineral nutrients that remain in the water solution.

Biological Nutrient Removal (Removal of Dissolved Inorganic Material). The original secondary-treatment systems as previously discussed were designed and operated in a manner to simply maximize biological digestion,

because elimination of organic material and its resulting BOD were considered the prime objectives. Ending up with a nutrient-rich discharge was not recognized as a problem. Today, with increased knowledge of the problem of cultural eutrophication, secondary activated-sludge systems have been added and are being modified and operated in a manner that achieves nutrient removal as well

(a)

(b)

▲ **FIGURE 18–12** *Trickling filters for secondary treatment.* (a) The water from primary clarifiers is sprinkled onto and trickles through a bed of rocks 6–8 feet deep. (b) Various bacteria and other detritus feeders adhering to the rocks consume and digest organic matter as it trickles by in the water. The water is collected at the bottom of the filters and goes on for final clarification and disinfection.

as detritus oxidation, a process known as **biological nutrient removal (BNR)**.

Recall that in the natural nitrogen cycle (Fig. 3–18), various bacteria convert nutrient forms of nitrogen (ammonia and nitrate) back to non-nutritive nitrogen gas in the atmosphere. This process is called **denitrification**. For biological removal of nitrogen, then, the activated-sludge system is partitioned into zones, and the environment in each zone is controlled in a manner to promote the denitrifying process (Fig. 18–13).

With respect to phosphate, in an environment that is rich in oxygen but relatively lacking in food—the environment of zone 3—bacteria take up phosphate from solution and store it in their bodies. Thus, phosphate is removed as the excess organisms are removed from the system (see again Fig. 18–13). These organisms, with the phosphate they contain, are added to and treated with the raw sludge, ultimately producing a more nutrient-rich treated sludge product.

As an alternative to BNR, there are various chemical processes that are used. One is simply to pass the effluent from standard secondary treatment through a filter of lime, which causes the phosphate to precipitate out as insoluble calcium phosphate. Another is to treat the effluent with ferric chloride, which produces insoluble ferric phosphate, or organic polymer, which produces a floc. With these methods, the phosphorus is removed during sludge removal.

Final Cleansing and Disinfection. With or without biological nutrient removal, the wastewater is subjected to a final clarification and disinfection. Although few pathogens survive the combined stages of treatment, public-health rigors still demand that the water be disinfected before discharge into natural waterways. The most widely used disinfecting agent is chlorine gas because it is both effective and relatively inexpensive. But this treatment also introduces chlorine into natural waterways, and even minute levels of chlorine can harm aquatic animals. Chlorine gas is also dangerous to work with. More commonly, sodium hypochlorite (Chlorox) provides a safer way of adding chlorine to achieve disinfection.

Other disinfecting techniques are coming into use. One alternative disinfecting agent is ozone gas, which is extremely effective in killing microorganisms and breaks down to oxygen gas in the process, which actually improves water quality. But, because ozone is unstable and hence explosive, it must be generated at the point of use, a step that demands considerable capital investment and energy. Another disinfection technique is to pass the effluent through an array of ultraviolet lights mounted in the water. (Standard fluorescent lights without the white coating emit ultraviolet.) The ultraviolet radiation kills microorganisms but does not otherwise affect the water.

After these steps of treatment, the wastewater has a lower organic and nutrient content than most bodies of water into which it is discharged; BOD values are commonly 10 to 20 ppm, as opposed to 200 ppm in the incoming sewage. In other words, discharge of such water actually contributes toward improving water quality in the receiving body. In water-short areas, there is every reason to believe that the treated water itself, with little additional treatment, could be recycled into the municipal water-supply system.

Bear in mind that the techniques we have described here are state-of-the-art. Many cities, even in the developed world, are still operating with antiquated systems that provide lower-quality treatment. A few coastal cities still persist in discharging sewage with only primary treatment directly into the ocean, although in the United States this practice has almost disappeared (see the "Ethics" essay, p. 450).

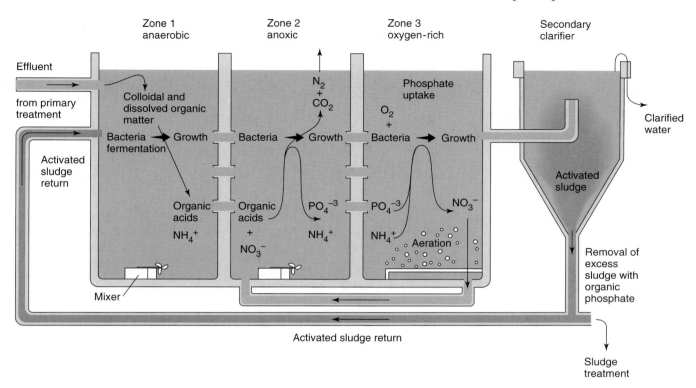

▲ **FIGURE 18–13** *Biological nutrient removal (BNR).* The secondary treatment—the activated sludge process—may be modified to remove nitrogen and phosphate while at the same time breaking down organic matter. For this BNR process, the aeration tank is partitioned into three zones, only the third of which is aerated. As seen in the diagram, ammonium (NH_4^+) is converted to nitrate (NO_3^-) in zone 3. Recycled to zone 2, which is without oxygen gas (anoxic), the nitrate supplies the oxygen for cell respiration. It is converted to nitrogen gas in the process and is released into the atmosphere. Phosphate is taken up by bacteria in zone 3 and removed with the excess sludge.

Sludge Treatment

Recall the particulate organic matter that settled out or floated to the surface of sewage water in primary treatment. This material forms the bulk of *raw sludge*, although there may be additions, as we noted, of excesses from activated sludge and BNR systems. As such, raw sludge is a gray, foul-smelling, syrupy liquid with a water content of 97% to 98%. Also, the potential for the presence of pathogens is significant because raw sludge includes material directly from toilets. Indeed, it is considered a biologically hazardous material. However, as nutrient-rich organic material, it has the potential to be used as organic fertilizer if it is suitably treated to kill pathogens and if it does not contain other toxic contaminants.

Several methods for treating sludge and converting it into organic fertilizer are commonly used: (1) *anaerobic digestion*, (2) *composting*, and (3) *pasteurization*. Since this is a developing industry, it is unclear which method will prove most cost-effective and environmentally acceptable over the long run. Also, none of these methods is capable of removing toxic substances such as heavy metals and non-biodegradable synthetic organic compounds. The presence of such toxins can preclude the use of sludge as fertilizer.

Anaerobic Digestion. Anaerobic digestion is a process of allowing bacteria to feed on the detritus in the *absence of oxygen*. The raw sludge is put into large air-tight tanks called **sludge digesters** (Fig. 18–14). In the absence of oxygen, a consortium of anaerobic bacteria accomplishes the breakdown of organic matter. The end products of this decomposition are carbon dioxide, methane, and water. Thus, a major byproduct of the anaerobic processes is **biogas**, a gaseous mixture that is about two-thirds *methane*. The other one-third is made up of carbon dioxide and various foul-smelling organic compounds that give sewage its characteristic odor. Because of its methane content, biogas is flammable and can be used for fuel. In fact, it is commonly collected and burned to heat the sludge digesters because the bacteria working on the sludge do best when maintained at about 104°F (38°C).

After four to six weeks, anaerobic digestion is more or less complete, and what remains is called **treated sludge** or **biosolids**. It consists of the remaining organic material, which is now a relatively stable, nutrient-rich, humus-like material in water suspension. Pathogens have been largely if not entirely eliminated, so they no longer present any significant health hazard.

Such treated sludge makes an excellent organic fertilizer. It may be applied directly to lawns and agricultural fields in the liquid state in which it comes from the digesters, providing the benefit of both the humus and the nutrient-rich

ETHICS

CLEANING UP THE FLOW

Coastal cities have a reputation for dragging their heels with regard to sewage treatment. With the enormous potential of an ocean for tidal mixing and flushing, they became accustomed to using the dilution solution for disposing their sewage. The city of Boston, Massachusetts, presents a case in point. For decades, Boston sewage—370 million gallons each day—received only primary treatment at a site on Deer Island before being discharged into Boston Harbor. Making matters worse, scum and sludge from the treatment process were also released to the harbor. The harbor had such a well-deserved reputation for gross pollution that the Republican presidential candidate George Bush used it to demolish the environmental record of Michael Dukakis (his opponent and governor of Massachusetts) in the 1988 presidential campaign.

Under federal government pressure for years to clean up the harbor, the state established the Massachusetts Water Resources Authority (MWRA) in 1984 to oversee the cleanup. The MWRA mapped out a plan for the cleanup that conforms with the federally mandated full implementation by 1999. The first milestone occurred in December 1991: The flow of raw sludge was stopped, and all sludge now goes to a pelletizing plant to be turned into fertilizer, marketed as Bay State Fertilizer. The second milestone took place in August 1997 when the sewage began receiving secondary treatment. A key component of the plan is

a 9.5-mile-long, 26-foot-diameter tunnel that will carry treated effluent well out of Boston Harbor and into Massachusetts Bay, where diffusers will release the effluent into bay water. By December 1999, the secondary treatment phase of the plant will be in full operation and the program will be fully implemented.

The facility, when finished, will be one of the largest in the world, able to treat 1.3 billion gallons per day of wastewater. The project will cost $3.7 billion, shared by federal, state, and local communities. It will serve 2.5 million people from 43 communities and 5500 businesses. Costs for a typical household have increased by 300% since 1986 and now average around $675 a year. However, this is well below earlier projections as a result of significant federal funding.

Boston Harbor itself has shown its approval of this project. Harbor porpoises and seals are now seen in unprecedented numbers, even in the inner harbor, where effluents had been released for years. Eight miles of recreational beaches are now swimmable all but a few days of the year. A recreational fishery for bluefish, striped bass, cod, and winter flounder is flourishing. Commercial lobster and shellfish are harvested at a value of more than $10 million a year. Further improvements are planned to eliminate the combined sewer/storm drain overflows at an additional $450 million. When this is accomplished, Boston Harbor will be one of the cleanest municipal water bodies in the world.

The project has not been without controversy, however. The 9.5-mile outfall pipe

has stirred up a firestorm of criticism from citizens and advocacy groups on Cape Cod. Their concern centers on the impact of the effluent on water quality in Cape Cod Bay, which adjoins Massachusetts Bay to the southeast. The opponents would prefer that the outfall remain in Boston Harbor. They claim that because of the location of the outfall pipe, coastal-water circulation may tend to carry the high-nutrient effluent into Cape Cod Bay. The Cape Codders cite numerous studies that show that nitrogen acts as a limiting factor in many coastal marine systems, and they fear that the added nitrogen from the Boston treatment plant will create conditions leading to blooms of algae, especially those that cause red tide. Red tides have been a recurring problem in recent years up and down the northeastern coast, often closing down shellfish harvesting because of the dangers of paralytic shellfish poisoning.

The MWRA claims that not only will the new treatment system obviously benefit Boston Harbor, it will also improve water quality in Massachusetts and Cape Cod bays. They point out that the system as currently operating is already polluting those bays, since the effluent is eventually carried out of the harbor by tidal action. On the surface, this conflict appears to be another case of NIMBY (Not In My Back Yard). One municipality is attempting to resolve its environmental problem in an apparently effective fashion, but there are risks that the solution will create another problem many miles away. What do you think?

---

water. Alternatively, the sludge may be "dewatered" by means of belt presses, where the sludge is passed between rollers that squeeze out most of the water (Figure 18–15a) and leave the organic material as a semisolid **sludge cake** (Fig. 18–15b). The sludge cake is easy to stockpile, distribute, and spread on fields with traditional manure spreaders.

Composting. Another process sometimes used to treat sewage sludge is **composting**. Raw sludge is mixed with wood chips or other water-absorbing material to reduce the water content. It is then placed in **windrows**, long narrow piles that allow air circulation and convenient turning with machinery.

Bacteria and other decomposers break down the organic material to rich, humus-like material.

Pasteurization. Raw sludge may be dewatered and the resulting sludge cake put through drying ovens that operate like oversized laundry dryers. In the dryers, the sludge is **pasteurized**, that is, heated sufficiently to kill any pathogens (exactly the same process that makes milk safe to drink). The product is dry, odorless organic pellets. Milwaukee, Wisconsin, which has a particularly rich sludge, resulting from the brewing industry, has been using this process for over 60 years. The city bags and sells the pellets throughout the country as an organic fertilizer under the trade name Milorganite.

◀ **FIGURE 18–14** *Anaerobic sludge digesters.* Bacterial digestion of raw sludge in the absence of oxygen in these tanks leads to the production of methane gas, which is tapped from the tops of the tanks, and humus-like organic matter that can be used as a soil conditioner. The "egg shape" of the tanks facilitates mixing and digestion.

Alternative Treatment Systems

Individual Septic Systems. Despite the expansion of sewage-collection systems, in rural and suburban areas there always will be countless homes not connected to a municipal system. For such homes, individual septic systems are required. The traditional and still most common system is the septic tank and a drain field (Fig. 18–16). Wastewater flows into the tank, where particulate organic material settles to the bottom. The tank acts like a primary clarifier in a municipal system. Water containing colloidal and dissolved organic material as well as the dissolved nutrients flows into the drain field and gradually percolates into the soil. Organic material that settles in the

tank is digested by bacteria, but accumulations still must be pumped out every three to five years. Soil bacteria decompose the colloidal and dissolved organic material that comes through the drain field. Some people establish successful vegetable gardens over septic drain fields, thus exercising the sound principle of recycling the nutrients. Prerequisites for this traditional system are suitable land area for the drain field and subsoil that allows sufficient percolation of water.

Using Effluents for Irrigation. The nutrient-rich water coming from the standard secondary treatment process is beneficial for growing plants. The problem is that we don't want to put it into waterways, where it will stimulate the growth of undesirable

▼ **FIGURE 18–15** *Dewatering treated sludge.* (a) Sludge (98% water) may be "dewatered" by means of belt presses, as shown here. The liquid sludge is run between canvas belts going over and between rollers so that much of the water is pressed out. (b) The resulting "sludge cake" is a semisolid humus-like material that may be used as an organic fertilizer.

(a)

(b)

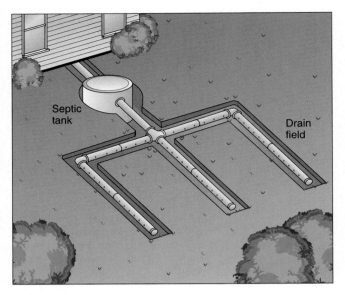

▲ **FIGURE 18–16** *Septic tank treatment.* Sewage treatment for a private home, using a septic tank and drain field. The pipes and the tank normally are buried underground. They are shown uncovered here only for illustration. (Source: USDA, Soil Conservation Service.)

algae. But why not use it for irrigating plants we do want to grow? It is a way of completing the nutrient cycle. Indeed, this concept has been put into practice in a considerable number of locations as a practical alternative to upgrading the treatment to remove nutrients. Again, there is the prerequisite that effluents not be contaminated with toxic materials.

For example, the nutrient-rich effluent from standard secondary treatment from St. Petersburg, Florida, was causing cultural eutrophication in Tampa Bay. Now St. Petersburg uses the effluent to irrigate 4000 acres (1600 hectares) of urban open space, from parks and residential lawns to a golf course. Revenues from the water sales help offset operating costs. Bakersfield, California, receives a $30,000 annual income from a 5000-acre (2000-ha) farm irrigated with its treated effluent. Clayton County, Georgia, is irrigating 2500 acres (1000 ha) of woodland with partially treated sewage. Hundreds of other similar projects are under way around the country.

A number of developing countries irrigate croplands with raw (untreated) sewage effluents. This practice is getting the bad with the good. The crops respond well, but parasites and disease organisms can easily be transferred to farm workers and consumers. Therefore, it is important to emphasize that only treated effluents should be used for irrigation.

Reconstructed Wetland Systems. It is also possible to utilize the nutrient-absorbing capacity of wetlands where suitable areas and climatic conditions exist. The project may be part of a wetlands recovery program, or artificial wetlands may be constructed.

As one example of the former, in the 1960s and 1970s much of the land around Orlando, which was originally wetlands, was drained and converted to cattle pasture. At the time, Orlando was discharging into the James River 13 million gallons (50 million liters) per day of nutrient-rich effluent following secondary treatment. Through the Orlando Easterly Wetlands Reclamation Project, 1200 acres (480 ha) of pastureland has been converted back to wetlands. The project involved scooping soil from and building berms (mounds of earth) around pastures to create a chain of shallow lakes and ponds. In addition, 1.2 million wetland plants ranging from bulrushes and cattails to various trees were planted. The effluent entering the upper end now percolates through the wetland for about 30 days before entering the James River virtually pure. Thus, the project has re-created a wildlife habitat. Wetland systems can be designed for small as well as large areas and are becoming an increasingly popular alternative for small communities. The key to success for such systems is to ensure that the systems are kept in balance and not loaded beyond their ability to handle inputs.

18.4 Public Policy

Responsibility for oversight of the health of the nation's waters rests with the Environmental Protection Agency. The EPA, however, can only develop regulations if Congress gives the authority to do so. Hence, the foundation for public policy must be the laws passed by Congress. Some major legislative milestones for protecting the nation's waters are shown in Table 18-3. The landmark legislation is the **Clean Water Act of 1972**, which gave the EPA jurisdiction and for the first time required permits for all point-source pollutant discharges. This act and subsequent amendments have provided $69 billion to help cities build treatment plants to meet the federal requirement for secondary treatment of all sewage. To this have been added more than $25 billion of state and local funds. However, the EPA estimates that an additional $140 billion will be needed over the next 20 years to meet funding needs for all eligible municipal wastewater treatment systems.

Reauthorization of the Clean Water Act is long overdue. Programs have been funded on a year-to-year basis, and Congress has been hung up on a debate over whether requirements should be strengthened or weakened, whether additional mandates should be subjected to a cost/benefit analysis, and whether regulatory relief should be provided to industries, states, cities, and individuals who are required to take actions to comply with the regulations. Recently, impatient with congressional inaction to reauthorize and strengthen the Clean Water Act, Vice President Gore directed federal agencies to develop a Clean Water Initiative to improve and strengthen our water pollution control efforts. An action plan was subsequently developed in 1998, and almost 100 actions were identified as part of the initiative. Most are existing activities and have been funded by way of appropriations to the federal agencies involved (EPA, USDA, Interior, etc.), with funding well below the President's request of $2.2 billion. New initiatives include a national response for harmful algal blooms (HABs—see the "Earth

Table 18-3 Legislative Milestones in Protecting the Nation's Waters

1899—Rivers and Harbors Act: First federal legislation protecting the nation's waters to promote commerce.

1948—Water Pollution Control Act: The federal government provides technical assistance and funds to state and local governments to promote efforts to protect water quality.

1972—The Clean Water Act: Comprehensive federal program created to achieve the goal of protecting and restoring the physical, chemical, and biological integrity of the nation's waters. Requires permits for any pollution discharge; strengthens water quality standards; encourages use of best achievable pollution control technology; provided billions of dollars for construction of sewage-treatment plants.

1972—Marine Protection, Research and Sanctuaries Act: Prevents unacceptable dumping in oceans.

1977—Clean Water Act amendments: Strengthened controls on toxic pollutants and allowed states to assume responsibility for federal programs.

1987—Water Quality Act: Major amendments to the Clean Water Act; created a revolving loan fund to provide ongoing support for construction of treatment plants; addressed regional pollution with a watershed approach; addressed nonpoint source pollution in Section 319 of the act, with requirements for states to assess the problem, develop and implement plans, with $400 million in grants available.

Watch" essay, p. 442), more attention to the assessment of watersheds in pollution abatement, and greater response to the problems associated with animal-feeding operations.

The EPA has identified nonpoint-source pollution as the nation's number one water pollution problem, with construction of new wastewater treatment facilities not far behind. Much progress has been made in the 27 years since the enactment of the Clean Water Act. The number of people in the United States served by adequate sewage-treatment plants has more than doubled, from 85 million to 173 million. Soil erosion has been reduced by 1 billion tons annually. And two-thirds of the nation's waterways are safe for fishing and swimming, a doubling over the 1972 situation.

Many of the nation's most heavily used rivers, lakes, and bays have been cleaned up and restored; for example, the Androscoggin River in Maine, Boston Harbor, the Upper Mississippi River, the South Platte River, the Upper Arkansas River, Lake Erie, the Illinois River, and the Delaware River. Fish now swim in rivers once so polluted that only bacteria and sludge worms could survive. Significantly, a 70% increase in bottom vegetation has been achieved since the mid-1980s in the Chesapeake Bay. Clearly, a national sense of stewardship has been applied to the rivers, lakes, and bays that are our heritage from a previous generation, and public policy has been enacted and billions of dollars spent to bring our waters back from a polluted state.

ENVIRONMENT ON THE WEB

CHALLENGES IN WATER QUALITY IMPROVEMENT

Thirty years ago, eutrophication of lakes and streams was one of our central water-quality concerns. The evidence of eutrophication could be seen in the excessive aquatic plant and algal growth, nighttime oxygen depletion, and fish kills. So great was public concern about the issue that it prompted a continent-wide effort to reduce phosphorus and nitrogen concentrations in natural waters. Key elements of this strategy were the reformulation of detergents to reduce phosphorus concentrations, and improved sewage treatment at publicly owned sewage-treatment plants throughout the United States and Canada.

Today, point sources of nutrients—industrial and municipal discharges—are strictly controlled through a variety of regulatory programs. Although nutrient levels and fish kills had declined significantly by the 1980s, eutrophication problems continue to persist in many systems. In part, this phenomenon is thought to be due to the re-release of older deposits of nutrients from lake and river sediments. A simpler explanation may be that some key sources—especially nonpoint sources—remain uncontrolled.

Nonpoint sources of nutrients are numerous in our society. They include diffuse runoff of rainwater and associated contaminants from urban lawns, roofs, roads, and parking lots, and drainage from agricultural fields, feedlots, and manure storage areas. These sources have long been known to contribute to water pollution. Yet compared to the costly technologies usually needed to clean industrial or sewage wastewater, nonpoint source controls are usually relatively inexpensive and freely available.

Why, then, has progress in controlling urban and agricultural diffuse source pollution been slow? Central challenges include the problem of who "owns" the pollution and who therefore should pay for clean up. Selection of appropriate remedial actions is also difficult, because the most effective approaches often require people to

change the way they behave; for instance, by fertilizing lawns or fields less frequently or by "stooping and scooping" after pets.

Because a combination of point and nonpoint sources of pollution creates observed water quality problems such as eutrophication, water quality improvement therefore means finding the most cost-effective combination of point- and nonpoint source controls. Where human attitudes and behavior patterns slow implementation, the "best" strategy may also involve a lengthy process of public education and citizen involvement. However, although public education programs can improve the success of nonpoint source controls, they can be time-consuming solutions.

Web Explorations

The Environment on the Web activity for this essay describes the use of computer simulation models to test alternative management strategies for water quality improvement. Go to the Environment on the Web activity (select Chapter 18 at http://www.prenhall.com/nebel) and learn for yourself:

1. how modeling can be used to answer different key environmental management questions;
2. how computer models can be used to assess and plan watershed restoration projects;
3. about the role of watershed management in water quality improvement; and
4. about the differences between physical and computer simulation models.

A suggested time frame for each of these exercises is 10–30 minutes.

REVIEW QUESTIONS

1. Define pollution, pollutant, nonbiodegradable, point source, and nonpoint source of pollutants.
2. Discuss each of the five categories of water pollutants and the problems they cause.
3. Describe and compare submerged aquatic vegetation (SAV) and phytoplankton. Where and how does each get nutrients and light?
4. Explain the difference between oligotrophic and eutrophic. Describe the entire process of eutrophication.
5. Distinguish between natural and cultural eutrophication.
6. Give six ways in which the symptoms of eutrophication can be addressed. What are the benefits and limitations of each?

7. Give a brief history of humans' handling of sewage wastes as the understanding of risks and potential benefits have been gained.
8. What is the water content of, and what are the four categories of, pollutants present in raw sewage?
9. Name and describe the facility and the process used to remove debris, grit, particulate organic matter, colloidal and dissolved organic matter, and dissolved nutrients.
10. Why is secondary treatment also called biological treatment? What is the principle involved? What are the two alternative techniques used?
11. What are the principles involved in, and what is accomplished by, biological nutrient removal? Where do nitrogen and phosphate go in the process?

12. Raw sludge is a byproduct of what treatment processes? What is its chemical and physical nature?

13. Name and describe three methods of treating raw sludge, and give the end product(s) that may be produced from each method.

14. Describe alternative methods for treating raw sewage.

15. How may the sewage from individual homes be handled in the absence of municipal collection systems?

16. What are some of the important issues in public policy relating to sewage treatment?

THINKING ENVIRONMENTALLY

1. A large number of fish are suddenly floating dead on a lake. You are called in to investigate the problem. You find an abundance of phytoplankton and no evidence of toxic dumping. Suggest a reason for the fish kill.

2. A number of regions in the United States have banned the use of phosphate-containing detergents in recent years. What harm is caused by such detergents, and what is hoped to be achieved by such bans?

3. Describe how planting trees on an eroding hillside may protect aquatic life in an estuary many miles away.

4. Arrange a tour to the sewage-treatment plant that serves your community. Contrast it with what is described in this chapter. Is the water being purified or handled in a way that will prevent cultural eutrophication? Are sludges being converted to and used as fertilizer? What improvements, if any, are in order? How can you help promote such improvements?

5. Suppose a new community of several thousand persons is going to be built in Arizona (warm, dry climate). You are called in as a consultant to design a complete sewage system, including collection, treatment, and use or disposal of byproducts. Write an essay describing the system you recommend and give a rationale for the choices involved.

6. Some people favor the use of oceans as a dumping ground for sewage. Do you support or oppose this alternative? Defend your position.

WEB REFERENCES

On-line resources for this chapter can be found on the World Wide Web at: **http://www.prenhall.com/nebel**. Click on Chapter 18 on the chapter selector.

19

Municipal Solid Waste: Disposal and Recovery

〰〰〰〰〰

Key Issues and Questions

1. Two hundred and ten million tons of municipal solid waste (MSW) are disposed of annually in the United States. What are the components of MSW, and how is this waste handled?

2. Fifty-five percent of MSW is disposed of in landfills. What are the problems with landfills, and how are these addressed in newer landfill sites?

3. Seventeen percent of MSW is combusted, mostly in waste-to-energy (WTE) combustion facilities. What are the advantages and disadvantages of WTE combustion?

4. The best solution to solid waste problems is to reduce waste at its source. How can the total volume of refuse be reduced?

5. More than 75% of MSW is recyclable. What role is recycling playing in waste management, and how is recycling best promoted?

6. Although most management of MSW occurs at the local level, federal regulations concerning MSW are increasing. What regulations have affected the management of municipal solid waste?

7. Much more can be done to move MSW management in a more sustainable direction. What are some recommendations to improve MSW management?

Danehy Park features four soccer fields, four softball fields, three basketball courts, a baseball diamond, two playgrounds, and a two-mile jogging trail. The 55-acre park, opened in 1990, is located close to a heavily populated area of North Cambridge, Massachusetts, and is in almost constant use in good weather (Fig. 19–1). An unusual feature of the park is a big red light in the public rest room that warns users to vacate the park if the light goes on. The light is tied in to an elaborate venting system that prevents methane from building up below ground. The system is necessary because Danehy Park is built on the former city dump. Aside from the presence of the light and a few minor settling problems on soccer fields that have interfered with water drainage, a newcomer would never know that this was once a blight on the neighborhood—an open, burning dump in the 1950s and then a "sanitary landfill" that was closed in 1972. The park increased Cambridge's open space by 20% when it was created.

Danehy Park is a lesson in land use that is being learned only slowly in the United States—and, for that matter, in the rest of the world. The lesson is that if we could more effectively picture what 50 years can bring in the way of change, we could solve one of the most contentious problems facing local communities—what to do with municipal waste. As old dumps and landfills were closed because of environmental concerns, they created a temporary problem identified as the "solid waste crisis" in the 1970s and 1980s. Now many of those dumps and landfills are being converted into parks, golf courses, and nature preserves.

In some ways, the solid waste crisis is still with us, as we will see when we examine the problems of interstate trash movement. It is commonly said that we are running out of space to put our trash and garbage. That is a misconception; actually, we are running out of disposal space because of a fundamentally irresponsible feature of our modern society: We are happy to purchase the goods displayed so

A bull-dozer compactor works over a mountain or trash in a municipal landfill.

457

▶ FIGURE 19–1 *Landfills to playing fields.* Thomas W. Danehy Park in Cambridge, Massachusetts, a former landfill that has been recycled into a recreational park. In addition to the playing fields, the park features a half-mile "glassphalt" pathway (built with recycled glass and asphalt), shown in the foreground.

prominently in our malls and advertised in the media, but we are reluctant to accept the consequences of getting rid of them responsibly.

This chapter is about solid waste issues. We examine current patterns of disposal—landfills, combustion, and recycling—and look for solutions to our solid waste problems that are sustainable. The ideal would be to imitate natural ecosystems and reuse everything. Recall the second principle of sustainability: *Ecosystems dispose of wastes and replenish nutrients by recycling all elements.* Some solutions do well in conforming to this principle, whereas others do not—and possibly cannot.

19.1   The Solid Waste Problem

The focus of this chapter is **municipal solid waste** (MSW), defined as the total of all the materials thrown away from homes and commercial establishments (commonly called trash, refuse, or garbage). Municipal solid waste is often viewed as *pollution*, but it is increasingly becoming a *resource* as we learn to recycle many of the materials found in the waste stream.

Disposing of Municipal Solid Waste

Over the years, the amount of MSW generated in the United States has grown steadily, in part because of increasing population but more so because of changing lifestyles and the increasing use of disposable materials and excessive packaging. MSW now amounts to somewhat over 4 pounds (2 kg) per person per day. At the current (1999) U.S. population of 273 million, that is enough waste to fill 80,000 garbage trucks each day, a total of 210 million tons (191 million metric tons) per year. The *problem* can be simply stated: *We have to dispose of this stuff in the most effective and efficient way, while protecting human and environmental health.*

The refuse generated by municipalities is a mixture of materials from households and small businesses, with proportions as shown in Fig. 19–2. However, the proportions vary greatly, depending on the generator (commercial versus residential), the neighborhood (affluent versus poor), and the time of year (during certain seasons, yard wastes, such as grass clippings and raked leaves, add greatly to the solid waste burden, often equaling all the other categories combined). It is fair to say that little attention is given to what people throw away in the trash; even if there are restrictions and prohibitions, these can be bypassed with careful packing of the trash containers. Thus, many nasty substances—paint, used motor oil, small batteries, and so on—are discarded with the feeling that they are gone forever.

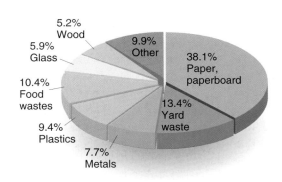

▲ FIGURE 19–2 *U.S. municipal solid waste composition.* The composition of municipal solid waste in 1996 in the United States. (Source: Data from EPA, Office of Solid Waste, "Characterization of Municipal Solid Waste in the United States," 1997 edition.)

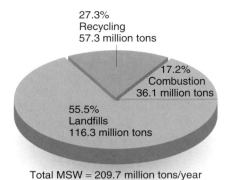

27.3%
Recycling
57.3 million tons

17.2%
Combustion
36.1 million tons

55.5%
Landfills
116.3 million tons

Total MSW = 209.7 million tons/year

▲ **FIGURE 19–3** *U.S. municipal solid waste disposal.*
Disposal of solid waste to landfills, combustion, and recycling, 1996.
(Source: Data from EPA, Office of Solid Waste, "Characterization of
Municipal Solid Waste in the United States," 1997 edition.)

Customarily, local governments have assumed the responsibility for collecting and disposing of MSW. The local jurisdiction may own the trucks and employ workers, or it may contract with a private firm to provide the collection service. Alternatively, some municipalities have opted for putting all trash collection and disposal in the private sector. The collectors bill each home by volume and weight of trash. This system allows competition among collectors and gives homeowners a strong incentive to reduce the volume of trash they produce. The MSW that is collected is then disposed of in a variety of ways, and it is at the point of disposal that state and federal regulations begin to apply.

Until the 1960s, most MSW was disposed of in open, burning dumps. The waste was burned to reduce its volume and lengthen the life span of the dump site, but refuse does not burn well. Smoldering dumps produced clouds of smoke that could be seen from miles away, smelled bad, and created a breeding ground for flies and rats. Some cities turned to incinerators, or combustion facilities, as they are called today—huge furnaces in which high temperatures allow the waste to burn more completely than in open dumps. Without controls, however, incinerators were also prime sources of air pollution. Public objection and air pollution laws forced the phaseout of open dumps and many incinerators during the 1960s and early 1970s. Open dumps were then converted to landfills.

In the United States in 1996, 56% of MSW was disposed of in approximately 2400 municipal landfills, 27% was recovered for recycling and composting, and the remainder (17%) was combusted (Fig. 19–3). Over the last 10 years, the landfill component has been declining, recycling has shown a steady increase, and combustion has remained fairly constant. The pattern is different in countries where population densities are higher and there is less open space for landfills. High-density Japan, for instance, combusts over 60% of its trash and recycles more than 20%. Most Western European countries also deposit less than half of their municipal waste in landfills and combust most of the rest.

Landfills

In a **landfill**, the waste is put on or in the ground and covered with earth. Because there is no burning and because each day's fill is covered with a few inches of earth, air pollution and vermin populations are kept down. Unfortunately, aside from those concerns and the minimizing of cost, no other factors were given real consideration when the first landfills were opened. Municipal waste managers generally had no understanding of or interest in ecology, the water cycle, or what products decomposing wastes would generate, and they had no regulations to guide them. Therefore, in general, any cheap, conveniently located piece of land on the outskirts of town became the site for a landfill. This site was frequently a natural gully or ravine, an abandoned stone quarry, a section of wetlands, or a previous dump (Fig. 19–4). Once the municipality acquired the land, dumping commenced without any precautions. After the site was full, it would be covered with earth and ignored; only recently have landfills been seen as a valuable open-space resource.

Problems of Landfills. Landfills are subjected to biological and physical factors in the environment and will undergo change over time as a consequence of the operation of those factors on the waste that is deposited. Several of the changes are undesirable, because they present the following problems if not dealt with effectively:

- leachate generation and groundwater contamination
- methane production
- incomplete decomposition
- settling

▲ **FIGURE 19–4** *Landfill siting problem.* A landfill in Staten Island, New York, encroaching on a wetland.

Leachate Generation and Groundwater Contamination. The most serious problem by far is groundwater contamination. Recall that as water percolates through any material, various chemicals in the material may dissolve in the water and get carried along in a process called *leaching.* The water with various pollutants in it is called *leachate.* As water percolates through MSW, a noxious leachate is generated that consists of residues of decomposing organic matter combined with iron, mercury, lead, zinc, and other metals from rusting cans, discarded batteries, and appliances—generously "spiced" with paints, pesticides, cleaning fluids, newspaper inks, and other chemicals. The nature of the landfill site and the absence of precautionary measures noted earlier funnel this "witches' brew" directly into groundwater aquifers.

All states have some municipal landfills that are or soon will be contaminating groundwater, but Florida has some unique problems. Flat and with vast areas of wetlands, much of the state is only a few feet above sea level and rests on water-saturated limestone. No matter where Florida's landfills were located, they were either in wetlands or just a few feet above the water table. The result? More than 150 former municipal landfill sites in Florida are now on the Superfund list. (Superfund is the federal program to clean up sites that are in imminent danger of jeopardizing human health through groundwater contamination.) All landfills in Florida now have state-of-the-art landfill liners.

Methane Production. Because it is about two-thirds organic material, MSW is potentially subject to natural decomposition. However, buried wastes do not have access to oxygen. Therefore, their decomposition is anaerobic, and a major byproduct of this process is *biogas,* which is about two-thirds methane and the rest hydrogen and carbon dioxide, a highly flammable mixture (see Chapter 18). Produced deep in a landfill, biogas may seep horizontally through the soil and rock, enter basements, and even cause explosions if it accumulates and is ignited. Over 20 homes at distances up to 1000 feet from landfills have been destroyed, and some deaths have occurred as a result of such explosions. Also, gases seeping to the surface kill vegetation by poisoning the roots. Without vegetation, erosion occurs, exposing the unsightly waste.

A number of cities have exploited the problem by installing "gas wells" in old and existing landfills. The wells tap the biogas, and the methane is purified and used as fuel. There are now 102 commercial landfill gas facilities in the United States; the largest, in Sunnyvale, California, generates enough electricity to power 100,000 homes. In 1996, commercial landfill gas produced the energy equivalent of 5 million tons of wood.

Incomplete Decomposition. The commonly used plastics in MSW are resistant to natural decomposition because of their molecular structure. Chemically, they are polymers of petroleum-based compounds that microbes are unable to digest. Biodegradable plastic polymers have been developed from such sources as corn starch, cellulose, lactic acid, soybean protein, amides, and others. For example, the German company Bayer AG has recently announced plans to market a 100% biodegradable plastic structured as a polyester amide

polymer. Plastic film made from the polymer degrades completely in 70 days, according to company tests. So far, however, none of the biodegradable plastics has seen common usage in consumer products.

A team of "archaeologists" from the University of Arizona, led by William Rathje, has been carrying out research on old landfills. Their research has shown that even materials formerly assumed to be biodegradable—newspapers, wood, and so on—are degraded only slowly, if at all, in landfills. In one landfill, 30-year-old newspapers were recovered in a readable state; layers of telephone directories, practically intact, were found marking each year. Since paper materials are 38% of MSW, this is a serious matter. The reason paper and other organic materials decompose so slowly is the lack of suitable amounts of moisture; the more water percolating through a landfill, the better the biodegradation of paper materials. However, the more percolation there is, the more toxic leachate is produced!

Settling. Finally, waste settles as it compacts and decomposes. Luckily, this eventuality was recognized from the beginning, so buildings have never been put on landfills. Settling presents a problem where landfills have been converted to playgrounds and golf courses, though, because it creates shallow depressions (and sometimes deep holes) that collect and hold water. This process can be addressed by continual monitoring of the facility and the use of fill to restore a level surface.

Improving Landfills. Recognizing the foregoing problems, the EPA upgraded siting and construction requirements for new landfills. Under current regulations:

- New landfills are sited on high ground, well above the water table. Often, the top of an existing hill is bulldozed off to supply a source of cover dirt and at the same time to create a floor that is above the water table.

- The floor is first contoured so that water will drain into a tile leachate-collection system. The floor is then covered with a plastic liner and at least two feet of compacted soil. On top of this is a layer of coarse gravel and a layer of porous earth. With this design, any leachate percolating through the fill will encounter the gravel layer and then move through that layer into the leachate collection system. The clay layer or plastic liner prevents leachate from ever entering the groundwater. Collected leachate can be treated as necessary.

- Layer upon layer of refuse is positioned such that the fill is built up in the shape of a pyramid. Finally, it is capped with a layer of clay and a layer of topsoil and then seeded. The clay-topsoil cap and the pyramidal shape help the landfill to shed water. In this way, water infiltration into the fill is minimized, and less leachate is formed.

- Finally, the entire site is surrounded by a series of groundwater-monitoring wells that are checked periodically, and such checking must go on indefinitely.

These design features are summarized in Fig. 19–5. Most landfills currently in operation have the improved

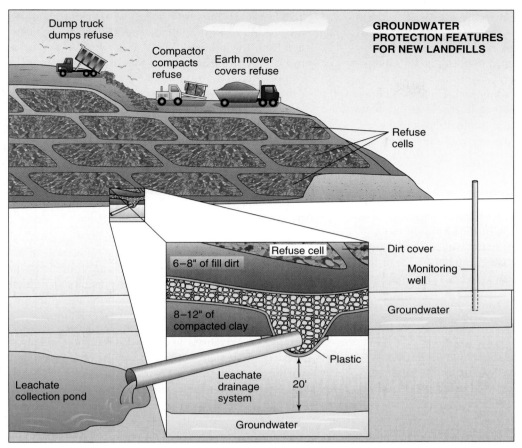

Dump truck
dumps refuse

Compactor
compacts
refuse

Earth mover
covers refuse

**GROUNDWATER
PROTECTION FEATURES
FOR NEW LANDFILLS**

Refuse
cells

Refuse cell

6–8" of fill dirt

8–12" of
compacted clay

Dirt cover

Monitoring
well

Groundwater

Plastic

20'

Leachate
collection pond

Leachate
drainage
system

Groundwater

◀ **FIGURE 19–5** *Features of
a modern landfill.* The landfill
is sited on a high location well
above the water table. The bottom
is sealed with compacted soil and a
plastic liner, overlaid by a rock or
gravel layer with pipes to drain the
leachate. Refuse is built up in
layers as the amount generated
each day is covered with soil, so
that the completed fill has a
pyramidal shape that sheds water.
The fill is provided with wells for
monitoring the groundwater.

technologies, which protect both human health and the environment.

Although the regulations protect groundwater, the landfill pyramids may well last as long as the Egyptian pyramids. (They are not likely to become tourist attractions, however!) And if they do break down, they become a threat to the groundwater; therefore, the need for monitoring remains. Nonetheless, the creative siting and construction of landfills has the potential to address some very significant future needs, as we have seen: The abandoned landfill can become an attractive golf course, recreational facility, or wildlife preserve.

Siting New Landfills. Many states in the northeastern United States are facing less than five years of remaining capacity for their landfills. This is due to the one problem associated with landfills that gets more attention than any other: **siting**. It is not that landfills take up enormous amounts of land; one landfill occupying 121 acres (Central Landfill) serves the entire state of Rhode Island, and Fresh Kill, the largest landfill in the world, serves much of New York City and its environs and is only 2400 acres in area. But as old landfills are closed, it has become increasingly difficult to locate land for the new ones needed to take their place.

People in residential communities (where MSW is generated) invariably reject proposals to site landfills anywhere near where they live. And those who already live close to existing landfills are anxious to close them down. Weary of the

odor and heavy truck traffic, Staten Island, New York, residents have pressured the city of New York to shut down the huge Fresh Kill landfill, the recipient of 13,000 tons of garbage per day. The city has promised to phase out the landfill by the end of 2001, but where the city's waste will go is still not clear.

With spreading urbanization, there are few suburban areas not already dotted with residential developments. Any site selection, then, is met with protests and legal suits. This problem has been repeated in so many parts of the country (and globe!) that it has given rise to several inventive acronyms: **LULU, NIMBY**, and **NIMTOO**. LULU means "Locally unwanted land use"; NIMBY is "Not in my backyard"; and *NIMTOO* is "Not in my term of office." The applications of these attitudes to the landfill siting problem are obvious.

The siting problem has some undesirable consequences. First, it drives up the costs of waste disposal, as alternatives to local landfills are invariably more expensive. Second, it leads to the inefficient and equally objectionable practice of long-distance transfer of trash, as waste generators look for private landfills anxious to receive trash. Very often, this transfer occurs across state and even national lines, leading to very real resentment and opposition on the part of citizens of the recipient state or nation. Such opposition has resulted in numerous state efforts to restrict or prohibit landfills from receiving out-of-state trash. These efforts have invariably been struck down by the courts as unconstitutional interference with interstate commerce. Five states—Illinois, Indiana, Pennsylvania, Ohio, and Virginia—import more than 1 million tons of MSW a year (Fig. 19–6). New Jersey, New York, and Missouri lead the exporting states, with New York far in the lead, at 3.8 million tons a year.

One positive impact of the siting problem is to encourage residents to reduce their waste and recycle as much as possible; this will be explored later. Another effect of the siting problem is to stimulate the use of combustion as an option for waste disposal.

Combustion: Waste to Energy

Because it has a high organic content, refuse (especially the plastic portion) can be burned. Currently, more than 130 combustion facilities are operating in the United States, burning about 36 million tons of waste annually—17% of the waste stream.

Advantages of Combustion. Combustion of MSW has some advantages:

- Combustion can reduce the weight of trash by over 70% and the volume by 90%, thus greatly extending the life of a landfill (which is still required to receive the ash).

- Toxic or hazardous substances are concentrated into two streams of ash, which are easier to handle and control than the original MSW. The *fly ash* (captured from the combustion gases by air pollution control equipment) contains most of the toxic substances and can be safely put into the landfill. The *bottom ash* (from the bottom of the boiler) can be used as fill in some construction sites and roadbeds. Some combustion facilities process the bottom ash further to recover metals and then convert the remainder into concrete blocks.

- No changes are needed in trash collection procedures or people's behavior; trash is hauled to a combustion facility instead of the landfill.

- Practically all modern combustion facilities are designed to generate electricity, which is sold to offset some of the costs of disposal. The newer facilities are equipped with scrubbers and filters or electrostatic precipitators, which remove acid gases and particles to bring the emissions into compliance with Clean Air Act regulations.

Simple incineration is an outmoded method of combustion; it accomplishes a reduction in volume of the trash, but without the recovery of energy. When burned, unsorted MSW releases about 35% as much energy as coal, pound for pound.

▶ **FIGURE 19–6** *Trash on the move.* Net import and export of municipal solid waste, by state, in 1992.

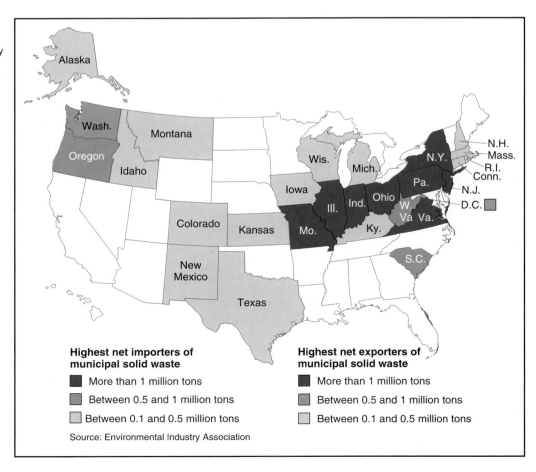

Highest net importers of municipal solid waste
- More than 1 million tons
- Between 0.5 and 1 million tons
- Between 0.1 and 0.5 million tons

Highest net exporters of municipal solid waste
- More than 1 million tons
- Between 0.5 and 1 million tons
- Between 0.1 and 0.5 million tons

Source: Environmental Industry Association

Most of the currently operating combustion facilities (110 of 130) are **waste-to-energy (WTE)** facilities equipped with modern emission-control technology. Many of these facilities add **resource recovery** to their waste processing, in which many materials are separated and recovered before (and sometimes after) combustion. Because the sale of the electricity offsets some of the costs of MSW disposal, the combustion facilities are usually able to compete successfully with landfills for MSW.

Drawbacks of Combustion. Combustion has some drawbacks, however:

- Trash does not burn cleanly. Despite being equipped with air pollution control devices, exhaust stacks emit toxic fumes into the air as burning oxidizes and vaporizes the assortment of metals, plastics, and hazardous materials that inevitably end up as municipal waste.

- Combustion facilities are expensive to build, and their siting has the same problem as the landfills: No one wants to live near one.

- Combustion ash is often loaded with metals and other hazardous substances and must be disposed of in secure landfills.

- To justify the cost of its operation, the combustion facility must have a continuing supply of MSW. For that reason, the facility enters into long-term agreements with municipalities, and these agreements can lessen the flexibility of the community's solid waste management options.

- Even if the combustion facility generates electricity, the process wastes both energy and materials, unless it is augmented with recycling and recovery. A number of combustion facilities compete directly with recycling for burnable materials such as newspapers and represent a major impediment to recycling in some municipalities.

An Operating Facility. Let us look at the operation of a typical modern WTE facility. The facility might serve a number of communities or a larger metropolitan area. Servicing a population of a million or more, the plant receives about 3000 tons of MSW per day. The waste comes in by rail and truck, and the communities pay tipping fees (the costs assessed at the disposal site) ranging from $15 to $100 per ton. Waste processing is efficient: Overall, about 80% of the MSW is burned for energy, 12% is recovered, and 8% is landfilled. The process, pictured in Fig. 19–7, is as follows:

1. Incoming waste is first inspected, and obvious recyclable and bulky materials are removed.

2. Waste is then pushed onto conveyers that feed shredders capable of reducing the width of waste particles to 6 inches or less.

3. Strong magnets remove about two-thirds of ferrous metals for recycling before combustion.

4. The waste is then blown into boilers, where light materials burn in suspension and heavier materials burn on a moving grate.

5. Water circulated through the walls of the boilers produce steam, which drives turbines for generating electricity.

6. After the waste is burned, the bottom ash is conveyed to a processing facility, where further separation of metals may occur in a process that recovers brass, aluminum, gold, copper, and iron. In some facilities, this process nets $1000 a day in coins alone!

7. Combustion gases are passed through a lime-based spray dryer/absorber to neutralize sulfur dioxide and other noxious gases and then through electrostatic precipitators that remove particles. The resulting waste stream is significantly lower in pollutants than would be emitted by an energy-equivalent utility based on coal or oil combustion.

8. The fly ash and bottom ash residues are put into landfills.

An appreciation for the impact of such a facility can be gained from looking at the outcome of a year's operation. In a year, 1 million tons of MSW is processed, 40,000 tons of metal is recycled, and 570,000 megawatt-hours of electricity is generated—the equivalent of more than 60 million gallons of fuel oil and enough electricity to power 65,000 homes. All this comes from stuff that people have thrown away!

The advantages of combustion over landfilling are obvious; the major disadvantage is that some recycling and composting opportunities are missed. This drawback can be addressed at the collection point, as we will see when we examine the recycling option. In any event, substantial recovery of recyclable materials—and energy—certainly can be accomplished with a modern WTE facility.

Costs of Municipal Solid Waste Disposal

The costs of disposing of MSW are becoming prohibitive. Increasing costs are not just a result of the new design features of landfills; more and more, they reflect the expenses of acquiring a site and providing transportation. Tipping fees now exceed $100 a ton at some landfills, and the waste collector must recover this cost, as well as transportation costs to the site. Tipping fees at some combustion facilities are no better; one WTE combustion facility in North Andover, Massachusetts, recently planned to raise its tipping fee from $92 to $240 per ton by 2000 to cover the cost of adding a mandated stack scrubber.

Getting rid of all trash is getting more expensive, and one sad consequence of this is illegal dumping. Some towns are charging up to $1.50 a bag for MSW disposal; it now costs $1 or more to get rid of an automobile tire and $30 or more to dispose of a refrigerator. As a result, tires, refrigerators, yard waste, car parts, and construction waste are appearing all over the landscape. Institutions and apartments that operate with dumpsters often must put padlocks on them in order to prevent their unauthorized use by people trying to avoid disposal costs. Many states have established a corps of environmental police to track down midnight dumpers and bring them to justice.

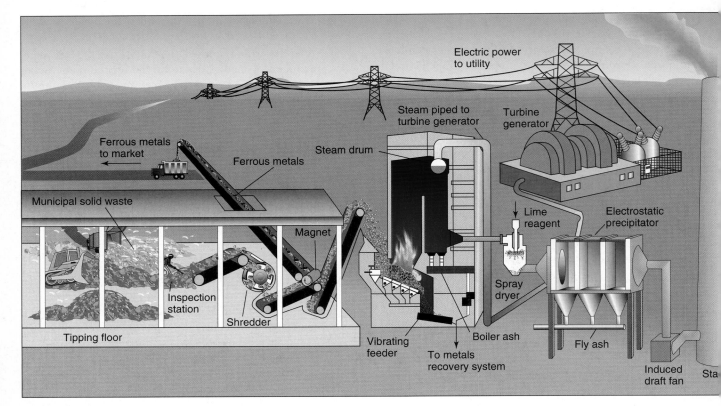

▲ **FIGURE 19–7** *Waste-to-energy combustion facility.* Schematic flow for the separation of materials and combustion in a typical modern waste-to-energy combustion facility.

19.2   Solutions

Source Reduction

The best strategy of all is to reduce waste at its source. This accomplishes two goals: It reduces the amount of waste that must be managed, and it conserves resources. It is noteworthy that the amount of waste per capita in the United States peaked at 4.5 lbs in 1990, has leveled off, and is now at 4.3 lbs. (We still lead the world in this dubious statistic, however.) This is a signal that some changes in lifestyle may be occurring that have an impact on waste generation. We cite two recent developments:

1. Lightening the weight of many items has reduced the amount of materials used in manufacturing. Steel cans are 60% lighter than they used to be; disposable diapers contain 50% less paper pulp, due to absorbent-gel technology; and aluminum cans contain only two-thirds as much aluminum as they did 10 years ago.
2. The Information Age may be having an impact on paper use: Electronic communication, data transfers, and advertising are increasingly performed on personal computers, and the number of homes with personal computers continues to rise. For example, there were 1.7 trillion e-mail messages sent in 1996, growing at a rate of 50% per year. The growth of the World Wide Web as a medium for information transfer has been phenomenal.

An option that thus far has received too little attention is the potential for reducing the volume of waste by keeping products in use longer. Reusing items in their existing capacity is an efficient form of waste reduction. The use of returnable versus nonreturnable beverage containers is a prime case in point.

Returnable vs. Nonreturnable Bottles. For many years, most beer and soft drinks were marketed by local bottlers and breweries in returnable bottles that required a deposit. Trucks delivered filled bottles to retailers and picked up empties to be cleaned and refilled. This procedure is efficient when the distance between producer and retailer is relatively short. As the distance increases, however, transportation costs become prohibitive because the consumer pays for hauling the bottles, as well as for the beverage. In the late 1950s, distributors, bent on expanding markets and growth, observed that transportation costs could be greatly reduced if they used lightweight containers that could be thrown out rather than shipped back. Thus, no-deposit, nonreturnable bottles and cans were introduced. The throwaway container is also an obvious winner for its manufacturers, who profit by each bottle or can they produce.

Through massive advertising campaigns promoting national brands and the convenience of throwaways, a handful of national distributors gained dominance during the 1950s and 1960s and countless local breweries and bottlers were driven out of business—and with them went the returnable bottle. At the same time, nonreturnable bottle and can manufacturing grew into a multibillion-dollar industry.

The average person drinks about a quart of liquid each day. Given that there are 273 million Americans, this daily consumption amounts to some 26 billion gallons of liquid per year. Most of this volume is packaged in single-serving containers that are used once and then thrown away.

Beverages in nonreturnable containers and those in returnable containers appear to be priced competitively on the market shelf, but this equality is seen to evaporate when one looks at the hidden costs of single-use containers. Nonreturnable containers constitute 6% of the solid waste stream in the United States and about 50% of the nonburnable portion; they also constitute about 90% of the nonbiodegradable portion of roadside litter. Broken bottles along roads, beaches, and parklands are responsible for innumerable cuts and other injuries, not to mention flat tires. Both the mining of the materials and the process used to manufacture these beverage containers create pollution. All of these are hidden costs that do not appear on the price tag. However, we pay them with taxes to clean up litter, as well as with our injuries, flat tires, environmental degradation, and so on.

In an attempt to reverse the trend, environmental and consumer groups have promoted *bottle laws*—laws that facilitate the recycling or reuse of beverage containers. Such laws generally call for a deposit on all beverage containers, both returnables and throwaways. Retailers are required to accept the used containers and pass them along for recycling or reuse.

Bottle laws have been proposed in virtually every state legislature over the last decade. Nevertheless, in every case, the proposals have met with fierce opposition from the beverage and container industries and certain other special-interest groups. The reason for their opposition is obvious—economic loss to their operations—but the arguments they put forth are more subtle. The container industry contends that bottle laws will result in loss of jobs and higher beverage costs for the consumer. They also claim that consumers will not return the bottles and that the amount of litter will not diminish. In a 1996 media blitz, these same arguments were used in defeating an expansion of Oregon's successive bottle bill.

In most cases, the industry's well-financed lobbying efforts have successfully defeated bottle laws. However, some states—10, as of 1998—have adopted bottle laws of varying types despite industry opposition (Table 19-1). Their experience has proved the beverage and bottle industry's arguments false. More jobs are gained than lost, costs to the consumer have not risen, a high percentage of bottles are returned, and there is a marked reduction in the can and bottle portion of litter.

A final measure of the success of bottle laws is continued public approval. Despite industry efforts to repeal bottle laws, no state that has one has repealed it. Repeated attempts have been made to get a national bottle law through Congress—to date, unsuccessfully. Opponents (the same ones that oppose state-level bottle bills) argue that such a law will threaten the newly won successes in curbside recycling, with some justification: Beverage containers represent the most important

TABLE 19-1 States that Have Passed Bottle Laws			
State	**Year Passed**	**State**	**Year Passed**
Oregon	1972	Iowa	1978
Vermont	1973	Massachusetts	1978
Maine	1976	Delaware	1982
Michigan	1976	New York	1983
Connecticut	1972	California	1991

source of revenue in curbside recycling. However, since curbside recycling currently reaches only half of the U.S. population, a national bottle law will undoubtedly recover a greater proportion of beverage containers. (States with bottle laws report 80% to 97% rates of return of containers.) A national bottle law would be labor intensive, employing tens of thousands of workers and creating countless new jobs.

Other Measures

Resale. Whenever items are reused rather than thrown away, the effect is a reduction in waste and better conservation of resources. In this respect, it is encouraging to see the growing popularity of yard sales, flea markets, consignment clothing stores and other not-new markets (Fig. 19–8).

Elimination of Junk Mail. We are all on bulk-mailing lists, because these lists are sold and shared widely; this guarantees that each of us will receive increasing volumes of advertising. To stay off such lists, simply inform mail-order companies and other organizations involved that you do not want your name and address shared. You can write to the Mail Preference Service, Direct Marketing Association, P.O. Box 9008, Farmingdale, NY 11735-9008. This is a no-charge service that provides a master list of people who do not wish to receive mass advertising; it works as long as marketers choose to consult it.

An examination of our domestic wastes and their disposal reveals that a mammoth stream of material flows in one direction, from our resource base to disposal sites. Just as natural ecosystems depend on recycling nutrients, if we are to move in the direction of sustainability, we will also ultimately need to learn to recycle our wastes. There is strong evidence that we are moving in this direction.

The Recycling Solution

In addition to reuse, recycling is another obvious solution to the solid waste problem. More than 75% of MSW is recyclable material. There are two levels of recycling: **primary** and **secondary**. *Primary recycling* is a process in which the original waste material is made back into the same material—for example, newspapers recycled to make newsprint. In

▲ **FIGURE 19–8** *Waste reduction by reuse.* Volunteers preparing items for sale in a secondhand store. By reselling used clothes, such stores are reducing the volume of discarded waste.

secondary recycling, waste materials are made into different products, which may or may not be recyclable—for example, cardboard from waste newspapers.

The primary items from MSW currently being heavily recycled are cans (aluminum and steel), bottles, plastic containers, newspapers, and yard wastes. Yet, there are many alternatives for reprocessing various components of refuse, and people are coming up with new ideas and techniques all the time. A few of the major established techniques, together with current percentages of their recovery by recycling, are as follows:

- Paper (41% recovery) can be repulped and reprocessed into recycled paper, cardboard, and other paper products; finely ground and sold as cellulose insulation; or shredded and composted.
- Glass (26% recovery) can be crushed, remelted, and made into new containers or crushed and used as a substitute for gravel or sand in construction materials such as concrete and asphalt.
- Some forms of plastic (5.4% recovery) can be remelted and fabricated into carpet fiber, outdoor wearing apparel, irrigation drainage tiles, and sheet plastic (Fig. 19–9).
- Metals can be remelted and refabricated. Making aluminum (38% recovery) from scrap aluminum saves up to 90% of the energy required to make aluminum from virgin ore. In addition, these ores are imported and are part of the mounting U.S. trade deficit. National recycling of aluminum saves energy, creates jobs, and reduces the trade deficit.
- Food wastes and yard wastes (leaves, grass, and plant trimmings—39% recovery) can be composted to produce a humus soil conditioner.

- Textiles (12% recovery) can be shredded and used to strengthen recycled paper products.
- Old tires (19% recovery) can be remelted or shredded and incorporated into highway asphalt or burned in special combustion facilities.

Recycling is both an environmental and an economic issue. Many people are motivated to recycle because of environmental concern, but the use of recycled materials is strongly driven by economic factors.

Municipal Recycling. Recycling is probably the most direct and obvious way most people can become involved in environmental issues. The appeal of recycling is obvious: If you recycle, you save some natural resources from being used (trees, in the case of paper) and you prevent landfills from becoming "landfulls."

Recycling's popularity is well established; only five states lack specific recycling goals. EPA sources report that only 6.7% of MSW was recycled in 1960 versus 27% in 1996. There is a great diversity of approaches to recycling in municipalities, from recycling centers requiring residents to drive miles to recycle to curbside recycling with sophisticated separation (Fig. 19–10). The most successful programs have the following characteristics:

▲ **FIGURE 19–9** *Plastic recycling.* Bales of plastic bottles in a recycling center.

◀ **FIGURE 19–10** *Curbside recycling.* Curbside pickup by waste hauler using a special truck. This approach is reaching over 50% of the U.S. population.

1. There is a strong incentive to recycle, in the form of direct charges for general trash and no charge for recycled goods.
2. Recycling is not optional; mandatory regulations are in place, with warnings and sanctions for violations.
3. Residential recycling is curbside, with free recycling bins distributed to households. (Curbside recycling has risen rapidly; over 50% of the U.S. populace is now served with curbside pickup.)
4. Recycling goals are ambitious yet clear and feasible. Some percent of the waste stream is targeted, and progress is followed and communicated.
5. A concerted effort is made to involve local industries in the recycling process.
6. The municipality employs an experienced and committed recycling coordinator.

Nevertheless, municipalities experience very different recycling rates. New York City, for example, has made great progress—from recycling only 5% to 30% in four years—by mandating curbside recycling for all 3 million households. Boston, on the other hand, with only 12% recycling, is holding Massachusetts back in its goal to achieve 46% by 2000. Worcester, farther to the west in the state, has achieved an outstanding 55% recycling rate.

Recycling has its critics, who base their arguments primarily on economics. If the costs of recycling (from pickup to disposal of recycled components) are compared with the costs of combustion or placing waste in landfills, recycling frequently comes out second best. Markets for recyclable materials fluctuate wildly, and residents often end up subsidizing the recycling effort. The shortfall between recycling costs and market value range from $20 to $135 a ton for some recyclables (items such as green glass and colored plastic bottles). In some cases (newspaper, glass), prices are controlled by the industry of origin. Competition between landfills and combustion facilities often lowers tipping fees, creating an even greater disincentive to recycle. Thus, critics of current recycling practices argue that unless recycling pays for itself through the sale of recovered materials, it should not be done. However, the critics ignore the subsidies that make the use of virgin materials less expensive than recycled materials. Also, garbage collection is big business, and those involved see recycling as cutting their trash business.

In spite of these economic considerations, the demonstrated support for recycling programs is strong; experience has shown that at least two-thirds of households will recycle if presented with a curbside pickup program. Even when people understand that it costs more to recycle, there is strong support for the effort.

Paper Recycling. By far the most important recycled item is newspapers, because of their predominance in the waste stream. It is a simple matter to tie up or bag household newspapers, and the amount recovered by recycling is increasing dramatically. Currently, more newspaper is being recycled than discarded (54%). Since more than 25% of the trees harvested in the United States are used to make paper, recycling paper obviously saves trees. Depending on the size and type of trees, a 1-meter stack of newspapers equals the amount of pulp from one tree.

There is often some confusion in what is meant by "recycled paper." The key is the amount of *postconsumer* recycled paper in a given product. Much paper is "wasted" in manufacturing processes, and this is routinely recovered and rerouted back into processing and called "recycled paper." Thus, the total recycled content of a paper can be 50%, although the actual postconsumer amount recovered by recycling might be only 10% of the total paper.

As with the recycling of other materials, paper recycling is highly dependent on the existence of a market for the recycled products. This involves cost considerations, from the point of collection of the wastepaper to the point of use by consumers. For example, there are all types of paper, and recycling mills want specific grades; gross mixtures have little market appeal. So the processor of the recycled paper must perform some separation and packaging for specific uses.

After the wastepaper is incorporated into a final product, the market becomes a critical factor. Is there a demand for recycled paper? Several years ago, much of the recycled paper was rough and off-color. In order to guarantee a market for the product, many organizations—including state and federal agencies—required purchasing and using paper with recycled content. The technology for producing high-quality paper from recycled stock has improved greatly, to the point where it is virtually impossible to distinguish it from "virgin" paper. Although the mandates for "closing the loop," as the practice of creating markets for recycled products is called, are important, they may soon disappear as paper manufacturers make the use of recovered paper a standard practice.

The market for used newspapers has fluctuated greatly over the past several decades. During the late 1980s, the market was saturated and municipalities often had to pay to get rid of newspapers. In 1995, discarded papers were so valuable—up to $160 a ton—thieves were stealing them off the sidewalks before the recycling trucks could pick them up. A year later, as more recycling programs came on line, the market collapsed again and many cities once more paid as much as $25 a ton to have the newspapers hauled away—this is still less expensive than the cost of landfilling.

There is a lively international trade as well in used paper (Fig. 19–11). Forest-poor countries in Europe and Asia purchase wastepaper from the United States and other industrial countries in the Northern Hemisphere, where there

continues to be a surplus of such paper. The largest importer is Taiwan (almost 2 million tons per year). Taiwan also boasts the highest percentage of reused paper content in its paper: 98%. The United States is at a much lower 33%.

Plastics Recycling. Plastics have a bad reputation in the environmental debate for many reasons. First, they have many uses that involve rapid throughput—for example, packaging, bottling, the manufacture of disposable diapers, and incorporation into a host of cheap consumer goods. Second, plastic production has increased 10% a year for the past three decades. Third, plastics are conspicuous in MSW and litter—ironically, most trash is disposed of in plastic bags manufactured just for that purpose. Finally, plastics do not decompose in the environment, because no microbes (or any other organisms) are able to digest them. (Imagine plastics in landfills delighting archaeologists hundreds of years from now.) Therefore, the possibility of recycling at least some of the plastic in products—liquid containers—interests the environmental community.

If you look on the bottom of a plastic container, you will see a number (or some letters) inside the little triangle of arrows that has come to represent recycling (Fig. 19–12). These numbers are used to differentiate between and sort the many kinds of plastic polymers—not all of which are recyclable. The two recyclable plastics most commonly in use are high-density polyethylene (HDPE), code 2, and polyethylene terephthalate (PETE), code 1. In the recycling process, the plastics must be melted down and poured into molds, and unfortunately, some contaminants from the original containers may carry over. This makes it difficult to reuse the plastic for food containers; thus, the uses for the recycled plastic are necessarily restricted. However, some new uses for recycled plastic are appearing, with more in the development stage: PETE, for example, is turned into carpet fiber and fill for outdoor wearing apparel, and HDPE becomes irrigation drainage tiles, sheet plastic, and, appropriately, recycling bins.

Critics of plastic recycling point out that if the process were purely market driven, it wouldn't happen. Recovering the plastic is more costly than starting from scratch—that is, beginning with petroleum derivatives—and manufacturers

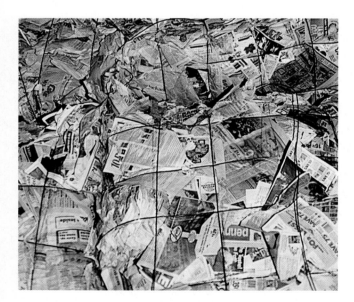

▲ **FIGURE 19–11** *Wastepaper exports.* Bales of wastepaper being loaded for shipment overseas.

▲ **FIGURE 19–12** *Recycling symbol.* This symbol representing high-density polyethylene is on the bottom of a plastic bottle.

are getting involved mainly because environmentally concerned consumers demand it. Industry-supported critics also point out that plastics in landfills create no toxic leachate or dangerous biogas; also, plastics in combustion facilities burn wonderfully hot and leave almost no ash. However, as oil reserves continue to dwindle, the petroleum corporations will find it increasingly difficult to promote a once-through waste stream for plastics. Perhaps we can conclude that even though plastic recycling has its problems, continued attention to plastics in MSW will keep us moving in a sustainable direction.

Composting

An increasingly popular way of treating yard waste and food scraps (currently 13% of MSW) is composting, especially as more and more states are banning yard waste from MSW collections. Recall that composting involves the natural biological decomposition (rotting) of organic matter in the presence of air. Composting can be carried out in a backyard by homeowners, or it can be promoted by municipalities as a way of dealing with a large fraction of MSW. On a large scale, composting involves laying down the wastes in windrows up to 12 feet high and 24 feet wide. The microbes and detritus feeders (worms, grubs, etc.) will decompose the organic matter in the compost heap and greatly reduce the volume of waste. The process is speeded by aeration, which is accomplished by turning the material with large mechanical forks.

Composting has clear economic and environmental benefits over putting waste in landfills and over combustion if the yard waste can be collected and processed independently of collecting trash. The end product is a residue of humuslike material, which can be used as an organic fertilizer and soil builder. Composting is one method of treating sewage sludge, as described in Chapter 18.

19.3 Public Policy and Waste Management

Management of MSW was entirely under the control of local governments until quite recently. It is still largely under local management, but state and federal legislation addresses some aspects of MSW disposal. Regulation at higher levels has taken some of the disarray out of the management scene, but towns and cities still have the basic responsibility for managing their MSW.

The Regulatory Perspective

At the federal level, the following legislation has been passed:

- The first attempt of Congress to address the problem was the Solid Waste Disposal Act of 1965. The legislation gave jurisdiction over solid waste to the Bureau of Solid Waste Management, but the agency's mandate was basically financial and technical rather than regulatory.

- With the creation of the EPA in 1970, the Resource Recovery Act of 1970 gave jurisdiction to the EPA and directed attention to recycling programs and other ways of recovering resources in MSW. The act also encouraged the states to develop some kind of waste management program.

- The passage of the Resource Conservation and Recovery Act (RCRA) of 1976 saw a more regulatory ("command and control") approach to MSW, as the EPA was given power to close local dumps and set regulations for landfills. Combustion facilities were covered by air pollution and hazardous waste regulations, again under the EPA's jurisdiction. The RCRA also required the states to develop comprehensive solid waste management plans.

- The Superfund Act of 1980 (described in Chapter 20) addressed abandoned hazardous waste sites throughout the country, many of which (41%) are old landfills.

- The Hazardous and Solid Waste Amendments of 1984 gave the EPA greater responsibility to set solid waste criteria for all hazardous waste facilities. Since even household waste must be assumed to contain some hazardous substances, this meant that the EPA had to determine all landfill and combustion criteria more closely.

Largely in response to these federal mandates, the states have been putting pressure on local governments to develop integrated waste management plans (see next section), with goals for recycling, source reduction, and landfill performance. State solid waste management is comprehensive and ambitious in many states, particularly in the Northeast, where landfills have been closing. States are setting goals and policies for recycling, one significant manifestation of which is the *materials recovery facility* (see the "Earth Watch" essay, p. 470), where processing of waste allows for improved recovery of materials for recycling. There are now more than 360 of these facilities, spread throughout the country, capable of processing 30,000 tons of waste per day.

Integrated Waste Management

It is not necessary to fasten onto just a single method of waste handling. Source reduction, waste-to-energy combustion, recycling, recovery facilities, landfills, and composting all have roles to play in waste management. Different combinations of these options will work in different regions of the country. A system having several alternatives in operation at the same time is called *integrated waste management*. Let us examine the available MSW management options, with a view toward developing recommendations that make environmental sense — that is, toward moving in a more sustainable direction.

Waste Reduction. Many observers have called the late twentieth-century United States a "throwaway society." We probably deserve that label for setting the world record for per capita waste production, along with per capita energy consumption

EARTH WATCH

REGIONALIZED RECYCLING

Currently, most recycling is on a town-by-town basis. Either the towns or their waste contractors must find markets for the recycled goods: cans, bottles, newspapers, and plastics. This problem has kept many municipalities from getting into recycling. The solution? A regionalized materials recovery facility (MRF), referred to in the trade as a "murf." Here's how the state-owned MRF in Springfield, Massachusetts, works.

Basic sorting takes place when waste is collected, either by curbside collection or by town recycling stations (sites where townspeople can bring wastes to be recycled). The waste is then trucked to the MRF and handled on three tracks—one for metal cans and glass containers, another for paper products, and a third for plastics. The materials are moved through the facility by escalators and conveyor belts, tended by workers who inspect and do further sorting. The objective of the process is to prepare materials for the recycled goods market. Glass is sorted by color, cleaned, crushed into small pebbles, and then shipped to glass companies, where it replaces the raw materials that go into glass manufacture—

sand and soda ash—and saves substantially on energy costs. Cans are sorted, flattened, and sent either to detinning plants or to aluminum processing facilities. Paper is sorted, baled, and sent to reprocessing mills. Plastics are sorted into four categories, depending on their color and type of polymer, and then sold.

The facility's clear advantages are its economy of scale and its ability to produce a high-quality end product for the recycled materials market. Towns know where to bring their waste, and they quickly become familiar with the requirements for initial sorting during collection and transfer.

Currently, there is only one MRF in Massachusetts, built with state funds at a cost of $5 million and operated by the state Department of Environmental Protection. Several more are on the drawing board. After its first year in operation, the Springfield MRF was declared a success. It took in 42,886 tons of materials from 91 towns, representing a population of 750,000. Depending on the community, the process has diverted 22% to 50% of the waste stream from landfills and incinerators and saved the towns millions of dollars in disposal costs. The facility has attracted recycled wastes

from Connecticut and New York while those states are setting up their own regional recycling centers.

Nevertheless, there were some unanticipated problems at the Springfield facility during its first year of operation. Several towns reduced their trash so much that they are not fulfilling their contractual obligations to a regional combustion facility. Landfill revenues have declined, forcing landfill operators to lower their fees in order to attract more haulers. These appear to be temporary problems that will be worked out in time, however. The state is committed to recycling almost half of MSW and combusting most of the rest by the year 2000. In theory, landfills will operate mainly as recipients of combustion ash and materials that can't be either recycled or burned.

Very likely, the future of regional recycling will be in the private sector. In 1996, 363 MRFs were operating in the United States, some with the high technology of magnetic pulleys, optical sensors, and air sorters, and many with simple manual sorting. These facilities have a daily capacity of 29,000 tons. All indications point to such facilities as increasing in the future.

(Fig. 19–13). How can we turn this around? True management of MSW begins in the home (or dorm room). Materialistic lifestyles, affluence, and overconsumption can pack our homes and our trash cans with stuff we simply do not need; indeed, some observers see this as a malady called *affluenza* (see the "Ethics essay," p. 472).

Several incentives could be put in place on the governmental level to encourage waste reduction. One initiative to reduce waste as part of living more sustainably is the Global Action Plan (explored in depth in Chapter 24). This international program is designed to enable people to change the way they live; families enrolled in the program generate 42% less waste, among other things. Another possibility would be for the government to offer incentives and opportunities to organizations and businesses, encouraging them to do what they can to bring about a decrease in the per capita MSW. For example, the EPA sponsors the WasteWise program, targeting the reduction of MSW by establishing partnerships with not only local governments and organizations but multinational corporations as well. This is a voluntary partnership

that allows the partners to design their own solid waste reduction programs. In this growing program, 1997 accomplishments of the partners include eliminating more than 816,000 tons of waste, avoiding $86 million in costs, and purchasing nearly $3 million of recycled-content products.

One more important step in waste reduction would be to stop subsidizing garbage disposal. Instead of paying for trash collection and disposal through local taxes, communities would levy curbside charges for all unsorted MSW— say, $1 or more per container (Fig. 19–14). In actuality this practice is on the increase, and communities adopting it report reductions of 25% to 45% in the MSW stream. Recycling in Seattle, Washington, led to increasing the MSW that was recycled from 24% in 1988 to the current 48% after a per-bag fee was initiated.

Another policy for bringing about waste reduction would be to establish a program of extended product responsibility (EPR), a concept that involves assigning some responsibility for reducing the environmental impact of a product at each stage of its "life cycle," especially at the end.

◀ **FIGURE 19–13** *The throwaway society.* The production of junk and accelerated obsolescence!

For example, Hewlett-Packard and Xerox make it easy for customers to return spent copier cartridges, and they then recycle the components of the cartridges. EPR originated in Western Europe; Germany, the Netherlands, and Sweden have well-developed regulations that require take-back of many items, promote reuse of products, and promote manufacture of more durable products.

Waste Disposal. No human society can avoid generating some waste. There will always be MSW, no matter how much we reduce, reuse, and recycle it. Integrated waste management will require providing several options for waste disposal. Most experts see an increase in the percentage of waste going to WTE combustion facilities, a

▲ **FIGURE 19–14** *Pay-as-you-throw trash pickup.* Consumers purchase the empty bags for a set price, and only these bags are picked up by the trash hauler. Next to the bags are three recycling containers.

decrease in the percentage going to landfills, and a continued rise in recycling.

It will be important to break the gridlock at local and state levels on the siting of new landfills and WTE combustion facilities. Landfills will still be needed, although they should last longer than in the past. Policymakers have long known about the landfill shortage but have opted for short-term solutions with the lowest political cost. One result of this is the long-distance hauling of MSW by truck and rail, described earlier. If regions and municipalities are required to handle their own trash locally, they will find places to site landfills, regardless of NIMBY considerations. Of course, new landfills must use the best technologies: proper lining, leachate collection and treatment, groundwater monitoring, biogas collecting, and final capping. If these are employed, a landfill will not be a health hazard. Also, sites should be selected with a view to some future use that is attractive.

Another policy goal should be to encourage more WTE combustion of MSW. Again, the technologies employed should be the best available, and again, if these are used, there should be no significant threat to human health. This option seems to be the best way to deal with mixed waste that is nonrecyclable. Much of the energy content of the waste is converted to electricity, metals are recovered, and the waste is greatly reduced in volume.

Recycling and Reuse. Recycling is certainly the wave of the future; it should not, however, be pursued in lieu of waste reduction and reuse. The move toward more durable goods is an overlooked and underutilized option.

Many states have passed mandatory recycling laws. Such laws require that each region or municipality, under threat of loss of state funds, recycle a certain percentage of its refuse by a certain date. Massachusetts, for example, has adopted a goal of recycling 46% (it is currently 35%) of all of its MSW by the year 2000.

ETHICS

AFFLUENZA: DO YOU HAVE IT?

First, take this sample quiz to see if you have the bug.

1. I'm willing to pay more for a T-shirt if it has a cool corporate logo on it.
2. I'm willing to work at a job I hate so I can buy lots of stuff.
3. I usually make just the minimum payment on my credit cards.
4. When I'm feeling blue, I like to go shopping and treat myself.
5. I spend much more time shopping each month than I do being involved in my community.
6. I'd rather be shopping right now.
7. I'm running out of room to store my stuff.

Give yourself two points for true and one point for false. If you scored 10 or above, you may have a full-blown case of "affluenza."

Just what is affluenza? Jessie H. O'Neill, who coined the term, defines it as a "dysfunctional relationship with wealth or money." Symptoms of affluenza include a love of shopping, a glut of stuff in your home or dorm room that you really don't need or use, and a dissatisfaction with what you have. It can strike anyone, regardless of economic status. According to two programs produced about it by the Public Broadcasting System, *Affluenza* and *Escape from Affluenza*, affluenza is a plague of materialism and overconsumption that is so pervasive that it actually characterizes our modern society. Recall that the United States leads the world in per capita waste generation. This is a symptom of a society-wide problem, according to O'Neill. It begins at an early age, as children are bombarded by TV commercials urging them to get the latest toys. Peer pressure makes it worse, as children are "forced" to wear only the approved apparel. Every fad that comes along must be accommodated. Eventually, the conditioned children grow

up, carry credit cards, and drive, and then affluenza takes on more serious consequences. Adults caught by the disease acquire so much stuff they have to rent a self-storage bin to hold things they can't possibly part with. Bankruptcy and credit card overloads are commonplace, a consequence of people's inability to control their spending.

How do you escape from affluenza? There are several steps you can take. First, admit you have a problem. Then, begin to take small withdrawal steps. Before you buy something, ask yourself: Do I need it? Could I borrow it from a friend or relative? How many hours do I have to work to pay for it? Another suggested step is to avoid those recreational shopping trips to the mall. Take a walk or play ball with some kids instead. Become an advertising critic. Make a budget. These are a few of the many possible pathways for escape from affluenza. As you do these things, you may be amazed at how challenging and rewarding it can be to live more simply. And, it might be said, more sustainably.

Banning the disposal of recyclable items in landfills and combustion facilities makes very good sense. Many states have incorporated this regulation into their management program; for example, Massachusetts phased out yard waste in 1991, metals and glass in 1992, and recyclable papers and plastic in 1994. Landfill- and combustion-facility operators are supposed to conduct random truck searches and are authorized to turn back trucks with significant amounts of banned items.

Enacting a national bottle-deposit law would make a giant stride toward the reuse and recycling of beverage containers and would also greatly reduce roadside litter in states lacking bottle laws.

Finally, closing the "recycling loop" remains a significant action to be taken by governments to encourage recycling. A number of states have opted for one or more of the following approaches: (1) minimum postconsumer levels of recycled content for newsprint and glass containers; (2) requirements stating that purchases of goods include recycled products even if they are more expensive than "virgin" products; (3) requirements that all packaging be reusable or made (at least partly) of recycled materials; (4) tax credits or incentives that encourage the use of recycled materials in manufacturing; (5) assistance in the development of recycling markets.

ENVIRONMENT ON THE WEB

DIFFICULTIES OF WASTE STREAM COMPOSITION

Like every other species, humans generate waste. On average, people in the United States and Canada generate 3 to 6 pounds of solid waste per person per day—an enormous difference from the 20 to 30 pounds per person *per year* that is typical of less developed countries. One of the most challenging aspects of managing all this waste is the composition of the municipal solid waste stream.

Typically, municipal waste is made up of about 40 to 50% paper, wood, and plastics (combustibles), about 10% metal and glass (noncombustibles), about 20 to 40% food wastes, yard wastes, and other compostable materials, and small proportions of inert solids, ash, sewage sludge, and bulky items such as furniture and appliances. Composition of a waste stream, however, can vary from community to community. In some places, the proportion of construction wastes—scrap wood, wallboard, plaster, concrete, and similar materials—may be high, while in other communities, the waste stream contains a high proportion of yard and food wastes, which are easily composted by homeowners. Regardless, most communities must contend with highly variable waste streams—for instance, very high compostable materials during some times of year (such as fall), or certain days of the week, but lower at other times. This uncertainty makes it difficult for municipalities to optimize treatment and disposal methods, even within the same region.

An integrated waste management strategy therefore needs to accommodate the full range of materials that can be expected in the waste stream, including provisions for waste collection, transportation, and disposal. Increasingly, municipalities are faced with rising difficulty and costs in all areas of waste management and, as a result, are turning to alternatives that encourage residents to reduce, reuse, and recycle rather than just discard their wastes.

Web Explorations

The Environment on the Web activity for this essay discusses the "pay-as-you-throw" (PAYT) policy that requires residents to pay a fee for each unit of garbage they generate. Go to the Environment on the Web activity (select Chapter 19 at http://www.prenhall.com/nebel) and learn for yourself:

1. about communities that use pay-as-you-throw policies;
2. how communities make the decision to adopt user-pay waste management policies; and
3. about illegal diversion and enforcement of pay-as-you-throw systems.

A suggested time frame for each of these exercises is 10–30 minutes.

REVIEW QUESTIONS

1. List the components of municipal solid waste (MSW).
2. Trace the historical development of refuse disposal. What method of disposal is now most common in the U.S.?
3. What are the major costs and limitations of placing waste in landfills?
4. Outline the EPA's latest regulations for the construction of new landfills.
5. Why are landfills not a sustainable option for solid waste disposal?
6. What are the advantages and disadvantages of combustion?
7. How can the volume of waste in the U.S. be reduced?
8. Define primary and secondary recycling.
9. Discuss the attributes of successful recycling programs.
10. Which waste materials may be composted?
11. What laws has the federal government adopted to control solid waste disposal?
12. What is integrated waste management? Name some options for solid waste management.

THINKING ENVIRONMENTALLY

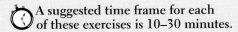

1. Compile a list of all the plastic items you used and threw away this week. How can you reduce this list?
2. How and where does your school dispose of solid waste? Is a recycling program in place? How well does it work?
3. Suppose your town planned to build a combustion facility or landfill near your home. Outline your concerns, and explain your decision for or against this site.
4. Does your state have a bottle bill? If not, what has prevented the bill from being adopted?

WEB REFERENCES

On-line resources for this chapter can be found on the World Wide Web at: **http://www.prenhall.com/nebel**. Click on Chapter 19 on the chapter selector.

Hazardous Chemicals: Pollution and Prevention

〜〜〜〜〜〜〜

Key Issues and Questions

1. Hazardous materials (HAZMATs) are chemicals that present a hazard or risk. What are the four properties that characterize HAZMATs? Why are heavy metals and synthetic organic compounds the two categories of chemicals that present the worst toxic hazards?

2. Before 1970, chemical wastes were disposed of indiscriminately. What were the consequences of unregulated land disposal?

3. One of the most daunting tasks our society faces is cleaning up the thousands of existing toxic waste sites. How does the Superfund Act of 1980 address this task?

4. A number of laws have been passed to protect the public and the environment from present toxic wastes. What roles do the Clean Air Act, the Clean Water Act, RCRA, EPCRA, and TSCA play in accomplishing this objective?

5. A future direction for pollution is one of pollution avoidance rather than pollution control. What is the distinction between avoidance and control? How has the EPA changed its strategies in order to bring about more cooperation from states and industries to accomplish pollution avoidance?

Homes in Black Creek Village, New York, are selling well. Over 200 older homes have been renovated and sold to eager buyers, and there is a waiting list for a few homes still scheduled for renovation. A developer is building new homes in the village, in full view of a chain-link fence with a "Danger—No Trespassing" sign. One retiree, whose backyard ends at the fence, likes living there because "It's a nice, quiet neighborhood." Twenty years ago it wasn't at all quiet in the neighborhood, which was known as **Love Canal**. The area was occupied by a school and a number of houses, all of which were perched on top of a chemical waste dump that had been filled over and developed. The surface of the dump began to collapse, exposing barrels of chemical wastes. Fumes and chemicals began seeping into cellars. Ominously, people began reporting serious health problems, including birth defects and miscarriages. An aroused neighborhood, led by citizen-activist Lois Gibbs, began demanding that the state do something about the problem. Following confirmation in 1978 of the contamination by toxic chemicals, President Carter signed an emergency declaration to relocate hundreds of residents. The state and federal governments closed the school and demolished a number of the homes nearest the site (Fig. 20–1). In all, more than 800 families moved out of the area because of the fear of damage to their health from the toxic chemicals.

The lack of public policy for dealing with the disposal of hazardous chemicals contributed to the occurrence at Love Canal. Hooker Chemical and Plastics Company had purchased an abandoned canal near Niagara Falls in 1942 and proceeded to fill it with hazardous wastes—an estimated 17,000 tons. Hooker covered the canal over with a clay cap and sold it to the school board for a small sum, reportedly after warning the board that there were chemicals buried on the property. Subsequent construction penetrated the clay cap, and rain seeped in and leached chemicals in all directions. The result was that the U.S. public became aware of hazardous chemicals for the first time; other deposits of chemical wastes were revealed to the public, and toxic and hazardous wastes soon became the number one environmental concern. Congress responded with the Superfund

Abandoned hazardous chemicals. Environmental Protection Agency workers sample the contents of illegally discarded drums of chemical wastes.

▶ **FIGURE 20–1** *Love Canal.* These are a few of the nearly 100 homes that the state was forced to buy and later destroy as a result of contamination from Love Canal. The area has now been cleaned up and redeveloped with new homes.

Act in 1980, giving responsibility for cleaning up the wastes to the chemical industries that generated them. Hooker Chemical's parent company, Occidental Petroleum, eventually spent more than $233 million on the cleanup and subsequent law suits.

Over time, a number of "chemical disasters" have made front-page headlines. Such tragedies have convinced the public that significant dangers are associated with the manufacture, use, and disposal of some chemicals. But very few would advocate giving up the advantages of the innumerable products that derive from our modern chemical industry. Better to learn to handle and dispose of chemicals in ways that minimize the risks. Thus, over the past 25 years regulations surrounding the production, transport, use, and disposal of chemicals have mushroomed. From having almost no controls at all, the chemical industry is now among the most thoroughly regulated of all industries, and everyone using or handling chemicals that are deemed hazardous is affected by the regulations in one way or another.

This chapter will focus on hazardous chemicals, explaining the nature of the risks involved; provide an overview of laws and regulations that have come into play to protect human and environmental health and their strengths and weaknesses; and point out future directions toward reducing the chemical hazards at less cost.

20.1 The Nature of Chemical Hazards: HAZMATs

A chemical that presents a certain hazard or risk is known as a **hazardous material (HAZMAT)**. The EPA categorizes substances on the basis of the following hazardous properties:

- **Ignitability**: Substances that catch fire readily (e.g., gasoline and alcohol)
- **Corrosivity**: Substances that corrode storage tanks and equipment (e.g., acids)
- **Reactivity**: Substances that are chemically unstable and which may explode or create toxic fumes when mixed with water (e.g., explosives, elemental phosphorus (not phosphate), and concentrated sulfuric acid)
- **Toxicity**: Substances that are injurious to health when ingested or inhaled (e.g., chlorine, ammonia, pesticides, and formaldehyde)

Radioactive materials, discussed in Chapter 14, are probably the most hazardous of all, and are treated as an entirely separate category.

Sources of Chemicals Entering the Environment

To understand how HAZMATs enter our environment, we need to look at various aspects of how people in our society live and work. First, consider that the materials making up almost everything we use, from the shampoo and toothpaste in the morning to the TV set we watch in the evening, are products of chemical technology. Our use constitutes only one step in the **total product life cycle**, a term that considers all steps from obtaining raw materials to final disposal of the product. Implicit in our use of hair spray, for example, is that raw materials were obtained and various chemicals were produced to make both the hair spray and its container. Inevitably, there are chemical wastes and byproducts in the production processes. In addition, consider the risks of accidents or spills occurring in the manufacturing process and in the transportation of the raw materials, the finished product, or the wastes. Finally, consider: What are the risks of breathing the hair spray? What happens to the container you throw into the trash, still holding some of the hair spray when the propellent is used up?

Multiply these steps by all the hundreds of thousands of products used by billions of people, in homes, factories, and business, and you can appreciate the magnitude of the situation. At every stage—from mining raw materials through manufacturing, use, and final disposal—various chemical products

and byproducts enter the environment, with potential consequences for both human and environmental health (Fig. 20–2).

The use of many products—pesticides, fertilizers, and road salts, for example—implies their direct introduction into the environment. Or, the intended use may entail a fraction of the material going into the environment—the evaporation of solvents from paints and adhesives, for example. Then there are the product life cycles of materials that are used tangentially to the desired item. Consider lubricants, solvents, cleaning fluids, cooling fluids, and so on, with whatever contaminants they may contain. Likewise, there are the product life cycles of the gasoline, coal, or other fuels that are consumed for energy. Again, in addition to the unavoidable wastes produced, there is in every case the potential for accidental releases, ranging from minor leaks in storage tanks to supertanker wrecks such as the *Exxon Valdez*, which in 1989 spilled 11 million gallons of crude oil into Prince William Sound in Alaska.

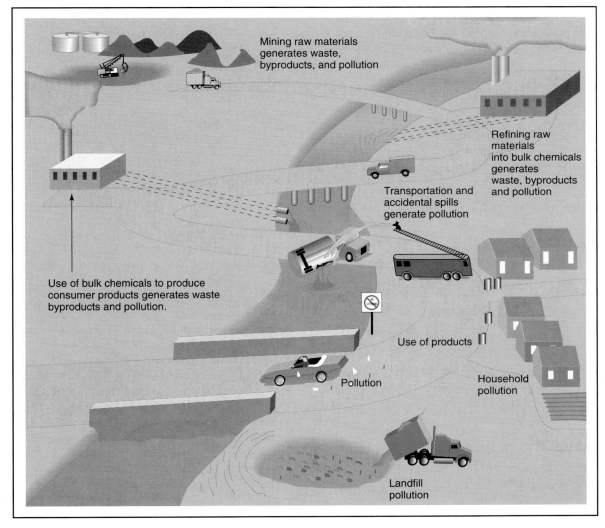

▲ **FIGURE 20–2** *Total product life cycle.* The life cycle of a product begins with the obtaining of raw materials and ends with final discard of the used product. At each step—or in transportation between steps—wastes, byproducts, or the product itself may enter the environment, causing pollution and creating various risks to human and environmental health.

Introductions of chemicals into the environment potentially occur in every sector, from major industrial plants to small shops and individual homes. Whereas single events involving large amounts of one chemical may constitute a disaster and make headlines, the total amounts entering the environment from millions of homes and businesses is far greater and presents a much greater health risk to society. Some idea of the quantities involved can be obtained from the *Toxics Release Inventory (TRI)*. The **Emergency Planning and Community Right-to-Know Act (EPCRA) of 1986** requires industries to report the locations and quantities of toxic chemicals stored on site to state and local governments, and to report releases of toxic chemicals to the environment; the TRI does not cover small businesses like dry-cleaning establishments and gas stations or household hazardous wastes (HHW). The TRI is disseminated on the Internet and provides an annual record of releases of more than 600 designated chemicals. The TRI for 1996 reveals the following information:

- Releases of toxic chemicals to the air: 722,000 tons.
- Releases of toxic chemicals to the water: 86,500 tons.
- Total environmental releases: 1,066,000 tons.
- Total production-related hazardous wastes: 11,702,500 tons.

Add to this the estimated 1.6 million tons each year of HHW, and an unknown quantity coming from small businesses, and you have some idea of the dimensions of this problem. The good news is that over the nine years since the TRI has been required, the quantities of virtually all categories have been reduced by half or more.

The Threat from Toxic Chemicals

All toxic chemicals, by definition, are hazards that pose a potential risk to humans (see Chapter 16). Fortunately, a large portion of the chemicals introduced into the environment are gradually broken down and assimilated by natural processes. Therefore, once these chemicals are diluted sufficiently, they pose no long-term human or environmental risk, even though they may be highly toxic in acute doses (high-level, short-term exposures).

There are two major classes of chemicals that do not readily degrade in the environment, however: (1) *heavy metals and their compounds* and (2) *synthetic organics*. Again, if sufficiently diluted in air or water, most of these compounds do not pose a hazard. Yet there are some notable exceptions.

Heavy Metals. The most dangerous heavy metals are lead, mercury, arsenic, cadmium, tin, chromium, zinc, and copper. These metals are widely used in industry, particularly in metal-working or metal-plating shops and in such products as batteries and electronics. They are also used in certain pesticides and medicines. In addition, because heavy-metal compounds can have brilliant colors, they are used in paint pigments, glazes, inks, and dyes. Thus, heavy metals may enter the environment wherever any of these products are produced, used, and ultimately discarded.

Heavy metals are extremely toxic because, as ions or in certain compounds, they are soluble in water and may be readily absorbed into the body, where they tend to combine with and inhibit the functioning of particular vital enzymes. Even very small amounts can have severe physiological or neurological consequences. The mental retardation caused by lead poisoning and the insanity and crippling birth defects caused by mercury are particularly well-known examples.

Synthetic Organics. Synthetic organic compounds are the chemical basis for all plastics, synthetic fibers, synthetic rubber, modern paintlike coatings, solvents, pesticides, wood preservatives, and hundreds of other products. Because of their chemical structure, many synthetic organics are resistant to biodegradation. This is an important part of what makes many such compounds useful; we wouldn't want fungi and bacteria attacking and rotting our tires, and paints and wood preservatives function only insofar as they are both nonbiodegradable and toxic to decomposer organisms.

These compounds are toxic because they are often readily absorbed into the body. There they interact with particular enzymes, but their nonbiodegradability prevents them from being broken down or processed further. When a person ingests a sufficiently high dose, the effect may be acute poisoning and death. With low doses over extended periods, however, the effects are insidious and can be mutagenic (mutation-causing), carcinogenic (cancer-causing), or teratogenic (birth defect-causing). They may cause serious liver and kidney dysfunction, sterility, and numerous other physiological and neurological problems.

A particularly troublesome class of synthetic organics is the **halogenated hydrocarbons**, organic compounds in which one or more of the hydrogen atoms have been replaced by atoms of chlorine, bromine, fluorine, or iodine. These four elements are classed as *halogens*, hence the name *halogenated hydrocarbons* (Fig. 20–3). Of the halogenated hydrocarbons, the **chlorinated hydrocarbons** (also called **organic chlorides**) are by far the most common. Organic chlorides are widely used in plastics (polyvinyl chloride), pesticides (DDT, Kepone, and Mirex), solvents (carbon tetrachlorophenol, tetrachloroethylene), electrical insulation (polychlorinated biphenyls), and many other products.

We can consider one of these compounds, tetrachloroethylene (Fig. 20–3), as a simple case study. This substance, also called perchloroethylene, or PERC, is colorless and nonflammable and is the major substance in dry-cleaning fluid. It is an effective solvent and finds uses in all kinds of industrial cleaning operations. It can also be found in shoe polish and typewriter correction fluid. PERC evaporates readily when exposed to air, but when in soil it can enter groundwater readily because it does not bind to the soil particles. Human exposure can occur in the workplace,

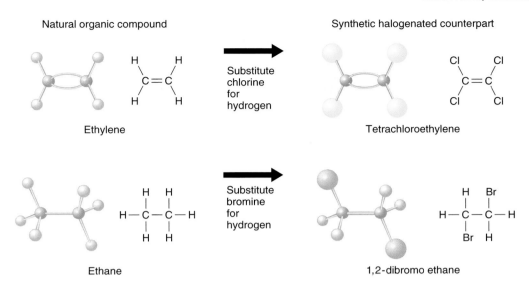

▲ **FIGURE 20–3** *Halogenated hydrocarbons.* These are organic (carbon-based) compounds in which one or more hydrogen atoms has been replaced by halogen atoms (chlorine, fluorine, bromine, or iodine). Such compounds are particularly hazardous to health because they are nonbiodegradable and they tend to bioaccumulate. Shown here are tetrachloroethylene and 1,2-dibromo ethane.

especially in connection with dry-cleaning operations. It can also occur if people use products containing PERC, like dry-cleaned garments. It enters the body most readily when breathed in with contaminated air. Breathing PERC for short periods of time can bring on dizziness, fatigue, incoordination, and unconsciousness. Over longer periods of time, PERC can cause liver and kidney damage. Laboratory studies also show PERC to be carcinogenic. It is not known to cause any environmental harm at levels normally found in the environment, except that it does add to the burden of volatile organic chemicals in causing photochemical smog. PERC is produced in large amounts in the United States, and according to the EPA's Toxic Release Inventory, almost 8 million pounds of PERC was released to the environment in 1996 from industrial sources, mostly by evaporation. Other *chlorinated hydrocarbons* and their health effects are listed in Table 20-1.

Involvement with Food Chains

The trait that makes heavy metals and nonbiodegradable synthetic organics particularly hazardous is their tendency to accumulate in organisms. We discussed the phenomena of bioaccumulation and biomagnification in connection with DDT in Chapter 17.

A tragic episode in the early 1970s, known as the Minamata disease, revealed the potential for biomagnification of mercury and other heavy metals. The disease is named for a small fishing village in Japan where the episode occurred. In the mid-1950s, cats in Minamata began to show spastic movements, followed by partial paralysis, coma, and death. At first this was thought to be a syndrome peculiar to

cats, and little attention was paid to it. However, when the same symptoms began to occur in people, concern escalated quickly. Additional symptoms, such as mental retardation, insanity, and birth defects were also observed. Scientists and health experts eventually diagnosed the cause as acute mercury poisoning.

A chemical company near the village was discharging wastes containing mercury into a river that flowed into the bay where the Minamata villagers fished. The mercury, which settled with detritus, was first absorbed and bioaccumulated by bacteria and then biomagnified as it passed up the food chain through fish to cats or to humans. Cats had suffered first and most severely because they fed almost exclusively on the remains of fish. By the time the situation was brought under control, some 50 persons had died and 150 had suffered serious bone and nerve damage. Even now, the tragedy lives on in the crippled bodies and retarded minds of some Minamata descendants.

20.2 A History of Mismanagement

In the past, chemical wastes of all kinds were disposed of as expediently as possible. From the dawn of the Industrial Age to the relatively recent past, it was common practice to exhaust all combustion fumes up smokestacks, vent all evaporating materials and solvents into the air, and flush all waste liquids and contaminated wash water into sewer systems or directly into natural waterways. Many human health problems occurred, but they either were not recognized as being caused by the pollution or were accepted as the "price of progress." Indeed, much of our understanding regarding

TABLE 20-1 Toxic Synthetic Organic Compounds Frequently Found in Chemical Wastes

Chemical	Mutations	Carcinogenic	Birth Defects	Still births	Nervous Disorders	Liver Disease	Kidney Disease	Lung Disease
Benzene	X	X	X	X				
Dichlorobenzene	X			X	X	X		
Hexachlorobenzene	X	X	X	X	X			
Chloroform	X	X	X		X			
Carbon tetrachloride	X		X	X	X	X		
Chloroethylene (vinyl chloride)	X	X			X	X		X
Dichloroethylene	X	X		X	X	X	X	
Tetrachloroethylene		X			X	X	X	
Trichloroethylene	X	X			X	X		
Heptachlor	X	X		X	X	X		
Polychlorinated biphenyls (PCBs)	X	X	X	X	X	X		
Tetrachlorodibenzo dioxin	X	X	X	X	X	X		
Toluene	X				X	X		
Chlorotoluene	X	X						
Xylene			X	X	X			

*Known Health Effects — Determined from human exposures or tests on experimental animals.

Source: Adapted from S. Epstein, L. Brown, and C. Pope, *Hazardous Waste in America.* Copyright © 1982 by Samuel S. Epstein, M.D., Lester O. Brown, and Carl Pope. Reprinted with permission of Sierra Club Books.

human health effects of hazardous materials is derived from those uncontrolled exposures. For example, the expression "mad as a hatter" comes from the fact that people who made hats in the 1800s frequently became insane. The insanity, it was later found, was caused by poisoning from the mercury used in the production process.

In the 1950s, as production expanded and synthetic organics came into widespread use in the developed countries, many streams and rivers essentially became open chemical sewers, as well as sewers for human waste. These waters were not only devoid of life, they were themselves hazardous. For example, in the 1960s the Cuyahoga River, which flows through Cleveland, Ohio, carried so much flammable material that it actually caught fire and destroyed seven bridges before it burned itself out (Fig. 20–4). Worsening pollution (both chemical and sewage) and increasing recognition of adverse health effects finally created a degree of public outrage that pushed Congress to pass the Clean Air Act of 1970 and the Clean Water Act of 1972. These acts set standards for allowable emissions into air and water and timetables for reaching those standards.

The Clean Air and Clean Water acts and their subsequent amendments remain cornerstones of environmental legislation. However, the passage of these acts in the early 1970s left an enormous loophole. If you can't vent wastes into the atmosphere or flush them into waterways, what do you do with them? Industry turned to *land disposal*, which was essentially unregulated at the time, as an expedient alternative. Indiscriminate air and water disposal became indiscriminate land disposal. Thus, in retrospect we see that the Clean Air and Clean Water acts, for all their benefits in improving air and water quality, also succeeded in transferring pollutants from one part of the environment to another.

◀ **FIGURE 20–4** *Cuyahoga River on fire.* Prior to laws and regulations curtailing pollution, all manner of wastes were indiscriminately discharged. In the 1960s the Cuyahoga River actually caught fire. Incidents such as this contributed to public outrage that led to the passage of the Clean Water Act of 1972.

Methods of Land Disposal

In the early 1970s, there were three primary land-disposal methods: (1) deep-well injection, (2) surface impoundments, and (3) landfills. With conscientious implementation of safeguards, each of these methods has some merit. Without adequate regulations or enforcement, however, contamination of groundwater is virtually inevitable.

Deep-Well Injection. Deep-well injection involves drilling a "well" into dry, porous material below groundwater (Fig. 20–5). In theory, hazardous waste liquids pumped into the well soak into the porous material and remain isolated indefinitely. In practice, however, it is almost impossible to guarantee that injected wastes will never escape and contaminate groundwater. For this reason, deep-well injection is a less used disposal source today. A number of commercial deep-well facilities accept wastes like chemical cleaning solutions, saline wastes, metals, cyanides, and corrosive materials. The total amount of deep-well injection has declined over the years, from 685 million tons in 1988 to 102 million tons in 1996.

Surface Impoundments. Surface impoundments are simple excavated depressions ("ponds") into which liquid wastes are drained and held. They were the least expensive and hence most widely used way to dispose of large amounts of water carrying relatively small amounts of chemical wastes. As waste is discharged into the pond, solid wastes settle and accumulate while water evaporates (Fig. 20–6). If the pond bottom is well sealed and if evaporation equals input, impoundments may receive wastes indefinitely. However, inadequate seals may allow wastes to percolate into groundwater, exceptional storms may cause overflows, and

volatile materials can evaporate into the atmosphere, adding to air pollution problems. Where surface impoundments are used today, they are seen as strictly temporary sites for hazardous-waste storage.

Landfills. When hazardous wastes are in a concentrated liquid or solid form, they are commonly put into drums and buried in landfills. If a landfill is properly lined, supplied with a system to remove leachate, provided with monitoring wells and properly capped, it may be reasonably safe and is referred to as a **secure landfill** (Fig. 20–7, p. 484). The various barriers are potentially subject to damage or deterioration, however, requiring adequate surveillance and monitoring systems to prevent leakage.

Because early land disposal was not regulated, in numerous instances not even the most rudimentary precautions were taken. In many cases, deep wells were injecting wastes directly into groundwater, abandoned quarries were sometimes used as landfills with no additional precautions being taken, and surface impoundments frequently had no seals or liners whatsoever. Even worse, considerable amounts of waste failed to get to any disposal facility at all.

Midnight Dumping and Orphan Sites. The need for alternative methods to dispose of waste chemicals created an opportunity for a new enterprise—waste disposal. Many reputable businesses entered the field, but—again in the absence of regulations—there were also disreputable operators. As stacks of drums filled with hazardous wastes "mysteriously" appeared in abandoned warehouses, vacant lots, or municipal landfills, it became clear that some operators simply were pocketing the disposal fee and then unloading the wastes in any available location, frequently under cover of darkness, an activity termed

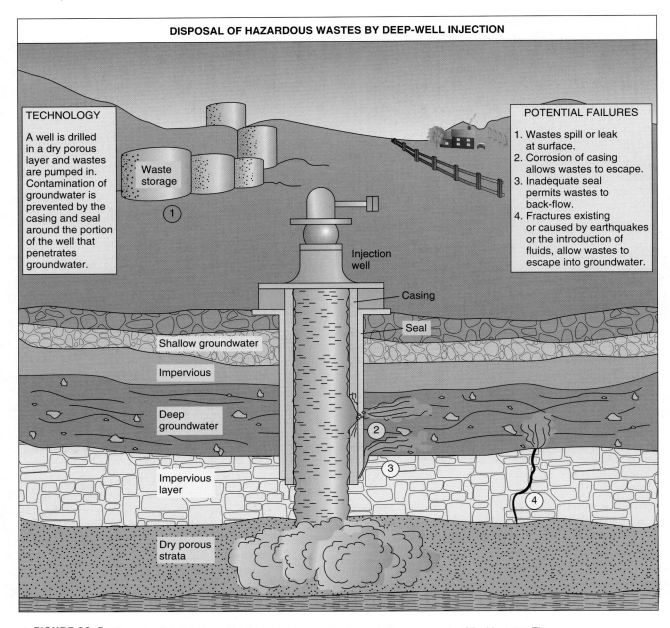

DISPOSAL OF HAZARDOUS WASTES BY DEEP-WELL INJECTION

TECHNOLOGY

A well is drilled in a dry porous layer and wastes are pumped in. Contamination of groundwater is prevented by the casing and seal around the portion of the well that penetrates groundwater.

POTENTIAL FAILURES

1. Wastes spill or leak at surface.
2. Corrosion of casing allows wastes to escape.
3. Inadequate seal permits wastes to back-flow.
4. Fractures existing or caused by earthquakes or the introduction of fluids, allow wastes to escape into groundwater.

Waste storage

Injection well

Casing

Seal

Shallow groundwater

Impervious

Deep groundwater

Impervious layer

Dry porous strata

▲ **FIGURE 20–5** *Deep-well injection.* This technique is used for disposal of large amounts of liquid wastes. The concept is that toxic wastes may be drained into dry, porous strata below ground, where they may reside harmlessly "forever." However, as the figure shows, failures can occur and allow the liquid wastes to contaminate groundwater. (Adapted with permission from Environmental Action, 1525 New Hampshire Ave., N.W., Washington, D.C. 20036.)

midnight dumping (Fig. 20–8). Authorities trying to locate the individuals responsible would find that they had gone out of business and were nowhere to be found.

Some companies or individuals simply stored wastes on their own properties and then went out of business, abandoning the property and the wastes. These became known as **orphan sites**—hazardous-waste sites without a responsible party to do cleanup. As drums containing hazardous chemicals corroded and leaked, there was great danger of reactive chemicals combining and causing explosions and fires. One of the most famous abandoned sites is the Valley of the Drums, in Kentucky (Fig. 20–9).

Scope of the Mismanagement Problem

The mounting problem of unregulated land disposal of hazardous wastes was brought vividly to public attention by the episode at Love Canal, which became a battle cry of citizens concerned about hazardous wastes. As both government and independent researchers began surveying the extent of the problem, bad disposal practices were found to be rampant. The World Resources Institute estimated that in the United States in the early 1980s, there existed 75,000 active industrial landfill sites, along with 180,000 surface impoundments and 200 other special facilities that were or

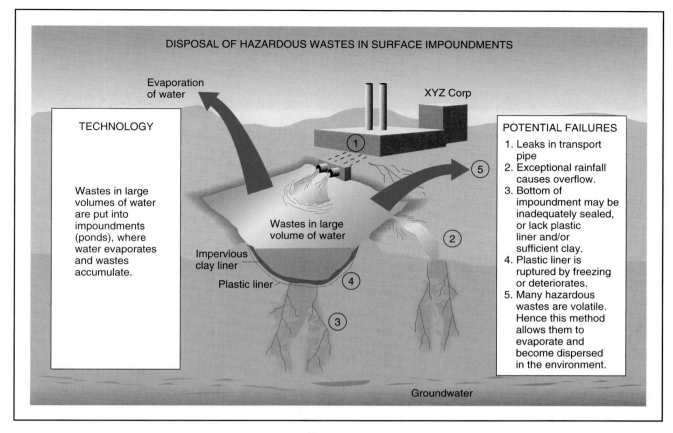

DISPOSAL OF HAZARDOUS WASTES IN SURFACE IMPOUNDMENTS

Evaporation
of water

XYZ Corp

TECHNOLOGY

Wastes in large
volumes of water
are put into
impoundments
(ponds), where
water evaporates
and wastes
accumulate.

Wastes in large
volume of water

Impervious
clay liner

Plastic liner

Groundwater

POTENTIAL FAILURES

1. Leaks in transport
   pipe
2. Exceptional rainfall
   causes overflow.
3. Bottom of
   impoundment may be
   inadequately sealed,
   or lack plastic
   liner and/or
   sufficient clay.
4. Plastic liner is
   ruptured by freezing
   or deteriorates.
5. Many hazardous
   wastes are volatile.
   Hence this method
   allows them to
   evaporate and
   become dispersed
   in the environment.

▲ **FIGURE 20–6** *Surface impoundment.* This is an inexpensive technique for disposal of large amounts of lightly contaminated liquid wastes. The supposition is that only water leaves the impoundment, by evaporation, while wastes remain and accumulate in the impoundment indefinitely. In reality, this method is subject to failure at a number of points and is limited today to use as a short-term measure.

could be possible sources of groundwater contamination. In most cases, the contaminated area was relatively small, 200 acres (80 ha) or less, but in total the problem was immense and affected every state in the country. As studies and tests proceeded, thousands of individual wells and some major municipal wells were closed because they were contaminated with toxic chemicals.

Many cases of private-well contamination caused by careless disposal were discovered only after people experienced "unexplainable" illnesses over prolonged periods. While these incidents did not receive the media attention of Love Canal, they were nevertheless devastating to the people involved.

Even more important than the damage already done was the recognition that this is an ongoing problem. As we have seen, more than 13 million tons of hazardous wastes are generated each year in the United States. The problems concerning toxic chemical wastes can be divided into three areas:

- cleaning up the "messes" already created, especially where they threaten drinking-water supplies;
- regulating the handling and disposal of wastes currently being produced so as to protect public and environmental health; and

- looking toward future solutions, such as reducing the quantity of hazardous waste produced.

We shall address each of these areas in the following sections.

20.3 Cleaning Up the Mess

A major public health threat from land-disposed toxic wastes is contamination of groundwater that is subsequently used for drinking. The first priority is to assure that people have safe water. The second priority is to clean up or isolate the source of pollution so that further contamination does not occur.

Assuring Safe Drinking Water

To protect the public from the risk of toxic chemicals contaminating drinking-water supplies, Congress passed the **Safe Drinking Water Act of 1974**. Under this act, the EPA sets national standards to protect the public health, including allowable levels of various contaminants. If any known contaminants are found to exceed maximum contaminant levels (MCL), the water supply is closed until adequate purification procedures or other alternatives are adopted. The

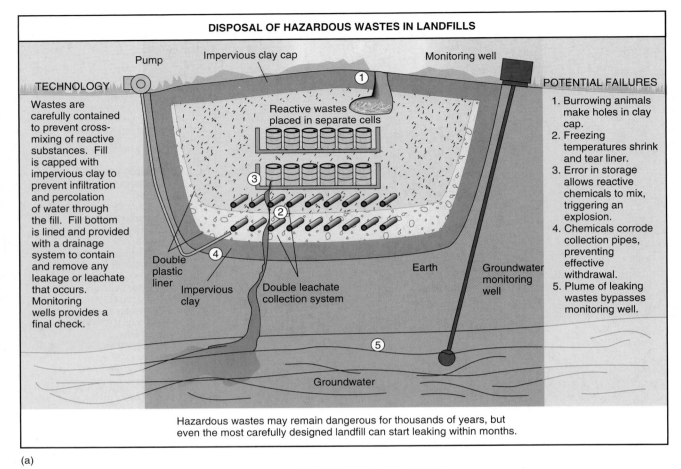

DISPOSAL OF HAZARDOUS WASTES IN LANDFILLS

TECHNOLOGY

Wastes are carefully contained to prevent cross-mixing of reactive substances. Fill is capped with impervious clay to prevent infiltration and percolation of water through the fill. Fill bottom is lined and provided with a drainage system to contain and remove any leakage or leachate that occurs. Monitoring wells provides a final check.

Pump   Impervious clay cap   Monitoring well

Reactive wastes placed in separate cells

① ② ③ ④ ⑤

Double plastic liner   Impervious clay   Double leachate collection system   Earth   Groundwater monitoring well

POTENTIAL FAILURES

1. Burrowing animals make holes in clay cap.
2. Freezing temperatures shrink and tear liner.
3. Error in storage allows reactive chemicals to mix, triggering an explosion.
4. Chemicals corrode collection pipes, preventing effective withdrawal.
5. Plume of leaking wastes bypasses monitoring well.

Groundwater

Hazardous wastes may remain dangerous for thousands of years, but even the most carefully designed landfill can start leaking within months.

(a)

(b)

▲ **FIGURE 20–7** *Secure landfill.* Landfilling is a technique widely used for disposal of concentrated chemical wastes. (a) Precautionary measures to make this method safe are listed. Before the 1980s, these measures frequently were not taken. Though they are taken today, potentials for failure remain. (b) An operating hazardous-waste landfill in Alabama. (Adapted from an illustration by Rick Farrell. All rights reserved.)

act was amended in 1986 to require the EPA to set MCLs for 83 contaminants and to give the EPA jurisdiction over groundwater. Public water agencies were required to monitor drinking water to be sure that it met the standards.

Although its major objectives have been accomplished, there has been serious debate over some impacts of the Safe Drinking Water Act (SDWA) and its amendments. Most communities never found toxic chemicals to the degree that requires remedial action. Therefore, many municipalities wanted to see the act amended because they felt that monitoring was too costly relative to the risks involved, especially for small towns supplying only a few hundred homes.

▲ **FIGURE 20–8** *Midnight dumping.* Hazardous wastes were often left by unscrupulous haulers on remote or unoccupied properties.

These and other concerns led the Congress to amend the SDWA in 1996. The amendments provide small municipalities with greater flexibility in water treatment and monitoring and require the EPA to assess risks, costs, and benefits before proposing regulations. EPA was given more time for processing regulations for the lengthening list of contaminants. Also, states and municipalities are now required to oversee protection of the sources of public water supplies by considering susceptibility of watersheds and other sources to contamination.

Because of concerns about public drinking water, many people have turned to drinking bottled water to avoid presumed contamination of municipal water supplies. Bottled water is considered a food and, as such, is regulated by the

Food and Drug Administration (FDA). However, FDA standards for bottled water specify only that it be as safe as tap water. Ironically, consumers pay for a product for which there are no guarantees that it is any better than tap water. Indeed, numerous samples of bottled water in one study in 1991 showed higher levels of contamination than did municipal supplies.

Groundwater Remediation

If dumps, leaking storage tanks, or spills of toxic materials have contaminated groundwater and such groundwater is threatening water supplies, all is not lost. **Groundwater remediation** is a developing and growing technology. Techniques involve drilling wells, pumping out the contaminated groundwater, purifying it, and reinjecting the purified water back into the ground or discharging it into surface waters (Fig. 20–10). Of course, cleaning up the source of the contamination is mandatory. If, however, the contamination is extensive, remediation may not be possible, and the groundwater must be considered unfit for use as drinking water.

Superfund for Toxic Sites

Probably the most monumental task we are facing is the cleanup of the tens of thousands of toxic sites resulting from the history of mismanaged toxic materials disposal. Where facilities were still operating, pressures were brought to bear on the operators to clean up the sites. Many operators who could not afford to do so, however, simply declared bankruptcy, and the sites joined those already abandoned.

The Comprehensive Environmental Response, Compensation, and Liability Act of 1980 (CERCLA), popularly

◀ **FIGURE 20–9** *Orphan waste site – "Valley of the Drums."* Thousands of drums of waste chemicals, many of them toxic, were unloaded near Louisville, Kentucky, around 1975 and left to "rot," seriously threatening the surrounding environment, waterways, and aquifers.

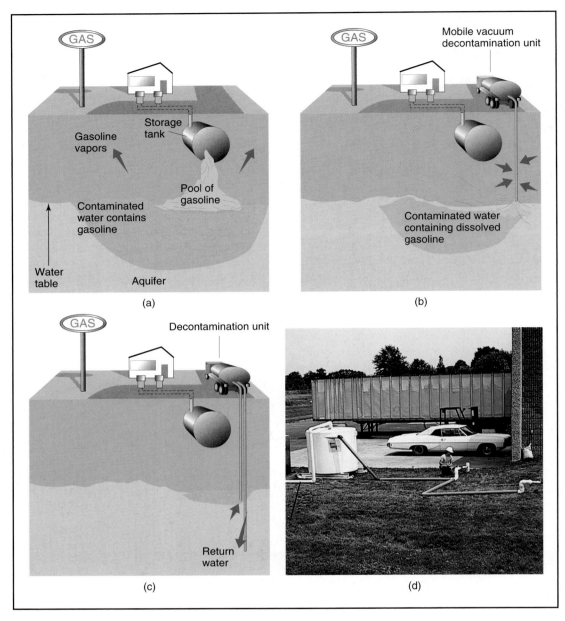

▲ FIGURE 20–10 *Groundwater remediation.* (a) Typical subsurface contamination from a leaking fuel tank at a gas station. (b) After the leak has been repaired, the vacuum-extraction process causes gasoline and residual hydrocarbons in the soil and on the water table to evaporate and then removes the vapors, preventing further contamination of the groundwater. (c) Contaminated groundwater is pumped out, treated, and returned to the ground. (d) Photo of remediation system.

known as Superfund, initiated a major federal program aimed at cleaning up such abandoned chemical waste sites. Through a tax on chemical raw materials, this legislation provides a fund for the identification of abandoned chemical waste sites, protection of groundwater near the site, remediation of groundwater if it has been contaminated, and cleanup of the site. Funding for the program has steadily grown from about $300 million per year in the early 1980s to about $2 billion per year in the 1990s. But the cleanup job is nowhere near complete. It is estimated that it will go on for at least another 30 years at a total cost of some $300 billion.

Such is the magnitude of the problem; Superfund, one of the EPA's largest ongoing programs, works as follows.

Setting Priorities. It should be obvious that resources are not sufficient to clean up all sites at once. Therefore, a system for setting priorities has been developed.

- As sites are identified—note that many abandoned sites had long since been forgotten—they are first analyzed in terms of current and potential threat to groundwater supplies by taking samples of the waste, determining waste characteristics, and testing groundwater around the site for contamination. If it is determined that no immediate threat exists, nothing more may be done.

- If a threat to human health does exist, the most expedient measures are taken immediately to protect the public. These measures may include digging a deep trench and installing a concrete dike around the site and recapping it with impervious layers of plastic and clay to prevent infiltration. Thus, the wastes are isolated, at least for the short term. If the situation has gone past the threat stage and contaminated groundwater is or will be reaching wells, the remediation procedures discussed above are begun immediately.

- The worst sites (those presenting the most immediate and severe threats) are put on a **National Priorities List (NPL)** and scheduled for total cleanup. A site on this list is reevaluated in terms of the most cost-effective method of cleanup, and finally cleanup begins.

Cleanup Technology. If the chemical wastes are contained in drums, they can be picked up, treated, and managed in proper fashion. The bigger problem is the soil (often millions of tons) contaminated by leakage. One procedure is to excavate contaminated soil and run it through an incinerator or kiln assembled on the site to burn off chemicals (Fig. 20–11). (Synthetic organic chemicals can be broken down by incineration, and heavy metals are converted to insoluble, stable oxides.) Another method is to drill a ring of injection wells around the site and a suction well in the center. Water containing a harmless detergent is injected into the injection wells and drawn to the suction well, cleansing the soil

along the way. The withdrawn water is treated to remove the pollutants and is then reused for injection.

Another rapidly developing cleanup technology is *bioremediation*. In many cases, the soil is contaminated with toxic organic compounds that are biodegradable. The problem is that they do not degrade, because the soil lacks organisms or oxygen or both. In **bioremediation**, oxygen and organisms are injected into contaminated zones. The organisms feed on and eliminate the pollutants (as in secondary sewage treatment) and then die when the pollutants are gone. Considerable research is being done to find and develop microbes that will break down certain kinds of wastes more readily. Bioremediation may be used in place of detergents to decontaminate soil. It is a rapidly expanding and developing technology of waste cleanup applicable to leaking storage tanks and spills as well as to waste-disposal sites.

Where the soil contaminants are heavy metals and nonbiodegradable organic compounds, *phytoremediation* has been employed with some success. **Phytoremediation** uses plants to accomplish a number of desirable steps in cleanups: stabilizing the soil, preventing further movement of contaminants by erosion, and extracting the contaminants by direct uptake from the soil. After full growth, the plants are removed and treated as toxic waste products. However, this is a slow process and can only be used at sites where contaminant types or concentrations are not toxic to plants.

Evaluating Superfund. Of the literally hundreds of thousands of various waste sites, over 40,000 have been deemed serious enough to be given Superfund status. Over time, the EPA judged that about 28,000 of these sites did not pose a significant public health or environmental threat and were assigned to the category "No Further Removal Action Planned," or NFRAP. Still, over 12,000 sites remain on the active list. It is assumed that this figure now includes most if not all of the sites in the United States that pose a significant risk. Interestingly, some of the worst that have come to light are on military bases, the result of what many characterize as totally heedless and unconscionable discard of toxic materials from military operations.

To date (1999), 1372 have been deemed to present threats serious enough to be put on the NPL; additions are made to the list as sites from the master list are assessed, and subtractions are made when remediation on a site is complete. Since 1980, when CERCLA went into effect, about 520 sites have received all the necessary cleanup-related construction, and 178 have been crossed off the list. Cleanup of sites takes an average of 12 years, at a mean cost of $30 million. Total cleanup often takes so long because groundwater remediation is a slow process. The other NPL sites are in various stages of analysis, remediation, and construction. (An example of a Superfund site and its progress toward cleanup is given in the "Earth Watch" essay, p. 488.)

CERCLA is based on the principle that "the polluter pays." However, the "history" of a waste-disposal site may go

▼ **FIGURE 20–11** *Toxic waste mobile incinerator.* The EPA is cleaning up some Superfund sites by running contaminated materials through this incinerator, which is erected at the site.

EARTH WATCH

THE CASE OF THE OBEE ROAD NPL SITE

The following is but one example of the more than 1300 sites on the EPA's Superfund National Priorities List.

Conditions at listing (January 1987): The Obee Road Site consisted of a plume of contaminated groundwater in the vicinity of Obee Road in the eastern section of Hutchinson, Reno County, Kansas. The Kansas Department of Health and Environment (KDHE) has been investigating the area since July 1983. At that time, the state detected volatile organic chemicals, including benzene, trans-1,2-dichloroethylene, chlorobenzene, 1,1-dichloroethylene, tetrachloroethylene, trichloroethylene, vinyl chloride, and toluene, in wells drawing on a shallow aquifer. An estimated 1900 residents of suburban Obeeville obtained drinking water from private wells in the aquifer.

To protect public health, Hutchinson connected the homes of the Obeeville residents and a school that was drawing water from a contaminated well, to the Reno County Rural Water District.

Preliminary work by the state identified the source of the contamination as the former Hutchinson City landfill, which is located at the eastern edge of what is now the Hutchinson Municipal Airport. Before it was closed in 1968, the landfill had accepted unknown quantities of liquid wastes and sludges from local industries, as well as solvents from small metal-finishing operations at local aircraft plants. The Department of Defense (DOD), which owned or maintained the airport until 1963, may also have disposed of solvents in the landfill. No records have been found for the design, use, or closure of the landfill, or for the wastes that were delivered there.

Status as of 1998: Further testing and analysis showed that the original landfill had "stabilized." That is, leaching had already run its course, and there was little likelihood of further contamination from the landfill. Therefore, excavation or other cleanup of the landfill has been deemed unnecessary, but monitoring of the site to detect any further leakage will continue for the foreseeable future. EPA's risk assessment for the site indicates that there are no present risks to human health or the environment. Remedial action selected includes restricting access and preventing future development of the landfill, which is now covered with vegetation. Eight wells have been installed, and groundwater monitoring is to occur for a five-year period to ascertain that no further groundwater contamination is occurring. Estimated total cost for assessment and remedial action is $88,000 to $100,000.

---

back 50 years or more, and users may have included everything from schools, hospitals, and small businesses to large corporations, all of which mount legal defenses to disclaim responsibility. Where liability is difficult to track down or the responsible parties are unable to pay, the Superfund trust fund kicks in. However, only about 25% of the cleanup costs have been paid for with public funds to date.

Over the years, the EPA has gained a great deal of experience in dealing with Superfund sites. The technology has become quite sophisticated, and many hazardous-waste remediation companies have become established. A great deal of progress has been made, and interestingly, although hazardous waste was once a top environmental concern, the American public now places it only eighth among a host of other environmental problems, according to a 1997 Roper survey.

Nevertheless, critics of the program are skeptical about its success, citing the high costs and slow progress. Industries complain about the "gold-plated cleanups" required by EPA regulations and claim that they are unfairly blamed for pollution that reaches back to activities that were legal before the enactment of CERCLA. Many feel that overly stringent standards of cleanup are costing large sums of money without providing any additional benefit to public health. (This problem of finding a suitable balance between costs and benefits will be explored in more depth in Chapter 23.) Many of these concerns have been reflected in attempts to reauthorize CERCLA by the last two Congresses, but bipartisan support for the proposed legislation has been lacking, and the law continues to operate on previously established regulations.

Brownfields. One highly successful recent Superfund development is the *brownfields* program. **Brownfields** are "abandoned, idled, or under-used industrial and commercial facilities where expansion or redevelopment is complicated by real or perceived environmental contamination" (EPA definition). It has been estimated that environmental hazards impair the value of $2 trillion of real estate in the United States. Federal and state initiatives are forging ahead with legislation and regulations that clear the way for cleanup and development of such properties by limiting liability and providing incentives to developers. Many brownfield sites lie in economically disadvantaged communities; their rehabilitation contributes jobs and exchanges a functional facility for an unsightly blight on the neighborhood (see the "Ethics" essay, p. 491). For example, the city of Chicago recently acquired and cleaned up a former bus barn in west Chicago that was becoming an illegal indoor garbage dump. The Illinois EPA cleared the site for reuse, and it was acquired by Scott Peterson Meats, which erected a $5.2 million smokehouse that employs 100 workers. Frequently, the rehabilitation of brownfield sites provides industries and municipalities with centrally located, prime land for facilities that would otherwise have been carved out of suburban or "greenfield" lands (land occupied by natural ecosystems).

20.4   Management of New Wastes

The production of chemical wastes is and will continue to be an ongoing phenomenon as long as modern societies persist. Therefore, human and environmental health can be protected only if we have management procedures to handle and dispose of wastes safely so that we will not be creating more and more Superfund sites. We have already noted that the Clean Water Act and the Clean Air Act limit discharges into water and air. When problems regarding disposal of wastes on land became evident in the mid-1970s, Congress passed the **Resource Conservation and Recovery Act of 1976 (RCRA)** in order to control land disposal. Thus, a company producing hazardous wastes in the United States today is under the regulations of these three major environmental acts.

It is important to understand these laws in somewhat more detail. First, let us briefly review their enactment and administration. The general process is that the U.S. Congress writes the legislation, and after it is passed, it becomes the EPA's responsibility to administer and enforce. Typically, the legislation requires the EPA to set particular standards and write rules regarding how certain objectives should be met. This requirement often puts the EPA at the center of controversies, because almost invariably environmentalists feel standards and regulations are too weak and industry feels they are too stringent. The whole package of law, standards, and regulations is passed on to respective agencies in each of the states to administer to local businesses and industries, but the EPA still oversees state activities. States may also apply their own standards and regulations if they are at least as strong as federal standards.

Checking for compliance with all the standards and regulations necessitates a tremendous amount of collecting and analyzing of samples, as well as inspection and enforcement. Consequently, the EPA and related agencies provide an important source of jobs for people interested in environmental improvement.

The Clean Air and Water Acts

The Clean Air Act of 1970, the Clean Water Act of 1972, and their various amendments make up the basic legislation limiting discharges into the air or water. More will be said about the Clean Air Act in Chapter 22. Under the Clean Water Act (see Chapter 18) any firm, including facilities such as sewage-treatment plants, discharging more than a certain volume into natural waterways must have a **discharge permit** (NPDES permit). You can't require a firm to stop polluting instantaneously, short of closing it down. The discharge permits are a means of seeing who is discharging what. Establishments with discharge permits are required to report all discharges of the substances covered by the Toxics Release Inventory. The renewal of the permits then is made contingent on reducing pollutants to meet certain standards within certain time periods. Standards are being made continually stricter as technologies for pollution control improve. Some manufacturing firms discharging wastewater into municipal sewer systems are required to pretreat such water to remove any pollutant that cannot be removed by the sewage-treatment plants, namely nonbiodegradable organics and heavy metals.

Nevertheless, even this restriction does not end all water pollution. Amounts of wastes are still legally discharged under permits, and though they may be a low percentage of total emissions, they still add up to large numbers, as we have seen. Moreover, small firms, homes, and farms are exempt from regulation and contribute an unknown quantity of toxics to air and water. Still further, a great deal of pollution comes from nonpoint sources such as urban and farm runoff.

The Resource Conservation and Recovery Act (RCRA)

RCRA (1976) and its subsequent amendments is the cornerstone legislation designed to prevent unsafe or illegal disposal of all solid wastes on land. RCRA has three main features. First, RCRA requires that all disposal facilities such as landfills be *permitted.* The permitting process requires that they have all the safety features described in Fig. 20–7a, including monitoring wells. This demand caused most old facilities to shut down—many subsequently became Superfund sites—and new high-quality landfills with safety measures to be constructed.

Second, RCRA requires that toxic wastes destined for landfills be pretreated to convert them to forms that will not leach. Such treatment now commonly includes biodegradation or incineration in various kinds of facilities, including cement kilns (Fig. 20–12). Biodegradation refers to the use of systems similar to secondary sewage treatment, as discussed in Chapter 18, and perhaps new species of bacteria capable of breaking down the synthetic organics. If treatment is thorough, there may be little or nothing to landfill. This is the ultimate objective.

For whatever is still going to disposal facilities, the third major feature of RCRA is to require "cradle-to-grave" tracking of all hazardous wastes. The generator (the company where the wastes originated) must fill out a form detailing the exact kind and amounts of waste generated. Persons transporting the waste, who are also required to be permitted, and operating the disposal facility must each sign the form, vouching that amounts of waste transferred are accurate and copies go to the EPA. All phases are subject to unannounced EPA inspections. The generator remains responsible for any waste "lost" along the way or any inaccuracies in reporting. You can see how this provision of RCRA assures that generators will deal only with responsible parties and curtails midnight dumping.

Reduction of Accidents and Accidental Exposures

A significant risk to personal and public health lies in exposures that occur as a result of leaks, accidents, and misguided use of hazardous chemicals in the home or workplace. A considerable number of laws bear on reducing the probability of

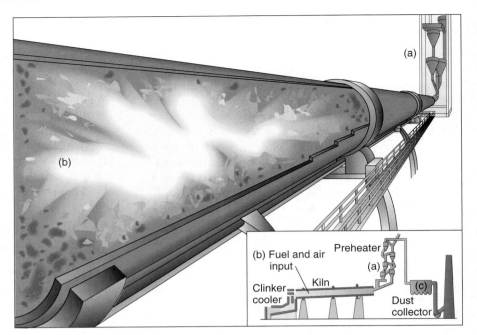

▶ **FIGURE 20–12** *Cement kiln to destroy hazardous wastes.* A cement kiln is a huge, rotating "pipe," typically 15 feet in diameter and 230 feet long, mounted on an incline. (a) Solid wastes fed in with the raw materials are fully incinerated and made to react with cement compounds as they gradually tumble toward the combustion chamber. (b) Flammable, liquid wastes added with the fuel and air impart fuel value as they are burned. (c) Waste dust is trapped and recycled into the kiln. (Redrawn with permission. Southdown Inc. Houston, TX 77002.)

accidents and on minimizing exposure of both workers and the public should accidents occur. Some examples follow.

Leaking Underground Storage Tanks—UST Legislation. One consequence of our automobile-based transportation system is millions of underground fuel-storage tanks at service stations. Putting such tanks underground greatly diminishes the risk of explosions and fires, but it also hides leaks. Underground storage tanks have a life expectancy of about 20 years before they may spring leaks. Many thousands of tanks are crossing this threshold each year and springing leaks. (By 1996, states reported over 300,000 of these failures.) Without monitoring, small leaks can go undetected until nearby residents begin to smell fuel-tainted water flowing from their faucets. **Underground Storage Tank (UST)** regulations, part of RCRA, now require strict monitoring of fuel supplies, tanks, and piping so that leaks may be detected early. When leaks are detected, remediation must begin within 72 hours. Rules now require all USTs to be upgraded with tank interior lining and cathodic protection (to retard electrolytic corrosion of the steel), and new tanks to be provided with the same protection if they are steel. Many service stations are turning to fiberglass tanks, which are noncorrodible.

Department of Transportation Regulations. Transport is an area particularly prone to accidents. As modern society uses increasing amounts and kinds of hazardous materials, the stage is set for accidents to become widescale disasters. To reduce this risk, **Department of Transportation Regulations (DOT Regs)** specify kinds of containers and methods of packing to be used in the transport of various hazardous materials. Such regulations are intended to reduce the risk of spills, fires, or poisonous fumes that could be generated from mixing certain chemicals in the case of an accident.

In addition, DOT Regs require that every individual container and the outside of a truck or railcar carry a standard placard identifying the hazards (flammability, corrosiveness, potential for poisonous fumes, and so on) of the material inside (Fig. 20–13). Such placards enable police and firefighters to identify the potential hazard and respond

▲ **FIGURE 20–13** *HAZMAT placards.* These are examples of some of the placards that are mandatory on trucks and railcars carrying hazardous materials. Numbers in place of the word on the placard, or on an additional orange panel, will identify the specific material. Placards alert workers, police, and firefighters to the kinds of hazards they face in the case of accidents.

ETHICS

ENVIRONMENTAL JUSTICE AND HAZARDOUS WASTE

Item: The largest commercial hazardous-waste landfill in the United States is located in Emelle, Alabama. This landfill receives wastes from Superfund sites and every state in the continental United States.

Item: Kettleman, California, has been selected as the site for the state's first commercial hazardous-waste incinerator.

Item: A Choctaw reservation in Philadelphia, Mississippi, was targeted to become the home of a 466-acre hazardous-waste landfill.

These three places have two things in common: They are considered to be appropriate locations for hazardous-waste disposal, and they are predominantly populated by people of color. African Americans make up 90% of Emelle's population; Latinos represent 95% of Kettleman's population; and the Choctaw reservation is entirely Native American. At issue is the question of *environmental justice*. The EPA defines **environmental justice** as "the fair treatment and meaningful involvement of all people regardless of race, color, national origin, or income with respect to the development, implementation, and enforcement of environmental laws, regulations, and policies. Fair treatment means that no group of people, including racial, ethnic, or socioeconomic group should bear a disproportionate share of the negative environmental consequences resulting from industrial, municipal, and commercial operations or the execution of federal, state, local, and tribal programs and policies."

Several recent studies have documented the fact that all across the United States,

waste sites and other hazardous facilities are more likely than not to be located in towns and neighborhoods where the majority of residents is non-Caucasian. These same towns and neighborhoods are also less affluent, a further element of environmental injustice. The wastes involved are primarily generated by affluent industries and the affluent majority, but somehow the wastes tend to end up well away from where they were generated and in the backyards of people of color. It seems fair to assume that the siting of hazardous facilities is a matter of political power, and those with the power would like to have the sites well away from their own backyards.

Within the non-Caucasian communities there are signs of a groundswell of grassroots concern about environmental racism and equity. The success of minorities in pushing for civil rights in recent decades has set a precedent that is beginning to empower people of color in seizing the initiative in this newly emerging arena. As a start, activists are turning their attention to cleaning up their own communities. Calling themselves the Toxic Avengers, a group of African American and Latino students in Brooklyn, New York, has taken on projects like battling the Radiac Research Corporation's toxic waste facility in the neighborhood, starting a recycling program, and conducting neighborhood workshops on fighting pollution. In Los Angeles, Watchdog is another grassroots, multiracial, working-class group that has successfully worked toward amending the regulations of the South Coast Air Quality Management District so that the regulations reflect special concerns of minorities and low-income workers. The

Good Road Coalition is an alliance of grassroots groups that successfully blocked the proposal by a Connecticut company to build a 6000-acre landfill on the Rosebud Sioux reservation in South Dakota.

The federal administration has taken this problem seriously. President Clinton issued Executive Order 12898 in 1994, focusing federal agency attention on environmental justice. The EPA's response has been to establish an Environmental Justice (EJ) program, put in place early in 1998, to capture the intent of Executive Order 12898 and to further a number of strategies already established by the EPA's Office of Solid Waste and Emergency Response. For example, in connection with the EPA's Brownfields Initiative, 78 communities are putting abandoned properties back into productive use. Many of the EJ efforts are directed toward addressing justice concerns before they become problems. In 1997, the EPA rejected a permit for a plastics plant in Convent, Louisiana; the plant was to be built in a predominantly minority community. The chemical company decided later to build the plant in an industrial area near Baton Rouge.

What is still uncertain about the environmental justice movement is the response of the Caucasian majority in the United States. Opposition to the EPA's EJ program has mounted from states, industries, and major cities; they claim that the EJ standards will impede economic progress. At times, the minorities themselves have opposed the EPA on the grounds that their low-income communities need the jobs that the industrial facilities will bring. What do you think?

---

appropriately in case of an accident. You may also see highway HAZMAT signs restricting truckers with hazardous materials to particular routes or lanes or to using a highway only during certain hours.

Worker protection—OSHA, Worker's Right to Know. In the past, it was not uncommon for industries to require workers to perform jobs that entailed exposure to hazardous materials without informing the workers of the hazards involved. This situation is now addressed by certain amendments to the **Occupational Safety and Health Act (OSHA)** known

as the **hazard communication standard,** or **"worker's right to know."** Basically, the law requires businesses and industries to make information regarding hazardous materials and suitable protective equipment available. One form taken by this information is the **Material Safety Data Sheets (MSDS),** which must accompany the shipping, storage, and handling of over 600 chemicals. These sheets contain information about the reactivity and toxicity of the chemicals, with precautions to follow when using the chemical. Notably, however, the responsibility to read the information and exercise proper precautions remains with the worker.

Community Protection and Emergency Preparedness— SARA, Title III. In 1984, an accident at a Union Carbide pesticide plant in Bhopal, India, caused the release of some 30–40 tons of an extremely toxic gas, methyl isocyanate (MIC). An estimated 600,000 people in communities surrounding the plant were exposed to the deadly fumes. The official death toll stands at 2500, but unofficial estimates start at 7000 and range much higher. At least 50,000 people suffered various degrees of visual impairment, respiratory problems, and other injuries from the exposure. Ironically, most of those deaths and injuries could have been avoided if people had known the simple protection. MIC is very soluble in water; thus, a wet towel over the head would have greatly reduced exposure, and showers would have alleviated aftereffects. Unfortunately, neither people nor medical authorities had any idea of the chemical they were confronted with, much less the protection or treatment.

In view of the Bhopal disaster, Congress passed legislation to address the problem. For expediency in passage, it was added as Title III to a bill reauthorizing Superfund, the Superfund Amendments and Reauthorization Act of 1986 (SARA). SARA, Title III, is better known as the **Emergency Planning and Community Right-to-Know Act (EPCRA),** which we have already seen in connection with the Toxics Release Inventory.

EPCRA requires companies that handle in excess of 5 tons of any hazardous material to provide a "complete accounting" of storage sites, feed hoppers, and so on. This information goes to a Local Emergency Planning Committee, one of which is also required in every governmental jurisdiction. This committee is made up of officials representing local fire and police departments, hospitals, and any other groups that might be involved in case of an emergency, as well as the executive officers of the companies in question.

The task of the committee is to draw up scenarios for possible accidents involving the chemicals on site and to have a contingency plan for every case. This means everything from having firefighters trained and properly equipped to fight particular kinds of chemical fires, to having hospitals stocked with medicines to treat exposures to the particular chemical(s) to implementing procedures. Thus, there can be an immediate and appropriate response to any kind of accident. Together with the TRI information, the act enables communities to draw up a chemical profile of their local area and encourages them to initiate pollution prevention activities and risk-reduction analyses.

The Toxic Substances Control Act (TSCA). In the past, new synthetic organic compounds were introduced for a specific purpose without any testing of potential side effects. For example, in the 1960s a new compound known as TRIS was found to be a very effective flame retardant and was widely used in children's sleepwear. It was only later discovered that TRIS is a potent carcinogen. Treated sleepwear was immediately withdrawn from the market. It is not known how many children (if any) developed cancer from TRIS, but the warning from this and other such cases is obvious.

Congress responded by passing the **Toxic Substances Control Act of 1976 (TSCA).** TSCA requires that before

manufacturing a new chemical in bulk, manufacturers must submit a "premanufacturing report" to the EPA in which the environmental impacts of the substance are assessed, including those that may derive from its ultimate disposal, and it is indicated whether it is a carcinogen. Depending on the results of the assessment, uses may be restricted or a product may be kept off the market altogether.

Laws applying to hazardous wastes are summarized in Fig. 20–14. It should be noted that nongovernmental consumer advocate groups have been a major force behind the passage of these laws as well as regulations requiring ingredient labels on all products. Citizen action does make a difference.

20.5 Looking Toward the Future

Essentially everyone, environmentalists as well as their critics, agree that pollution control has become an enormously complex and costly business. Consider the costs to business of installing and operating all the pollution-control or treatment equipment. Then reflect on the costs to federal, state, and local governments for all the inspection, monitoring, and enforcement of all the regulations. The EPA estimates that the current yearly cost of complying with federally mandated pollution-control and cleanup programs is over $100 billion. Isn't there a less costly way to accomplish the same objectives of protecting human and environmental health?

Too Many or Too Few Regulations?

Considering all the complexities and costs of pollution control, it is not hard to see why some critics feel that business is overregulated. Nor is it hard to see why some U.S. companies have moved operations to developing countries where regulations are less strict. Yet which laws or regulations would you abandon? The environmental nightmare of toxic wastes that has come to light in the former U.S.S.R. shows what can happen in the absence of regulations (see the "Global Perspective" essay, p. 494).

Further, regulation is far from complete. One area of concern is the legally permitted discharges. The low concentrations that standards still allow in discharges add up. Although the figure is gradually going down, there still are more than 2 million tons of toxic chemicals legally discharged into the environment annually. Do the regulations need to be made even stricter?

A second area of concern is the exemption of companies that produce less than 100 kilograms (220 pounds) of hazardous waste per month, a category that homeowners and farmers fall into as well. Items such as batteries and unemptied pesticide containers go into the trash and end up in municipal landfills, most of which don't have protective measures for toxic wastes. Can we regulate such activities without creating a police state?

In short, pollution control seems to have come to a regulatory impasse. However, a new approach is developing that may provide the best of both worlds, reducing environmental

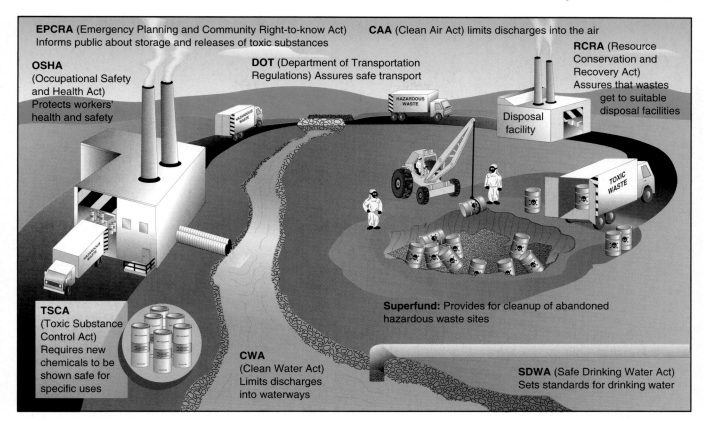

▲ **FIGURE 20–14** *Major hazardous-waste laws.* Summary of the major laws pertaining to the protection of workers, the public, and the environment from hazardous materials.

pollution while at the same time reducing costs and regulatory complexity. The concept is called *pollution avoidance* or *pollution prevention.*

Pollution Avoidance for a Sustainable Society

First, let us be clear regarding the distinction between *pollution control* and *pollution prevention.* Pollution control refers to adding on a filter or other device at the "end of the pipe" to prevent pollutants from entering the environment. Disposal of the captured pollutants still has to be dealt with, and this entails more regulation and control. **Pollution avoidance**, on the other hand, refers to changing the production process or the materials used, or both, so that the harmful pollutants won't be produced in the first place. Take a simple example: Adding a catalytic converter to the exhaust pipe of your car is a case of pollution control. Redesigning the engine so that less pollution is produced, or switching to an electric car, is a case of pollution avoidance.

Pollution prevention often translates as better product or materials management—that is, less wastage. Thus, pollution prevention often creates a cost savings. For example, Exxon Chemical Company added simple "floating roofs" to storage tanks of its most volatile chemicals, reducing evaporative emissions by 90% and gaining a savings of $200,000 per year. That paid for the cost of the roofs in six months. As

another example, Carrier, Inc., a maker of air conditioners, installed equipment for more precise metal cutting. According to a spokesperson for the company, the new equipment greatly reduces cutting wastes, eliminates the need for a toxic cutting lubricant, and leads to the production of a better product—all results that make the company more competitive. These are examples of *minimization* and *elimination* of pollution.

A second angle on pollution avoidance is *substitution:* finding nonhazardous substitutes for hazardous materials. For example, the dry-cleaning industry, which uses large quantities of toxic organic chemicals, is exploring the use of water-based or **wet-cleaning**. Preliminary results from some entrepreneurs indicate that wet-cleaning is quite comparable in cost and performance to dry cleaning; there are now more than 100 wet-cleaners in North America, and the number is growing. Clairol switched from water to foam balls for flushing pipes in hair-product production, reducing wastewater by 70% and saving $250,000 per year. Water-based inks and paints are being substituted for those that contained synthetic organic solvents, and inks and dyes based on biodegradable organic compounds are being substituted for those based on heavy metals.

A third approach is *reuse:* to clean up and recycle solvents and lubricants. Some military bases, for example, have been able to distill solvents and reuse them instead of discarding them into the environment.

GLOBAL PERSPECTIVE

A TOXIC WASTELAND

Given the magnitude of the toxic waste problem in the United States, it is tempting to think that things might be better in a society characterized by central planning on the part of the regime in power, that environmental concerns might more easily be addressed at the highest levels of power. Recent events have put such thinking to rest. Now that the shrouds of secrecy have been lifted from the former Communist countries of Eastern Europe and the republics of the former U.S.S.R., it is apparent that central planning has been responsible for the worst kinds of environmental pollution imaginable. Pollutants were emitted from the stacks of industry and power plants with no controls, ruining thousands of square miles of forests and creating untold health problems; untreated sewage from cities fouled the rivers, destroying fish and rendering the water unfit even for industrial uses; heavy metals and toxic chemicals were poured untreated into the Baltic Sea, turning the bottom into a marine desert (see map).

The truth is, the leaders of Communist regimes almost completely ignored environmental concerns in the interest of promoting economic production. In the words of Murray Feshback and Alfred Friendly, Jr., in their book *Ecocide in the U.S.S.R.*, "No other great industrial civilization so systematically and so long poisoned its air, land, water and people....And no advanced society faced such a bleak political and economic reckoning with so few resources to invest toward recovery." The people of Eastern Europe and the former U.S.S.R. are now faced with both a bankrupt environment and a bankrupt economy, and they are looking to the Western democracies for help in both areas.

In environmental affairs in the United States and other democratic societies, grassroots action has always preceded change in public policy. The progress in solving the hazardous-waste problem—the laws passed by Congress and the implementation and enforcement of those laws—has been a direct result of grassroots pressure both on recalcitrant leaders and through electing new leaders with environmental values. All the evidence indicates that without this grassroots pressure—citizens demanding a clean, healthful environment—political leaders are likely to promote economic and business concerns that benefit their own in-group and will ignore environmental responsibilities.

How are the environmental grassroots movements in the countries of Eastern Europe and the former U.S.S.R. doing? Grassroots movements are never encouraged in totalitarian societies; they are seen as subversive elements. After the nuclear accident at Chornobyl' in 1986, environmentalism achieved some legitimacy in the U.S.S.R., grassroots movements were tolerated, and some protests were successful in bringing about limited environmental changes. But now that political processes are open, environmental grassroots movements have declined, as people have become more concerned with basic economic survival. There is currently a fatalism in the response of people living in the polluted former Communist lands. They know their environment is unhealthy, but they still feel powerless to do anything about it. Thus, experience demonstrates that it is in liberal democracies, where citizens have the freedom to organize and to demand a better environment, that pollution control and prevention regulations are passed and implemented. Of course, the freedom to organize and act is only half the answer; people must take advantage of that freedom. We can only hope that the newly freed people of the former Communist nations and other peoples of the world, including ourselves, will take advantage of freedom and act to create a better environment.

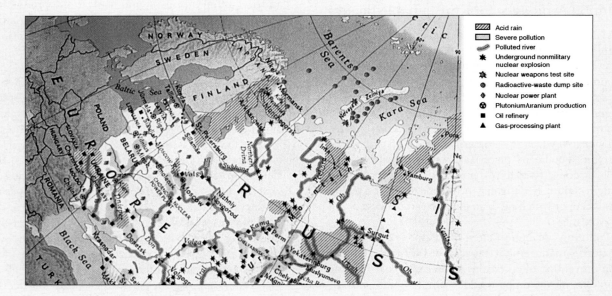

▨	Acid rain
▢	Severe pollution
〰	Polluted river
✳	Underground nonmilitary nuclear explosion
✲	Nuclear weapons test site
◉	Radioactive-waste dump site
◆	Nuclear power plant
⊗	Plutonium/uranium production
■	Oil refinery
▲	Gas-processing plant

(Reprinted from M. Edwards, "Lethal legacy," *National Geographic* [August 1994]:80–81.)

Perhaps the most profound feature of pollution prevention is that it is putting environmentalist groups, government regulators, and industry representatives around the same table. Pollution control, by its very nature, tends to create an adversarial relationship—"us against them"—the regulators against the regulated (industry). Over the years, this adversarial relationship has cost billions in lawyer fees and court costs of parties suing one another. Now we are beginning to find citizens and government workers side by side with company executives looking for ways to reduce both pollution and costs through reaching consensus.

Industry is also acting unilaterally to clean up its act. The Chemical Manufacturers Association, a trade group of chemical industry executives, has initiated what they call a **Responsible Care Program**. The aim of this program is to go beyond what is required by the law and commit members to an ongoing, long-term reduction of toxic releases and of all hazardous wastes generated by their industry.

The government's involvement in pollution prevention is crucial. The EPA has recently been changing its image from the strict "command and control" regulatory agency to one of providing leadership in ways that encourage states, industries, and consumers to cooperate to find better solutions to toxic waste problems. The EPA has initiated programs such as

- Common Sense Initiative—EPA and industries sharing information about industrial processes and environmental goals;
- Project XL—EPA encouraging industries or individuals to test new pollution-prevention and -control technologies; and
- Design for the Environment—a voluntary program to bring environmental considerations into the design and redesign of technical processes and products.

Because of its oversight responsibilities, the EPA has often had contentious relationships with state agencies that are implementing environmental programs. To address this problem, the EPA has established the National Environmental Performance Partnership System (NEPPS), designed to give states more flexibility and to encourage the EPA and state offices to work together to accomplish pollution-control objectives.

Finally, pollution avoidance can also be applied to the individual consumer. So far as you are able to reduce or avoid using products containing harmful chemicals, you are preventing those amounts of chemicals from going into the environment. You are also reducing the byproducts resulting from producing those chemicals. The average American home contains as much as 100 pounds of HHW. These are substances like paints, stains and varnishes, batteries, pesticides, motor oil, oven cleaners, etc. They need to be stored safely, used responsibly, and disposed of properly if they are not to pose a risk of injury or death and a threat to environmental health. (Many communities hold regular HHW collection days; if yours doesn't, you could initiate one!)

Increasingly, companies are gradually beginning to produce and market **green products**, a term used for products that are more environmentally benign than their traditional counterparts. How fast and to what degree these green products displace or replace traditional products will in large part depend on how we behave as consumers. In other words, consumerism can be an extremely potent force.

In conclusion, we should see that there are four ways to address the problems of chemical pollution: (1) pollution prevention, (2) recycling, (3) treatment (breaking down or converting the material to harmless products), and (4) safe disposal. The first three of these methods promote a minimum of waste in harmony with the second principle of sustainability. There is every reason to believe that we can have the benefits of modern technology without destroying the sustainability of our environment with pollution.

ENVIRONMENT ON THE WEB

DEFINING HAZARDOUS

Effective management of hazardous wastes requires that we be able to define those materials that are of most concern and to develop approaches for handling, treating, and disposing of those materials that are different from those used for solid wastes. As discussed earlier in this chapter, the EPA determines whether a substance is hazardous according to four categories: ignitability, corrosivity, reactivity, and toxicity. These categories are simple to understand and make sense to us—for instance, we can imagine that a highly ignitable substance, for instance, is of much greater concern than an inert material. However, because there is no absolute definition for any of these terms used to categorize hazardous materials, the practical limits on each category are essentially arbitrary and subject to debate.

An example of this debate can be found in the definition of toxicity. The EPA defines a toxic material as one that is injurious to health when ingested or inhaled, but the term "injurious" is a vague one. In the past, our understanding of human health injury centered on acute and visible damage such as death or obvious illness. More recently, we have become concerned about chronic effects such as cancers, miscarriages, and persistent respiratory distress. Still, it is possible to broaden the definition of "toxic" to include changes in an individual's perception—for instance, the quality of peripheral vision. Gait changes, typical of the effects of some heavy metals on human health, could also be considered a measure of toxic impact. Or perhaps changes in cell structure, such as tiny chromosomal changes, could be considered evidence of toxicity, even if no impacts on human health are apparent.

Intergenerational impacts could also be part of a definition of toxicity. Some substances, such as diethyl stilbestrol (DES, a drug given to pregnant mothers to prevent miscarriages and other pregnancy problems), for example, have few or no impacts on the current generation but more significant or very serious impacts on the children or grandchildren of those exposed to the substance. (In the case of diethyl stilbestrol, there are no impacts on the mothers. However, DES has been linked to a rare vaginal cancer in daughters and genital problems in sons born to mothers who used DES during pregnancy.)

There is little consensus about how toxicity can or should be defined, and different jurisdictions evaluate toxicity differently. As is the case with many other environmental issues, our working definition of this term, and indeed of the category of substance we currently consider "hazardous," reflects our society's current best understanding of the problem. As our knowledge of the nature and range of impacts grows, our definition of "hazardous wastes"—and the regulatory framework for their management—will also evolve.

Web Explorations

The Environment on the Web activity for this essay describes the challenges in managing hazardous wastes. Go to the Environment on the Web activity (select Chapter 20 at http:// www.prenhall.com/nebel) and learn for yourself:

1. how the definition of "hazardous" and "waste" can affect regulatory approaches to managing waste;
2. about the Basel Convention on the international trade and disposal of hazardous wastes; and
3. about the strengths and weaknesses of the Basel Ban on hazardous wastes.

A suggested time frame for each of these exercises is 10–30 minutes.

REVIEW QUESTIONS

1. The EPA divides hazardous chemicals into what four categories?
2. Define what is meant by "total product life cycle," and describe the many stages at which pollutants may enter the environment.
3. What are the two classes of chemicals that pose the most serious long-term toxic risk, and how do they affect food chains?
4. How were chemical wastes generally disposed of before 1970?
5. What two laws pertaining to the disposal of hazardous wastes were passed in the early 1970s? Describe how the passage of the laws shifted pollution from one part of the environment to another.
6. Describe three methods of land disposal that were used in the 1970s. What was a common denominator of all of them in terms of pollution?
7. What laws exist to protect the public against contaminated drinking water? Against new chemicals having dangerous side effects?
8. Can contaminated groundwater be cleaned up? How?
9. What law was passed to cope with the problem of abandoned hazardous waste sites? What are the main features of the legislation?
10. What law was passed to ensure safe land disposal of hazardous wastes? What are the main features of the legislation?

11. What laws exist to protect the public against exposures resulting from accidents? What are the main features of the legislation?

12. What methods of hazardous waste disposal or treatment may be used more in the future? Used less?

13. What is the distinction between pollution control, pollution prevention, and pollution avoidance?

14. Discuss how environmentalists and businesspeople can work together to reduce pollution. How likely is it that they will cooperate?

THINKING ENVIRONMENTALLY

1. Do a product life-cycle study for all the materials used in both manufacturing and running your car. List all the pollutants produced at all stages.

2. Before the 1970s, it was not illegal to dispose of hazardous chemicals in unlined pits, and many companies did so. Should they be held responsible today for the contamination those wastes are causing, or should the government (taxpayers) pay for the cleanup? Give a rationale for your position.

3. Our knowledge regarding bioaccumulation of toxic chemicals and their long-term health effects in humans is very limited. Why?

4. Suppose there is a proposal to build an incineration facility near your community. It is proposed that hazardous wastes currently being landfilled at the same location will be disposed of in the new facility. Would you support or oppose the proposal? Give a rationale for your position.

5. Do you favor cutting back on regulations pertaining to hazardous chemicals to help balance the federal budget? What laws or regulations, if any, would you cut back? Defend your answer.

WEB REFERENCES

On-line resources for this chapter can be found on the World Wide Web at: **http://www.prenhall.com/nebel**. Click on Chapter 20 on the chapter selector.

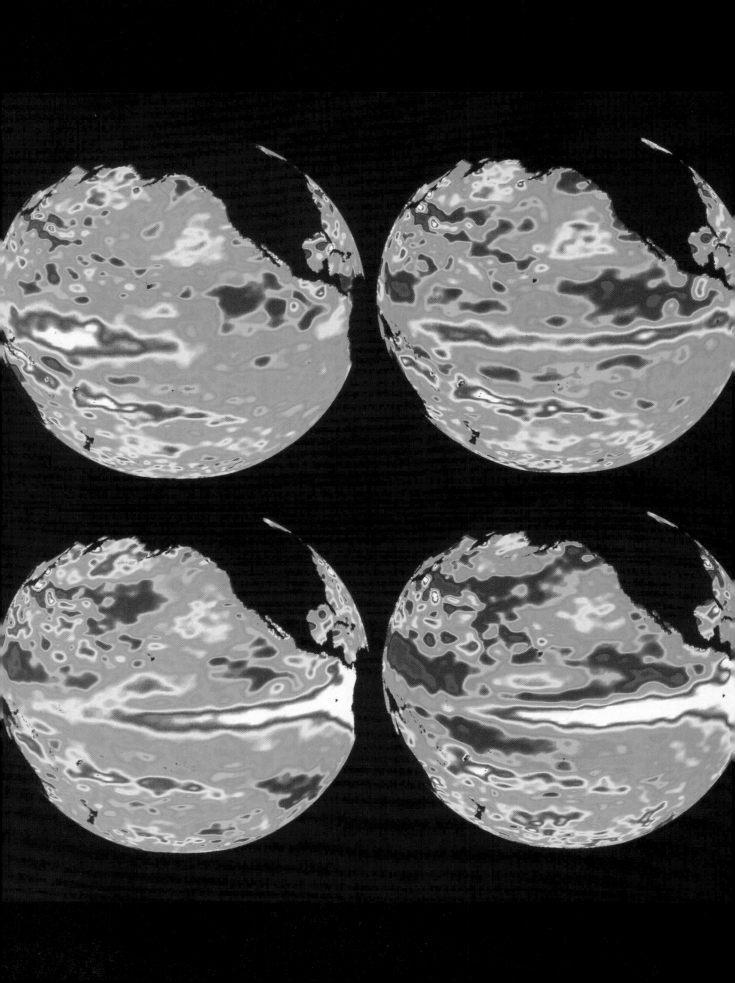

The Atmosphere: Climate, Climate Change, and Ozone Depletion

Key Issues and Questions

1. The atmosphere is the site and source of our weather. How is the atmosphere structured, and how does it function to bring us the major features of our weather?

2. There are ways of investigating the past to reveal features of past climates. What does the past tell us about climate change?

3. The oceans and atmosphere are closely linked in creating climate. What is the ocean Conveyor, and how could the functioning of this system change to bring about a major climate change?

4. The interaction of solar radiation with atmospheric gases controls the balance of warming and cooling of Earth. How have atmospheric gases from human activities affected this balance?

5. Greenhouse gases are increasing in the troposphere. What is the probability that these gases will bring on global warming? If it occurs, what will be the major effects of global warming?

6. Global warming poses a serious threat to the world's climate. How have the countries of the world responded to the threat?

7. The stratospheric ozone shield is vital in protecting life from damaging ultraviolet radiation. How is ozone both formed and destroyed in the stratosphere?

8. Chlorofluorocarbons (CFCs) and other gases destroy stratospheric ozone. What evidence confirms these destructive processes?

9. Continued ozone destruction represents a threat to life on Earth. What is the global community doing about slowing or reversing ozone loss?

El Niño has become a household name. Beginning in April 1997 and extending through the spring of 1998, this climatic event has linked the world together. In California and Oregon, unusually severe storms battered the coastline, causing major coastal erosion and flooding rivers (Fig. 21–1). On the Eastern seaboard and Gulf of Mexico, residents relaxed through a hurricane season that was the mildest in many years. Rainfall at least five times the normal deluged East Africa, often a region of drought. Fires blackened 1400 square miles of drought-impacted forests in Indonesia, creating huge clouds of smoke that blanketed much of Southeast Asia. In New Guinea, the lack of rain brought crop failures, necessitating a massive effort in food relief in order to prevent famine. Record crop harvests were enjoyed in India, Australia, and Argentina. Unusual rainfall in California and Florida triggered lush growth of vegetation. Even after El Niño was officially declared over, in May 1998, the impact of El Niño continued, with wildfires sweeping through woods and suburbs in Florida, fueled by the lush vegetation that had dried during several months of dry weather. The estimates of global damage are as high as $30 billion, and at least 2000 deaths have been traced to the 1997–1998 El Niño. It has been a lesson in global climate the world will not soon forget, as atmosphere and oceans teamed up to produce a reminder that we live at the mercy of a system we do not control.

What caused this incredible shift in weather over so much of the globe? Briefly, a major shift in atmospheric pressure over the central equatorial Pacific Ocean led to a reversal of the

Development of the 1997 El Niño. Satellite images (from 25 April, 25 May, 25 June and 25 Sept.) demonstrate the development of warm water in the eastern Pacific over time (white is warmest, red is next). The El Niño is the area moving eastward along the equator across the Pacific.

trade winds that normally blow from an easterly direction. Warm water spread toward the east, patterns in precipitation and evaporation were impacted, the jet streams shifted from their normal courses, and the system was sustained for more than a year. Meteorologists are quite good at explaining what an El Niño is and where and how it will impact the different continents and oceans, but they are still unable to explain why it happens. What we do know, however, is that there have been many El Niños in the past, but none so intense as this latest one. In the past 12 years, there have been five El Niños, a frequency that is unprecedented. If such changes in major weather patterns persist, we will in fact be experiencing a climate change. This latest El Niño has revealed to people everywhere that the atmosphere, oceans, and land are linked together and that when normal patterns are disrupted, the climate on the whole Earth can be affected. Could it be that the warming trend now evident in global temperatures is responsible for triggering this latest intense El Niño? Are we in for a major climate change? Indeed, what does control the climate? We will seek answers to these questions as we investigate the atmosphere, how it is structured, and how it brings us our weather and climate. Then we will consider the evidence for global climate change, finishing with a look at what is happening to ozone in the upper atmosphere.

(a)

(b)

(c)

(d)

▲ **FIGURE 21–1** *Impacts of El Niño.* (a) Landslides on the California coast. (b) Food relief in New Guinea. (c) Smoke and fires in Indonesia. (d) Flooding in Kenya.

21.1 Atmosphere and Weather

Atmospheric Structure

As we learned in Chapter 3, the atmosphere is a collection of gases that gravity holds in a thin envelope around Earth. The gases within the troposphere are responsible for moderating the flow of energy to Earth and are involved with the biogeochemical cycling of many elements and compounds: oxygen, nitrogen, carbon, sulfur, and water, to name the most crucial ones. The lowest layer, the **troposphere**, extends up to 10 miles (16 km) in the tropics and 5 miles (8 km) in higher latitudes. This layer contains practically all of the water vapor and clouds; it is the site and source of our weather. Except for local temperature inversions, the troposphere gets colder with altitude (Fig. 21–2). Air masses in this layer are well mixed vertically; pollutants can reach the top within a few days. Substances entering the troposphere—including pollutants—may be changed chemically and washed back to Earth's surface by precipitation (Chapter 22). Capping the troposphere is the **tropopause**.

Above the tropopause is the **stratosphere**, a layer within which temperature *increases* with altitude, up to about 40 miles above the surface of Earth. The temperature increases here primarily because the stratosphere contains ozone (O_3), a form of oxygen that absorbs high-energy radiation emitted by the Sun. Because there is little vertical mixing of air masses in the stratosphere and no precipitation from it, substances that enter it can remain there for a long time. Beyond the stratosphere are two more layers, the *mesosphere* and the *thermosphere*, where ozone concentration declines and only small amounts of oxygen and nitrogen are found. Because none of the reactions we are concerned with occur in the mesosphere or thermosphere, we shall not discuss those two layers. Table 21-1 summarizes the characteristics of the troposphere and stratosphere.

Weather

The day-to-day variations in temperature, air pressure, wind, humidity, and precipitation—all mediated by the atmosphere—constitute our **weather**. Climate is the result of long-term weather patterns in a region. The scientific study of the atmosphere, both of weather and climate, is **meteorology**. It is fair to think of the atmosphere-ocean system as an enormous weather engine, fueled by the Sun and strongly affected by the rotation of Earth. Solar radiation enters the atmosphere and then takes a number of possible courses (Fig. 21–3); some is reflected by clouds and Earth's surfaces, but most is absorbed by the atmosphere, oceans, and land, which are heated in the process. The land and oceans can then radiate some of their heat back upward as infrared energy. Some of this back-radiated heat is transferred to the atmosphere. Thus, air masses will become warmer at the surface of Earth and will tend to expand, becoming lighter. This lighter air will rise, creating vertical air currents; on a large scale, it creates the major *convection currents* we encountered in Chapter 9 (Fig. 9–5). Of course, air must flow to replace the rising

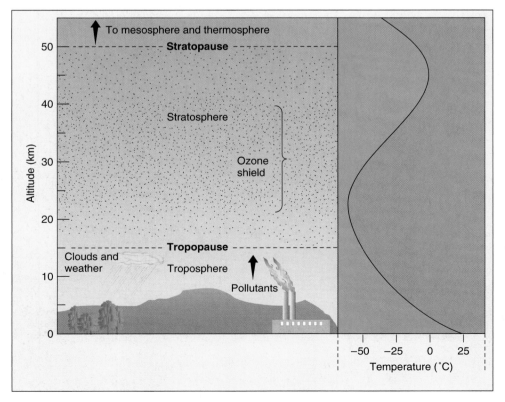

◀ **FIGURE 21–2** *Structure and temperature profile of the atmosphere.* The left-hand plot shows the layers of the atmosphere, while the plot on the right shows the vertical temperature profile.

TABLE 21-1 Characteristics of Troposphere and Stratosphere

Troposphere	Stratosphere
Extent: Ground level to 10 miles (16 km)	Extent: 10 miles to 40 miles (16 km to 65 km)
Temperature normally decreases with altitude to -70°F (-59°C)	Temperature increases with altitude to +32°F (0°C)
Much vertical mixing	Little vertical mixing, slow diffusion exchange of gases with troposphere
Substances entering may be washed back to Earth	Substances entering remain unless attacked by sunlight or other chemicals
All weather and climate take place	Isolated from the troposphere by the tropopause

warm air, and this leads to horizontal airflows, or wind. The ultimate source of this horizontal flow is cooler air that is sinking, and the combination is the Hadley cell (Fig. 9–5). As we saw in Chapter 9, these major flows of air create regions of high rainfall (equatorial), deserts (25° to 35° north and south of the equator), and horizontal winds (trade winds).

On a smaller scale, **convection currents** bring us the day-to-day changes in our weather as they move in a general pattern of west to east. Weather reports inform us of regions of high and low pressure, but where do these come from? Rising air (due to solar heating) creates high pressure up in the atmosphere, leaving behind a region of lower pressure close to Earth. Conversely, once the moist, high-pressure air has cooled through radiation to space and loss through condensation (generating our precipitation), the air then flows horizontally toward regions of sinking cool, dry air (where the pressure is lower). There the air is warmed at the surface and creates a region of higher pressure (Fig. 21–4). These differences in pressure will lead to air flows, which are the winds we experience. As Fig. 21–4 shows, the winds will tend to flow from high-pressure regions toward low-pressure regions.

▶ **FIGURE 21–3 Solar energy balance.** Much of the incoming radiation from the Sun is reflected back to space (30%), but the remainder is absorbed by the oceans, land, and atmosphere (70%), where it creates our weather and fuels photosynthesis. Eventually this absorbed energy is back radiated to space as infrared energy.

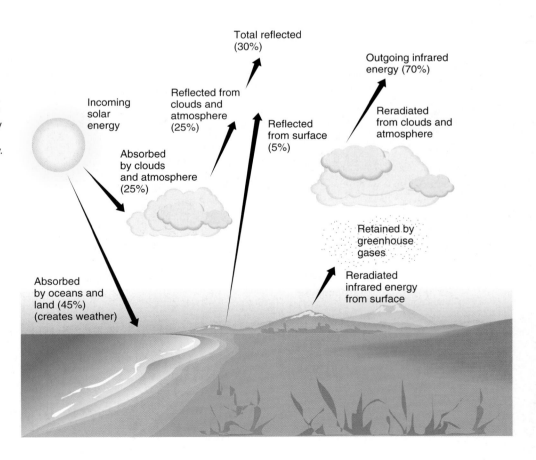

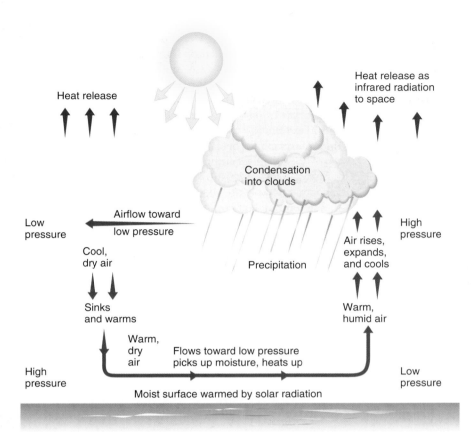

◀ **FIGURE 21-4** *A convection cell.*
Driven by solar energy, these cells produce the main components of our weather as evaporation and condensation occur in rising air and precipitation results, followed by the sinking of dry air. Horizontal winds are generated in the process.

The larger scale air movements of Hadley cells are influenced by Earth's rotation from west to east. Thus, we have the trade winds over the oceans and the general flow of weather from west to east. Higher in the troposphere, Earth's rotation and air-pressure gradients generate veritable rivers of air, called **jet streams**, which flow eastward at speeds of over 300 mph and which undergo considerable meandering. The jet streams are able to steer major air masses in the lower troposphere. One example is the polar jet stream, which steers cold air masses into North America when it dips downward in latitude.

Air masses of different temperature and pressures will meet at boundaries we call **fronts**, and these are regions of rapid weather change. Other movements of air masses due to differences in pressure and temperature include hurricanes and typhoons and the local but very destructive phenomena of tornadoes. Finally, we also have major seasonal airflows, the **monsoons**, which often represent a reversal of previous wind patterns. These are created by major differences in cooling and heating between oceans and continents. The summer monsoons of the Indian subcontinent are notorious for the beneficial rains they bring as well as the frequently devastating floods that can occur when the rains are heavy. When we put together the general atmospheric circulation patterns and the resulting precipitation, then add the wind and weather systems generating it and mix this with the rotation of Earth along with the tilt of

Earth on its axis, which brings us the seasons, we have the general patterns of weather that characterize different regions of the world. These patterns are referred to as the region's **climate**.

21.2 Climate

Earlier (Chapter 2) we described climate as the description of the average temperature and precipitation expected throughout a typical year in a given region. Recall that the different temperature and moisture regimes in different parts of the world "created" different major ecosystem types that we call biomes. These biomes represent the adaptations of plants, animals, and microbes to the prevailing weather patterns, or climate, of this region. Of course, the temperature and precipitation patterns are actually caused by other forces, the major determinants of weather previously outlined. Humans have shown that they can adjust to practically any climate (short of the brutal conditions on high mountains or burning deserts), but this is not true of the other inhabitants of the particular regions we occupy. If other living organisms in a region are adapted to a particular climate, then a change in the climate represents a major threat to the structure and function of the existing ecosystems. The subject of climate change is such a burning issue today because we depend on these other organisms for a host of vital goods and services without which we could not survive (Chapter 3). If the climate changes, can these ecosystems change with it in such a way as not to interrupt the vital support they provide us? How rapidly can organisms and ecosystems adapt to changes in the climate? How rapidly do climates change? One way to answer these questions is to look into the past, where there may be records of climate changes.

Climates in the Past

Searching the past for evidence of climate change has become a major scientific enterprise, one that becomes more difficult the further into the past we try to search. Systematic records of the factors making up weather—temperature, precipitation, storms, and so forth—have been kept for little more than a hundred years. Nevertheless, these can already inform us that the climate is far from constant. The record of surface temperatures, from weather stations around the world and from literally millions of observations of sea-surface temperatures, tells an interesting story (Fig. 21–5): Since 1855, global average temperature has shown periods of cooling and warming, but in general has increased 0.6°C.

Using indirect methods, it is possible to extend observations on climatic changes much further back in time. By

delving into historical accounts, it appears that from 1100 to 1300 A.D. the Northern Hemisphere enjoyed a medieval warming period, followed by the "Little Ice Age," between 1400 and 1850 A.D. Some work done quite recently on ice cores has provided a startling view of a global climate that oscillates according to several cycles, as well as some evidence that remarkable changes can occur in the climate within as little as a few decades. Ice cores in Greenland and the Antarctic have been analyzed for ice thickness, gas content—specifically, carbon dioxide and methane (two greenhouse gases)—and **isotope** data. Oxygen and hydrogen isotopes—isotopes are alternate chemical configurations due to different nuclear components—will behave differently at different temperatures when condensed in clouds and incorporated into ice.

The record based on these analyses indicates that Earth's climate has oscillated between ice ages and warm periods (Fig. 21–6); during major ice ages, huge amounts of water were tied up in glaciers and ice sheets, and the sea level was lower by as much as 400 feet (120 m). The most likely explanation for these major oscillations is the existence of known variations in Earth's orbit, such that in different modes of orbital configuration the distribution of solar radiation over different continents and latitudes varies substantially. These oscillations take place according to several periods, called

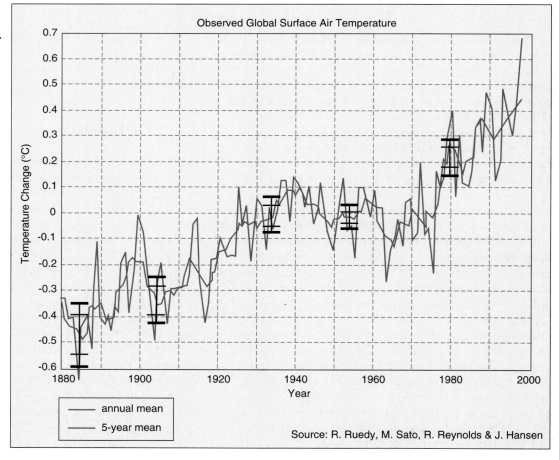

▶ **FIGURE 21–5**
Annual mean global surface temperatures. The baseline, or zero point, is the 1950–1980 average temperature; the red line represents the five-year running average. The warming trend since 1970 is conspicuous.

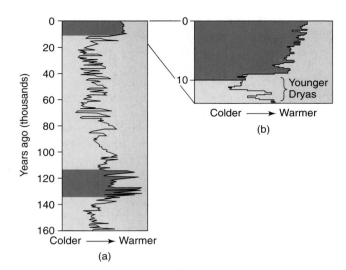

▲ **FIGURE 21–6** *Past climates, as determined from an ice core.* (a) Temperature patterns of the last 160,000 years, demonstrating climatic oscillations. (b) Higher resolution of the last 12,000 years. The "Younger Dryas" cold spell occurred at the start of this record. Note how rapidly this cold spell dissipated, at around 10,700 years before present. (After Christopherson, Robert, *Geosystems.* 3rd ed., 1997.)

Milankovitch cycles (after the Serbian scientist who first described them): 100,000, 41,000, and 23,000 years. However, superimposed on these major oscillations is a record of rapid climatic fluctuations during periods of glaciation and warmer times (Fig. 21–6b). One such rapid change occurred toward the end of the last ice age, called the *Younger Dryas* event (Dryas is a genus of Arctic flower). Earth had been warming up for 6000 years and then plunged again into 1500 years of cold weather. At the end of this event, 10,700 years ago, Arctic temperatures rose 7°C in 50 years! The impact on living systems must have been enormous. One possible explanation for this warming that has been discarded is variations in solar output; there is simply no evidence that the Sun has changed over the last million years, let alone undergone

major changes in just a few decades. In search of other explanations for these rapid changes, scientists have narrowed the field to the link between the atmosphere and the ocean system. We return now to this link, having already seen the crucial role played by the oceans in El Niño events.

Ocean and Atmosphere

Earth is mostly a water planet, covered more than two-thirds by oceans. Since we all live on land, it is hard for us to imagine that the oceans play a dominant role in determining our climate. Yet the oceans are the major source of water for the hydrologic cycle and the main source of heat to the atmosphere. As we saw in Chapter 9, the evaporation of ocean water supplies the atmosphere with water vapor, and when water vapor condenses in the atmosphere (Fig. 9–5), it supplies the atmosphere with heat (latent heat of condensation). The oceans also play a vital role in climate because of their innate heat capacity—the ability to absorb energy when water is heated. Indeed, the entire heat capacity of the atmosphere is equal to just the top 3 m of ocean water. The well-known ameliorating effect of oceans on the climate of coastal land areas is a consequence of this property. Finally, through the movement of ocean currents, the oceans are highly important as conveyors of heat. Enormous quantities of heat are moved laterally through water, from hot equatorial regions to higher latitudes (Fig. 21–7); an example is the Gulf Stream that keeps Western Europe warm. Thus, because of the ocean's innate heat capacity and its role as conveyor of heat, any discussion of climate must examine the links between the atmosphere and the oceans.

A crucial concept to our concerns here is the thermohaline circulation pattern dominating oceanic currents; thermohaline refers to the contributions that temperature and salinity make to the density of seawater. This **Conveyor** system (Fig. 21–8) acts as a giant, complex conveyor belt, where water masses move according to density from surface to deep oceans and back again. A key area is the high latitude North Atlantic, where salty water from the Gulf Stream moves

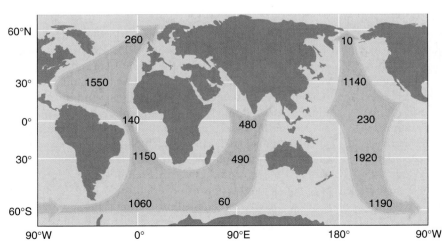

◀ **FIGURE 21–7** *Heat transport by the oceans.* These estimates are in units of terawatts (10^{12} W). Note the connections between the oceans, indicating that some heat transported by the North Atlantic originates in the Pacific.

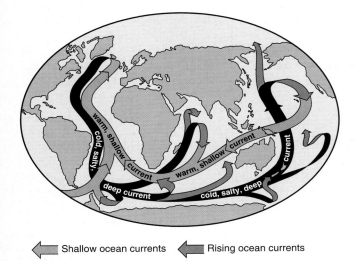

| Shallow ocean currents | Rising ocean currents |
| Deep ocean currents | Descending ocean currents |

▲ **FIGURE 21–8** *The oceanic Conveyor system.* Salty water flowing to the North Atlantic is cooled and sinks, forming the North Atlantic Deep Water. This deep flow extends southward and is joined by Antarctic water, where it extends into the Indian and Pacific oceans. Surface currents then proceed in the opposite direction, returning the water to the North Atlantic. (Adapted from Mackenzie, Fred T., *Our Changing Planet*, 2nd ed., 1998. Reprinted by permission of Prentice-Hall, Inc., Upper Saddle River, NJ.)

northward on the surface and is cooled by Arctic air currents. Cooling increases the density, and the water sinks to depths of up to 4000 m—the North Atlantic Deep Water (NADW). This deep water spreads southward through the Atlantic to the southern tip of Africa, where it is joined by cold Antarctic waters and spreads northward into the Indian and Pacific oceans as deep currents. The currents gradually slow down and warm, becoming less dense and welling up to the surface, where they are further warmed and begin a movement of surface waters back again toward the North Atlantic. Indeed, it is this movement that transfers enormous quantities of heat toward Europe, providing a climate that is much warmer than the high latitudes there would suggest. This system operates over a period of about 1000 years for one complete cycle and is vital to the maintenance of current climatic conditions.

Recent evidence has indicated that the Conveyor system has been interrupted in the past; when that happens, the climate is severely changed. One mechanism that could accomplish such a shutdown is the appearance of unusual quantities of fresh water in the North Atlantic, diminishing the density of the water and therefore preventing much of the massive sinking that normally occurs there. There is evidence in marine sediments of the periodic invasion of icebergs from the Polar ice cap that supplied huge amounts of fresh water as they melted. The evidence indicates that these invasions coincided with rapid cooling events as recorded in ice cores and suggests that the Conveyor system shifted southward, with deep water forming near Bermuda instead of Greenland. When this occurred, a major climate cooling happened within a few decades. Although it is not clear how, the normal Conveyor pattern returned and brought about

another abrupt change, this time warming conditions in the North Atlantic. The Younger Dryas event is believed to have involved just such a shift in the Conveyor system.

It is significant that one of the likely consequences of extended global warming is increased precipitation over the North Atlantic. If such a pattern is sustained, it could lead to a breakdown in the normal operation of the Conveyor and a rapid reversal of climatic conditions into a new ice age, a possibility that oceanographer Wallace Broecker has called "The Achilles' heel of our climate system!"

21.3 Global Climate Change

Geochemist James W. C. White, writing in the July 15, 1993, issue of the journal *Nature*, said it well: "If the earth came with an operating manual, the chapter on climate might begin with a caveat that the system has been adjusted at the factory for optimum comfort, so don't touch the dials." Many scientists are convinced that we are madly twirling the dials of Earth's "operating system" as we release pollutants into the atmosphere that put the planet in danger of unprecedented warming. Skeptics ask for more evidence of global warming before taking what will surely be costly steps to prevent it. Are we now on the threshold of a major human-caused change in the climate? Let us look at the evidence.

The Earth as a Greenhouse

Warming Processes. You are familiar with the way the interior of a car heats up when the car is sitting in the Sun with the windows closed. This heating occurs because sunlight comes in through the windows and is absorbed by the seats and other interior objects. In being absorbed by the objects, the light energy is converted to heat energy, which is given off in the form of infrared radiation. Unlike sunlight, infrared radiation is blocked by glass and so cannot leave the car. The trapped heat energy causes the interior air temperature to rise. This is the same phenomenon that keeps a greenhouse warmer than the surrounding environment.

On a global scale, carbon dioxide, water vapor, and other gases in the atmosphere play a role analogous to the glass in a greenhouse. Therefore, they are called **greenhouse gases**. Light energy comes through the atmosphere and is absorbed by Earth and converted to heat energy at the planet's surface. The infrared heat energy radiates back upward through the atmosphere and into space. The greenhouse gases naturally present in the troposphere absorb some of the infrared radiation and reradiate it back toward the surface; other gases (N_2 and O_2) in the troposphere do not. This greenhouse effect was first recognized in 1827 by French scientist Jean-Baptiste Fourier and is now firmly established. The greenhouse gases are like a heat blanket, insulating Earth and delaying the loss of infrared to space (Fig. 21–9). Without this insulation, average surface temperatures on Earth would be 21°C

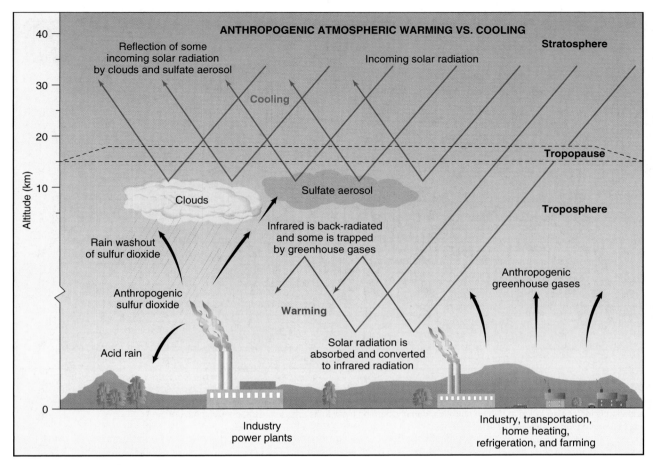

ANTHROPOGENIC ATMOSPHERIC WARMING VS. COOLING

▲ **FIGURE 21–9** *Global warming and cooling.* Anthropogenic factors involved in atmospheric warming and cooling are illustrated. Some gases promote global warming; others, such as sulfur dioxide, lead to cooling.

colder, and life as we know it would be impossible. Therefore, our global climate is dependent on Earth's concentrations of greenhouse gases. If these concentrations increase or decrease significantly, our climate will change accordingly.

Global Cooling. Earth's atmosphere is also subject to cooling factors. For example, on average, clouds cover 50% of Earth's surface and reflect some 21% of solar radiation away to space. This reflection of sunlight is called the **planetary albedo**, and it contributes to overall cooling. (More accurately, it prevents warming from occurring.) This is especially true for *low* clouds. The effect can be readily appreciated if you think of how you are comforted on a hot day when a large cloud passes over the Sun. The radiation that was heating you and your surroundings is suddenly intercepted higher up in the atmosphere, and you feel cooler. However, *high* clouds have a greenhouse effect, absorbing some of the infrared radiation and emitting some infrared themselves. Overall, the net impact of clouds is judged to be a slightly cooling effect.

Volcanic activity can also lead to planetary cooling. When Mount Pinatubo in the Philippines erupted in 1991, some 20 million tons of particles and aerosols entered the atmosphere and contributed to a significant drop in global temperature as

radiation was reflected and scattered away. This cooling effect was global, and it lasted until the volcanic debris was finally cleansed from the atmosphere by chemical change and deposition—a process that took several years.

Finally, climatologists have recently found that anthropogenic sulfate aerosols (from ground-level pollution) play a significant role in canceling out some of the warming from greenhouse gases. Sulfur dioxide from industrial sources enters the atmosphere and reacts with compounds there to form a high-level aerosol, a sulfate haze. This haze reflects and scatters some sunlight and also contributes to the formation of clouds, with a concomitant increase in planetary albedo. The mean residence time of sulfates forming the aerosol is about a week; thus, the aerosol does not increase over time, as the greenhouse gases do.

Global atmospheric temperatures are a balance of the effects of factors leading to cooling and warming. The net result varies, depending on global location. As we will see, this balance contributes much uncertainty to our predictions of what will happen in the future as greenhouse gases continue to increase. Fig. 21–9 shows a summary of the anthropogenic factors that interact to influence the temperature at any given location.

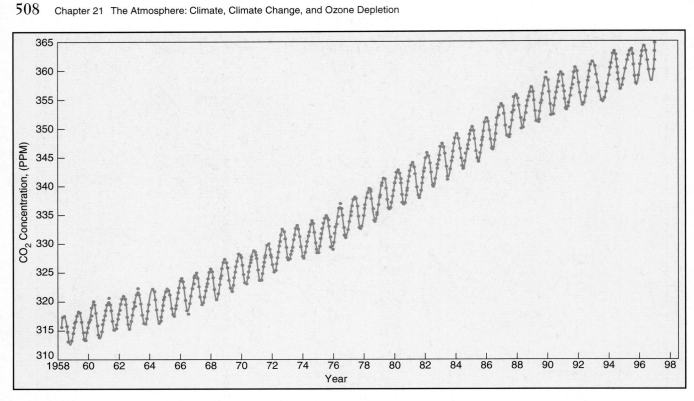

▲ FIGURE 21–10 *Atmospheric carbon dioxide concentrations.* The concentration of carbon dioxide in the atmosphere fluctuates between winter and summer because of seasonal variation in photosynthesis. The average concentration is increasing owing to human activities, namely, burning fossil fuels and deforestation. (All measurements made at Mauna Loa Observatory, Hawaii.)

The Carbon Dioxide Story

In an article published in 1938 entitled, "The Artificial Production of Carbon Dioxide and Its Influence on Temperature," scientist G. Callendar reasoned that human beings' use of fossil fuels had the potential to increase atmospheric carbon dioxide concentrations. If that were to happen, the climate could change. This suggestion was largely forgotten until 1958, when Charles Keeling began measuring carbon dioxide levels on Mauna Loa, in Hawaii. Measurements there have continued to be recorded, and they reveal a striking increase in atmospheric levels of carbon dioxide (Fig. 21–10). The concentrations increased exponentially until the energy crisis in the mid-1970s, then rose linearly for two decades and resumed their exponential increase in the mid-1990s. The data also reveal an annual oscillation of 5 ppm, which reflects seasonal changes of photosynthesis and respiration in terrestrial ecosystems in the Northern Hemisphere. When respiration predominates (late fall through spring), levels rise; when photosynthesis predominates (late spring through early fall), levels fall. Most salient, however, is the relentless rise in carbon dioxide levels, ranging in recent years from 0.8 to 1.7 ppm per year. Carbon dioxide levels are now (in 1999) near 370 ppm, 35% higher than they were before the Industrial Revolution. Thus, our insulating blanket is thicker, and there is reason to expect that this will have a warming effect.

As Callendar suggested long ago, the obvious place to look for the source of increasing carbon dioxide levels is our use of fossil fuels. Every kilogram of fossil fuel burned results in the production of about 3 kg of carbon dioxide. (The mass triples because each carbon atom in the fuel picks up two oxygen atoms in the course of burning and becoming CO_2.) Currently, 6.5 billion tons of fossil fuel carbon are burned each year, adding about 24 billion tons of carbon dioxide to the atmosphere. The vast majority of this and past fossil-fuel carbon dioxide has come from the industrialized countries (Fig. 21–11).

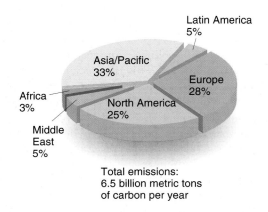

Total emissions:
6.5 billion metric tons
of carbon per year

▲ FIGURE 21–11 *Sources of carbon dioxide emissions from fossil-fuel burning.* Total emissions in 1996 were approximately 22 billion metric tons of carbon dioxide. (Source: Data from World Energy Council.)

Careful calculations show that if all the carbon dioxide emitted from burning fossil fuels accumulated in the atmosphere, the concentration would rise by at least 3 ppm per year, not the 1.7 ppm or less charted in Fig. 21–10. Fortunately, the oceans are undersaturated with carbon dioxide, and there is now broad agreement that the oceans serve as a *sink* for much of the carbon dioxide emitted. There are limitations to the ocean's ability to absorb carbon dioxide, however, because only the top 300 m of the ocean exchanges gases with the atmosphere. (As mentioned earlier, the deep ocean layers do mix with the upper layers, but the mixing time is over 1000 years.)

The seasonal swings that are obvious in Fig. 21–10 show that the biota can influence atmospheric carbon dioxide levels. Can some of the "lost" carbon dioxide be traced to net biological uptake? Can we hope that the forests will take up increasing amounts of carbon dioxide as our emissions continue to rise? The evidence suggests that forests are now serving as a net *source* of carbon dioxide, not a sink, due to deforestation. It is estimated that the burning of forest trees is adding 1.6 billion tons annually to the 6.5 billion tons of carbon already coming from industrial processes. The net loss of forests, a serious biological concern in itself, therefore is also a cause for alarm in the context of global warming.

Other Greenhouse Gases

Several other gases also absorb infrared radiation and add to the insulating effect of carbon dioxide (Table 21-2). Some of these gases also have anthropogenic sources and are increasing in concentration, raising the concern that future warming will extend well beyond the calculated effects of carbon dioxide alone.

Water Vapor. Water vapor absorbs infrared energy and is a powerful greenhouse gas. However, its concentration in the troposphere is quite variable. Through evaporation and precipitation, water undergoes rapid turnover in the lower atmosphere, and water vapor does not tend to accumulate over time. It does, however, appear to be a major factor in what has been called the "supergreenhouse effect" in the tropical Pacific Ocean. As it traps energy that has been radiated back to the atmosphere, the high concentration of water vapor contributes significantly to the heating of the ocean surface and lower atmosphere in the tropical Pacific. This informs us that if temperatures over the land and oceans rise, evaporation increases and water vapor concentration rises, which in turn causes even more warming.

Methane. Methane (CH_4) is a product of microbial fermentative reactions (its main natural source is from wetlands) and is also emitted from coal mines, gas pipelines, and oil wells. Because methane is generated in the stomachs of ruminants, animal husbandry is thought to be responsible for much of the increase in the gas appearing in the troposphere. Although methane is gradually destroyed in reactions with other gases in the atmosphere, it is being added to the atmosphere faster than it can be broken down. The concentration of atmospheric methane has doubled since the Industrial Revolution, as revealed in core samples taken from glacial ice.

Nitrous Oxide. Nitrous oxide (N_2O) levels are also on the increase. Sources of the gas include biomass burning and the use of chemical fertilizers; lesser quantities come from fossil fuel burning. The emission of nitrous oxide is particularly unwelcome because its long residence time (120 years) will be a problem in not only the troposphere, where it contributes to warming, but also the stratosphere, where it contributes to the destruction of ozone.

CFCs and Other Halocarbons. Emissions of halocarbons are entirely anthropogenic in origin and are increasing more rapidly than those of any other greenhouse gas. Like nitrous oxide, halocarbons are long lived and contribute both to global warming in the troposphere and to ozone destruction in the stratosphere. Used as refrigerants, solvents, and fire retardants, halocarbons have a much greater capacity (10,000 times) for absorbing infrared radiation than does carbon dioxide. Although the rate of production of chlorofluorocarbons

Gas	Average Concentration 100 years ago (ppb)[1]	Approximate Current Concentration (ppb)	Average Residence Time in Atmosphere (years)
Carbon dioxide (CO_2)	290,000	370,000	100
Methane (CH_4)	900	1700	12
Nitrous oxide (N_2O)	285	310	120
Chlorofluorocarbons and halocarbons	0	3	60-100

TABLE 21-2 Anthropogenic Greenhouse Gases in the Atmosphere

[1] Parts per billion.

(CFCs) has declined since the Montreal Accord of 1987, these gases are very stable and will continue to accumulate for a few more years.

Together, these other anthropogenic greenhouse gases are estimated to trap about 60% as much infrared radiation as carbon dioxide does. Although the tropospheric concentrations of some of these gases are rising, it is hoped that they will gradually decline in importance because of steps being taken to reduce them, leaving CO_2 as the primary greenhouse gas to cope with in the future.

Amount of Warming and Its Probable Effects

It is certain that levels of carbon dioxide and the other greenhouse gases are increasing in the troposphere as a result of human activities. How can we be *certain*, however, that the increase in greenhouse gases will bring on a permanent increase in Earth's average temperature? The short answer is that we can't be sure—yet. Many variables affect global temperatures, including the reflection of solar radiation by cloud cover, minor changes in the Sun's intensity, airborne particulate matter from volcanic activity, and sulfate aerosol. However, as Fig. 21–5 indicates, global temperatures are on the rise, and they coincide with the observed rise in greenhouse gases. A more informed answer to the question is that all the evidence examined so far points to the *strong probability* that, as levels of greenhouse gases increase in the troposphere, global temperatures will indeed rise and, as a consequence, the climate will undergo major changes.

The Contributions of Modeling. Weather forecasting employs powerful computers capable of handling large amounts of atmospheric data and applying appropriate mathematical equations involved with the processes taking place in the at-

mosphere, oceans, and land. Forecasts of conditions for 72 hours or more have become quite accurate. Even though the computers used have greatly increased in computing power in recent years, the somewhat chaotic behavior of weather parameters prevents more long-term forecasting.

Modeling the climate is an essential strategy for exploring the potential impacts of rising greenhouse gases. Climatologists employ the same powerful computers used for weather forecasting and have combined global atmospheric circulation patterns with ocean circulation and cloud-radiation feedback to produce general circulation models (GCMs) capable of simulating long-term climatic conditions—especially, temperatures. Fourteen centers around the world are now intensely engaged in running models coupling atmosphere, oceans, and land. The latest information on the cooling effects of sulfates has been incorporated in current simulations, giving modelers more confidence in their results (Fig. 21–12). Many models have now successfully simulated past climates, based on known differences in Earth's orbit and ocean-atmosphere coupling. The most complex models, requiring supercomputers, recently scored a major triumph in predicting the intensity and locations of the latest El Niño.

The main purpose of the models is to project the future global climate. Typically, a model will be run for a few simulated decades under standard conditions and then repeated with a doubled greenhouse gas concentration to determine the climate change "caused" by the doubling. The projections indicate that if the concentration of greenhouse gases were to double, Earth would warm up between 1.5° and 4.5°C, with most estimates around 2.5°C. This doesn't sound like much, but consider that temperatures were only 5°C cooler during the last ice age, 18,000 years ago. At that time, ice sheets up to a mile thick stretched across North America from New York through the Great Lakes states and all of Canada. Figure 21–13

▶ **FIGURE 21–12** *Computer-simulated global warming.* Model results with and without the effects of sulfate aerosol are compared with the actual temperature record.

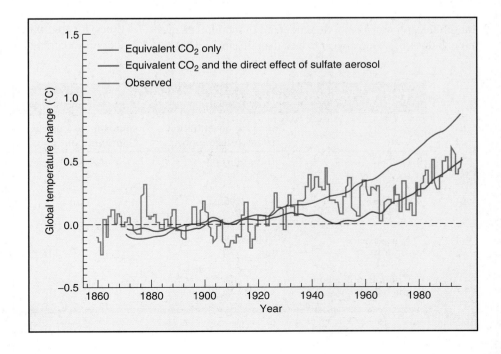

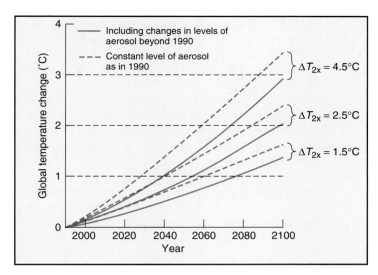

◀ **FIGURE 21–13** *Future global mean temperature changes.* The changes shown are the result of the "business-as-usual" scenario, as presented by the IPCC. Results show the response of temperature to a doubling of greenhouse gas emissions over preindustrial levels, with and without corrections for sulfate aerosol emissions. The range of responses (1.5°C to 4.5°C) indicates the significant uncertainty involved in such extrapolations.

gives a model result based on doubling greenhouse gas concentrations under what has been called a "business-as-usual" scenario. This scenario assumes that no strong controls will be placed on emissions throughout the twenty-first century.

More recently, some modelers have been extending their models beyond the traditional doubling scenario. After all, these scientists reason, carbon dioxide doubling is only the beginning; there is no reason to believe that we will stop generating greenhouse gases when that level is reached. As the models simulate time into the future, the uncertainties in the models' outcomes that can be traced to variations in cloud behavior and sulfate aerosol are all dwarfed by the relentless impact of rising greenhouse gas concentrations. The models indicate that if carbon dioxide builds up to four or more times its preindustrial level, it will promote a rise in global temperature of 7°C and a rise in sea level of about 2 m. Although this development is a century or more away, a change of this magnitude would clearly have catastrophic effects on life on Earth, as it would occur over a period of decades.

Impacts of Future Warming. Rising global temperatures are linked to two major impacts: *regional climatic changes* and a *rise in sea level*. Both of these effects show up in all of the models and are expected to become very evident in a matter of a few decades. Responding to these changes will likely involve unprecedented and costly adjustments. The impact on natural ecosystems could be highly destabilizing.

Warming will seriously affect rainfall and agriculture. The present-day difference between temperatures at the poles and those at the equator is a major driving force for atmospheric circulation. Greater heating at the poles than at the equator will reduce this force, changing atmospheric circulation patterns as well as rainfall distribution patterns. Some regions will be drier, and some will be wetter. The frequency of droughts will increase. Further changes in climate could be initiated by alterations in the global oceanic circulation, which could be brought on by greenhouse warming. Oceanographers fear that as the ocean's Conveyor goes into a stall, sudden shifts in land temperature may occur.

Another expected impact of global warming is weather *change*. The following changes have already been observed: 1) Winters in Europe have been warmer and wetter during the last decade as a result of differences in North Atlantic circulation. 2) As noted, El Niño events are becoming more frequent and more intense. 3) Thunderstorms, windstorms, and hurricanes have been more frequent and more severe in recent years, a fact well noted by the insurance industry, which has become interested in global-warming issues as huge sums for weather-related losses accrued in the 1990s.

The greatest difficulty for the agricultural community in coping with climatic change, however, is not knowing what to expect. Already, farmers lose an average of one in five crops because of unfavorable weather. As the climate shifts, the vagaries of weather will become more pronounced and crop losses are likely to increase. However, farmers are capable of rapidly switching crops and land uses, so the impact may not be as bad as some observers think.

With global warming, sea level will rise because of two factors: thermal expansion as ocean waters warm, and melting of glaciers and ice fields. Sea level is already on the rise at the annual rate of 2 mm. The warming associated with a doubling of greenhouse gas levels is likely to be more pronounced in polar regions (as much as 10°C) and less pronounced in equatorial regions (1 to 2°C). The warming at the poles will have a tremendous effect when the ice melts, because so much water is stored in the world's remaining ice— enough to raise sea level by 75 m. The area of greatest concern is the Antarctic, which holds most of the world's ice. West Antarctica's ice sheet is not greatly elevated above sea level; in fact, most of it rests on land below sea level. The Ross and Weddell Seas extend outward from this ice field, and the landward fringes of these seas are covered with thick floating shelves of ice. Currently, the rate of melting and of buildup seem to be about equal, and many climate models indicate that ice will build up even more in the Antarctic with increasing greenhouse gases.

There is great uncertainty about the magnitude of the rise in sea level; projections for the next century range from 15 to

95 cm (6 to 37 in.), most likely 50 cm. Even the lowest estimated rise, however, will flood many coastal areas and make them much more prone to damage from storms, forcing people to abandon properties and migrate inland. For many of the small oceanic nations, a rise in sea level would mean obliteration, not just alteration. The highest estimated rise would cause disaster for most coastal cities, which are home to half the world's population and its business and commerce. Moreover, the estimate of 15 to 95 cm extends only through the next century; the impact will certainly be greater beyond the year 2100. Are inland cities and communities ready to accommodate the billions of people that will be displaced? Are we ready to build dikes or modify all ports to accommodate the higher sea level?

Coping with Global Warming

Is Global Warming Here? Some would deny that preparing for the disastrous effects of global warming is necessary. They point out that Earth has a great capacity for recovering from harm and that false assumptions can skew the results of computer models. What kind of evidence would convince most people that global warming was indeed occurring? For one thing, average annual temperatures would be expected to rise. There is so much natural variation in weather from year to year that trends are not always apparent, but we would expect to see a gradual warming trend in global temperatures (while remaining aware that local temperatures do not necessarily follow globally averaged ones).

It is a fact that fourteen of the hottest years on record have occurred since 1980 (Fig. 21–5). This is a revealing piece of evidence, especially since there is a likely explanation for cooler temperatures during the remaining five years: When Mount Pinatubo erupted in the Philippines in 1991, climatologists using their GCMs predicted that it would produce a temporary cooling trend—and they were correct. After debris from Mount Pinatubo cleared from the atmosphere, we returned to the warming trend that was so evident in the 1980s. Both 1997 and 1998 set new records for global temperatures, adding further compelling evidence for global warming.

Another significant impact of global warming is the retreat of glaciers. The World Glacier Monitoring Service has followed the melting and slow retreat of glaciers all over the world. The response time of glaciers to climatic trends is slow—10 to 50 years—but the data collected clearly show a wide-scale recession of glaciers over the past 100 years, which is consistent with an increase in greenhouse gases.

Ironically, the sulfate aerosol in the industrialized regions of the Northern Hemisphere appears to be canceling out greenhouse warming over those regions. Figure 21–14 shows the global energy balance for summer conditions and

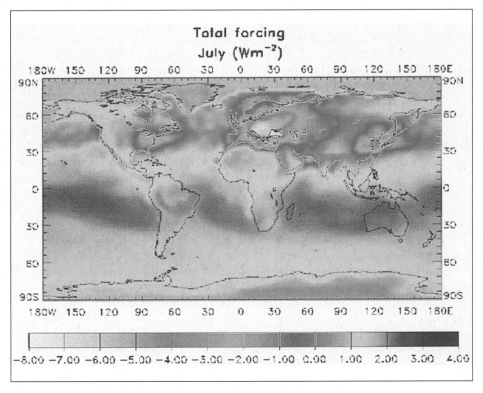

▲ **FIGURE 21–14** *Global map of energy balance in midsummer in the Northern Hemisphere.* The map shows the regional cooling that occurs in the industrialized North as a result of sulfate aerosols intercepting radiation and stimulating the formation of clouds. The map is color coded to show which regions receive more or less energy from sunlight, compared with the global average. White areas represent a large drop in incoming solar energy, blue a slight drop, green no change, yellow a slight increase, and red a major increase.

ETHICS

STEWARDSHIP OF THE ATMOSPHERE

The atmosphere is a global commons—a resource used by all countries yet not owned by any. Countries are free to draw from it to meet their needs and to add to it to get rid of pollutants. Because the air in the troposphere moves so quickly, what one country does to it can easily affect others downwind. The stratosphere is less turbulent, but inevitably it is affected by many gases that enter the lower atmosphere—as we see with the depletion of the ozone layer. Management of a global commons presents a difficult problem; treaties must be made between countries, and compliance must be assured if the global commons is to be protected. Some progress has been made in forging international agreements to curb acid emissions, which represent one form of transboundary air pollution. The Montreal Protocol and its amendments represent the most far-reaching and potentially effective international effort to protect a global resource: the stratospheric ozone layer.

Yet, taking action on ozone-depleting chemicals is child's play compared with what lies ahead to take action on global climate change. Perhaps the most challenging aspect of this action is that what we do about it now will do little to affect the present generation. It will likely take several generations for greenhouse gas concentrations to come to a doubling of the preindustrial levels, with almost certain severe impacts. So our dilemma is this: Shall we take action *now* to restrain our use of fossil fuels in order to benefit *future* generations? Increasing numbers of people are answering this question with a definite *yes*! For example, more than 2000 economists agreed that there are many policies to reduce greenhouse gas emissions for which total benefits outweighed the costs. Over 1500 scientists, including 104 Nobel laureates, signed a statement urging government leaders attending the 1997 Kyoto conference to "act immediately to prevent the potentially devastating consequences of human-induced global warming." In fact, these people are assuming the role of stewards of the atmosphere and are concerned about what they are passing on to their children and grandchildren.

One scientist whose efforts have been crucial to the work of the IPCC (Intergovernmental Panel on Climate Change) is British meteorologist Sir John Houghton. Houghton is co-chair of the IPCC Working Group I and chief editor of their published report, *Climate Change 1995: The Science of Climate Change*. He also has published a succinct and very readable book on climate change, *Global Warming: The Complete Briefing*, currently in its second edition. Chapter 8 of his book, entitled "Why Should We Be Concerned," addresses the question of human responsibility for caring for the planet. He points out that current human exploitation of the natural world is unsustainable and that going "back to nature" or hoping for some "technical fix" to correct the damage of global warming are not reasonable options. Houghton speaks of "a basic instinct that we wish to see our children and our grandchildren well set up in the world and wish to pass on to them some of our most treasured possessions," in particular "an earth that is well looked after and which does not pose to them more difficult problems than those we have had to face." He states that "we have no right to act as if there is no tomorrow."

Houghton makes the point that action on environmental problems depends not only on our understanding of how the world works but also on how we value it. He examines our responsibility to care for all living things and presents perspectives on this view from a number of the world's religions. Houghton then speaks of the concept of stewardship—one of the continuing themes of this text—that is derived from the Judaeo-Christian Old Testament. He recommends as a model for stewardship the biblical picture of the Garden of Eden, where the first humans were placed in the garden "to work and take care of it," as Genesis puts it. He states, "… at the basis of this stewardship should be a principle extending what has traditionally been considered wrong—or in religious parlance as sin—to include unwarranted pollution of the environment or lack of care for it."

Stewardship of the atmosphere suggests that the world's nations act on the Precautionary Principle and begin to curb greenhouse gas emissions now and more deeply in the future. What do you think, and what will you do?

Sir John Houghton (left) and text co-author Richard T. Wright (right) discuss global climate change at a conference in Cambridge, UK, in 1998.

indicates clearly just where cooling effects of the aerosol occur—over the very regions most responsible for greenhouse gas emissions. However, the aerosol cooling effect is temporary. As the industrialized nations continue to reduce sulfate emissions (because of acid rain and impacts on human health) and greenhouse gases continue to build up, the Northern Hemisphere will likely experience its own share of warming.

Political Developments. The world's industries and transportation networks are so locked into the use of fossil fuels that massive emissions of carbon dioxide and other greenhouse

TABLE 21-3 IPCC Consensus on Global Climate Change

Issue	Statement	Consensus
Basic characteristics	Fundamental physics of the greenhouse effect	Virtually certain*
	Added greenhouse gases add heat energy	Virtually certain
	Greenhouse gases increasing because of human activity	Virtually certain
	Significant reduction of uncertainty will require a decade or more	Virtually certain
	Full recovery will require many centuries	Virtually certain
Projected effects by mid-21st century	Reduction of sea ice	Very probable**
	Arctic winter surface warming	Very probable
	Rise in global sea level	Very probable
	Local details of climatic change	Uncertain***
	Increases in number of tropical storms	Uncertain

* Nearly unanimous agreement among scientists; no credible alternative.
** Roughly 9 out of 10 chance.
*** Roughly 2 out of 3 chance.
See: Houghton, John T., ed., et al., *Climate Change 1995: The Science of Climate Change*, Working Group I; Watson, Robert T., ed., et al., *Impacts, Adaptations, and Mitigation of Climate Change*, Working Group II; Bruce, James P., ed., et al., *Economic and Social Dimensions of Climate Change*, Working Group III. All three IPCC volumes published by Cambridge University Press, New York, 1996.

gases seem sure to continue for the foreseeable future. It is possible, however, to lessen the rate at which emissions are added to the atmosphere and eventually to bring about a sustainable balance—although no one thinks that this will be easy. The means for accomplishing this goal lie in the political arena.

In its latest report, the Intergovernmental Panel on Climate Change (IPCC), which includes over 1000 scientists from all the member countries of the United Nations, has affirmed that human-produced emissions of greenhouse gases are producing a warming trend (see the "Ethics" essay, p. 513). Thus, there is strong consensus in the scientific community on the serious threat of global climate change. The IPCC consensus is summarized in Table 21-3.

A variety of steps have been suggested to combat global climate change, with the goal of *stabilizing the greenhouse gas content of the atmosphere*:

- Place a worldwide cap on carbon dioxide emissions by limiting the use of fossil fuels in industry and transportation.
- Encourage the development of nuclear power, but only if cost-effectiveness, reliability, spent fuel, and high-level waste issues are resolved.
- Stop the loss of tropical forests and encourage the planting of trees over vast areas now suffering from deforestation.
- Make energy conservation rules much more stringent. (Tighten building codes to require more insulation, use energy-efficient lighting, and so forth.)
- Reduce the amount of fuels used in transportation by raising mileage standards, encouraging car pooling, stimulating mass transit in urban areas, and imposing increasingly stiff carbon taxes on fuels.
- Invest in and deploy known renewable energy technologies: wind power, solar collectors, solar thermal, photovoltaics, tidal power, and geothermal, among others.
- Make every effort to slow growth of the human population;

a continually growing population will inevitably produce increasing amounts of greenhouse gas emissions.

All of these steps move societies in the direction of sustainability and have many benefits beyond their impacts on global climate change. For example, investing in more efficient use and production of energy reduces acid deposition, is economically sound, lowers the harmful health effects of air pollution, lowers our dependency on foreign oil—and, of course, reduces carbon dioxide emissions.

Skeptics about global warming can be found even within the scientific community. They stress the uncertainties of the global climate models and, in particular, emphasize that there is much we do not know about the role of the oceans, the clouds, the biota, and the chemistry of the atmosphere. In the absence of "convincing evidence," many think that the threat of global warming may well be overplayed. Why take enormously costly steps, they ask, to prevent something that may never happen? In response to charges like this, the 1992 Earth Summit introduced the **"Precautionary Principle"** as part of the Rio Declaration. This states: "In order to protect the environment, the precautionary approach shall be widely applied by States according to their capabilities. Where there are threats of serious or irreversible damage, lack of full scientific certainty shall not be used as a reason for postponing cost-effective measures to prevent environmental degradation." Employing this principle is like taking out insurance; we are taking measures to avoid a highly costly but uncertain situation.

Is the world taking global warming seriously? One of the five documents signed by heads of state at the United Nations Conference on Environment and Development (UNCED) Earth Summit in Rio de Janeiro in 1992 was the **Framework Convention on Climate Change (FCCC)**. This convention agreed to aim at reducing greenhouse gas emissions to 1990 levels by the year 2000, but countries were to achieve this goal by voluntary means. Five years later, it was obvious that the voluntary approach was failing. All the developed countries except the European Union countries increased their greenhouse gas emissions by 7 to 9% in the ensuing five years. The developing countries increased theirs by 25%.

Prompted by a coalition of island nations (whose very existence is threatened by global climate change), the third Conference of Parties to the FCCC met in Kyoto, Japan, in December 1997 to craft a binding agreement on reducing greenhouse gas emissions. In Kyoto, 38 industrialized countries agreed to reduce emissions of six greenhouse gases below 1990 levels, to be achieved by 2010. The percent reductions varied (Table 21-4). At the conference, the developing countries refused to agree to any reductions, arguing that the developed countries had created the problem and that it was only fair that they continue on their path to development as the developed countries did, energized by fossil fuels. Each country must ratify the accord, which signifies compliance. In the U.S., this is far from easy. To achieve a 7% reduction in emissions below 1990 levels by 2010 actually means a reduction of close to 25% below what we would be generating in the absence of the Kyoto accord. There is fierce opposition to such cuts from the fuel industry, automotive companies, electric utilities, and labor unions. Although the Administration favors the accord, Congress is opposed.

Assuming the United States *does* ratify the agreement, however, we will not actually be reducing emissions until close to 2010, and by that time, the developing countries will have increased their emissions such that the total world production of greenhouse gas emissions will have *increased* by 30%. Further, these reductions crafted at Kyoto are far less than what is needed to stabilize atmospheric concentrations of greenhouse gases. The IPCC calculates that it would take emission reductions of at least 60% worldwide to stabilize greenhouse gas concentrations at today's levels. Thus, the greenhouse gases will continue to rise. Nevertheless, Kyoto was an important first step toward the stabilization of the atmosphere.

So are we in fact on the threshold of a major human-caused change in the climate? Whatever the outcome, it is certain that we are conducting an enormous global experiment and our children and their descendants will be living with the consequences.

Let us now turn our attention to ozone depletion, an atmospheric problem the public often confuses with global warming.

21.4 Depletion of the Ozone Layer

In another major atmospheric challenge traced to human technology, the work of environmental scientists again has uncovered a serious problem for which there was no prior warning. Our knowledge of the workings of the atmosphere has been appallingly poor, and one consequence of that lack of understanding is the strong possibility that ultraviolet radiation will increase in intensity all over Earth. As with global warming, however, there are skeptics who are not convinced that it is a serious problem. Let us examine the evidence.

Radiation and Importance of the Shield

Radiation from the Sun includes *ultraviolet (UV) radiation*, along with visible light. Radiation is the emission of electromagnetic waves of a whole range of energy and wavelengths (Fig. 21–15). Visible light is that part of the electromagnetic spectrum that can be detected by the eye's photoreceptors. UV radiation is like visible light radiation except that UV wavelengths are slightly shorter than the wavelengths of violet light, which are the shortest wavelengths visible to the human eye. It is important to distinguish between UVB and UVA radiation. UVB wavelengths range from 280 to 320 nanometers (0.28 to 0.32 mm), UVA from 320 to 400 nanometers. Since energy is inversely related to wavelength, UVB is more energetic and therefore more dangerous.

On penetrating the atmosphere and being absorbed by biological tissues, UV radiation damages protein and DNA molecules at the surfaces of all living things. (This is what occurs when you get a sunburn.) If the full amount of ultraviolet radiation falling on the stratosphere reached Earth's surface, it is doubtful that any life could survive. We are spared more damaging effects from ultraviolet rays because most UVB radiation (over 99%) is absorbed by ozone in the stratosphere (Fig. 21–2). For that reason, stratospheric ozone is commonly referred to as the **ozone shield**.

However, even the small amount (less than 1%) that does reach us is responsible for all the sunburns and more than 700,000 cases of skin cancer and precancerous ailments per year in North America, as well as for untold damage to plant crops and other life-forms (see the "Global Perspective" essay, p. 517).

TABLE 21-4 Kyoto Emission Allowances and 5-year Emission Changes

Country or Region	Allowances % (1990–2010)	Observed % (1990–1995)
European Union	-8	-1
Australia	+8	+8
Canada	-6	+9
Japan	-6	+8
Norway	+1	+9
United States	-7	+7
Russian Federation	0	n.a.
Eastern Europe	-1	-29
Developing countries	–	+25

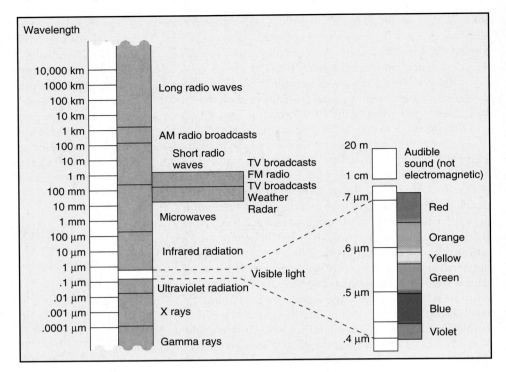

▶ **FIGURE 21–15** *The electromagnetic spectrum.* Ultraviolet, visible light, infrared, and many other forms of radiation are just different wavelengths of the electromagnetic spectrum.

Formation and Breakdown of the Shield

The need to maintain the ozone shield is obvious. However, certain anthropogenic pollutants are causing it to break down. Ozone in the stratosphere is a product of UV radiation acting on oxygen (O_2) molecules. The high-energy UV radiation first causes some molecular oxygen (O_2) to split apart into free oxygen (O) atoms, and these atoms combine with the molecular oxygen to form ozone as follows:

$$O_2 + UVB \longrightarrow O + O \quad (1)$$
$$O + O_2 \longrightarrow O_3 \quad (2)$$

Not all of the molecular oxygen is converted to ozone, however, because free oxygen atoms may also combine with ozone molecules to form two oxygen molecules:

$$O + O_3 \longrightarrow O_2 + O_2 \quad (3)$$

Finally, when ozone absorbs UVB, it is converted back to free oxygen and molecular oxygen:

$$O_3 + UVB \longrightarrow O + O_2 \quad (4)$$

Thus, the amount of ozone in the stratosphere is dynamic; there is an equilibrium due to the continual cycle of reactions of formation (Eqs. 1 and 2) and reactions of destruction (Eqs. 3 and 4). Because of seasonal changes in solar radiation, ozone concentration in the Northern Hemisphere is highest in summer and lowest in winter. Also, in general, ozone concentrations are highest at the equator and diminish as latitude increases—again, a function of higher overall amounts of solar radiation. However, as we have learned in recent years, the presence of other chemicals in the stratosphere can upset the normal ozone equilibrium and promote unsustainable reactions there.

Halogens in the Atmosphere. Chlorofluorocarbons (CFCs) are a type of halogenated hydrocarbon. (See Chapter 20.) They are nonreactive, nonflammable, nontoxic organic molecules in which both chlorine and fluorine atoms replace some of the hydrogen atoms. At room temperature, CFCs are gases under normal (atmospheric) pressure, but they liquefy under modest pressure, giving off heat in the process and becoming cold. When they revaporize, they reabsorb the heat and become hot. These attributes led to the widespread use of CFCs (over a million tons per year in the 1980s) for the following:

- In refrigerators, air conditioners, and heat pumps as the heat-transfer fluid
- In the production of plastic foams
- By the electronics industry for cleaning computer parts, which must be meticulously purified
- As the pressurizing agent for aerosol cans

All of these uses led to the release of CFCs into the atmosphere, where they mixed with the normal atmospheric gases and eventually reached the stratosphere. In 1974, chemists Sherwood Rowland and Mario Molina published a classic paper (for which they were awarded the Nobel Prize in 1995) concluding that CFCs could be damaging the stratospheric ozone layer through the release of chlorine atoms and, as a result, would increase UV radiation

GLOBAL PERSPECTIVE

COPING WITH UV RADIATION

People living in Chile and Australia are no strangers to the effects of the thinning ozone shield. UV alerts are given in those parts of the Southern Hemisphere during their spring months as lobes of ozone-depleted stratospheric air move outward from the Antarctic. At times, the ozone above these regions can be less than half its normal concentration, and the consequence is, of course, greater intensities of UV radiation reaching Earth's surface. It is now predicted that one out of every three Australians will develop serious and perhaps fatal skin cancer in his or her lifetime.

A thinning ozone layer has now appeared above the Northern Hemisphere and is getting serious attention. Concerned over the rising incidence in the United States of skin cancer and cataract surgery, the EPA, together with the National Oceanic and Atmospheric Administration and the Centers for Disease Control, has initiated a new "UV Index." The index is in the form of daily forecasts of UV exposure, issued by the Weather Service for 58 cities (see inset table). Satellite measurements of stratospheric ozone are combined with other weather patterns. The goal of the index is to remind people of the dangers of UV radiation and to prompt them to take appropriate action to avoid cancer, premature skin aging, eye damage, cataracts, and blindness. For at least the next two decades, the dangers will be very noticeable, ranging from a greater incidence of sunburn to a likely epidemic of malignant melanoma.

Less obvious, but no less important, are the chronic effects of exposure to the Sun.

The "normal aging" of skin is now known to be largely the result of damage from the Sun: coarse wrinkling, yellowing, and the development of irregular patches of heavily pigmented and unpigmented skin. This can happen even in the absence of episodes of sunburn. Further, a large proportion of the million cataract operations performed annually in the United States can be traced to UV exposure, another of the chronic effects of this kind of radiation.

The most serious impact of chronic exposure to the Sun is skin cancer, which occurs in some 700,000 new cases each year in the United States. Three types of skin cancer are traced to UV exposure: basal cell carcinoma (BCC), squamous cell carcinoma (SCC), and melanoma. Most of the skin cancers are BCC (75% to 90%), slow growing, and therefore treatable. SCC accounts for about 20% of skin cancers. This form can also be cured if treated early. However, it metastasizes (spreads away from the source) more readily than BCC and thus is potentially fatal. Melanoma is the most deadly

form of skin cancer, as it easily metastasizes; it is also the form of cancer that causes the greatest mortality. Melanoma is often traced to occasional sunburns during childhood or adolescence or to sunburns in people who normally stay out of the Sun.

What is an appropriate response to this disturbing information? First, know when UV intensity is greatest (the hours around midday) and take precautions accordingly. Protect your eyes with sunglasses and a hat, and apply sunscreen with a Sunscreen Protection Factor (SPF) rating of 15 or higher during such times. Think twice before considering sunbathing or tanning beds, and never proceed without sunscreen that is strong enough to prevent sunburn. Always protect children with sunscreen; their years of potential exposure are many, and their skin is easily burned. If you detect a patch of skin or a mole that is changing in size or color or that is red and fails to heal, consult a dermatologist for treatment.

UV Index (EPA)

Exposure Category Index Value		Minutes to Burn for "Never Tans" (most susceptible)	Minutes to burn for "Rarely Burns" (least susceptible)
Minimal	0–2	30 minutes	>120 minutes
Low	4	15 minutes	75 minutes
Moderate	6	10 minutes	50 minutes
High	8	7.5 minutes	35 minutes
Very high	10	6 minutes	30 minutes
	15	<4 minutes	20 minutes

and cause more skin cancer. Rowland and Molina reasoned that although CFCs would be stable in the troposphere (they have been found to last 60 to 100 years there), in the stratosphere they would be subjected to intense UV radiation, which would break them apart, releasing free chlorine atoms as follows:

$$CFCl_3 + UV \longrightarrow Cl + CFCl_2 \qquad (5)$$

Ultimately, all of the chlorine of a CFC molecule would be released as a result of further photochemical breakdown. The free chlorine atoms would then attack stratospheric ozone to form chlorine monoxide (ClO) and molecular oxygen:

$$Cl + O_3 \longrightarrow ClO + O_2 \qquad (6)$$

Furthermore, two molecules of chlorine monoxide may react to release more chlorine and an oxygen molecule:

$$ClO + ClO \longrightarrow 2\,Cl + O_2 \qquad (7)$$

Reactions 6 and 7 are called the **chlorine cycle**, because chlorine is continuously regenerated as it reacts with ozone.

Thus, chlorine acts as a **catalyst**, a chemical that promotes a chemical reaction without itself being used up by the reaction. Because every chlorine atom in the stratosphere can last from 40 to 100 years, it has the potential to cause the breakdown of 100,000 molecules of ozone. Thus, CFCs are judged to be damaging because they act as transport agents that continuously move chlorine atoms into the stratosphere. The damage can persist because the chlorine atoms are removed from the stratosphere only very slowly. Figure 21–16 shows the basic processes of ozone formation and destruction, including recent refinements of those processes that will be explained shortly.

After studying the evidence, the EPA became convinced that CFCs were a threat and, in 1978, banned their use in aerosol cans in the U.S. Manufacturers quickly switched to nondamaging substitutes, such as butane, and things were quiet for several years. CFCs continued to be used in applications other than aerosols, however, and critics demanded more convincing evidence of their harmfulness.

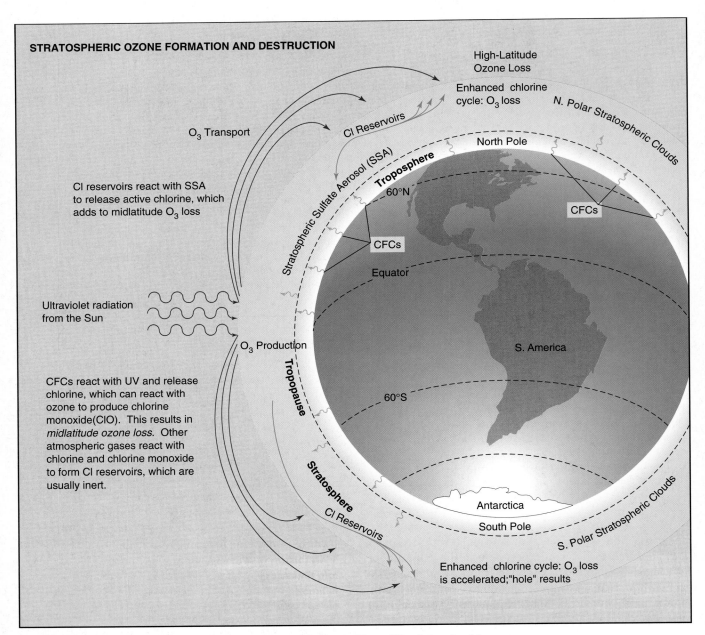

▲ **FIGURE 21–16** *Stratospheric ozone formation and destruction.* UV radiation stimulates ozone production at the lower latitudes, and ozone-rich air migrates to high latitudes. At the same time, chlorofluorocarbons and other compounds carry halogens into the stratosphere, where they are broken down by UV and release chlorine and bromine. Ozone is subject to high-latitude loss during winter, as the chlorine cycle is enhanced by the polar stratospheric clouds. Midlatitude losses occur as chlorine reservoirs are stimulated to release chlorine by reacting with stratospheric sulfate aerosol.

EARTH WATCH

ATMOSPHERIC TROUBLE FROM AIR TRAFFIC

Researchers have recently found unusually high concentrations of nitrogen oxides at altitudes between 5 and 9 miles above sea level. The source of these oxides is mainly air traffic: The normal cruising altitude for commercial airlines is 5–8 miles (9–13 km). This height is close enough to the tropopause to enable the gases to drift into the stratosphere. Recent work has shown that in the stratosphere, nitrogen dioxide may break down ozone, adding to the impact of CFC emissions. Further, nitrogen oxides and water vapor contribute to the formation of the stratospheric clouds implicated in the formation of the Antarctic ozone "hole." Studies indicate that ozone depletion over

the Northern Hemisphere thus may be twice as high as what would occur with CFCs alone. Measurements above Europe have shown a 3% decrease in ozone between 6 and 10 miles up and an 8% loss between 10 and 13 miles. These data support the conclusion that much of the measured ozone losses above the temperate zone in the Northern Hemisphere can be attributed to air traffic.

Of course, the bulk of the nitrogen oxides emitted from aircraft remain in the high troposphere, where they contribute to tropospheric ozone formation and global warming. Substances released so high in the troposphere tend to remain there at least 100 times longer than those released at ground level, due to less active cloud for-

mation and less precipitation at those heights. When these substances finally do descend, they contribute to acid deposition. Thus, *one* activity—high-altitude air traffic—leads to depletion of the ozone layer, global warming, *and* acid deposition!

What can be done? Cruising altitudes should be kept below the tropopause, which can be as low as 6 miles during the winter. Plans to develop supersonic aircraft that cruise higher than 9 miles up should be scrapped forever. And air traffic should be subjected to the same scrutiny as all other forms of transportation, with attention given to the development of jet engines that emit only low levels of nitrogen oxides as well as to limitations on the volume of traffic.

Atmospheric scientists reason that any substance carrying reactive halogens to the stratosphere has the potential to deplete ozone. These substances include halons, methyl chloroform, carbon tetrafluoride, and methyl bromide. Because of its extensive use as a soil fumigant and pesticide, methyl bromide is thought to cause as much as 10% of current stratospheric ozone loss (bromine is 40 times as potent as chlorine in ozone destruction).

The Ozone "Hole." In the fall of 1985, some British atmospheric scientists working in Antarctica reported a gaping "hole" (actually, a thinning of one area) in the stratospheric ozone layer over the South Pole (Fig. 21–17). There, in an area the size of the United States, ozone levels were 50% lower than normal. The "hole" would have been discovered earlier by NASA satellites monitoring ozone levels, except that computers were programmed to reject data showing a drop as large as 30% as instrument "hiccups." Scientists had assumed that the loss of ozone, if it occurred, would be slow, gradual, and uniform over the whole planet. The "hole" came as a surprise, and if it had occurred anywhere but over the South Pole, the UV damage would have been extensive.

News of the ozone "hole" stimulated an enormous scientific research effort. A unique set of conditions is responsible for the ozone "hole." In the summer, gases such as nitrogen dioxide and methane react with chlorine monoxide and chlorine to trap the chlorine, forming the so-called **chlorine reservoirs** (Fig. 21–16), preventing much ozone depletion.

When the Antarctic winter arrives in June, it creates a vortex (like a whirlpool) in the stratosphere, which confines

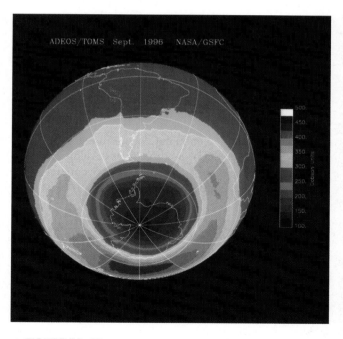

▲ **FIGURE 21–17** *The Antarctic ozone hole.* TOMS (Total Ozone Mapping Spectrometer) instruments aboard two new satellites recorded the largest ever ozone hole over Antarctica during September and October 1996. NASA's *Earth Probe* satellite and Japan's *Advanced Earth-Orbiting Satellite* (ADEOS) are the instrument platforms. The 1996 "hole" exceeded 25.9 million km² (10 million square miles). This is an area larger than Canada, the United States, and Mexico combined. The color scale represents ozone concentrations in Dobson units (D.U.), with blues and purples for amounts of less than 200. Measurements dropped as low as 111 Dobson units in 1996. (A normal winter ozone value for the Antarctic is 300 D.U.)

stratospheric gases within a ring of air circulating around the Antarctic. The extremely cold temperatures of the Antarctic winter cause the small amounts of moisture and other chemicals present in the stratosphere to form the South Polar stratospheric clouds. During winter, the cloud particles provide surfaces on which chemical reactions occur and release molecular chlorine (Cl_2) from the chlorine reservoirs.

When sunlight returns to the Antarctic in the spring, the Sun's warmth breaks up the clouds; UV light then attacks molecular chlorine, releasing free chlorine and initiating the chlorine cycle, which rapidly destroys ozone. By November, the beginning of the Antarctic summer, the vortex breaks down and ozone-rich air returns to the area. By this time, however, ozone-poor air has spread all over the Southern Hemisphere. Shifting patches of ozone-depleted air have caused UV radiation increases of 20% above normal in Australia. Television stations there now report daily UV readings and warnings for Australians to stay out of the Sun. Based on current data, estimates indicate that in Queensland, where the ozone shield is thinnest, three out of four Australians are expected to develop skin cancer. The "hole" has reappeared every spring in Antarctica (our autumn) and has been intensifying—exactly what would be expected with increasing levels of CFC. In 1997, Antarctic ozone depletion was intense (15 to 40% below normal) and grew to almost 14 million square miles (an area larger than all of Europe and North America combined) before it dissipated.

Because severe ozone depletion occurs under polar conditions, observers have been keeping a close watch on the Arctic. To date, although a winter ozone loss occurs over the Arctic, no "hole" has developed. The higher temperatures and weaker vortex formation there have so far prevented the severe losses that have become routine in the Antarctic. However, spring ozone depletion does occur in the Arctic, most recently a 25% reduction. If a hole does develop, its consequences could be severe: Arctic air spreads over much of the densely populated Northern Hemisphere.

Controversy over Ozone Depletion. Skeptics about the CFCs' role in ozone depletion claim, among other things, that probably there has always been an Antarctic ozone "hole," but we didn't use the right instruments to look for it. They also claim that there is little evidence of a thinning ozone layer and almost no evidence of ground-level UVB increases. One of the most persistent claims of the skeptics is that volcanoes have been ejecting enormous amounts of chlorine into the stratosphere for eons, with no apparent impact on the ozone shield. This chlorine, the claim goes, dwarfs any chlorine carried up into the stratosphere by CFCs.

Careful scrutiny reveals that this is a false argument. Whereas volcanoes do inject hydrochloric acid into the atmosphere, very little of it reaches the stratosphere, because of the speed with which it is washed out of the atmosphere by precipitation. Furthermore, the volcanic production was highly exaggerated because of a serious calculation error by one of the skeptics, Dixy Lee Ray, former director of the Atomic Energy Commission. Despite this and other serious mistakes and flaws, the talk-radio circuit continues to repeat them.

Climatologists have, in fact, recently found evidence for ozone loss traced to volcanic eruption—but not from volcanic chlorine. Enormous volcanic eruptions, such as the 1991 Mount Pinatubo eruption, inject sulfur particles as high as the lower stratosphere. There they form a sulfate aerosol (the stratospheric sulfate aerosol pictured in Fig. 21–16) that reacts in complex ways with the chlorine reservoir chemicals normally tying up stratospheric chlorine. Active chlorine is released, which is likely one cause of ozone loss in midlatitudes.

More Ozone Depletion. Ozone losses have not been confined to Earth's polar regions, although they are most spectacular there. A worldwide network of ozone-measuring stations sends data to the World Ozone Data Center in Toronto, Canada. Reports from the center reveal a loss rate of 2–4% per decade since 1979 over midlatitudes in the Northern Hemisphere. Ozone loss is expected to peak at about the year 2001, when, hopefully, the chlorine and bromine concentrations in the stratosphere will start to decline as a consequence of the international agreements that have been forged.

Is the ozone loss significant to our future? The EPA has calculated that the ozone losses of the 1980s will eventually have caused 12 million people in the United States to develop skin cancers over their lifetime and that 93,000 of these cancers will be fatal. Americans are estimated to develop more than 900,000 new cases of skin cancer a year in the 1990s. The ozone losses of the 1990s are expected to allow more UVB radiation than ever to reach Earth. Studies have confirmed increased UVB levels, especially at higher latitudes.

Coming to Grips with Ozone Depletion

The dramatic pictures of the growing "hole" in the ozone layer (Figure 21–17) have galvanized a response around the world. In spite of the skepticism in the United States, scientists and politicians here and in other countries have achieved legislation designed to avert a UV disaster.

International Agreements. Under its Environmental Program, the United Nations convened a meeting in Montreal, Canada, in 1987 to address ozone depletion. Member nations reached an agreement, known as the **Montreal Protocol**, to scale CFC production back 50% by 2000; to date, 140 countries (including the United States) have signed the agreement (see "Global Perspective" essay, p. 521).

The Montreal protocol was written even before CFCs were so clearly implicated in driving the destruction of ozone and before the threat to Arctic and temperate-zone ozone was recognized. Because ozone losses during the late 1980s were greater than expected, an amendment to the protocol was adopted in June 1990. The amendment requires participating nations to completely phase out the

GLOBAL PERSPECTIVE

LESSONS FROM THE OZONE TREATY

The 1987 Montreal Protocol and its amendments are a remarkable and encouraging development in human affairs. Most of the countries of the world have agreed to take steps that are economically costly in order to protect a global resource—the ozone shield. This agreement was reached on the basis of research from the scientific sector, and it was a response not to an immediate problem, but to one that is expected to occur in the future. The protocol is a statement that today's generation actually has some concern about future generations. Some important lessons from the Protocol should not be missed:

- It is significant that such broad-based cooperation can be forthcoming from the collection of nations that are most accustomed to putting national self-interest at the top of their agendas.

- It is quite apparent that scientists must continue to play a role in the crafting of policy. It was their work that first drew attention to the ozone losses and the global-warming complex. The development of consensus among scientists likewise has been remarkable.

- Governments must be bold enough to act even while scientific certainty is lacking. By its very nature, the scientific method cannot provide certainty short of an actual observation. At some point, delaying a decision becomes an immoral act. The costs of such a delay may be enormous and far overshadow the relatively low cost of mitigation.

- Political leadership from nongovernmental organizations and a few countries were crucial in bringing about the Montreal Protocol; such leadership may be just as essential in moving to a legally binding treaty on global warming.

- The United Nations, through its Environment Program, played a crucial role in the ozone accord. It has already been important in the negotiations on global warming and will likely continue to play the role of a catalyst for action.

- And Sherwood Rowland, Mario Molina, and Paul Crutzen were awarded the 1995 Nobel Prize for chemistry for their discoveries and subsequent fight in the political arena. The Nobel committee said, "... the 3 researchers contributed to our salvation from a global environmental problem that could have catastrophic consequences."

major chemicals destroying the ozone layer by 2000 in developed countries and by 2010 in developing countries. In the face of evidence that ozone depletion was accelerating even more, another amendment to the protocol was adopted, in November 1992, moving the target date for the complete phaseout of CFCs to 1996. Timetables for phasing out all of the suspected ozone-depleting halogens were shortened in the 1992 meeting.

Unfortunately, even with the ban, quantities of CFCs are still being manufactured, in particular in Russia, India, and China. A lively black market has developed for CFCs, and by 1998, the Justice Department had charged 62 individuals and seven businesses for smuggling CFCs into the U.S., leading to fines of $58 million, significant jail time, and recovery of tons of the banned chemicals.

Action in the United States. The United States was the leader in the production and use of CFCs and other ozone-depleting chemicals, with du Pont Chemical Company being the major producer. Following 15 years of resistance, du Pont pledged in 1988 to phase out CFC production by 2000. In late 1991, a spokesperson announced that in response to new data on ozone loss, the company would accelerate its phaseout by three to five years. Many of the large corporate users of CFCs (AT&T, IBM, and Northern Telcom, for example) completely phased out their CFC use by 1994. Also, du Pont spoke in opposition to three bills introduced in September 1995 in the 104th Congress that were designed to terminate U.S. participation and compliance with the CFC-banning protocols.

The Clean Air Act of 1990 also addresses this problem in Title VI, "Protecting Stratospheric Ozone." Title VI is a comprehensive program that restricts the production, use, emissions, and disposal of an entire family of chemicals identified as ozone depleting. For example, the program calls for a phaseout schedule for the hydrochlorofluorocarbons (HCFCs), a family of chemicals being used as less damaging substitutes for CFCs until nonchlorine substitutes are available. Halons—used in chemical fire extinguishers—were banned in 1994. The act also regulates the servicing of refrigeration and air-conditioning units.

January 1, 1996, has come and gone, and in most of the industrialized countries, CFCs are no longer being produced. Substitutes for CFCs are available and, in some cases, are even less expensive than the CFCs. The most commonly used ones are HCFCs, which still contain some chlorine and are scheduled for a gradual phaseout. The most promising substitutes are HFCs—hydrofluorocarbons, which contain no chlorine and are judged to have no ozone-depleting potential.

Final Thoughts. The ozone story is a remarkable episode in human history. From the first warnings in 1974 that something might be amiss in the stratosphere because of a practically inert and highly useful industrial chemical, through the development of the Montreal Protocol and the final steps of CFC phaseout that are still occurring, the world has shown it can respond collectively and effectively to a clearly perceived threat. The scientific community has played a crucial role in this episode, first alerting the world and then plunging into intense research programs to ascertain the validity of the threat. Scientists continue to influence the political process that has forged a response to the threat.

Although skeptics still stress the uncertainties in our understanding of ozone loss and its consequences, the strong consensus in the scientific community convinced the world's political leaders that action was clearly needed. This is a most encouraging development as we look forward to the creation of a sustainable society.

ENVIRONMENT ON THE WEB

METHANE: THE "OTHER" GREENHOUSE GAS

Over the last hundred years, the average temperature of Earth's surface has increased about 0.6°C. There is now persuasive evidence that some—perhaps most—of this increase is attributable to the actions of humans. Human activities are thought to have a significant influence on levels of carbon dioxide, methane, nitrous oxide, and, to a much smaller extent, anthropogenic chemicals such as chlorofluorocarbons. Of these greenhouse gases, carbon dioxide seems to make the greatest contribution to global warming, but the role of methane (CH_4) is also significant and often overlooked. Atmospheric methane levels have almost doubled since the early 1800s and now account for about 20% of the overall global-warming effect.

Methane is produced as anaerobic bacteria break down organic wastes. Sources of methane therefore occur where abundant organic material is present but oxygen levels are low, such as in wetlands (the term "swamp gas" refers in part to methane) and where organic wastes are stored or burned.

Scientists have few accurate measurements of methane emissions to the atmosphere, so available estimates are only approximate. Of the global total of 500 million tons/year, the most important sources are thought to be wetland systems (120 million tons/year), fossil fuel extraction (100 million tons/year), human and animal waste management activities (80 million tons/year), landfill degradation (40 million tons/year), deforestation and biomass burning (40 million tons/year), and agriculture (160 million tons/year).

Agriculture's contribution of methane per year includes 60 million tons/year from rice paddies and 100 million tons/year from enteric (intestinal) fermentation in the guts of ruminant animals such as cows, sheep, and water buffalo. It seems incredible that processes such as cattle digestion could have an impact on the vast mass of Earth's atmosphere, currently estimated at almost 6×10^{15} tons. But

it is precisely because the quantities of trace gases like methane are so small that modest emissions—and modest controls—can alter their proportions in important ways. The 12-year lifetime of methane in the atmosphere is short compared to other greenhouse gases, so the impact of emission reductions could be apparent in only a few years.

Web Explorations

Such reductions, like those mentioned above, may be well worth consideration. Molecule for molecule, methane has 21 times more warming effect than carbon dioxide in the atmosphere. Methane levels are also rising eight times faster than those of carbon dioxide. Programs, like those described in the Environment on the Web activity for this essay, developed in partnership with agriculture and other industrial sectors, are helping methane producers increase the efficiency—and productivity—of their operations while reducing their impact on the atmospheric environment. Go to the Environment on the Web activity (select Chapter 21 at http://www.prenhall.com/nebel) and learn for yourself:

1. why varying the quality of livestock feed can affect methane production;
2. how governments are preparing for potential methane increases expected from their expanding agriculture base; and
3. about the Kyoto Agreement and the rift separating many developing nations from their industrialized counterparts.

A suggested time frame for each of these exercises is 10–30 minutes.

REVIEW QUESTIONS

1. What are the important characteristics of the troposphere? Of the stratosphere?
2. Differentiate between weather and climate.
3. How does solar radiation create our weather patterns?
4. What has been discovered about global climate trends in the past? What is the contemporary trend showing?
5. What is the Conveyor system? How does it work? Why is it important?
6. How do greenhouse gases work? What is planetary albedo?
7. Which of the greenhouse gases are the most significant contributors to global warming? Describe the heat-trapping effects of carbon dioxide.
8. What is modeling? How can it help our efforts to improve the condition of the atmosphere?
9. What are four possible impacts of rising global temperature?

10. What steps could be taken to stabilize the greenhouse gas content of the atmosphere?

11. What does the "Precautionary Principle" of the 1992 Earth Summit state?

12. Trace the political history of the Framework Convention on Climate Change (FCCC) from 1992 to present.

13. How is the ozone shield formed? What causes its breakdown?

14. Describe four sources of CFCs entering the stratosphere. How do CFCs affect the concentration of ozone in the stratosphere and contribute to the formation of the ozone "hole"?

15. Discuss several efforts that are currently under way to protect our ozone shield.

THINKING ENVIRONMENTALLY

1. Compile a list of ways you are producing carbon dioxide. What steps could you take to decrease this production?

2. In your opinion, does the greenhouse effect exist? What about ozone depletion? Consider the "Precautionary Principle" introduced at the 1992 Earth Summit. What might be the short- and long-term economic effects of using this approach to solve environmental problems such as those stated above?

3. How would a ban on HCFC production change the way you live? Give your reasoning for or against such a ban.

WEB REFERENCES

On-line resources for this chapter can be found on the World Wide Web at: **http://www.prenhall.com/nebel**. Click on chapter 21 within the chapter selector.

22

Atmospheric Pollution

^^^^^^^^^

Key Issues and Questions

1. Understanding air pollution begins with understanding normal atmospheric function. How is the atmosphere normally cleansed?

2. Ever since the Industrial Revolution, unhealthy smogs have plagued human cities. How are industrial and photochemical smogs generated?

3. The variety and effects of the major air pollutants have now been identified. What are the eight major air pollutants and their most serious effects?

4. Much is known now about the origins and chemistry of air pollutants. Where do primary pollutants originate, and how do they form secondary pollutants?

5. Acid deposition impacts natural ecosystems as well as industrial centers. What is acid deposition, and what are its most significant effects on the environment?

6. Public policy now identifies standards for air pollutants based on their impacts on human health. What are the existing U.S. air pollution standards?

7. The most recent legislation addressing air pollution is the Clean Air Act of 1990. What are the main provisions of this act, and how are they related to former legislation?

8. Coping with acid deposition requires both technological and political change. What is being done in the United States to reduce acid deposition?

9. Further improvements in air quality may require rethinking how we structure our society. What further steps could be taken to improve air quality and our way of life?

On Tuesday morning, October 26, 1944, the people of Donora, Pennsylvania (population 13,000), awoke to a dense fog (Fig. 22–1). Donora lies in an industrialized valley along the Monongahela River. On the outskirts of town a sooty sign read, DONORA: NEXT TO YOURS, THE BEST TOWN IN THE U.S.A. The town had a large steel mill, which used high-sulfur coke, and a zinc-reduction plant that roasted ores having a high sulfur content. At the time, most homes in the area were heated with coal. At first the fog did not seem unusual. Most of Donora's fogs lifted by noon, as the Sun warmed the upper atmosphere and then the land. This one didn't lift for five days.

Through Wednesday and Thursday, the air began to smell of sulfur dioxide—an acrid, penetrating odor. By Friday morning, the town's physicians began to get calls from people in trouble. At first, the calls were from elderly citizens and those with asthmatic conditions. They were having difficulty breathing. The calls continued into Friday afternoon; people young and old were complaining of stomach pain, headaches, nausea, and choking. Work at the mills went on, however. The first deaths occurred on Saturday morning. By 10 A.M. one mortician had nine bodies; two other morticians each had one. On Sunday morning the mills were shut down. Even so, the owners were certain that their plants had nothing to do with the trouble. Mercifully, the rain came on Sunday and cleared the air—but not before more than 6000 townspeople were stricken and 20 had died. The cause? Eventually it was determined to be a combination of polluting gases and particles, a thermal inversion in the lower atmosphere, and a stagnant weather system that together gave a new meaning to the term *air pollution*.

Smog in Mexico City. Because of its geographic location and automobile traffic, Mexico City has some of the worst air pollution anywhere.

▶ **FIGURE 22–1** *Donora, Pennsylvania.* Donora was the scene of a major air pollution disaster in 1944.

With the advent of the Industrial Revolution, the mixture of gases and particles in our atmosphere began changing rapidly, and the effects on natural ecosystems and human health have proved to be dramatic and serious. In the previous chapter we encountered the structure and function of the atmosphere and looked at its links to climate, climate change, and ozone depletion. Here we will consider what happens when we add pollutants to the air.

22.1   Air Pollution Essentials

Pollutants and Atmospheric Cleansing

For a long time we have known that the atmosphere contains numerous **gases**: major constituents are N_2 (nitrogen) at 78.08%, O_2 (oxygen) at 20.95%, Argon at 0.93%, CO_2 (carbon dioxide) at 0.03%, and water vapor ranging from 0 to 4%. Smaller amounts of at least 40 "trace gases" are normally present, including ozone, helium, hydrogen, nitrogen oxides, sulfur dioxide, and neon. In addition, there are also aerosols present in the atmosphere—microscopic liquid and solid particles originating primarily from land and water surfaces and carried up into the atmosphere, like dust, carbon particles, pollen, sea salts, and microorganisms.

Air pollutants are substances in the atmosphere—*certain* gases and aerosols—that have harmful effects. We must make a distinction between natural and anthropogenic air pollutants, however. For millions of years, volcanoes, fires, and dust storms have sent smoke, gases, and particles into the atmosphere. Coniferous trees and other plants emit volatile organic compounds into the air around them. However, there are mechanisms in the biosphere that remove, assimilate, and recycle these natural pollutants. First, the pollutants disperse and become diluted in the atmosphere. Then, as shown in Fig. 22–2, a naturally occurring cleanser, the **hydroxyl radical** (OH), oxidizes many of them to products that are harmless or that can be brought down to the ground or water by precipitation. Microorganisms in the soil further convert some of these products into harmless compounds. These processes hold natural pollutants below toxic levels (except in the immediate area of a source, such as around an erupting volcano). Many of the pollutants oxidized by the hydroxyl radical are of concern because human activities have raised their concentrations far above normal levels.

Recent studies have shown that the hydroxyl radical plays the key role in the removal of anthropogenic pollutants from the atmosphere. Thus, highly reactive hydrocarbons are oxidized within an hour of their appearance in the atmosphere; nitrogen oxides (NO_x) are converted to nitric acid (HNO_3) within a day. It takes months, however, for less reactive substances like carbon monoxide (CO) to be oxidized by hydroxyl. It appears that the atmospheric levels of hydroxyl are in turn determined by the levels of anthropogenic pollutant gases—thus, hydroxyl's cleansing power is "used up" when high concentrations of pollutants are oxidized, and the pollutants are able to build up to damaging levels. On the other hand, as Fig. 22–2 shows, photochemical breakdown of tropospheric ozone is the major source of the hydroxyl radical, and as we will see, higher levels of ozone result from higher concentrations of other polluting gases.

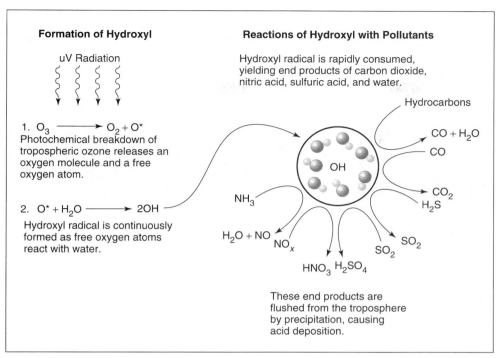

Formation of Hydroxyl

uV Radiation

1. $O_3 \longrightarrow O_2 + O^*$
Photochemical breakdown of tropospheric ozone releases an oxygen molecule and a free oxygen atom.

2. $O^* + H_2O \longrightarrow 2OH$
Hydroxyl radical is continuously formed as free oxygen atoms react with water.

Reactions of Hydroxyl with Pollutants

Hydroxyl radical is rapidly consumed, yielding end products of carbon dioxide, nitric acid, sulfuric acid, and water.

Hydrocarbons

$CO + H_2O$

CO

OH

CO_2
H_2S

NH_3

$H_2O + NO$
NO_x

HNO_3 H_2SO_4

SO_2
SO_2

These end products are flushed from the troposphere by precipitation, causing acid deposition.

◀ **FIGURE 22–2** *The hydroxyl Radical.* This is a simplified model of atmospheric cleansing by the hydroxyl radical. The first step is the photochemical destruction of ozone, which is the major process leading to ozone breakdown in the troposphere. The second step produces hydroxyl that reacts rapidly with many pollutants, converting them to substances that are either less harmful or can be brought down to Earth in precipitation.

Organisms are able to deal with certain levels of pollutants without suffering ill effects. The pollutant level below which no ill effects are observed is called the **threshold level**. Above this level, the effect of a pollutant depends on both its concentration and the duration of exposure to it. Higher levels may be tolerated if the exposure time is short. Thus, for any given pollutant, the threshold level is high for shorter exposures but gets lower as exposure time increases (Fig. 22–3). It is not the absolute amount of pollutant, but rather the *dose* that is important. As you learned in Chapter 2, dose is defined as concentration multiplied by time of exposure.

Three factors determine the level of air pollution:

- Amount of the pollutants entering the air
- Amount of space into which the pollutants are dispersed
- Mechanisms that remove pollutants from the air

The Appearance of Smogs

Down through the centuries, the practice of venting combustion and other fumes into the atmosphere remained the natural way to avoid their obviously noxious effects. With the Industrial Revolution of the 1800s came crowded cities and the use of coal for heating and energy, and air pollution began in earnest. In *Hard Times*, Charles Dickens commented on a typical scene: "Coketown lay shrouded in a haze of its own, which appeared impervious to the sun's rays. You only knew the town was there because there could be no such sulky blotch upon the prospect without a town." This shrouding haze became known as industrial smog (a combination of *sm*oke and f*og*!), an irritating, grayish mixture of soot, sulfurous compounds, and water vapor (Fig. 22–4a). This kind of smog continues to be found wherever industries

are concentrated and where coal is the primary energy source. The Donora incident is a classic example. Currently, industrial smog can be found in the cities of China, Korea, and a number of Eastern European countries.

After World War II, the booming economy in the United States led to the creation of vast suburbs and a mushrooming use of cars for commuting to and from work. As other fossil fuels were used in place of coal for heating, industry, and transportation, industrial smog was largely replaced by another kind of smog. Increasingly, Los Angeles and other cities served by huge freeway systems began to be enshrouded daily in a

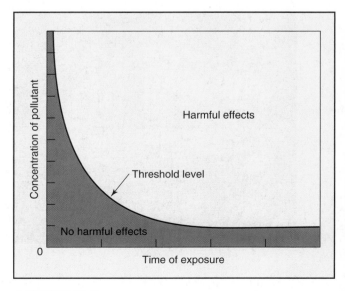

▲ **FIGURE 22–3** *The threshold level.* The threshold level diminishes with increasing exposure time. It differs for each pollutant.

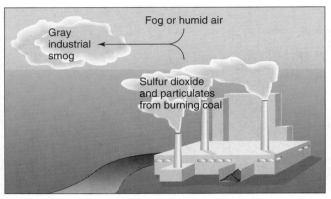

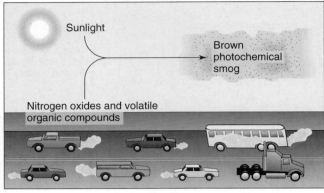

(a) Industrial smog   (b) Photochemical smog

▲ **FIGURE 22–4** *Industrial and photochemical smog.* (a) Industrial smog, or gray smog, occurs where coal is burned and the atmosphere is humid. (b) Photochemical smog, or brown haze, occurs where sunlight acts on vehicle pollutants.

brownish, irritating haze clearly different from the more familiar industrial smog (Fig. 22–5). Weather conditions were usually warm and sunny rather than cool and foggy, and typically the new haze would arise during the morning commute and only begin to dissipate by the time of the evening commute. This **photochemical smog**, as it came to be called, is produced when several pollutants from automobile exhaust—nitrogen oxides and volatile organic carbon compounds—are acted on by sunlight (Fig. 22–4b).

Certain weather conditions intensify levels of both industrial and photochemical smogs. The most significant of these conditions is a *temperature inversion*. Under normal conditions, daytime air temperature is highest near the ground because sunlight strikes Earth, and the absorbed heat radiates to

the air near the surface. The warm air near the ground rises, carrying pollutants upward and dispersing them at higher altitudes (Fig. 22–6a). At night, when the sun is no longer heating Earth, the currents cease. This condition of cooler air below and warmer air above is called a **temperature inversion** (Fig. 22–6b). Inversions are usually very short-lived, as the next morning's Sun begins the process anew and any pollutants that accumulated overnight are carried up and away. During cloudy weather, however, the Sun may not be strong enough to break up the inversion for hours or even days. Or a mass of high-pressure air may move in and sit above the cool surface air, trapping it. Topography can also intensify smogs as the airflow is affected by nearby mountain ranges or cool ocean air. Los Angeles has both!

(a)   (b)

▲ **FIGURE 22–5** *Los Angeles air.* A typical episode of photochemical smog. (a) Early in the morning, the air is clear. (b) Midmorning of the same day, the air is very hazy with smog. These photographs are close to the same view over Los Angeles.

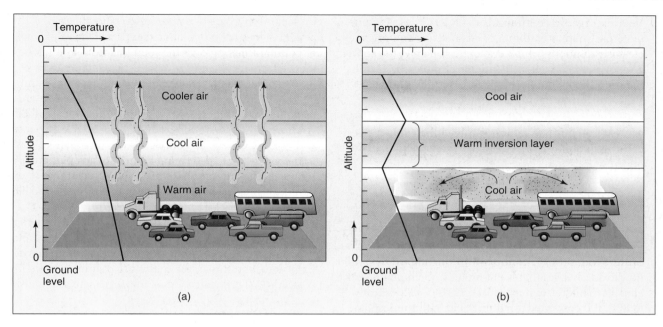

▲ **FIGURE 22–6** *Temperature inversion.* A temperature inversion may cause episodes of high concentrations of air pollutants. (a) Normally, air temperatures are highest at ground level and decrease at higher elevations. (b) A temperature inversion is a situation in which a layer of warmer air overlies cooler air at ground level.

When such long-term temperature inversions occur, pollutants can build up to dangerous levels, prompting local health officials to urge people with breathing problems to stay indoors. For many people, smog causes headaches, nausea, and eye and throat irritation; it may aggravate preexisting respiratory conditions such as asthma and emphysema. In some industrial cities (Donora, Pennsylvania, for example), air pollution has reached such high levels under severe temperature inversions that mortalities have increased significantly. These cases became known as *air pollution disasters*. London experienced repeated episodes of inversion-related disasters in the mid-1900s; there were 4000 pollution-related deaths during one episode in 1952.

The effects of air pollution are not limited to people. In recent years, many species of trees and other vegetation in and near cities began to die back, and farmers near cities suffered damage to or even total destruction of their crops because of air pollution. A conspicuous acceleration in the rate of metal corrosion and the deterioration of rubber, fabrics, and other materials was also noted.

22.2 Major Air Pollutants and Their Impact

Major Pollutants

The following eight pollutants have been identified as most widespread and serious:

1. **Suspended particulate matter.** This is a complex mixture of solid particles and aerosols (liquid particles) suspended

in the air. We see these particles as dust, smoke, and haze (Fig. 22–7). Particulates may carry any or all of the other pollutants dissolved in or adsorbed to their surfaces. Because they can be inhaled, the particles impair many respiratory functions, especially in individuals with chronic respiratory problems.

FIGURE 22–7 *Industrial pollution.* Because the stack of this wood-products mill in New Richmond (Quebec Province, Canada) lacks electrostatic precipitators, the plume contains substantial amounts of suspended particulate matter that can be seen from a great distance.

2. **Volatile organic compounds (VOCs).** These include materials such as gasoline, paint solvents, and organic cleaning solutions, which evaporate and enter the air in a vapor state, as well as fragments of molecules resulting from the incomplete oxidation of fuels and wastes. VOCs are prime agents of ozone formation.

3. **Carbon monoxide (CO).** This is an invisible, odorless gas that is highly poisonous to air-breathing animals because of its ability to block the delivery of oxygen to the organs and tissues.

4. **Nitrogen oxides (NO_x).** These include several nitrogen-oxygen compounds, all gases. They are converted to nitric acid in the atmosphere and are a major source of acid deposition. Nitrogen dioxide is a lung irritant that can lead to acute respiratory disease in children.

5. **Sulfur oxides (SO_x), mainly sulfur dioxide (SO_2).** Sulfur dioxide is a gas that is poisonous to both plants and animals. Children and the elderly are especially sensitive to SO_2. It is converted to sulfuric acid in the atmosphere and is also a major source of acid deposition.

6. **Lead and other heavy metals.** Lead is very dangerous at low concentrations and can lead to brain damage and death. It accumulates in the body and impairs many tissues and organs.

7. **Ozone and other photochemical oxidants.** Recall that ozone in the upper atmosphere (stratosphere) must be preserved to shield us from ultraviolet radiation (Chapter 21). However, ozone is also highly toxic to both plants and animals; it damages lung tissue and is implicated in many lung disorders. Therefore, ground-level (tropospheric) ozone is a serious pollutant. The case of ozone emphasizes the fact that "a pollutant is a chemical out of place."

8. **Air toxics and radon.** Toxic chemicals in the air include carcinogenic chemicals, radioactive materials, and other chemicals (such as asbestos, vinyl chloride, and benzene) that are emitted as pollutants but are not included in the preceding list of conventional pollutants. The Clean Air Act identifies 188 hazardous air pollutants in this category, many of which are known human carcinogens. Radon is a radioactive gas produced by natural processes within Earth. All radioactive substances have the potential to be damaging to any living matter they contact.

Adverse Effects of Air Pollution on Humans, Plants, and Materials

It is important to recognize that air pollution is not a single entity but an alphabet soup of the foregoing materials mixed with the normal constituents of air. Further, the amount of each pollutant present varies greatly, depending on proximity to the source and various conditions of wind and weather. As a result, we are exposed to a mixture that varies in makeup and concentration from day to day—even from hour to hour—and from place to place. Consequently, the effects we feel or observe are rarely, if ever, the effects of a single pollutant; they are the combined impact of the whole mixture of pollutants acting over the total life span, and frequently these effects are *synergistic*—that is, two or more factors combine to produce an effect greater than their simple sum.

For example, both plants and animals may be so stressed by pollution that they become more vulnerable to other environmental factors, such as drought or attack by parasites and disease. Given the complexity of this situation, it is extremely difficult to determine the role of any particular pollutant in causing an observed result. Nevertheless, some significant progress has been made in linking cause and effect.

Effects on Human Health. Humans breathe 30 lb (14 kg) of air into their lungs each day. Although some of the symptoms of pollution that people suffer involve the moist surfaces of the eyes, nose, and throat, the major site of impact is the lungs (Fig. 22–8). Three categories of impact can be distinguished:

1. **Chronic:** Pollutants cause the gradual deterioration of a variety of physiological functions over a period of years.
2. **Acute:** Pollutants bring on life-threatening reactions within a period of hours or days.
3. **Carcinogenic:** Pollutants initiate changes within cells that lead to uncontrolled growth and division (cancer).

Chronic effects. Almost everyone living in areas of urban air pollution suffers from chronic effects. Long-term exposure to sulfur dioxide can lead to bronchitis (inflammation of the bronchi). Chronic inhalation of ozone and particulates can cause inflammation and, ultimately, fibrosis of the lungs, a scarring that permanently impairs lung function. Carbon monoxide reduces the capacity of the blood to carry oxygen, and extended exposure to low levels of carbon monoxide can contribute to heart disease. Chronic exposure to nitrogen oxides is known to impair the immune system, leaving the lungs open to attack by bacteria and viruses.

Those most sensitive to air pollution are small children, asthmatics, people with chronic pulmonary and heart disease, and the elderly. Asthma, an immune system disorder characterized by impaired breathing caused by the constriction of air passageways, is usually brought on by contact with allergens. Many air pollutants increase the severity of asthma; the 58% increase in the incidence of asthma in the United States since 1970 is thought to be largely due to the synergistic effects of air pollutants. Studies have shown that asthmatics are extremely sensitive to levels of sulfur dioxide that are well below concentrations found in polluted cities.

Air quality is poor in countries where industrial air pollution is uncontrolled (Eastern Europe and China) and where motor-vehicle use is on the rise (many "megacities" in developing countries). The toll on human health and mortality in these regions can be severe. For example, the rate of infant mortality from respiratory disease in Eastern Europe is nearly 20 times that in North America. The chronic effects of air

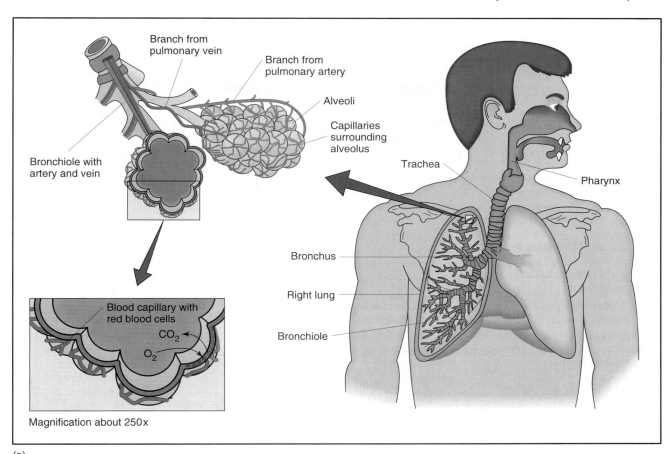

(a)

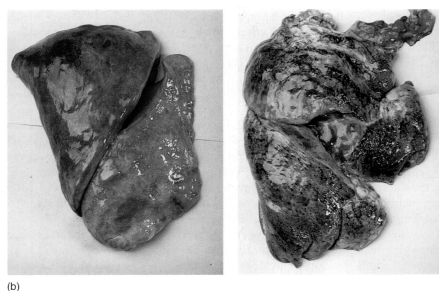

(b)

▲ **FIGURE 22–8** *The respiratory system.* (a) In the lungs, air passages branch and rebranch and finally end in millions of tiny sacs called alveoli. Alveoli are surrounded by blood capillaries. As blood passes through these capillaries, oxygen from inhaled air diffuses into the blood from the alveoli. Carbon dioxide diffuses in the reverse direction and leaves the body in the exhaled air. (b) On the left, normal lung tissue; on the right, lung tissue from a person who suffered from emphysema, a chronic lung disease in which some of the structure of the lungs has broken down. Cigarette smoking and heavy air pollution are associated with the development of emphysema and other chronic lung diseases.

GLOBAL PERSPECTIVE

MEXICO CITY: LIFE IN A GAS CHAMBER

"I'm getting out of here one of these days," said one resident of Mexico City. "The ecologists are right. We live in a gas chamber." By all accounts, Mexico City residents have the worst air in the world. The city exceeds safety limits for ozone more than 300 days a year, with values sometimes almost 400% over norms set by the World Health Organization. Smog emergencies are frequent occurrences. The city's 3 million vehicles (a large proportion are old and in disrepair) and 60,000 industries pour more than 5 million tons of pollutants into the air every day. Because it is nestled in a natural bowl, Mexico City experiences thermal inversions every day during the winter. One study showed that 40% of the city's young children suffer chronic respiratory sickness during the winter months. The pollution is estimated to cost $11.5 million a day in health costs and lost industrial production; an estimated 500,000 residents suffer from some form of smog-related health problem. Writer Carlos Fuentes has nicknamed the metropolis "Makesicko City"!

Mexico City's problems are indicative of a general dilemma throughout the developing world. Jobs and industrial activity, vital for a nation's economy, depend on the use of motor vehicles and on the industrial plants surrounding the city. Many automobile owners are absolutely dependent on their cars but can ill afford to keep them repaired in order to reduce exhaust emissions. Some people see cars as status symbols and will acquire one as soon as possible—often a "clunker" discarded from the developed world. Industries in Mexico are not regulated as they are in the United States, and lack the technology and capital to control emissions.

The Mexican government has initiated an air-monitoring system and, more recently, mandated emissions tests and catalytic converters for automobiles. The city has adopted a "Day Without a Car" policy: Drivers are given decals that permit driving only on specific days. Residents, however, get around the rule by purchasing another used car for use on alternate days. New air pollution laws are beginning to be enforced: One major refinery responsible for 4% of the air-quality problem was shut down by then President Salinas in the spring of 1991, and 80 other industries were temporarily or permanently closed for violating the new laws. New subway lines are being built, and less polluting buses are being added to the fleet. Lead, sulfur dioxide, and carbon monoxide levels are declining. By the mid-1990s, levels of air pollution were declining by 10% a year, but a rising city population and automobile fleet threaten to cancel the gains made. In the meantime, Mexico City's 18 million inhabitants are part of an unintentional experiment in demonstrating the health effects of severe air pollution.

---

pollution are reflected in morbidity and mortality data from Poland, a nation with perhaps the worst air in the world. Life expectancy for Polish men is lower than it was 20 years ago, and 30% to 45% of students are below international norms for height, weight, and other health indicators. Acute respiratory disease is responsible for some 4 million deaths a year of children under five in developing countries, second only to infant diarrhea in its impact on mortality.

One of the heavy-metal pollutants, lead, deserves special attention in this discussion. For several decades, lead poisoning has been recognized as a cause of mental retardation. Researchers once thought that eating paint chips containing lead was the main source of lead contamination in humans. In the early 1980s, however, elevated lead levels in blood were shown to be much more widespread than previously expected, and they were present in adults as well as children. Learning disabilities in children, as well as high blood pressure in adults, were correlated with high levels of lead in the blood. The major source of this widespread contamination was traced to leaded gasoline. (The lead in exhaust fumes may be inhaled directly or may settle on food, water, or any number of items that are put in the mouth.) This knowledge led the EPA to mandate the elimination of leaded gasoline, completed at the end of 1996. The result has been a dramatic reduction in lead concentrations in the environment.

Acute effects. In severe cases, air pollution reaches levels that cause death, although it should be noted that such deaths usually occur among people already suffering from severe respiratory or heart disease or both. The gases present in air pollution are known to be lethal in high concentrations, but such concentrations only occur in cases of accidental poisoning. Therefore, deaths attributed to air pollution are not the direct result of simple poisoning. However, intense air pollution puts an additional stress on the body, and if a person is already in a weakened condition (for example, elderly or asthmatic), this additional stress may be fatal. Most of the deaths in air pollution disasters reflect the acute effects of air pollution.

Carcinogenic effects. The heavy-metal and organic constituents of air pollution include many chemicals known to be carcinogenic in high doses. According to industrial reporting required by the EPA, 3.7 million tons of hazardous air pollutants (air toxics) are released annually into the air in the United States. The presence of trace amounts of these chemicals in air may be responsible for a significant portion of the cancer observed in humans. The **Clean Air Act of 1990** says that once an air pollutant is shown by laboratory studies to have mutagenic or carcinogenic properties, it may not be emitted if there is a lifetime risk of cancer greater than 1 in 1 million to the most exposed individual in the population.

▲ **FIGURE 22–9** *Air pollution damage.* The countryside around this Butte, Montana, smelter is devastated by the toxic pollutants from the industrial processes.

In some cases, exposure to a pollutant can be linked directly to cancer and other health problems by way of epidemiological evidence. One pollutant that clearly and indisputably is correlated with cancer and other disorders is benzene. This organic chemical is present in motor fuels. It is also used as a solvent for fatty substances and in the manufacture of detergents, explosives, and pharmaceuticals. Environmentally, benzene is found in emissions from fossil fuel combustion: motor-vehicle exhaust and burning coal and oil. It is also present in tobacco smoke, which accounts for half of the public's exposure to benzene. EPA has classified benzene as a Group A, known human carcinogen, linked to leukemia in persons encountering benzene through occupational exposure. Chronic exposure can also lead to numerous blood disorders and damage to the immune system.

Effects on Agriculture and Forests. Experiments show that plants are considerably more sensitive to gaseous air pollutants than are humans. Before emissions were controlled, it was common to see wide areas of totally barren land or severely damaged vegetation downwind from smelters or coal-burning power plants (Fig. 22–9). The pollutant responsible was usually sulfur dioxide.

The dying off of vegetation in large urban areas and the damage to crops, orchards, and forests downwind of urban centers are mainly caused by exposure to ozone and other photochemical oxidants. Estimates of crop damage by ozone range from $2 billion to $6 billion per year, with an estimated $1 billion in California alone. For example, California summer crops, like cotton and grapes, were calculated to experience losses of yield from 20% to 30% between 1989 and 1992. Mean ozone levels during this time were 0.085 ppm, with some eight-hour maxima as high as 0.135 ppm (clean air ozone levels range from 0.02 to 0.05 ppm). It is significant that much of the world's grain production occurs in regions that receive enough ozone pollution to reduce crop yields. The same parts of the world that produce 60% of the world's food—North America, Europe, and eastern Asia—also produce 60% of the world's air pollution.

The negative impact of air pollution on wild plants and forest trees may be even greater than on agricultural crops. Significant damage to valuable ponderosa and Jeffrey pines occurred along the entire western foothills of the Sierra Nevada in California. U.S. Forest Service studies have concluded that ozone from the nearby Central Valley and San Francisco–Oakland metropolitan region was responsible for this damage. Heavy metals, ozone, and acids carried in by clouds from the Midwest were implicated in the death of red spruce in Vermont's Green Mountains (Fig. 22–10). During the 1970s in the San Bernardino Mountains of California, an area that receives air pollution from Los Angeles, 50% of the trees died in some areas. As a result of air pollution controls, those same areas have shown significant improvement in tree growth in recent years, an encouraging sign.

Forests under stress from pollution are more susceptible to damage by insects and other pathogens than are unstressed forests. For example, the deaths of the ponderosa and Jeffrey pines in the Sierra Nevada were attributed to western pine beetles, which invade trees weakened by ozone. If widespread pollution worsens, reductions in tree growth and survival could occur with disastrous suddenness as threshold levels of more and more species are exceeded. As we will see, acid deposition from air pollution also has a decided impact on the growth of forest trees downwind of major urban areas.

▲ **FIGURE 22–10** *Pollution effect on forests.* Heavy metals and acids from distant urban areas were responsible for the death of red spruce trees in Vermont's Green Mountains.

▲ **FIGURE 22–11** *Pollution effect on monuments.* The corrosive effects of acids from air pollutants are dissolving away the features of many monuments and statues, as seen in the faces of these statues in Brooklyn, New York.

Effects on Materials and Aesthetics. Walls, windows, and other exposed surfaces turn gray and dingy as particulates settle on them. Paint and fabrics deteriorate more rapidly, and the sidewalls of tires and other rubber products become hard and checkered with cracks because of oxidation by ozone. Metal corrosion is dramatically increased by sulfur dioxide and acids derived from sulfur and nitrogen oxides, as are weathering and deterioration of stonework (Fig. 22–11). These and other effects of air pollutants on materials increase the costs for cleaning or replacing them by hundreds of millions of dollars a year. Many of the materials damaged are irreplaceable.

A clear blue sky and good visibility are matters of health, but they also carry significant aesthetic value and can have a deep psychological impact on people. Can a value be put on these benefits? Many of us spend thousands of dollars and hundreds of hours commuting long distances to work so that we can live in a less polluted environment than the one in which we work. Ironically, the resulting traffic and congestion cause much of the very pollution we are trying to escape. Then we travel to national parks and wilderness areas and too often find that the visibility in these natural areas is impaired by what has been called "regional haze," coming from particulates and gases originating hundreds of miles away. In a surprise move, EPA has proposed regulations to reduce regional haze, with the goal of a 10% improvement in visibility every 10 to 15 years during the next century. This may be the first time the government has attempted to regulate pollutants for aesthetic reasons!

Having considered the identity of the major air pollutants and their impacts, we now turn our attention to their origins.

22.3 Pollutant Sources

In large measure, air pollutants are direct and indirect byproducts of the burning of coal, gasoline, other liquid fuels, and refuse (wastepaper, plaster, and so on). These fuels and wastes are organic compounds. With complete combustion, the byproducts of burning them are carbon dioxide and water vapor, as shown by the following equation for the combustion of methane:

$$CH_4 + 2O_2 \rightarrow CO_2 + 2H_2O$$

Unfortunately, oxidation is seldom complete, and substances far more complex than methane are involved.

The first six of the major pollutants listed earlier in this chapter (**particulates, VOCs, CO, NO_x, SO_x, and lead**) are called **primary pollutants** because they are the direct products of combustion and evaporation. Some of the primary pollutants may undergo further reactions in the atmosphere and produce additional undesirable compounds. The latter products are called **secondary pollutants.** Sources of air pollution are summarized in Fig. 22–12. Notice that while industrial processes are the major source of particulates, transportation and fuel combustion account for the lion's share of the other pollutants. Strategies for control of air pollutants are different for these different sources.

Primary Pollutants

When fuels and wastes are burned, particles consisting mainly of carbon are emitted into the air; these are the particulates we see as soot and smoke. In addition, various unburned fragments of fuel molecules remain; these are the VOC emissions. Incompletely oxidized carbon is carbon monoxide (CO), in contrast to completely oxidized carbon, which is carbon dioxide (CO_2). Combustion takes place in the air, which is 78% nitrogen and 21% oxygen. At high combustion temperatures, some of the nitrogen gas is oxidized to form the gas nitric oxide (NO). In the air, nitric oxide immediately reacts with additional oxygen to form nitrogen dioxide (NO_2), nitrogen tetroxide (N_2O_4), or both. (These compounds are collectively referred to as the **nitrogen oxides**.) Nitrogen dioxide absorbs light and is largely responsible for the brownish color of photochemical smog.

In addition to their organic matter content, fuels and refuse contain impurities and additives, and these substances, too, are emitted into the air during burning. Coal, for example, contains from 0.2% to 5.5% sulfur. In combustion, this sulfur is oxidized, giving rise to the gas sulfur dioxide (SO_2). Coal may contain heavy-metal impurities, and refuse, of course, contains an endless array of "impurities."

According to EPA data, 1996 emissions in the United States of the first five primary pollutants amounted to 154 million tons. By comparison, in 1970, when the first Clean Air Act became law, these same five pollutants totaled 244

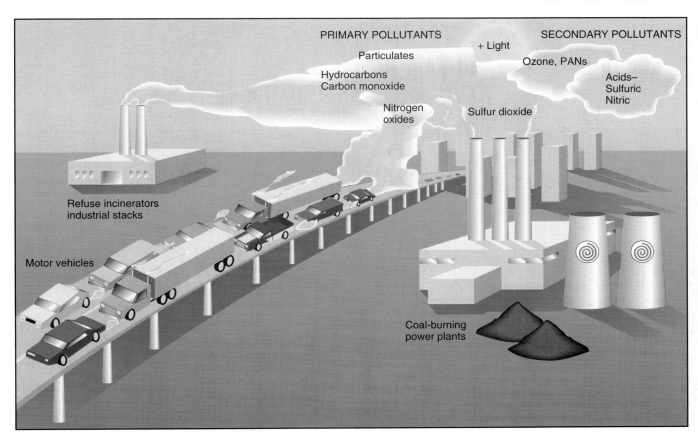

▲ FIGURE 22–12 *The prime sources of the major air pollutants.*

million tons. The relative amounts emitted in the United States, as well as their major sources, are illustrated in Fig. 22–13. Ten-year trends in the emissions of these primary pollutants are represented in Fig. 22–14. Those emissions tied most directly to transportation (CO, VOC) show consistent long-term improvement, while SO_2 shows some improvement since 1987 but has risen 3% in the last year. NO_x has shown a 5% *increase* in the last 10 years, due primarily to an increase in emissions from power plants. The trends clearly reflect the accomplishments as well as the shortcomings of Clean Air Act regulations, as we shall see.

The sixth type of primary pollutant—lead and other heavy metals—is discussed separately because the quantities emitted are far less than the levels for the first five. Before the EPA-directed phaseout, lead was added to gasoline as an inexpensive way to prevent engine knock. Emitted with the exhaust from gasoline-burning vehicles, the lead remained airborne and traveled great distances before settling as far away as the glaciers of Greenland. Since the phaseout, concentrations of lead in the air of cities in the United States have shown a remarkable decline: Between 1977 and 1996, ambient lead concentrations have declined 97%. At the same time, levels of lead in children's blood have dropped greatly. Lead concentrations in Greenland ice have also decreased significantly. All these declines indicate that lead restrictions in the United States and a few other Northern Hemisphere nations have had a global impact. Nonetheless,

lead is still used in some U.S. manufacturing processes and is subject to few restrictions elsewhere in the world.

As with lead, the concentrations of toxic chemicals and radon in the air are small compared with those of the other primary pollutants. Some of the air's toxic compounds—benzene, for example—originate with transportation fuels. Most, however, are traced to industries and small businesses. Radon, on the other hand, is produced by the spontaneous breakdown of fissionable material in rocks and soils. Radon escapes naturally to the surface and seeps into buildings through cracks in foundations and basement floors, sometimes collecting in the structures.

Secondary Pollutants

Ozone and numerous reactive organic compounds are formed as a result of chemical reactions between nitrogen oxides and volatile organic carbons. Because sunlight provides the energy necessary to propel these reactions, these products are known collectively as **photochemical oxidants**. In preindustrial times, ozone concentrations ranged from 10 to 15 ppb (parts per billion), well below harmful levels. Summer concentrations of ozone in unpolluted air in North America now range from 20 to 50 ppb. Polluted air often contains ozone concentrations of 150 ppb or more, considered very unhealthy if encountered for extended periods of

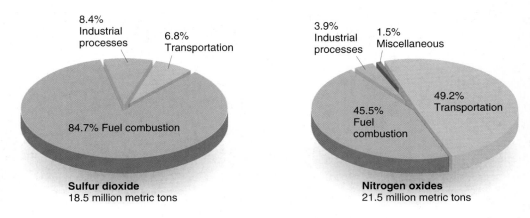

Sulfur dioxide
18.5 million metric tons

Nitrogen oxides
21.5 million metric tons

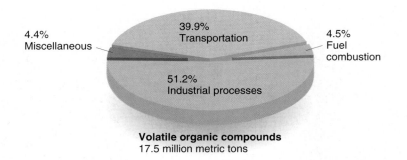

Volatile organic compounds
17.5 million metric tons

▶ **FIGURE 22–13** *Current emissions in the United States of five primary air pollutants, by source, for 1996.* Fuel combustion refers to fuels burned for electrical power generation and for space heating. Note especially the different contributions by transportation and fuel combustion, the two major sources of air pollutants. PM-10 particulate matter percentages are for non-fugitive dust sources.

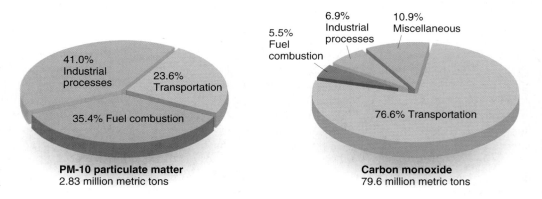

PM-10 particulate matter
2.83 million metric tons

Carbon monoxide
79.6 million metric tons

▶ **FIGURE 22–14** *Ten-year trends in emissions of six criteria pollutants since 1987.* Improvements in Carbon monoxide and VOC emissions reflect Clean Air Act successes. Nitrogen oxides have not improved, because little attention has been paid to them.

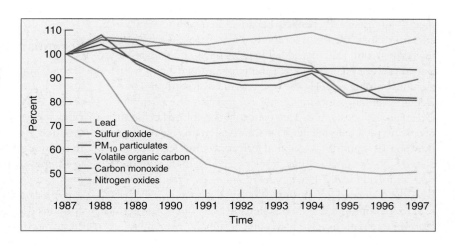

time. The good news is that EPA data indicate ambient ozone levels, as measured at almost 400 sites across the United States, have been declining at a rate of about 1% per year since 1987.

The major reactions in the formation of ozone and other photochemical oxidants are shown in Fig. 22–15. Nitrogen dioxide absorbs light energy and splits to form nitric oxide and atomic oxygen. The atomic oxygen rapidly combines with oxygen gas to form ozone. If other factors are not involved, ozone and nitric oxide then react to form nitrogen dioxide and oxygen gas. A steady-state concentration of ozone results, and there is no appreciable accumulation of ozone (Fig. 22–15a). When volatile organic compounds are present, however, the nitric oxide reacts with them instead of with the ozone, causing several serious problems. First, the reaction between nitric oxide and VOCs leads to highly reactive and damaging compounds known as peroxyacetyl nitrates, or PANs (Fig. 22–15b). Second, numerous aldehyde and ketone compounds are produced as the VOCs are oxidized by atomic oxygen, and these compounds are also noxious. Finally, with the nitric oxide tied up in this way, the ozone tends to accumulate. Because of the complex air chemistry involved, ozone concentrations usually peak 30 to 100 miles (50 to 160 km) downwind of urban centers where the primary pollutants were generated. As a result, ozone

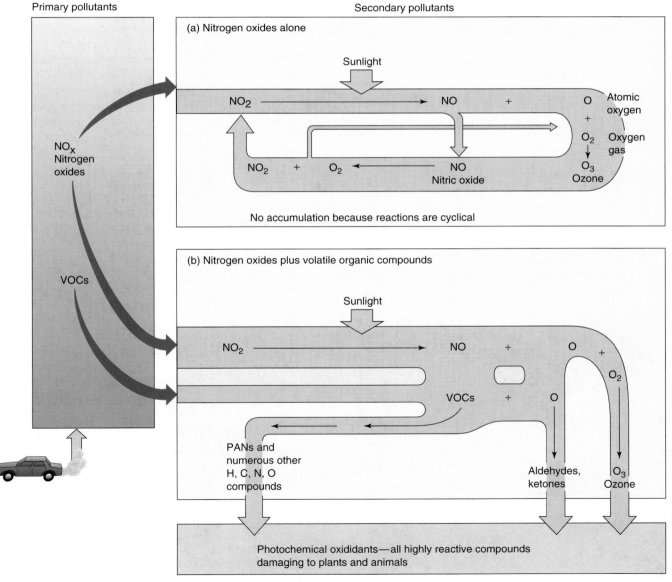

FIGURE 22–15 *Formation of ozone and other photochemical oxidants.* (a) Nitrogen oxides, by themselves, would not cause ozone and other oxidants to reach damaging levels, because reactions involving them are cyclic. (b) When VOCs are also present, however, reactions occur that lead to the accumulation of numerous damaging compounds—most significantly, ozone, the most injurious.

concentrations above the health standard in the United States are often found in rural and wilderness areas.

Particulates take on additional potency when they adsorb other pollutants onto their surfaces. (*Ad*sorption means that the pollutants simply adhere to a surface, like flies sticking to flypaper. By contrast, *ab*sorption means "soaking in.") When inhaled, many of these contaminated particulates are fine enough to penetrate deeply into the lungs, releasing the adsorbed pollutants onto the moist lung surfaces, where they often remain trapped for life.

In a sense, **sulfuric** and **nitric acids** can also be considered secondary pollutants, since they are products of sulfur dioxide and nitrogen oxides reacting with atmospheric moisture and oxidants such as hydroxyl. They are the sources of the well-known atmospheric pollution known commonly as acid rain (technically referred to as acid deposition). We now turn our attention to this important and widespread problem.

22.4 Acid Deposition

Acid precipitation refers to any precipitation—rain, fog, mist, or snow—that is more acidic than usual. Because dry acidic particles are also found in the atmosphere, the combination of precipitation and dry particle fallout is called **acid deposition**. In the late 1960s, Swedish scientist Svante Oden first documented the acidification of lakes in Scandinavia and traced it to air pollutants originating in other parts of Europe and Great Britain. Since then, careful monitoring has shown that broad areas of North America, as well as most of Europe and other industrialized regions of the world, are regularly experiencing precipitation that is between 10 and 1000 times more acidic than usual. This is affecting ecosystems in diverse ways, as illustrated in Fig. 22–16.

To understand the full extent of the problem, first we must understand some principles about acids and how we measure their concentration.

Acids and Bases

Acidic properties (sour taste, corrosiveness) are due to the presence of hydrogen ions (H^+, a hydrogen atom without its electron), which are highly reactive. Therefore, an **acid** is any chemical that releases hydrogen ions when dissolved in water. The chemical formulas of a few common acids are shown in Table 22-1. Note that all of them ionize—that is, their components separate—to give hydrogen ions plus a negative ion. The higher the concentration of hydrogen ions in a solution, the more acidic the solution.

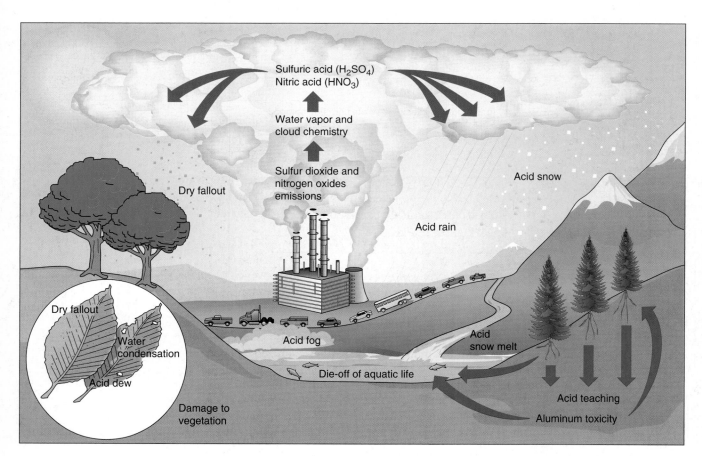

▲ **FIGURE 22–16** *Acid deposition.* Emissions of sulfur dioxide and nitrogen oxides react with the hydroxyl radicals and water vapor in the atmosphere to form their respective acids, which come back down either as dry acid deposition or, mixed with water, acid precipitation. Various effects of acid deposition are noted.

TABLE 22-1 Common Acids and Bases

Acid	Formula	Yields	H^+ Ion(s)	Plus	Negative Ion	
Hydrochloric acid	HCl	→	H^+	+	Cl^-	Chloride
Sulfuric acid	H_2SO_4	→	$2H^+$	+	SO_4^{2-}	Sulfate
Nitric acid	HNO_3	→	H^+	+	NO_3^-	Nitrate
Phosphoric acid	H_3PO_4	→	$3H^+$	+	PO_4^{3-}	Phosphate
Acetic acid	CH_3COOH	→	H^+	+	CH_3COO^-	Acetate
Carbonic acid	H_2CO_3	→	H^+	+	HCO_3^-	Bicarbonate

Base	Formula	Yields	OH^- Ion(s)	Plus	Positive Ion	
Sodium hydroxide	NaOH	→	OH^-	+	Na^+	Sodium ion
Potassium hydroxide	KOH	→	OH^-	+	K^+	Potassium ion
Calcium hydroxide	$Ca(OH)_2$	→	$2OH^-$	+	Ca^{2+}	Calcium ion
Ammonium hydroxide	NH_4OH	→	OH^-	+	NH_4^+	Ammonium

A **base** is any chemical that releases hydroxide ions (OH^-, oxygen-hydrogen groups with an extra electron) when dissolved in water (see Table 22-1). The bitter taste and caustic properties of all alkaline, or basic, solutions are due to the presence of hydroxide ions.

The concentration of hydrogen ions is expressed as **pH**. The pH scale goes from 0 (highly acidic) through 7 (neutral) to 14 (highly basic) (Fig. 22–17). Each of the numbers on the scale represents the negative logarithm (power of 10) of the hydrogen ion concentration, expressed in grams per liter. For example, to say that a solution has a pH of 1 means that the concentration of hydrogen ions in the solution is 10^{-1} g/L (0.1 g/L); pH = 2 means that the hydrogen ion concentration is 10^{-2} g/L, and so on.

At pH = 7, the hydrogen ion concentration is 10^{-7} (0.0000001) g/L, and the hydroxide ion concentration is also 10^{-7} g/L. This is the neutral point, where small but equal amounts of hydrogen ions and hydroxide ions are present in pure water. The pH numbers above 7 continue to express the negative exponent of hydrogen ion concentration, but they also represent an increase in hydroxide ion concentration. Because solutions of pH greater than 7 contain higher concentrations of hydroxide ions than of hydrogen ions, they are called basic solutions.

We use the same negative logarithm arrangement to express the concentration of hydroxide ions: **pOH**. There is a reciprocal relationship between pH and pOH: As the pH of a given solution goes up, its pOH goes down, and vice versa. In fact, we can state this as a rule: For any aqueous solution, pH + pOH = 14. For example, pH = 13 means that the hydrogen ion concentration is 10^{-13} (a decimal followed by 12 zeros and a 1) g/L. However, the hydroxide ion concentration of this same solution is 10^{-1} g/L, so the pOH = 1. In a neutral solution, equal amounts of hydrogen ions and hydroxide ions are present; thus, pH = 7 and pOH = 7; and pH + pOH = 14.

Since numbers on the pH scale represent powers of 10, there is a *tenfold difference* between each unit and the next. For example, pH 5 is 10 times as acidic (has 10 times as many H^+ ions) as pH 6, pH 4 is 10 times as acidic as pH 5, and so on.

One easy way to measure pH is with indicator paper, which is available from any laboratory supply house. This paper contains pigments that change color when they are wetted with an acidic or a basic solution. The pH is determined by dipping a strip of indicator paper in the solution and matching the color

◀ **FIGURE 22–17** *The pH scale.*

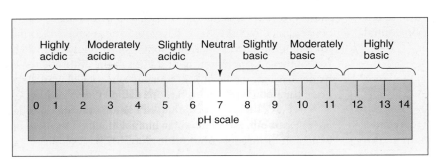

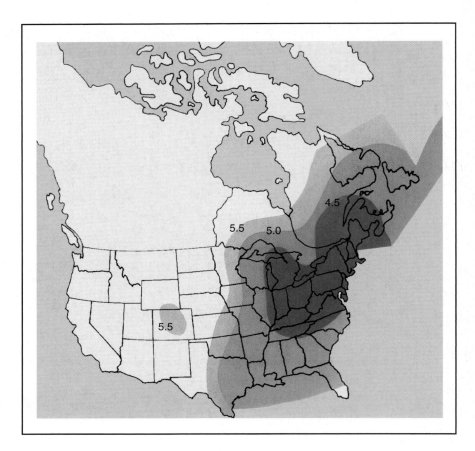

▶ **FIGURE 22–18 *Acid deposition in North America.*** Monitoring the pH of deposition now reveals that acid deposition is occurring over most of the eastern United States and Canada. It is especially severe in the northeastern United States and eastern Canada.

of the wet paper with a color chart provided with the paper. Where accuracy and precision are important, however, it is necessary to use electronic instruments for measuring pH.

Extent and Potency of Acid Precipitation

In the absence of any pollution, rainfall is normally slightly acidic, with a pH of 5.6, because carbon dioxide in the air readily dissolves in and combines with water to produce an acid called carbonic acid. **Acid precipitation**, then, is any precipitation with a pH of 5.5 or less.

Unfortunately, acid precipitation is now the norm over most of the industrialized world. As Fig. 22–18 shows, the pH of rain and snowfall over a large portion of eastern North America is typically about 4.5. Many areas in this region regularly receive precipitation having a pH of 4.0 and, occasionally, as low as 3.0. Fogs and dews can be even more acidic; in mountain forests east of Los Angeles, scientists found fog water of pH 2.8—almost 1000 times more acidic than usual—dripping from pine needles. Acid precipitation has been heavy in Europe, from the British Isles to central Russia. Also, it is now well documented in Japan.

Sources of Acid Deposition

Chemical analysis of acid precipitation in eastern North America and Europe reveals the presence of two acids, **sulfuric acid** (H_2SO_4) and **nitric acid** (HNO_3), in a ratio

of about two to one. (In the western states and provinces, nitric acid predominates, principally formed from tailpipe emission.) As we have seen, burning fuels produce sulfur dioxide and nitrogen oxides, and so the source of the acid deposition problem is evident. These oxides enter the troposphere in large quantities from both anthropogenic and natural sources. Once in the troposphere, they are oxidized by hydroxyl radicals (Fig. 22–2) to sulfuric and nitric acids, which dissolve readily in water or adsorb to particles and are brought down to Earth in acid deposition. This usually occurs within a week of the oxides' entering the atmosphere.

We must recognize that *natural sources* contribute substantial quantities of pollutants to the air: 50 to 70 million tons per year of sulfur (from volcanoes, sea spray, and microbial processes) and 30 to 40 million tons per year of nitrogen oxides (from lightning, burning of biomass, and microbial processes). *Anthropogenic sources* are estimated at 100 to 130 million tons per year of sulfur dioxide and 60 to 70 million tons per year of nitrogen oxides. The vital difference between these two sources is that anthropogenic oxides are strongly concentrated in industrialized regions, whereas the emissions from natural sources are spread out over the globe and are a part of the global environment. Levels of the anthropogenic oxides have increased at least fourfold since 1900, while levels of the natural emissions have remained fairly constant.

As Fig. 22–13 indicates, 18.5 million tons of sulfur dioxide are released into the air annually in the United States, 85% of which are from fuel combustion (mostly from coal-burning power plants). Some 21.5 million tons of nitrogen oxides are released annually, 49% traced to transportation emissions and 46% to fuel combustion at fixed sites. For the eastern United States, the source of over 50% of the acid deposition has been identified as the tall stacks of 50 huge coal-burning power plants (Fig. 22–19). These tall stacks were built to alleviate local sulfur dioxide pollution at ground level. The unfortunate result of the "dilution solution" is that emitting sulfur dioxide and nitrogen oxides high in the air simply provides more opportunity for them to spread hundreds of miles from their source.

Effects of Acid Deposition

Acid deposition has been recognized as a problem in and around industrial centers for over 100 years. Its impact on ecosystems, however, was noted only about 35 years ago, when anglers started noticing sharp declines in fish populations in many lakes in Sweden, Ontario, and the Adirondack Mountains of upper New York State. Since that time, as ecological damage has continued to spread, studies have revealed many ways in which acid deposition alters and may destroy ecosystems.

Impact on Aquatic Ecosystems. The pH of an environment is extremely critical because it affects the function of virtually all enzymes, hormones, and other proteins in the bodies of all organisms living in that environment. Ordinarily, organisms are able to regulate their internal pH; however, consistently low environmental pH often overcomes this regulatory ability in many life forms. Most freshwater lakes, ponds, and streams have a natural pH in the range of 6 to 8, and organisms are adapted accordingly. The eggs, sperm, and developing young of these organisms are especially sensitive to changes in pH. Most are severely stressed, and many die if the environmental pH shifts as little as one unit from the optimum.

As such aquatic ecosystems are acidified, there is a rapid die-off of virtually all higher organisms, either because the acidified water kills them or because it keeps them from reproducing. Figure 22–16 illustrates the fact that acid precipitation may leach aluminum and various heavy metals from the soil as the water percolates through it. Normally, the presence of these elements in the soil does not pose a problem, because they are bound in insoluble mineral compounds and therefore are not absorbed by organisms. As these compounds are dissolved by low-pH water, however, the metals are freed;

(a)

(b)

▲ **FIGURE 22–19** *Midwestern coal-burning power plant.* (a) Standard smokestacks of this coal-burning power plant were replaced by new 1000-foot (330-m) stacks to aid in the dispersion of pollutants into the atmosphere. The taller stacks alleviate local air pollution problems but create more widespread distribution of acid-generating pollutants. (b) Locations of the 50 largest sulfur dioxide emitters, all of which are utility coal-burning power plants.

they may then be absorbed and are highly toxic to both plants and animals. For example, mercury tends to accumulate in fish as lake waters become more acidic. Mercury levels are so high in the Great Lakes that many bordering states advise against eating fish caught in these waters.

In Norway and Sweden, the fish have died in at least 6500 lakes and seven Atlantic salmon rivers. In Ontario, Canada, approximately 1200 lakes now harbor no life. In the Adirondacks, a favorite recreational region for New Yorkers, more than 200 lakes are without fish, and many are devoid of all life save resistant algae and bacteria. More than 190 New England lakes and rivers are suffering from recent acidification. The physical appearance of such lakes is deceiving. From the surface they are clear and blue, the outward signs of a healthy condition. However, a view under the surface is eerie: In spite of ample light shimmering through the clear water, there is not a sign of life in them.

As Fig. 22–18 indicates, wide regions of the United States receive roughly equal amounts of acid precipitation, yet not all areas have acidified lakes. Apparently, many areas remain healthy, whereas others have become acidified to the point of becoming lifeless. How is this possible? The key lies in the system's *buffering capacity*. Despite the addition of acid to it, a system may be protected from pH change by a **buffer**—a substance that, when present in a solution, has a large capacity to absorb hydrogen ions and thus hold the pH relatively constant.

Limestone ($CaCO_3$) is a natural buffer (Fig. 22–20) that protects lakes from the effects of acid precipitation in many areas of the North American continent. Lakes and streams receiving their water from rain and melted snow that has percolated through soils derived from limestone will contain dissolved limestone. The regions sensitive to acid precipitation are those containing much granitic rock that does not yield good buffers. For these areas, the most critical time of year is the spring thaw, when accumulated winter snow melts. If the thaw is sudden, streams, rivers, and lakes are hit with what has been called "acid shock," as accumulated acids send pH levels plummeting in a sudden burst of meltwater. Making matters worse, the acid shock often coincides with the times of spawning and egg laying in aquatic animals, when they are at their most vulnerable.

Any buffer has limited capacity. Limestone, for instance, is used up by the buffering reaction and so is no longer available to react with more added hydrogen ions (Fig. 22–20). Ecosystems that have already acidified and collapsed are those that had very little buffering capacity. Those that remain healthy have greater buffering capacity. However, many of these still healthy lakes are gradually losing their buffering capacity as more and more acid continues to be deposited. In time, many healthy lakes will join the growing number of dead lakes if acid precipitation is not curbed.

Impact on Forests. Along with dying lakes, the decline of forests has been conspicuous. From the Green Mountains of Vermont to the San Bernardino Mountains of California, the die-off of forest trees in the 1980s caused great concern. Red spruce forests are especially vulnerable; in New England, 1.3 million acres of high-elevation forests have been devastated. Commonly, the damaged trees lost needles and often fell prey to insects and diseases before they succumbed.

Much of the damage of acid precipitation to forests is due to chemical interactions within the forest soils. Sustained acid precipitation at first adds nitrogen and sulfur to the soils, which stimulate tree growth. In time, these chemicals impact the soils by leaching out large quantities of the buffering chemicals, usually calcium and magnesium salts. When these buffering salts no longer neutralize the acid rain, aluminum ions, which are toxic, are then dissolved from minerals in the soil. The combination of aluminum and the increasing scarcity of calcium that is essential to plant growth leads to reduced tree growth. Research at the Hubbard Brook Experimental Forest in the White Mountains of New Hampshire has shown a marked reduction in calcium and magnesium in the forest soils from the 1960s on, which is reflected in the calcium of tree rings over the same time period. The net result of these changes has been a severe decline in forest growth.

Declining forests are a serious problem in some parts of Europe, and the evidence indicates that the same kinds of soil-chemical exchanges are occurring there. Because of the variations in the buffering capacity of soils and the differing amounts of sulfur and nitrogen brought in by acid precipitation, forests are affected to varying degrees. Some forests continue to grow, whereas others are in decline. One result of the sustained acidification is sometimes a gradual shift toward more acid-tolerant species. In New England, the balsam fir is moving in to replace the dead spruce.

Impact on Humans and Their Artifacts. One of the more noticeable effects of acid deposition is the deterioration of artifacts. Limestone and marble (which is a form of limestone) are favored materials for the outsides of buildings and for monuments (collectively called **artifacts**). The reaction between acid and limestone is causing these structures to

▶ **FIGURE 22–20** *Buffering.* Acids may be neutralized by certain nonbasic compounds called *buffers.* A buffer such as limestone (calcium carbonate) reacts with hydrogen ions as shown. Hence, the pH remains close to neutral despite the additional acid.

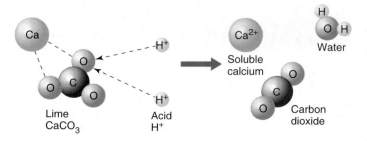

erode at a tremendously accelerated pace. Monuments and buildings that have stood for hundreds or even thousands of years with little change are now dissolving and crumbling away, as was shown in Fig. 22–11. The corrosion of buildings, monuments, and outdoor equipment by acid precipitation costs billions of dollars for replacement and repair each year in the United States. A 37-foot bronze Buddha in Kamakura, Japan, is slowly dissolving away as acid rain from Korea and China bathes the statue in rain that is more acidic than that recorded in the United States.

Whereas the decay of such artifacts is a tragic loss in itself, it should also stand as a grim reminder of how we are dissolving away the buffering capacity of ecosystems. In addition, some officials are concerned that acid precipitation's mobilization of aluminum and other toxic elements may result in the contamination of both surface water and groundwater. Increased acidity in water also mobilizes lead from the pipes used in some old plumbing systems and from the solder used in modern copper systems.

As the song says, "What goes up must come down." The sulfur and nitrogen oxides pumped into the troposphere at the rate of 40 million tons per year in the United States come down as acid deposition, generally to the east of their origin, because of the way weather systems flow. The deposits cross national boundaries: Canada receives half of its acid deposition from the United States, and Scandinavia gets it mostly from Great Britain and other Western European nations. Japan receives the windborne pollution from widespread coal burning in Korea and China. There is now a broad scientific consensus that the problem of acid deposition must be addressed at national and international levels. We turn now to the efforts being made to curb air pollution.

22.5 Bringing Air Pollution Under Control

By the 1960s, it was obvious that pollutants produced by humans were overloading natural cleansing processes in the atmosphere. The unrestricted discharge of pollutants into the atmosphere could no longer be tolerated.

Under grassroots pressure from citizens, the U.S. Congress passed the **Clean Air Act of 1970**. Together with amendments passed in 1977 and 1990, this law, administered by the EPA, represents the foundation of U.S. air pollution control efforts. It calls for identifying the most widespread pollutants, setting **ambient standards**—levels that need to be achieved to protect environmental and human health—and establishing control methods and timetables to meet these goals.

The Clean Air Act of 1970 mandated the setting of these standards for four of the primary pollutants—particulates, sulfur dioxide, carbon monoxide, and nitrogen oxides—and for the secondary pollutant ozone. At the time, these five pollutants were recognized as the most widespread and objectionable; today, with the recent addition of lead, they are known as the *criteria pollutants* and are

TABLE 22-2 National Ambient Air-Quality Standards for Criteria Pollutants

Pollutant	Averaging Time*	Primary Standard
PM_{10} particulates	1 year	$50 \ \mu g/m^3$
	24 hours	$150 \ \mu g/m^3$
$PM_{2.5}$ particulates$^\Psi$	1 year	$15 \ \mu g/m^3$
	24 hours	$65 \ \mu g/m^3$
Sulfur dioxide	1 year	0.03 ppm
	24 hours	0.14 ppm
	3 hours	0.5 ppm
Carbon monoxide	8 hours	9 ppm
	1 hour	35 ppm
Nitrogen oxides	1 year	0.05 ppm
Ozone	8 hours	0.08 ppm
Lead	3 months	$1.5 \ mg/m^3$

*Averaging time is the period over which concentrations are measured and averaged.
$^\Psi PM_{2.5}$ is the particulate fraction having a diameter smaller than or equal to 2.5 micrometers. It supplements the existing PM_{10} standard.
Source: EPA National Air Quality and Emissions Trends Report, 1996.

covered by the **National Ambient Air Quality Standards (NAAQS)** (Table 22-2). The **primary standard** for each pollutant is based on the highest level that can be tolerated by humans without noticeable ill effects, minus a 10% to 50% margin of safety. For many of the pollutants, long-term and short-term levels are set. The short-term levels are designed to protect against acute effects, while the long-term standards are designed to protect against chronic effects.

Table 22-2 reflects two standards announced by EPA in July 1997: the new $PM^{2.5}$ category, and a revision of the ozone standard from 0.12 ppm for one hour to 0.08 ppm for eight hours. These changes were established in response to health studies that indicated that substantial improvement in lung function and early death prevention could be accomplished by the changes. Cost/benefit estimates indicated that the health benefits far outweighed the costs of compliance (estimated at $9.7 billion per year). The new ozone standards are also believed to prevent a substantial amount of the damage to vegetation by ozone. Both of these changes were met with strong opposition from industry groups and state and local officials, and there is the possibility that Congress may consider legislation to rescind the new standards.

In addition, **National Emission Standards for Hazardous Air Pollutants (NESHAPS)** have been issued for eight toxic substances: arsenic, asbestos, benzene, beryllium, coke oven emissions, mercury, radionuclides, and vinyl chloride. The Clean Air Act of 1990 greatly extended this section of the EPA's regulatory work by specifically naming 188 toxic air pollutants for EPA to track and regulate.

EARTH WATCH

PORTLAND TAKES A RIGHT TURN

For many of us, the American lifestyle includes spending time in the automobile each day going to and from work. Vehicle miles per year in the United States have increased much more rapidly than population, and too often the only response of state and city governments is to build more lanes on the expressways.

Portland, Oregon, had its share of expressways 20 years ago but took a different tack when faced with the prospect of more and more commuters on the road. In response to an Oregon land-use law, Portland threw away its plans for more expressways and instead built a light rail system, Metropolitan Area Express (MAX). This public transportation system now carries the equivalent of two lanes of traffic on all the roads feeding into downtown Portland. The result? Smoggy days have declined from 100 to 0 per year, the downtown area has added 30,000 jobs with no increase in automobile traffic, and Portland's economy has prospered.

This has clearly been a situation where everyone wins. MAX has made a major contribution to the clean air and continued economic success of downtown Portland. The Portland solution seems so sensible that you have to ask why it is the exception and not the rule.

Control Strategies

The basic strategy of the 1970 Clean Air Act was to regulate air pollution so that the criteria pollutants remained below the primary standard levels. This approach is called **command-and-control** because industry was given regulations to achieve a set limit on each pollutant, to be accomplished by specific control equipment. The assumption was that human and environmental health could be significantly improved by a reduction in the output of pollutants. If a particular region was in violation for a given pollutant, a local government agency would track down the source(s) and order reductions in emissions until the area came into compliance.

Unfortunately, this strategy proved difficult to implement. Most of the regulatory responsibility fell on the states and cities, which were often unable or unwilling to enforce control. Many areas violated the standards. Even today, after 30 years of legislated air pollution control, many metropolitan areas still consistently fail to meet the standards. The good news is that total air pollutants have been reduced by some 37% during a time when both population and economic activity have substantially increased. However, we are still sending huge quantities of pollutants into the atmosphere.

The Clean Air Act of 1990 addresses these failures by more directly targeting specific pollutants and more aggressively demanding compliance, through such means as the imposition of sanctions. As with the earlier law, the states do much of the work in carrying out the mandates of the 1990 act. Each state must develop a *State Implementation Plan (SIP)* that must go through a process of public comment before being submitted to the EPA for approval. The SIP is designed to reduce emissions of every NAAQS pollutant whose control standard (Table 22-2) has not been attained. One major change is a permit application process (already in place for the release of pollutants into waterways). Polluters must apply for a permit that identifies the kinds of pollutants they release, the quantities of those pollutants, and the steps they are taking to reduce pollution. Permit fees provide funds the states can use to support their air pollution control activities. The act also provides more flexibility than the earlier command-and-control approach by allowing polluters to choose the most cost-effective way to accomplish the goals. In addition, the law uses a market system to allocate pollution among different utilities.

Reducing Particulates. Prior to the 1970s, the major sources of particulates were industrial stacks and the open burning of refuse. The Clean Air Act of 1970 mandated the phaseout of open burning of refuse and required that particulates from industrial stacks be reduced to "no visible emissions."

The alternative generally taken to dispose of refuse was landfilling, a solution that has created its own set of environmental problems (discussed in Chapter 19). To reduce stack emissions, many industries were required to install filters, electrostatic precipitators, and other devices. Unfortunately, the solid wastes removed from exhaust gases frequently contain heavy metals and other toxic substances (Chapter 20). Although these measures have markedly reduced the levels of particulates since the 1970s (see Fig. 22–14), particulates continue to be released from steel mills, power plants, cement plants, smelters, construction sites, diesel engines, and so on.

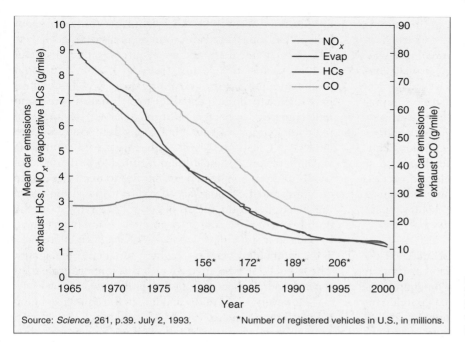

Source: *Science,* 261, p.39. July 2, 1993.   *Number of registered vehicles in U.S., in millions.

◀ **FIGURE 22–21** *Automobile emissions trends.* Average car emissions from vehicles, in grams per vehicle mile traveled, from 1965 to (estimated) 2000. Note numbers of vehicles on the road in the United States for 1980 to 1995.

Wood-burning stoves and wood and grass fires also contribute to the particulate load, making regulation even more difficult.

Under the Clean Air Act of 1990, the regions of the United States that have failed to attain the required levels must submit *attainment plans.* These plans must be based on **reasonably available control technology (RACT)** measures, and offending regions must convince the EPA that the standards are reached within a certain time frame.

Also, the EPA added the new standard for particulates (**PM$_{2.5}$**) based on information that smaller particulate matter (less than 2.5 micrometers in diameter) has the greatest effect on health because the finer particulate matter tends to originate from combustion processes and from atmospheric chemical reactions between pollutants. The coarser material (**PM$_{10}$**) often comes from windblown dust, and although it is still regulated, it is not considered as significant in health problems as the finer particles.

Limiting Pollutants from Motor Vehicles. Cars, trucks, and buses release nearly half of the pollutants that foul our air. Vehicle exhaust sends out the VOCs, carbon monoxide, and nitrogen oxides that lead to ground-level ozone and PANs. Additional VOCs come from the evaporation of gasoline and oil vapors from fuel tanks and engine systems.

The Clean Air Act of 1970 mandated a 90% reduction in these emissions by 1975. This timing proved to be unrealistic, but enough improvements have been made over the years that a new car today emits 75% less pollutants than did pre-1970 cars (Fig. 22–21). This is fortunate, because driving in the United States has been increasing much more than the population has: Between 1970 and 1990, the number of vehicle miles has doubled from 1 trillion to 2 trillion miles per year and between 1980 and 1995, the number of vehicles on the road increased over 30% (see inset data on Fig 22–21).

It is hard to imagine what the air would be like without the improvements mandated by the Clean Air Act.

The reductions in automobile emissions have been achieved with a general reduction in the size of passenger vehicles, along with a considerable array of pollution control devices. These include the computerized control of fuel mixture and ignition timing, allowing more complete combustion of fuel and decreasing VOC emissions. To this day, however, the most significant control device on cars is the **catalytic converter** (Fig. 22–22). As exhaust passes through this device, a chemical catalyst made of platinum-coated beads oxidizes most of the VOCs to carbon dioxide and water.

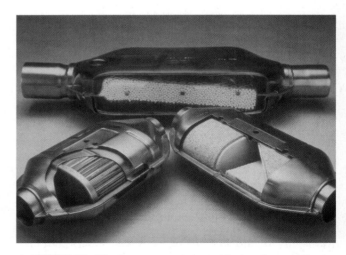

▲ **FIGURE 22–22** *Cutaways of three kinds of catalytic converters.* Top: ceramic pellets. Left: metal foil plates. Right: ceramic monolith substrates. The catalysts are platinum, palladium, and rhodium—metals that promote the oxidation of residual VOC and carbon monoxide to CO$_2$ and water.

The catalytic converter also oxidizes most of the carbon monoxide to carbon dioxide. Although newer converters reduce nitrogen oxides as well, the reduction is not impressive.

Despite these efforts, the continuing failure to meet standards in many regions of the United States, as well as the aesthetically poor air quality in many cities, made it clear during the 1980s that further legislation was in order. In preparing for the Clean Air Act of 1990, the EPA considered three options to improve the air in our cities: (1) Tighten emissions standards further; (2) encourage the development and use of cleaner burning fuels; (3) and persuade people to drive less. The new law addressed the first two options but made only a token gesture toward promoting less driving. Following are the highlights of the sections on motor vehicles and fuels:

1. New cars sold in 1994 and thereafter must emit 30% less VOCs and 60% less nitrogen oxides than cars sold in 1990. Emission-control equipment must function properly (as represented in warranties) to 100,000 miles; on 1990 cars, this equipment need work only for 50,000 miles. Buses and trucks must meet more stringent standards. Also, the EPA is given authority to control emissions from all nonroad engines that contribute to air pollution, including lawn and garden equipment, motorboats, off-road vehicles, and farm equipment.

2. Starting in 1992 in the 38 cities with continuing carbon monoxide problems, oxygen has been added to gasoline in the form of alcohols and ethers to stimulate more complete combustion and to cut down on carbon monoxide emissions. Beginning in 1995, special "clean" fuels are required for the nine cities having the worst air pollution.

3. Before 1990, only a few states and cities required that vehicle inspection stations be capable of measuring emissions accurately. The new act requires more than 40 metropolitan areas to initiate inspection and maintenance programs while other cities improve their current programs. The result is expected to be cost effective in reducing air pollution from cars and trucks.

4. Employers in cities with high smog levels will be required to take steps to increase the number of their employees who car pool or use mass transit.

Two factors now affect fuel efficiency and consumption rates. First, the elimination of federal speed limits in early 1996 reduced fuel efficiencies because of higher speeds. Second, the sale of sport utility vehicles (SUVs) has surged since 1990. SUVs now represent 6% of all cars and trucks on the road. In city traffic, these vehicles get 12 to 16 mpg and no higher than 22 mpg on the highway. Remember: As fuel efficiency decreases, tailpipe emissions increase.

Managing Ozone. Because ozone is a secondary pollutant, the only way to control ozone levels is to address the compounds that lead to its formation. For a long time it was assumed that the best way to reduce ozone levels was to reduce emissions of VOCs. The steps pertaining to motor vehicles in

the Clean Air Act of 1990 addresses only the sources of about half of VOC emissions. **Point sources** (industries) account for 30% of such emissions, and **area sources** (numerous small emitters such as dry cleaners, print shops, and household products) represent the remaining 20%. RACT measures have already been mandated for many point sources, and much progress has been made through EPA, state, and local regulatory efforts to reduce emissions from these sources. Under the Clean Air Act of 1990, regions with continuing violation of ozone standards are required to further control VOC emissions from both point and area sources. The degree of control and the timing vary according to the seriousness of the violation, but the means of control will range from the prohibition of some consumer products, to annual fees paid by point sources ($5000 per ton of VOCs), to sanctions such as the withholding of federal highway funds and other grants. The last of these is an effective means of forcing recalcitrant states to address their air pollution problems.

However, recent understanding of the complex chemical reactions involving NO_x, VOCs, and oxygen has thrown some uncertainty into this strategy of emphasizing a reduction in VOCs. The problem is that *both* NO_x and VOC concentrations are crucial to the generation of ozone. Either one or the other can become the rate-limiting species in the reaction that forms ozone (see Fig. 22–15b). Thus, as the ratio of VOCs to NO_x changes, the concentration of NO_x can become the controlling chemical factor because of an excess of VOCs. This happens more commonly in air pollution episodes that occur over a period of days than in the daily photochemical smog of the urban city. The EPA has recently proposed a rule to address this problem; the rule sets a budget for emissions of NO_x from 22 states east of the Mississippi, designed to greatly reduce the regional transport of ozone.

Controlling Toxic Chemicals in the Air. By EPA estimates, the total amount of toxic substances emitted into the air in the United States is around 3.7 million tons annually. Cancer and other adverse health effects, environmental contamination, and catastrophic chemical accidents are the major concerns associated with this category of pollutants.

Under the Clean Air Act of 1990, Congress identified 188 toxic pollutants. It directed the EPA to identify major sources of these pollutants and to develop **Maximum Achievable Control Technology (MACT)** standards. Besides control technologies, the setting of these standards includes options for substituting nontoxic chemicals, giving industry some flexibility in meeting goals. State and local air pollution authorities will be responsible for seeing that industrial plants achieve the goals. The program should reduce emissions by at least 20% by the year 2005 at an estimated cost to the industry of $7 billion.

Coping with Acid Deposition

Ways to Reduce Acid-Forming Emissions. Scientists have calculated that a 50% reduction in current acid-causing emissions in the United States would effectively prevent further acidification of the environment. This reduction may not correct the already bad situations, but together with natural buffering processes, it is estimated to be capable of preventing further environmental deterioration. Because we know that about 50% of acid-producing emissions come from the tall stacks of coal-burning plants that generate electricity, control strategies are centered on these sources. Six main strategies have been proposed: (1) coal washing, (2) fluidized bed combustion, (3) fuel switching, (4) scrubbers, (5) alternative power plants, and (6) reductions in the consumption of electricity.

Economic factors do not favor the first two strategies. Coal washing to remove sulfur is costly, both economically and environmentally; large amounts of polluted water would be the outcome. In fluidized bed combustion, coal is burned in a mixture of sand and lime, churned by air forced in from underneath. The sulfur in the coal combines with the lime during combustion and is removed with the ash that is formed. When new plants are built, this may be the preferred method of emissions control, but installing the technology in existing plants would entail tearing down and rebuilding a major portion of each plant.

Oil is more expensive than coal, and switching to oil as a fuel for generating electricity would increase the dependence of the United States on foreign oil. There is, however, low-sulfur coal available in central Appalachia and the western United States, and—despite the higher cost of the coal and its transportation—switching to this source represents an attractive strategy for power plants seeking to reduce their sulfur dioxide emissions without changing their technology.

Scrubbers are "liquid filters" that put exhaust fumes through a spray of water containing lime (Fig. 22–23). The sulfur dioxide reacts with the lime and is precipitated as calcium sulfate ($CaSO_4$). Scrubbers have been required for all major power plants and smelters built in the United States since 1977, but the law has not required retrofitting the large, older plants in the Midwest that are the source of much of the acid rain plaguing the East and Canada. Plants in Europe and Canada and a few in the United States have demonstrated that scrubbers can be added to existing power plants relatively quickly; the scrubbers are highly effective in controlling emissions and are not prohibitively expensive.

The last two strategies in the foregoing list require major shifts in the energy policy of the United States. Currently, nuclear power represents the only alternative technology for generating electricity. It accounts for 21% of electrical power generation in the United States. However, as Chapter 14 explains, the future of nuclear energy is in serious question because of concerns about safety and the storage of nuclear waste. Reducing the consumption of electricity (conservation) makes very good sense and has already been implemented in some applications, such as appliances and insulation. Conservation and energy-efficiency strategies represent a relatively untapped potential for future reductions in air pollution, although putting it to work will involve major changes in many sectors of U.S. economic and social life, as explained in Chapter 13.

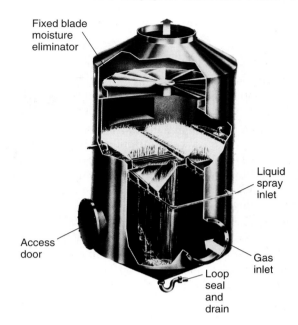

Fixed blade moisture eliminator

Liquid spray inlet

Access door

Gas inlet

Loop seal and drain

▲ **FIGURE 22–23** *Scrubbers.* Sulfur dioxide may be removed from flue gases by passing the flue gas through either a spray of lime slurry in a venturi scrubber or a sodium hydroxide or sodium carbonate solution in an impingment scrubber. The sulfur dioxide reacts with the alkali in the scrubbing water to form sulfite and sulfate salts. The device shown here is a cutaway view of the SLY IMPINJET gas scrubber.

Political Developments. Although evidence of the link between power-plant emissions and acid deposition was well established by the early 1980s, no legislative action was taken until 1990. The problem was one of different regional interests. Western Pennsylvania and the states of the Ohio River valley, where older coal-burning power plants produce most of the electrical power, argued that controlling their sulfur dioxide emissions would make electricity in the region unaffordable. Throughout the 1980s, a coalition of politicians from these states, fossil-fuel corporation representatives of high-sulfur coal producers, and representatives from the electric power industry effectively blocked all attempts at passing legislation that would take action on acid deposition.

On the other side of the question were New York and New England, as well as most of the environmental and scientific community, which argued that it was both possible and necessary to address acid deposition and that the best way to do it was to control sulfur dioxide emissions. Also, since 70% of Canada's acid-deposition problem came from the United States, diplomatic pressure toward a resolution was applied.

Title IV of the Clean Air Act of 1990. The outcome of the controversy is now history. However, we lost two decades of action because of political delays. **Title IV** of this act is the first law in our history to address the acid-deposition problem, by mandating reductions in both sulfur dioxide and nitrogen oxide levels. The major provisions of Title IV are as follows:

1. By the year 2000, total sulfur dioxide emissions must be reduced 10 million tons below 1980 levels. This is the 50% reduction called for by scientists, and it involves the setting of a permanent cap of 8.9 million tons on such emissions.

2. The reduction will be implemented gradually. By 2000, the combined sulfur dioxide output of all the country's power plants must not exceed the permanent cap of 8.9 million tons. The utilities are required to install equipment that closely monitors their emissions of acid-generating gases.

3. In a major departure from the command-and-control approach described in Chapter 15, Title IV authorizes the EPA to use a free-market approach to regulation. Each plant is granted **emission allowances** based on formulas in the legislation. The penalty for exceeding these allowances is severe: a fee of $2000 per ton and the requirement that the utility must compensate for the excess emissions the next year. A utility may not emit more sulfur dioxide than allowances permit, but if it emits less, the difference in allowance may be sold, so that another utility may purchase the first plant's difference in place of reducing its own emissions. Judged highly successful, the sulfur dioxide emissions market is currently worth $900 million a year!

4. In the future, new utilities will not receive allowances; they must buy into the system by purchasing existing allowances. Thus, there will be a finite number of allowances in existence.

5. Nitrogen oxide emissions must be reduced by 2 million tons by the year 2000. This is to be accomplished by regulating the boilers used by the utilities and by mandating the continuous monitoring of emissions.

Industry's Response to Title IV. The utilities industry has responded to the new law with three actions: (1) Many of the utilities are switching to low-sulfur coal. This will enable the targeted high-capacity utilities to achieve compliance rapidly. (2) At least 15 power plants are adding scrubbers during the first few years of compliance efforts. Many more will be forced to scrub in order to reach final compliance by 2000. Since the technology for high-efficiency scrubbing is well established, many plants will upgrade their existing scrubbers. (3) Many of the utilities are trading in emissions allowances. At current prices, the purchase of allowances often represents a less costly way to achieve compliance than by purchasing low-sulfur coal or adding a scrubber. A typical transaction might involve a trade in the rights to emit, say, 10,000 tons of sulfur dioxide at a cost of $110 a ton. This approach has created so many efficiencies it has cut compliance costs to 10% of what was expected.

In the wake of the passage of the Clean Air Act of 1990, Canada and the United States have signed a treaty providing that Canada will cut its sulfur dioxide emissions by half and cap them at 3.2 million tons by the year 2000. As a result of several treaties in Europe, sulfur dioxide emissions there are down 40% from 1980 levels and should decline another 18% by 2010 if the signatories fulfill their treaty obligations. The Europeans are also targeting nitrogen oxide emissions, with the goal of a 30% reduction in emissions by 1999.

The main target of these agreements is the remaining healthy aquatic and forested ecosystems, which will hopefully be protected from future acid deposition. In addition, it is hoped that ecosystems already harmed will be able to recover from current damage and that we will be back on the road to a sustainable interaction with the atmosphere.

Unfortunately, acid rain still falls on eastern forests, although the acid levels have dropped slightly. The reason? Very little was done in the Clean Air Act amendments to curb emissions of nitrogen oxides. Recall that the NO_x emissions have actually increased by 5% over the last 10 years. Scientists and politicians from the East are exerting pressure to amend the Clean Air Act to call for more drastic cuts in NO_x emissions. The EPA's recent ruling on curbing regional emissions of NO_x should help, but most observers believe that much more must be done before the problem of acid deposition begins to disappear. Industry representatives object that such measures would lead to economic disaster; they said the same thing about sulfur dioxide reductions, and they were wrong.

22.6 Taking Stock

Without question, measures taken to reduce air pollution carry an economic cost. The United States sends a clear signal that human health and environmental quality rank highly in the development and implementation of public policy by investing heavily in pollution control. Also, there is strong citizen support for this investment. According to a recent *Times Mirror* poll, 89% of Americans believe that both a strong economy and environmental protection can be accomplished, while only 22% believe that the economy is more important than the environment. Other polls indicate that most Americans are ready to pay more for cleaner gasoline. Currently, pollution control costs are estimated at $125 billion annually, about one fourth of which represents efforts to reduce air pollution. The EPA estimates that after being fully implemented in 2005, the new Clean Air Act will cost an additional $25 billion. Although this is an enormous sum, it represents less than 1% of the estimated $7 trillion economy for that year.

Some critics, especially economists, have charged that air pollution controls are not cost effective; that is, the benefits are not nearly as great as the costs. They see lost opportunities for economic growth and tend to disregard the *avoided costs* (from improved health) despite a recent analysis of the Clean Air Act that conservatively found direct benefits totaling $6.8 trillion—a net benefit of $6.4 trillion to society (see "Earth Watch" essay, p. 549). In addition, pollution control has become a major industry. Providing over 2 million jobs, it is a significant part of the national economy.

EARTH WATCH

THE CLEAN AIR ACT BRINGS A WINDFALL

The Clean Air Act (1970, 1977, and 1990) has been the subject of open political warfare between those who think its cost has been too high for industry, taxpayers, labor, and consumers and those who think the health and environmental benefits were justified. Compliance has affected patterns of industrial production, employment, and capital investment. Although these expenditures must be viewed as investments that have generated benefits and opportunities, the dislocation in some regions was severe: such as reductions in high-sulfur coal mining and cutbacks in polluting industries such as steel. A need developed for a real cost-benefit analysis.

In 1990, Congress requested the EPA answer the question: How do the overall health, welfare, ecological, and economic benefits of Clean Air Act programs compare with the costs of these programs? In response, the EPA performed the most exhaustive cost-benefit analysis of public policy ever attempted. Here is what the EPA reported in a 1996 study:

- The total direct cost to implement the Clean Air Act for all federal, state, and local rules from 1970 to 1990 was $436 billion (in 1990-value dollars). This cost was borne by businesses, consumers, and government entities in the form of higher prices for many goods, services, and some utilities.

- The mean estimate of direct benefits from the Clean Air Act from 1970 to 1990 was $6.8 trillion.

- Therefore, the net benefit of the Clean Air Act has been $6.4 trillion!

"The finding is overwhelming. The benefits far exceed the costs of the Clean Air Act in the first 20 years," said Richard Morgenstern, associate administrator for policy planning and evaluation at the EPA. Further, the report states that "all benefits may be significantly underestimated due to the exclusion of large numbers of benefits from the monetized benefit estimate."

The benefits to society, directly and indirectly, have been widespread across the entire population. Here is a summary:

- Reduced air pollution (described in this chapter)
- Improved human health: each year, 79,000 lives saved, 18,000 fewer heart attacks, 10,000 fewer strokes, 13,000 fewer hypertension cases, and 15 million fewer respiratory cases.

- "Avoided cost": improved health has meant less debilitating disease, hospitalization, special care, and medicines.

- Less lead to harm children: In 1990, 220,000 tons of lead were not burned in gasoline, because of Clean Air Act measures. Exposure to lead impairs the cognitive development of children; therefore, the huge reductions in lead produced a benefit of retained IQ and the possibility of a more productive, less dependent life.

- Lower cancer rates
- Less acid deposition

The EPA study result should encourage us in our hopes for a more sustainable future. Society knew what to do, took action despite disruptive efforts by special-interest and political partisans, and reaped about $16 in benefits for every $1 invested to control air pollution.

(Source: Adapted from R. Christopherson, Geosystems, An Introduction to Physical Geography, 3rd ed., by permission of Prentice Hall, Inc., 1997, pp. 646–648.)

Future Directions

In the early 1970s, many cities were experiencing more than 100 days per year of pollution in the "unhealthful" range or above; Los Angeles had about 300 such days per year. The original goal of air pollution control was to achieve "good" air quality on all days by 1975. This goal was not met; however, marked progress has occurred for most of the primary air pollutants (Fig. 22–14). Certainly, the quality of the air is much better than if controls had not been initiated, and the total avoided cost is significant.

Nevertheless, the backsliding of the last few years is troubling. From 1970 to 1990, pollution control devices and greater fuel efficiency reduced a typical car's emissions by 75%. As old, polluting cars were replaced by new, "cleaner" cars over this period, emission levels decreased despite increasing numbers of cars and miles driven. Today, this reduction by replacement has leveled out, and the increasing number of cars and steady erosion of fuel economy are the dominant factors in preventing further progress. The Clean Air Act addresses this problem primarily by mandating increased emission controls and cleaner burning fuels. Whether these measures will be successful in further reducing air pollution remains to be seen.

A more general approach that would benefit the whole country would be to increase the fuel efficiency standard for cars (known as corporate average fuel economy, CAFÉ, standards), which is now locked at 27.5 mpg. Intense lobbying by the automobile and petroleum industries prevented more stringent standards of fuel efficiencies from being included in the Clean Air Act of 1990. (Fuel efficiency standards were actually rolled back in the 1980s!) In 1995, the actual average fuel economy of all vehicles on U.S. highways was 24.5 mpg. The SUV, classified as light trucks, feature average efficiencies of only 18 mpg! Available yet neglected technologies could bring the average mpg to 60. For example, two-stroke engines (which fire on every thrust of the piston instead of on every other thrust, as in the current four-stroke engine) would cut engine weight in half; and lightweight cars with smaller engines give mileage in the range of 80 to

100 mpg. Increasing the fuel efficiency of vehicles would proportionately reduce emissions. It would also address the problems associated with global climate change (Chapter 21) and with future shortages of crude oil (Chapter 13).

California, which by all measures has the greatest problem with photochemical smog and vehicular pollutants, has taken a completely different approach to reduce pollution from vehicles. Starting in 1998, California law requires that 2% of all vehicles sold in the state must be "emission free"—in other words, powered by electricity. The required percentage will rise to 10% by 2003. Similar laws have been enacted in nine northeastern states and the District of Columbia. In response to these developments, automakers are reluctantly devoting increasing resources to the development of affordable electric cars. Entrepreneurial firms such as U.S. Electricar of California and Solectria of Massachusetts are forging ahead of the big three automakers and already have electric cars on the market (Fig. 22–24a). These cars are considerably lighter than conventional cars and quite limited in range (traveling 50 to 100 miles before needing recharging). They also lack such amenities as air-conditioning and other power accessories. An interesting option is the hybrid electric vehicle (HEV), which is an electric car with a small internal combustion engine equipped with an electric generator to charge the batteries. HEVs are being developed by several foreign manufacturers; Toyota has recently begun marketing the Prius in Japan, a five-passenger sedan that gets 66 mpg, reaches speeds of 100 mph, and produces one-tenth the pollution of a comparable gasoline car (Fig. 22–24b).

Electrical cars or hybrids may well be the wave of the future, especially for short trips at low speeds, which waste energy and cause much pollution. Of course, switching from gasoline to electrical power will transfer the site of pollution emission from the moving vehicle to the power plant—a trade-off with uncertain consequences, unless we make strides in renewable energy sources. It can be argued that all this emphasis on how vehicles burn fuel is just tinkering with the mechanics of the problem. To achieve lasting progress, we need to address the fundamental way we have organized society and industry. Consider that we pour billions of dollars into expanding and improving our highways, build malls beside every major highway corridor, and build homes in bedroom communities that are scores of miles from major employment centers; but we put relatively little money into improving mass transit systems, so prominent in the first half of the century. It makes great sense, in the coming years, to fashion legislation and policies that would guide new development and redevelopment of depreciated urban areas to reduce average commuting distances and times and to facilitate greater use of mass-transit systems. At the same time, we can make development of efficient mass-transit systems and fuel-efficient vehicles matters of national priority.

The President's Council on Sustainable Development in its 1996 report said, "The federal government should redirect policies that encourage low-density sprawl to foster investment in existing communities. It should encourage shifts in transportation toward transit and expansion of transit options rather than new highway construction." The revolution in information technology, which has spawned telecommuting and greater use of the home office, is an encouraging step in the right direction. These measures will do much to move our society toward sustainability.

(a)

(b)

▲ FIGURE 22–24 *Electric cars.* (a) GM's Impact electric car. (b) Toyota's Prius, a hybrid electric vehicle.

ENVIRONMENT ON THE WEB

THE COSTS OF GROUND LEVEL OZONE

Chapter 21 emphasized the important role of stratospheric ozone (O^3) as a shield against ultraviolet radiation. However, when ozone occurs at ground level, it is much less welcome, damaging valuable crops and affecting human health. Currently, ozone accounts for $1 to $2 billion in crop damage every year in the United States, and tens of millions of dollars a year in Canada.

Tropospheric (ground-level) ozone—sometimes known as photochemical smog—is formed when nitrogen compounds in the atmosphere react with volatile organic compounds (Fig. 22–15). Ozone is most likely to be formed in valleys and low-lying basins when winds are calm and warm, and sunny conditions prevail. As a result, it is usually found at highest concentrations during the summer growing season. Although the precursors of ozone arise primarily in urban areas, drifting air masses can carry ozone far into rural and wilderness areas.

Tropospheric ozone damages crops by affecting vegetation in two ways. First, as a powerful oxidant, it causes thinning and discoloration of leaf tissues. Once the leaf is weakened, it becomes more vulnerable to insect pests and disease. Ozone damage can also promote premature tissue aging, leaf loss, and reduced pollen and seed production. Available evidence suggests that ozone can reduce plant growth by up to 40%. Consequently, ozone's second impact is reduced plant growth.

Overall, a farmer with ozone-damaged crops can expect up to a 20% loss in yields from the affected area. In some regions, ozone damage now outweighs pests and disease as the most serious threat to farm income. This level of damage is not surprising since ozone concentrations of 70 to 200 parts per billion (ppb) are commonly found downwind of major cities—concentrations as low as 40 ppb are enough to injure plants.

Through its tissue-damaging effects, ozone also endangers valuable timber stands and fragile wilderness ecosystems. As a component of urban smog, ozone impairs the aesthetics of those systems and creates secondary impacts on urban and wilderness habitats. Such damage is already apparent in urban trees and in parks downwind of major cities around the world.

However, the human health impacts of ozone are also significant. Ozone is a major cause of human respiratory distress and asthma (especially in children), causing such respiratory illnesses in as many as a million individuals each year. As a result, environmental managers need a way of tracking and reporting air quality in a way that is easily understood by the public and by decision makers. In the case of ozone, for instance, regulators need to alert the public when smog levels have reached a point that is likely to cause human health problems or endanger valuable agricultural crops. To do this, agencies use simple language and a few key "indicator" parameters to convey air quality in easily understood terms.

Web Explorations

The Environment on the Web activity for this essay discusses some of the challenges in choosing, and using, suitable indicators of air quality. Go to the Environment on the Web activity (select Chapter 22 at http://www.prenhall.com/nebel) and learn for yourself:

1. how air quality indicators are used to monitor urban pollution;
2. how pollution results can vary between cities within a region; and
3. how monitoring a site location can influence pollution monitoring results.

A suggested time frame for each of these exercises is 10–30 minutes.

REVIEW QUESTIONS

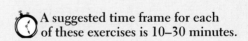

1. Give three examples of both natural and anthropogenic air pollutants.
2. What naturally occurring cleanser helps to remove pollutants from the atmosphere? What molecule (that we studied in Chapter 21) is the major source for this cleanser?
3. What is a pollutant threshold level?
4. What three factors determine the level of air pollution?
5. Describe the origin of industrial and photochemical smog. What are the differences in cause and appearance of each?
6. What are the major air pollutants and their most serious effects?
7. What impact does air pollution have on human health? Give the three categories of impact and distinguish between them.
8. Describe the negative effects of pollutants on crops, forests, and other materials. Which pollutants are mainly responsible for these effects?
9. Where do primary pollutants originate? How do they form secondary pollutants?
10. What is the difference between an acid and a base? What is the pH scale? The pOH scale? How are they related?
11. What two major acids are involved in acid deposition? Where does each come from?
12. How can a shift in environmental pH affect aquatic ecosystems? In what other ecosystems can acid deposition be observed? What are its effects?
13. Discuss ways in which the Clean Air Act of 1990 addresses the failures of former legislation.
14. What technological and political changes are taking place in the United States to reduce acid deposition?
15. What can we do in the future—as individuals and as a nation—to further improve air quality?

THINKING ENVIRONMENTALLY

1. Motor vehicles release close to half the pollutants that dirty our air. What alternatives might be introduced to encourage a decrease in our use of automobiles and other vehicles with internal combustion engines?

2. How might traditional patterns of life change if cleaner air is not achieved? Write a short essay describing the possibilities.

3. What arguments have the utility industries presented to delay action on acid deposition? Evaluate the validity of their concerns.

4. The Swedish government has made efforts to attack the symptoms of acid deposition directly by trying to neutralize acidic lakes with lime. Discuss the pros and cons of this method.

MAKING A DIFFERENCE

PART FIVE: Chapters 16, 17, 18, 19, 20, 21, 22

1. If you are a smoker, enlist in a stop smoking program. Insist that others do not smoke in your presence or in your home or workplace.

2. Monitor the hazards in your own life and take steps to reduce behavior ranked as risky by the experts. Be discriminating as you hear about hazards and risks. Especially be aware of the media's tendency to exploit the outrage element of risks.

3. Survey the lakes and ponds in your area and determine whether cultural eutrophication is occurring. If you find cases, follow up by consulting a local environment group and get involved in combating the problem.

4. Use phosphate-free detergents for all washing. Information will be printed on the container (often in very fine print!).

5. Explore sewage treatment in your community. To what extent is the sewage treated? Where does the effluent go? The sludge? Are there problems, alternatives that are ecologically more sound? Consult local environment groups for help.

6. Investigate how municipal solid waste is handled in your school or community and what the future plans are for waste management. Promote curbside recycling if it is not happening, and investigate the possibility of adopting a system that charges for each container of unrecycled waste.

7. Consider your consumption patterns. Live as simply as possible so as to minimize your impact on the environment; buy and use durable products, and minimize your use of disposables.

8. Read labels and become informed about the potentially hazardous materials you use in the home and workplace. Wherever possible, use nontoxic substitutes. If toxic substances are used, ensure that they are disposed of properly.

9. Encourage your community to set up a hazardous waste collection center where people can take leftover hazardous materials.

10. Reduce your contribution to air pollution by amending your lifestyle to drive fewer miles: Arrange to live near your workplace, car pool, use public transport, bicycle to work or school, avoid unnecessary trips.

11. Properly maintain all fuel-burning equipment to burn efficiently (oil burner, gas heater, lawnmower, outboard motor, automobile, etc.).

12. Keep informed on how the Clean Air Act of 1990 is being implemented and enforced. If there are attempts to weaken it, protest to your congresspersons. Ask them to promote new standards to increase the average vehicle fuel efficiencies.

13. Be an advocate of energy conservation on all fronts in supporting the steps we can take to combat global climate change. Insist on full U.S. participation in global efforts to curb greenhouse gas emissions like the Kyoto Accord of 1997.

14. Make sure anyone servicing a refrigerator or air-conditioning unit for you will recapture and recycle the CFCs if the system needs to be opened.

WEB REFERENCES

On-line resources for this chapter can be found on the World Wide Web at: **http://www.prenhall.com/nebel**. Click on Chapter 22 on the chapter selector.

PART SIX

Toward a Sustainable Future

How will we shape a sustainable future? In previous chapters we have uncovered many issues that need to be addressed and resolved by human society if we are indeed to make our way to a sustainable relationship with the environment. We are convinced that a key to this mission can be found in environmental public policy and individual lifestyles.

In Chapter 23 we highlight economics and public policy as they relate to the environment. We examine how a nation can call on its wealth to address societal problems and what three factors constitute the wealth of nations: produced assets, natural capital, and human resources. In the last chapter (Chapter 24), we explore one of the most difficult problems faced by every nation today, which is promoting eminently livable and sustainable communities, particularly in the many crowded cities of the developing world. We look at examples of cities moving in a sustainable direction. We also look at some of the work being done by organizations grappling with the question of sustainability. Finally, we look at the vital but difficult task of adopting lifestyles that embrace some of the stewardship values capable of shaping a sustainable future. The outcome is uncertain, but we know enough now to be able to point in a sustainable and stewardly direction.

A street in downtown Nelson, British Columbia, Canada.

Sustainable Communities
and Lifestyles

〰〰〰〰〰〰〰〰

Key Issues and Questions

1. Urban trends since World War II are marked by exurban migration, urban sprawl, urban blight, and a growing dependence on cars. Describe these trends and how they are all related.

2. Urban sprawl is at the root of many environmental problems. List and describe these problems.

3. Urban sprawl is giving way to Smart Growth in many communities. What is being done to reign in urban sprawl?

4. Exurban migration set into motion a vicious cycle of urban decline. Describe the factors involved in this cycle. What is economic exclusion?

5. Many cities around the world are attractive and "livable." What are the characteristics that make a city livable?

6. The U.N.'s Sustainable Cities Program and the work of the U.S. Council on Sustainable Development are working toward establishing sustainable communities. Describe these programs.

7. Environmental decision-making is strongly influenced by values. What are some values that may be employed by people to promote stewardship and sustainability?

8. Changes in lifestyles are vital to sustainability. What are the most important changes? How can people best become involved in working toward building a sustainable society?

Waitakere City (opposite) is the sixth largest city in New Zealand, with a population of 136,000 and a land area of 39,134 hectares (96,660 acres). Waitakere City has been experiencing the typical problems of urban growth: urban sprawl, traffic congestion, and pressure on natural areas within the city. In addition, the city reflects a diverse ethnic makeup—a distinctly European ethnic majority and minorities of several cultural groups, but especially the native Maori. This city formally adopted *Agenda 21* in 1993 and was New Zealand's first eco-city.

Adopting the *Agenda 21* goal to protect the indigenous people and cultures of countries while meeting other goals of sustainable development, Waitakere City focused on Maori principles and values first and then on New Zealand legislation and *Agenda 21* principles to develop the guidelines for the city's efforts. Using the Maori perspective of *tangata whenua*, or "people of the land," which incorporates ecological and stewardship principles into the relationship of people and land,

the city has developed a strategic plan called Greenprint. In this plan the Maori are to be included in decision-making by the city council, and Maori principles are to serve as guidelines in the city's ongoing growth and development. The Greenprint plan makes specific reference to the principles of *taonga* (a community's treasures such as forests, rivers, animals, humans, and the health of the environment) and *kaitiaki* (those selected by a tribe to be stewards of the taonga). With input from the Maori, the city council has developed a Green Network Plan with the objective of "protecting, restoring and enhancing the natural environment across the whole city, while increasing human enjoyment and appreciation of that environment. This goal places human beings within the natural world, as stewards of its natural and spiritual health." The whole community is enlisted in "greening the city": restoring native plants, providing corridors for wildlife, providing access to the parks and reserves that will maintain people's safety (from crime). Because the city's population is growing, Green-

Waitakere City, New Zealand. This city, New Zealand's first eco-city, covers a large area in Aukland, New Zealand. Shown here is one of the more developed areas, called Titirangi.

print will be continually challenged by issues related to urban sprawl, like loss of habitat and the impacts of exotic species and pets on native wildlife. Waitakere City has, however, accepted the challenge of becoming an eco-city and has embarked on a course of action that may well make it a world leader in the development of sustainable cities.

On the other side of the world, with the aid of Isles, Inc., citizens of Trenton, New Jersey, have made remarkable progress in rehabilitating their city. Trenton is the capital of New Jersey, with a multicultural population of 88,600. Isles, Inc., was founded in 1981 by nearby Princeton University personnel but quickly evolved into a locally owned and controlled development organization. Trenton was suffering from the all too common problems of a decaying urban center, with ethnic segregation and urban blight. To date, Isles has fostered the following changes: Vacant land has been transformed into community garden sites now providing $150,000 profits annually; over 90 units of affordable housing have been created, with more than 100 still in progress; numerous unemployed young people have been given job training in connection with the affordable housing projects; environmental education programs have been provided to 7000 students in seven schools; and the city, with a grant from the EPA, is working on

redeveloping four ugly "brownfield" sites (see Chapter 20). Isles, with an annual budget of over $1 million, is now an effective organization that has learned how to work with city and environmental groups to mobilize community action in working together to accomplish the renovation of an aging city. Like Waitakere City, Trenton has established a course of action and has found help in its efforts toward renewal.

This tale of two cities is tied together by a common resurgence of attention to urban living and a desire on the part of city residents to take charge of their own communities and reverse the trends that have been leading to a decline in the livability of their cities. It is significant that the direction they take is toward sustainability. These city residents recognize that things need to change, and they are becoming agents of changes that involve individual lifestyles and institutional redirection. One city is coping with sprawl, and the other is tackling urban blight.

Our objective in this chapter is first to show how these two problems—urban sprawl and urban blight—are related and how they can be remedied. Then we shall consider the global and national efforts toward forging sustainable cities and communities. Finally, we look at some ethical concerns and personal lifestyles as we consider the urgent needs of people and the environment in entering a new millennium.

24.1   Urban Sprawl

The dominant and most environmentally provocative feature of life in the United States is our near-total dependence on cars. We have developed an urban-suburban layout in which the locations where we live, work, go to school, and shop are widely separated. Thus, we have come to rely on private cars, which we drive an average of 200 miles (300 km) a week simply to meet our everyday needs. We shall refer to this lifestyle that revolves around going everywhere by car as a *car-dependent lifestyle*.

This far-flung urban-suburban network of residential areas, shopping malls, industrial parks, and other facilities loosely laced together by multilane highways is referred to as **urban sprawl**. The word *sprawl* is used because the perimeters of the city have simply been extended outward into the countryside, one development after the next, with little plan as to where the expansion is going and no notion as to where it will stop. That the sprawl is continuing is all too evident. Almost everywhere we go near urban areas, we are confronted by farms and natural areas giving way to new developments, new highways being constructed, and old roadways being upgraded and expanded (Fig. 24–1).

The Origins of Urban Sprawl

Until the end of World War II, a relatively small percentage of people owned cars. Cities had developed in ways that allowed people to meet their needs by means of the transportation

available—mainly walking. Every few blocks had a small grocery, a pharmacy, and other stores, as well as professional offices integrated with residences. Often, buildings had a store at the street level and residences above (Fig. 24–2). Schools were scattered throughout the city, as were city parks for outdoor recreation or relaxation. Thus, walking distances were generally short, and bicycling was even more convenient.

For more specialized needs, people boarded public transportation—electric trolleys, cable cars, and buses—from neighborhoods to the "downtown" area, where big department stores, specialty shops, and offices were located. Public transportation did not change the compact structure, because people still needed to walk to the transit line. At the outer ends of the transit lines, cities gave way abruptly to farms, which provided most of the food for the city, and open country. The small towns and villages surrounding cities, the original suburbs, were compact for the same reasons and mainly served farmers in the immediate area. At the end of World War II, this pattern began to change dramatically.

Despite the many advantages of urban life, many people found cities less-than-pleasant places to live. Especially in industrial cities, poor housing, inadequate sewage systems, inadequate refuse collection, pollution from home furnaces and industry, and generally congested, noisy conditions were much the norm. A decrease in services during World War II aggravated those problems. Hence, many people had the desire—the "American Dream"—to live in their own house on their own piece of land away from the city.

◀ **FIGURE 24–1** *New highway construction.* The hallmark of development over the past several decades has been the construction of new highways. New highways, however, have only fed the environmentally damaging car-dependent lifestyle.

The Automobile. Because of the massive production of war materiel there was a shortage of consumer goods during World War II. As a result, when the war ended, there was a pent-up demand for consumer goods. Also, many civilians and returning veterans had accumulated considerable savings during the war. When mass production of cars commenced at the end of the war, people flocked to buy them. With private cars, people were no longer restricted to living within walking distance of their workplaces or of transit lines.

They could move out of their cramped city apartments and into homes of their own outside the city. Their cars would allow them to drive back and forth easily to their jobs, shopping, recreation, and so on.

Developers responded quickly to the demand for private homes. They bought farmlands and natural areas outside the cities and put up houses. The government aided this trend by providing low-interest mortgages through the Veterans Administration and the Federal Housing

◀ **FIGURE 24–2** *The integrated city.* Before the widespread use of automobiles, cities had an integrated structure. A wide variety of small stores and offices on ground floors, with residences on upper floors, placed everyday needs within walking distance.

Administration, and interest payments on mortgages were made tax-deductible (rent continued to be non-tax-deductible). Property taxes in the suburbs were much lower than in the city. These financial factors meant that for the first time in history, making monthly payments on one's own home in the suburbs was cheaper than paying rent for equivalent or less living space in the city.

Thus, the mushrooming development around cities did not proceed according to any plan; rather, it happened wherever developers could acquire land (Fig. 24–3). In fact, planning was more than merely neglected; it was actively prevented by the fact that cities were surrounded by a maze of more or less autonomous local jurisdictions (towns, villages, townships, counties). No governing body existed to devise, much less enforce, an overall plan. Local governments were quickly thrown into a catch-up role of trying to provide schools, sewers, water systems, and other public facilities—and most of all, roads—to accommodate the uncontrolled growth.

Highways. The influx of commuters into previously rural areas soon resulted in traffic congestion, creating a need for new and larger roads. To raise money to build and expand highways, Congress passed the Highway Revenue Act of 1956, which created the **Highway Trust Fund**. This legislation placed a tax on gasoline and earmarked the revenues to be used exclusively for building new roads.

▲ **FIGURE 24–3** *Urban sprawl.* Prime agricultural land is being sacrificed for development around most U.S. cities and towns. Location of developments is determined more by the availability of land than by any overall urban planning.

It is evident how the Highway Trust Fund perpetuates development. A new highway not only alleviates existing congestion, it also encourages development at farther locations because time, not distance, is the limiting factor for commuters. (When people describe how far they live from work, it is almost always put in terms of minutes, not distance.) The average person is willing to spend 20 to 40 minutes each way on a daily commute. Given this time limit, people who have to walk would have to live within 1 or 2 miles (1.5–3.0 km) of their work. With cars and expressways, people can live 20 to 40 miles (30–60 km) from work and still get there in the same amount of time.

Therefore, new highways that were intended to reduce congestion actually fostered development of open land and commuting by more drivers from distant locations (Fig. 24–4). Soon traffic conditions became as congested as ever. Average commuting *distance* has doubled since 1960, but average commuting *time* has remained about the same. The increase in commuting distance, however, requires more fuel, which generates more money for the Highway Trust Fund. Thus, the whole process is repeated in a continuing cycle (Fig. 24–5). As a result, government policy, far from curtailing urban sprawl, became a party to supporting and promoting it.

Residential developments are followed (or sometimes led) by shopping malls and industrial parks and office complexes. These commercial centers are often situated in such a way that the only possible access is by car; this has only changed the direction of commuting. Whereas in the early days of suburban sprawl the major traffic flow was into and out of cities, it is now between suburban centers. Multilane highways connecting suburban centers are perpetually congested with traffic going in *both* directions.

In broad perspective, then, urban sprawl is a process of **exurban migration**—that is, a relocation of residences, shopping areas, and workplaces from their traditional spots in the city to outlying areas. Population growth, although it has occurred, has played a relatively minor part in the development of urban sprawl. The populations of many U.S. cities, excluding the suburbs, have actually declined in the last several decades as a result of exurban migration (Table 24-1). Exurban migration is continuing in a leapfrog fashion as people from older suburbs move to **exurbs** (communities farther from cities than suburbs).

The "love affair" with cars is not just a U.S. phenomenon. Around the world, in both developed and developing countries, people aspire to own cars and adopt the car-dependent lifestyle. Consequently, urban sprawl is occurring around many developing world cities as people become affluent enough to own cars. Love of the car-dependent lifestyle, however, is not sufficient to make it sustainable. In fact, this lifestyle is at the crux of a number of the unsustainable trends described in earlier chapters.

Environmental Impacts of Urban Sprawl

The environmental impacts of urban sprawl can be put into the following categories:

◀ **FIGURE 24–4** *The new highway-traffic congestion cycle.* (a) Growing traffic congestion creates the need for new and upgraded highways. (b) However, upgraded highways encourage development at more distant locations, thus creating more traffic congestion and hence the need for more highways.

(a)

(b)

- depletion of energy resources
- air pollution
- water pollution and degradation of water resources
- loss of landscapes and wildlife
- loss of agricultural land

Depletion of Energy Resources. Shifting to a car-dependent lifestyle has entailed an ever increasing demand for energy resources. From 1945 to 1980, U.S. oil consumption nearly quadrupled while population grew by just 60%, a nearly threefold increase in per capita consumption. In addition,

individual suburban homes require 1.5 to 2 times more energy for heating and cooling than comparable attached city dwellings. We have amply described the costs and risks of becoming increasingly dependent on foreign sources of oil for fuel (see Chapter 13).

Air Pollution. Despite improvements in pollution control, many cities still fail to meet desired air-quality standards (see Chapter 22). Vehicles are responsible for an estimated 80% of the air pollution in metropolitan regions. Likewise, the threefold increase in per capita energy consumption exacerbates the potential for global warming as burning fossil fuels

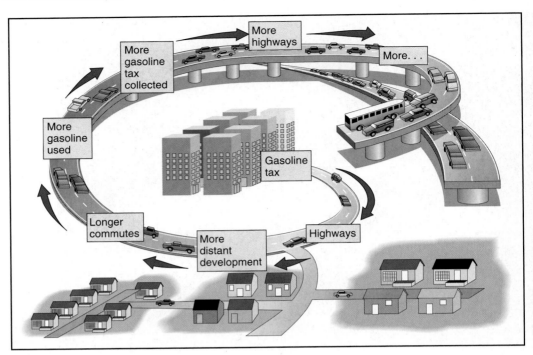

▶ **FIGURE 24–5** *The development cycle spawned by the Highway Trust Fund.*

produce carbon dioxide. Automobiles likewise are implicated in depletion of the stratospheric ozone shield because a major source of CFCs entering the atmosphere is that which escapes from vehicle air conditioners (see Chapter 21).

Water Pollution and Degradation of Water Resources. All the highways, parking lots, driveways, and other paved areas associated with urban sprawl lead to a substantial increase in runoff, as well as decreased infiltration over large regions. Even suburban lawns increase runoff because they are more compacted than the original soils. Recall the consequences of increasing runoff—increased flooding, streambank erosion, and decreased water quality, to name a few—and the consequences of decreasing infiltration—depletion of groundwater, drying of springs and waterways, saltwater encroachment, and others (see Chapter 9). Also, water quality

TABLE 24-1 Decline in Population of American Cities 1950–1996

City	Population (thousands) 1950	1970	1996	Percent Change 1950–1996
Baltimore, MD	950	905	675	-29
Boston, MA	801	641	558	-30
Buffalo, NY	580	463	311	-46
Cleveland, OH	915	751	498	-46
Detroit, MI	1850	1514	1000	-46
Louisville, KY	369	362	261	-29
Minneapolis, MN	522	434	359	-31
Philadelphia, PA	2072	1949	1478	-29
St. Louis, MO	857	662	352	-59
Washington, D.C.	802	757	543	-32

(Source: Data from Statistical Abstracts of the United States, U.S. Bureau of the Census, 1998.)

is degraded by the runoff of fertilizer, chemicals, crankcase oil (the oil that drips from engines), pet droppings, and so on.

Loss of Landscapes and Wildlife. New highways are frequently routed through parklands or along stream valleys because such open areas provide the least expensive rights of way. The result is the sacrifice of aesthetic, recreational, and wildlife values in the very places where they are most important: metropolitan areas. The excess individuals of every species displaced by development are inevitably assigned to death even if similar habitat exists, because nearby ecosystems are inevitably already occupied. Of course, if no similar habitat exists, the result is self-evident.

Furthermore, many species of wildlife need certain minimum unbroken areas to maintain viable populations, and unbroken "lanes" of habitat through which to migrate. Wildlife biologists are finding marked declines in countless species ranging from birds to amphibians as highways and new developments fragment natural areas. Also, expanded and improved highways lead to increasing numbers of roadkills. Today, much more wildlife is killed by vehicles than by hunters.

Loss of Agricultural Land. Finally, and perhaps most serious in the long run, is the loss of prime agricultural land. In the United States alone, sprawling development has eaten up over 4,250,000 acres of prime agricultural land since 1982. For the time being, it is considered that the United States has adequate agricultural land and can tolerate this loss. However, most of the food for cities used to be locally grown on small, diversified family farms surrounding the city. With most of these farms turned into housing developments, it is estimated that food now travels an average of 1000 miles (1500 km) from where it is produced—mostly on huge commercial farms—to where it is eaten. The loss is not just the locally grown produce but also the social interactions and ties with the farm community.

The loss of agricultural land to development is a worldwide phenomenon, and in countries having no land to spare, the consequences are becoming significant. For example, China became food-independent with the Green Revolution. However, with its push toward modernization and industrialization, it is experiencing mounting losses of prime agricultural land to development. This loss of agricultural land, coupled with growing population and a growing demand for meat, is making China dependent on imports again.

These environmental impacts resulting from or exacerbated by urban sprawl and development are summarized in Fig. 24–6.

Reining In Urban Sprawl: Smart Growth

The obvious way to curtail urban sprawl might appear to be to simply pass laws that preclude development of exurban properties. Indeed, a number of European countries and Japan do have such laws. However, laws cannot be passed without the support of the public. In the United States the strong sentiment toward the right of a property owner to develop property as he or she feels fit has made such restrictions, with few exceptions, politically impossible to pass or financially impossible to implement.

Yet, many outdated laws and regulations that exacerbate urban sprawl might well be changed. Examples include archaic zoning laws that require such things as large lot sizes, detached housing, and the separation of residential and commercial or industrial areas. Most such laws were passed by

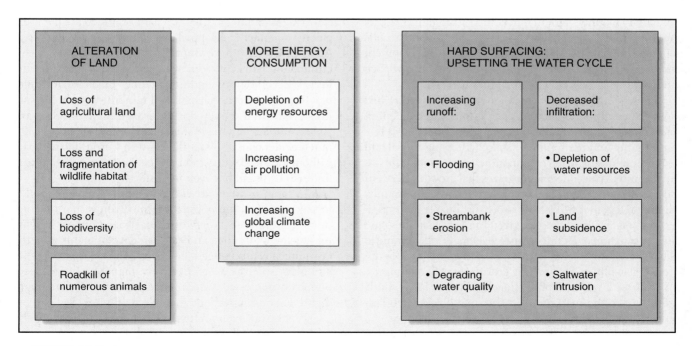

▲ **FIGURE 24–6** *A summary of the environmental impacts of urban sprawl.*

pressures from the first exurban migrants in the 1940s and 1950s as a means of preventing "undesirable elements" from invading the new exurban neighborhoods. By demanding such separation and low-density residential areas, however, these regulations now assure a car-dependent lifestyle. Mass transit (bus or rail lines), a much promoted idea, is unworkable where residential densities are low because there are not enough riders between any two locations to make it viable. It makes no sense, economically or ecologically, to run a 40-passenger bus with only a few passengers.

Furthermore, the requirement for separation of different functions led architects and developers themselves to focus on specialized aspects—homes, shopping malls, industrial parks—while government transportation departments focused on providing highways. Lost in the process was any sense of creating integrated communities for people.

As reported in the Worldwatch Institute's study, titled "Shaping Cities: The Environmental and Human Dimensions," a key to developing sustainable communities is changing zoning laws to allow stores, light industries, professional offices, and higher-density housing to be integrated. A second point stressed is the need to provide affordable housing around industrial parks and shopping malls for the people who work and shop in those places. Such housing need not be unattractive or lacking in privacy. A third point emphasized is the need to provide safe bicycle and pedestrian access between residential areas and workplaces. It is assumed that employers will provide a convenient, safe place to keep a bicycle; at present, they seldom do, although they do spend a great deal of money providing parking spaces for automobiles. These steps would allow more people to walk or bicycle to their jobs and would reduce the space needed for highways and parking. The effectiveness of these steps in reining in urban sprawl is illustrated by the following example. Fairfax County, Virginia, next to Washington, D.C., is now almost 100% covered by sprawl-type development. It is estimated that if Fairfax had *allowed* the kind of development described above, the county would still be 70% open space and would be far less congested and less polluted.

There is considerable indication that what is prescribed above is beginning to occur. For example, the concept of *Smart Growth* has recently emerged. This concept involves communities and metropolitan areas addressing sprawl and purposely choosing to develop in more environmentally sustainable ways. The concept recognizes that growth will occur and focuses on economic, environmental, and community values that together will lead to more sustainable communities. Zoning laws are being changed in many localities, and a new generation of architects and developers is beginning to focus on creating integrated communities as opposed to disassociated facilities. By directing growth to particular places, Smart Growth provides protection to sensitive lands. Early indications in the marketplace—the final determining factor—are that the new communities are a great success.

In addition, there have been repeated attempts to break the cycle created by the Highway Trust Fund by allowing revenues to be used for purposes other than construction of more highways. The first significant inroad was made in 1991 with the passage of the Intermodal Surface Transportation Efficiency Act (ISTEA), a bill environmentalists had been promoting for many years. Under this act almost half of the money levied by the Highway Trust Fund is now eligible for use on other modes of transportation, including cycling and walking as well as mass-transit facilities. The legislation also requires states to hire bicycle and pedestrian coordinators to oversee programs and to establish long-range pedestrian-bicycle plans. The act also requires that states spend a total of $3 billion on "transportation enhancements," which include pedestrian and bicycle facilities. The objectives of ISTEA were significantly reinforced and extended with passage of the Transportation Equity Act for the Twenty-first Century (TEA 21) in 1998. TEA 21 authorizes a total of $218 billion for federal highway and mass-transit programs for the next five years and provides a major increase over ISTEA funding for programs that address the environmental impacts of surface transportation.

24.2 Urban Blight

Now let us take a look at the other end of the exurban migration, the city from which people are moving. Here we find that exurban migration is the major factor underlying the *urban decay*, or *blight*, that has occurred over the last 50 years.

Economic and Ethnic Segregation

Historically, U.S. cities have included people with a wide diversity of economic and ethnic backgrounds. However, moving to the suburbs required some degree of affluence—at least the ability to manage a down payment and mortgage on a home and the ability to buy and drive a car. Therefore, exurban migration, for the most part, excluded the poor, the elderly, and the handicapped. The poor were largely African Americans and other minorities because a history of discrimination had kept them from education and well-paying jobs. Discriminatory lending practices by banks and sales practices by real estate agents kept minorities out of the suburbs even when money was not a factor. Civil rights laws passed in the 1960s made such practices illegal, but there is much evidence that they still persist in various forms. Even without the racial segregation, the economic segregation continues to exist and has even intensified.

In short, exurban migration and urban sprawl have led to segregation of the population into groups sharing common economic, social, and cultural backgrounds. Moving into the new suburban and exurban developments were the economically advantaged, whereas many areas of the cities and older suburbs were effectively abandoned to the economically depressed, who in large part were ethnic minorities, the handicapped, and the elderly (Fig. 24–7).

People moving from rural areas to urban regions are also split according to their economic and ethnic status. Likewise, new immigrants are segregated along the same lines.

(a)

(b)

◀ **FIGURE 24–7** *Segregation by exurban migration.* Exurban migration, the driving force behind urban sprawl, has also led to the segregation of the population along economic and racial lines. In (a) an area of suburbia and (b) an area of inner-city Baltimore, you see the contrast.

The Vicious Cycle of Urban Blight

Affluent people moving to the suburbs set into motion a vicious cycle of exurban migration and urban blight, which still continues. To understand how this downward cycle occurs, we must note several points concerning local governments. First, local governments (usually city or county, though the particular entities vary from state to state) are responsible for providing public schools, maintenance of local roads, police and fire protection, refuse collection and disposal, public water and sewers, welfare services, libraries, and local parks. Second, a local government's major source of revenue to pay for these services is local property taxes, a

yearly tax proportional to the market value of the property and home or other buildings on it. If property values increase or decrease, the property taxes are adjusted accordingly. Third, in most cases, the central city is a governmental jurisdiction separate from the surrounding suburbs.

Because of these three characteristics of local government, exurban migration has the following consequences: By the economics of supply and demand, property values in the suburbs escalate with the influx of affluent newcomers. Thus, suburban jurisdictions enjoy increasing tax revenue with which they can improve and expand local services. At the same time, property values in the city decline because of decreasing demand. Also, property taxes create a powerful disincentive toward maintaining property. Often, landlords allow property to deteriorate to reduce their taxes while keeping rents high to maximize their income. Many properties end up being abandoned by the owner for nonpayment of taxes. The city "inherits" such abandoned properties, but they are a liability for the city rather than a source of revenue (Fig. 24–8). The declining tax revenue resulting from declining property values is referred to as an **eroding**, or **declining**, **tax base** and it has been a serious handicap for most U.S. cities since the exurban migration started in the late 1940s. Adding to the problem is the fact that many of those remaining in the city are disadvantaged people requiring public assistance of one sort or another. Thus, cities bear a disproportionate burden of welfare obligations as well as the declining tax revenues.

The eroding tax base forces city governments to cut local services or increase the tax rate or, generally, both. Hence, the property tax on a home in a city is often two to three times greater than on a comparably priced home in the suburbs, a difference that amounts to $2000 to $3000 per year in the tax bill on an average-priced home. At the same time, schools, refuse collection, street repair, libraries, parks and other services and facilities deteriorate from neglect. This situation of increasing taxes and deteriorating services causes more people to leave the city and become part of urban sprawl even if the car-dependent lifestyle is not their primary choice. However, it is still only the relatively more affluent individuals who can make this move and thus the whole process of exurban migration, declining tax base, and deteriorating real estate, worsening schools and other services and facilities is perpetuated in a vicious cycle

(Fig. 24–9). This downward spiral of conditions is referred to as **urban blight** or **urban decay**.

Economic Exclusion of the Inner City

The exurban migration includes more than just individuals and families. The exiting of affluent people from the city removes the purchasing power necessary to support stores, professionals, and other businesses. Faced with declining business, these enterprises are forced either to go out of business or to move to the new locations in the suburbs. Either way, vacant storefronts are the result, and people remaining in the city lose convenient access to goods, services, and, most important, *jobs*, because each business also represents employment opportunities.

Unemployment rates run at 50% or more in depressed areas of inner cities and what jobs do exist are mostly at the minimum wage. The new jobs being created are in the shopping malls and new industrial parks in the outlying areas that are largely inaccessible for lack of public transportation. Therefore, people remaining in the inner city are not only poor from the outset, but the cycle of exurban migration and urban decay has now led to their **economic exclusion** from the mainstream. As a result, drug dealing, crime, violence, and other forms of deviant social behavior are widespread and worsening in such areas.

To be sure, in the last two decades, many cities have redeveloped core areas. Shops, restaurants, hotels, convention and entertainment centers, office buildings, residences, and places for walking and relaxing are all combining to bring a new spirit of life and hope, as well as the practical assets of

▶ **FIGURE 24–8** *Abandoned buildings.* Many cities' ills are because of too *little* population, not too *much*. When the exurban migration causes population to drop below a certain level, businesses are no longer supported. The result is abandonment, the bane of urban blight. Shown here is a section of Baltimore only a few blocks from the redeveloped Inner Harbor shown in Fig. 24–10.

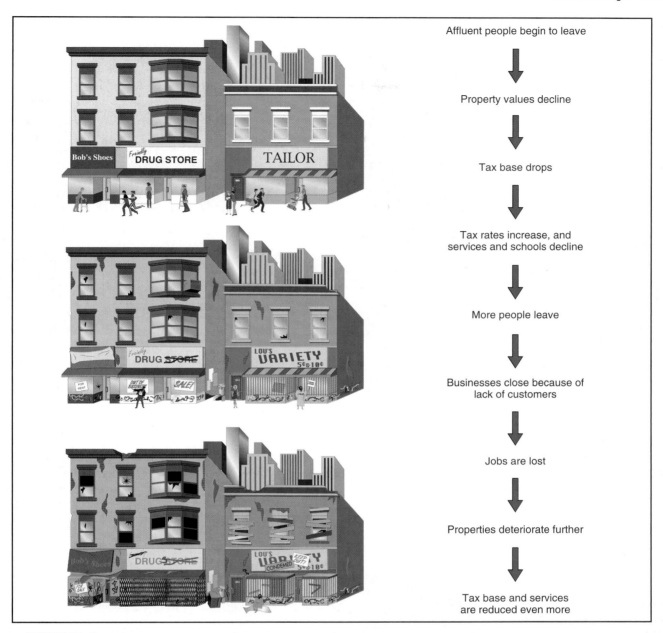

Affluent people begin to leave

Property values decline

Tax base drops

Tax rates increase, and services and schools decline

More people leave

Businesses close because of lack of customers

Jobs are lost

Properties deteriorate further

Tax base and services are reduced even more

▲ FIGURE 24–9 *The process of urban decay.* The vicious cycle of exurban migration and urban decay.

taxes and employment, back into cities (Fig. 24–10). This is an encouraging new trend, so far as it goes. However, walk a few blocks in any direction from the sparkling, revitalized core of most U.S. cities and you will find urban blight continuing unabated.

What Makes Cities Livable?

More and more people in all walks of life are recognizing that the environmental consequences of urban sprawl and the social consequences of urban blight are two sides of the same coin. There is increasing awareness that a sustainable future will depend on both reining in urban sprawl and revitalizing cities. The only possible way to sustain the global population is by having viable resource-efficient cities, leaving the countryside for agriculture and natural ecosystems. The key word is *viable*, which means livable. No one wants to or should be required to live in the conditions that have come to typify urban blight.

Livability is a general index based on people's response to the question, "Do you like living here, or would you rather live somewhere else?" Crime, pollution, recreational, cultural, and professional opportunities, and many other social and environmental factors are summed up in the subjective answer.

Though many people assume that the social ills of the city are an outcome of high population densities, crime rates and other problems in U.S. cities have climbed while

▶ **FIGURE 24–10** *Inner Harbor, Baltimore.* High population densities help *make* a city, not destroy it. Redeveloping key areas of cities with heterogeneous mixtures of office buildings, shopping, restaurants, residences, and recreational facilities, all of which attract people, has helped to bring new life back into cities.

populations dwindled as a result of the exurban migration. As reported in the Worldwatch Institute's "Shaping Cities":

> There is no scientific evidence of a link between population density per se and social ills….Copenhagen and Vienna—two cities widely associated with urban charm and livability—each have relatively high "activity density" (the number of residents and jobs in the city), 47 and 72 people per hectare (1 hectare = 2.5 acres) respectively. By contrast, low-density cities such as Phoenix (13 people per hectare) often are dominated by unwelcoming, car-oriented commercial strips and vast expanses of concrete and asphalt.

Renowned city planner William Whyte goes even further. His studies show that wherever urban population density drops below a certain point, there are no longer enough people to support restaurants, department stores, and so on, and the area "dies" as a result.

Looking at livable cities around the world, we find that the common denominators are (1) maintaining a high population density; (2) preserving a heterogeneity of residences, businesses, stores, and shops; and (3) keeping layouts on a human dimension, where people can incidentally meet, visit, or conduct business over coffee at a sidewalk café, or stroll on

▶ **FIGURE 24–11** *Livable cities.* The key to livable cities is having a heterogeneity of residences, businesses, and stores, and in keeping layouts on a human dimension where people can incidentally meet or conduct business over coffee in a sidewalk café or stroll on a promenade through an open area. In short, the space is designed for and devoted to people. Shown here is Quincy Market, Boston.

◀ **FIGURE 24–12** *Car-centered cities.* In contrast to what is required for livability, city development of the last several decades has focused on moving and parking cars and creating homogeneity. Note the sterility of parking lots surrounding buildings in the foreground of this photo of Los Angeles.

a promenade through an open area. In short, the space is designed for and devoted to people (Fig. 24–11). In contrast, development of the past 50 years has focused on accommodating automobiles and traffic. Two-thirds of the land in cities that have grown up in the era of the automobile is devoted to moving, parking, or servicing cars, and such space is essentially alien to the human psyche (Fig. 24–12). William Whyte remarked, "It is difficult to design space that will not attract people. What is remarkable is how often this has been accomplished."

The world's most livable cities are not those with "perfect" auto access between all points; they are ones that have taken measures to reduce outward sprawl, reduce automobile traffic, and improve access by foot and bicycle in conjunction with mass transit. For example, Geneva, Switzerland, prohibits automobile parking at workplaces in the city's center, forcing commuters to use the excellent public transportation system. Copenhagen bans all on-street parking in the downtown core. Paris has removed 200,000 parking places in the downtown area. Curitiba, Brazil, with a population of 1.5 million, is cited as the most livable city in all of Latin America. The achievement of Curitiba is due almost entirely to the efforts of Jaime Lerner, who, serving as mayor since the early 1970s, has guided development in terms of mass transit rather than cars. The space saved by not building highways and parking lots has been put into parks and shady walkways, causing green area per inhabitant to increase from 4.5 square feet (0.5 m^2) in 1970 to 450 square feet (50 m^2) today.

In Tokyo, millions of people ride bicycles (Fig. 24–13), either all the way to work or to stations from which they catch fast, efficient subways or the "bullet train" to their destination. By sharply restricting development outside certain city limits, Japan has maintained population densities within cities and along metropolitan corridors that ensure the viability of commuter trains. Japan's cities have maintained a heterogeneous

urban structure that mixes small shops, professional offices, and residences in such a way that a large portion of the population meets their needs without cars. In maintaining an economically active city, it is probably no coincidence that street crime, vagrancy, and begging are virtually unknown in the vast expanse of Tokyo despite the seeming congestion.

▲ **FIGURE 24–13** *Bicycles in Tokyo.* Most people in Tokyo do not own automobiles but ride bicycles or walk to stations, from which they take fast, efficient, inexpensive subways to reach their destinations.

▶ **FIGURE 24–14** *Portland, Oregon.* By reducing traffic, Portland, Oregon, was able to convert a former expressway and a huge parking lot into the now-renowned Tom McCall Waterfront Park. Due to such planning, Portland is now ranked among the world's most livable cities.

Portland, Oregon, is one of the few U.S. cities that have taken more than token steps to curtail automobile use. The first step was to encircle the city with an urban growth boundary, a line outside of which new development was prohibited. Thus, compact growth rather than sprawl was ensured. Second, an efficient light rail and bus system was built (see the "Earth Watch" essay, p. 544), which now carries 43% of all commuters to downtown jobs. (In most U.S. cities, only 10% to 25% of commuters ride public transit systems.) By reducing traffic, Portland was able to convert a former expressway and huge parking lot into the now-renowned Tom McCall Waterfront Park (Fig. 24–14). Portland is now ranked among the world's most livable cities.

Livable Equals Sustainable. The same factors that underlie livability also lead to sustainability. The reduction of auto traffic and greater reliance on foot and public transportation reduce energy consumption and pollution. Urban heterogeneity can facilitate recycling of materials. Housing can be retrofitted with passive solar heating of space and hot water. Use of landscaping can provide cooling, as described in Chapter 15. A number of cities are developing vacant or cleared areas into garden plots (Fig. 24–15), and rooftop hydroponic gardens are becoming popular. Such gardens will not make cities agriculturally self-sufficient, of course, but they add to livability, provide an avenue for recycling compost and nutrients removed from sewage, give a source of fresh vegetables, and have the potential of providing income for many unskilled workers (see the "Earth Watch" essay, p. 591). If urban sprawl is curbed, relatively close-in farms could provide most of the remaining food needs for the city.

This discussion of urban issues may seem to have strayed far from nature and environmental science. However, there is a close connection—the decay of our cities is hastening the breakdown of our larger environment. As the growing human population spreads outward from the old cities, we are encroaching on all other life. Thus, without the creation of sustainable human communities, there is little chance for sustainability of the rest of the biosphere.

By making urban areas more appealing and economically stable, we can not only improve the lives of those who choose to remain or must remain in cities but also spare surrounding areas. Our parks, wildernesses, and farms will not be replaced by exurbs and shopping malls but saved for future

▲ **FIGURE 24–15** *Urban gardens.* Urban garden plots on formerly vacant lots in Philadelphia. Urban gardening is becoming recognized as having sociological, economic, environmental, and aesthetic benefits.

EARTH WATCH

ROOFTOP GARDENS FOR LIVABILITY AND FOOD

For years, workers at ECHO (Educational Concerns for Hunger Organization), a nonprofit organization helping to meet food needs of the poor, have been developing techniques for growing vegetables on rooftops at ECHO's model facilities in Fort Myers, Florida. Even though it would be possible to use hydroponic technology to accomplish this, ECHO staff decided to avoid traditional hydroponics because of the dependence on water or air pumps to keep roots aerated—technology that could be expensive and energy-dependent. Instead, they found that they could grow luxuriant vegetables using a variety of methods that were adapted to local conditions.

ECHO workers found that they could grow almost any vegetable in a shallow (3- to 4-inch) layer of compost or slightly decomposed refuse. No soil was used, and the beds were kept shallow to keep weight down

(rooftops are often too weak to support heavy weights). In one developing country, application of this technology—on a school and hospital rooftop in Port-au-Prince, Haiti—were utilized for raising high-nutrient crops to feed students and patients. Water and nutrients were fed to the vegetables with a wick made from a sheet of polyester cloth. The cloth was spread out on the cement roof, vegetable plants were spaced out on the cloth, and a few inches of pine needles were used to fill in around the plants for lightweight support and protection against drying. A 5-gallon bucket was then filled with a complete, soluble fertilizer, a 3/8-inch hole was drilled in the lid, and the bucket was turned upside down on a clear section of the cloth. Water and nutrients simply wicked out to the vegetables through the cloth. The workers were even able to raise field corn with this system!

Another popular method uses old automobile tires, which are all too common

almost everywhere. The tire garden consists of tires, plastic garbage bags, and soil or compost. To prepare a tire for planting, the rim is cut on the top side of the tire, a plastic sheet is placed in the inside bottom, and the cut piece from the rim is turned upside down and fit into the bottom rim, holding the plastic firmly in place. Several inches of soil, seeds, water, and fertilizer complete the "garden." Tire gardens are easily moved to maximize shade or sun exposure or to prevent theft of the produce.

The lessons to be learned are straightforward: Plants really need light, water, and nutrients, but, surprisingly, little soil. These needs can be met in virtually any developing-world city, by the literally acres of flat "land" in the form of rooftops in the city. This kind of urban "farming" also has the potential to rehabilitate cities in the industrialized countries, giving people the opportunity to raise some of their own food and enjoy living in the company of green plants.

---

generations. In the last decade, some remarkable developments have begun to accelerate the movement toward building sustainable communities, and we turn to them now.

24.3 Moving Toward Sustainable Communities

We noted in Chapter 1 that the 1992 U.N. Conference on Environment and Development (UNCED) created the Commission on Sustainable Development. This new body was given the responsibility of monitoring and reporting on the implementation of the agreements that were made at UNCED, in particular, those that related to sustainability. According to *Agenda 21*, one of the agreements signed at UNCED, countries agreed to work on sustainable development strategies and action plans that put the country on a track toward sustainability. Some 100 countries have now developed National Environmental Action Plans or sustainable development strategies that lay out public policy priorities. Most of these countries are now actively engaged in transforming the policies into action plans, and we will investigate two of the outcomes of these activities: the Sustainable Cities Program and the work of the U.S. Council on Sustainable Development. We will also see, however, that many of the objectives of these programs have already been

adopted in the United States in an ongoing movement called the "Sustainable Communities Movement."

Sustainable Cities

A joint facility of the U.N. Environment Program and the U.N. Center for Human Settlements, the Sustainable Cities Program (SCP) is designed to foster the planning and management needed to move cities in the developing countries toward the objective of sustainability. The program defines a sustainable city as one "where achievements in social, economic and physical development are made to last." The cities are especially in focus because they are absorbing two-thirds of population growth in the developing countries and in the process are experiencing serious environmental degradation in and around the growing urban centers. The program is viewed as a "capacity building program," meaning that it is intended to help cities and countries mobilize resources and capabilities mostly available within the cities by building management capacity. The approaches fostered by the program are thus "bottom up" in nature, calling for the involvement of people at all economic and social levels and reconciling their interests where conflicts are evident. Common principles seen in sustainable development theory are put into practice—principles like social equity, economic efficiency, and environmental planning and management. Working groups specific to unique city issues are the core element in the program.

A number of SCP cities have been identified and are working on implementing the program, including Madras (India), Dar es Salaam (Tanzania), Accra (Ghana), Shenyang (China), Concepción (Chile), and many others (Fig. 24–16). The process involves a preparatory phase, moving then into demonstration activities at the level of the whole city, and then extending to replications in other cities of the country. The following are the processes in focus at the outset of a city's involvement:

- clarification of environmental issues
- agreement on joint strategies
- coordination of action plans
- implementation of technical support and capital investment programs
- establishment of a continuing environmental planning and management process

▼ **FIGURE 24–16** *Dar es Salaam, Tanzania.* The first Sustainable Cities Program demonstration city, Dar es Salaam is growing at a rate of 8% per year and is struggling to cope with deteriorating environmental conditions cause by its rapid growth.

Funding for the demonstration phase comes primarily from the city itself, but additional aid often comes from partnership with other countries. Denmark, Canada, France, the United States, and Italy have provided support and expertise at this level of involvement. The World Bank and other U.N. agencies are often participants in the SCP activities; indeed, the program's approach is quite conducive to collaboration between NGOs and other countries and the host country and city. Although this program is in its early stages, it is a very promising approach to sustainable development in exactly the places where it is most needed.

Sustainable Communities

Revitalizing urban economies and rehabilitating cities requires coordinated efforts on the part of all sectors of society. In fact, such efforts are beginning to occur. What is termed a "sustainable communities movement" is taking root in cities around the United States. Chattanooga, Tennessee, is now regarded as a prototype of what can occur with such a movement.

Twenty-five years ago Chattanooga, which straddles the Tennessee River, was a decaying industrial city with high levels of pollution. Employment was falling, and residents were fleeing to the suburbs, leaving abandoned properties and increasing crime. Then, **Chattanooga Venture**, a nonprofit organization, launched *Vision 2000*, the first step of which was to bring people from all walks of life together to build a consensus about what the city *could* be like. Literally thousands of ideas were gradually distilled into 223 specific projects. Then, with the cooperation of all sectors—including government, business, lending institutions, and average persons—work on the projects began, providing construction employment for more than 7300 people and permanent employment for 1380 and investing more than $800 million. Among the projects were the following:

- With the support of the Clean Air and Clean Water acts, local government clamped down hard on industries to control pollution.
- Low-cost housing was renovated.
- A new industry to build pollution-control equipment was spawned, as was another industry to build electric buses, which are now serving the city without noise or pollution.
- A recycling center employing mentally handicapped adults to separate materials was built.
- An urban greenway demonstration farm, which schoolchildren may visit, was created.
- A zero-emissions industrial park utilizing pollution-avoidance principles was built.
- The river was cleaned up, and parks were created along the riverbanks.

- Theaters, museums, and a freshwater aquarium were renovated or built.
- All facilities and a renovated business district were made pedestrian-friendly and accessible.

With these and numerous other projects, many of them still ongoing, Chattanooga has moved its reputation from one of the worst to one of the best places to live (Fig. 24–17). Chattanooga Venture has developed a step-by-step guide for community groups to assist them in similar efforts to build sustainable communities; the city's experience is being modeled in cities throughout the United States as well as internationally. The city recently sponsored *Revision 2000*, with over 2600 participants taking up where *Vision 2000* left off. *Revision 2000* identified an additional 27 goals and more than 120 recommendations for further improving Chattanooga. The city's experience demonstrates that visioning and change must occur and continue for success in creating and maintaining sustainable community development.

President's Council on Sustainable Development

What role, if any, should the government play in the development of sustainable communities? In the United States, the President's Council on Sustainable Development has the responsibility of guiding the country toward greater sustainability. We have seen examples of the council's accomplishments in earlier chapters; here we give an overview of the council's work.

President Clinton created the council in 1993 to find ways to "meet the needs of the present without jeopardizing the future." Made of members from the administration, the business community, and the environmental community, the council met for three years and delivered its major report in 1996, "Sustainable America: A New Consensus." The council began its work by first developing a "We Believe Statement" that helped to forge a cooperative spirit among members. One of the beliefs relates to change and the nature of sustainability: "Change is inevitable and necessary for the sake of future generations and for ourselves. We can choose a course for change that will lead to the mutually reinforcing goals of economic growth, environmental protection, and social equity." These three goals, which appear in Fig. 1–10 as the key to sustainable development, give structure to the 186-page report, as it goes on to present a series of 10 more specific national goals for achieving sustainable development (Table 24-2). The national goals are then given closer attention in the report in the form of policy recommendations, specific actions, and indicators of progress.

The introduction to the report is a remarkable challenge and call for change. Major global changes affecting everyone are brought into focus, such as continuing population growth, major impacts on natural resources, the link between economic growth and increasing global pollution, and the globalization of information and economic activity. The council cited the need to move "from conflict to collaboration," meaning that businesses, the federal government, communities, and citizens must learn to work together in forging public policies. The guide for accomplishing a sustainable America, in the eyes of the council, needs to be a stewardship ethic. "Principles of stewardship help define appropriate human interaction with the natural world. Stewardship ... is a set of values that applies to a variety of decisions. It provides moral standards that cannot be imposed but can be taught,

◀ **FIGURE 24–17** *Chattanooga, Tennessee.* With numerous projects to reduce pollution and traffic and to provide attractions and amenities for people, Chattanooga, Tennessee, has changed its reputation from one of the worst to one of the best places to live in the United States.

TABLE 24-2 National Goals Toward Sustainable Development.

Goal 1: Health and the Environment. Ensure that every person enjoys the benefit of clean air, clean water, and a healthy environment at home, at work, and at play.

Goal 2: Economic Prosperity. Sustain a healthy U.S. economy that grows sufficiently to create meaningful jobs, reduce poverty, and provide the opportunity for a high quality of life for all in an increasingly competitive world.

Goal 3. Equity. Ensure that all Americans are afforded justice and have the opportunity to achieve economic, environmental, and social well-being.

Goal 4. Conservation of Nature. Use, conserve, protect, and restore natural resources—land, air, water, and biodiversity—in ways that help ensure long-term social, economic, and environmental benefits for ourselves and future generations.

Goal 5. Stewardship. Create a widely held ethic of stewardship that strongly encourages individuals, institutions, and corporations to take full responsibility for the economic, environmental, and social consequences of their actions.

Goal 6. Sustainable Communities. Encourage people to work together to create healthy communities where natural and historic resources are preserved, jobs are available, sprawl is contained, neighborhoods are secure, education is lifelong, transportation and health care are accessible, and all citizens have opportunities to improve the quality of their lives.

Goal 7. Civic Engagement. Create full opportunity for citizens, businesses, and communities to participate in and influence the natural resource, environmental, and economic decisions that affect them.

Goal 8. Population. Move toward stabilization of U.S. population.

Goal 9. International Responsibility. Take a leadership role in the development and implementation of global sustainable development policies, standards of conduct, and trade and foreign policies that further the achievement of sustainability.

Goal 10. Education. Ensure that all Americans have equal access to education and lifelong learning opportunities that will prepare them for meaningful work, a high quality of life, and an understanding of the concepts involved in sustainable development.

Source: President's Council on Sustainable Development. "Sustainable Development: A New Consensus." (Washington, D.C.: U.S. Government Printing Office, 1996): 12–13.

---

encouraged, and reinforced.... Principles of stewardship can illuminate complex policy choices and guide individuals toward the common good." The report also points to the need for individual responsibility as a counter to apathy and inaction. The common good is good for individuals also, and for sustainable development to be successful, people must work together and also be held accountable for their actions.

The work of the council continues on in the form of special task forces and reports. One of these is "Sustainable Communities," published in 1997 as the report of the Sustainable Communities Task Force. The report includes a number of case studies, including that of Chattanooga, a set of community profiles that highlight community actions and action groups from each of the 50 states, and a number of examples of states providing leadership on sustainable development. In so doing, the task force has captured the best of the sustainable communities movement and has extended the work of the President's council in producing a set of goals and recommendations to be implemented by federal, state, and local governments, businesses, and especially, communities and the people in them.

The most recent work of the President's council was in sponsoring the National Town Meeting (NTM) for a Sustainable America in Detroit, from May 2–5, 1999. This event featured examples of sustainability from around the country and emphasized building individual and institutional capacity so that the best sustainability practices can be replicated elsewhere.

24.4 Epilogue

Sustainability, stewardship, and sound science—these are the themes that have kept our eyes on the basics of how we must live on our planet. As we pointed out in Chapter 1, sustainability is the practical goal that our interactions with the natural world should be working toward; stewardship is the ethical and moral framework that informs our public and private actions; sound science is the basis for our understanding of how the world works and how human systems interact with it. How are we doing?

Our Dilemma

Is sustainable development going to work to reduce the desperate poverty and lowered life expectancy of the billions who live on less than $2 a day? Will we reduce our use of fossil fuels to halt global warming and the rising sea level? Can we hope to feed the four billion more who will join us by midcentury and simultaneously reduce the malnutrition that still plagues 800 million? Our intention in this book has been to avoid dwelling on the bad news and instead to raise the hope—indeed, the certainty—that all environmental problems *can* be addressed successfully. Throughout the text, we have pointed to policies and possibilities that can help move human affairs in a sustainable direction.

As Nobel Laureate Henry Kendall said, environmental problems at root are human, not scientific or technical. Even though our scientific understanding is not complete and our grasp of sustainable development is still tentative, we know enough to be able to act decisively in most circumstances. Therefore, it is human decisions both at the personal and societal level that can bring about change, and it is these decisions that define our stewardship relationship with Earth.

However, making decisions is not a simple task. Decision-making is affected by our personal values and needs, and competing values and needs exist at every turn. One group wins, another loses. If we stop all destruction of the rain forests, the farmer who needs land to farm will not eat. If we force a halt to the harvesting of shrimp because of concerns about the by catch, thousands of shrimpers will be out of work. Making decisions means considering the competing needs and values and reaching the best conclusion in the face of the numbing complexity of demands and circumstances. It is the nature of our many dilemmas that "business as usual" resolves only a very small fraction of them. Even though public policies at the national and international level are absolutely essential elements of our success in turning things around, these in the end are the outcome of decisions by very human people. And even if thoughtful decisions are made by those in power, they will fail unless they are supported by the people who are affected by them.

What compels people to act to promote the common good? We can identify a number of values that can be the basis of stewardship and help us in our ethical decisions, decisions involving both public policies and personal lifestyles. For example, *compassion* for those less well off, those suffering from malnutrition or other forms of deprivation, can become a major element of our foreign policy, energizing our country's efforts in promoting sustainable development in the countries most in trouble. It can also motivate young people to spend two years of their life in the Peace Corps or a hunger relief agency. A *concern for justice* is another crucial value. Just policies can become the norm for international economic relations, and a concern for justice can also move people to protest the placement of hazardous facilities in communities of color. Another stewardly value is *honesty*, or a concern for the truth—in the sense of keeping the laws of the land and in openly examining issues from different perspectives. Other important values for stewardship are *frugality*—using no more than necessary of Earth's resources; *humility*—instead of demanding our rights, being willing to share and even defer to others; *neighborliness*—being concerned for other members of our communities, such that we do not engage in activities that are harmful to their welfare, and instead make positive contributions to their welfare.

To achieve sustainability, we will have to couple values like these with the knowledge that can come from our understanding of how the natural world works and what is

happening to it as a result of human activities. For this to work, people must be willing to grapple with scientific evidence and basic concepts and develop a respect for the consensus that scientists have reached in most areas of environmental concern. It has been our objective to convey this consensus to you in this text. Recall also that sustainable solutions have to be economically feasible, socially desirable, and ecologically viable (Fig. 1–10). Again, we have tried to emphasize all three of these components as we point to the political decisions that are shaping our future in so many of our environmental concerns.

The good news is that many people are engaging environmental problems, working with governments and industries, bringing relief to people in need, and demonstrating exactly the kinds of decisions that are needed for effective stewardship (see "Ethics" essay, p. 596). Thus, traditional environmentalist organizations as well as people from other sectors of society are working together in the many issues that we have highlighted in this text, bringing about policy changes and also *changes in personal lifestyles*. We close with this very important topic.

Lifestyle Changes

What are some of the lifestyle changes that are needed, and are they in fact occurring? We should be encouraged and inspired by the literally millions of people in all walks of life who are acutely aware of the problems and who are making outstanding efforts to bring about solutions. Every pathway toward solutions that we have mentioned represents the work of thousands of dedicated professionals and volunteers ranging from scientists and engineers to businesspeople, lawyers, and public servants. Indeed, we are all involved whether we recognize it or not. Simply by our existence on the planet, everything we do—the car we drive, the products we use, the wastes we throw away, virtually every choice we make and action we take—has a certain environmental impact and a certain consequence for the future. Therefore, it is not a matter of having an effect, but of what and how great that effect will be. It is a matter of each of us asking ourselves, Will I be part of the problem or part of the solution? The outcome will depend on how each of us responds to the challenges ahead.

There are a number of levels on which we may participate to work toward a sustainable society:

- individual lifestyle changes
- political involvement
- membership and participation in nongovernmental environmental organizations
- volunteer work
- career choices

Lifestyle changes may involve such things as switching to a more fuel-efficient car or using a bicycle for short errands; recycling paper, cans, and bottles; retrofitting your home with solar energy; starting a backyard garden and composting and

ETHICS

THE TANGIER ISLAND COVENANT

Chesapeake Bay, the largest estuary in the United States, is famous for its blue crabs and the unique "watermen" who fish for the crabs and other marine animals. Watermen are independent fishers who have lived on the bay for generations and fiercely defend their way of life. In recent years the watermen of Tangier, a small island on the Virginia part of the eastern shore of the bay, have been in constant battle with the Chesapeake Bay Foundation and the state regulators over crab and oyster laws restricting when and where they could harvest. They have reputedly skirted restrictions by keeping undersized crabs and dredging clams and oysters out of season. It has also been alleged that the watermen have thrown their trash overboard, causing the shores of the island to be littered with debris.

In 1997 the island community of Tangier became part of an unusual but significant process called "values-based stewardship." "Values-based stewardship" is a form of conflict resolution that requires working within the value system of the community to effectively bring about sustained changes in behavior toward the environment.

Susan Drake, a doctoral candidate in anthropology at the University of Wisconsin, came to the island to carry out her thesis research on how conflicts are dealt with in homogeneous communities like Tangier. Originally attracted to the project because of the highly religious community present in Tangier—the two churches on the island are the center of life and authority there—Drake knew the watermen considered

themselves part of a rich bounty in an environment created by God. After working with watermen on their boats and spending time in their churches and with their families, Drake became concerned about her perceived lack of a connection between what the God-fearing watermen believed and what they actually did in their profession.

Convinced that she might be able to intervene in a positive way, Drake requested permission by her thesis committee to engage in "participatory action research," where the researcher may enter the conflict situation and attempt to bring about a resolution. Working within the value system of the community—in this case, a value system based on faith—Drake addressed a joint meeting of the two churches on the island one Sunday morning. Referring to an image of Jesus standing behind a young fisherman at the wheel of a boat in rough seas that almost all Tangier watermen displayed in their boat cabins, Drake addressed the issues of dumping trash overboard and skirting the laws. She showed the picture to the congregation and then put paper blinders over the eyes of Jesus as she listed the ways in which the watermen were disobeying the law. The day of her sermon coincided with a very high tide that gathered the debris from around the shores of the island and washed it up into the streets and people's yards.

The meeting ended with a call by Drake to the watermen to adopt a covenant she had drafted earlier with the help of some of the island's women. Fifty-six watermen went to the altar and promised to abide by the following covenant: "The Watermen's

Stewardship Covenant is a covenant among all watermen regardless of their profession of religious faith. As watermen we agree to: (1) be good stewards of God's Creation by setting a high standard of obedience to civil laws (fishery, boat and pollution laws), and (2) commit to brotherly accountability. If any person who has committed to this Covenant is overtaken in any trespass against this covenant, we agree to spiritually restore such a one in a spirit of gentleness. We also agree to fly a red ribbon on the antennas of our boats to signify that we are part of the Covenant."

The next day, and up to the present, scores of watermen began flying red ribbons, taking trash bags out on their boats, and fishing according to the laws. Trucks began to pick up the trash on the island early in the morning. The two churches proceeded to work with the entire island community to draft a plan for the future sustainability of the island and their way of life and form the "Tangier Watermen's Stewardship for the Chesapeake" to implement the plan. The plan is far-reaching and involves the women of the island lobbying the state legislature and opening up lines of communication with scientists and government regulators involved in overseeing the bay's health.

Values-based stewardship is a call to life-changing stewardship put within the value system of a community. In this case, putting the call within the context of the faith of the watermen resulted in the establishment of structures within the community that show great promise in helping to bring the entire island community toward a sustainable future.

recycling food and garden wastes into your soil; putting in time at a soup kitchen to help the homeless; choosing low-impact recreation such as canoeing rather than speedboating; living closer to your workplace, and any number of additional things.

Political involvement ranges from supporting and voting for particular candidates to expressing your support for particular legislation through letters or phone calls. Membership in nongovernmental environmental organizations can enhance both lifestyle change and political involvement. As a member of an environmental organization, you will receive and may help disseminate information, making you and others more aware of particular environmental

problems and things you can do to help. Specifically, you will be informed regarding environmentally significant legislation so that you may focus your political efforts at the most effective time and place. Also, your membership and contribution serve to support lobbying efforts of the organization. A lobbyist representing only him- or herself has relatively little impact on the legislators. On the other hand, if the lobbyist represents a million-member organization that can follow up with that many phone calls and letters (and ultimately votes), the impact is considerable. Finally, where enforcement of existing law has been the weak link, some organizations, such as Public Interest Research Groups, the Natural

Resources Defense Council, and Environmental Defense Fund, have been very influential in bringing polluters or the government to court to see that the law is upheld. Again, this can be done only with the support of members.

Another form of involvement is volunteer organizations. Many effective actions that care for people and the environment are carried out by groups that depend on volunteer labor. Political organizations and virtually all nongovernmental organizations are highly dependent on volunteers. Many helping organizations, like those dedicated to alleviating hunger and to building homes for needy families (Fig. 24–18) are adept at mobilizing volunteers to accomplish some vital tasks that often fill in the gaps left behind by inadequate public policies.

Finally, you may choose to devote your career to implementing solutions to environmental problems. Environmental careers go far beyond the traditional occupation of wildlife or park management. There are any number of lawyers, journalists, teachers, research scientists, engineers, medical personnel, agricultural extension workers, and others focusing their talents and training on environmental issues or hazards. There are innumerable business and job opportunities in pollution control, recycling, waste management, ecological restoration, city planning, environmental monitoring and analysis, nonchemical pest control, production and marketing of organically grown produce, and so on. Some developers concentrate on rehabilitation and reversal of urban blight as opposed to contributing more to urban sprawl. Some engineers are working on the development of pollution-free vehicles to help solve the photochemical smog dilemma of our cities. Indeed, it is difficult to think of a vocation that cannot be focused on promoting solutions to environmental problems.

We are entering a new millennium, and it is not just in terms of the calendar flipping from 1999 to 2000. We are engaged in an environmental revolution, a major shift in world view and practice from seeing nature as resources to be exploited to seeing nature as life's supporting structure that demands our stewardship. You are living in the early stages of this revolution, and we invite you to be one of the many who will make it happen.

▲ **FIGURE 24–18** *Habitat for Humanity volunteers.* By using charitable contributions and volunteer labor, Habitat for Humanity creates quality homes—over 70,000 built throughout the world by 1998—that needy families can afford to buy with no-interest mortgages, enabling the money to be recycled to build more homes. Prospective home buyers participate in the construction and may gain valuable skills in the process.

"For civilization as a whole, the faith that is so essential to restore the balance now missing in our relationship to the earth is the faith that we do have a future. We can believe in that future and work to achieve it and preserve it, or we can whirl blindly on, behaving as if one day there will be no children to inherit our legacy. The choice is ours; the earth is in the balance."

Al Gore, *Earth in the Balance*

ENVIRONMENT ON THE WEB

INTEGRATIVE FRAMEWORK FOR ENVIRONMENTAL MANAGEMENT

In the mid-1980s, David Crombie, a popular former mayor of Toronto, Ontario, was appointed Head of a Royal Commission on the Future of the Toronto Waterfront. Crombie, affectionately termed "Toronto's tiny perfect mayor," was to lead the Commission in developing proposals to restructure Toronto's decaying waterfront area. The waterfront was typical of many older industrial cities, containing a mix of industrial uses, open space, and railway lands. An elevated highway blocked the view of Lake Ontario from the city and created a physical obstacle for citizens wanting to access the shoreline areas.

Crombie tackled the task with his normal wit and energy but soon found himself coming to a surprising conclusion. The waterfront, he declared, could not be managed without consideration of the entire watershed area. A study of the entire watershed would mean detailed examination of several major river basins, the entire city of Metropolitan Toronto, many outlying communities and agricultural areas, transportation corridors, and areas of parkland and green space. Because the current condition of those lands was affected by their human uses—and users—Crombie required that the Commission also examine community values relating to the waterfront, recreational, cultural, and other social issues, and the economics of water and waterfront use. In short, David Crombie and the Royal Commission had concluded that effective planning for sustainable use of one watershed component—the narrow waterfront strip—required the examination of that component in the context of the larger watershed system.

The work of the Royal Commission, now a permanent body called the Waterfront Regeneration Trust, was important not because of its scientific sophistication, or even because of the complexity of the issues it considered. The work was remarkable because it integrated biophysical, social, and economic factors in a framework that gave high regard to the values and opinions of average citizens. As the Commission proceeded with its investigations, holding workshops and open houses, listening to regulators, scientists, and citizens, it gradually educated watershed residents about watershed systems—and the implications for those systems of individual actions.

Web Explorations

The Environment on the Web activity for this essay describes the process of integrated watershed management as an example of an integrative framework for sustainable development. Go to the Environment on the Web activities (select Chapter 24 at http:// www.prenhall.com/nebel) and learn for yourself:

1. how to plan a successful watershed management plan;
2. how to fund an integrated management plan; and
3. about an example of integrative management in another field.

A suggested time frame for each of these exercises is 10–30 minutes.

REVIEW QUESTIONS

1. What problems do Trenton and Waitakere City illustrate, and how are they organizing to deal with those problems?
2. How did the structure of cities begin to change after World War II? What factors were responsible for the change?
3. Why are the terms car-dependent and urban sprawl used to describe our current suburban lifestyle and urban layout?
4. What federal laws and policies tend to support urban sprawl?
5. Describe five categories of environmental consequences stemming from the car-dependent lifestyle and urban sprawl.
6. What is Smart Growth? How does it address urban sprawl?
7. What services do local governments provide, and what is their prime source of revenue for these services?
8. What is meant by "erosion of a city's tax base"? Why does it follow from exurban migration? What are the results?
9. How do exurban migration, urban sprawl, and urban decay become a vicious cycle?
10. What is meant by "economic exclusion"? How is it related to the problems of crime and poverty in cities?
11. What are three common denominators of livable cities? What are the connections between livability and sustainability?
12. How does the U.N. Sustainable Cities Program work?
13. What is the Sustainable Communities Program? How does Chattanooga illustrate the program?
14. How is the President's Council on Sustainable Development promoting sustainability in cities and suburbs?
15. What are some stewardship values, and what role could they play in fostering sustainability?
16. Give five levels on which people can participate in working toward a sustainable future.

THINKING ENVIRONMENTALLY

1. Interview an older person (65 years or older) about what his or her city or community was like before 1950 in terms of meeting people's needs for shopping, recreation, getting to school, work, and so on. Was it necessary to have a car?

2. Do a study of your region. What aspects of urban sprawl and urban blight are evident? What environmental and social problems are evident? Are they still going on? Are efforts being made to correct them?

3. Identify nongovernmental organizations in your region that are working toward reining in urban sprawl or working toward preserving or bettering the city. What specific projects are under way in these areas? What roles are local governments playing in the process? How can you become involved?

4. Suppose you are a planner or developer. Design a community that is people-oriented and that integrates the principles of sustainability and self-sufficiency. Consider all the fundamental aspects: food, water, energy, sewage, solid waste, and how people get to school, shopping, work, and recreation.

MAKING A DIFFERENCE

PART SIX: Chapters 23 and 24

1. Be aware of the factors that influence environmental public policy and become involved to assure fairness and consistency in the application of laws and regulations passed by Congress.

2. Use the available Web resources to locate specific regulations developed by the EPA or other regulatory agencies, and research the details of how the agency conforms to Executive Order 12866 in developing the economic impact of the regulations.

3. Investigate land-use planning and policies in your region. Is urban sprawl or urban blight a problem? Become involved in zoning issues, and support zoning and policies that will support rather than hinder ecologically sound land use (Smart Growth).

4. Follow up on the work of the President's Council on Sustainable Development to find out what is happening in your state or city to promote sustainable development.

5. Contact local conservation organizations or land trusts and offer your help in protecting and preserving open land in your region.

6. Find out if there is a Habitat for Humanity project or group in your area and, if there is, volunteer some time in support of their work.

WEB REFERENCES

On-line resources for this chapter can be found on the World Wide Web at: **http://www.prenhall.com/nebel**. Click on Chapter 24 on the chapter selector.

Study Guide

Environmental Science
Seventh Edition

CONTENTS

CHAPTER 1

INTRODUCTION: SUSTAINABILITY, STEWARDSHIP AND SOUND SCIENCE

1.   Characterize how differently people in developing and industrialized countries are tied to their environment.

      a. Developing countries:_____

      b. Developed countries: _____

2.   The historic residents of Easter Island learned that civilization collapses when:

      a. _____

      b. _____

      c. _____

3.   What four steps need to be taken to prevent a global version of the collapse of civilization on Easter Island?

      a. _____

b. _____

c. _____

d. _____

The Global Environmental Picture

4. What global trends represent views of the interactions of human and natural ecosystems?

a. _____   b. _____

c. _____   d. _____

Population Growth

5. Give the world population figures in:

a. 1995 _____   b. 2050 _____

6. Given the rapid human population growth rate and expected global population size by 2050, discuss the growing conflict between "global quality of life" and "American lifestyle".

_____

_____

Degradation of Soils

7. What are the indicators of global soil degradation?

a. _____   b. _____

c. _____   d. _____

e. _____

Global Atmospheric Changes

8. Give one example of an air pollution problem with global dimensions.

_____

9. The carbon dioxide level in the atmosphere has grown from _____ parts

per million (ppm) in 1900 to _____ ppm today.

10. Explain the expected relationship between atmospheric carbon dioxide and global warming.

_____

_____

Loss of Biodiversity

11. List three ways in which habitat alteration is taking place.

    a. _____ b. _____

    c. _____

12. What three factors are the greatest contributors to loss of biodiversity?

    a. _____ b. _____

    c. _____

13. Give three reasons why the loss of biodiversity is so critical.

    a. _____

    b. _____

    c. _____

Three Unifying Themes

14. List three interrelated themes that may be applicable to changing or giving direction to the interaction between human and natural ecosystems.

    a. _____

    b. _____

    c. _____

15. The current trends of human uses of the world's natural resources (Is, Is Not) sustainable.

16. Imagine you are in the commercial fishing (e.g., tuna) business. Fisheries biologists have estimated that the fishery consists of 10K tuna of which 8K are the surplus resulting from the reproducing component of the population. In order to maintain a **sustainable yield** how many tuna should be harvested? _____

17. Justify your answer to question #16.

    _____

    _____

18. List three characteristics of a sustainable society.

    a. _____

    b. _____

c. _____

Sustainable Development

19. Apply the meaning of sustainable development to the disciplines of:

    a. Economics: _____

    b. Sociology: _____

    c. Ecology: _____

20. List four dimensions to sustainable development.

    a. _____

    b. _____

    c. _____

    d. _____

Stewardship

21. Describe three ways in which the stewardship ethic is demonstrated.

    a. _____

    b. _____

    c. _____

Justice and Equity

22. What is the major problem being addressed by the environmental justice movement?

    _____

23. How is environmental racism demonstrated? _____

    _____

24. Is it (true or false) that the perpetrators of environmental racism come only from outside developing countries?

25. List four conditions (local and global) that contribute to poverty in developing countries?

    a. _____   b. _____

c. _____   d. _____

26. Is it (true or false) that failure of debtor nations to repay their loans will not hurt the economies of developed nations?

27. Trace the conservation attitude of Americans during the following time periods by indicating whether the attitude was positive [+] or negative [-].

    a. [   ] late nineteenth century
    b. [   ] scientific era after World War I
    c. [   ] depression years (1930 to 1936)
    d. [   ] during World War II
    e. [   ] after World War II until the 1960s
    f. [   ] from the mid 1960s to 1980
    g. [   ] 1980 to now

28 Identify one environmental problem that developed during the post-World War II era in the

    a. Air: _____

    b. Waters: _____

    c. Wildlife: _____

29. What "cause and affect" relationship did Rachel Carson portray in her book titled: *Silent Spring*?

    _____

30. Why does the position of citizens aligned with the environmental movement include wildlife?

    _____

31. Identify four environmental organizations that grew out of the environmental movement.

    a. _____   b. _____

    c. _____   d. _____

32. List four successes associated with the environmental movement.

    a. _____

    b. _____

    c. _____

    d. _____

33. What are the two major points of contention in the spotted owl debate?

a. _____ vs. b. _____

34. Is it (true or false) that polluters represent one group of people?

35. Indicate whether the following statements are [+] or are not [-] examples of the policy recommendations of the President's Council on Sustainable Development.

[   ] restoration of biodiversity of terrestrial and aquatic ecosystems
[   ] identification of the real stakeholders in the process
[   ] enhancing agricultural productivity
[   ] coordination of stakeholders to achieve specific goals
[   ] habitat restoration
[   ] partnership development

Sound Science

36. Scientific information is based on repeated _____.

37. The art of compiling scientific information is a series of steps called the _____ method.

38. List the four basic assumptions that form the unquestioned foundation of the scientific method.

a. _____

b. _____

c. _____

d. _____

39. The scientific method begins with:

a. observations, b. developing hypotheses, c. experimentation, d. formulation of theories.

40. In science, the only observations accepted are those made with the basic five senses, namely _____, _____, _____, _____, and _____.

41. Before it is accepted, an observation must be confirmed or _____.

42. Indicate whether the following statements are [+] or are not [-] correct.

[   ] All observations are facts.
[   ] All facts are verified observations.

6

43.   The purpose of a scientific experiment is to:

        a. test a hypothesis, b. prove that a hypothesis is correct, c. confuse the public
        d. use fancy instruments, e. have fun at public expense

44.   A hypothesis that cannot be tested is:

        a. correct, b. incorrect, c. useful, d. useless

45.   To give definitive results, an experiment must involve a minimum of _____ groups.

46.   The two groups are called the _____ group and the _____
        group.

47.   The _____ group is subjected to treatment or condition which the scientist
        believes is responsible for the observed effect.

48.   The control group:
        a. must be an exact replicate in every respect,
        b. must be the same except for the factor being tested,
        c. is any other group

49.   A group of plants is grown on a medium without potassium (a chemical treatment). The plants
        do poorly. The result of this experiment:

        a. proves that potassium is required for plant growth.
        b. proves that potassium is not required for plant growth.
        c. proves nothing because there is no control.
        d. proves nothing because other factors could be responsible for the poor growth.
        e. Both c and d are correct.

50.   A concept that provides a logical explanation for a certain body of facts is called a

        _____.

51.   Theories are:

        a. tested in much the same way as are hypotheses. (True, False)
        b. subjected to if..then reasoning. (True, False)
        c. used to predict outcomes. (True, False)
        d. subject to change. (True, False)

52.   When we observe that certain events always occur in a precisely predictable way (e.g., an
        unsupported object always falls) we define such behavior as a _____ or
        natural _____.

53. Principles and natural laws are:

    a. the same as hypotheses.
    b. the same as theories.
    c. descriptions of the way certain things are always observed to behave.
    d. laws passed by a council of scientists.

54. Give two examples of natural laws.

    a._____   b._____

55. List three uses of instruments in conducting the process of science.

    a. _____

    b. _____

    c. _____

56. Scientists do not use instruments to:

    a. increase their power of observation.
    b. control conditions in various experiments.
    c. quantify observations.
    d. avoid making observations.

57. Indicate whether the following statements are true [+] or false [-] concerning the process of science.

    [ ] There are no controversies or arguments among scientists.
    [ ] Progress in science can be slow.
    [ ] We are continually confronted by new observations.
    [ ] Some observed phenomena may not lend themselves to simple experiments.
    [ ] Science is incapable of providing absolute proof for any theory.
    [ ] The process of science can be used to test value judgments.
    [ ] The validity of science is based on the ability to do experiments.

58. Describe the scientific community.

    _____

    _____

59. Give two examples of junk science.

    a. _____

    b. _____

VOCABULARY DRILL

Directions: Match each term with the list of definitions and examples given in the tables below. Term definitions and examples might be used more than once.

Term	Definition	Example	Term	Definition	Example
biodiversity			habitat alteration		
stewardship			sustainable yields		
environmental racism			sustainable society		
environmental movement			sustainable development		
environmentalist			sustainable		

VOCABULARY DEFINITIONS

a. anyone concerned with reducing pollution and protecting wildlife

b. placement of waste sites and other hazardous facilities in towns and neighborhoods where the majority of residents are not white

c. habitat changes that may or may not be reversible

d. the total assemblage of all plants, animals, microbes that inhabit the earth

e. militant citizenry demanding curtailment and cleanup of pollution and protection of pristine environments

f. continued indefinitely without depleting any of the material or energy resources to keep it running

g. harvest of renewable resources that does not exceed replacement rates

h. continues without depleting its resource base or producing pollutants

i. progress that meets the needs of the present without compromising the needs of future generations

j. an ethic that provides a guide to actions taken to benefit the natural world and other people

VOCABULARY EXAMPLES

a. Rachel Carson	f. urbanization
b. Earth Day 1970	g. American indians
c. clear-cut logging	h. Our Common Future
d. global flora and fauna	i. saving the spotted owl
e. suburbs of Chattanooga, TN	j. cashing in on the interest only

SELF-TEST

1. Rank (1 = first event to 5 = last event) the following chain of events in the order they would occur leading to the collapse of a civilization.

   Events   Rank

   a. loss of topsoil   [   ]
   b. population increase beyond food growing capacity [   ]
   c. deforestation   [   ]
   d. soil erosion   [   ]
   e. disparity between haves and have nots   [   ]

2. Explain why the Easter Island population could have been more susceptible to over population and resource depletion than continental populations?

   _____

   _____

3. During which of the following time periods did Americans have a positive attitude toward conservation?

   a. scientific era after World War I   b. depression years
   c. during World War II   d. after World War II until the 1960s

4. In general, a positive conservation attitude by the majority of Americans is initially stimulated by:

   a. public education   b. environmental crisis
   c. love of nature   d. federal laws

5. Which of the following was not a post-World War II environmental problem in America?

   a. air pollution   b. decline of wildlife populations
   c. water pollution   d. over population

6. Indicate whether the following are gains [+] or losses [-] in the war for the environment.

   a. [   ] population growth   b. [   ] pollution control
   c. [   ] environmental laws   d. [   ] biodiversity

7. Check the most important initial battle to win in the war for the environment.

   a. [   ] population growth   b. [   ] pollution control
   c. [   ] environmental laws   d. [   ] biodiversity

8. During the spotted owl controversy in the Northwest, the debate was over the
   _____ versus _____.

9. Match the evidence with list of global events indicating that environmentalism is losing the war to save our planet.

Evidence

[   ] habitat conversions
[   ] CO2 emissions
[   ] desertification
[   ] 10 billion
[   ] commercial exploitation
[   ] urban development

Events

1. population growth
2. atmospheric changes
3. soil degradation
4. loss of biodiversity

10. Any movement toward sustainable development will require public:

a. understanding and awareness of the problems.
b. adoption of new values.
c. demands for change.
d. all of the above

11. Below is a list of world view assumptions. Check all those that apply to anyone who can be called an environmentalist.

[   ] Natural resources are to be exploited for the advantage of humans.
[   ] Natural resources are essentially infinite.
[   ] We need to stay on the same course of human uses of natural resources as in the past.
[   ] It is alright to spend both the interest and the capital of our natural resource inheritance.
[   ] Natural resources are to be protected and maintained.
[   ] Natural resources are limited by the regenerative capacities of the natural environment.
[   ] We need to change the ways humans use natural resources.
[   ] We need to live only on the interest of our natural resource inheritance.

12. Any actions that promote a sustainable human use of natural resources will require public:

a. understanding and awareness of the problems.
b. adoption of new values.
c. demands for change.
d. all of the above

True [+] or False [-]

13.   [   ] American,s attitudes toward conservation has been consistent throughout the 20th century.
14.   [   ] Environmentalists pretty much dictate the future use of global resources.
15.   [   ] Air and water in America is cleaner now than in the 1960s.
16.   [   ] The outcomes of global trends that are not sustainable are predictable.
17.   [   ] The outcomes of establishing a sustainable society are predictable.
18.   [   ] The Wise Use and Environmental Movements have compatible goals.
19.   [   ] The present American lifestyle is compatible with a sustainable society.
20.   [   ] The majority of Americans support the views of contemporary environmentalists.

21. The scientific method begins with:

a. observations, b. developing hypotheses, c. experimentation, d. formulation of theories

22. Hypotheses are confirmed through:

a. observations, b. experiments, c. authority figures, d. technology

23. Experiments must include:

a. experimental groups, b. control groups, c. controlled environments, d. all of these

24. The purpose of the control group is to:

a. show that conditions can be controlled.
b. show that factors other than the test factors are not responsible for the observed effect.
c. prevent contamination of the experimental group.
d. verify the beliefs of authority figures.

25. Theories are:

a. tested in much the same way as hypotheses.
b. subjected to if...then reasoning.
c. subject to change.
d. all of the above.

26. Events that always occur in a precisely predictable way are called:

a. observations, b. hypotheses, c. theories, d. laws

27. Scientists do not use instruments to:

a. increase their power of observation.
b. control conditions in various experiments.
c. quantify observations.
d. avoid making observations.

28. Which of the following is not an accurate statement concerning the process of science?

a. The process of science may lead to answers and more questions.
b. The validity of science is based on the ability to do experiments.
c. There are no controversies or arguments among scientists.
d. Progress in science can be slow.

29. Which of the following is not an accurate statement concerning science?

a. Science can test value judgments.
b. Science can test one-time events.
c. Science can give us answers to all questions.
d. All of the above are not accurate.

ANSWERS TO STUDY GUIDE QUESTIONS

1. closely tied to the land because their survival depends on it; nearly totally isolated from natural world. 2. society fails to care for the environment and sustain it, population increases beyond food growing capacity, disparity between haves and have nots widens; 3. understand how the natural world works, understand how human systems interact with natural systems, assess the status and trends of crucial natural systems, promote and follow a long-term, sustainable relationship with the natural world; 4. population growth, atmospheric changes, soil degradation, loss of biodiversity; 5. 6 billion, 10 billion; 6. American=s consumption of the world=s energy and natural resources to produce more possessions is having negative global environmental effects; 7. erosion, depleted water supplies, desertification, urban development, salinization; 8. global warming; 9. 280, 360; 10. CO_2 traps heat escaping from the earth raising the temperature of the earth=s atmosphere; 11. habitat conversion, pollution, commercial exploitation; 12. habitat alteration, exploitation, pollution; 13. genetic derivatives of all agricultural plants and animals, derivatives of medicinal drugs and chemicals, maintain the stability of natural ecosystems; 14. sustainability, stewardship, sound science; 15. Is not; 16. 8K; 17. The 8K represents the surplus. Leave the 2K alone to produce another surplus; 18. continues generation after generation, does not exceed sustainable yields, does not pollute beyond nature=s capacity to absorb; 19. concerned with growth, efficiency, and maximum use; focus on human needs and concepts like equity; concern for preserving integrity of ecosystems, maintaining carrying capacity and dealing with pollution; 20. environmental, social, economic, political; 21. recognize that a trust has been given, responsible care for something not owned, desire to pass something on to future generations; 22. environmental racism; 23. placement of waste site and hazardous facilities in nonwhite neighborhoods; 24. false; 25. wealthy elite horde and squander the resources of developing countries, unjust economic practices of wealthy industrialized countries, patterns of international trade, international debt; 26. false; 27. +, -, +, -, -, +, -; 28. murky and irritating, fowled with raw sewage, decline or extinction; 29. pesticide (DDT) use and high bird mortality; 30. wildlife are the indicators of environmental conditions; 31. Environmental Defense Fund, Greenpeace, Natural Resource Defense Council, Zero Population Growth; 32. Environmental Protection Agency, environmental laws, species saved from extinction, pollution abatement; 11. law, engineering, economics, science; 33. environment vs jobs; 34. false; 35. +, -, +, +, +, +; 36. observations; 37. scientific; 38. we perceive reality with our five basic senses; objective reality functions according to certain basic principles and laws; explainable cause and effect; we have the tools and capabilities to understand the basic principles and natural laws; 39. a; 40. sight, smell, taste, sound, touch; 41. rejected; 42. -, +; 43. a; 44. d; 45. 2; 46. experimental, control; 47. experimental; 48. b; 49. e; 50. theory; 51. all true; 52. principle, law; 53. c; 54. Law of Gravity, Conservation of Matter; 55. increase power of observation, control experimental conditions, quantify observations; 56. d; 57. -, +, +, +, +, -, +; 58. A collective body of scientists, working in a given field who, because of their competence and experience, establish what is sound science and what is not; 59. presentation of selective results, political distortions of scientific works, publications in quasi-scientific journals.

ANSWERS TO SELF-TEST

1. 4, 1, 2, 3, 5 (may be debate on where the chain starts); 2. Population had no where else to go, had no import capabilities, land mass small and isolated; 3. b; 4. b; 5. d; 6. -, +, +, -; 7. Population growth; 8. environment vs. jobs; 9. 4, 2, 3, 1, 4, 3; 10. All of the above; 11. Checks of the last four and no others are the characteristics of an environmentalist. If you checked a blend of the first and second group of four, you are confused! 12. d; 13. -; 14. -; 15. +; 16. +; 17. +; 18 -; 18. 9. -; 20. + (depending on your opinion of the American public); 21. a; 22. b; 23. d; 24. b; 25. d; 26. d; 27. d; 28. c; 29. d.

VOCABULARY DRILL ANSWERS					
Term	Definition	Example	Term	Definition	Example
biodiversity	d	d	habitat alteration	c	c, f
stewardship	j	i	sustainable yields	g	j
environmental racism	b	e	sustainable society	h	g
environmental movement	e	b	sustainable development	i	h
environmentalist	a	a	sustainable	f	f

CHAPTER 2

ECOSYSTEMS: UNITS OF STABILITY

What Are Ecosystems?

1.   All **biotic communities** consist of _____, _____, and _____ communities.

2.   Identify the following community components as **abiotic** [A] or **biotic** [B].

    a. [   ] plants   b. [   ] animals   c. [   ] temperature

    d. [   ] microbes   e. [   ] rainfall   f. [   ] wind

3.   An organism can be designated a distinct **species** if it can _____ with others of its own kind and produce _____ offspring.

4.   A group of bullfrogs (*Rana catesbiana*) is the same farm pond constitutes a
_____.

5.   An **ecosystem** may be defined as a grouping of _____, _____,
and _____ interacting with _____ and their
_____.

6.   Furthermore, the interrelationships are such that the grouping may be _____.

7.   List six examples of major kinds of ecosystems.

a. _____   b. _____   c. _____

d. _____   e. _____   f. _____

8.   Each ecosystem is characterized by distinctive plant _____.

9.   Identify the **ecotone** in Fig. 2-2? _____

10.   The major **biome** of the eastern United States is _____.

11.   The major biome of the central United States is _____.

12.   The major biome of the southwestern United States is _____.

13.   The distribution of major biomes globally is the consequence of _____.

14.   Is it (True or False) that what occurs in one ecosystem or biome may affect other ecosystems or
biomes?

15.   Give two ways that terrestrial ecosystems remain connected even though they may be far apart
(e.g., coniferous forests in the northwest with southwest grasslands).

a. _____

b. _____

16.   All creatures on earth and the environments they live in constitute one huge ecosystem called the
_____.

The Structure of Ecosystems

17.   The two basic structural components of all ecosystems are the _____
communities and _____ environmental factors.

Trophic Categories

18.   All the organisms in an ecosystem can be assigned to one of three categories. These three
categories are _____, _____, and _____.

19. Producers include all plants that do what? _____

20. In photosynthesis, plants use _____ energy to convert
_____ gas and _____ into _____.

21. The word **in**organic refers to substances such as _____.

22. The chemicals that make up the bodies of organisms are referred to as _____.

23. In short, producers convert (Inorganic, Organic) chemicals into (Inorganic, Organic) chemicals.

24. The energy for this production comes from _____.

25. Organisms which can produce all their own organic materials from inorganic raw materials in the environment are called (Autotrophs, Heterotrophs). Give an example. _____

26. Organisms which must consume organic material as a source of nutrients and energy are called (Autotrophs, Heterotrophs). Give an example. _____

27. Is it (True or False) that **all** plants are autotrophs? If you said false, give an example of an exception. _____

28. List three organisms that could be classified as consumers.

    a. _____ b. _____ c. _____

29. Consumers that feed directly on producers are called **primary consumers** or
_____.

30. List two organisms that are primary consumers.

    a. _____ b. _____

31. Consumers that feed on other consumers are called secondary consumers or
_____ or _____.

32. Indicate whether the following list of animals would be **predators** [PR] or **prey** [PY] in **predator/prey** relationships.

    a. [   ] wolf   b. [   ] coyote   c. [   ] quail
    d. [   ] bear   e. [   ] deer   f. [   ] salmon

33. Give an example of:

    a. internal parasite: _____   b. external parasite: _____

    c. parasite host: _____

34. All dead organic matter (dead leaves, twigs, bodies of animals, fecal matter) is called
_____.

35. Give three examples of **detritus feeders.**

a. _____, b. _____, c. _____

36. **Decomposers** consist of _____ and _____.

Trophic Relationships: Food Chains, Food Webs, and Trophic Levels

37. Construct a **food chain** using a grassland community that consists of [a] big bluestem grass,
[b] grasshoppers, [c] frogs, [d] snakes, and [e] owls. First consider what each organism
eats. For example:

a. grasshoppers eat: _____   b. frogs eat: _____

c. snakes eat: _____   d. owls eat: _____

e. Next, place each animal in a sequence of **feeding relationships** [a to e] that demonstrates a
food chain.

[ ] , [ ] , [ ] , [ ] , [ ]

38. Let's add some more organisms to the above grassland community including: fescue grass, berry
bushes, mice, white-tailed deer, humans, rabbits, foxes, red-tailed hawk, ground squirrels, and
sunflowers. Construct a **food web** in the below diagram by placing each species in its proper
trophic level and then drawing lines of feeding relationships (see Fig. 2-13a for an example of
this exercise). The species to use in this exercise are as follows: a. bluestem grass,
b.grasshopper, c. frogs, d. snakes, e. owls, f. fescue grass, g. berry bushes, h. mice, i. deer,
j.humans, k. rabbits, l. foxes, m. hawks, n. ground squirrels, and o. sunflowers.

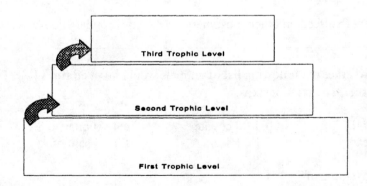

18

39.   Indicate whether the following categories of organisms occupy the [a] first, [b] second, [c] third, or [d] more than one trophic level.

a. [   ] producers   b. [   ] consumers   c. [   ] decomposers

d. [   ] carnivores   e. [   ] herbivores   f. [   ] predator

g. [   ] prey   h. [   ] omnivores   i. [   ] parasites

j. [   ] autotrophs   k. [   ] heterotrophs

40.   How is **biomass** determined?

_____

_____

41.   In an ecosystem the biomass of herbivores is always (Greater, Less) than the biomass of producers, and the biomass of carnivores is always (Greater, Less) than the biomass of herbivores.

42.   Diagram the biomass relationship for three trophic levels.

Nonfeeding Relationships

43.   Describe an example of **mutualism**.

_____

44.   How does your example of mutualism demonstrate the concept of **symbiosis**?

_____

45.   Where an organism lives is called its (habitat or niche). How an organism lives is called its (habitat or niche).

46.   Give an example of how interspecies niche competition is reduced in:

a. space: _____

b. time: _____

Abiotic Factors

47.   Three examples of abiotic factors are:

a. _____   b. _____   c. _____

48.   Label the below figure with the following terms (see Fig. 2-19).

a. optimum, b. range of tolerance, c. limits of tolerance, d. zones of stress

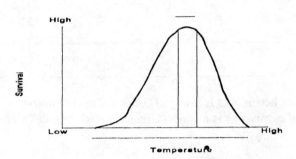

49.   As the amount or level of any factor increases or decreases from the optimum, the plant or animal is increasingly _____.

50.   If any factor increases or decreases beyond an organism's limit of tolerance, what occurs? _____.

51.   Is it (True or False) that each species has an optimum zone of stress, and limits of tolerance with respect to every factor?

52.   How many **limiting factors** need to be beyond a plant's limit of tolerance to preclude its growth? _____

53.   Give two examples of limiting factors.

a. _____   b. _____

Global Biomes

The Role of Climate

54.   The definition of **climate** involves both _____ and_____.

55.   _____ is the main climatic factor responsible for the separation of ecosystems into forests, grasslands, and deserts.

56.   Even though it takes 30 inches or more of rainfall to support a forest ecosystem, _____ will determine the kind of forest.

57.   Give an example of an ecosystem that would contain **permafrost**. _____

Microclimate and Other Abiotic Factors

58.   A localized area within an ecosystem that has moisture and temperature conditions different from the overall climate of the area is called a _____.

Biotic Factors

59.   The biotic factor that limits grass from taking over high rainfall regions is

       a. tall trees b. herbivores c. parasites d. insects

Physical Barriers

60.   The spread of species into other ecosystems may be prevented by natural physical barriers such as _____, _____, and _____.

61.   The spread of species into other ecosystems may be prevented by physical barriers produced by humans such as _____, _____, and _____.

Implications for Humans

Three Revolutions

62.   As hunter-gatherers, humans were much like: a. herbivores, b. carnivores, c. omnivores

63.   With _____, humans created their own distinctive ecosystem apart from natural ecosystems.

64.   List five ways humans have overcome the usual limiting factors that prevent the spread of our species.

       a. _____

       b. _____

       c. _____

       d. _____

       e. _____

65.   It is (True or False) that the human ecosystem is upsetting and destroying other ecosystems.

VOCABULARY DRILL

Directions: Match each term with the list of definitions and examples given in the tables below. Term definitions and examples may be used more than once.

Term	Definition	Example(s)	Term	Definition	Example(s)
abiotic			heterotroph		
abiotic factors			host		
autotroph			inorganic molecule		
biomass pyramid			limiting factors		
biomass			limits of tolerance		
biome			microclimate		
biosphere			mutualism		
biota			niche		
biotic community			omnivore		
biotic factors			optimum		
biotic structure			organic molecule		
carnivore			parasites		
chlorophyll			parasitism		
climate			photosynthesis		
consumers			population		
decomposer			predator		
detritus			prey		
detritus feeders			primary consumer		
ecologist			producers		
ecology			range of tolerance		
ecosystem			secondary consumer		
ecotone			species		
food chain			symbiotic		
food web			synergistic effect		
habitat			trophic levels		
herbivore			zone of stress		

VOCABULARY DEFINITIONS

a. grouping or assemblage of living organisms in an ecosystem

b. assemblage of nonliving physical or chemical components in an ecosystem

c. specific kinds of plants, animals, or microbes that interbreed and produce fertile offspring

d. number of individuals that make up an interbreeding, reproducing group within a given area

e. a grouping of plants, animals, and microbes interacting with each other and their environment in such a way to perpetuate the grouping

f. study of ecosystems and interactions

g. people who study ecosystems and interactions

h. the transitional area where one ecosystem blends into another

i. total combined dry weight of all organisms at a trophic level

j. all the species and ecosystems combined

k. the way different categories of organisms fit together in an ecosystem

l. organisms that produce their own organic material from inorganic constituents

m. organisms that must feed on complex organic material to obtain energy and nutrients

n. the process of converting sunlight energy and CO_2 to sugar and O_2

o. molecule that plants use to capture light energy for photosynthesis

p. constructed in large part from carbon and hydrogen atoms

q. constructed in large part from elements other than carbon and hydrogen

r. dead or partially digested plant or animal material

s. primary detritus feeders

t. feed directly on producers

u. feed on primary consumers

v. meat-eating secondary consumer

w. plant and meat-eating consumer

x. animal that attacks, kills, and feeds on another animal

y. animal killed and eaten by a predator

z. a predator that feeds off its prey for a long time typically without killing it

aa. plant or animal fed upon by a parasite

bb. pathways where one organism is eaten by a second, which is eaten by a third, and so on

cc. a way of demonstrating the interrelatedness of food chains

dd. feeding levels within a food chain or web

ee. shape of biomass potential at each trophic level

ff. a symbiotic relationship between two organisms in which both derive benefit

VOCABULARY DEFINITIONS

gg. living together
hh. a symbiotic relationship between organisms in which one benefits and the other is harmed
ii. plant community and physical environment where an organism lives
jj. what an organism feeds on, where and when it feeds, where it finds shelter and nesting sites
kk. physical and chemical components of ecosystems
ll. a certain abiotic level where organisms survive best
mm. abiotic level between optimal range and high and low level of tolerance
nn. high and low ranges of tolerance to abiotic levels
oo. any factor that limits the growth, reproduction, and survival of organisms
pp. greater effect of two factors interacting together than individually
qq. description of average temperature and precipitation each day throughout the year
rr. conditions in specific localized areas
ss. limiting factors caused by other species
tt. grouping of related ecosystems into major kinds of ecosystems
uu. the entire span of abiotic values that allow any growth at all
vv. organisms that feed primarily on dead, decaying, or partially digested organic matter

VOCABULARY EXAMPLES	
a. plants, animals, microbes	q. salt (NaCl)
b. CO2, rainfall, pH, O2, salinity, temperature	r. dead leaves
c. *Rana catesbiana* (bullfrogs)	s. bacteria or fungi
d. *All Rana catesbiana* in a farm pond	t. deer or rabbit
e. Fig 2-2	u. wolf or owl
f. what you are doing now	v. pigs and bears
g. your teacher	w. Fig 2-9
h. where grassland meet forest	x. Fig 2-13a
i. Fig. 2-14	y. Figs. 2-16 and 2-17
j. Earth	z. Fig. 2-8 or Fig. 2-9
k. producers, consumers, decomposers	aa. Fig. 2-19
l. green plants	bb. chemical A → effect B, chemical C → effect D, chemical A + B → effect E
m. herbivores, carnivores, omnivores	cc. Fig. 2-20
n. $6CO_2 + 6H2O \rightarrow C_6H_{12}O_6 + 6O_2$	dd. Fig. 2-22
o. green organelle in plant tissue	ee. Fig. 2-4
p. glucose ($C_6H_{12}O_6$)	

An ecosystem is best defined as a grouping of 1. _____ ,
2. _____ and 3. _____ interacting with 4. _____
and their 5. _____ in such a way that the grouping is perpetuated.

Ecosystems evolve in sequential patterns involving plant, animal, and abiotic components. What is the component sequence in ecosystem development?

6. First Component: _____   7. Second Component: _____

8. Third Component: _____

9. A major biome in the central United States is a :

a. deciduous forest, b. grassland, c. desert, d. coniferous forest

10. Which of the following is not correctly matched?

a. oak tree - producer
b. squirrel - consumer
c. mushroom - detritus feeder
d. fungi - decomposer

11. The three major biotic components of ecosystem structure are

a. producers, herbivores, and carnivores
b. producers, consumers, and decomposers
c. plants, animals, and climate
d. consumers, detritus feeders, and decomposers

Below is a typical food chain illustration. Answer the series of questions below the illustration by choosing level 1, 2, 3, 4, 5, or ALL.

Level 5: OWL
Level 4: SNAKE
Level 3: FROG
Level 2: GRASSHOPPER
Level 1: GRASS

At which level:

12. _____ is the most transferable energy available?

13. _____ would one find the least biomass?

14. _____ is a primary consumer located?

15. _____ is a producer located?

16. _____ could a white-tailed deer be a representative?

17. _____ is sunlight the original energy source?

18. _____ would one find decomposers?

19. _____ would one find parasites?

20. The biomass relationship of a food chain can only be depicted as a (a. box, b. rectangle, c. circle, d. pyramid)

21. Explain your answer choice to #20.

_____

22. Which of the following is not an example of an abiotic factor?

   a. water, b. light, c. dead log, d. temperature

23. The points at which a factor becomes so high or low as to threaten an organism's survival is referred to as

   a. range of optimums, b. range of limits, c. range of tolerance, d. limits of tolerance

24. How many factors need to be beyond an organism's limits of tolerance to cause stress?

   a. one, b. five, c. ten, d. hundreds

   Which limiting factor (a. temperature, b. rainfall, c. pH, d. soil type) is the best determinant of:

25. _____ deserts, grasslands, and forests?

26. _____ kind of forest?

27. _____ kind of desert?

   Identify those conditions that do [+] and do not [-] represent the human system and then indicate whether the listed conditions are [+] or are not [-] compatible with natural ecosystems.

Human System Conditions	Compatibility with natural ecosystems
28. [ ] Recycle wastes	32. [ ]
29. [ ] Practice population control	33. [ ]
30. [ ] Control limiting factors	34. [ ]
31. [ ] Causing rapid abiotic/biotic changes	35. [ ]

27

ANSWERS TO STUDY GUIDE QUESTIONS

1. plants, animals, microbe; 2. B, B, A, B, A, A; 3. reproduce, viable; 4. population; 5. plants, animals, microbes, each other, environment; 6. perpetuated; 7. deciduous forest, grassland, deserts, coniferous forest, tundra, tropical rain forest; 8. communities; 9. grassland/forest; 10. deciduous forest; 11. grassland; 12. desert; 13. climate; 14. true; 15. flow of rivers, migration of animals; 16. biosphere; 17. biotic, abiotic; 18. producers, consumers, decomposers/detritus feeders; 19. carry on photosynthesis; 20. light, carbon dioxide, water, glucose; 21. air, rocks, or water; 22. organic; 23. organic, inorganic; 24. light; 25. autotrophs, green plants; 26. heterotrophs, animals and fungi; 27. false, Indian pipe; 28. humans, fish, worms; 29. herbivores; 30. deer, mice; 31. carnivores, omnivores; 32. PR, PR, PY, PR, PY, PY; 33. tapeworm, tick, any living organism; 34. detritus; 35. earthworms, clams, crayfish; 36. fungi, bacteria; 37. grass, grasshoppers, frogs, snakes, a, b, c, d, e; 38. First trophic level: a, f, g, o, Second trophic level: b, h, i, k, n, Third trophic level: c, d, e, j, l, m, Example of food web relationship would be to link all the Second trophic level organisms with all the First trophic level organisms; 39. a, d, d, c, b, c, b, d, d, a, d; 40. total dry weight of all organisms occupying a trophic level; 41. less, less; 42. illustrations should be a pyramid; 43. relationship between bees and flowers; 44. bees and flowers require a close union to accomplish life functions; 45. habitat, niche; 46. birds feeding in different parts of a tree, day versus night-time feeders on same food source; 47. wind, temperature, precipitation; 48. see Fig. 2-19; 49. stressed; 50. death; 51. true; 52. one; 53. light, temperature, water; 54. temperature, precipitation; 55. water; 56. temperature; 57. Alpine tundra; 58. microclimate; 59. a; 60. ocean, desert, mountain range; 61. roadways, dams, cities; 62. c; 63. agriculture; 64. producing abundant food, creating reservoirs and distributing water, overcoming predation and disease, constructing our own habitats, overcoming competition with other species; 65. true

---

ANSWERS TO SELF-TEST

1. plants; 2. animals; 3. microbes; 4. each other; 5. environment; 6. abiotics; 7. plants; 8. animals; 9. b; 10. c; 11. b; 12. 1; 13. 5; 14. 2; 15. 1; 16. 2; 17. 1; 18. all; 19. all; 20. d; 21. energy loss from one level to the next decreases potential for biomass production at higher levels; 22. c; 23. d; 24. a; 25. b; 26. a; 27. a; 28. -; 29. -; 30. +; 31. +; 32. +; 33. + 34. -; 35. -.

Term	Definition	Example(s)	Term	Definition	Example(s)
abiotic	b	b	heterotroph	m	c,m,s,t,u,v,w
abiotic factors	kk	b	host	aa	a,c,g,k,l,m,s,t, u,v,w
autotroph	l	l	inorganic molecule	g	q
biomass pyramid	ee	x	limiting factors	oo	b
biomass	i	i	limits of tolerance	uu	aa
biome	tt	ee	microclimate	rr	dd
biosphere	j	j	mutualism	ff	y
biota	a	a	niche	jj	z
biotic community	a	a	omnivore	w	v
biotic factors	ss	a	optimum	ll	aa
biotic structure	k	k	organic molecule	p	p
carnivore	v	u	parasites	z	w
chlorophyll	o	o	parasitism	hh	w
climate	qq	cc	photosynthesis	n	n
consumers	m	m	population	d	d
decomposer	s	s	predator	x	u
detritus	r	r	prey	y	c,t,v
detritus feeders	vv	s	primary consumer	t	t
ecologist	g	g	producers	l	l
ecology	f	f	range of tolerance	uu	aa
ecosystem	e	e	secondary consumer	u	u
ecotone	h	e,h	species	c	c
food chain	bb	x	symbiotic	gg	y
food web	cc	x	synergistic effect	pp	bb
habitat	ii	e	trophic levels	dd	x
herbivore	t	t	zone of stress	mm	aa

VOCABULARY DRILL ANSWERS

CHAPTER 3

ECOSYSTEMS: HOW THEY WORK

Matter, Energy and Life
    Matter in Living and Nonliving Systems
    Energy Considerations
    Energy Changes in Organisms and Ecosystems

Principles of Ecosystem Function
    Energy Flow in Ecosystems
    Biogeochemical Cycles
        The Carbon Cycle
        The Phosphorus Cycle
        The Nitrogen Cycle

Implications for Humans
    Sustainability
    Value
    Managing Ecosystems

Vocabulary Drill
    Vocabulary Definitions
    Vocabulary Examples

Self-Test

Answers to Study Guide Questions

Answers to Self-Test

Vocabulary Drill Answers

Matter, Energy and Life

1.   **Matter**, in a chemical sense, refers to all _____, _____, and _____.

2.   **Atoms** or **elements** are the basic building blocks of all _____.

3.   How many different kinds of elements occur naturally? _____

4.   Only _____ of all the known elements are used to produce living tissues (see Table 3-2, iodine is missing).

31

5. Indicate whether the following statements are true [+] or false [-] concerning the **Law of Conservation of Matter**.

    [ ] atoms cannot be created
    [ ] atoms can be destroyed
    [ ] atoms can be converted into others
    [ ] atoms can be rearranged to form different molecules

Matter in Living and Nonliving Systems

6. Is it (True or False) that a **molecule** = a **compound**? Is it (True or False) that a compound = a molecule?

7. Indicate whether each of the following is an element [E] or compound [C].

    a. [ ] oxygen b. [ ] carbon dioxide c. [ ] water   d. [ ] nitrogen
    e. [ ] ozone   f. [ ]   phosphorus   g. [ ] carbon h. [ ] protein
    i. [ ] sugar   j. [ ] hydrogen

8. For each of the following elements, essential in living organisms, give a compound or molecule which contains it and where it is found in the abiotic environment. (see Table 3-1).

Element	Compound Containing It	Environmental Location
carbon	_____	_____
hydrogen	_____	_____
oxygen	_____	_____
nitrogen	_____	_____
phosphorus	_____	_____

9. Match the list of elements and molecules.

Elements	Molecules
a. N. _____	1. glucose
b. C. _____	2. proteins
c. H. _____	3. starch
d. O. _____	4. fats
e. P. _____	5. nucleic acids
f. S. _____	6. all of the above

10. Is it (True or False) that all **organic** compounds are found in living systems.

Energy Considerations

11. **Matter** is defined as anything that occupies _____ and has _____.

12. From the most to least dense states, matter can exist in the forms of
_____, _____ and _____.

13. **Energy** is defined as: _____

14. List three forms of energy. _____, _____, and
_____

15. Indicate whether the following statements about energy are true [+] or false [-].

[ ] Energy occupies space.
[ ] Energy can be weighed.
[ ] Kinetic energy can be transformed into potential energy.
[ ] Potential energy can be transformed into kinetic energy.
[ ] Energy can be recycled.
[ ] Energy can be created.
[ ] Energy can be converted from one form to another.
[ ] There is some loss in each energy conversion.
[ ] Energy can be measured.

16. Heat, light, and motion are forms of (Kinetic, Potential) energy.

17. Glucose, a compressed spring, and fossil fuels are forms of (Kinetic, Potential) energy.

18. The high potential energy in fuels is commonly referred to as _____ energy.

19. A _____ is the unit measure of the amount of heat required to raise the
temperature of 1 gram of water 1 degree Celsius.

20. Indicate whether the following statements pertain to the [a] **First** or [b] **Second Laws of Thermodynamics**.

[ ] Energy cannot be created or destroyed.
[ ] Energy can be converted from one form to another.
[ ] Energy conversions always result in net losses of useful energy.
[ ] Heat is the final energy conversion form.

Energy Changes in Organisms and Ecosystems

21. Identify whether the level of potential energy in the following types of molecules is high
[H] or low [L].

[ ] inorganic   [ ] organic   [ ] coal
[ ] water   [ ] sugar   [ ] wood

22. In most ecosystems, the primary producers are plants that carry on _____.

23. Light energy for photosynthesis is absorbed by the _____ molecule.

24. The efficiency of photosynthesis in converting light energy to chemical energy is at best _____ percent efficient.

25. The primary product of photosynthesis is _____.

26. List three ways plants use the glucose produced during photosynthesis.

    a. _____

    b. _____

    c. _____

27. The basic source of energy and raw materials for animal growth and developement is

    _____

28. **Cell respiration** is effectively the _____ process of photosynthesis in terms of what is consumed and released.

29. Indicate whether each of the following refers to photosynthesis [P], cell respiration [R], both [B], or neither [N].

    [   ] releases oxygen
    [   ] stores energy
    [   ] releases carbon dioxide
    [   ] consumes carbon dioxide
    [   ] releases energy
    [   ] produces sugar
    [   ] consumes sugar
    [   ] consumes oxygen

30. Place an [x] by any of the following organisms that carry on cell respiration for at least part of their energy needs.

    [   ] consumers [   ] decomposers [   ] plants [   ] fungi [   ] bacteria

31. The lack of certain nutrients in your diet may lead to various diseases caused by

    _____.

32. The point of a balanced diet is that it supplies both _____ and
    _____ in adequate but not excessive amounts.

33. Trace the three pathways of food eaten by any consumer.

    a. _____

b. _____

c. _____

34.   Is it (True or False) that fecal waste is no longer a source of nutrients and energy?

35.   Give two examples of detritus feeders that require a mutualistic relationship with other organims to digest **cellulose**.

a. _____   b. _____

Principles of Ecosystem Function

Energy Flow in Ecosystems

36.   What three factors account for energy loss from one trophic level to another?

a. _____

b. _____

c. _____

37.   There is an estimated _____ % loss of energy as it moves from one trophic level to the next.

38.   Biomass (Increases, Decreases) at each successive trophic level.

39.   The energy available to producers is (Greater, Less) than that available to herbivores, therefore, the biomass of producers will be (Greater, Less) than that of herbivores.

40.   The energy available to herbivores is (Greater, Less) than that available to carnivores, therefore, the biomass of herbivores will be (Greater, Less) than that of carnivores.

41.   Indicate whether the following terms pertain to the first [1], second [2], third [3] or fourth [4] principle of ecosystem sustainability (See Table 3-1).

[   ] biomass [   ] recycling [   ] solar energy [   ] nonpolluting [   ] carbon

[   ] food chains [   ] ammonium and nitrates

42.   Explain why the illustration of total biomass at different trophic levels is a pyramid.

_____

_____

43.   Eventually all of the energy escapes as _____ energy.

44. Is it (true or false) that energy can be recycled?

45. *For sustainability, ecosystems use sunlight as their source of energy* is the (First, Second, Third) basic principle of ecosystem sustainability.

46. Identify two benefits derived from solar energy.

    a. _____ b. _____

Biogeochemical Cycles

47. *For sustainability, ecosystems dispose of wastes and replenish nutrients by recycling all elements* is the (First, Second, Third) basic principle of ecosystem sustainability.

48. Natural ecosystems do not harm themselves with their own waste products because these products are _____.

The Carbon Cycle

49. The reservoir of carbon for the carbon cycle is _____ present in the _____.

50. Carbon dioxide is first incorporated into sugar by the process of _____ and then it may be passed to other organic compounds in other organisms through food chains.

51. At any point on the food chain an organism may use the carbon compounds in cell respiration to meet its energy needs. When this occurs, the carbon atoms are released as _____.

52. The amount of atmospheric CO_2 is being increased greatly by the burning of _____.

53. Complete the below diagram of the CARBON CYCLE by putting in all arrows of interrelationships between organisms and labeling the empty boxes (see Fig 3-16 for an example of this exercise).

The Phosphorus Cycle

54. The reservoir of the element phosphorus exists in various rock and soil
    _____.

55. Plants absorb phosphate from the _____ or _____
    solution.

56. Phosphate is first assimilated into organic compounds as _____
    phosphate and then passed to various other organisms through the food chain.

57. At any point on the food chain, an organism may use organic phosphorus in respiration.
    When this occurs, the phosphate is released back to the environment in _____.

58. Humans disrupt the phosphorus cycle by _____ and
    _____ .

59. Complete the below diagram of the PHOSPHORUS CYCLE by putting in all arrows of
    interrelationships between organisms and labeling the empty boxes (see Fig 3-17 for an
    example of this exercise).

The Nitrogen Cycle

60. The main reservoir of N_2 is in the air which is about _____ percent nitrogen gas.

61. Is it (True, False) that plants cannot assimilate nitrogen gas from the atmosphere?

62. To be assimilated by higher plants, nitrogen must be present as _____ or _____ ions.

63. A number of bacteria and certain blue green algae can convert nitrogen gas to the ammonium form, a process called nitrogen _____.

64. The bacteria have a mutualistic relationship with the _____ family of plants.

65. Once fixed, nitrogen may be passed down food chains, excreted as _____ and recycled as the mineral nutrient called _____.

66. Nitrogen eventually returns back to the atmosphere because certain bacteria convert the nitrate and ammonia compounds back to _____.

67.   Complete the below diagram of the NITROGEN CYCLE by putting in all arrows of interrelationships between organisms and labeling the empty boxes (see Fig 3-18).

68.   Is it (True or False) that all natural terrestrial ecosystems are dependent on the presence of nitrogen fixing organisms and legumes?

69.   The human agricultural system bypasses nitrogen fixation by using fertilizers containing _____ or _____.

Implications for Humans

Sustainability

70.   Indicate whether the following human actions are a violation of the first [1] of second [2] principle of ecosystem sustainability.

        [   ] lack of recycling
        [   ] excessive use of fossil fuels
        [   ] feeding largely on the third trophic level
        [   ] excessive use of fertilizers

[   ] use of nuclear power
[   ] using agricultural land to produce meat
[   ] destruction of tropical and other rainforests
[   ] nutrient overcharge into marine and freshwater ecosystems
[   ] production and use of nonbiodegradable compounds

Value

71.   **Natural capital** is identified by the _____ and _____
        represented in natural ecosystems.

72.   List five of the 17 major ecosystem goods and services.

        a. _____   b. _____

        c. _____   d. _____

        e. _____

73.   What is the estimated total value to human welfare for one year of natural ecosystem goods
        and services? _____ trillion dollars

Managing Ecosystems

74.   What is the goal of ecosystem management?

        _____

75.   What does the goal of ecosystem management mean?

        _____

VOCABULARY DRILL

Directions: Match each term with the list of definitions and examples given in the tables below. Term definitions and examples may be used more than once.

Term	Definition	Example(s)	Term	Definition	Example(s)
anaerobic			malnutrition		
atoms			matter		
cellulose			mineral		
chemical energy			mixture		
compound			molecule		
elements			natural organics		
energy			organic molecule		
entropy			organic phosphate		
fermentation			overgrazing		
First Law of Thermodynamics			oxidation		
inorganic molecule			potential energy		
kinetic energy			Second Law of Thermodynamics		
Law of Conservation of Matter			synthetic organics		

VOCABULARY DEFINITIONS
a. anything that occupies space and has mass
b. basic building blocks of all matter
c. any two or more atoms bound together
d. any two or more different atoms bound together
e. no chemical bonding between molecules involved
f. any hard crystalline material of a given composition
g. carbon-based molecules which make up the tissue of living organisms
h. molecules with neither carbon-carbon nor carbon-hydrogen bonds
i. compounds produced by living organisms
j. compounds produced by humans in test tubes

VOCABULARY DEFINITIONS

k. the ability to move matter
l. energy in action or motion
m. energy in storage
n. potential energy contained in fuels
o. energy is neither created nor destroyed by may be converted from one form to another
p. in any energy conversion, you will end up with less usable energy that you started with
q. disorder
r. systems will go spontaneously in one direction only: toward increasing entropy
s. process of oxidizing glucose in the presence of oxygen into carbon dioxide and water
t. process of breaking down molecules
u. absence of one or more nutrients in the diet
v. stored form of glucose in plant cell walls
w. partial oxidation of glucose that can occur in the absence of oxygen
x. oxygen-free environment
y. phosphate bound in organic compounds
z. a form of nonsustainable harvest
aa. atoms are not created or destroyed in chemical reactions

VOCABULARY EXAMPLES

a. gases, liquids, solids	j. burning match	s. alcohol
b. NCHOPS	k. a potatoe	t. deep space
c. O_2	l. gasoline	u. nucleotides
d. CH_4	m. You can't get something for nothing.	v. clear-cut logging
e. muddy water	n. You can't even break even.	w. $C_6H_{12}O_6 + 6O_2 \rightarrow 6CO_2 + 6H_2O$
f. diamond	o. decomposition	
g. NaCl	p. digestion	
h. plastics	q. vitamin deficiency	
i. walking	r. fiber	

SELF-TEST

1.   Identify 17 elements out of the 108 (Table C-1) that are essential in the structure or function of living organisms? (Your text gave you 16 in Table 3-2.)

The essential elements of life include:_____

_____

       If a nonessential element becomes incorporated into a living system (e.g., yourself); what are three potential impacts of this "element insult" on the living system?

2. _____

3. _____

4. _____

       Identify from the examples of organization of matter on the right the most simple and the most complex levels.

5.   [   ] Most simple   a. organisms
                                     b. proteins, carbohydrates, fats
6.   [   ] Most complex   c. atoms
                                     d. cells

7.   What is the carbon source for photosynthesis? _____

8.   What is the energy source for photosynthesis? _____
9.   What is the carbon source for cellular respiration? _____

10.   What is the energy source for cellular respiration? _____

       Use the below chemical equations to answer the next series of questions.

       a. $6CO_2 + 6H_2O \rightarrow C_6H_{12}O_6 + 6O_2$

       b. $C_6H_{12}O_6 + 6O_2 \rightarrow 6CO_2 + 6H_2O$

       Which equation:

11.   [   ] is the chemical formula for photosynthesis?

12.   [   ] is the chemical formula for cellular respiration?

13.   [   ] shows glucose as an end product?

14.   [   ] shows the raw materials for photosynthesis as the end product?

15.   [   ] uses a kinetic energy source?

16.   [   ] uses a potential energy source?

17.   How does equation [a] demonstrate the Law of Conservation of Matter?

_____   _____

_____   _____

Associate the below terms with [a] First Law or [b] Second Law of Thermodynamics.

18.   [   ] entropy

19.   [   ] loss

20.   [   ] change

21.   In every energy conversion there is:

        a. some energy is converted to heat   b. heat energy is lost
        c. lost heat may be recaptured and reused   d. a and b

22.   Consumers must feed on preexisting organic material to obtain:

        a. nutrients   b. energy   c. both a and b
        d. neither a or b, consumers make their own nutrients and energy
23.   Which of the following is not recycled in natural ecosystems

        a. nitrogen   b. carbon   c. energy   d. phosphorus

24.   The carbon atoms in the food you eat will

        a. be exhaled as carbon dioxide b. become part of your body tissues
        c. pass through as fecal wastes   d. a, b, and c

25.   In the phosphorus cycle, soil phosphate first enters the ecosystem through:

        a. plants   b. herbivores   c. carnivores   d. decomposers

26.   Nitrogen fixation refers to:

        a. releasing nitrogen into the air b. converting it to a chemical form plants can use
        c. repairing broken molecules   d. applying fertilizer

27.   The nitrogen cycle relies on

        a. mutualism   b. commensalism   c. parasitism   d. synergism

        Indicate whether the carbon [C], nitrogen [N], phosphorus [P], or more than one [M]
cycle has been negatively effected by the following human actions.

28.   [   ] abandonment of rotation cropping

29.   [   ] heavy reliance on fossil fuels

30.   [   ] discharges of household detergents into natural waterways

31.   What LAW explains why it is impossible to have a greater biomass at the highest level in a
        food chain?

        a. Law of Conservation of Matter,   b. First or,   c. Second Law of Thermodynamics

32.   How does this law (your response to item #31) explain why it is impossible to have a greater
        biomass at the highest level in a food chain.

        _____

        _____

33.   Explain why it makes sense to nurture a recycling mentality in the human system?

        _____

ANSWERS TO STUDY GUIDE QUESTIONS

1. gases, liquids, solids; 2. life; 3. 92; 4. 17; 5. +, +, -, +; 6. true, false; 7. E, C, C, E, C, E, E, C, C, E; 8. (carbon dioxide, water), (water, water), (oxygen gas, air), (nitrogen gas, air), (phosphate ion, dissolved in water or in rock or soil minerals); 9. 2 and 5, 6, 6, 6, 4 and 5, 2 and 5; 10. false; 11. space, mass; 12. solids, liquids, gas; 13. ability to move matter; 14. kinetic, potential, chemical; 15. -, -, +, +, -, -, +, +, +; 16. kinetic; 17. potential; 18. chemical; 19. calorie; 20. a, a, b, b; 21. L, H, H, L, H, H; 22. photosynthesis; 23. chlorophyll; 24. 2-5%; 25. glucose; 26. raw materials for production of other organic molecules, plant cell respiration, stored for future use; 27. glucose; 28. reverse; 29. P, B, R, P, B, P, B, B; 30. all are marked; 31. malnutrition; 32. calorie, nutrients; 33. 60% used to produce energy, body growth and maintenance, undigested residue passes out as wastes; 34. false; 35. termites, cows; 36. much of the preceding trophic level is standing biomass and is not consumed, much of what is consumed is used for energy, some of what is consumed is undigested and passes through the organism; 37. 90; 38. decreases; 39. greater, greater; 40. greater, greater; 41. 3, 1, 2, 2, 1, 3, 1; 42. because of energy loss from lower to higher trophic levels; 43. heat; 44. false; 45. first; 46. nonpolluting, nondepleting; 47. second; 48. recycled; 49. carbon dioxide, air; 50. photosynthesis; 51. carbon dioxide; 52. fossil fuels; 53. see Fig. 3-16; 54. minerals; 55. soil, water; 56. organic; 57. urine and other wastes; 58. cutting down rain forests, phosphate fertilizer runoff from agricultural lands; 59. see Fig. 3-17; 60. 78; 61. true; 62. ammonium, nitrates; 63. fixation; 64. legume; 65. ammonium, nitrates; 66. nitrogen gas; 67. see Fig. 3-18; 68. true; 69. ammonium, nitrate; 70. 1, 2, 2, 1, 2, 1, 2, 1, 1, 1; 71. goods, services; 72. see Table 3-3; 73. 33; 74. manage ecosystems to assure their sustainability; 75. maintaining ecosystems such that they continue to provide the same level of goods and services indefinitely into the future.

ANSWERS TO SELF-TEST

1. B, Ca, C, Cl, Cu, H, I, Fe, Mg, Mn, Mo, N, O, P, K, Na, S, Zn; 2. death; 3. the element might be stored in body fat or organs; 4. the element may be discharged with other wastes; 5. c; 6. a; 7. carbon dioxide; 8. sunlight; 9. glucose; 10. glucose; 11. a; 12. b; 13. a; 14. b; 15. a; 16. b; 17. the same number of C, H, and O atoms that go into the reaction come out at the end of the reaction; 18. b; 19. b; 20. a; 21. d; 22. c; 23. c; 24. d; 25. a; 26. b; 27. a; 28. N; 29. C; 30. P; 31. c; 32. energy loss from one level to the next decreases potential for biomass production at higher levels; 33. because all nonrenewable resources are available in fixed amounts and recycling converts fixed to infinite

Term	Definition	Example (s)	Term	Definition	Example (s)
anaerobic	x	t	malnutrition	u	q
atoms	b	b	matter	a	a
cellulose	v	r	mineral	f	f
chemical energy	n	l	mixture	e	e
compound	d	d	molecule	c	c, d
elements	b	b	natural organics	i	d
energy	k	i	organic molecule	g	d
entropy	q	o	organic phosphate	y	u
fermentation	w	s	overgrazing	z	v
First Law of Thermodynamics	o	m	oxidation	t	i, j, p, s, w
inorganic molecule	h	g	potential energy	m	h, k, l, r, s
kinetic energy	l	i, j	Second Law of Thermodynamics	p, r	n
Law of Conservation of Matter	aa	w	synthetic organics	j	l

VOCABULARY DRILL ANSWERS

CHAPTER 4

ECOSYSTEMS: POPULATIONS AND SUCCESSION

Population Dynamics

1.   A balance between births and deaths is called a **population** _____.

Population Growth Curves

2.   A population that continues to grow exponentially until it exhausts essential resources and then has a precipitous dieoff demonstrates the _____ curve of population growth.

3. A population that grows exponentially and then exhibits a leveling off demonstrates the _____ curve of population growth.

4. Animal populations subjected to human disturbances usually demonstrate the _____ curve of population growth.

5. The _____ curve of population growth is an indicator of ecosystem nonsustainability.

Biotic Potential versus Environmental Resistance

6. Indicate whether the following factors are examples of **biotic potential** [BP] or **environmental resistance** [ER].

   a. [   ] adverse weather conditions   b. [   ] ability to cope with weather conditions

   c. [   ] predators   d. [   ] defense mechanisms

   e. [   ] competitors   f. [   ] reproductive rate

7. If biotic potential is higher than environmental resistance, the population will (Increase, Decrease).

8. Population explosions occur when biotic potential is (Higher, Lower) than environmental resistance.

9. When a population is more or less stable in terms of numbers, biotic potential is (Higher Than, Lower Than, Equal To) environmental resistance.

10. Is it (True or False) that there is a finite number of plants or animals that can be supported by an ecosystem?

11. Describe two types of **reproductive strategies** used by plants or animals.

    a. _____

    b. _____

12. Is it (True or False) that the dynamic balances which exist between biotic potential and environmental resistance may be easily upset by humans?

13. Is it (True or False) that when balances are upset, there are population explosions of some species and extinctions of other species?

Density Dependence and Critical Numbers

14. Does environmental resistance (Increase or Decrease) as **population density** increases?

15. How are the concepts of **critical numbers** and **endangered** species related?

    _____

Mechanisms of Population Equilibrium

Predator-Prey and Host-Parasite Dynamics

16.   Explain how Fig. 4-5 represents an example of *predator-prey balance*.

        _____

17.   If a herbivore population is not controlled by natural enemies, it is likely that the herbivore
        population will (Increase, Decrease) and the vegetation it feeds on will (Increase, Decrease).

18.   Which of the following groups of organism requires a high degree of adaptation to maintain
        a balance between environmental resistance and biotic potential? (Check all that apply.)

        [　] predator   [　] prey   [　] parasite   [　] host

19.   Is it (True or False) that a natural enemy can be so effective that it causes the extinction of
        its prey or host? If you said True, give an example _____

20.   If individuals are stressed by environmental resistance, indicate whether they would [+] or
        would not [-] exhibit the following characteristics.

        [　] less able to compete successfully with other species
        [　] more vulnerable to attack by predators
        [　] more vulnerable to attack by parasitic and disease organisms
        [　] more likely to attempt to disperse from the area

Introduced Species

21.   Explain how ecosystems got out of balance with the following introductions of exotics.

        a. rabbits in Australia: _____

        b. chestnut blight in America: _____

        c. domestic cats on islands: _____

        d. zebra mussel in Great Lakes: _____

22.   Give an example of an introduction of an exotic that did not affect the balance.

        _____

23.   List the two conditions that enable introduced species to thrive in their new ecosystems.

        a. _____

        b. _____

Territoriality

24.   Is it (True or False) that effective territorial behavior by an animal results in priority use of those resource available in a specific area?

25.   List three options for survival available to individuals unable to claim a territory.

   a. _____   b. _____

   c. _____

Plant-Herbivore Dynamics

26.   Indicate whether the following activities will maintain [+] or upset [-] the plant-herbivore balance.

   [   ] introducing goats on islands
   [   ] killing off herbivore predators
   [   ] confining herbivores in a localized area
   [   ] allowing the herbivores to exceed the carrying capacity of the habitat
   [   ] perpetuating monocultures

Competition Between Plant Species

27.   Identify three factors that prevent one plant species from driving out others in competion.

   a. _____

   b. _____

   c. _____

28.   Indicate whether the following conditions are [+] or not [-] examples of **balanced herbivory**.

   [   ] plant species that demonstrate unique adaptations to specific microclimates
   [   ] introducing plant species that out compete native species
   [   ] introducing herbivores that out compete native grazing animals
   [   ] introducing carnivores that out compete native carnivores
   [   ] introducing carnivores that prey species do not recognize
   [   ] development of rubber plantations

Disturbance and Succession

29.   Is it (True or False) that changes in ecosystem structure and/or function, over time, are inevitable.

30.   How is your answer to #29 consistent with or a contradiction to the *Equillibrium Theory*?

       _____

       _____

Ecological Succession

31.   Is it (True or False) that there is an end point in the process of **ecological succession**?

32.   List three examples of ecological succession.

       a. _____   b. _____   c. _____

33.   In what order (1 to 4) would the following plant communities appear during primary
       succession?

       [ ] trees   [ ] mosses   [ ] shrubs   [ ] larger plants

34.   How do mosses change the conditions of the area so that it can support larger plants?

       _____

35.   Suppose that a section of eastern deciduous forest was cleared for agriculture and later
       abandoned. In what order (1 to 4) would the following species invade the area?

       [ ] deciduous trees   [ ] crabgrass   [ ] pine trees   [ ] grasses

36.   What prevents hardwood species of the deciduous forest from reinvading the area
       immediately?

       _____

37.   The end result of aquatic succession is a _____ ecosystem.

Disturbance and Biodiversity

38.   What role does biodiversity play in ecological succession?

       _____

Fire and Succession

39.   In which of the below ecosystems can fire be used as a management tool? (Check all that
       apply.)

       [ ] grasslands   [ ] tropical rainforests   [ ] redwood trees   [ ] pine forests

40. Indicate whether the following events are [+] or are not [-] the results of controlled ground fires.

    [   ] removal of dead litter
    [   ] permanent harm of mature trees
    [   ] seed release of some tree species
    [   ] destruction of wood-boring insects
    [   ] crown fires
    [   ] loss of human lives and property

Non-equilibrium Systems

41. List three ways disturbances contribute to ecosystem structure.

    a. _____   b. _____

    c. _____

42. Sustainability of ecosystems is dependent on maintaining constant _____ and _____ conditions.

The Fourth Principle of Ecosystem Sustainability

43. Write the fourth principle of ecosystem sustainability.

    _____

Implications for Humans

44. The human population is exploding because we have effectively

    a. increased our biotic potential. (True or False)
    b. decreased our environmental resistance. (True or False)

45. The human population growth pattern presently resembles a _____ curve.

46. List three ways in which humans are causing the upset of natural ecosystems.

    a. _____

    b. _____

    c. _____

VOCABULARY DRILL

Directions: Match each term with the list of definitions and examples given in the tables below. Term definitions and examples may be used more than once.

Term	Definition	Example(s)	Term	Definition	Example(s)
aquatic succession			host specific		
balanced herbivory			monoculture		
biotic potential			natural enemies		
carrying capacity			population explosion		
climax ecosystem			population density		
critical number			primary succession		
ecological restoration			recruitment		
ecological succession			replacement level		
environmental resistance			reproductive strategies		
exponential increase			secondary succession		
fire climax			territoriality		

VOCABULARY DEFINITIONS
a. number of offspring a species may reproduce under ideal conditions
b. number of offspring that survive and reproduce
c. the way species achieve recruitment in their populations
d. a doubling of each generation
e. the result of exponential increase
f. biotic and abiotic factors that tend to decrease population numbers
g. number of offspring required to replace breeding population
h. number of individuals per unit area
i. the minimum population base (number) of a species
j. predators and parasites that control population size
k. defending territory against encroachment of others of the same species
l. maximum population a given habitat will support without the habitat being degraded over time
m. growth of a single species over a wide area

VOCABULARY DEFINITIONS

n. parasites that attack only one species and close relatives
o. balance among competing plant populations maintained by herbivores
p. the orderly transition from one biotic community to another
q. a stage of equilibrium between all species and the physical environment
r. succession in an area that has not been previously occupied
s. succession in an area that has been previously occupied
t. replacement of aquatic by terrestrial communities
u. ecosystems that depend on the recurrence of fire to maintain existing balance
v. reclaiming natural ecosystems

VOCABULARY EXAMPLES

a. live births, eggs laid, seeds in plants
b. seeds that sprout and produce more seeds
c. produce massive numbers of young
d. $2 \rightarrow 4 \rightarrow 8 \rightarrow 16 \rightarrow 32$
e. lack of food, space, and mates
f. number of offspring required to replace breeding population
g. 2 deer/acre
h. endangered species
i. predators and parasites
j. scent marking
k. upper limit on S-shaped population growth curve
l. a field of cotton
m. host specific: dog tape worm
n. the opposite of monoculture
o. bare rock $\rightarrow$ moss $\rightarrow$ herbaceous plants $\rightarrow$ shrubs $\rightarrow$ trees
p. deciduous forest
q. deciduous forest $\rightarrow$ monoculture $\rightarrow$ grassland $\rightarrow$ shrubs $\rightarrow$ pine forest $\rightarrow$ deciduous forest
r. lake $\rightarrow$ marsh $\rightarrow$ grassland
s. grasslands and pine forests
t. wetlands

SELF-TEST

Indicate whether the following factors are examples of biotic potential [BP] or environmental resistance [ER].

1.   [   ] lack of suitable habitat
2.   [   ] disease
3.   [   ] reproductive rate
4.   [   ] ability to migrate
5.   [   ] + adaptability
6.   [   ] + competition

Identify whether the following traits are characteristic of [J] or [S] population growth curves.

7.   [   ] biotic potential = environmental resistance

8.   [   ] biotic potential > environmental resistance

9.   [   ] population increases and dies off in rapid sequence

10.   [   ] violates the Fourth Principle of Ecosystem Sustainability

11.   [   ] the human population growth curve is an example

12.   What concept concerning population numbers is the direct opposite of carrying capacity

_____

13.   In what order (1-4) would the following plant communities appear during primary succession?

        [   ] trees   [   ] mosses   [   ] grasses   [   ] shrubs

14.   In what order (1-4) would the following plant communities appear during secondary succession?

        [   ] grasses   [   ] pine trees   [   ] herbaceous plants   [   ] deciduous trees

15.   What ecological factors define primary versus secondary succession?

_____

16.   In which of the following ecosystems can fire not be used as a management tool?

        a. grasslands   b. coniferous forests   c. hardwood forests   d. a and b

17.   Which of the following is not an intended function of ground fires?

        a. removal of dead litter b. seed release of some tree species
        c. starting crown fires   d. destruction of wood-boring insects

18.   Aggressive defense of habitat is called

       a. biotic potential   b. succession   c. territoriality   d. the niche

19.   The process whereby the growth of one community causes changes that make the
       environment more favorable to a second community and less favorable to the first is called

       a. biotic potential   b. succession   c. territoriality   d. the niche

20.   Which of the following is not a characteristic of climax ecosystems?

       a. have maximum species diversity
       b. have different dominant plant forms
       c. are self-perpetuating
       d. will maintain themselves regardless of climatic, abiotic, or biotic changes

Which of the following human actions will [+] and will not [-] cause upsets of natural ecosystems?
(Note: the following list expands on that given in the text)

21.   [   ] increasing population size

22.   [   ] decreasing the gross national product

23.   [   ] supporting conservation organizations

24.   [   ] introducing species from one ecosystem to another

25.   [   ] eliminating natural predators

ANSWERS TO STUDY GUIDE QUESTIONS

1 equillibrium; 2. J; 3. S; 4. J; 5. J; 6. ER, BP, ER, BP, ER, BP; 7. increase; 8. higher; 9. equal to; 10. true; 11. have massive numbers of offspring, low reproductive rate with long-term care of young; 12. true; 13. true; 14. increase; 15. species become endangered when they reach critical numbers; 16. figure shows predictable rises and falls in populations of wolves and moose; 17. increase, decrease; 18. all should be checked; 19. true, humans; 20. all +; 21. rabbit population exploded and overgrazed grassland, wiped out American chestnut, overly effective predators, endemic species displacement; 22. ring-necked pheasant; 23. lack of competition with endemic species, no predators; 24. true; 25. live to maturity, die, migrate; 26. all -; 26. true; 27. microclimates or microhabitats, single species cannot use all the resources in the area, balanced herbivory; 28. +,-,-,-,-,-; 29. true; 30. it is not consistent because nothing stays the same forever; 31. true; 32. primary, secondary, aquatic; 33. 4, 1, 3, 2; 34. provide first layer of humus; 35. 4, 1, 3, 2; 36. temperature too hot; 37. terrestrial; 38. provides populations for invasions of new habitats; 39. grasslands, redwoods, pine forests; 40. +,-,+,+,-,-; 41. remove organisms, reduce populations, create opportunities for other species to colonize; 42. abiotic, biotic; 43. For sustainability, biodiversity is maintained; 44. true, true; 45. J; 46. introducing species from one ecosystem to another, eliminating natural predators, losing biodiversity, altering abiotic factors, misunderstanding the role of fire.

ANSWERS TO SELF-TEST

1. ER; 2. ER; 3. BP; 4. BP; 5. BP; 6. ER; 7. S; 8. J; 9. J; 10. J; 11. J; 12. critical number; 13. 4, 1, 3, 2; 14. 1, 3, 2, 4; 15. a suitable soil base to start with; 16. c; 17. c; 18. c; 19. b; 20. d; 21. -; 22. +; 23. +; 24. -; 25. -.

VOCABULARY DRILL ANSWERS

Term	Definition	Example(s)	Term	Definition	Example(s)
aquatic succession	t	r	host specific	n	m
balanced herbivory	o	n	monoculture	m	l
biotic potential	a	a	natural enemies	j	i, m
carrying capacity	l	k	population explosion	e	d
climax ecosystem	q	p	population density	h	g
critical number	i	h	primary succession	r	o
ecological restoration	v	t	recruitment	b	b
ecological succession	p	o, q, r	replacement level	g	f
environmental resistance	f	e	reproductive strategies	c	c
exponential increase	d	d	secondary succession	s	q, r
fire climax	u	s	territoriality	k	j

CHAPTER 6

THE HUMAN POPULATION: DEMOGRAPHICS

The Population Explosion and Its Cause

The Explosion

1.   Chart the progress of human population growth from 1830 to that projected for 2046.

Year	Billions of People	Years to reach population
a. 1830	_____	_____

Year	Billions of People	Years to reach population
b. 1930	_____	_____
c. 1960	_____	_____

Year	Billions of People	Years to reach population
b. 1930	_____	_____
c. 1960	_____	_____
d. 1975	_____	_____
e. 1987	_____	_____
f. 1998	_____	_____
g. 2009	_____	_____
h. 2020	_____	_____
I. 2033	_____	_____
j. 2046	_____	_____

Reasons for the Explosion

2.   Why was the human population in a dynamic balance prior to the 1800s?

      _____

3.   List four factors that increased the recruitment rate of the human population after the
      1800s.

      a. _____   b. _____

      c. _____   d. _____

4.   Human population growth is expected to level off at _____ billion.

Different Worlds

Rich Nations and Poor Nations

5.   The countries of the world fall into three major economic categories (A. high-income,
      highly developed; B. middle-income, moderately developed; and C. low-income).
      Identify these categories with examples of countries or regions that are in each category
      (see Fig. 6-4).

      [   ] United States, Japan, Western Europe
      [   ] Latin America, East Asia
      [   ] East and Central Africa, India, China

6.   Developed countries have about _____ percent of the Earth's population, but they
      control about _____ percent of the world's wealth.

7.   Developing countries have _____ percent of the Earth's population, but control
      only _____ percent of the Earth's wealth.

8.   List six conditions of "absolute poverty."

      a. _____   b. _____   c. _____

      d. _____   e. _____   f. _____

Population Growth in Rich and Poor Nations

9.   Identify the countries that have the shortest and longest doubling times (see Table 6-3).

      a. _____   b. _____

10.   Over the past few decades, the fertility rates in developed countries have (Increased, Decreased) and in developing countries have (Increased, Decreased).

Different Populations Present Different Problems

11.   Match the characteristics below with [a] developed or [b] developing countries.

      [   ] high total fertility rates
      [   ] low total fertility rates
      [   ] short doubling times
      [   ] high consumptive lifestyles
      [   ] long doubling times
      [   ] annual rates of increase $\geq$ 2%
      [   ] annual rates of increase < 1%
      [   ] population growth is most intense
      [   ] poverty is most intense
      [   ] environmental degradation is most intense
      [   ] eating high on the food chain
      [   ] demand for tropical hardwoods
      [   ] demand for exotic wildlife

12.   The negative impacts of high consumptive lifestyles may be moderated by a factor called _____ regard.

13.   What three conditions must be met in order to achieve a sustainable population?

      a. _____   b. _____

      c. _____

Environmental and Social Impacts of Growing Populations and Affluence

The Growing Populations of Developing Nations

14.   List five ways people in developing nations are attempting to resolve population growth and land availability.

      a. _____ b. _____

c. _____ d. _____

e. _____

15. Indicate whether the following are [+] or are not [-] environmental conditions resulting from attempts to increase agriculture in developing countries.

[   ] decrease in per capita agricultural land
[   ] attempt agriculture on marginal land
[   ] soil erosion
[   ] loss of topsoil
[   ] water pollution
[   ] loss of soil productivity
[   ] deforestation

16. What is the fundamental problem with attempting to open new lands for agriculture in developing countries? _____

17. Indicate whether the following are [+] or are not [-] environmental/social conditions resulting from people in developing countries attempting to seek a better life by moving from the country to cities.

[   ] high population densities
[   ] lack of utilities and social services
[   ] polluted water
[   ] disease
[   ] high crime rates

18. Identify two sources of illicit income that people in developing countries might turn to in order to earn enough money to survive.

a. _____ b. _____

19. Do the majority of their customers come from (Developed, Developing) countries?

20. Many people in developing countries see (Emigration, Immigration) as their best hope for a brighter future.

21. Developed countries are placing (More, Fewer) restrictions on immigration.

22. Is it (True or False) that women in developing countries have access to an adequate social welfare system?

23. Is it (True or False) that cities in developing countries have high numbers of stray or abandoned children?

Effects of Increasing Affluence

24. Is (True or False) that the affects of affluence are all negative?

25. The United States contains _____ percent of the Earth's population and is responsible for _____ percent of the global emissions of carbon dioxide.

26. Explain why the cities in developing countries are more polluted (air and water) than cities in developed countries?

_____

27. TAKE A STAND! What is your position [a. agree, b. neutral, c. disagree] concerning the following population issues?

[   ] Further population growth will be beneficial.
[   ] Our technology will solve any problems associated with population growth.
[   ] Even at a population of 12 billion, we can all enjoy an affluenct lifestyle.

Dynamics of Population Growth

Population Profiles

28. Population profiles can be used to project future:

a. _____   b. _____

29. Indicate whether the following fertility rates will lead to an increase [+] decrease [-], or stability [0] in population growth.

[   ] Fertility rate < 2
[   ] Fertility rate = 2
[   ] Fertility rate > 2

30. Which of the above fertility rates is known as replacement fertility? _____

31. Match the below shapes of population profiles with the fertility rates given in question #29.

a. Column-shaped: Fertility rate = _____
b. Pyramid-shaped: Fertility rate = _____
c. Upside down pyramid: Fertility rate = _____

32. The population pyramid of Denmark would be _____ shaped.

33. The population pyramid of Kenya would be _____ shaped.

Population Projections

34. Identify two population projections provided in population profiles.

a. _____   b. _____

Based on the population profiles for Denmark and Kenya, predict whether MORE or LESS of the following goods and services will be needed over the next 20 years.

Service	Denmark	Kenya
Schools	35. _____	39. _____
Facilities for old people	36. _____	40. _____
New housing units	37. _____	41. _____
Utility services	38. _____	42. _____

Population Momentum

43.   Given the population momentum of Kenya their populations will continue to grow for _____ to _____ years after achieving replacement fertility levels.

The Demographic Transition

44.   What has been the observed trend in economic development and fertility rates?
_____

45.   Subtracting the crude death rate from the crude birth rate (CBR-CDR) and then dividing by 10 gives the annual rate of _____.

46.   Doubling time for a population may be found by dividing the annual rate of increase into (7, 70, 700).

47. Calculate the annual rate of increase and doubling time for the following countries.				
Country	Crude Birth Rate	Crude Death Rate	Rate of Increase	Doubling Time
Kenya	46	7		
Mexico	29	6		
United States	17	9		
Denmark	12	12		

48.   The goal of each country should be to have a (Long, Short) doubling time.

49.   The demographic transition refers to changes in infant _____ and _____ rates which tend to parallel development.

50.   There are (One, Two, Three, Four) major phases in the demographic transition?

51. In the table below, indicate whether the variables given after each phase are high [+], declining [-], or stable [0].

Phase	Birth Rate	Death Rate	Population
I	[　]	[　]	[　]
II	[　]	[　]	[　]
III	[　]	[　]	[　]
IV	[　]	[　]	[　]

52. Developed nations are in phase _____ .

53. Developing nations are in phase _____ .

VOCABULARY DRILL

Directions: Match each term with the list of definitions and examples given in the tables below. Term definitions and examples might be used more than once.

Term	Definition	Example(s)	Term	Definition	Example(s)
absolute poverty			doubling time		
age structure			environmental regard		
crude birth rate			natural increase		
crude death rate			population profile		
demographer			population momentum		
demographic transition			replacement fertility		
developed countries			total fertility rate		
developing countries					

VOCABULARY DEFINITIONS

a. fertility rate that will just replace the population of parents

b. collecting, compiling, and presenting information regarding populations

c. continued population growth after fertility rate = replacement rate

d. the gradual shift from primitive to modern condition that is correlated with development

e. number of births per 1000 of the population per year

f. number of deaths per 1000 of the population per year

g. relative proportion of people in each age group

h. net increase or loss of population per year

i. the number of years it will take a population growing at a constant percent per year to double

j. factors that moderate environmental impacts of affluent lifestyles

k. average number of children each woman has over her lifetime

l. a condition of life so limited by malnutrition, iliteracy, disease, high infant mortality, and low life expectancy as to be beneath any defintion of human decency.

m. high income nations

n. middle to low income nations

o. the number of people at each age for a given population

VOCABULARY EXAMPLES

a. 90% of the people in developing countries	g. (CBR - CDR) ÷ 10
b. Table 6-2	h. Fig. 6-11
c. Fig. 6-18	i. 50 to 60 years in Kenya
d. Kenya	j. 2.0
e. 18 to 1155 years	k. 1.3 to 6.7
f. recycling + conservation + pollution control	l. United States

SELF-TEST

1.   Draw a graph showing the change in human population size on the Y axis from historical times through the present and projected into the future on the X axis.

Y axis

X axis

Identify the following countries or regions as [a] developed or [b] developing countries.

2.   [ ] Latin America   3. [ ] Japan   4. [ ] Central Africa   5. [ ] United States

6.   Developed countries have about _____ percent of the Earth's population but control about _____ percent of the Earth's wealth.

7.   Calculate the people to resource ratio for developed countries in units of resource per person. _____

8.   Developing countries have about _____ percent of the Earth's population but control only _____ percent of the Earth's wealth.

9.   Calculate the people to resource ratio for developing countries in units of resource per person. _____

Assume that people living in HDCs reduced their resource consumption by 50% and donated the additional resource to LDCs.

10.   Calculate the new people to resource ratio for developed countries in units of resource per person. _____

11.   Calculate the new people to resource ratio for developing countries in units of resource per person. _____

Explain the results of your calculations in questions 10 and 11 in terms of:

12.   Impact on developed countries: _____

13.   Impact on developing countries: _____

69

Calculate the annual rate of increase and doubling time for the countries listed below.				
Country	CBR	CDR	Rate of Increase	Doubling Time
14. Developed	14	9		
15. Developing	34	10		

Indicate whether the following consequences of exploding population are the results of high fertility rates in [a] developing countries, the affluence of [b] developed countries, or [c] both.

16.   [   ] population pressure in the countryside   17. [   ] pressures on forests and
                                                              endangered species

18.   [   ] global pollution   19. [   ] high emigration rates

20.   Give one example of how population profiles can be used.

_____

21.   In the phases of demographic transition, having stable birth and death rates is generally seen in Phase (I, II, III, IV).

22.   A population profile shows

        a. standard of living.   b. causes of death.

        c. numbers of people in each 5-year age group.   d. factors that control fertility rates.

23.   The most significant factor(s) that determine how fast a population grows is (are)

        a. age structure of the population.   b. total fertility rate.

        c. infant and childhood mortality rates.   d. all of the above

24.   The population profile of a developed country has the shape of a

        a. column.   b. pyramid.   c. upside down pyramid.   d. football on its point.

25.   Why would Kenya have to wait 50 years to experience a population decrease?

_____

ANSWERS TO STUDY GUIDE QUESTIONS

1. 1, since the beginning of human history; 2, 100; 3, 30; 4, 15; 5, 12; 6, 11; 7, 11; 8, 11; 9, 13; 10, 13; 2. low recruitment; 3. cause of diseases recognized, improvements in nutrition, discovery of penecillin, improvements in medicine; 4. 12; 5. A, B, C; 6. 20, 80; 7. 80, 20; 8. malnutrition, illiteracy, disease, squalid surroundings, high infant mortality, low life expectancy; 9. Kenya, Italy; 10. decreased, increased; 11. b, a, b, a, a, b, a, b, b, b, a, a, a; 12. environmental; 13. stabalize population, decrease consumptive lifestyles, increase stewardly care; 14. subdivide farms, open new land for agriculture, move to cities, engange in illicit activities, emigrate to other countries; 15. all +; 16. no suitable land left to open for agriculture; 17. all +; 18. drugs and poaching; 19. developed; 20. emigration; 21. more; 22. false; 23. true; 24. false; 25. 4.5, 25%; 26. developed countries mine the resources of developing countries and leave their wastes in developing countries; 27. answer depends on your point of view; 28. population growth and demand for goods and services; 29. -, 0, +; 30. =2; 31. = 2, > 2, < 2; 32. column; 33. pyramid; 34. births and deaths; 35. less; 36. more; 37. less; 38. less; 39. more; 40. less; 41. more; 42. more; 43. 50 to 60; 44. inverse relationship between economic development and fertility rates; 45. increase; 46. 70; 47. see Table 6-2; 48. long; 49. births and deaths; 50. four; 51. +, +, +; +, -, +; -, -, 0; 0, 0, 0; 52. IV; 53. III.

ANSWERS TO SELF-TEST

1. draw a J-shaped curve; 2. L; 3. H; 4. L; 5. H; 6. 25, 80; 7. 3; 8. 75, 20; 9. 0.26; 10. 1.6; 11. 0.8; 12. considerable change in quality of life; 13. very small change in resource availabilty; 14. 2.4, 30; 15. 0.5, 137; 16. L; 17. B; 18. B; 19. L; 20. make projections on future population trends; 21. IV; 22. c; 23. d; 24. a; 25. population momentum.

Term	Definition	Example(s)	Term	Definition	Example(s)
absolute poverty	l	a	doubling time	i	e
age structure	g	h	environmental regard	j	f
crude birth rate	e	b	natural increase	h	g
crude death rate	f	b	population profile	o	h
demography	b	h	population momentum	c	i
demographic transition	d	c	replacement fertility	a	j
developed countries	m	d	total fertility rate	k	k
developing countries	n	l			

VOCABULARY DRILL ANSWERS

CHAPTER 7

ADDRESSING THE POPULATION PROBLEM

Reassessing the Demographic Transition

1.   Interpret Figure 7-1 in terms of what Thomas
     Malthus pointed out in 1798.

     _____

     _____

     _____

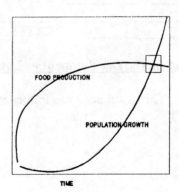

2. What two considerations were not included in the Malthusian interpretation of Figure 7-1?

   a. _____ b. _____

3. What are two basic schools of thought on the issue of population and development?

   a. _____ b. _____

Factors Influencing Family Size

4. Indicate whether the following social and economic factors would **most** determine the family size of [A] developing, [B] developed, [C] both, or [D] neither countries.

   [   ] children: an economic asset or liability
   [   ] old-age security
   [   ] importance of education
   [   ] status of women
   [   ] religious beliefs
   [   ] infant and childhood mortality rates
   [   ] availability of contraceptives

Conclusions

5. Which of the following social events prevented a population explosion in industrialized societies? (Check all that apply.)

   [   ] reduction in infant and childhood mortality
   [   ] availability of jobs in the cities
   [   ] lower fertility rates
   [   ] strict control of immigration

Development

6. What is the expected sequence of events promoting development in low-income nations?

   BETTER JOBS → BETTER _____ → HIGHER STANDARD OF _____ → EXPANDED _____ → FOSTERING _____, _____, AND _____ → LOWER _____ RATES.

Promoting the Development of Low-Income Countries

7. The world bank has influenced the course of development in poor nations for the past _____ years.

Past Successes and Failures of the World Bank

8.   Score the world bank's successes [+] and failures [-] in promoting development in low-income nations using the following list.

    [   ] increased GNP for some countries
    [   ] social progress in some countries
    [   ] declining fertility rates in some countries
    [   ] achievement of replacement level fertility rates in some countries
    [   ] the number of people living in absolute poverty
    [   ] electric generating facility in India
    [   ] large cattle operations in Latin America
    [   ] mechanized plantations for growing cash crops

9.   Indicate whether the following are [+] or are not [-] the working conditions in industrial plants in developing nations.

    [   ] union-protected workforce
    [   ] low wages and no fringe benefits
    [   ] enforced pollution regulations
    [   ] enforced child labor laws

10.   Indicate whether the following are [+] or are not [-] benefits of World Bank loans applied to agriculture in developing countries.

    [   ] higher production per acre
    [   ] crop diversity
    [   ] year-round food availability
    [   ] direct farmer-to-consumer exchange
    [   ] maintains the traditions of subsistence farming

The Debt Crisis

11.   Is the national debt obligation of developing nations (Growing, Shrinking)?

12.   What happens when a developing nation cannot even pay the interest on its loan?

    _____

13.   What are three strategies used by developing nations to make even partial payment on their debt?

    a. _____   b. _____

    c. _____

14.   How is the debt crisis in developing nations "the essence of nonsustainability"?

    _____

15. Give three examples of the large-scale centralized projects funded by the World Bank.

a. _____   b. _____

c. _____

16. Is it (True or False) that large-scale centralized projects have alleviated absolute poverty conditions?

World Bank Reform

17. List two major differences in how the World Bank does business in developing countries today compared to the past.

a. _____

b. _____

A New Direction for Development - Social Modernization

18. Identify those factors that are are characteristics of the process of social modernization. (Check all that apply.)

[ ] good health and nutrition for mothers and children
[ ] old-age security apart from children
[ ] educational opportunities for women
[ ] availability of contraceptives
[ ] career opportunities
[ ] rights of women
[ ] improving resource management

Education

19. List two benefits of providing more education in developing countries.

a. _____

b. _____

Improving Health

20. Health care in developing countries means: (Check all that apply.)

[ ] bypass surgery
[ ] good nutrition and hygiene
[ ] proper treatment of infections
[ ] AIDS prevention and care

Family Planning

21.   Indicate whether the following practices are [+] or are not [-] characteristic of family planning services and information.

[   ] Counseling on reproduction, hazards of STDs, and contraceptive techniques
[   ] Supplying contraceptives
[   ] Counseling on pre- and post-natal health of mother and child
[   ] Counseling on the health advantages of spacing children
[   ] Promoting as many abortions as possible
[   ] Nursing a baby for as long as possible
[   ] How to avoid unwanted or high-risk pregnancies

22.   Which of the following elements are confounding the process of helping women in developing countries to lower their fertility? (Check all that apply)

[   ] Inadequate government funding
[   ] Lack of access to and cost of contraceptives
[   ] Inadequate sources of information on contraceptives

Enhancing Income

23.   What three factors preclude most people in developing countries from obtaining a loan?

a. _____   b. _____

c. _____

24.   What is the Grameen Bank doing to provide start-up money for people in developing countries?
_____

25.   Who are the preferred customers of the Grameen Bank? _____

26.   Projects funded by the Grameen Bank:

a. do not upset the existing social structure. (True or False)
b. involve a high degree of training or skill. (True or False)
c. utilize local resources. (True or False)
d. utilize centralized workplaces. (True or False)
e. allow individuals to develop self-reliance. (True or False)

Improving Resource Management

27.   Is it (True, False) that people in developing countries lack the basic skill in resource management.

Putting it All Together

28.   What are the expected sequence of events from breaking the cycle of poverty, high fertility, and environmental degradation in low-income nations?

BETTER HEALTH → ECONOMIC _____ → BETTER _____ → DELAYED _____ → FEWER _____

The Cairo Conference

29.   State the conceptual framework of all goals established in the 1994 ICPD Program of Action.

_____

_____

30.   What population comonents (men, women, children) are particularly targeted in the 1994 ICPD Program of Action.

31.   How will the Program of Action be funded?

_____

NO VOCABULARY DRILL FOR THIS CHAPTER!

SELF-TEST

Indicate whether the following social events have happened in [a] developed, [b] developing, or [c] both countries.

1.   [   ] reduction in infant and childhood mortality
2.   [   ] availability of jobs in the cities
3.   [   ] lower fertility rates

Indicate whether the following statements are true [+] or false [-] concerning the majority of project funded by the World Bank.

4.   [   ] relatively easy to administer, measure, monitor, and to demonstrate end products
5.   [   ] result in substantial increases in the gross national product
6.   [   ] result in substantial increases in wealth for all residents of the developing country
7.   [   ] aggravate poverty
8.   [   ] require the use of modern machinery and technology
9.   [   ] increases diversity in crops and methods of farming
10.   [   ] promote nonsustainable agriculture
11.   [   ] increase the debt crisis in developing countries

12. Explain why developing nations are experiencing a population explosion and developed nations are not.

_____

Give three reasons why U.S. firms are moving their production facilities to developing nations.

13. _____

14. _____

15. _____

16. Explain how the movement of U.S. firms to developing nations has the potential of doing more harm than good for the people.

_____

_____

Indicate whether the following are [+] or are not [-] characteristics of World Bank loans applied to agriculture in developing countries.

17. [   ] cash crops   18. [   ] higher production per acre   19. [   ] use of large tractors

20. [   ] combat absolute poverty

21. The utilization of local resources without upsetting the social structure is a characteristic of projects funded by the (World, Grameen) bank?

List the three factors that are **most** directly correlated with low fertility rates.

22. _____

23. _____

24. _____

25. Family planning services do not provide

   a. counseling on reproduction and contraceptive techniques.
   b. contraceptives.
   c. unlimited abortions.
   d. counseling on the health advantages of spacing children.

ANSWERS TO STUDY GUIDE QUESTIONS

1. population growth will exceed food-producing capacity of nations; 2. agricultural technology, declining fertility rates; 3. controlling population, economic development; 4. A, A, B, B, D, A, B; 5. check first three, last one blank; 6. incomes; living; markets; trade, cooperation, peace; fertility; 7. 50; 8. -, +, +, -, -, -, -, -; 9. -, +, -, -; 10. all -; 11. growing; 12. unpaid interest is added to principal and country shrinks deeper into debt; 13. cash crops, austerity measures, exploitation of natural resources; 14. they liquidate capital assets to raise cash for short term needs; 15. electric power plants, transportation (e.g., super highways), mechanized agriculture; 16. false; 17. environmental sensitivity, focus on the needs of the poor; 18. all checked; 19. empowers people to get more information, enhances economic strength of country; 20. check all but the first box; 21. +, +, +, +, -, +, +; 22. all checked; 23. credit risks, loans too small, status of women; 24. micro lending; 25. women; 26. true, false, true, false, true; 27. true; 28. productivity, education, marriage, children; 29. to create an economic, social, and cultural environment in which all people, regardless of age, race, or gender, can equitably share a state of well-being; 30. women and children; 31. 7% of each nation's GNP.

ANSWERS TO SELF-TEST

1. C; 2. A; 3. A; 4. +; 5. -; 6. -; 7. +; 8. +; 9. -; 10. +; 11. +; 12. the people of developing nations were unable to enhance their quality of life by moving to the city and finding jobs which led to high fertility rates and a population explosion; 13. low wages; 14. weak child labor laws; 15. weak environmental laws; 16. exploits the poor by keeping them poor with low wages and social and environmental degradation; 17. +; 18. -; 19. +; 20. -; 21. Grumman; 22. availability of education for women; 23. availability of contraceptives; 24. good health and nutrition for mothers and children; 25. c.

CHAPTER 9

WATER: HYDROLOGIC CYCLE AND HUMAN USE

Water - a Vital Resource

The Hydrologic Cycle
    Evaporation, Condensation, and Purification
    Precipitation
    Water Over and Through the Ground
    Summary of the Water Cycle

Human Impacts on the Water Cycle

Sources and Uses of Fresh Water

Overdrawing Water Resources
    Consequences of Overdrawing Surface Waters
    Consequences of Overdrawing Groundwater

Obtaining More Water

Using Less Water
    Irrigation
    Municipal Systems

Desalting Sea Water

Storm Water
    Mismanagement and Its Consequences
    Improving Storm Water Management
    Water Stewardship

Vocabulary Drill
    Vocabulary Definitions
    Vocabulary Examples

Self-Test

Answers to Study Guide Questions

Answers to Self-Test

Vocabulary Drill Answers

Water - a Vital Resource

1.   Water covers _____ percent of the Earth's surface.

2.   _____ percent of the Earth's water is saltwater.

3.   Eighty-seven percent of the fresh water is bound up in:

        a. _____   b. _____ and c. _____

4.   Only _____ percent is accessible fresh water.

5.   Water supply shortages loom in at least _____ countries in 1997.

The Hydrologic Cycle

6.   The Earth's **water cycle** is also called the _____ cycle.

7.   Water enters the atmosphere through _____ and _____.

8.   Water returns from the atmosphere through _____ and _____.

Evaporation, Condensation, and Purification

9.   The three physical states of water are _____, _____, and _____.

10.   These three states are the result of different degrees of _____ bonding between water molecules.

11.   What are the two forces that determine the different degrees of interaction between water molecules?

        a. _____   b. _____

12.   Hydrogen bonding tends to (Attract, Separate) water molecules.

13.   Kinetic energy tends to (Attract, Separate) water molecules.

14.   Below freezing, (Hydrogen Bonding, Kinetic Energy) is the dominant force determining the association between water molecules.

15.   Above freezing, (Hydrogen Bonding, Kinetic Energy) is the dominant force determining the association between water molecules.

16.   During **evaporation**, kinetic energy is (High, Low) when compared to hydrogen bonding.

17.   During **condensation**, kinetic energy is (High, Low) when compared to hydrogen bonding.

18. Evaporation and condensation are two water purification processes similar to
_____.

19. Is it (True or False) that all the natural freshwater on earth comes from this distillation process?

20. Water vapor enters the atmosphere by way of:

a. _____ from all water and moist surfaces.

b. _____ which is evaporation through the leaves of plants.

21. The amount of water vapor held in the air at various temperatures is called _____ humidity.

22. Warm air holds (More or Less) water than cold air.

Precipitation

23. Rising air results in (High, Low) precipitation whereas descending air results in (High, Low) precipitation.

24. Air rises over equatorial regions and descends over subequatorial regions (see Figure 9-6), therefore:

a. rainfall is (High, Low) in equatorial regions.
b. rainfall is (High, Low) in subequatorial regions.

25. The **rainshadow** (see Figure 9-7) refers to a region of (High, Low) precipitation.

Water Over and Through the Ground

26. When precipitation hits the ground, what two alternative pathways might it follow?

a. _____   b. _____

27. All ponds, lakes, streams, and rivers are referred to as _____ waters.

28. Water that infiltrates may follow either of two pathways including:

a. _____ or b. _____ water.

29. Plants draw mainly from _____ water.

30. Water that percolates down through the soil eventually comes to an
_____ layer and accumulates filling all the empty pores and spaces.
This accumulated water is now called _____ water and its upper
surface is called the _____ table.

31. The layers of porous material through which groundwater moves are called an
_____.

32. Where water actually enters an aquifer is called the _____ area.

33. Natural exits of groundwater to the surface are _____ or _____.

34. Springs and seeps feed streams and rivers becoming part of _____ water.

Summary of the Water Cycle

35. Summarize the below events in the water cycle by indicating whether they primarily occur as part of the [a] surface runoff loop, [b] evaporation-transpiration loop, or [c] groundwater loop (see Figure 9-9).

    [   ] springs
    [   ] water table
    [   ] transpiration
    [   ] percolation
    [   ] surface runoff
    [   ] cloud formation
    [   ] water vapor
    [   ] capillary water
    [   ] gravitational water
    [   ] aquifers
    [   ] ground water
    [   ] infiltration

36. The two filters for water that infiltrates through the ground are _____ and _____.

37. Match the terms on the left with the water descriptions on the right (see Table 9-1).

    Water Terms   Water Descriptions

    [   ] Fresh   a. a mixture of fresh and salt water
    [   ] Salt   b. salt concentration < 0.01%
    [   ] Brackish   c. salt concentration at least 3%
    [   ] Hard   d. high mineral content
    [   ] Soft   e. low mineral content

Human Impacts on the Water Cycle

38. List the three main impacts that humans have on the water cycle.

    a. _____

    b. _____

    c. _____

Sources and Uses of Fresh Water

39.   The two primary human concerns regarding fresh water are _____ and
        _____ water.

40.   What proportion of the Earth's water resources are used for:

        a. irrigation _____% b. industry _____% c. human consumption _____%

41.   The major source of fresh water is: a. ground water. b. surface water.

42.   Center pivot irrigation systems use huge amounts of (Surface, Ground) water and may use as
        much as (1, 10, 100) thousand gallons per minute.

Overdrawing Water Resources

Consequences of Overdrawing Surface Waters

43.   No more than _____ percent of a river's average flow can be taken risking shortfalls.

44.   However, the demand on some rivers exceeds _____ percent of the average flow.

45.   The consequences of overdrawing surface waters include:

        a. wetlands along and at the mouth of the river drying up. (True or False)
        b. estuaries at the mouth of the river becoming increasingly salty. (True or False)
        c. dieoffs of large populations of wildlife. (True or False)

Consequences of Overdrawing Groundwater

46.   Depletion of ground water will very likely result in

        a. cutbacks in agricultural production. (True or False)
        b. diminishing surface water. (True or False)
        c. adverse effects on stream and river ecosystems. (True or False)
        d. land subsidence. (True or False)
        e. formation of sinkholes. (True or False)
        f. saltwater intrusion into wells in coastal regions. (True or False)
        g. required cutbacks in municipal consumption. (True or False)

Obtaining More Water

47.   It may be possible to increase water supplies:

        a. at modest cost. (True, False)
        b. without severe ecological impacts. (True, False)

48. Future water shortages will

   a. become increasingly common. (True or False)
   b. be of longer duration. (True or False)
   c. affect increasingly large areas and numbers of people. (True or False)
   d. go away. (True or False)

Using Less Water

49. The greatest demands on U.S. fresh water (Consumptive, Nonconsumptive) uses are particularly in (see Table 9-2):

   a. agriculture,   b. electrical power production, c. industry   d. homes

Irrigation

50. About _____ percent of irrigation water is wasted through evaporation.

51. A water conserving alternative to surface and sprinkling irrigation is _____ irrigation.

Municipal Systems

52. Water consumption in modern homes averages around _____ gallons per person per day.

53. The most effective conservation technique would be to use (More, The Same, Less) water.

54. An example of recycling water is _____ water.

Desalting Sea Water

55. What are the two technologies used for desalinization:

   a. _____   b. _____

56. What is the cost per 1,000 gallons of desalinized water? $_____

57. Farmers currently pay $_____ per 1,000 gallons of fresh water for irrigation.

Storm Water

58. As land is developed, natural soil surfaces are replaced by various hard surfaces. By this action

   a. infiltration is (Increased or Decreased).
   b. runoff is (Increased or Decreased).

Mismanagement and Its Consequences

59. List three consequences of stormwater mismanagement.

 a. _____ b. _____ c. _____

60. As a result of the change in infiltration and runoff, predict whether each of the following will increase [+], decrease [-], or remain the same [0].

 [   ] amount of groundwater recharge   [   ] sediment deposition in the stream channel
 [   ] level of water table   [   ] breadth of stream channel
 [   ] amount of flow from springs   [   ] depth of stream channel
 [   ] stream flow between rains   [   ] pollution of stream channel
 [   ] stream flow during rains   [   ] frequency of flooding
 [   ] stream bank erosion   [   ] heights of floods

61. List the categories of pollutants that enter streams and rivers from surface runoff.

 a. _____ b. _____ c. _____

 d. _____ e. _____ f. _____

 g. _____

62. To alleviate the problem of urban flooding , streams may be **channelized**. In this process

 a. a channel is dug out. (True or False)
 b. curves are smoothed out. (True or False)
 c. the channel is lined with concrete or rock. (True or False)
 d. a natural water flow is restored. (True or False)
 e. a natural stream ecology is restored. (True or False)

Improving Storm Water Management

63. Describe one technique for improving storm water management (see Fig. 9-27).

 _____

Water Stewardship

64. Current trends in water use (Are, Are Not) sustainable.

65. List four stakeholders in future water use decisions.

 a. _____ b. _____

 c. _____ d. _____

VOCABULARY DRILL

Directions: Match each term with the list of definitions and examples given in the tables below. Term definitions and examples might be used more than once.

Term	Definition	Example(s)	Term	Definition	Example(s)
aquifer			polluted water		
capillary water			precipitation		
channelization			rain shadow		
condensation			recharge area		
consumptive water use			relative humidity		
desalinization			saltwater		
drip irrigation			saltwater intrusion		
evaporation			sinkhole		
evapotranspiration			spring		
fresh water			storm water management		
gravitational water			surface waters		
gray water			surface runoff		
groundwater			transpiration		
humidity			water purification		
hydrological cycle			water cycle		
infiltration			water table		
land subsidence			water vapor		
nonconsumptive water use			water quality		
percolation			water quantity		
physical states of water			watershed		
			xeroscaping		

VOCABULARY DEFINITIONS

a. amount of fresh water

b. purity of fresh water

c. water molecules entering the atmosphere

d. water conditions given different temperatures

e. water molecules rejoining by hydrogen bonding

f. salt content less than 100 ppm

g. water molecules in the air

h. amount of water vapor in the air

i. amount of water as a percentage of what air can hold at that temperature

j. salt content more than 100 ppm

k. containing elements or particulate matter that is beyond the norm

l. process of removing unwanted elements or particulate matter

m. dry region downwind of a mountain

n. all the land area that contributes water to a particular stream or rive

o. non enclosed water impoundments

p. water held in the soil

q. water that soaks into the ground

r. water rising to the atmosphere through evaporation or transpiration and leaving it through condensation and precipitation

s. water movement in the gaseous state promoted by Earth's heat or through plants

t. the effect of condensation and cooling temperatures

u. movement of excess noninfiltrated water

v. accumulation of water in an impervious layer of rock or dense clay

w. water subjected to the pull of gravity

x. movement through the soil beyond capillary water

y. the upper surface of accumulated groundwater

z. where water enters an aquifer

aa. layers of porous material through which groundwater moves

bb. where water exits the ground as a significant flow from a relatively small opening

cc. lost for further human use

dd. results of a collapsed underground cavern drained of it supporting groundwater

ee. gradual settling of land when the water table drops

VOCABULARY DEFINITIONS

ff. movement of saltwater into a freshwater aquifer
gg. remains available for further human use
hh. a system of irrigation that applies small steady amount of water to the plant roots
ii. water that has been used for other purposes and reused again
jj. change in the structural characteristics of stream beds
kk. landscaping with desert species that require no additional watering
ll. converting salt into fresh water
mm. movement of water through plants
nn. procedures used to offset the destructive effects of storm water runoff

VOCABULARY EXAMPLES

a. 0.4%	p. rivulets, streams
b. nonpolluted and < 100ppm salt	q. Ogallala Aquifer
c. water in gaseous state	r. infiltrated water not available to plant roots
d. gas, liquid, solid	s. western Nebraska grasslands
e. water going from gaseous to liquid state	t. irrigation
f. water vapor + temperature	u. Fig. 9-18
g. oceans, seas	v. San Joaquin Valley in California
h. acid rain	w. Fig. 9-19
i. desalinization	x. gray water
j. Death Valley	y. Fig. 9-20
k. Missouri River Basin	z. washing machine, shower water
l. lakes, rivers	aa. Fig. 9-26
m. water available to plant roots	bb. cacti, mesquite bush
n. Fig. 9-4	cc. Middle East - Saudi Arabia
o. rain, snow	dd. Fig. 9-27

1-15. Illustrate the hydrological cycle including all filtration loops and relevant terms. For example, give yourself one point for each correct representation of: **surface runoff loop, evaporation-transpiration loop, groundwater loop, water vapor, condensation, cloud formation, transpiration, precipitation, evaporation, infiltration, percolation, water table, groundwater, springs, surface runoff.**

Impervious Rock

Indicate whether the depletion of groundwater will [+] or will not [-] lead to:

16. [   ] increased groundwater use in agriculture.
17. [   ] subsidence.
18. [   ] saltwater intrusion.
19. [   ] increased municipal consumption.
20. [   ] decreased flow through seeps and springs.
21. [   ] adverse effects on stream and river ecosystems.

As land is developed, natural soil surfaces are replaced by various hard surfaces. As a result of this change indicate whether the following will increase [+], decrease [-], or remain the same [O].

22. [   ] amount of groundwater recharge
23. [   ] level of water table
24. [   ] amount of spring flow

25.   [   ] stream flow during rains
26.   [   ] frequency of flooding
27.   [   ] pollution of stream channel

        Explain how human activities have adversely effected the following filtration loops in the hydrological cycle.

28.   Surface runoff loop: _____

29.   Evapotranspiration loop: _____

30.   Groundwater loop: _____

ANSWERS TO STUDY GUIDE QUESTIONS

1. 71; 2. 97; 3. ice, groundwater, atmosphere; 4. 0.4; 5. 26; 6. hydrological; 7. evaporation, transpiration; 8. condensation, precipitation; 9. gas, liquid, solid; 10. hydrogen; 11. kinetic energy, hydrogen bonding; 12. attract; 13. separate; 14. hydrogen bonding; 15. kinetic energy; 16. high; 17. low; 18. distillation; 19. true; 20. evaporation, transpiration; 21. relative; 22. more; 23. high, low; 24. high, low; 25. low; 26. infiltration, runoff; 27. surface; 28. capillary, gravitational; 29. capillary; 30. impervious, ground, water; 31. aquifer; 32. recharge; 33. springs, seeps; 34. surface; 35. c, c, b, b, a, b, b, b, c, c, c, b; 36. soil, porous rock; 37. b, c, a, d, e; 38. changing earth's surface, pollution, over withdrawals; 39. supply, purity; 40. 70, 23, 8; 41. a; 42. ground, 10; 43. 30; 44. 90; 45. all true; 46. all true; 47. all false; 48. true, true, true, false; 49. consumptive, a; 50. 60; 51. drip; 52. 100; 53. less; 54. gray; 55. micro filtration/reverse osmosis, distillation; 56. $3; 57. $0.02; 58. decreased, increased; 59. flooding, streambank erosion, increased pollution; 60. -, -, -, -, +, +, +, +, +, -, +, +, +; 61. fertilizer, "cides," bacteria, salt, toxic chemicals, oil and grease, litter; 62. true, true, true, false, false; 63. see Fig. 9-27; 64. are not; 65. natural ecosystem, industry, agriculture, domestic.

ANSWERS TO SELF-TEST

1-15. (see Figure 11-3); 16. -; 17. +; 18. +; 19. -; 20. +; 21. +; 22. -; 23. -; 24. -; 25. +; 26. +; 27. +; 28. large quantities and different types of pollutants are part of urban runoff; 29. air pollutants become mixed with water molecules during condensation and precipitation (e.g., acid rain); 30. depletion of groundwater and pollutants in soil that are carried into ground water.

Term	Definition	Example(s)	Term	Definition	Example(s)
aquifer	aa	n, q	polluted water	k	h
capillary water	p	m	precipitation	t	n, o
channelization	jj	aa	rain shadow	m	j
condensation	e	e	recharge area	z	s
consumptive water use	cc	t	relative humidity	i	f
desalinization	ll	cc	saltwater	j	g
drip irrigation	hh	y	saltwater intrusion	ff	w
evaporation	c	c, n	sinkhole	dd	u
evapotranspiration	s	n	spring	bb	n
fresh water	f	b	Storm water management	nn	dd
gravitational water	w	n, r	surface waters	o	l
gray water	ii	z	surface runoff	v	n, p
groundwater	v	n, q	transpiration	mm	n
humidity	h	c	water purification	l	i
hydrological cycle	r	n	water cycle	r	n
infiltration	q	n	water table	y	n, q
land subsidence	ee	v	water vapor	g	c
nonconsumptive water use	gg	x	water quality	b	b
percolation	x	n	water quantity	a	a
physical states of water	d	d	watershed	n	k
			xeroscaping	kk	bb

VOCABULARY DRILL ANSWERS

94

CHAPTER 10

THE PRODUCTION AND DISTRIBUTION OF FOOD

Crops and Animals: Major Patterns of Food Production
  The Development of Modern Industrialized Agriculture
  The Transformation of Traditional Agriculture
  Subsistence Agriculture in the Developing World
  Animal Farming and its Consequences
  Prospects for Increasing Food Production
  The Promise of Biotechnology

Food Distribution and Trade
  Patterns in Food Trade
  Levels of Responsibility in Supplying Food

Hunger, Malnutrition, and Famine
  Nutrition vs. Hunger
  Extent and Consequences of Hunger
  Root Cause of Hunger
  Famine...
  Food Aid

Building Sustainability Into The Food Arena
  Sustainable Agriculture
  Final Thoughts on Hunger

Vocabulary Drill
  Vocabulary Definitions
  Vocabulary Examples

Self-Test

Answers to Study Guide Questions

Answers to Self-Test

Vocabulary Drill Answers

Crops and Animals: Major Patterns of Food Production

The Development of Modern Industrialized Agriculture

1. Indicate whether the following practices were [+] or were not [-] common agricultural practices 150 years ago.

    [   ] bringing additional land into cultivation
    [   ] increasing use of fertilizer
    [   ] increasing use of irrigation
    [   ] increasing use of chemical pesticides
    [   ] substituting new genetic varieties of grains
    [   ] rotating crops
    [   ] growing many different kinds of crops
    [   ] recycling animal wastes

The Transformation of Traditional Agriculture

2. What other revolution contributed to a revolution in agriculture? _____

3. Today, less than _____ percent of the U.S. workforce produces enough food for the nation's needs.

4. Indicate whether the following practices were [+] or were not [-] characteristics of the agricultural revolution that increased food production so dramatically in the last 40 years.

    [   ] bringing additional land into cultivation
    [   ] increasing use of fertilizer
    [   ] increasing use of irrigation
    [   ] increasing use of chemical pesticides
    [   ] substituting new genetic varieties of grains
    [   ] rotating crops
    [   ] growing many different kinds of crops
    [   ] recycling animal wastes

5. Which of the following agriculture practices will [+] and will not [-] increase food production in the **next** 40 years.

    [   ] bringing additional land into cultivation
    [   ] increasing use of fertilizer
    [   ] increasing use of irrigation
    [   ] increasing use of chemical pesticides
    [   ] substituting new genetic varieties of grains
    [   ] rotating crops
    [   ] growing many different kinds of crops
    [   ] recycling animal wastes

6.	The shift from animal labor to machinery has led to:

	a. high dependency on fossil fuel energy. (True or False)
	b. soil compaction. (True or False)
	c. recycling of animal wastes. (True or False)

7.	Any future attempts to bring additional land under cultivation will probably result in:

	a. use of marginal, highly erodible lands. (True or False)
	b. loss of forests and wetlands. (True or False)

8.	Adding more fertilizer than the optimum required leads to:

	a. ever increasing productivity. (True or False)
	b. increased plant vulnerability to pests. (True or False)
	c. pollution. (True or False)

9.	The use of pesticides in agriculture:

	a. provided better pest control. (True or False)
	b. increased crop yields. (True or False)
	c. led to pest resistance. (True or False)
	d. caused adverse side effects to human and environmental health. (True or False)

10.	Even though irrigated acreage increased about 2.6 times from 1950 to 1980, irrigation is increasing at a slower pace due to:

	a. lack of new water sources. (True or False)
	b. development of drought resistant crops. (True or False)
	c. waterlogging and accumulation of salts in the soil. (True or False)

11.	The new varieties of wheat and rice produced in the 1960s required more
	_____ and _____ but gave yields _____ to
	_____ when compared to traditional varieties.

12.	The high world wide production of wheat and rice resulting from the new genetic varieties
	was hailed as the _____ revolution.

13.	Identify which of the following aspects of the Green Revolution were benefits [B] or
	drawbacks [D] for the people of developing nations.

	[ ] closed the gap between food production and food needs in some countries
	[ ] the growth and production limits of "high-yielding" varieties of wheat and rice
	[ ] heavy reliance on irrigation
	[ ] heavy reliance on fertilizers
	[ ] impact on farm laborers and small landowners
	[ ] impact on culturally-specific crops

Subsistence Agriculture in the Developing World

14. Using the terms **MORE** or **LESS**, complete the following statements about subsistence farming. *Subsistence farming in developing countries:*

is _____ labor-intensive, requires _____ technology, utilizes _____ marginally productive land, and involves the clearing of _____ tropical forests.

Animal Farming and its Consequences

15. What proportion of the world's crop lands are used to feed animals? _____

16. In the United States, _____ percent of the grain crop goes to animals.

17. Indicate whether the following are [+] or are not [-] environmental consequences of animal farming.

[   ] overgrazing
[   ] deforestation of tropical rain forests
[   ] increases of atmospheric carbon dioxide

Prospects for Increasing Food Production

18. List two prospects for increasing food production in the future.

a. _____

b. _____

19. List two factors that will limit the ability to increase crop yields.

a. _____   b. _____

The Promise of Biotechnology

20. Explain genetic engineering in terms of three results.

a. _____

b. _____

c. _____

21. The products of genetic engineering:

a. will be affordable to farmers in developing countries. (True or False).
b. are considered safe by all consumers. (True or False).
c. will be part of agriculture as we know it in the 21st. century. (True or False).
d. will keep food production in pace with population growth. (True or False).

Food Distribution and Trade

Patterns in Food Trade

22.   The world region recognized as the major source of exportable grains is

_____ _____.

23.   Identify whether the 1990 grain import (I) export (E) ratio of each of the following countries
is
[a] I = E, [b] I < E, or [c] I > E

[   ] North America [   ] Asia [   ] Latin America   [   ] Africa [   ]Mexico

24.   For those countries that import more grain than they export, what has been the primary cause
for the trade imbalance?

_____

Levels of Responsibility in Supplying Food

25.   Identify the three major levels of responsibility for meeting food needs.

a. _____   b. _____   c. _____

26.   Indicate whether the following policies focus on meeting the food needs of [a] family, [b]
nation, or [c] globe (see Fig. 10-9).

[   ] effective safety net
[   ] food aid for famine relief
[   ] employment security
[   ] effective family planning
[   ] fair trade
[   ] just land distribution

Hunger, Malnutrition, and Famine

Nutrition vs. Hunger

27.   Indicate whether the following are conditions predominantly found in [a] developing or [b
] developed countries.

[   ] heart disease and high blood pressure
[   ] malnutrition and undernutrition

Extent and Consequences of Hunger

28.   At least _____ percent of the people on the Earth suffer from the effects of hunger and
malnutrition.

29. In what order (1 to 3) are the consequences of malnutrition and hunger reflected in a population? _____ men _____ children _____ women

Root Cause of Hunger

30. The root cause of hunger is _____.

31. Is it (True or False) that our planet can produce enough food for everyone alive today?

32. Food surpluses enter the cash economy and flow in the direction of _____ demand not of _____ need.

33. Using the answer to item #32, is it MORE or LESS probable that:

    _____ pet cats will be well fed.
    _____ undernourished children will be well fed.

Famine

34. The two primary causes of famine are _____ and _____.

35. Indicate whether the primary famine in the following countries was due to [a] drought, [b] war, [c] both or [d] government incompetence.

    [   ] Sahel region of West Africa
    [   ] Mozambique
    [   ] Sudan
    [   ] North Korea

36. Explain FEWS in terms of:

    What it means: _____

    What it does: _____

Food Aid

37. Indicate whether the following statements are true [+] or false [-] concerning the overall effects of providing food aid to countries in need. Providing food:

    [   ] actually alleviates chronic hunger in developing countries.
    [   ] undercuts prices for domestically produced food sold in local markets.
    [   ] causes the local economy to deteriorate.
    [   ] contributes to environmental and ecological deterioration.

38. Food aid to developing countries usually (Disrupts, Stimulates) the local economy and leads to a/an (Increase, Decrease) in local food production.

39. Food aid tends to (Improve, Aggravate) the problem of chronic hunger.

Building Sustainability Into The Food Arena

40.   Is it (True or False) that the agricultural practices of the past 40 years are sustainable?

Sustainable Agriculture

41.   What was nonsustainable about the agricultural practices of the following civilizations?

        a. Sumerians: _____

        b. Greeks: _____

42.   What is the goal of sustainable agriculture?

        _____

        _____

43.   Which principle of ecosystem sustainability (1, 2, 3, or 4) do the following agricultural practices promote?

        [   ] use of green manures
        [   ] poly cropping
        [   ] growing gardens rather than grass
        [   ] recognizing that rangelands have a carrying capacity

Final Thoughts on Hunger

44.   In order to alleviate world hunger, humankind needs to develop new (a. technologies, b. forums of political and social action).

45.   In order to alleviate world hunger, humankind needs to distribute food based on (a. need, b. ability to pay).

VOCABULARY DRILL

Directions: Match each term with the list of definitions and examples given in the tables below. Term definitions and examples might be used more than once.

Term	Definition	Example(s)	Term	Definition	Example(s)
subsistence farming			sustainable agriculture		
food security			organic farming		
hunger			green manures		
malnutrition			green revolution		
undernutrition			cash crops		
absolute poverty			exports		
food aid			high-yielding		
famine			imports		

VOCABULARY DEFINITIONS

a. lack of sufficient income to meet most basic biological needs
b. crops sold by a country
c. products sold by a country
d. severe shortage of food accompanied by significant increase in death rate
e. distribution of food where famine is evident
f. ability to meet the nutritional needs of all family members
g. a remarkable increase in crop production
h. crop residue
i. crops ability to produce beyond the norm through genetic manipulation
j. lack of basic food required for energy and meeting nutritional needs
k. products purchased by a country
l. lack of essential nutrients
m. regular addition of crop residues and animal manures to build up soil humus
n. farming for household needs and small cash crops
o. maintain agricultural production without degrading the environment
p. lack of adequate food energy (calories)

VOCABULARY EXAMPLES	
a. 1.2 billion people	g. lawn clippings
b. coffee, cocaine, cocoa	h. Fig. 10-13
c. Fig. 10-12	i. Fig. 10-5
d. Egypt	j. applying the four principles of ecosystem sustainability
e. the family level of responsibility	k. 786 million people
f. wheat, corn, rice	

SELF-TEST

Which of the following factors will enable humankind to increase agricultural production in the **next** 40 years? (Check all that apply.)

1. [   ] bringing additional land into cultivation

2. [   ] increasing use of fertilizer

3. [   ] increasing use of irrigation

4. [   ] increasing use of chemical pesticides

5. [   ] substituting new genetic varieties of grains

Briefly explain your responses to items 1 through 5 above.

6. Bringing additional land into cultivation:_____

_____

7. Increasing use of fertilizer: _____

_____

8. Increasing use of irrigation:_____

_____

9. Increasing use of chemical pesticides: _____

_____

10. Substituting new genetic varieties of grains: _____

_____

Identify [+] those characteristics of agricultural that will be absolutely necessary to initiate future Green Revolutions in developing countries.

11.   [   ] reliance on "high-yielding" varieties of wheat and rice

12.   [   ] polycropping techniques

13.   [   ] application of the principles of ecosystem sustainability

14.   [   ] more mobilization of the human work force in food production

15.   [   ] utilization of culturally-specific crops

16.   [   ] reliance on irrigation and fertilizers

17.   There appears to be a direct relationship between increased reliance on food imports and _____ growth in some developing countries.

Formulation of policies to meet the food needs of the global population are required at the levels of:

18. _____   19. _____   20. _____

21.   Explain how the global priorities on food distribution make it more probable that a pet cat rather than a starving child will be well fed.

        _____

        _____

22.   Explain, from the agribusiness side of the issue, why food aid to developing countries does not solve but rather aggravates their hunger problems.

        _____

        _____

ANSWERS TO STUDY GUIDE QUESTIONS

1. -, -, -, -, -, +, +, +; 2. industrial; 3. 3; 4. +, +, +, +, +, -, -, -; 5. -, -, -, -, -, +, +, +; 6. true, true, false; 7. true, true; 8. false, true, true; 9. all true; 10. true, false, true; 11. fertilizer, water, double, triple; 12. green; 13. B, D, D, D, D, D; 14. more, less, more, more; 15. 25%; 16. 70; 17. all +; 18. continue to increase crop yields, grow more food instead of cash crops; 19. loss of cropland through soil erosion, climate changes; 20. crossbreeds of genetically different plants, incorporation of desired traits into crop lines and animals, cloning of domestic animals; 21. false, false, true, true; 22. North America; 23. b, c, c, c, c; 24. population growth surpassed food production; 25. family, nation, globe; 26. b, c, a, a, c, b; 27. a, b; 28. 20; 29. 3, 1, 2; 30. poverty; 31. true; 32. economic, nutritional; 33. more, less; 34. war, poverty; 35. a, b, b, c; 36. famine early warning system, measures trends in vegetation and rainfall in Africa in order to predict famines; 37. -, +, +, +; 38. disrupts, decreases; 39. aggravate; 40. false; 41. intense irrigation that led to salinization of soil, overgrazing and deforestation; 42. producing enough food to feed the human population without ruining any part of the supporting system or degrading the environment; 43. 1, 4, 2, 3; 44. b; 45. a.

ANSWERS TO SELF-TEST

1 through 5. all blank, cannot rely on any of these methods to increase agricultural production in the next 40 years. 6. there is not additional tillable land left to bring into agriculture; 7. the optimum level is already being used, using more will not increase production; 8. ground table water reserves are nearly depleted, no other large supplies of water left; 9. pest resistance capabilities and adverse effects on humans and the environment preclude continued use of pesticides; 10. there are no new laboratory designed "high-yielding" varieties to use; 11. blank; 12. +; 13. +; 14. +; 15. +; 16. blank; 17. population; 18. family; 19. nation; 20. global; 21. surpluses are distributed based on economic priorities rather than the nutritional needs of people, i.e., more money can be made from turning surpluses into cat food than feeding starving people; 22. free food aid undercuts the peasant farmer's ability and willingness to grow crops and free food is not distributed equitably or to those who need it most.

Term	Definition	Example(s)	Term	Definition	Example(s)
			VOCABULARY DRILL ANSWERS		
subsistence farming	n	i	sustainable agriculture	o	j
food security	f	e	organic farming	m	h
hunger	j	h	green manures	h	g
malnutrition	l	h	green revolution	g	f
undernutrition	p	h, k	cash crops	b	b
absolute poverty	a	a, h	exports	c	b
food aid	e	d	high-yielding	i	f
famine	d	c	imports	k	b

CHAPTER 11

WILD SPECIES: BIODIVERSITY AND PROTECTION

Value of Wild Species
    Biological Wealth
    Two Kinds of Values
    Sources for Agriculture, Forestry, Aquaculture, and Animal Husbandry
    Sources for Medicine
    Recreational, Aesthetic, and Scientific Value
    Intrinsic Value

Saving Wild Species
    Game Animals in the United States
    The Endangered Species Act

Biodiversity
    The Decline of Biodiversity
    Reasons for the Decline
    Consequences of Losing Biodiversity
    International Steps to Protect Biodiversity
    Stewardship Concerns

Vocabulary Drill
    Vocabulary Definitions
    Vocabulary Examples

Self-Test

Answers to Study Guide Questions

Answers to Self-Test

Vocabulary Drill Answers

Value of Wild Species

1.   What is the estimated worth of the goods and services provided by wild species?
            _____ billion

Biological Wealth

2.   We presently know of the existence of _____ million species and estimate that at least
            _____ to _____ million additional species have not been studied.

3.   How many plant and animal species have become extinct in the United States? _____

Two Kinds of Values

4. _____ value exists if a species' existence or use benefits some other entity.

5. _____ value considers the species value for its own sake.

6. List the four values of natural species.

   a. _____

   b. _____

   c. _____

   d. _____

Sources for Agriculture, Forestry, Aquaculture, and Animal Husbandry

7. Indicate whether the following statements pertain to [a] wild populations of plants or animals or [b] cultivated (**cultivar**) populations of plants or animals.

   [   ] highly adaptable to changing environmental conditions
   [   ] highly nonadaptable to changing environmental conditions
   [   ] have numerous traits for resistance to parasites
   [   ] lack genetic **vigor**
   [   ] can only survive under highly controlled environmental conditions
   [   ] have a high degree of genetic diversity in the gene pool of the population
   [   ] have a low degree of genetic diversity in the gene pool of the population
   [   ] represents a reservoir of genetic material commonly called a **genetic bank**

8. Of the estimated 7,000 plant species existing in nature, how many have been used in agriculture? _____

9. What is the significance of the winged bean (Fig. 11-2) in regard to the answer to the previous question?

   _____

Sources for Medicine

10. Of what value is the chemical called vincristine?

    _____

11. Vincristine is an extract from the _____.

12. Of what value is the chemical called capoten?

    _____

13. Capoten is an extract from the _____.

14. Is it (True or False) that there is the potential of discovering innumerable drugs in natural plants, animals, and microbes?

Recreational, Aesthetic, and Scientific Value

15. One of the largest income-generating enterprises in many developing countries is

    _____.

16. As leisure time (Increases, Decreases), larger portions of the economy become connected to supporting activities related to the natural environment.

17. Pollution (Increases, Decreases) the commercial benefits of the natural environment.

18. The contribution of wild plants and animal species to the U.S. economy was calculated to be more than $ _____ billion.

19. Indicate whether the following are [a] recreational, [b] aesthetic, or [c] scientific values of the biota of natural ecosystems.

    [   ] hunting
    [   ] sport fishing
    [   ] hiking
    [   ] camping
    [   ] being in a forest rather than a city dump
    [   ] study of ecology
    [   ] just knowing that certain plant and animal species are alive

Intrinsic Value

20. Intrinsic value arguments center on _____ rights and _____.

Saving Wild Species

Game Animals in the United States

21. Was it (Commercial or Regulated) hunting that caused the extinction or near extinction of some game animals?

22. Restrictions on hunting that represent wildlife management are

    a. requiring a hunting license. (True or False)
    b. limiting the type of weapon used in the harvest. (True or False)
    c. hunting within specified seasons. (True or False)
    d. setting size and/or sex limits on harvested animals. (True, False)
    e. limiting the number that can be harvested. (True or False)
    f. prohibiting commercial hunting. (True or False)

23.   Which of the following do [+] and do not [-] represent contemporary wildlife problems.

        [   ] near extinction of the wild turkey
        [   ] road killed animals
        [   ] urban wildlife, e.g., opossums, skunks, raccoons
        [   ] lack of natural predators
        [   ] wildlife as vectors for certain diseases
        [   ] cougar predation on humans
        [   ] urban canada geese populations
        [   ] pets predation by coyotes

The Endangered Species Act

24.   Snowy egrets were heavily exploited in the 1800s for their (a. meat b. eggs c. nests
        d. feathers) that were used in (a. fancy restaurants b. ladies hats c. curiosity shops d.
        grocery stores).

25.   The two states that were first to pass laws protecting plumed birds were
        _____ and _____.

26.   The law that forbids interstate commerce in illegally killed wildlife is called the
        _____ Act.

27.   The law that protects species from extinction is called the _____
        _____ Act of _____.

28.   Indicate whether the following provisions are strengths [+] or weaknesses [-] of the
        Endangered Species Act.

        [   ] the need for *official recognition*
        [   ] stiff fines for the killing, trapping, uprooting, or commerce of endangered species
        [   ] controls on government development projects in critical habitats
        [   ] enforcement of the act
        [   ] recovery programs
        [   ] taxonomic status of some animals protected by the act

29.   Identify two endangered bird species that were "recovered."

        a. _____   b. _____

30.   Explain the connection between sandhill and whooping cranes.

        _____

31.   What are two major concerns of opponents of the ESA that surfaced in the spotted owl
        controversy?

        a. _____   b. _____

Biodiversity

32. Over geological time, the net balance between _____ and _____ has favored the gradual (Accumulation, Loss) of species.

33. The number of plant and animal species on the earth today is estimated at _____ million with the greatest species richness found in _____ plants and _____.

34. Which country has at least 5% of all living species? _____

The Decline of Biodiversity

35. How many species of animals _____ and plants _____ have become extinct since 1600?

36. Globally, how many species of animals _____ and plants _____ are in danger of becoming extinct?

37. In what type of ecosystem can one find the greatest biodiversity? _____

38. List four sources of evidence that there is a decline of biodiversity occurring.

    a. _____ b. _____

    c. _____ d. _____

39. Give two reasons why species on oceanic islands are more susceptible to extinction than continental species.

    a. _____

    b. _____

Reasons for the Decline

40. List five factors that represent evidence of or contribute to the decline in biodiversity.

    a. _____

    b. _____

    c. _____

    d. _____

    e. _____

41. Match the following examples of factors that represent evidence of or contribute to the decline of biodiversity with those you listed [a through e] in the previous question.

[   ] habitat fragmentation
[   ] clear cut logging
[   ] expansion of the human population over the globe
[   ] the brown tree snake
[   ] the greenhouse effect
[   ] prospects of huge immediate profits
[   ] the growing fad for exotic pets, fish, reptiles, birds, and house plants
[   ] habitat conversion
[   ] endocrine disrupters
[   ] black bear gall bladders

42. Indicate whether the following examples of habitat alteration represent [a] conversion, [b] fragmentation, or [c] simplification

[   ] stream channelization
[   ] shopping malls
[   ] highway development

43. What is the graphic relationship (see Figure 11-12) between human population size and species survival worldwide? _____

44. As human population increases, species survival worldwide will (Increase, Decrease).

45. The greatest catastrophe to hit natural biota in 60 million years will be

a. acid rain   b. human population growth   c. global warming   d. urbanization

46. Species adapt very (Quickly, Slowly) to environmental change.

47. Global warming is predicted to occur very (Quickly, Slowly).

48. Exotic species are responsible for _____ percent of all animal extinctions since 1600.

49. Usually, the introduction of exotic species into a country (Increases, Decreases) biodiversity.

50. Usually, an exotic species is a (More, Less) effective competitor for resources.

51. Give three examples of exotic species introduced into the United States.

a. _____   b. _____   c. _____

52. As certain plants and animals become increasingly rare, the price people are willing to pay will (Increase, Decrease).

53. Would you predict that exploiters of natural biota would (Encourage, Discourage) public education on poaching and black market trade?

54. Below is a chain of events that perpetuates overuse of natural biota. Which link in the chain needs to be removed to most effectively curtail overuse?

   a. hunter b. trading post c. taxidermist d. tourist shop e. consumer tourist

55. Give an example of human use of the following wildlife or plant species or their parts.

   Small songbirds: _____

   Tropical hardwoods: _____

   Bear gall bladders: _____

   Parrots: _____

Consequences of Losing Biodiversity

56. Explain why **keystone species** are such an important group to consider as a consequence of losing biodiversity?

   _____

International Steps to Protect Biodiversity

57. List two international steps that have been taken to protect biodiversity.

   a. _____   b. _____

58. Which of the above steps

   a. focuses on trade in wildlife and wildlife parts? _____

   b. focuses on conserving biological diversity worldwide? _____

   c. does not yet have political support in the U.S.? _____

Stewardship Concerns

59. List four steps that must be taken to protect our biological wealth.

   a. _____

   b. _____

   c. _____

   d. _____

VOCABULARY DRILL

Directions: Match each term with the list of definitions and examples given in the tables below. Term definitions and examples might be used more than once.

Term	Definition	Example(s)	Term	Definition	Example(s)
anthropocentric			fragmentation		
biodiversity			genetic bank		
biological wealth			instrumental value		
biota			intrinsic value		
conversion			keystone species		
cultivar			simplification		
ecotourism			speciation		
endangered species			threatened species		
exotic species			vigor		
extinction					

VOCABULARY DEFINITIONS

a. reduced to a point where it is eminent danger of extinction

b. judged to be in jeopardy but not on brink of extinction

c. assemblage of all living organisms

d. assemblage of all living organisms and their ecosystems

e. existence benefits some other entity

f. an entity that derives instrumental value from biota

g. something that has value for its own sake

h. tolerance to adverse conditions

i. cultivated variety of a wild strain

j. source of all genetic traits of wild strains of plants and animals

k. visiting a place in order to observe unique ecological sites

l. the creation of new species

m. the disappearance of species

n. human development of natural areas

o. process of cutting natural areas into separated parcels

p. diminishing abiotic and biotic habitat diversity

q. species introduced into an area from somewhere else

r. a species whose role is absolutely vital for the survival of many other species in the ecosystem

VOCABULARY EXAMPLES	
a. 1.4 million species	i. housing subdivisions
b. channelization	j. humans
c. competitiveness	k. Table 11-1
d. corn, wheat, rice	l. Fig. 11-7
e. Darwin's finches	m. large predators
f. Fig. 11-1	n. cats
g. Galapagos Islands	o. sources for agriculture
h. highways	p. just knowing that it exists

SELF-TEST

Indicate whether the following species have basically [a] instrumental or [b] intrinsic value.

1.   [   ] tuna
2.   [   ] butterflies
3.   [   ] rosy periwinkle
4.   [   ] hummingbirds

Match the examples on the left with the values of natural species on the right.

Examples

5.   [   ] cultivars
6.   [   ] captone
7.   [   ] bird watching
8.   [   ] ecotourism
9.   [   ] just knowing the species exists

Values

a. intrinsic
b. commercial
c. recreational
d. medicinal
e. agricultural

10.   The intrinsic value of wild species is rooted in _____ rights and _____ beliefs.

11.   Biodiversity is measured by the _____ of different species in an ecosystem.
12.   Which of the following types of ecosystems has the greatest biodiversity?

        a. grassland   b. tropical rain forest   c. urban development   d. cornfield

Indicate whether the following activities conserve [+] or destroy [-] biodiversity.

13.   [   ] conversion of habitat
14.   [   ] pollution
15.   [   ] introduction of exotics
16.   [   ] expansion of human population globally

17.   [   ] overuse
18.   [   ] regulated hunting
19.   [   ] recovery plans
20.   [   ] ecosystem management

Complete the statements 21 and 22 using the terms MORE or LESS.

21.   More expansion of the human population leads to _____ pollution, _____ use
        of wildlife species, and _____ biodiversity.

22.   More introduction of exotics leads to _____ competition for resources, _____
        predation on endemic prey species, and _____ biodiversity.

23.   The consequence(s) of species extinctions is (are):

        a. cascading loss of other species.
        b. loss of nature's services.
        c. loss of genetic banks.
        d. all of the above.

24.   It was (Commercial, Regulated) hunting that caused the extinction or near extinction of some
        wild animals.

Indicate whether the following statements pertain to [a] ESA, [b] CITES, [c] Convention on
Biological Diversity, or [d] more than one of these.

25.   [   ] focuses on the trade of wildlife or their parts
26.   [   ] uses recovery plans to re-establish species numbers
27.   [   ] emphasizes sovereignty of a nation over its biodiversity

ANSWERS TO STUDY GUIDE QUESTIONS

1. 33; 2. 1.4, 3 to 30; 3. 500; 4. instrumental; 5. intrinsic; 6. agriculture, medicine, aesthetics and recreation, intrinsic; 7. a, b, a, b, b, a, b, a; 8. 30; 9. a wild species that can become an alternative human food source; 10. treatment for leukemia; 11. rosy periwinkle; 12. controls high blood pressure; 13. Brazilian pit viper; 14. true; 15. ecotourism; 16. increases; 17. decreases; 18. 200; 19. a, a, a, a, b, c, b; 20. animal, religion; 21. commercial; 22. all true; 23. -, all the rest +; 24. d, b; 25. Florida, Texas; 26. Lacey; 27. Endangered Species, 1973; 28. -, +, +, -, -, -; 29. wild turkey, bald eagle; 30. sandhill cranes are surrogate parents for whooping crane chicks; 31. property rights and jobs; 32. speciation, extinction, accumulation; 33. 1.75, flowering, insects; 34. Costa Rica; 35. 484, 654; 36. 5,400, 26,000; 37. tropical rainforest; 38. commercial fishing down, decline in waterfowl populations, disappearance of songbirds from some regions, overall decline in songbird populations; 39. small area limits population size, human intrusions most severe; 40. habitat alterations, human population growth, pollution, exotic introductions, overuse; 41. habitat alterations, habitat alteration or overuse, human population growth, exotics, pollution, overuse, overuse, habitat alterations, pollution, overuse; 42. c, a, b; 43. inverse; 44. decrease; 45. c; 46. slowly; 47. quickly; 48. 39; 49. decreases; 50. more; 51. house mouse, Norway rat, wild boar, donkey, horse, nutria, red fox, armadillo, pheasant, starling, house sparrow, honey bee, fire ant; 52. increase; 53. discourage; 54. e; 55. food, furniture, perceived medicinal value, exotic pets; 56. a species whose role is absolutely vital for the survival of many other species in an ecosystem; 57. CITES, Convention on Biological Diversity; 58. CITES, Convention, Convention; 59. reform policies that lead to declines in biodiversity, address the needs of people whose livelihood is derived from exploiting wild species, practice conservation at the landscape level, promote more research on biodiversity.

ANSWERS TO SELF-TEST

1. a; 2. b; 3. a; 4. b; 5. e; 6. d; 7. c; 8. b; 9. a; 10. animal, religious; 11. number; 12. b; 13. -; 14. -; 15. -; 16. -; 17. -; 18. +; 19. +; 20. +; 21. more, more, less; 22. more, more, less; 23. d; 24. commercial; 25. b; 26. a; 27. c.

VOCABULARY DRILL ANSWERS

Term	Definition	Example(s)	Term	Definition	Example(s)
anthropocentric	f	j	fragmentation	o	h
biodiversity	d	a	genetic bank	j	a
biological wealth	d	a	instrumental value	c	o
biota	c	a	intrinsic value	g	p
conversion	n	i	keystone species	r	m
cultivar	i	d	simplification	p	b
ecotourism	k	g	speciation	l	e
endangered species	a	l	threatened species	b	k
exotic species	q	n	vigor	h	c
extinction	m	f			

CHAPTER 13

ENERGY FROM FOSSIL FUELS

━━━

━━━

Energy Sources and Uses

Harnessing Energy Sources: An Overview

1. Throughout human history, the major energy source was _____ labor.

2. The primary limiting factor to machinery designs in the early 1700s was a _____ source.

3. The _____ engine launched the Industrial Revolution in the late 1700s.

4. The first major fuel for steam engines was (a. coal, b. wood, c. gasoline, d. electricity).

5. By the 1800s the major fuel for steam engines was (a. coal, b. wood, c. gasoline, d. electricity).

6. By 1920, coal provided _____ percent of all energy used in the United States.

7. Three 1800 technologies that provided an alternative to coal were the
_____ engine, _____ drilling technology, and the ability to
refine _____ oil.

8. Crude oil provides _____ percent of the total U.S. energy demand.

9. Coal provides _____ percent of the total U.S. energy demand.

10. Indicate whether the following attributes are [A] advantages, or [D] disadvantages of oil-
based fuels when compared to coal.

   [ ] portability   [ ] pollution   [ ] energy content   [ ] known reserves

Electrical Power Production

11. Electricity is a (Primary, Secondary) energy source.

12. One **primary energy source** for generating electrical power is (a. coal b. wood c. oil-based
fuels).

13. Indicate whether the following are advantages [A] or disadvantages [D] of electrical
power.

   [ ] pollution from secondary energy source
   [ ] pollution from primary energy sources
   [ ] habitat alterations
   [ ] environmental effects of mining and processing
   [ ] efficiency of secondary energy source

14. Thermal production of electricity is only _____ percent efficient.

15. _____ percent of the primary energy is lost in the form of _____ causing
_____ pollution in aquatic ecosystems.

16. Explain how the below forms of generating electricity represent "transferral of pollution
from one place to another."

   Coal-burning power plants: _____

   Hydroelectric power plants: _____

   Nuclear power plants: _____

Matching Sources to Uses

17. Identify the four major categories of energy use (see Figure 13-9).

   a. _____   b. _____

   c. _____   d. _____

18.   What percent of the total energy use is devoted to:

a. transportation _____ %
b. industry _____%
c. residential _____ %
d. electrical power generation _____ %

19.   Match the dominant primary energy sources with list of secondary energy uses.

a. oil-based fuels   b. natural gas   c. coal   d. nuclear power and other sources
e. more than one of these energy source is used for the stated purpose

[ ] transportation
[ ] industrial processes
[ ] space heating and cooling
[ ] generation of electrical power

20.   The energy source that supplies over 40 percent of our total energy and virtually 100 percent of our energy for transportation is (a. oil-based fuels, b. natural gas, c. coal, d. nuclear power and other sources).

21.   All nuclear power and virtually all coal is used to generate _____ power.

Reserves of Crude Oil

22.   We are now dependent on foreign sources for over _____ % of our crude oil

23.   List four economic, political, or ecological costs of foreign oil dependence.

a. _____   b. _____

c. _____   d. _____

How Fossil Fuels Are Formed

24.   Three examples of fossil fuels are:

a. _____   b. _____   c. _____

25.   Fossil fuels are accumulations of organic matter from rapid _____ activity that occurred at various times in the Earth's history.

26.   Indicate whether the following statements are true [+] or false [-] concerning fossil fuels.

[ ] Supplies are limited.
[ ] The significant accumulations of organic matter that produced them can happen today.
[ ] We are using them far faster than they can be produced.

Crude Oil Reserves vs. Production

27.   Amounts of crude oil that are estimated to exist in the earth are called **estimated**
       _____ .

28.   Oil fields that have been located and measured are called _____ **reserves**.

29.   The maximum annual production from a given oil field is limited to about _____
       percent of the (Proven, Remaining) reserves.

Assume that a field has 100 million barrels of recoverable reserves. Given a maximum production of
10 percent, how much can be withdrawn in

30.   Year 1: _____ barrels leaving _____ barrels

31.   Year 2: _____ barrels leaving _____ barrels

32.   It is (True or False) that production from a given field may continue indefinitely.

33.   It is (True or False) that as maximum production goes down, the price per barrel will
       increase.

Declining U.S. Reserves and Increasing Importation

34.   List two revelations of the "Hubbart" curve applied to U.S. oil production.

       a. _____

       b. _____

The Oil Crisis of the 1970s

35.   Indicate whether the following events have increased [+] or decreased [-] in the United
       States since the 1970s.

       [   ] consumption of fuels derived from oil
       [   ] discoveries of new oil in the United States
       [   ] production of oil in the United States
       [   ] the gap between production and consumption
       [   ] United States dependence on foreign oil

36.   **OPEC** stands for

       _____

37.   The OPEC nations created an oil "crisis" in the 1970s by (Increasing, Suspending) oil
       exports.

38.   Through its ability to create world oil shortages, OPEC (Increased, Decreased) the price of
       crude oil.

Adjusting to Higher Prices

39. Indicate whether increasing the price of crude oil caused the following events to increase [+] or decrease [-] in the United States.

[ ] the rate of exploratory drilling and discovery of new oil
[ ] renewed production from old oil fields
[ ] efforts toward fuel conservation
[ ] consumption
[ ] development of alternative energy sources
[ ] dependence on foreign oil

40. It is (True, False) that the results of the oil "crisis" were not shortages but an oil glut.

41. As a result of oil surpluses, the price of a barrel of crude oil (Increased, Decreased).

42. As a result of oil surpluses, the price of a gallon of gas at the gas station

a. increased   b. decreased   c. stayed at the same level as during the crisis.

Victims of Our Success

43. Indicate whether the collapse in oil prices has caused the following events to increase [+] or decrease [-] in the United States.

[ ] the rate of exploratory drilling and discovery of new oil
[ ] renewed production from old oil fields
[ ] efforts toward fuel conservation
[ ] consumption
[ ] development of alternative energy sources
[ ] dependence on foreign oil

44. U.S. dependence on foreign oil in 1994 has grown to _____ percent (see Fig. 13-14).

45. It is (True or False) that OPEC oil reserves will last forever.

Problems of Growing U.S. Dependency on Foreign Oil

46. List three problems resulting from U.S. Dependency on foreign oil.

a. _____ b. _____

c. _____

47. The cost for foreign crude oil represents about _____ % of the U.S. balance trade deficit.

48. Compare the economic impact of using

    oil produced at home _____

    foreign oil _____

49. The Persian Gulf War was fought primarily to

    a. free the people of Kuwait.
    b. protect Kuwait oil fields from Saddamm Hussein
    c. drive oil prices up.
    d. force OPEC nations to come to terms on oil prices.

50. Military intervention (Increased, Decreased) the price per barrel of oil.

51. Our actual cost of a barrel of oil from Persian Gulf sources is $14 + $_____ for military support services or $_____ a barrel.

52. The last major oil discovery in the U.S. was in _____ in 1968.

53. The Middle East possesses _____ % of the world's proven oil reserves.

54. List three ways we can reduce our dependency on foreign oil.

    a. _____

    b. _____

    c. _____

Alternative Fossil Fuels

55. Unlike oil, the U.S. is rich in supplies other fossil fuels including:

    a. _____   b. _____   c. _____

56. Identify the fossil fuel (a. natural gas, b. coal, c. synfuels, d. oil shales, e. oil sands) that ranks the highest in each of the following categories.

    [ ] air pollution
    [ ] cost of extraction
    [ ] proven reserves
    [ ] greenhouse effect
    [ ] retrofitting to existing transportation technologies
    [ ] habitat alteration
    [ ] cost competition with current oil prices

Sustainable Energy Options

57. Which fossil fuel produces the least CO_2 emissions? _____

58.   Identify two options to combat future oil shortages.

     a. _____   b. _____

Conservation

59.   A major preparation for future oil shortages is to (Increase, Decrease) oil consumption.

60.   One way oil consumption can be decreased is through _____.

61.   It is (True or False) that between 1970 and 1980 the U.S. economy expanded without a corresponding rise in energy consumption.

62.   At least _____ percent of the conservation reserve remains to be tapped.

63.   List four ways to increase the conservation reserve.

     a. _____

     b. _____

     c. _____

     d. _____

Development of Non-Fossil Fuel Energy Sources

64.   Identify two major pathways for developing alternative energy sources.

     a. _____

     b. _____

NO VOCABULARY DRILL FOR THIS CHAPTER!

SELF-TEST

1.   Rank (1 to 5) the following fuel types in their order of historical use by human society.

     [ ] coal   [ ] wood   [ ] electricity   [ ] gasoline   [ ] natural gas

Classify each energy type in terms of (a. primary b. secondary) source and (a. renewable b. nonrenewable).

Energy Types	Primary or Secondary	Renewable or Nonrenewable
2. wood	[ ]	[ ]

Energy Types	Primary or Secondary	Renewable or Nonrenewable
3. coal	[ ]	[ ]
4. oil-based	[ ]	[ ]
5. natural gas	[ ]	[ ]
6. electricity	[ ]	[ ]
7. nuclear	[ ]	[ ]
8. synfuels	[ ]	[ ]
9. solar	[ ]	[ ]

Classify each energy type in terms of primary end uses including (a. transportation b. industry c. residential d. electric power generation e. not used).

10. [ ] wood
11. [ ] coal
12. [ ] oil-based fuels
13. [ ] natural gas
14. [ ] electricity
15. [ ] nuclear fuels
16. [ ] synfuels
17. [ ] solar energy

Indicate which energy types have **all three advantages**: low cost of extraction, low pollution, and high known reserves. (Check all that apply.)

18. [ ] wood   22. [ ] electricity
19. [ ] coal   23. [ ] nuclear fuels
20. [ ] oil-based fuels 24. [ ] synfuels
21. [ ] natural gas   25. [ ] solar energy

26. Which of the above energy types is a primary source, renewable, not used, and has all three advantages listed? _____

27. Explain why this energy source is not used.
_____
_____

Define "fossil" fuel and give two examples.

28. "Fossil" fuel:
_____

Two examples are: 29. _____ and 30. _____

Explain why fossil fuels are "nonrenewable" and "nonrecyclable."

31. Nonrenewable: _____

32. Nonrecyclable: _____

33.   Explain why electric heat systems represent a **double** energy waste:

_____

Indicate whether the following actions will increase [+], decrease [-], or have no impact [0] on future supplies of oil-based fuels.

34.   [   ] searching for new reserves
35.   [   ] taking OPEC oil by force
36.   [   ] increasing the price of gasoline at the pump
37.   [   ] increaseing use of mass transit systems
38.   [   ] continuing consumption at present rates
39.   [   ] car pooling
40.   [   ] increasing engine fuel efficiency
41.   [   ] increasing the beginning driving age to 18

42.   Identify an oil field that has a potential production of 6 million barrels per day, is three times the size of the Alaskan field, is inexhaustible, and its exploitation will not adversely effect the environment. This oil field is called the _____ reserve.

ANSWERS TO STUDY GUIDE QUESTIONS

1. human; 2. power; 3. steam; 4. b; 5. a; 6. 80; 7. internal combustion, oil, crude; 8. 41; 9. 39; 10. A, A, A, D; 11. secondary; 12. a; 13. A, D, D, D, D; 14. 35; 15. 65, heat, thermal; 16. major source of acid rain, flooding of terrestrial habits from dams, disposal of nuclear wastes; 17. transportation, industry, residential, electrical power generation; 18. 28, 23, 13, 36; 19. a, e, e, c; 20. a; 21. electrical; 22. 50; 23. trade imbalances, military actions, pollution of the oceans, coastal oil spills; 24. coal, crude oil, natural gas; 25. photosynthetic; 26. +, -, +; 27. reserves; 28. proven; 29. 10, remaining; 30. 10 million, 90 million; 31. 9 million, 81 million; 32. false; 33. true; 34. U.S. oil production peaked in 1970, will become increasingly dependent on foreign oil; 35. I, D, D, I, I; 36. Organization of Petroleum Exporting Countries; 37. suspending; 38. increased; 39. I, I, I, D, I, D; 40. true; 41. decreased; 42. c; 43. D, D, D, I, D, I; 44. 50; 45. false; 46. costs of purchases, supply disruption, resource limitations, 47. 59; 48. provides jobs that produce other goods and services, money leaves U.S. as another form of foreign aid; 49. b; 50. increase; 51. 80, 94; 52. Alaska; 53. 64; 54. use other fossil fuel sources, reduce our energy demand, develop alternatives to fossil fuels; 55. natural gas, coal, oil shales and tar sands; 56. b, c, b, b, a, a or b, e; 57. natural gas; 58. conservation, develop nonfossil fuel sources; 59. decrease; 60. conservation; 61. true; 62. 80; 63. increase fuel efficiency in cars, cogeneration, use florescent lights, increase home insulation; 64. nuclear and solar power.

ANSWERS TO SELF-TEST

1. 2, 1, 4, 3, 5; 2. a, a; 3. a, b; 4. a, b; 5. a, b; 6. b, b; 7. a, b; 8. a, b; 9. a, a; 10. e; 11. d; 12. a; 13. b or c; 14. b or c; 15. d; 16. e; 17. e; 18 to 24. not checked; 25. checked; 26. solar; 27. a multitude of political, economic, and cultural reasons; 28. derived from ancient photosynthetic processes; 29. coal; 30. oil; 31. cannot be made again in the short term; 32. cannot be used again once burned; 33. burn coal to produce heat to turn a turbine to produce electricity to produce heat for a home; 34. O; 35. O; 36. +; 37. +; 38. -; 39. +; 40. +; 41. +; 42. conservation.

CHAPTER 15

RENEWABLE ENERGY

Principles of Solar Energy

1. Solar energy is a (Potential, Kinetic) form of energy originating from (Chemical, Thermonuclear) reactions in the sun.

2. Solar energy is a (Renewable, Nonrenewable) form of energy.

Putting Solar Energy to Work

3.   The four main hurdles to overcome in using solar energy are:

     a. _____   b. _____ _____

     c. _____   d. _____

4.   List three direct uses of solar energy.

     a. _____   b. _____

     c. _____

Solar Heating of Water
Solar Space Heating

5.   Indicate whether the below components are required of **active** [A], **passive** [P], or both
     [B] types of solar heating systems (see Figs. 15-6 to 15-11).

     [   ] **flat-plate solar collector**
     [   ] water pump
     [   ] heat storage facility
     [   ] expansion tank
     [   ] blowers
     [   ] back-up system
     [   ] heat exchanger
     [   ] valves
     [   ] improved insulation
     [   ] appropriate landscaping
     [   ] earth berms

6.   Indicate whether the below components are requirements of solar hot water [W], space
     [S], or both [B] types of solar heating systems (see Figs. 15-6 to 15-11).

     [   ] **flat-plate solar collector**
     [   ] water pump
     [   ] heat storage facility
     [   ] expansion tank
     [   ] blowers
     [   ] back-up system
     [   ] heat exchanger
     [   ] valves
     [   ] improved insulation
     [   ] appropriate landscaping
     [   ] earth berms

7.   The most economical means of solar heating is an/a (Active, Passive) system.

8.　　A well-designed passive solar home can reduce energy bills by _____ percent with added construction costs of _____ to _____ percent.

9.　　It is (True or False) that partial conversion to solar energy will make a large amount of natural gas available for fueling vehicles.

10.　　About _____ percent of the total U.S. energy budget is used for heating buildings and water.

11.　　The chief barrier to more wide-spread use of passive solar designs is _____.

Solar Production of Electricity

12.　　List two technologies that convert solar energy into electricity.

　　a. _____　　　　　b. _____

13.　　A device that converts light directly to electricity is the _____ cell.

14.　　The photovoltaic or solar cell consists of two layers of

　　a. water　b. electrons　c. atoms　d. mylar film.

15.　　The top layer contains atoms with (Additional, Missing) electrons in the outer orbital.

16.　　The bottom layer contains atoms with (Additional, Missing) electrons in the outer orbital.

17.　　The top layer will readily (Lose, Accept) an electron.

18.　　The bottom layer will readily (Lose, Accept) an electron.

19.　　The kinetic energy of sunlight forces a movement of electrons from the top to the bottom layer creating an electrical imbalance or _____.

20.　　Solar cell technology:

　　a. is becoming more cost-effective with increased use. (True or False)
　　b. will not wear out. (True or False)
　　c. is already competitive in costs per watt output. (True or False)
　　d. has the potential to provide power for anything from a watch to a home. (True or False)

21.　　Solar trough collectors convert _____ percent of incoming sunlight to electric power at a cost of _____ cents per kilowatt-hour.

22.　　The economic feasibility of the solar trough collector is enhanced by (More, Less) environmental pollution when compared to coal-fired facilities.

23.　　Power towers use mirrors to convert solar energy to _____ which runs a _____ that creates electricity.

24. The dish/engine system has demonstrated efficiencies of _____%.

The Promise of Solar Energy

25. List four disadvantages of using solar energy technologies.

    a. _____   b. _____

    c. _____   d. _____

26. Identify three hidden costs in the use of traditional energy sources.

    a. _____   b. _____

    c. _____

27. What is the best strategy for integrating solar energy technology with existing energy consumption patterns?

    _____

28. About _____% of our electrical power is generated by coal-burning and nuclear power plants.

29. Complete the following statements with MORE or LESS

    "The use of solar electrical power will:"

    a. create _____ reliance on coal or nuclear power.
    b. lead to significantly _____ acid rain.
    c. produce _____ electrical power for villages in developing countries.

Solar Production of Hydrogen -The Fuel of the Future

30. Indicate whether the following are benefits [+] or limitations [-] of hydrogen power.

    [   ] substitute for natural gas
    [   ] substitute for gasoline
    [   ] pollution factor
    [   ] production technology
    [   ] modification of existing transportation technologies
    [   ] production costs
    [   ] national distribution system

Indirect Solar Energy

31. List four indirect forms of solar energy.

    a. _____   b. _____

c. _____   d. _____

Hydropower

32.   The earliest form of hydropower was the use of (Natural, Artificial) water falls.

33.   The contemporary form of hydropower is the use of (Natural, Artificial) water falls.

34.   Artificial waterfalls are created by building huge _____.

35.   The artificial waterfalls of huge dams generate _____ **power**.

36.   Indicate whether the following are benefits [+] or drawbacks [-] of hydroelectric power.

      [   ] level of pollution generated
      [   ] level of environmental degradation
      [   ] amount of total energy produced
      [   ] geographical distribution of energy produced
      [   ] ecological impacts above and below the dam

37.   Only _____ percent of the nation's rivers remain free-flowing.

38.   Many of the remaining free-flowing rivers are protected by the

      _____.

39.   The Nam Theun Two Dam will displace _____ people and destroy
      _____ square miles of sensitive environments.

40.   The Three Gorges Dam will displace _____ people.

Wind Power

41.   The earliest form of harnessing wind power were wind (Mills, Turbines).

42.   The contemporary form of harnessing wind power is wind (Mills, Turbines).

43.   Indicate whether the following are benefits [+] or drawbacks [-] of wind power.

      [   ] the size limitations on wind turbines
      [   ] the number of megawatts of electricity produced
      [   ] the level of pollution generated
      [   ] the level of environmental degradation
      [   ] geographical distribution of energy produced
      [   ] cost-effectiveness
      [   ] aesthetics

Biomass Energy

44.   List four examples of biomass energy or bioconversion.

a. _____   b. _____

c. _____   d. _____

45.   Indicate whether the following are benefits [+] or drawbacks [-] of biomass energy.

[   ] availability of the biomass resource
[   ] access to the biomass resource
[   ] public acceptance and utilization of biomass energy
[   ] past history of human harvests within a maximum sustained yield

46.   Methane gas is (Direct, Indirect) use of biomass as fuel.

47.   It is (True or False) that one can get more electrical power from cows than from nuclear power.

48.   Alcohol production is a (Direct, Indirect) use of biomass as fuel.

49.   Alcohol production utilizes (Fermentation, Distillation, Both).

Additional Renewable Energy Options

50.   List two additional sustainable energy options.

a. _____   b. _____

Geothermal Energy

51.   Indicate whether the following are benefits [+] or drawbacks [-] of geothermal energy.

[   ] if accessible, it is an everlasting energy resource
[   ] the consistency in number of megawatts of electricity produced
[   ] the level of pollution generated
[   ] the level of environmental degradation
[   ] geographical distribution of energy produced
[   ] cost-effectiveness
[   ] technology required for extraction

Tidal Power

52.   Tidal power converts the energy of (Incoming, Outgoing, Both) tides.

53.   Indicate whether the following are benefits [+] or drawbacks [-] of tidal power.

[   ] if accessible, it is an everlasting energy resource
[   ] the consistency in number of megawatts of electricity produced
[   ] the level of pollution generated

[   ] the level of environmental degradation
[   ] geographical distribution of energy produced
[   ] cost-effectiveness
[   ] technology required for extraction

Ocean Thermal Energy Conversion

54.   Indicate whether the following are benefits [+] or drawbacks [-] of **OTEC**.

[   ] technology
[   ] cost-effectiveness
[   ] environmental consequences

Policy for a Sustainable Energy Future

55.   Indicate whether the following are [+] or are not [-] aspects of the existing U.S. energy policy.

[   ] mandates to increase fuel efficiency of cars
[   ] exploitation of public lands for fossil fuel reserves
[   ] subsidies to producers of solar energy or solar energy products
[   ] continued research on and development of renewable energy resources
[   ] continued research on and development of alternative energy technologies
[   ] advocacy of energy conservation
[   ] public education on solar and other renewable energy sources

56.   Indicate whether the following energy policies represent "business as usual" [B] or "transition to the future" [T].

[   ] tax subsidies for solar energy technology production and use
[   ] mandates to increase fuel efficiency of cars
[   ] depletion allowances
[   ] military support to assure access to oil in Mid East
[   ] damage to human health
[   ] savings of $600 million in electric bills
[   ] cars that get 80 miles per gallon
[   ] climate change research
[   ] deregulation of the electrical industry
[   ] net metering
[   ] carbon tax

VOCABULARY DRILL

Directions: Match each term with the list of definitions and examples given in the tables below. Term definitions and examples might be used more than once.

Term	Definition	Example(s)	Term	Definition	Example(s)
active systems			OTEC		
bioconversion			passive systems		
biomass energy			power gird		
carbon tax			PV cell		
electrolysis			solar trough		
flat-plate collectors			wind farms		
fuel cells			wind turbine		
geothermal energy					

VOCABULARY DEFINITIONS

a. broad box with a glass top and a black bottom with imbedded water tubes
b. moves water or air with pumps or blowers
c. moves water or air with natural convection currents or gravity
d. cell that converts sunlight energy into electricity
e. a network of power lines taking power from generating stations to customers
f. long trough-shaped reflectors tilted toward the sun
g. disassociation of H_2O using electricity
h. devices in which H is recombined with O_2 chemically to produce electric potential
i. wind-driven generator
j. collections of thousands of wind-driven generators
k. deriving energy from present-day photosynthesis
l. concept of using ocean's temperature differences to produce power
m. naturally heated groundwater
n. tax on fuels according to the amount of CO_2 produced during consumption

VOCABULARY EXAMPLES	
a. burning wood	h. Hawaii
b. Fig. 15-12	I. Fig. 15-9
c. Fig. 15-19	j. pumps, valves, blowers
d. Fig. 15-6	k. European countries
e. Fig. 15-23	l. Fig. 15-15
f. overhead electrical lines	m. Fig. 15-20
g. H and O_2 bubbles	

SELF-TEST

1. The biggest problem connected with using solar energy is

   a. there is not enough energy inherent in sunlight to be worth its use.
   b. there is no good means of collecting sunlight energy.
   c. the means of storing it are bulky and expensive.
   d. there are no good ways of converting solar energy into profits.

   Match the below terms or concepts with examples of solar and other renewable energy sources.

Terms and Concepts	Examples
2. [ ] flat plate collector	a. solar energy
3. [ ] active solar system	b. additional renewable energy options
4. [ ] passive solar system	c. indirect solar energy
5. [ ] photovoltaic cells	
6. [ ] solar trough collector	
7. [ ] power tower	
8. [ ] hydrogen production	
9. [ ] hydropower	
10. [ ] wind turbine	
11. [ ] bioconversion	
12. [ ] OTEC	
13. [ ] geysers	
14. [ ] twice-daily rise and fall	

15. The most cost-effective method of using solar energy for home use is through

   a. passive solar systems.
   b. active solar systems.
   c. photovoltaic solar systems.
   d. selling your excess power back to the utility companies.

16. Which of the following is not a direct means of using solar energy?

   a. Direct conversion to electrical power.
   b. Heating homes and other buildings.
   c. Production of flammable hydrogen gas.
   d. Production of biomass for conversion to flammable liquids.

17. Solar cell technology

   a. cannot become cost effective even with increased use.
   b. will wear out.
   c. is competitive with nuclear power in costs per watt output.
   d. has severe limitations in power output.

18. A major drawback of solar production of hydrogen is

   a. its ability to replace our reliance on natural gas.
   b. its ability to replace gasoline in cars.
   c. its waste products.
   d. the ability to produce free hydrogen atoms.

19. Hydroelectric power cannot provide a large percentage of power in the future because of the

   a. level of thermal pollution generated.
   b. amount of energy produced.
   c. limited geographic distribution of energy produced.
   d. level of air and water pollution generated.

20. A major drawback of wind power is the

   a. amount of energy produced.
   b. size limitations of wind turbines.
   c. level of environmental degradation.
   d. cost-effectiveness.

21. The major drawback that geothermal energy, tidal and wave power all have in common is

   a. the number of megawatts of electricity that could be produced is low.
   b. the technology required for energy conversion is not cost-effective.
   c. the level of environmental degradation.
   d. the limitations on long-term dependency as an energy source.

22. In view of the nature of U.S. energy problems and associated environmental problems, which of the following areas should be given a higher priority in research and development funding?

   a. fusion power b. solar power c. breeder reactors d. utilization of coal

ANSWERS TO STUDY GUIDE QUESTIONS

1. kinetic, thermonuclear; 2. renewable; 3. collection, conversion, storage, cost-effectiveness; 4. heating, electricity, fuel; 5. B, A, A, A, A, B, A, A, B, B, P; 6. W, W, B, W, B, B, W, W, S, S, S; 7. passive; 8. 75, 5, 9. true; 10. 25; 11. ignorance; 12. photovoltaic cells, solar trough collectors; 13. solar; 14. b; 15. missing; 16. additional; 17. accept; 18. accept; 19. current; 20. true, false, false, true; 21. 22, 8; 22. less; 23. steam, turbogenerators; 24. 30; 25. expensive technologies, only works during the day, requires back-up energy source or storage battery, some climates not sunny enough; 26. air pollution, strip-mining, nuclear waste; 27. use it during the daytime hours to meet peak energy demands and use traditional energy sources in the evening hours; 28. 77; 29. less, less, more; 30. all true; 31. wind, water, biomass, ocean thermal; 32. natural; 33. artificial; 34. dams; 35. hydroelectric; 36. +, -, -, -, -; 37. 2; 38. Wild and Scenic Rivers Act of 1968; 39. 4,500, 2,485; 40. 1.9 million; 41. mills; 42. turbines; 43. -, +, +, +, +, +, -; 44. fire wood, burning MSW, methane production, alcohol production; 45. +, -, +, -; 46. indirect; 47. true; 48. indirect; 49. both; 50. geothermal, tidal; 51. -, -, -, -, -, -, -; 52. both; 53. +, +, +, -, -, -, -; 54. -, -, +; 55. -, +, -, -, -, -, -; 56. T,T,B,B,B,T,T,T,T,T,T.

ANSWERS TO SELF-TEST

1. d; 2. a; 3. a; 4. a; 5. a; 6. a; 7. a; 8. a; 9. c; 10. c; 11. c; 12. c; 13. b; 14. b; 15. a; 16. d; 17. d; 18. d; 19. c; 20. b; 21. b; 22. b.

VOCABULARY DRILL ANSWERS

Term	Definition	Example(s)	Term	Definition	Example(s)
active systems	b	j	OTEC	l	h
bioconversion	k	a	passive systems	c	i
biomass energy	k	a	power gird	e	f
carbon tax	n	k	PV cell	d	b
electrolysis	g	g	solar trough	f	l
flat-plate collectors	a	d	wind farms	j	m
fuel cells	h	c	wind turbine	i	m
geothermal energy	m	e			

CHAPTER 16

ENVIRONMENTAL HAZARDS AND HUMAN HEALTH

Links Between Human Health and the Environment.
  The Picture of Health
  Environmental Hazards

Pathways of Risk
  The Risks of Being Poor
  The Cultural Risk of Smoking
  Risk and Infectious Diseases
  Toxic Risk Pathways

Risk Analysis
  Risk Analysis by the EPA
  Risk Management
  Risk Perception

Vocabulary Drill
  Vocabulary Definitions
  Vocabulary Examples

Self-Test

Answers to Study Guide Questions

Answers to Self-Test

Vocabulary Drill Answers

1.   The bacterium that causes Lyme disease comes from the _____ which can be transmitted to deer and humans by the primary **vector** called the _____.

2.   Explain the relationship between oak trees, gypsy moths, mice and Lyme disease.

   _____

   _____

3.   A study of the connections between hazards in the environment and human disease and death is called _____ **health.**

Links Between Human Health and the Environment

4.   In this chapter, **environment** means: _____

        _____ .

5.   In the context of environmental health, give three examples of **hazards**.

        a. _____   b. _____

        c. _____

6.   **Risk** is defined as: _____

        _____ .

The Picture of Health

7.   List four dimensions of the meaning of health.

        a. _____   b. _____

        c. _____   d. _____

8.   Human life expectancy has (Increased, Decreased) in the last 40 years.

9.   List three advances that caused a change in human life expectancy in the last 40 years.

        a. _____   b. _____

        c. _____

10.   Forty-three per cent of the mortality is developing countries is caused by
        _____ compared to _____ % in developed countries.

Environmental Hazards

11.   What are two ways to consider hazards to human health?

        a. _____   b. _____

12.   Match the hazards the left with their classifications on the right.

        Hazards   Classifications

        [   ] smoking   a. Cultural
        [   ] AIDS   b. Biological
        [   ] plague   c. Physical
        [   ] malaria   d. Chemical
        [   ] tornadoes

Hazards   Classifications
[   ] floods   a. Cultural
[   ] pesticides   b. Biological
[   ] cleaning agents   c. Physical
[   ] sunbathing   d. Chemical
[   ] uses vectors
[   ] building in a floodplain
[   ] consume harmful drugs
[   ] respiratory infections
[   ] leaded paint

13.   Explain the relationship between AIDS and tuberculosis.

_____

14.   Give the common pathogens that cause:

Diarrheal diseases: _____

Malaria: _____

15.   What insect is the vector of malaria? _____

16.   What part of the human anatomy is infected by malaria? _____

17.   List three symptoms of malaria.

a. _____   b. _____

c. _____

18.   What country has the greatest exposure to tornadoes? _____

19.   List four methods of exposure to chemical hazards.

a. _____   b. _____

c. _____   d. _____

20.   Toxicity = _____ of exposure + _____ of a toxic substance.

21.   List four potential effects of toxic chemicals on human health.

a. _____   b. _____

c. _____   d. _____

22.   Twenty-three per cent of all U.S. deaths are traced to _____.

23.   _____ % of cancers can be traced to environmental causes.

Pathways of Risk

The Risks of Being Poor

24.   Why is poverty the world's biggest killer?

        _____

        _____

25.   It is (True or False) that wealth = health.

26.   It is (True or False) that people in developed countries tend to live longer.

27.   List three reasons that explain the differences in time and cause of death between people in
        developed and developing countries.

        a. _____   b. _____

        c. _____

28.   How does equitable income distribution affect the health of a nation?

        _____

The Cultural Risk of Smoking

29.   Smoking is responsible for _____ % of the cancer deaths in the U.S.

30.   Tobacco related illness is estimated to cost $_____ billion/year in lost work capacity
        and health care expenses.

31.   Since 1964, the U.S. smoking population has dropped from 40% to _____ %.

32.   What is ETS? _____

33.   It is (True or False) that nicotine is addictive.

34.   It is (True or False) that the U.S. continues to subsidize the tobacco industry.

Risk and Infectious Diseases

35.   What does an epidemiologist do?

        _____

36.   List two major causes of food and water contamination.

        a. _____   b. _____

37.   It is (True or False) that major outbreaks of diarrheal diseases occur only in developing countries.

38.   In addition to malaria, list four more diseases known to be carried by mosquitoes.

a. _____   b. _____

c. _____   d. _____

39.   What major problem has developed in the campaign to control mosquito populations?

_____

40.   Indicate whether the following activities have increased [+] or decreased [-] the incidence of malaria.

[   ] deforestation
[   ] irrigation
[   ] creation of dams
[   ] high intensity agriculture

Toxic Risk Pathways

41.   It is (True or False) that air inside the home and workplace often contains much higher levels of pollutants than outdoor air.

42.   Factors contributing to indoor air pollution include

a. products used indoors. (True or False)
b. well insulated and sealed buildings. (True or False)
c. duration of exposure to indoor pollutants. (True or False)

43.   List five types or sources of indoor air pollutants (see Fig. 16-14).

a. _____   b. _____

c. _____   d. _____

e. _____

Risk Analysis

44.   Which of the following actions carry the greatest risk (see Table 16-3)?

a. being a policeman   b. drinking alcohol   c. smoking cigarettes

Risk Analysis by the EPA

45.   Risk analysis includes

        a. hazard assessment. (True or False)
        b. dose-response assessment. (True or False)
        c. exposure assessment. (True or False)
        d. risk characterization. (True or False)

46.   Epidemiological studies are used to make a cause and effect relationship between human health and (Past, Future) exposures to carcinogenic chemicals.

47.   Animal tests are used to make a cause and effect relationship between human health and (Past, Future) exposures to carcinogenic chemicals.

48.   List three objections that have been raised about animal testing.

        a. _____   b. _____

        c. _____

49.   Match the steps in risk analysis with the methods used to perform each step.

Step	Method
[ ] hazard assessment	a. final estimate of probability of fatal outcome from hazard
[ ] dose-response assessment	b. associates dose with incidence and severity of response
[ ] exposure assessment	c. uses epidemiological studies and animal tests
[ ] risk characterization	d. studies human groups with previous exposures

Risk Management

50.   List the two steps in **risk management**.

        a. _____

        b. _____

51.   Regulatory decisions are based on:

        a. cost-benefit analysis   b. risk-benefit analysis   c. public preferences   d. all of these

Risk Perception

52.   How many of the public's top 11 environmental concerns are also on the EPA's list of top 11 risks (see Table 16-4)? _____

53.   The public's **risk perception** is (the Same, Different) from that of EPA scientists.

54. List the outrage factors that influence the public's risk perception.

    a. _____   b. _____   c. _____

    d. _____   e. _____   f. _____

    g. _____

55. It is (Public Concern or Cost-benefit Analysis) the drives public policy.

56. It is (True or False) that public concern for human health overrides concern for major ecological risks.

VOCABULARY DRILL

Directions: Match each term with the list of definitions and examples given in the tables below. Term definitions and examples might be used more than once.

Term	Definition	Example(s)	Term	Definition	Example(s)
distributive justice			morbidity		
dose			mortality		
dose-response			outrage		
ecological risks			risk		
environmental health			risk analysis		
epidemiology			risk characterization		
exposure			risk management		
hazard			risk perception		
hazard assessment			vector		

VOCABULARY DEFINITIONS

a.	disease carrying agent
b.	connections between hazards in the environment and human disease and death
c.	anything that causes sickness, shorten life or contributes to human suffering
d.	incidence of disease in a population
e.	incidence of death in a population
f.	study of the presence, distribution and control of disease in populations
g.	probability of suffering from injury, disease, or death
h.	process of evaluating the risks associated with a particular hazard before taking the action
I.	concentrations of the chemical in a test
j.	incidence and severity of response when exposed to different doses of a chemical
k.	identifying subjects, origin of chemical, dose and length of exposure
l.	probability of a fatal outcome due to a hazard
m.	examining evidence linking a potential hazard to its harmful effects
n.	concerning nature's welfare
o.	intuitive judgement about risks
p.	additional public concerns other than fatalities
q.	risk characterization of a hazard followed by regulatory decisions
r.	making sure all stakeholders are involved in the decision-making process

VOCABULARY EXAMPLES

a.	mosquito	g.	involuntary risks
b.	Table 16-3	h.	Table 16-2
c.	tracking geographic distribution of HIV	i.	toxic chemicals and cancer
d.	everyone votes on the decision	j.	Table 16-1
e.	How many will die?	k.	Table 16-4
f.	Chernobyl, Russia		

SELF-TEST

Fill in the below table with the human disease(s) the list of organisms are known to cause or carry.

Organisms	Disease(s)
1. black-legged tick	
2. white-tailed deer	
3. mosquito	
4. *Salmonella*	
5. *Escherischia coli*	
6. *Borrelia burgdorferi*	
7. *Plasmodium*	
8. Rats	

Indicate whether the following are a. cultural, b. biological, c. physical, or d. chemical hazards.

9.   [   ] earth quakes
10.   [   ] cholera
11.   [   ] sunburn
12.   [   ] cleaning solutions

13.   In the U.S., about what percent of all deaths can be traced to cancer?

        a. 15   b. 25   c. 50   d. 75

14.   Why is poverty the world's greatest killer?

        _____

15.   It is (True or False) that people in developed countries live longer than people in developing countries. Explain why.

        _____

16.   The major single cause of cancer deaths in the U.S. is _____.

17.   The smoking population in the U.S. has (Increased, Decreased) over the past 40 years.

18.   It is (True or False) that there could be a resurgence of malaria in developing countries.

19.   List four indoor air pollutants

        a. _____   b. _____

        c. _____   d. _____

The below definitions are examples of [a] hazard, [b] risk, [c] outrage.

20.   [   ] People who have a choice will accept risk at a greater rate than those having no choice.
21.   [   ] the probability of suffering injury or other loss as a result of exposure to hazard.
22.   [   ] anything that has the potential of causing suffering due to exposure.

Match the steps in risk analysis with the methods used to perform each step.

Step	Method
23. [ ] hazard assessment	a. final estimate of probability of fatal outcome from hazard
24. [ ] dose-response assessment	b. associates dose with incidence and severity of response
25. [ ] exposure assessment	c. uses epidemiological studies and animal tests
26. [ ] risk characterization	d. studies human groups with previous exposures

ANSWERS TO STUDY GUIDE QUESTIONS

1. mouse, black-legged tick; 2. + moths ⇨ - oak leaves ⇨ - mice ⇨ - Lyme disease; 3. environmental; 4. the whole context of human life - physical, chemical, biological; 5. injury and disease, property damage, environmental decay; 6. probability of suffering from injury, disease or death; 7. physical, mental, spritual, emotional; 8. Increased; 9. social, medical, economic; 10. infectious diseases, 2; 11. lack of access to resources, exposure to a hazard; 12. a, a, b, b, c, c, d, d, a, b, c, a, b, d; 13. HIV infection weakens the immune system which allows TB to flourish; 14. Salmonella, Campytobacter, E. coli, Plasmodium; 15. mosquito; 16. red blood cells; 17. fever, chills, headaches, vomiting; 18. U.S.; 19. eating, drinking, breathing, direct use; 20. time, dose; 21. impairment of immune system, infertility, brain impairment, birth defects; 22. cancer; 23. 25; 24. the poor have no access to health care resources and have a greater exposure to hazards; 25. true; 26. true; 27. education, nutrition, modernization; 28. overall improvement of quality of life leads to healthier society; 29. 30; 30. 200; 31. 26; 32. second hand tobacco smoke; 33. true; 34. true; 35. traces geographic distribution of a disease, mode of transmission and consequences; 36. lack of sewage treatment and personal hygiene; 37. false; 38. yellow fever, dengue fever, elephantiasis, encephalitis; 39. mosquito and Plasmodium resistance to controls; 40. all +; 41. true; 42. all true; 43. see Fig. 16-14; 44. c; 45. all true; 46. past; 47. future; 48. rodents and humans have different responses to same dose, higher than normal doses given to rodents, people object to animal experiments; 49. a, b, c, d; 50. complete risk assessment, regulatory decision; 51. d; 52. 3; 53. different; 54. technological ignorance, lack of choice, memory of past hazards, overselling safety, morality, lack of control, fairness; 55. public concern; 56. true.

ANSWERS TO SELF-TEST

1. Lyme; 2. Lyme; 3. malaria, yellow fever, encephalitis, elephantiasis; 4. diarrheal disease; 5. diarrheal disease; 6. Lyme; 7. malaria; 8. bubonic plague; 9. c; 10. b; 11. a; 12. d; 13. b; 14. lack of access to health care resources and greater exposure to hazards; 15. true, people in developed countries have greater access to health care resources and less exposure to hazards; 16. smoking; 17. decreasd; 18. true; 19. b; 20. a; 21. c; 22. b; 23. d; 24. a.

VOCABULARY DRILL ANSWERS

Term	Definition	Example(s)	Term	Definition	Example(s)
distributive justice	r	d	Morbidity	d	j
dose	i	f	Mortality	e	j
dose-response	j	f	Outrage	p	g
ecological risks	n	h	risk	g	b
environmental health	b	j	risk analysis	h	k
epidemiology	f	c	risk characterization	l	e
exposure	k	f	risk management	q	j
hazard	c	b	risk perception	o	j
hazard assessment	m	b	vector	a	a

CHAPTER 17

PESTS AND PEST CONTROL

The Need for Pest Control

Defining Pests

1.   It is (True or False) that the term **pest** refers only to bugs.

Match the organisms on the left with their pest category on the right.

	Organism		Pest Category
2.	[ ] weeds		a. disease
3.	[ ] molds		b. annoyance
4.	[ ] coyotes		c. competitive feeder
5.	[ ] snails		d. predator
6.	[ ] bacteria		e. decomposer
7.	[ ] mosquitos		f. nutrient competitor

The Importance of Pest Control

8.   It is (True or False) that humankind need no longer rely on pest control.

9.   Crop losses to pests have (Increased or Decreased) since the 1950s.

10.   An estimated _____ percent of agricultural production are lost to pests now compared to _____ percent in 1950s.

11.   Indicate whether the following chemicals are meant to target (a. plants or b. animals).

[ ] pesticides   [ ] herbicides

Different Philosophies of Pest Control

12.   List the two philosophies of pest control.

a. _____   b. _____

13.   Which philosophy treats only the symptoms of pest outbreaks? _____

14.   Which philosophy stimulates ecosystem immunity to pest outbreaks? _____

15.   Which philosophy combines the two approaches? _____

Promises and Problems of the Chemical Approach

Development of Chemical Pesticides and Their Successes

16.   Indicate whether the following chemicals are meant to target: [a] all organisms, [b] mice and rats, [c] insects, or [d] fungi.

[ ] fungicides   [ ] rodenticides   [ ] insecticides   [ ] **biocides**

154

17. A common chemical pesticide used in the 1930s was

 a. cyanide b. arsenic c. sulfur d. DDT

18. Indicate whether the following are characteristics of [a] first-generation, [b] second-generation, or [c] both types of pesticides.

 [   ] consisted of toxic heavy metals
 [   ] are inorganic compounds
 [   ] are synthetic organic compounds
 [   ] are persistent
 [   ] promote pest resistance
 [   ] are broad spectrum in their effect
 [   ] examples are cyanide and arsenic
 [   ] an example is DDT (dichlorodiphenyltrichloroethane)
 [   ] stimulated Rachel Carson to write *Silent Spring*

Problems Stemming from Chemical Pesticide Use

19. List three problems associated with the use of synthetic organic pesticides.

 a. _____

 b. _____

 c. _____

20. Chemical pesticides gradually lose their _____.

21. Loss of effectiveness leads to two choices by growers, namely, use (More, Less) or use the (Same, New) pesticides.

22. The use of chemical pesticides on insects leads to selection for the (Sensitivity, Resistance) allele in the gene pool of the population.

23. Each generation of insects exposed to a chemical pesticide will have (More, Less) alleles for resistance in the gene pool of the population.

24. It is (True or False) that as a pest population becomes resistant to one pesticide it may, at the same time, become resistant to others to which it has not yet been exposed.

25. Match the following definitions with [a] resurgence or [b] secondary pest outbreak.

 [   ] Insects that were previously not a problem, become a pest category
 [   ] The return of the pest at higher and more severe levels

26. Pesticide treatments often have a (Less, Greater) effect on the natural enemies of pests.

27.   Which of the following statements are [+] or are not [-] explanations of why pesticides have a greater impact on natural enemies than upon the target pest?

[   ] Herbivore species are more resistant to the pesticides than are its predators.
[   ] Biomagnification gives the predator a higher dose of the pesticide.
[   ] Predatory species may be starved out due to a temporary lack of prey.

28.   How is the chemical approach contrary to basic ecological principles?

_____

_____

29.   How many million cases of acute occupational poisoning occur each year in developing countries? Between _____ and _____ million.

30.   List three ways children and families in developing countries come into contact with pesticides.

a. _____   b. _____

c. _____

31.   List three chronic effects of pesticide exposure on humans.

a. _____   b. _____

c. _____

32.   What were three sources of evidence that indicated DDT was the causative factor in reproductive failure in fish-eating birds?

a. DDT levels in fragile egg shells were (High, Low).
b. DDT (Promotes, Reduces) calcium metabolism.
c. DDT levels in fish-eating birds was (Higher, Lower).

33.   The highest DDT levels can be expected at the (Bottom, Top) of the food chain.

34.   DDT accumulates in the body _____ of humans and other animals.

35.   DDT was banned in the United States in the early 19_____s.

36.   It is (True or False) that when DDT was banned, pesticides were no longer used.

37.   About _____ million tons of pesticides are used globally.

38.   If less than one percent of the pesticides dumped on the environment come into contact with the pest organism, where does the remaining 99 percent of the pesticide go?

    a. Drifts into the air (True or False)
    b. Settles on surrounding ecosystems (True or False)
    c. Settles on bodies of water (True or False)
    d. Remains as residue on foods (True or False)
    e. Leaches from the soil into aquifers (True or False)
    f. Is captured and recycled by growers (True or False)

39.   Construct the last four steps in the pesticide treadmill below using [a] new and larger quantities of chemicals and [b] more resistance and secondary outbreaks.

    [a] Step 1 → [b] Step 2 → [   ] Step 3 → [   ] Step 4 → [   ] Step 5 → [   ] Step 6 → ?

Nonpersistent Pesticides: Are They the Answer?

40.   Assume that 1000 pounds of DDT was applied to a field in Nebraska in 1950. How much of the 1000 pounds will be left in:

    a. 1970 _____ b. 1990 _____ c. 2010 _____

    d. When will the DDT disappear? _____

41.   The half-life of nonpersistent pesticides is (a. days, b. weeks, c. years, d. a or b).

42.   Indicate whether the following statements are true [+] or false [-] concerning nonpersistent pesticides.

    [   ] stay in the environment for a long time
    [   ] are less toxic than DDT
    [   ] are broad spectrum pesticides
    [   ] do not cause resistance in pest species
    [   ] do not cause resurgence or secondary outbreaks

Alternative Pest Control Methods

43.   List the four alternative pest control methods.

    a. _____

    b. _____

    c. _____

    d. _____

157

Cultural Controls

44. Indicate whether the following control measures are in the category of [a] sanitation or [b] personal hygiene.

    [   ] water purification
    [   ] bathing
    [   ] wearing clean clothing
    [   ] proper and systematic disposal of garbage
    [   ] using clean cooking and eating utensils
    [   ] refrigeration, freezing, and canning foods

45. Indicate whether the following control measures are examples of

    a. Selection of what to grow and where to grow it
    b. Management of lawns and pastures
    c. Water and fertilizer management
    d. Timing of planting
    e. Destruction of crop residues
    f. Bordering crops with beneficial "weeds"
    g. Crop rotation
    h. Polyculture
    I. Quarantines

    [   ] prohibiting biological materials that carry pests from entering a country
    [   ] planting plants under their optimum conditions
    [   ] growing two or more species together or in alternate rows
    [   ] cutting grass no less than 3 inches
    [   ] alternating crops from one year to the next in a given field
    [   ] providing the ideal levels of water and fertilizer
    [   ] plowing under or burning the residue left after harvest
    [   ] eliminating plants that act as pest attractants and grow others that act as repellents
    [   ] delaying planting so most of the pest population starves before the plants are available

Control by Natural Enemies

46. Match the following examples of pests with the natural enemy that was used to control each pest.

Pest	Natural Enemy
[ ] scale insects	a. manatees
[ ] caterpillars	b. plant-eating insects
[ ] gypsy moths and Japanese beetles	c. bacteria
[ ] prickly pear cactus	d. parasitic wasps
[ ] water hyacinths	e. vedalia (ladybird) beetles
[ ] rabbits in Australia	

47. The first step in using natural enemies is _____ of the natural enemies that already exist.

48. It is (True or False) that finding natural enemies to pests generates as much profit for industry as do synthetic chemicals.

49. Natural enemies are a (Short, Long) term form of pest control.

50. Natural enemies are (More, Less) profitable in the long term than synthetic chemicals.

Genetic Control

51. List three forms of genetic controls.

   a. _____   b. _____

   c. _____

52. A chemical barrier means that the plant produces some chemical that is _____ or _____ to potential pests.

53. A chemical barrier was introduced into wheat plants that killed Hessian fly _____ .

54. Physical barriers are _____ traits that impede pest attacks.

55. An example of a physical barrier is _____ hairs on leaf surfaces.

56. It is (True or False) that pests cannot overcome genetic controls through resistance.

57. The sterile male procedure involves subjecting the (a. egg, b. larvae, c. pupa, d. adults) to just enough high energy radiation to render them sterile.

58. It is (True or False) that pests cannot develop resistance to the sterile male technique.

59. An example of application of the sterile male technique is the _____ fly.

60. What is a transgenic crop?_____

_____

61. Give three examples of transgenic crops.

   a. _____   b. _____

   c. _____

62. List three drawbacks in the use of genetically altered crops.

   a. _____   b. _____

   c. _____

Natural Chemical Control

63. Indicate whether the following definitions pertain to [a] hormones or [b] pheromones.

    [   ] A chemical secreted by one insect that influences the behavior of another.
    [   ] Chemicals that provide signals for development and metabolic functions.

64. The four aims of natural chemical control are to _____,
    _____, _____, and _____ an
    insect's own hormones or pheromones to disrupt its life cycle.

65. The two advantages of natural chemicals are that they are _____
    and _____ .

66. Juvenile hormone has a direct adverse affect on the (a. egg, b. larva, c. pupa, d. adult).

67. Sex attractant pheromones may be used in the _____ or
    _____ techniques.

Socioeconomic Issues In Pest Management

Pressures to Use Pesticides

68. Natural controls are aimed at (Management, Eradication) of pest populations.

69. Significant damage is implied when the cost of damage is considerably (Greater, Less) than
    the cost of pesticide application.

70. Indicate whether the following attitudes or actions are examples of:

    a. cosmetic spraying.   b. insurance spraying.   c. chemical company vested interest.

    [   ] consumer desire for unblemished produce
    [   ] the use of pesticides "just to be safe"
    [   ] ignorance of whether a "bug" is causing damage or is beneficial
    [   ] the "Snow White" syndrome
    [   ] promoting and exploiting the above attitudes

Integrated Pest Management

71. Integrated pest management (IPM) addresses all _____,
    _____, and _____ factors.

72. IPM is a (Short, Long) term approach.

73. IPM (Never, Sometimes) requires use of chemical pesticides.

74.   List five economic or environmental benefits derived from the Indonesian IPM program.

    a. _____

    b. _____

    c. _____

    d. _____

    e. _____

Organically Grown Food

75.   Control of pests in organic farming is likely to consist of :(Check all that apply.)

    [   ] nonpersistent pesticides
    [   ] natural controls
    [   ] cultural controls
    [   ] synthetic organic pesticides

Public Policy

76.   What are the three basic concerns in regulating the use of pesticides?

    a. _____

    b. _____

    c. _____

FIFRA

77.   What is the purpose of FIFRA? _____

78.   The registration procedure requires _____ to determine toxicity to animals.

79.   List four shortcomings of FIFRA.

    a. _____

    b. _____

    c. _____

    d. _____

80.   Indicate whether the following statements reflect FIFRA shortcomings related to [a] inadequate testing, [b] bans on a case-by-case basis, [c] pesticide exports, or [d] lack of public trust.

[   ] 1400 chemicals and 40,000 formulations.
[   ] 100 million pounds of pesticides shipped to less developed countries.
[   ] EDB found in hundreds of wells in Florida and California.
[   ] Using 100 to 200 mice over a one- to two-year period.
[   ] Substituting healthy mice for experimental mice that died.
[   ] Intense lobbying efforts by the chemical industry as opposed to public forums.

FQPA of 1996

81.   What is the purpose of the **Food Quality Protection Act** of 1996?

_____

_____

Pesticides in Developing Countries

82.   What is the annual export of pesticides to developing countries by the United States?
        _____ metric tons

83.   It is (True or False) that these pesticide exports do not include those banned in the U.S..

84.   Explain how PIC and FAO Code's of Conduct are protecting developing countries from imports of unwanted pesticides.

PIC: _____

FAO: _____

New Policy Needs

85.   What new pesticide regulatory rules are needed to help farmers jump off the pesticide treadmill? (Check all that apply.)

[   ] groundwater protection
[   ] food safety
[   ] measuring extent and dosage of pesticide exposures
[   ] worker protection
[   ] expanded pesticide testing requirements

86.   List one pesticide legislative reform that is being considered.

_____

VOCABULARY DRILL

Directions: Match each term with the list of definitions and examples given in the tables below. Term definitions and examples might be used more than once.

Term	Definition	Example(s)	Term	Definition	Example(s)
biomagnification			natural chemicals		
broad-spectrum			nonpersistent		
chemical barriers			organically grown		
cosmetic spraying			persistent		
cultural control			pest		
economic threshold			pesticide treadmill		
FIFRA			pesticide		
first-generation			physical barriers		
genetic control			resistance		
herbicide			resurgence		
insect life cycle			second-generation		
insurance spraying			secondary outbreaks		
IPM			sex attractant		
natural enemies					

VOCABULARY DEFINITIONS

a. any organism that has a negative impact on human health or economics

b. chemicals that kill animal pests

c. chemicals that kill plants

d. pesticides used before the 1930s

e. pesticides used during and after the 1930s

f. pesticides that do not select target species only

g. pesticides that stay in the ecosystem for a long time

h. pest organism's ability to develop genetic immunity to a pesticide

i. pest population explodes to higher and more severe levels

j. population explosions of pests that previously were of no concern

k. the process of accumulating higher and higher doses through the food chain

l. cycle of pesticide use, resistance, resurgence, pesticide use, etc.

m. pesticides that remain for only a short period of time in the ecosystem

VOCABULARY DEFINITIONS

n. when economic losses due to pest damage outweigh cost of applying pesticide
o. use of pesticides to control pests that harm only item's outward appearance
p. use of pesticides when no control need is evident
q. metamorphic stages in an insect's life
r. chemical substance the alters the reproductive behavior of the opposite sex of the same species
s. identifying and using the natural predator of the pest species
t. developing a genetic incompatibility between the pest and its host species
u. alteration of one or more environmental factors that attract pest species
v. alteration of one or more stages in a pest species' life cycle
w. substances that are lethal or repulsive to would-be pests
x. structural traits that imped the attack of a pest
y. involves the integration of several sequential pest control techniques
y. key legislation to control pesticides in the United States
z. crop production without the use of synthetic chemical pesticides or fertilizers

VOCABULARY EXAMPLES

a. flies, tick, mosquitos	m. "just to be safe"
b. DDT	n. Fig. 17-9
c. Round-up	o. pheromones
d. arsenic, cyanide	p. Fig. 17-11
e. Fig 17-4	q. sterile males
f. all pest species	r. Fig. 17-10
g. cotton mites	s. skunk odor
h. $0.001 \rightarrow 0.01 \rightarrow 0.10 \rightarrow 1.0$ ppm	t. hairy leaves
i. Fig. 17-6	u. Federal Insecticide, Fungicide, and Rodenticide Act
j. parathion	v. Integrated Pest Management
k. Fig. 17-15	w. Fig. 17-17

SELF-TEST

1.   Define the term **pest**.

_____

Identify whether the following characteristics pertain to [a] first generation, [b] second generation, [c] both first and second generation, [d] nonpersistent, or [e] all pesticides.

2.   [   ] are broad spectrum in their effect

3.   [   ] are harmful to humans and the environment

4.   [   ] cause pest resistance

5.   [   ] cause resurgence

6.   [   ] cause secondary outbreaks

If a pesticide has been present in an aquatic ecosystem for several years, match the pesticide levels given below with various parts of an aquatic ecosystem.

        Pesticide Levels: a. 0.02 ppm, b. 2000 ppm, c. 40-300 ppm, d. 5 ppm, e. 1600 ppm

7. [   ] water   8. [   ] fish-eating birds   9. [   ] algae   10. [   ] plant-eating fish

11.   Give the term that is used to explain your answers to 7 through
        10._____

What are three sources of evidence that implicated DDT as the causative factor in reproductive failure in fish-eating birds?

12.   _____

13.   _____

14.   _____

15.   Explain the association between **resistance** and **resurgence**.

        _____

        _____

List four ways of controlling pest species that do not include the use of inorganic or synthetic organic chemicals.

16.   _____   17.   _____

18. _____   19. _____

Match the following actions with one of the ways (16-19) of controlling pests that you listed above.

20. _____ using high energy radiation   21. _____ crop rotation

22. _____ use of pheromone   23. _____ using ladybug beetles

24. _____ chemical barriers   25. _____ disposal of garbage

26. _____ juvenile hormone   27. _____ destruction of crop residues

28. Ecological pest management seeks to control pests by

   a. manipulating natural factors affecting the host and pest relationship.
   b. letting nature take its course.
   c. using pesticides that are known to be ecologically sound.
   d. using biological controls only.

29. Which of the following shortcomings in FIFRA promotes heavy lobbying by the chemical industry?

   a. inadequate testing
   b. lack of public inpu.
   c. pesticide exports
   d. ban issued on a case-by-case basis when threats are proven

30. Cosmetic spraying is done to

   a. improve the appearance of produce.
   b. improve the taste.
   c. improve the storability.
   d. protect the public health.

ANSWERS TO STUDY GUIDE QUESTIONS

1. false; 2. f; 3. e, 4. d; 5. c; 6. a; 7. b; 8. false; 9. increased; 10. 37, 31; 11. b, a; 12. chemical technology, ecological pest management; 13. chemical technology; 14. ecological pest management; 15. integrated pest management; 16. e, c, d, a; 17. a; 18. a, a, b, c, c, c, a, b, b; 19. pest resistance, pest resurgence, adverse environmental and human health effects; 20. effectiveness; 21. more, new; 22. resistance; 23. more; 24. true; 25. b, a; 26. greater; 27. -, +, +; 28. assumes ecosystem is a static entity in which one species, the pest, can be eliminated; 29. 3.5 and 5; 30. aerial spraying, dumping of pesticide wastes, using pesticide containers for drinking water storage; 31. dermititis, cancers, neurological disorders, sterility, birth defects; 32. high, reduces, higher; 33. top; 34. fat; 35. 70; 36. false; 37. 3; 38. true, true, true, true, true, false; 39. a, b, a, b; 40. 500, 250, 125, never; 41. d; 42. -, -, +, -, -; 43. cultural, natural enemies, genetic, natural chemicals; 44. a, b, b, a, a, a,; 45. i, a, h, b, g, c, e, f, d; 46. e, d, c, b, a, c; 47. identification; 48. true; 49. long; 50. more; 51. chemical barriers, physical barriers, sterile males; 52. lethal, repulsive; 53. larva; 54. structural; 55. hooked; 56. false; 57. d; 58. true; 59. screw worm; 60. crops that have had the genes of other plant species, bacteria or viruses incorporated into their genome; 61. see Fig. 17-13; 62. not well suited to developing countries, possible development of superweeds, development of insect resistance to transgenic crops; 63. b, a; 64. isolate, identify, synthesize, use; 65. specific, nontoxic; 66. b; 67. trapping, confusion; 68. management; 69. greater; 70. a, b, b, a, c; 71. social, economic, ecological; 72. long; 73. sometimes; 74. savings on pesticide purchases, no investments in pesticide applicators, environmental health, fish in rice paddies, health benefits; 75. blank, check the rest; 76. intended use of pesticides and impacts on human and environmental health, training of pesticide applicators, protection of food supply; 77. law requiring manufacturers to register pesticides with the government before marketing them; 79. inadequate testing, bans on case-by-case basis, pesticide exports, lack of public input; 80. a, c, d, a, a, d; 81. establishes standards to protect food supply from becoming contaminated with high pesticide levels; 82. 200,000; 83. false; 84. exporting countries inform all potential importing countries of actions they have taken to ban or restrict use of pesticides, conditions of safe use of pesticides are addressed in detail; 85. all checked; 86. 50% reduction in pesticide use.

ANSWERS TO SELF-TEST

1. any organism that is noxious, destructive, or troublesome; 2 to 6. all e; 7. a; 8. b; 9. d; 10. c; 11. biomagnification; 12. DDT levels in fragile egg shells was high; 13. DDT is known to reduce calcium metabolism; 14. DDT levels in fish-eating birds was higher than other organisms in the food chain; 15. pests develop a genetic resistance to pesticides exemplified by surviving members of the population surviving and reproducing many more pests (resurgence) that are resistance; 16. natural enemies; 17. genetic controls; 18. natural chemicals; 19. cultural controls; 20. genetic control; 21. cultural control; 22. natural chemicals; 23. natural enemies; 24. genetic controls; 25. cultural control; 26. natural chemicals; 27. cultural control; 28. a; 29. b; 30. a.

VOCABULARY DRILL ANSWERS

Term	Definition	Example(s)	Term	Definition	Example(s)
biomagnification	k	h	natural chemicals	v	o
broad-spectrum	f	b	nonpersistent	m	j
chemical barriers	w	h	organically grown	z	w
cosmetic spraying	o	l	persistent	g	b
cultural control	u	r	pest	a	a
economic threshold	n	k	pesticide treadmill	l	i
FIFRA	y	u	pesticide	b	b
first-generation	d	d	physical barriers	x	t
genetic control	t	q	resistance	h	e
herbicide	c	c	resurgence	i	f
insect life cycle	q	n	second-generation	e	b
insurance spraying	p	m	secondary outbreaks	j	g
IPM	y	v	sex attractant	r	o
natural enemies	s	p			

CHAPTER 18

WATER: POLLUTION AND PREVENTION

---

1. Loss of the sea grasses in the Chesapeake bay was due to (a. toxic chemicals, b. herbicides, c. turbid water).

2. Using the terms MORE and LESS, compare the following characteristics of the bay ecology BEFORE and AFTER the massive die off of sea grasses.

Ecological Characteristics	Before	After
a. depth of light penetration	_____	_____
b. amount of sediment	_____	_____
c. amount of phytoplankton	_____	_____
d. dissolved oxygen	_____	_____
e. bacteria	_____	_____

Ecological Characteristics	Before	After
f. photosynthesis	_____	_____
g. fish populations	_____	_____

Water Pollution

3. The presence of a substance in the environment that causes undesirable environmental alterations is called _____.

Pollution Essentials

4. Match the sources of pollutants on the right with categories of pollutants on the left (see Figure 18-1).

Categories

[   ] toxic chemicals
[   ] nutrient oversupply
[   ] sediments
[   ] particulates
[   ] acid forming compounds
[   ] fuel combustion products
[   ] pesticides/herbicides

Sources

a. soil erosion and dredging
b. refuse burning
c. coal burning power plants
d. cars, trucks, buses, airplanes
e. leaching from lawns and agricultural fields
f. industrial discharges and disposal sites
g. fertilizer runoff from several sources

5. List four undesirable environmental alterations caused by pollution.

a. _____   b. _____

c. _____   d. _____

6. List the four actions required to reduce pollution problems.

a. _____

b. _____

c. _____

d. _____

7. What is one strategy that could be used to address pollution? _____

Water Pollution: Sources, Types

8. Identify the following are **point** [P] or **nonpoint** [NP] pollution sources.

[   ] atmospheric deposition
[   ] oil wells
[   ] sewage systems

[ ] agricultural runoff

9.   List two basic strategies that are employed to bring water pollution under control.

      a. _____   b. _____

10.   List four basic types of water pollutants.

      a. _____   b. _____

      c. _____   d. _____

11.   The major problem resulting from untreated sewage wastes is the spread of infectious
      _____.

12.   Disease causing bacteria, viruses, and other parasitic organisms that infect humans and other
      animals are called _____.

13.   List four diseases caused by the pathogens present in sewage (see Table 18-1).

      a. _____   b. _____

      c. _____   d. _____

14.   Transmission of pathogens in infected sewage waste may occur in what three ways?

      a. _____   b. _____

      c. _____

15.   It is (True or False) that serious and widespread cholera outbreaks are still occurring.

16.   Public health measures which prevent a disease cycle involve:

      a. _____

      b. _____

      c. _____

17.   Bacterial decomposition of organic matter in water will (Increase or Decrease) O_2
      concentration?

18.   A common water quality test that measures the amount of organic material, in terms of the
      amount of O_2 will be required to break it down biologically is the _____.

19.   Classify the below list of pollutants as **inorganic** [IN] or **organic** [O].

      [ ] heavy metals
      [ ] petroleum products

[ ] pesticides
[ ] acid precipitation

20.   The source of all sediments is _____ erosion.

21.   List seven major points of soil erosion and sediments.

a. _____   b. _____   c. _____

d. _____   e. _____   f. _____

g. _____

22.   It is (True or False) that streams, rivers, lakes, and estuaries support complex ecosystems based on many kinds of plant and animal organisms living on or attached to the bottom.

23.   Indicate whether the following types of damage to aquatic ecosystems is caused by [a] clay or [b] the bedload of sand and silt.

[ ] reduction of light penetration and rate of photosynthesis
[ ] smothers fish by clogging their gills
[ ] scours bottom-dwelling organisms clinging to rocks
[ ] fills in hiding and resting places for fish and crayfish
[ ] causes the stream to become more shallow

24.   _____ is considered the foremost pollution problem of streams and rivers.

25.   The impacts of sediments cost the United States $_____ billion each year.

26.   The most obvious point sources of excessive nutrients are _____ outfalls.

27.   The most notorious nonpoint source of excessive nutrients is _____ runoff.

28.   The most obvious environmental impact of excessive nutrients is _____.

Eutrophication

Different Kinds of Aquatic Plants

29.   Two kinds of aquatic plants are _____ plants which are rooted to the bottom and _____ which mostly consist of single cells or small groups of cells which grow at or near the water surface.

30.   The depth to which light penetrates in sufficient intensity to support photosynthesis is known as the _____ zone.

31.   Depending on the turbidity of the water, the euphotic zone may vary from nearly _____ feet in clear water down to only a few inches in very turbid water.

32. Phytoplankton can maintain itself near the surface where there is plenty of light, but they depend on (Surface, Bottom) nutrients.

33. Benthic plants get their nutrients from the (Surface, Bottom).

34. The balance between benthic plants and phytoplankton depends on the level of _____ in the water.

35. (Benthic Plants or Phytoplankton) depend on light penetration to the bottom to enable photosynthesis.

The Impacts of Nutrient Enrichment

36. Indicate whether the following properties increase [+] or decrease [-] given the listed conditions.

    [+] nutrient → [   ] phytoplankton → [   ] turbidity → [   ] SAV's

37. The term oligotrophic means nutrient (Rich, Poor).

38. The two filters of nitrogen and phosphate for natural waterways are:

    a. _____   b. _____

39. The major source of oxygen in aquatic ecosystems comes from the

    a. atmosphere b. benthic plants c. phytoplankton

40. Indicate whether the following conditions are LOW [-] or HIGH [+] in oligotrophic bodies of water. (You will need to read further into the chapter to answer all items.)

    a. [   ] dissolved oxygen   b. [   ] bacterial decomposition

    c. [   ] light penetration   d. [   ] benthic plants

    e. [   ] phytoplankton   f. [   ] water temperature

    g. [   ] water turbidity   h. [   ] nutrient concentrations

    i. [   ] species diversity   j. [   ] aesthetic qualities

    k. [   ] recreational qualities   l. [   ] **BOD**

    m. [   ] detritus decomposition   n. [   ] sediments

41. The term **eutrophic** means nutrient (Rich, Poor).

42.   Phytoplankton do not contribute significantly to dissolved oxygen because

      a. they do not carry on photosynthesis. (True or False)
      b. they do not produce oxygen in photosynthesis. (True or False)
      c. the oxygen they produce escapes from the surface into the atmosphere. (True or False)

43.   The amount of detritus (Increases or Decreases) in eutrophic water.

44.   The increase in detritus comes from the die back of (Benthic Plants, Phytoplankton).

45.   Attack of detritus by _____ uses so much _____ that bottom dwelling fish and shellfish suffocate.

46.   Decomposers (bacteria) can keep the dissolved oxygen at or near zero for as long as there is detritus to feed on because in the absence of oxygen they can carry on _____ respiration.

47.   Indicate whether the following conditions are LOW [-] or HIGH [+] in eutrophic bodies of water. (You will need to read further into the chapter to answer all items.)

    a. [   ] dissolved oxygen   b. [   ] bacterial decomposition

    c. [   ] light penetration   d. [   ] benthic plants

    e. [   ] phytoplankton   f. [   ] water temperature

    g. [   ] water turbidity   h. [   ] nutrient concentrations

    i. [   ] species diversity   j. [   ] aesthetic qualities

    k. [   ] recreational qualities   l. [   ] **BOD**

    m. [   ] detritus decomposition   n. [   ] sediments

48.   A eutrophic body of water is not technically "dead" because it still supports the abundant growth of _____ and _____.

49.   In shallow lakes and ponds, the (SAV, phytoplankton) are the indicators of eutrophication.

50.   It is (True or False) that eutrophication of natural bodies of water is part of the natural aging process.

51.   List three nutrient sources the cause cultural eutrophication.

    a. _____   b. _____   c. _____

Combating Eutrophication

52.   List the two general approaches to combating eutrophication.

     a. _____

     b. _____

53.   List four methods that have been used in attacking the symptoms of eutrophication and state
the major reason that each is less that successful or practical.

           Method   Shortcoming

     a. _____   _____

     b. _____   _____

     c. _____   _____

     d. _____   _____

54.   It is (True or False) that the real control of eutrophication requires decreasing nutrient and
sediment inputs.

55.   List three major sources of nutrients that end up in streams and rivers.

     a. _____

     b. _____

     c. _____

56.   List seven long-term strategies for controlling eutrophication.

     a. _____

     b. _____

     c. _____

     d. _____

     e. _____

     f. _____

     g. _____

57.   The key to controlling eutrophication is to reduce _____ inputs.

Sewage Management and Treatment

Development of Collection and Treatment Systems

58.   Prior to the late-1800s, the general means of disposing of human excrements was the
        _____ privy which often was located near a _____.

59.   What person showed the relation between sewage-borne bacteria and infectious diseases?
        _____

60.   The earliest forms of sewage disposal transferred it directly into _____
        drains.

61.   _____ societies initiated the system of flushing sewage into natural
        waterways.

62.   Around 1900, sewage treatment included _____ the wastewater and
        separating _____ and _____ water.

63.   It is (True or False) that there are no longer any cases of raw sewage overflowing with storm
        water into waterways.

The Pollutants in Raw Sewage

64.   The total of all the drainage from sinks, bath tubs, laundry, and toilets constitutes a mixture
        called raw _____.

65.   Raw sewage is actually _____ percent water and _____ percent pollutants.

66.   List the four major categories of pollutants in raw sewage and give an example of each.

            <u>Category</u>   <u>Example</u>

        a. _____   _____

        b. _____   _____

        c. _____   _____

        d. _____   _____

Removing the Pollutants From Sewage

67.   The two major sewage treatment problems in the United States are those of dealing with:

        a. _____ and b. _____

68. Match the below list of wastewater treatment technologies or products with [a] preliminary treatment, [b] primary treatment, [c] secondary treatment, [d] biological nutrient removal.

[   ] primary clarifiers
[   ] bar screen
[   ] grit settling tank
[   ] raw sludge
[   ] biological treatment
[   ] activated sludge system
[   ] activated sludge
[   ] rotating screens
[   ] course sand and gravel
[   ] removal of dissolved nutrients
[   ] uses decomposers and detritus feeders
[   ] settling of raw sludge
[   ] aeration

69. First, large pieces of debris are removed by passing the water through a _____ screen.

70. Coarse sand and other dirt particles are removed by passing the water through a _____ settling tank where the velocity of the water is (Increased, Decreased) and the particles are allowed to _____ out.

71. The large debris from the bar screen are taken out and _____.

72. Grit from the grit settling tank is put in _____.

73. In primary treatment

a. the water flows into very large tanks called _____ **clarifiers.**
b. the water flows through (Quickly, Slowly).
c. heavier particles of organic matter (Disperse, Settle Out).

74. What percentage of the organic material is removed by this settling process? _____

75. The material that settles out is called **raw** _____.

76. Secondary treatment is also called _____ treatment because decomposers and detritus feeders are used.

77. In the activated sludge system, a mixture of organisms is added to the water to be treated and it is passed through a tank where it is _____.

78. In the aeration tank, the organisms feed on _____ matter.

79. Following the aeration tank, the water is passed on to a secondary _____ tank where the organisms are returned to the _____ tank.

80. Through the activated sludge systems, _____ percent of the organic material is removed.

81. BNR utilizes aspects of the _____ and _____ cycles.

82. Indicate whether the following statements represent positive [+] or negative [-] attributes of using chlorine gas as the means of killing pathogenic bacteria in wastewater.

[   ] Cost
[   ] Toxicity during transportation and to fish.
[   ] Chlorinated hydrocarbons.

83. Two alternative technologies for killing pathogenic bacteria in wastewater are

a. _____ and b. _____

Sludge Treatment

84. Raw sludge is about _____ percent water and _____ percent organic matter.

85. Three sludge treatment methods are

a. _____   b. _____

c. _____

86. **Anaerobic digestion** of raw sludge in the absence of _____ produces the products of _____ gas and _____ **sludge.**

87. Composting basically produces _____.

88. Pasteurization and drying produce sludge _____.

Alternative Treatment Systems

89. List three alternative methods of extracting wastewater nutrients.

a. _____ b. _____

c. _____

Public Policy

90. List six indicators of progress in water pollution abatement since the enactment of the Clean Water Act 27 years ago.

a. _____ b. _____

c. _____ d. _____

e. _____ f. _____

VOCABULARY DRILL

Directions: Match each term with the list of definitions and examples given in the tables below. Term definitions and examples might be used more than once.

Term	Definition	Example(s)	Term	Definition	Example(s)
activated sludge			primary clarifiers		
bar screen			primary treatment		
benthic plants			phytoplankton		
biogas			raw sludge		
biological treatment			raw sewage		
BNR			sanitary sewers		
BOD			point-source		
chlorinated hydrocarbons			secondary treatment		
co-composting			settling tank		
composting toilet			sludge cake		
cultural eutrophication			pollutant		
emergent vegetation			pollution		
euphotic zone			red tides		
eutrophication			sediment trap		
non-point source			SAV		
nonbiodegradable			tidal wetlands		
nontidal wetlands			turbid		
oligotrophic			wetlands		
pasteurized			sludge digester		
pathogens			storm drains		
preliminary treatment			treated sludge		

VOCABULARY DEFINITIONS

a. human-caused addition of any material or energy in amounts that cause undesired alterations

b. material that causes pollution

c. resist attack and breakdown by detritus feeders and decomposers

d. nutrients rich

e. cloudy or murky

f. free-floating aquatic plants

g. deep rooted aquatic plants

h. aquatic plants that grow entirely under water

i. aquatic plants that grow partially under and above water

j. depth of adequate light for photosynthesis

k. nutrient poor

l. measure of pollution level by ppm oxygen required for decomposition

m. accelerated eutrophication caused by humans

n. land areas naturally covered by water for specified periods of time

o. coastal wetlands

p. inland wetlands

q. indicators of oceanic eutrophication

r. a fixed-point pollutant source

s. a diffuse pollutant source

t. a means of controlling erosion from construction sites

u. sewage with no treatment whatsoever

v. disease-causing organisms

w. system to collect runoff from precipitation

x. system to collect all wastewater

y. first stage in the removal of pollutants from sewage

z. devise to remove large objects from wastewater

aa. devise to remove sand and gravel from wastewater

bb. first stage in the removal of organic matter from wastewater

cc. devise that allows organic matter to settle out of wastewater

dd. material removed during primary treatment

ee. stage in wastewater treatment that uses natural decomposers and detritus feeders

ff. a mixture of sludge and detritus-feeding organisms

VOCABULARY DEFINITIONS
gg. stage in wastewater treatment that removes dissolved nutrients
hh. result of spontaneous reactions between chlorine and organic compounds
ii. devise that allows bacteria to feed on detritus in the absence of oxygen
jj. a major byproduct from the sludge digester
kk. a dehydrated form of treated sludge
ll. a mixture of raw sludge and shredded waste paper
mm. procedure for killing pathogens
nn. a waterless toilet

VOCABULARY EXAMPLES	
a. cattails, water sedges	p. methane
b. ppm dissolved O_2	q. Fig.18-14
c. lawn fertilizers and pet droppings	r. cancer-causing compounds
d. 1 inch to 600 feet	s. Fig. 8-15
e. Chesapeake Bay	t. Cilvus Multrum
f. urban runoff	u. Fig.18-13
g. synthetic organic compounds	v. Milorganite
h. swamps	w. bacteria and viruses
i. wetlands	x. fluid that enters the sewage treatment plant
j. filamentous or single-cell algae	y. plumbing from sinks, tubs, and toilets
k. sewage treatment plant	z. bacteria, protozoa, fungi
l. nitrate and phosphate	aa. Fig. 8-11
m. eutrophication	bb. street curb and gutter
n. marine phytoplankton	cc. nutrient-rich humus like material
o. hay bales on highway construction projects	dd. Florida Everglades

SELF-TEST

Indicate whether the following conditions are characteristic of [O] oligotrophic or [E] eutrophic bodies
of water.

1. [   ] high dissolved oxygen   8. [   ] high bacterial decomposition

2. [   ] low light penetration   9. [   ] abundant benthic plants

3. [   ] abundant phytoplankton   10. [   ] cool water temperature

4. [   ] low water turbidity   11. [   ] high nutrient concentrations

5. [   ] high species diversity   12. [   ] poor aesthetic qualities

6. [   ] high recreational qualities   13. [   ] low BOD

7. [   ] high detritus decomposition   14. [   ] low sediments

List four major sources of sediments and nutrients that end up in streams and rivers.

               <u>Sediments</u>   <u>Nutrients</u>

15. _____   16. _____

17. _____   18. _____

19. _____   20. _____

21. _____   22. _____

Explain how each of the following events or materials contribute to **eutrophication**.

23.   Sedimentation: _____

_____

24.   Nutrients: _____

_____

25.   Decomposition: _____

_____

26.   Which of the following approaches to combating eutrophication is getting at the root cause?

         a. chemical treatments   b. decreasing nutrient input   c. aeration   d. harvesting algae

27.   The foremost pollution problem of streams and rivers is

a. toxic wastes b. salinization c. sedimentation d. eutrophication

28.   Nutrients flow into aquatic ecosystems from

a. fertilizers from croplands.
b. animal wastes from feedlots.
c. detergents containing phosphate.
d. all of the above.

29.   Lands that are naturally covered by shallow water at certain times and more or less drained at others are called

a. grasslands b. wetlands c. rain forests d. estuaries

Indicate whether the following functions of wetlands will increase [+] or decrease [-] after the establishment of human "development" projects on or adjacent to wetlands.

30.   [   ] water purification.

31.   [   ] provision of habitat and food for waterfowl and other wildlife.

32.   [   ] groundwater recharge

33.   The two major problems resulting from untreated sewage wastes going into waterways are

a _____ and b. _____ .

34.   The primary categories of pollutants in raw sewage are

a. pathogens, detritus, and dissolved nutrients.
b. phosphates, nitrates, and sulfates.
c. heavy metals, synthetic organisms, and pesticides.
d. industrial, household, and municipal wastes.

35.   What proportion of raw sewage is actually pollutant? _____ %

36.   Match the below list of wastewater treatment technologies or products with [A] preliminary treatment, [B] primary treatment, [C] secondary treatment, or [D] BNR.

[   ] irrigation
[   ] bar screen
[   ] aeration
[   ] raw sludge
[   ] denitrifying bacteria
[   ] activated sludge

37.   Match each of the following facilities [a to d] with the functions it performs.

a. activated sludge system b. bar screen c. primary clarifier d. none of these

[   ] screens out large pieces of debris
[   ] enables microorganisms to digest organic matter present
[   ] allows 50 to 70 percent of the organic matter to settle out
[   ] removes most of the dissolved nutrients

38.   Most sewage treatment plants do not have

a. pretreatment b. primary treatment c. secondary treatment d. BNR

39.   Explain why conventional methods of wastewater treatment only partially solve the problem of water pollution.

_____

40.   The treated water from most sewage treatment plants contains

a. nutrients   b. bacteria   c. sediments   d. colloidal materials

184

ANSWERS TO STUDY GUIDE QUESTIONS

1. c; 2. more, less; less, more; less, more; more, less; less, more; more, less; more, less; 3. pollution; 4. f, g, a, b, c, d, e; 5. aesthetic, biological, human health, local to global; 6. identifying material or materials causing the pollution problem, identifying the source(s), develop and implement control strategies; 5. apply the second principle of ecosystem sustainability; 8. NP, P, NP, P; 9. reduce the sources, treat the water to remove the pollutants; 10. pathogens, organic wastes, chemical pollutants, sediments; 11. diseases; 12. pathogens; 13. see Table 18-1; 14. drinking, eating food, body contact; 15. true; 16. disinfection of public water supplies, improving personal hygiene and sanitation, sanitary collection and treatment of sewage wastes; 17. decrease; 18. biological oxygen demand (BOD test); 19. IN, O, O, IN; 20. soil; 21. croplands, overgrazed rangelands, deforested areas, construction sites, surface mining, gully erosion; 22. true; 23. a, a, b, b, b; 24. sediment; 25. 6; 26. sewage; 27. agricultural; 28. eutrophication; 29. benthic, phtoplankton; 30. euphotic; 31. 90; 32. surface; 33. bottom; 34. nutrient; 35. benthic plants; 36. +, +, -; 37. poor; 38. watersheds, wetlands; 39. b; 40. +, -, +, +, -, -, -, -, +, +, +, -, -, -; 41. poor; 42. false, false, true; 43. increases; 44. benthic plants; 45. decomposers, oxygen; 46. anaerobic; 47. -, +, -, -, +, +, +, +, -, -, -, +, +, +; 48. phytoplankton, fish; 49. SAV; 50. true; 51. sewage treatment plants, poor farming practices, urban runoff; 52. attack the symptoms, get at the root cause; 53. (chemical treatments, phytoplankton resistance), (aeration, expense), (harvesting algae, expense), (dredging, cost and no where to put material); 54. true; 55. agriculture/forestry, urban/suburban runoff, sewage effluents; 56. banning phosphate detergents and advanced sewage treatment, controlling agricultural and urban runoff, controlling sediments from construction and mining sites, controlling stream erosion, protecting wetlands, controlling air pollution; 57. nutrient; 58. outdoor, stream; 59. Pasteur; 60. storm; 61. western; 62. treating, waste, storm; 63. false; 64. sewage; 65. 99.9, 0.1; 66. (debris and grit, rags and plastic bags), (particulate organic material, food wastes), (colloidal or organic material, feces), (dissolved inorganic materials, nitrogen); 67. nutrients, sludge; 68. b, a, a, b, c, c, c, a, a, d, c, b, c; 69. bar; 70. grit, decreased, settles; 71. incinerated; 72. land fills; 73. primary, slowly, settles out; 74. 30 to 50; 75. sludge; 76. biological; 77. aerated; 78. organic; 79. clarifier, aeration; 80. 90 to 95; 81. nitrogen, phosphate; 82. +, -, -; 83. ozone, ultraviolet radiation; 84. 98, 2; 85. anaerobic, composting, pasteurization; 86. digester, methane, treated; 87. humus; 88. pellets; 89. septic systems, using effluents for irrigation, reconstructing wetlands; 90. number of people served by adequate sewage treatment plants; reduction in soil erosion; more waterways safe for fishing and swimming; rivers, lakes and bays cleaned up and restored; fish returned to rivers; increase in bottom vegetation in Chesapeake bay.

ANSWERS TO SELF-TEST

1. O; 2. E; 3. E, 4. O; 5. O; 6. O; 7. E; 8. E; 9. O; 10. O; 11. E; 12. E; 13. O; 14. O; 15. croplands; 16. overgrazed rangelands; 17. deforested areas; 18. construction sites; and surface mining, gully erosion; 19. fertilizer from cropland; 20. fertilizer from lawns and gardens; 21. animal wastes from feedlots; 22. pet wastes; and human excrement, phosphate detergents, acid precipitation; 23. decreases light penetration to benthic plants; 24. increases phytoplankton growth; 25. decreases oxygen concentration killing animals; 26. b; 27. c; 28. d; 29. b; 30. -; 31. -; 32. -; 33. disease, eutrophication; 34. a; 35. 0.1%; 36. d, a, b, b, d, c; 37. b, a, c, d; 38. d; 39. most do not remove dissolved nutrients; 40. a.

VOCABULARY DRILL ANSWERS

Term	Definition	Example(s)	Term	Definition	Example(s)
activated sludge	l	q	primary clarifiers	i	aa
bar screen	f	aa	primary treatment	h	aa
benthic plants	g	a	phytoplankton	f	j, n
biogas	p	p	raw sludge	j	k
biological treatment	k	z	raw sewage	a	l
BNR	m	u	sanitary sewers	d	y
BOD	l	b	point-source	r	k
chlorinated hydrocarbons	n	r	secondary treatment	k	aa
co-composting	r	s	settling tank	g	aa
composting toilet	t	t	sludge cake	q	s
cultural eutrophication	m	f	pollutant	b	l, n
emergent vegetation	i	a	pollution	a	m
euphotic zone	j	d	red tides	q	n
eutrophication	d	e	sediment trap	t	o
non-point source	s	f	SAV	h	a
nonbiodegradable	c	g	tidal wetlands	o	e, dd
nontidal wetlands	p	h	turbid	e	e
oligotrophic	k	i	wetlands	n	e, h, dd
pasteurized	s	v	sludge digester	o	aa
pathogens	b	w	storm drains	c	bb
preliminary treatment	e	aa	treated sludge	p	s, v

CHAPTER 19

MUNICIPAL SOLID WASTE: DISPOSAL AND RECOVERY

The Solid Waste Problem.
    Disposing of Municipal Solid Waste
    Landfills
    Combustion: Waste to Energy
    Costs of Municipal Solid Waste Disposal

Solutions
    Source Reduction
    Other Measures
    The Recycling Solution
    Composting

Public Policy and Waste Management
    The Regulatory Perspective
    Integrated Waste Management

Vocabulary Drill
    Vocabulary Definitions
    Vocabulary Examples

Self-Test

Answers to Study Guide Questions

Answers to Self-Test

Vocabulary Drill Answers

The Solid Waste Problem

Disposing of Municipal Solid Waste

1.   List four factors that have contributed to ever increasing amounts of MSW in the U.S.

    a. _____   b. _____

    c. _____   d. _____

2.   The three largest components of municipal solid waste are _____, _____, and _____.

3.	The proportions of the components of refuse change in terms of

	a. the generator b. the neighborhood c. time of year d. all of these factors

4.	Who assumes responsibility for collecting and disposing of municipal solid wastes?

	_____

5.	The early form of municipal solid waste disposal was open _____ which were later converted to _____.

6.	What proportions of the MSW in the U.S. are:

	a. dumped in landfills? _____ b. recycled? _____ c. combusted? _____

Landfills

7.	Indicate whether the following examples are related to the landfill problems of [a] leachate generation, [b] methane production, [c] incomplete decomposition, or [d] settling.

	[ ] explosions
	[ ] groundwater contamination
	[ ] development of shallow depressions
	[ ] "witches brew"
	[ ] biogas
	[ ] 30-year old newspapers

8.	Indicate whether the following landfill regulations address [a] leachate generation, [b] methane production, or [c] both a and b.

	[ ] siting landfills on high ground well above the water table
	[ ] contoured floors to drain water into tiles
	[ ] covering the floor of the landfill with 12 inches of impervious clay or a plastic liner
	[ ] a gravel layer that surrounds the entire fill
	[ ] shaping the refuse pile into a pyramid
	[ ] monitoring wells

9.	One of the most formidable problems facing city governments is _____.

10.	What is the implication of *NIMBY*?

	_____

11.	List two undesirable consequences of **siting**.

	a. _____ b. _____

12.	List two positive aspects of the siting problem.

	a. _____ b. _____

Combustion: Waste to Energy

13.   Indicate whether the following are advantages [A] or disadvantages [D] of combustion.

      [   ] 90% reduction in waste volume
      [   ] generation of electricity
      [   ] incinerator effluents
      [   ] cost of construction
      [   ] competition with recycling efforts
      [   ] WTE
      [   ] concrete blocks
      [   ] resource recovery

14.   Indicate whether the following are advantages [A] or disadvantages [D] of WTE facilities.

      [   ] tons of MSW processed
      [   ] fuel oil savings
      [   ] recycling opportunities
      [   ] resource recovery

Costs of Municipal Solid Waste Disposal

15.   Increasing costs of disposing of MSW are related to

      a. new design features of landfills. (True or False)
      b. acquiring landfill sites. (True or False)
      c. transportation to landfills. (True or False)
      d. **tipping fees** at landfills. (True or False)

16.   Acquiring a landfill site far enough away from suburbia automatically increases
      _____ costs.

17.   Indicate whether the following events will [+] or will not [-] happen because of difficulties in establishing new landfill locations.

      [   ] continued use of old landfills with inadequate safeguards
      [   ] more old landfills closed than can be replaced with new landfills
      [   ] a landscape covered with pyramids
      [   ] reduced refuse production by society

Solutions

Source Reduction

18.   The two largest contributors to waste volume are _____ products and excessive
      _____.

19.   _____ items in their existing capacity is the most efficient form of recycling.

20. Identify two methods of reusing items in their existing capacity.

a. _____ b. _____

21. Nonreturnable containers constitute what percent of

a. solid waste stream in the U.S.? _____
b. nonburnable portion of MSW? _____
c. roadside litter? _____

22. Larger profits are generated from nonreturnable containers through

a. reduced transportation costs. (True or False)
b. indefinite bottle production. (True or False)
c. eliminating the returnable bottle competition. (True or False)
d. opposing bottle laws. (True or False)

23. Bottle laws encourage the use of (Returnable, Nonreturnable) containers.

24. Bottle laws will (Decrease, Increase) jobs and (Decrease, Increase) litter.

25. How many states have bottle laws? _____

26. Who are the greatest opponents to bottle bills? _____

Other Measures

27. List two other measures to reduce the MSW stream.

a. _____ b. _____

The Recycling Solution

28. What percent of the MSW is recyclable? _____ %

29. Match the below types of refuse with the type of product that reprocessing will produce.

Refuse Type	Product
[ ] paper	a. compost
[ ] glass	b. refabrication without ore extraction
[ ] plastic	c. synthetic lumber
[ ] metals	d. substitute for gravel and sand
[ ] food and yard wastes	e. cellulose insulation
[ ] textiles	f. strengthening agent in recycled paper

30. List the six characteristics of an effective municipal recycling program.

   a. _____

   b. _____

   c. _____

   d. _____

   e. _____

   f. _____

31. What proportion of the MSW stream is newspaper? _____ %

32. It is (True or False) that more paper is being recycled than landfilled.

33. The market price for a ton of used newspaper is $_____.

34. A ton of recycled newspapers = _____ trees.

35. It is (True or False) that all plastic containers are recyclable.

36. List the products made from recycled plastic

   PETE: _____

   HDPE: _____

Composting

37. Composting produces a residue of decomposition called _____.

38. It is (True or False) that there are business opportunities in composting.

Public Policy and Waste Management

The Regulatory Perspective

39. Match the federal legislation with the role each plays in addressing MSW management.

   a. Solid Waste Disposal Act of 1965   b. Resource Recovery Act of 1970
   c. Resource Recovery Act of 1976   d. Superfund Act of 1980
   e. Hazardous and Solid Waste Amendments of 1984

   [   ] gave EPA responsibility to set solid-waste criteria for all hazardous-waste facilities
   [   ] gave jurisdiction over solid waste to Bureau of Solid Waste Management
   [   ] encouraged states to develop some kind of waste management program

[   ] gave EPA power to close local dumps and set regulations for landfills
[   ] addressed abandoned hazardous waste sites

Integrated Waste Management

40.   Integrated waste management uses (One, Several) recycling methods.

41.   List three recommendations for MSW management that offer the greatest opportunity to promote sustainability.

a. _____   b. _____

c. _____

VOCABULARY DRILL					
Directions:	Match each term with the list of definitions and examples given in the tables below. Term definitions and examples might be used more than once.				

Term	Definition	Example(s)	Term	Definition	Example(s)
landfill			resource recovery		
MSW			secondary recycling		
primary recycling			waste-to-energy		

VOCABULARY DEFINITIONS
a. surface repositories for MSW
b. conversion of MSW to electricity
c. separating and recovering materials from MSW
d. modern resource recovery facility
e. original waste material made back into the same material
f. total of all the materials thrown away from homes and commercial establishments
g. waste materials made into different products

VOCABULARY EXAMPLES	
a. Fig. 19-4	d. trash, refuse, garbage
b. Fig. 19-7	e. newspapers to cardboard
c. newspapers to newsprint	

SELF-TEST

1. Which of the following has not been a cause of increasing volumes of MSW?

    a. changing lifestyles b. disposable materials c. recycling d. increased gross national product

2. The three largest categories of municipal solid wastes are

    a. paper, cans, and bottles.
    b. paper, yard waste, metals
    c. plastics, paper, and metals
    d. food, paper, and plastics

3. The proportions of the components of refuse change in terms of

    a. the generator   b. the neighborhood   c. time of year   d. all of these

4. Most municipal solid waste in the United States is presently disposed of

    a. by burning it in a closed incinerator
    b. in landfills
    c. by barging it to sea and dumping it
    d. in open-burning dumps

5. The potential of groundwater contamination from landfills is primarily due to

    a. leachate   b. methane gas   c. settling   d. explosions

6. Newspaper into newsprint is an example of (Primary, Secondary) recycling.

7. Increasing costs of landfilling are related to

    a. new design features b. acquiring new sites c. transportation d. all of these

8. Even though establishing new landfill locations is extremely difficult, which of the following events will not occur in spite of this difficulty?

    a. continued use of old landfills with inadequate safeguards
    b. more old landfills closed than can be replaced with new landfills
    c. reduced refuse production by society
    d. a landscape covered with pyramids

9. Converting PETE into carpet fiber is an example of (Primary, Secondary) recycling?

10.   In terms of recycling potential, paper can be

       a. repulped and made into paper and paper products
       b. manufactured into cellulose insulation
       c. composted to make a nutrient-rich humus
       d. all of the above

11.   Which of the below examples of recycling laws places the burden mainly on the consumer
       rather than the government?

       a. mandatory recycling
       b. advance disposal fees
       c. bans on the disposal of certain items
       d. mandating government purchase of recycled materials

12.   Burning raw refuse does not eliminate which of the following problems of recycling?

       a. sorting   b. hidden costs   c. reprocessing   d. marketing

13.   Which of the following is not a potential value derived from incinerated refuse?

       a. metal salvage   b. returnable bottles   c. extended use of existing landfills
       d. fill dirt for construction sites

14.   Which of the following is not a potential value of "bottle bills"?

       a. reduced transportation costs
       b. indefinite bottle production
       c. increased jobs
       d. decreased litter

15.   Which of the following measures reduce the amount of material going into trash?

       a. buying less   b. decreased obsolescence   c. yard sales   d. all of these

ANSWERS TO STUDY GUIDE QUESTIONS

1. increasing population, changing lifestyles, disposable materials, excessive packaging; 2. paper, yard wastes, metals; 3. d; 4. government; 5. burning, landfills; 6. 75, 12, 13; 7. b, a, d, a, b, c; 8. all a; 9. siting; 10. no one wants a landfill near their home or neighborhood; 11. increased cost of waste disposal, long-distance transfer of MSW; 12. residents reduce waste output, recycling occurs; 13. A, A, D, D, D, A, A, A; 14. A, A, D, A; 15. all true; 16. transportation; 17. +, +, +, -; 18. disposable, packaging; 19. reusing; 20. returnable bottles, yard (garage) sales, salvation army contributions; 21. 6, 50, 90; 22. all true; 23. returnable; 24. increase, decrease; 25. 10; 26. industry; 27. resale, eliminating junk mail; 28. 75; 29. e, d, c, b, a, f; 30. strong incentive to recycle, recycling not optional, residential recycling is curbside, recycling goals are clear, involves local industries, municipality employs knowledgeable coordinator, 31. 13; 32. true; 33. 30; 34. 89; 35. false; 36. carpet fibers, irrigation drainage tiles; 37. humus; 38. true; 39. E, A, B, C, D; 40. several; 41. waste reduction, waste disposal, recycling and reuse.

ANSWERS TO SELF-TEST

1. c; 2. b; 3. d; 4. b; 5. a; 6. primary; 7. d; 8. c; 9. secondary; 10. d; 11. b; 12. b; 13. b; 14. b; 15. d

VOCABULARY DRILL ANSWERS					
Term	**Definition**	**Example(s)**	**Term**	**Definition**	**Example(s)**
Landfill	a	a	resource recovery	c	b
MSW	f	d	secondary recycling	g	e
primary recycling	e	c	waste-to-energy	b, d	b

CHAPTER 20

HAZARDOUS CHEMICALS: POLLUTION AND PREVENTION

The Nature of Chemical Hazards: HAZEMATs
    Sources of Chemicals Entering the Environment
    The Threat From Toxic Chemicals
    Involvement with Food Chains

A History of Mismanagement
    Methods of Land Disposal
    Scope of the Management Problem

Cleaning Up the Mess
    Assuring Safe Drinking Water
    Groundwater Remediation
    Superfund for Toxic Sites

Management of New Wastes
    The Clean Air and Water Acts
    The Resource Conservation and Recovery Act (RCRA)
    Reduction of Accidents and Accidental Exposures

Looking Toward the Future
    Too Many or Too Few Regulations?
    Pollution Avoidance for a Sustainable Society

Vocabulary Drill
    Vocabulary Definitions
    Vocabulary Examples

Self-Test

Answers to Study Guide Questions

Answers to Self-Test

Vocabulary Drill Answers

1. What types of chemical disasters occurred at the following locations?

    a. Cuyahoga River: _____

    b. Minimata, Japan: _____

    c. Bhopal, India: _____

    d. Niagra Falls, NY: _____

The Nature of Chemical Hazards: HAZEMATs

2. Match the examples of hazardous materials on the left with the EPA category on the right.

    [ ] gasoline   a. toxicity
    [ ] acids   b. reactivity
    [ ] explosives   c. corrosivity
    [ ] pesticides   d. ignitability

Sources of Chemicals Entering the Environment

3. List five environmental costs in the **product life cycle** of your shampoo.

    a. _____

    b. _____

    c. _____

    d. _____

    e. _____

4. It is (True or False) that the predominant source of chemicals entering the environment are from households.

5. What is the purpose of the **Toxics Release Inventory (TRI)**?

    _____

The Threat From Toxic Chemicals

6. The two major classes of chemicals that are particularly significant environmental pollutants are _____ and _____.

7.   Examples of heavy metals are:

a. _____   b. _____   c. _____

d. _____   e. _____   f. _____

g. _____   h. _____

8.   How are heavy metals used in industry?

_____

9.   Heavy metals are extremely toxic because they combine with and inhibit the functioning of _____ causing _____ and _____ effects.

10. Synthetic organic chemicals are the basis for the production of all _____, synthetic _____, synthetic _____, _____ coatings, _____, _____, and _____ preservatives.

11. Synthetic organic chemicals are (Biodegradable, Nonbiodegradable).

12. List three health effects from synthetic organics.

a. _____   b. _____   c. _____

13. A particularly dangerous group of synthetic organic compounds are the _____ **hydrocarbons**.

14. The main feature of halogenated hydrocarbons is that one or more hydrogen atoms have been substituted by an atom of _____, _____, _____ or _____.

15. The most common halogenated hydrocarbons are those containing chlorine. Such compounds are referred to as _____ **chlorides**.

16. Examples of chlorinated hydrocarbons include:

a. _____   b. _____   c. _____

d. _____   e. _____.

Involvement with Food Chains

17. Bioaccumulation occurs because toxic compounds are

    a. biodegradable. (True or False)
    b. nonbiodegradable. (True or False)
    c. readily absorbed by the body. (True or False)
    d. readily excreted from the body. (True or False)

18. Biomagnification demonstrates how each higher trophic level receives and accumulates a (higher, lower, the same) dose than the one before.

19. Mercury caused a biomagnification episode called _____ disease.

20. Develop the food chain that led to the biomagnification episode in Minimata, Japan. Match the following events [a to e] with the steps in the below biomagnification/food chain diagram. The events that led to the disaster were: a. mercury in detritus, b. biomagnification by fish, c. discharge of mercury wastes into river, d. biomagnification by bacteria, e. Minimata disease in humans.

```
┌──────────────────┐
│   Step 5 =        │
├──────────────────┤
│        ↑          │
├──────────────────┤
│   Step 4 =        │
├──────────────────┤
│        ↑          │
├──────────────────┤
│   Step 3 =        │
├──────────────────┤
│        ↑          │
├──────────────────┤
│   Step 2 =        │
├──────────────────┤
│        ↑          │
├──────────────────┤
│   Step 1 =        │
└──────────────────┘
```

21. Why was biomagnification of mercury in cats not a critical step in causing Minimata disease in humans?

    _____

A History of Mismanagement

22. Prior to the passing of environmental laws in the 60s and 70s, it was common practice to dispose of chemical wastes into the _____ or natural _____.

23. The public outcry against pollution in the 60s led the U.S. Congress to pass two major pieces of legislation called:

   a. _____ Date _____

   b. _____ Date _____

24. In the years since these laws were passed direct discharges of wastes into natural waterways and the air have (Decreased, Increased).

Methods of Land Disposal

25. Indicate whether the following statements describe the disposal technique of [a] deep well injection, [b] surface impoundments, or [c] landfills.

   [   ] concentrated wastes put into drums and buried
   [   ] a reverse well
   [   ] an open pond without an outlet
   [   ] volatile wastes may enter the atmosphere
   [   ] waste put into deep rock strata
   [   ] involves the removal of leachate

26. Indicate whether the following problems of improper disposal pertain to [a] deep well injection, [b] surface impoundments, [c] landfills, or [d] all three.

   [   ] absence of liners or leachate collection systems
   [   ] deposition of wastes above or into aquifers
   [   ] immersion of wastes into ground water
   [   ] lack of reliable monitoring systems to detect leakage

27. It is (True or False) that once wastes have seeped into groundwater, there is no practical way to remove them.

28. Small amounts of toxic chemicals in groundwater are extremely hazardous because of their known ability to _____ and cause serious _____

   _____ .

29. It is (True or False) that when land disposal facilities are properly sited, constructed, and maintained, they guarantee that wastes will never seep or leach into groundwater.

30. Two problems inherent in land disposal is that wastes _____ at disposal facilities and that they _____ there.

201

Scope of the Management Problem

31.   The use of designated disposal sites that have no precautions against ground pollution are examples of _____ land disposal. Give an example. _____

32.   There are an estimated _____ million tons of hazardous wastes generated each year in the United States.

33.   List the three major aspects to the toxic waste problem.

        a. _____

        b. _____

        c. _____

Cleaning Up the Mess

Assuring Safe Drinking Water

34.   A federal law that attempts to assure safe drinking water supplies is the _____ Date _____

35.   Indicate whether the following are [+] or are not [-] provisions of the Safe Drinking Water Act of 1974.

        [   ] setting standards regarding allowable levels of pollution
        [   ] monitoring municipal water supplies
        [   ] proper location and construction of injection wells
        [   ] systematic monitoring of groundwater and private wells
        [   ] determination of "safe" and "unsafe" levels for all chemicals
        [   ] provision of funds to compensate for illness resulting from contaminated wells

36.   Identify two emerging issues in the public debate over the Safe Drinking Water Act of 1974.

        a. _____

        b. _____

Groundwater Remediation

37.   List the four basic steps in groundwater remediation (see Fig. 20-10).

        a. _____   b. _____

        c. _____   d. _____

Superfund for Toxic Sites

38.   The **Comprehensive Environmental Response, Compensation, and Liability Act of 1980** is commonly known as the _____.

39.   How is this act funded? _____

40.   Funding from the program was about _____ million/year in the early 1980s to above _____ billion/year in the 1990s.

41.   It is (True or False) that all toxic waste sites will be cleaned up in the near future.

42.   What is the first step in the types of actions funded by the Superfund?

        _____

43.   What action is taken concerning the worst sites?

        _____

44.   The administrative unit for the Superfund is the _____.

45.   List four soil cleanup technologies.

        a. _____   b. _____

        c. _____   d. _____

46.   Indicate whether the following have been positive [+] or negative [-] attributes of the EPA's "administration" of the Superfund.

        [   ] placement of **NPL** sites on the Superfund list
        [   ] scope of action across all **NPL** sites in the United States
        [   ] development of technologies to clean-up **NPL** sites
        [   ] cost of cleaning-up **NPL** sites
        [   ] cost of administering the Superfund

47.   How do **brownfields** demonstrate an example of environmental racism?

        _____

Management of New Wastes

48.   Identify the general roles that the below government entities play in the management of toxic wastes.

        U.S. Congress: _____

        EPA: _____

The Clean Air and Water Acts

49. It is (True or False) that the Clean Water Act has stopped nearly all dumping of toxic wastes into natural waterways.

50. What is the purpose of **discharge permits**?

_____

The Resource Conservation and Recovery Act (RCRA)

51. The RCRA legislation required

    a. permitting of disposal sites. (True or False)
    b. pretreatment of wastes destined for landfills. (True or False)
    c. cradle to grave tracking of all hazardous wastes. (True or False)

52. Indicate whether the following are known [+] or anticipated [?] results of the RCRA legislation.

    [   ] the closure of many waste disposal facilities
    [   ] disposal of wastes in unregulated landfills
    [   ] legally permitted discharges
    [   ] new methods of midnight dumping
    [   ] tougher laws in some states
    [   ] shipment of hazardous wastes to other countries
    [   ] wholesale movement of some industries to other countries

Reduction of Accidents and Accidental Exposures

53. Match the four potential sources of accidental exposures to hazardous wastes with the safeguards or legislation that have been developed to prevent accidental exposures.

Sources of Exposure	Safeguards
[ ] leaking underground storage tanks	a. EPCRA
[ ] transportation accidents	b. SARA
[ ] workplace accidents	c. TSCA
[ ] new chemicals	d. UST
[ ] worker protection	e. OSHA
[ ] community protection	f. DOT Regs

Looking Toward the Future

54. The EPA estimates of annual costs of pollution control and clean-up programs is $_____ billion.

Too Many or Too Few Regulations?

55. There are _____ million tons of toxic chemicals legally discharged into the environment annually.

56. Two groups that are usually exempted from toxic waste regulation compliance are _____ and _____.

57. Why are these groups exempted? _____

Pollution Avoidance for a Sustainable Society

58. List four general approaches in pollution prevention programs.

   a. _____

   b. _____

   c. _____

   d. _____

VOCABULARY DRILL

Directions: Match each term with the list of definitions and examples given in the tables below. Term definitions and examples might be used more than once.

Term	Definition	Example(s)	Term	Definition	Example(s)
bioaccumulation			ground water remediation		
biomagnification			NPL		
bioremediation			orphan sites		
brown fields			organic chlorides		
halogenated hydrocarbons			pollution avoidance		
HAZMAT			pollution control		
midnight dumping			product life cycle		

VOCABULARY DEFINITIONS
a. a chemical that presents a hazard or a risk
b. cradle to grave product history
c. organic compounds in which one or more hydrogen atoms have been replaced by halogens
d. chlorinated hydrocarbons

VOCABULARY DEFINITIONS

e. occurs when harmless amounts of toxic waste received over long periods of time reach toxic levels
f. multiplying effect of toxic wastes through the food chain
g. illegal deposition of toxic wastes in any location available under the cover of darkness
h. a technique to clean contaminated groundwater
i. sites representing the most immediate and severe toxic waste threats
j. utilization of oxygen and organisms to clean-up toxic waste sites
k. adding filters to prevent environmental pollution
l. changing the production process so that harmful pollutants are not produced
m. hazardous waste sites without a responsible party to do cleanup
n. abandoned idustrial facilities where redevelopment complicated by possible environmental contamination

VOCABULARY EXAMPLES

a. electric cars	h. Fig. 20-2
b. biodegradable toxic organic compounds	i. military bases
c. catalytic converter	j. Minamata disease
d. Fig. 20-8	k. Fig. 20-10
e. $Hg \rightarrow$ detritus $\rightarrow$ bacteria $\rightarrow$ fish $\rightarrow$ humans	l. economically disadvantaged communities
f. gasoline, acids, explosives	m. Fig. 20-9
g. $CCl4$	

SELF-TEST

Give an example of a hazardous chemical that fits each of the categories listed below.

1. Toxicity: _____

2. Reactivity: _____

3. Corrosivity: _____

4. Ignitability: _____

Match the sources of chemicals entering the environment with an example.

	Sources	Examples
5.	[ ] entire products	a. tanker trucks
6.	[ ] fractional products	b. UST's
7.	[ ] waste byproducts	c. your discarded shampoo bottle
8.	[ ] household solvents	d. cleaning fluids
9.	[ ] unused portions	e. heavy metals
10.	[ ] leaks	f. paint solvents
11.	[ ] Accidental spills	g. Fertilizers

12.   The two categories of wastes that are particularly toxic are

        a. midnight dumping and inadequate landfills.
        b. heavy metals and synthetic organics.
        c. deicing salt and waste road oil.
        d. sewage sludge and transportation spills.

13.   _____ is the term defining organisms' inability to excrete heavy metals.

14.   _____ is the term defining the multiplying effect of concentrations of heavy metals in the food chain.

Indicate whether the following statements describe the disposal technique of [a] deep well injection, [b] surface impoundments, [c] landfills, or [d] all three

15.   [   ] concentrated wastes put in drums and buried
16.   [   ] a reverse well
17.   [   ] an open pond without an outlet
18.   [   ] volatile wastes may enter the atmosphere
19.   [   ] waste put into deep rock strata
20.   [   ] involves the removal of leachate
21.   [   ] absence of liners or leachate collection systems
22.   [   ] deposition of wastes above or into aquifers
23.   [   ] immersion of wastes into ground water
24.   [   ] lack of reliable monitoring systems to detect leakage

25.   The federal program aimed at identifying and cleaning up existing toxic waste sites is the

        a. Resources Conservation and Recovery Act of 1976
        b. Clean Water Act of 1974
        c. FIFRA
        d. Comprehensive Environmental Response, Compensation, and Liability Act of 1980

Which of the following legislative acts were designed to control disposal of chemical (toxic) wastes. (Check all that apply)

26.   [   ] Clean Air Act of 1970
27.   [   ] Clean Water Act of 1972
28.   [   ] Safe Drinking Water Act of 1974

29.	[ ] Comprehensive Environmental Response, Compensation, and Liability Act of 1980
30.	[ ] Resource Conservation and Recovery Act of 1976
31.	[ ] Toxic Substances Control Act of 1978

Distinguish the difference in pollution avoidance and control.

32.	Avoidance: _____

33.	Control: _____

ANSWERS TO STUDY GUIDE QUESTIONS

1. caught on fire, mercury poisoning of residents, accidental release of toxic chemicals on residents, toxic chemicals leaked from an old dump site; 2. d, c, b, a; 3. loss of raw materials, chemical wastes from production process, risks of accidents or spills in manufacturing process, used shampoo rinsed down the drain, containers and packaging thrown into landfill; 4. true; 5. requires industries to report storage and release of toxic chemicals; 6. heavy metals, nonbiodegradable synthetic organics; 7. lead, mercury, arsenic, cadmium, tin, chromium, zinc, copper; 8. metal-workings, batteries, electronics; 9. enzymes, physiological, neurological; 10. plastics, fibers, rubber, paint, solvents, pesticides, wood; 11. nonbiodegradable; 12. b, a, c; 13. halogenated; 14. chlorine, bromine, fluorine, iodine; 15. chlorinated; 16. plastics, pesticides, solvents, electric insulation, flame retardants; 17. false, true, true, false; 18. higher; 19. Minimata; 20. c, a, d, b, e; 21. cats not in the food chain leading to humans; 22. sewers, waterways; 23. Clean Air Act, 1970; Clean Water Act, 1972; 24. decreased; 25. c, a, b, b, a, c; 26. c, a, d, d; 27. true; 28. bioaccumulation, health hazards; 29. false; 30. arrive, stay; 31. unregulated, Love Canal; 32. 150; 33. assure safe water, regulating the handling and disposal of wastes, looking for future solutions; 34. Safe Drinking Water Act, 1974; 35. +, +, +, -, -, -; 36. cost of protection against a hazard that does not exist, no provision for systematic monitoring of private wells; 37. drilling wells, pumping out contaminated water, purifying water, reinjecting purified water; 38. Superfund; 39. tax on chemical raw materials; 40. 300, 2; 41. false; 42. setting priorities (triage); 43. placed on the NPL; 44. EPA; 45. incineration, bioremediation, detergent injection, phytoremediation; 46. +, -, -, -, -; 47. they occur in the most socioeconomic depressed areas of cities; 48. writes the legislation and passes laws, administers programs and enforces laws; 49. false; 50. provide a means of seeing who is discharging what; 51. all true; 52. all +; 53. d, f, e, c, e, a, b; 54. 115; 55. 2; 56. homeowners, farmers; 57. produce < 200 lbs hazardous mat; 58. better product and/or materials management, find nonhazardous substitute for hazardous material, clean and recyle solvents and lubricants, putting environmentalists and industrialists on the same team.

ANSWERS TO SELF-TEST

1. pesticides; 2. explosives; 3. acids; 4. gasoline; 5. g; 6. f; 7. e; 8. d; 9. c; 10. b; 11. a; 12. b; 13. bioaccumulation; 14. biomagnification; 15. c; 16. a; 17. b; 18. b; 19. a; 20. c; 21. c; 22. a; 23. d; 24. d; 25. d; 26 - 30. check; 31. blank; 32. does not produce wastes in the first place; 33. filters existing pollution.

VOCABULARY DRILL ANSWERS					
Term	**Definition**	**Example(s)**	**Term**	**Definition**	**Example(s)**
bioaccumulation	e	j	ground water remediation	h	k
biomagnification	f	e	NPL	i	i
bioremediation	j	b	organic chlorides	c, d	g
brownfields	n	l	orphan sites	m	m
halogenated hydrocarbons	c	g	pollution avoidance	l	a
HAZMAT	a	f	pollution control	k	c
midnight dumping	g	d	product life cycle	b	h

CHAPTER 21

THE ATMOSPHERE: CLIMATE, CLIMATE CHANGE AND OZONE DEPLETION

Atmosphere and Weather

Atmospheric Structure

1.   Indicate whether the following statements pertain to the [a] **troposphere**, [b] **tropopause**, or [c] **stratosphere**.

      [   ] contains ozone
      [   ] extends up 10 miles
      [   ] caps the troposphere
      [   ] gets colder with altitude

[   ] Temperature increases with altitude.
[   ] well mixed vertically
[   ] site and source of our weather
[   ] substance that enter remain there a long time
[   ] Pollutants can reach the top within a few days.

Weather

2.   What happens to most of the solar radiation that enters the Earth's atmosphere?

_____

3.   How are trade winds produced?

_____

4.   How are jet streams produced?

_____

5.   How are tornadoes produced?

_____

Climate

6.   The average temperature and precipitation conditions of an area produced different

_____

7.   It is (True or False) that humans are isolated from the effects of climate change on other organisms.

Climates in the Past

8.   At this time is the Earth in a global (Cooling or Warming) period?

9.   Explain the association between variations in the Earth=s orbit and changing climates.

_____

10.   Any explanations of rapid global climatic changes must consider the
_____ and the _____.

Ocean and Atmosphere

11.   What proportion of the Earth is covered by oceans? _____

12.   Oceans are the major source of:

       a. _____ and b. _____.

13. What two roles do the oceans play in heating the atmosphere?

a. _____ and b. _____

14. List the two abiotic factors that make the **Conveyor** system work.

a. _____ and b. _____

15. Water masses in the Conveyor system move according to _____.

16. Identify the following events in the Conveyor system in terms of More [M] or Less [L] density.

[   ] Gulf stream
[   ] North Atlantic Deep Water
[   ] Indian and Pacific deep currents

17. It takes the Conveyor system about _____ years to complete one cycle.

18. Which waters transfer enormous quantities of heat toward Europe.

a. Gulf Stream   b. North Atlantic Deep   c. Pacific deep currents

19. Does the addition of fresh water (Increase or Decrease) ocean water density.

20. Complete the below sequence of events by choosing the bracketed choices.

Extended global warming → [+ or -] melted polar ice → [+ or -] ocean water density → [northern or southern] shift in the Conveyor system → global [cooling or warming].

Global Climate Change

The Earth as a Greenhouse

21. Light energy is (Blocked, Absorbed) by greenhouse gases.

22. Light energy is converted to _____ energy in the form of infrared radiation.

23. When heat energy is trapped in an enclosed space and the temperature rises, this is known as the _____ effect.

24. What experimental evidence demonstrates a 60% increase of atmospheric CO_2 since the ice ages?

_____

25. Which of the following factors contribute to **planetary albedo**? (Check all that apply.)

[   ] atmospheric CO_2
[   ] low clouds

[   ] high clouds
[   ] sulfate aerosols
[   ] atmospheric volcanic ash

The Carbon Dioxide Story

26. CO_2 levels are now _____ percent higher than before the Industrial Revolution.

27.   As carbon dioxide concentrations increase, global temperatures will (Increase, Decrease).

28.   A major *sink* for atmospheric CO_2 is the _____.

29.   Every kilogram of fossil fuel that is burned results in the production of an additional _____ kilograms of carbon dioxide.

30.   Forests are a net (Source or Sink) of CO_2?

31.   A total of _____ billion tons of carbon dioxide is added annually to the Earth's atmosphere from burning fossil fuels and tropical rain forests.

Other Greenhouse Gases

32.   Identify the sources of four other greenhouse gases.

Other Green House Gases	Sources
a. Water Vapor	
b. Methane	
c. Nitrous Oxide	
d. CFC=s and other Halocarbons	

Amount of Warming and Its Probable Effects

33.   Identify four variables that affect global temperatures.

    a. _____   b. _____

    c. _____   d. _____

34.   What is the purpose of general circulation models?

    _____

35.   If concentrations of greenhouse gases were to double, the Earth would warm up between _____ and _____ degrees centigrade.

36. Earth's temperatures were only _____ degrees C cooler during the ice age.

37. Two predicted major impacts of global warming are

    a. _____   b. _____

38. Indicate whether the following are [+] or are not [-] probable effects of global warming.

    [   ] melting of polar ice caps
    [   ] flooding of coastal areas
    [   ] massive migrations of people inland
    [   ] alteration of rainfall patterns
    [   ] deserts becoming farmland and farmland becoming deserts
    [   ] significant losses in crop yields

Coping With Global Warming

39. List three sources of evidence that global warming might be here.

    a. _____   b. _____

    c. _____

40. Which of the following are strategies to reduce carbon dioxide emissions and the effects of global warming? (Check all that apply.)

    [   ] limiting the use of fossil fuels in industry and transportation
    [   ] adopt a wait-and-see attitude
    [   ] develop and adopt alternative energy sources
    [   ] encourage vast tree planting programs
    [   ] examine other possible causes of global warming
    [   ] make and enforce energy conservation rules
    [   ] rely on the government to develop the needed strategies
    [   ] adopt the **Precautionary Principle**

41. Why did the **Framework Convention on Climate Change (FCCC)** fail?

    _____

42. After adopting the FCCC, CO_2 emissions of member nations (Increased or Decreased)?

43. The IPCC calculates that it would take emission reductions of at least _____ % worldwide to stabilize greenhouse gas concentrations at today's levels. Is it highly (Likely or Unlikely) that this level of reduction will be achieved.

Depletion of the Ozone Layer

Radiation and Importance of the Shield

44. Ultraviolet light is responsible for thousands of cases of _____ cancer per year in the U.S.

45. _____ in the stratosphere prevents ultraviolet light from entering the Earth's atmosphere.

46. The presence of this chemical barrier to ultraviolet light is known as the _____ shield.

47. It is (True or False) that the ozone in the stratosphere and the ozone that is a serious pollutant on the Earth's surface are the same ozone.

Formation and Breakdown of the Shield

48. Complete the following reactions that form ozone.

    Reaction #1: _____ light + O_2 → _____ + _____

    Reaction #2: free _____ + O_2 → _____

    Reaction #3: free _____ + O_3 → _____ + _____

    Reaction #4: _____ light + O_3 → _____ + _____

49. Indicate whether O_3 concentrations are high [+] or low [-] given the below conditions.

    [   ] at the equator [   ] northern latitudes   [   ] winter [   ] summer

50. Identify four sources of halogens or chlorofluorocarbons (CFCs).

    a. _____   b. _____

    c. _____   d. _____

51. A _____ is a chemical which promotes a chemical reaction without itself being used up by the reaction.

52. Complete the following reactions that destroy ozone.

    Reaction # 5: $CFCl_3$ + _____ light → _____ + $CFCl_2$

    Reaction #6: Cl + _____ → _____ + O_2

    Reaction #7: ClO + _____ → 2 _____ + O_2

53. The source of chlorine entering the stratosphere is _____.

54.   Which of the above reactions releases Cl from CFCs (5, 6, 7)?

55.   Which of the above reactions generates more Cl (5, 6, 7)?

56.   Chlorine is a **catalyst** that destroys the product of reaction (1, 2, 3, 4)?

57.   Less ozone in the stratosphere will cause (More or Less) UV light penetration.

58.   It is (True or False) that CFCs will eventually be flushed out of the atmosphere.

59.   It is (True or False) that CFCs are water insoluble.

60.   A gaping hole in the ozone shield was discovered over the _____ pole.

61.   This hole represented a _____ % reduction in ozone levels.

62.   The ozone hole:

a. persists throughout the year. (True or False)
b. reappears every year over the South Pole. (True or False)
c. is becoming larger at each appearance. (True or False)
d. may cause the destruction of marine phytoplankton. (True or False)
e. could be become more destructive if located over the North Pole. (True or False)

63.   Where is the ozone shield the thinnest? _____

64.   What is the predicted public health impact in this area?

_____

65.   It is (True or False) that volcanic eruptions inject significant amounts of Cl into the stratosphere.

66.   It is (True or False) that the Cl in the stratosphere from volcanic eruptions equals that of anthropogenic sources.

More Ozone Depletion

67.   Indicate whether the following are [+] or are not [-] trends in the continuing saga of ozone depletion.

a. [   ] high ClO over both poles b. [   ] ozone holes over both poles
c. [   ] ground-level UVB increases   d. [   ] skin cancer increases

68.   Explain why ozone loss is expected to peak in 2001.

_____

Coming to Grips With Ozone Depletion

69. It is (True or False) that several nations and industries are scaling back CFCs production.

70. What was the purpose of the **Montreal Protocol**?

     _____

71. Is it (True or False) that there are CFC substitutes for industrial use?

72. U.S. chemical companies must halt all CFC production by December 31, 19_____.

73. What was the purpose of Title IV of the Clean Air Act of 1990?

     _____

VOCABULARY DRILL

Directions: Match each term with the list of definitions and examples given in the tables below. Term definitions and examples might be used more than once.

Term	Definition	Example(s)	Term	Definition	Example(s)
anthropogenic			fronts		
catalyst			greenhouse gases		
chlorine reservoirs			meteorology		
chlorine cycle			monsoons		
chlorfluorocarbon			ozone shield		
climate			planetary albedo		
convection currents			weather		

VOCABULARY DEFINITIONS

a. coming from human activities
b. day-to-day variations in temperature, air pressure, wind, humidity and precipitation
c. scientific study of weather and climate
d. result of long-term weather patterns in a region
e. vertical air currents resulting from rising warmer air
f. boundaries where air masses of different temperatures and pressures meet
g. atmospheric gases that play a role analogous to the glass in a greenhouse
h. a natural reflection of light that contributes to overall cooling
i ozone in the stratosphere

VOCABULARY DEFINITIONS
j. organic molecules in which Cl and Fl replace some of the hydrogens
k. continual regeneration of Cl as it interacts with ozone
l. promotes a chemical reaction without itself being used up in the reaction
m. temporary reactions that hold Cl out of the chlorine cycle
n. seasonal air flows the represent a reversal in previous wind patterns

VOCABULARY EXAMPLES	
a. biomes	g. halogenated hydrocarbons
b. low clouds	h. tornadoes
c. CFC's, CO_2, CH_4	i. Indian subcontinent summer weather
d. Cl	j. weather predictions
e. $ClO + ClO \rightarrow 2\,Cl + O_2$	k. NO_2, CH_4
f. Fig. 9-6	l. Fig. 21-17

SELF-TEST

1. The site and source of our weather resides in the

   a. troposphere b. stratosphere c. mesosphere d. thermosphere

2. That part of the atmospheric structure that blocks UV radiation is the

   a. troposphere b. stratosphere c. mesosphere d. thermosphere

3. The Earth's weather engine consists of

   a. atmosphere b. ocean c. solar radiation d. all of these

4. When air masses of different temperature and pressure meet, the weather could be a

   a. tornado b. monsoon c. typhoon d. any one of these

5. Temperature and moisture are the primary predictors of:
   a. climate b. biomes c. weather d. a and b

Identify two ways that oceans play a dominant role in determining our environment.

6. _____ and 7. _____

Identify the two major components of the **Conveyor** system.

8. _____ and 9. _____

Complete the below sequence of events by choosing the bracketed choices.

Extended global warming → [10. + or -] melted polar ice → [11. + or -] ocean water density → [12. northern or southern] shift in the Conveyor system → global [13. cooling or warming].

Identify the source and corresponding pollutant that cause each of the following environmental conditions.

Condition	Sources	Pollutant
Greenhouse Gases	14.	15.
	16.	17.
	18.	19.
	20.	21.
Depletion O3 Shield	22.	23.
	24.	25.

26.   Define **global warming** and apply the term to climatic change: _____

_____

Indicate whether the following environmental conditions are or might be the result of or result in [A] planetary albedo, [B] greenhouse effect or [C] ozone depletion.

27.   [   ] increased concentrations of UVB light hitting the Earth's surface
28.   [   ] melting of polar ice caps
29.   [   ] deserts becoming farmland and farmland becoming deserts
30.   [   ] global cooling
31.   [   ] higher incidence of skin cancer in humans
32.   [   ] volcanic eruptions
33.   [   ] flooding of coastal areas.

Identify whether the following pollution abatement procedures are [A] getting at the root cause or [B] attacking the symptoms.

34. _____ decrease fertilizer use 35. _____ banning aerosol sprays 36. _____ using solar energy

37. _____ recapture CFCs 38. _____ increase fuel efficiency 39. _____ increase fuel costs

40.   The ozone shield in the stratosphere protects Earth's biota from the damaging effects of

        a. CFCs b. UV light c. global warming d. planetary albedo

ANSWERS TO STUDY GUIDE QUESTIONS

1. c, a, b, a, c, a, a, c, a; 2. absorbed by atmosphere, oceans and land; 3. upward movements of air called convection currents; 4. Earth's rotation and air pressure gradients; 5. convergence of air masses of different temperatures and pressures; 6. climates; 7. false; 8. warming; 9. changes in the Earth=s orbit cause changes in the distribution of solar radiation over different continents and latitudes thus changin temperature and precipitation; 10. atmosphere, ocean; 11. 2/3; 12. water for hydrological cycle, heat to the atmosphere; 13. stores and conveys heat; 14. temperature and salinity; 15. density; 16. less, more, less; 17. 1000; 18. a; 19. decrease; 20. +, -, southern, cooling; 21. absorbed; 22. heat; 23. greenhouse; 24. CO_2 in gas bubbles trapped in glacier ice; 25. blank, check, blank, check, check; 26. 25; 27. increase; 28. ocean; 29. 3; 30. source; 31. 30.5; 32. water cycle, animal husbandry, chemical fertilizers, refrigerants; 33. cloud cover, solar intensity, volcanic activity; 34. project future global climate; 35. 1 to 5; 36. 5; 37. regional climatic changes, rise in sea level; 38. all +; 39. nine out of 14 hottest years, wide-scale recession of glaciers, sea level is rising; 40. check, blank, check, check, blank, check, blank, check; 41. relied on a voluntary approach; 42. increase; 43. 60; 44. skin; 45. ozone; 46. ozone; 47. false; 48. see Fig.21-16; 49. +, -, -, +; 50. refrigerants, plastic foams, electronics cleaners, aerosol cans; 51. catalyst; 52. see Fig.21-16; 53. CFCs; 54. 5; 55. 7; 56. 2; 57. more; 58. false; 59. true; 60. south; 61. 50; 62. false, the rest all true; 63. Queensland, Australia; 64. 3/4 Australians will develop skin cancer; 65. false; 66. false; 67. +, -, +, +; 68. anthropogenic sources of Cl and Br in stratosphere will start to decline; 69. true; 70. an agreement by 68 nations to scale back CFC production 50% by 2000; 71. true; 72. 1995; 73. restricts the production, use, emissions and disposal of a family of chemicals identified as ozone depleting.

ANSWERS TO SELF-TEST

1. a; 2. b; 3. d; 4. d; 5. d; 6.convey heat; 7. store heat; 8. salinity; 9. density; 10. +; 11. -; 12. Southern; 13. cooling; 14. coal; 15. carbon dioxide; 16. cattle; 17. methane gas; 18. refrigerants; 19. CFC's; 20. gasoline; 21. nitrous oxides; 22. refrigerants; 23. CFCs; 24. plastic foams; 25. electronics; 26. global warming is a process of gradual increase in Earth=s temperature which may cause a southern shift in the Gulfstream leading to another ice age; 27. C; 28. B; 29. B; 30. A; 31. C; 32. A; 33. B; 34. A; 35. A; 36. A; 37. A; 38. B; 39. B; 40. B.

	VOCABULARY DRILL ANSWERS				
Term	**Definition**	**Example(s)**	**Term**	**Definition**	**Example(s)**
anthropogenic	a	c	fronts	f	h
catalyst	l	d	greenhouse gases	g	c
chlorine reservoirs	m	k	meterology	c	j
chlorine cycle	k	e	monsoons	n	i
chlorfluorocarbon	j	g	ozone shield	i	l
climate	d	a	planetary albedo	h	b
convection currents	e	f	weather	b	h

CHAPTER 22

ATMOSPHERIC POLLUTION

Air Pollution Essentials
    Pollutants and Atmospheric Cleansing
    The Appearance of Smogs

Major Air Pollutants and Their Impact
    Major Pollutants
    Adverse Effects of Air Pollution on Humans, Plants, and Materials

Pollutant Sources
    Primary Pollutants
    Secondary Pollutants

Acid Deposition
    Acids and Bases
    Extent and Potency of Acid Deposition
    Sources of Acid Deposition
    Effects of Acid Deposition

Bringing Air Pollution Under Control
    Control Strategies
    Coping With Acid Precipitation

Taking Stock
    Future Directions

Vocabulary Drill
    Vocabulary Definitions
    Vocabulary Examples

Self-Test

Answers to Study Guide Questions

Answers to Self-Test

Vocabulary Drill Answers

Air Pollution Essentials

Pollutants and Atmospheric Cleansing

1. List two processes that hold natural pollutants below the toxic level.

    a. _____

    b. _____

2. It is (True or False) that organisms do have the capacity to deal with certain amounts or levels of pollutants without suffering ill effects.

3. Pollution effect depends on _____ and _____ of exposure.

4. Indicate whether the **threshold level** is high [+] or low [-] for:

    [   ] short exposure time   [   ] long exposure time

5. It is (True or False) that there are some compounds that have no threshold level.

6. Dose is defined as _____ multiplied by time of _____.

7. List the three factors that determine air pollution levels.

    a. _____   b. _____

    c. _____

The Appearance of Smogs

8. The discovery of _____ began human inputs of air pollutants.

9. In *Hard Times*, Charles Dickens referred to _____ **smog**.

10. Thousands of cars venting unprotected exhaust + sunlight + a mountain topography = _____ **smog**.

11. Normally temperature (Increases, Decreases) as elevation increases.

12. The warmer surface air _____ which tends to carry pollutants away.

13. Warmer surface air cannot rise when a layer of (Warmer, Cooler) air overlies warmer air, a situation that is referred to as a _____ **inversion**.

14. What is the highest recorded number of human deaths due to an air pollution disaster? _____

15. Air pollution may

    a. adversely affect human health. (True or False)
    b. cause damage to crops and forests. (True or False)
    c. increase the rate of corrosion and deterioration of materials. (True or False)

16. It is (True or False) that human illnesses and damage to crops and forests from air pollution has become increasingly commonplace.

Major Air Pollutants and Their Impact

Major Pollutants

17. List the eight pollutants or pollutant categories that are the most widespread and serious.

    a. _____ b. _____ c. _____

    d. _____ e. _____ f. _____

    g. _____ h. _____

18. In large measure, all of the above pollutants are direct or indirect products of fossil fuel _____.

Adverse Effects of Air Pollution on Humans, Plants, and Materials

19. Air pollution

    a. is the result of one chemical mixed with the normal constituents of air. (True or False)
    b. varies in time and place. (True or False)
    c. varies with environmental conditions. (True or False)

20. It is (Easy or Difficult) to determine the role of particular pollutants in causing an observed result.

21. Identify the three categories of air pollutant impacts on humans.

    a. _____ b. _____ c. _____

22. Serious adverse effects of air pollution on human health are seen

    a. mainly among smokers b. mainly among nonsmokers c. equally among both groups

23. It is (True or False) that smoking by itself greatly increases the risk of serious disease.

24. Learning disabilities in children and high blood pressure in adults are correlated with high levels of _____ in the blood.

25. The source of this widespread contamination is

a. people who smoke b. use of leaded gasoline c. high ozone levels d. pesticide use

26. Which country has the worst air in the world? _____

27. The EPA mandated the phase-out of leaded gasoline by the end of 1988. It is (True or False) that you can still buy leaded gasoline.

28. Air pollution

a. may destroy vegetation. (True or False)
b. may seriously retard the growth of crops and forests. (True or False)
c. may cause forests to become vulnerable to insect pests and disease. (True or False)

29. Indicate whether the following effects of air pollutants were [+] or were not [-] demonstrated in open-chamber experiments.

[   ] which pollutants cause the damage
[   ] the sensitivity level of plants to gaseous air pollutants
[   ] the degree to which air pollutants retard plant growth and development
[   ] the degree to which air pollutants reduce crop yields and profits

30. The most serious pollutant affecting agriculture and natural areas is _____.

31. It is (True or False) that crops and wild plants react in the same manner to air pollutants.

32. It is (True or False) that just a small increase in concentration or duration of exposure may push some plants beyond their ability to cope with an air pollutant.

33. Indicate whether the following conditions could [+] or could not [-] be an effect of air pollution.

[   ] grey and dingy walls and windows
[   ] deteriorated paint and fabrics
[   ] corrosion of metals
[   ] a bright, clear, blue sky
[   ] increased real estate values
[   ] decreased real estate values

Pollutant Sources

34. Air pollutants are direct or indirect by-products from burning _____, _____ and _____.

35. Oxidation of coal, gasoline, and refuse by burning (Is, Is Not) usually complete.

Primary Pollutants

36. Indicate which of the following are [+] or are not [-] primary pollutants from combustion.

[   ] sulfuric acid
[   ] hydrocarbon emissions
[   ] ozone
[   ] nitric oxide
[   ] sulfur dioxide
[   ] carbon monoxide
[   ] PANs
[   ] particulates
[   ] nitrogen oxides
[   ] lead
[   ] **photochemical oxidants**
[   ] volatile organic compounds (VOCs)

Secondary Pollutants

37. Indicate which of the following are [+] or are not [-] secondary pollutants from combustion.

[   ] sulfuric acid
[   ] hydrocarbon emissions
[   ] ozone
[   ] nitric oxide
[   ] sulfur dioxide
[   ] carbon monoxide
[   ] PANs
[   ] particulates
[   ] nitrogen oxides
[   ] lead
[   ] **photochemical oxidants**
[   ] volatile organic compounds (VOC's)

38. Identify the major source(s) of (see Figure 22-13):

Particulates: _____   All other air pollutants: _____

Acid Deposition

39. Industrialized regions of the world are regularly experiencing precipitation that is from 10 to _____ times more acid than normal.

40. List four forms of acid precipitation.

a. _____   b. _____

c. _____   d. _____

Acids and Bases

41. Indicate whether the following are properties of [a] acids, [b] bases, or [c] water.

    [   ] Any chemical that will release OH^- ions.
    [   ] Any chemical that will release H^+ ions.
    [   ] The combination of one OH^- and 2 H^+ ions.
    [   ] The balance or neutral point between acidic and basic solutions.

42. **pH** is a measurement of (Hydrogen, Hydroxyl) ions in solution.

43. On the pH scale, the number _____ represents the pH of pure water or neutral.

44. On the pH scale, numbers decreasing from 7 (e.g., 6, 5, 4, 3, 2, 1) represent increasing (Acidity, Alkalinity).

45. Each unit on the pH scale represents a factor of (1, 10, 100, 1000) in the acid concentration of a solution..

46. A solution of pH 5 is (1, 10, 100, 1000) times more acid than a solution of pH 6.

47 A solution of pH 4 is (1, 10, 100, 1000) times more acid than a solution of pH 6.

Extent and Potency of Acid Deposition

48. Acid precipitation is defined as any precipitation with a pH of _____ or less.

49. Give the pH values for the following examples.

    a. precipitation in eastern U.S.: from _____ to _____ (see Fig. 22-19)
    b. mountain forests east of Los Angeles: _____

Sources of Acid Deposition

50. About two thirds of the acid in acid precipitation is _____ acid and one third is _____ acid.

51. It is well known that sulfur and nitrogen are found in fossil _____.

52. Indicate whether the following statements pertain to anthropogenic [a] or natural [n] oxide emissions.

    [   ] produced by volcanic or lightening activity
    [   ] produced by human activities
    [   ] concentrated in industrial regions
    [   ] spread out over the globe

53. In the Eastern U.S., the source of over 50% of acid deposition are _____ burning power plants.

54. Power plants attempt to alleviate sulfur dioxide emission at ground level by building (Shorter, Taller) stacks that ultimately disperse the pollutant (Closer, Further) from the source.

Effects of Acid Deposition

55. pH controls all aspects of _____, _____, and _____.

56. Indicate whether the following conditions are [+] or are not [-] known effects of acid pH in aquatic ecosystems.

   [ ] alteration of plant and animal reproduction
   [ ] introduction of other toxic elements (e.g. aluminum) from soil leachate
   [ ] shift from eutrophic to oligotrophic conditions
   [ ] totally barren and lifeless aquatic ecosystems
   [ ] alterations in the food chains

57. Suppose two regions receive roughly equal amounts of acid precipitation. How is it possible for one region to have acidified lakes while the other region does not?

   _____

58. It is (True or False) that lakes can maintain their buffering capacity indefinitely.

59. Hydrogen ion concentrations in solution will (Increase, Decrease) in the presence of a buffer.

60. A mineral that is an important buffer in nature is _____, known chemically as calcium carbonate.

61. The amount of acid that a given amount of limestone can neutralize is known as its _____ capacity.

62. When an ecosystem loses its buffering capacity

   a. additional hydrogen ions will remain in solution. (True or False)
   b. the buffer is used up. (True or False)
   c. the ecosystem will become acidified. (True or False)

63. When the buffering capacity is exhausted, there is a (Rapid, Slow) drop in pH.

64. Indicate whether the below tree species are tolerant [T] or vulnerable [V] to acid rain.

   [ ] red spruce [ ] balsam fir   [ ] sugar maple

65. Indicate whether the following are direct [D] or indirect [I] effects of acid precipitation on forest ecosystems. (Not covered in the text directly, but think about it!)

[   ] leaching of nutrients
[   ] break down and release of aluminum into solution
[   ] rapid changes in soil chemistry
[   ] reduced growth and diebacks of plant and animal populations
[   ] increased plant vulnerability to natural enemies and drought
[   ] increased soil erosion
[   ] increased flooding
[   ] increased sedimentation of waterways

66. Indicate whether the following effects of acid precipitation on humans is related to [a] artifacts, [b] health, or [c] aesthetics.

[   ] mobilization of lead and other toxic elements
[   ] corrosion of limestone and marble
[   ] deterioration of lakes and forests

Bringing Air Pollution Under Control

67. The law that mandated setting standards for air pollutants is the _____.

68. The Clean Air Act mandated setting standards for

a. all air pollutants. (True or False)
b. five pollutants recognized as most widespread and objectionable. (True or False)

69. Explain the difference between **ambient** and **primary** standards.

_____

_____

70. List the five pollutants that are most widespread and objectionable.

a. _____   b. _____   c. _____

d. _____   e. _____

Control Strategies

71. Explain the **command-and-control** approach.

_____

_____

72.   Indicate the level, high [H] or low [L], of success of the command-and-control approach for the following air pollutants.

a. [   ] lead   b. [   ] ozone   c. [   ] particulate matter   d. [   ] SO2

73.   Lead emissions have been reduced by _____ percent in the past 25 years.

74.   Identify two devices that some industries have installed to reduce particulates.

a. _____   b. _____

75.   It is (True or False) that these devices remove all toxic substances.

76.   List four contemporary sources of particulates.

a. _____   b. _____

c. _____   d. _____

77.   How does the Clean Air Act of 1990 address the 83 regions of the U.S. that have failed to attain air particulates standards?

_____

78.   A new car today emits _____ percent less pollutants than pre-1970 cars.

79.   Explain how a **catalytic converter** works.

_____

_____

80.   List the three major considerations of the Clean Air Act of 1990 concerning motor vehicle pollutants.

a. _____

b. _____

c. _____

81.   Identify a **point** and **area** source of VOC emissions.

Point: _____   Area: _____

82.   List two VOC control strategies in the Clean Air Act of 1990.

a. _____

b. _____

83.   The estimated total toxic chemicals emitted into the air in the U.S. is around _____ metric tons annually.

84.   At least _____ toxic pollutants have been identified.

Coping With Acid Precipitation

85.   How much of a reduction in acid-causing emissions is required to prevent further acidification of the environment? _____ %

86.   Identify the major problem(s) associated with the adoption of any of the following ways to reduce acid-forming emissions.

[a] expense [b] major reconstruction of industrial facility [c] societal acceptance
[d] requires major shift in U.S. energy policy [e] still have the problem of where to put the "waste"

[   ] fuel switching
[   ] coal washing
[   ] fluidized bed combustion
[   ] scrubbers
[   ] alternative power plants
[   ] reductions in electricity consumption

87.   Indicate whether the following strategies for reducing acid-producing emissions are examples of [a] fuel switching, [b] coal washing, [c] fluidized bed combustion, [d] scrubbers, or [e] alternative power plants.

[   ] exhausting fumes through a spray of water containing lime
[   ] combustion of coal in a mixture of sand and lime
[   ] building more nuclear power plants
[   ] pulverize and chemically wash coal
[   ] using low-sulfur coals

88.   The circumstantial evidence linking power plant emissions to acid deposition is (Insufficient, Overwhelming).

89.   Utility companies have (More, Less) governmental support than does the public on the acid precipitation issue.

90.   The federal government's response to the overwhelming evidence on the cause and effect relationship of acid deposition was more _____.

91.   What made Title IV a unique addition to The Clean Air Act?

_____

92.   List the five provisions of Title IV.

        a. _____

        b. _____

        c. _____

        d. _____

        e. _____

93.   Which of the following have been industry's response to Title IV? (Check all that apply.)

        [   ] fuel switching
        [   ] coal washing
        [   ] scrubbers
        [   ] emissions allowance trading
        [   ] using low-sulfur coals

Taking Stock

94.   Currently, pollution control costs are estimated at $ _____ billion per year.

95.   It is (True or False) that pollution control is now a major industry.

Future Directions

96.   List four future improvements that can be made to reduce air pollution.

        a. _____   b. _____

        c. _____   d. _____

VOCABULARY DRILL

Directions: Match each term with the list of definitions and examples given in the tables below. Term definitions and examples might be used more than once.

Term	Definition	Example(s)	Term	Definition	Example(s)
acids			catalytic converter		
acid deposition			chronic		
acid precipitation			industrial smog		
acute			photochemical smog		
air pollutants			pH		
ambient standards			point sources		
area sources			primary standard		
artifacts			primary pollutants		
bases			secondary pollutants		
buffer			temperature inversion		
carcinogenic			threshold level		

VOCABULARY DEFINITIONS

a. precipitation that is more acidic than usual

b. concentration of H ions in solution

c. substances in the atmosphere that have harmful effects

d. pollutant level below which no ill effects are observed

e. an irritating grayish mixture of soot, sulfurous compounds and water vapor

f. air pollutant resulting from sunlight interacting with primary sources

g. effect of certain weather conditions on smog levels

h. air pollutant levels that need to be achieved to protect environmental and human health

i. gradual deterioration of a variety of physiological functions over a period of years

j. life-threatening reactions within a period of hours or days

k. cancer-causing

l. direct products of combustion and evaporation

m. products from reactions between primary pollutants and atmospheric conditions

n. the highest level of pollutant that can be tolerated by humans without noticeable ill effects.

VOCABULARY DEFINITIONS
o. a device that removes toxic substances from internal combustion engines
p. direct pollution sources
q. diffuse pollution sources
r. any precipitation that is 5.5 or less
s. substance that has a large capacity to absorb H ions and hold pH relatively constant
t. human-made objects
u. H ion > OH ion concentration
v. H ion < OH ion concentration

VOCABULARY EXAMPLES	
a. Fig. 22-14	k. household products
b. bronchitis, fibrosis of the lungs	l. industry
c. Fig. 22-22	m. Table 22-2
d. death	n. PM-10 for particulates
e. Fig. 22-6	o. $0 \rightarrow 7 \rightarrow 14$
f. Fig. 22-3	p. limestone
g. lungs	q. monuments
h. Fig. 22-2	r. pH > 7
I. Fig. 22-4	s. pH < 7
j. *Hard Times* by Charles Dickens	t. H_2SO_4

SELF-TEST

1. Which of the following statements is not accurate concerning the air pollution problem?

    a. Organisms have the capacity to deal with certain amounts and levels of pollution.
    b. There is a threshold pollution level that is tolerable.
    c. Air pollution has been with human society since the discovery of fire.
    d. The solution to pollution is dilution.

2. Unprotected car exhaust + sunlight + a mountain topography =

    a. a temperature inversion b. photochemical smog c. an air pollution episode d. acid rain

3. The condition of a warm air mass overlying a cool air mass is referred to as

    a. a temperature inversion b. photochemical smog c. an air pollution episode d. acid rain

4. Air pollution may

   a. adversely affect human health.
   b. cause damage to crops and forests.
   c. increase the rate of corrosion and deterioration of materials.
   d. cause all of the above effects.

5. Any chemical that releases H^+ ions is

   a. an acid   b. a base   c. water   d. all of these

6. Measurement of pH is actually a measure (hydrogen, hydroxyl) ions.

7. A solution of pH 5 (1, 10, 100, 1000) times more acid than a solution of pH 6.

8. About two thirds of the acid in acid precipitation is

   a. chromic   b. nitric   c. sulfuric   d. carboxylic acid

9. It is well known that the source(s) of acid precipitation is (are)

   a. sulfur dioxide in coal.
   b. nitrogen oxides in gasoline.
   c. coal fired power plants.
   d. all of these.

Identify four direct or indirect effects of air pollutants on pine trees and the forest ecosystem in general.

10. _____

11. _____

12. _____

13. _____

14. A major problem with the Clean Air Act of 1970 and its 1977 and 1990 amendments is

   a. identifying pollutants.
   b. demonstrating cause and effect relationships.
   c. determining the source of pollutants.
   d. developing and implementing controls.

15. Indicate whether the following air pollutants are Primary [P] or Secondary [S].

[   ] particulates   [   ] PANs
[   ] hydrocarbon emissions   [   ] VOCs
[   ] carbon monoxide   [   ] lead
[   ] nitric oxide   [   ] sulfuric acid
[   ] sulfur dioxide   [   ] nitric acid
[   ] ozone

16. A major change in the Clean Air Act of 1990 was:

a. identifying criteria pollutants.
b. imposition of sanctions on polluters.
c. allowing polluters to choose the most cost-effective means of pollution control.
d. all of the above.

17. Air pollution is known to have adverse effects on

a. human health   b. agricultural crops   c. forests   d. all of these

18. List three effects of air pollution on human health.

a. _____   b. _____

c. _____

19. Indicate whether the strategies described below were designed to control emissions of
a. particulates, b. pollutants from motor vehicles, c. sulfur dioxide and acids,
d. ozone, e. air toxics.

[   ] phasing out the use of leaded gasoline
[   ] "no visible emissions"
[   ] putting catalytic converters on car exhaust systems
[   ] installation of filters and electrostatic precipitators at industrial sites
[   ] tall smoke stacks
[   ] withholding federal highway funds

ANSWERS TO STUDY GUIDE QUESTIONS

1. disperse and dilute in atmosphere, oxidation by hydroxyl ions; 2. true; 3. concentration, time; 4.+, -; 5. true; 6. concentration, exposure; 7. amount, space, mechanisms of removal; 8. fire; 9. industrial; 10. photochemical; 11. decreases; 12. rises; 13. cooler, thermal; 14. 4,000; 15. all true; 16. true; 17. suspended particulate matter, volatile organic compounds, carbon monoxide, nitrogen oxides, sulfur oxides, heavy metals, ozone, air toxics; 18. combustion; 19. all true; 20. difficult; 21. chronic, acute, carcinogenic; 22. a; 23. true; 24. lead; 25. b; 26. Poland; 27. false; 28. all true; 29. all +; 30. ozone; 31. false; 32. true; 33. +, +, +, -, -, +; 34. coal, gasoline, refuse; 35. is not; 36. -, +, -, +, +, +, -, +, -, +, -, +; 37. the reverse of #36; 38. industry, fossil fuel combustion; 39. 1000; 40. rain, fog, mist, snow; 41. b, a, c, c; 42. hydrogen; 43. 7; 44. acidity; 45. 10; 46. 10; 47. 100; 48. 5.5; 49. 4 to 3, 2.8; 50. sulfuric, nitric; 51. fuels; 52. n, a, a, n; 53. coal; 54. taller, further; 55. enzymes, hormones, proteins; 56. +, +, -, +, +; 57. soil has a natural buffer; 58. false; 59. decrease; 60. limestone; 61. buffering; 62.

all true; 63. rapid; 64. V, T, T; 65. d, d, d, d, d, i, i, i; 66. b, a, c; 67. Clean Air Act; 68. false, true; 69. levels that need to be achieved to protect environmental and human health, level of human tolerance without suffering noticeable health effects; 70. particulates, SO_2, CO, NO, O_3; 71. regulate air pollution so criteria pollutants remain below primary standard level; 72. H, L, L, L; 73. 98; 74. filters, electrostatic precipitators; 75. false; 76. steel mills, power plants, smelters, construction sites; 77. region must submit plans based on reasonably available control strategies (RACT); 78. 75; 79. platinum coated beads oxidize VOC to CO_2 and H_2O, oxide CO to CO_2, and reduces some NO; 80. tighten emission standards, develop cleaner-burning fuel, drive less; 81. industry, household products; 82. prohibition of some consumer goods, withholding federal highway funds; 83. 600,000; 84. 189; 85. 60%; 86. d; a, e; a, b, e; e; d, e; c, d; 87. d, c, e, b, a; 88. overwhelming; 89. more; 90. study; 91. addressed acid rain and CFCs specifically; 92. reductions in SO2 levels, methods of implementing reductions, emissions allowances and trading, emissions purchasing, reductions in NO; 93. check, blank, check, check, check; 94. 125; 95. true; 96. increase fuel efficiency, emission-free vehicles, improving mass transit systems, reduce commuting distances.

ANSWERS TO SELF-TEST

1. d; 2. b; 3. a; 4. d; 5. a; 6. hydrogen; 7. 10; 8. c; 9. d; 10-13. see study guide question #65; 14. d; 15. P, P, P, P, P, S, S, S, P, P, S, S; 16. d; 17. d; 18. chronic, teratogenic, carcinogenic; 19. b, a, b, a, c, d.

VOCABULARY DRILL ANSWERS					
Term	Definition	Example(s)	Term	Definition	Example(s)
acids	u	s, t	catalytic converter	o	c
acid deposition	a	p	chronic	i	b
acid precipitation	r	p	industrial smog	e	j
acute	j	d	photochemical smog	f	i
air pollutants	c	h, i , j	pH	b	o
ambient standards	h	n	point sources	p	l
area sources	q	k	primary standard	n	m
artifacts	t	q	primary pollutants	l	a
bases	v	r	secondary pollutants	m	a
buffer	s	p	temperature inversion	g	e
carcinogenic	k	g	threshold level	d	f

CHAPTER 24

SUSTAINABLE COMMUNITIES AND LIFESTYLES

Urban Sprawl

The Origins of Urban Sprawl

1.   List three factors that contributed the most to urban sprawl.

   a. _____   b. _____

   c. _____

2.   Urban sprawl is a function of (Time, Distance).

3. Indicate whether the following factors enhanced [+] or had no effect on [-] urban sprawl.

    [   ] ability to purchase cars
    [   ] commuting time
    [   ] **Highway Trust Fund**
    [   ] commuting distance
    [   ] population growth
    [   ] the lack of planned development
    [   ] low rural tax base
    [   ] the environmental or social consequences
    [   ] Henry Ford
    [   ] low-interest mortgages
    [   ] Highway Revenue Act of 1956

4. How have average commuting distances and time changed since 1960?

    distance _____   time _____

5. Explain the difference between a **suburb** and **exurb**.

    _____

Environmental Impacts of Urban Sprawl

6. Indicate whether the following consequences of urban sprawl are related to

    a. depletion of energy resources
    b. air Pollution, acid rain, the greenhouse effect, and stratospheric ozone depletion
    c. degradation of water resources and water pollution
    d. negative impact on recreational and scenic areas and wildlife
    e. loss of agricultural land

    [   ] loss of 2.5 million acres per year
    [   ] highways through parks or along stream valleys
    [   ] decreases infiltration over massive areas
    [   ] increased numbers and densities of cars
    [   ] increased commuting miles
    [   ] increased road kills
    [   ] three-fold increase in per capita oil consumption

Reining In Urban Sprawl: Smart Growth

7. List three new considerations in city planning.

    a. _____

    b. _____

    c. _____

240

8.   Which of the following are [+] and are not [-] characteristics of *Smart Growth*?

[   ] disassociated homes, workplaces, recreation areas
[   ] planning for car transportation
[   ] protection of sensitive lands
[   ] community integration

9.   Explain the significance of the Intermodal Surface Transportation Efficiency Act of 1991 in promoting suburb sustainability.

_____

10.   What is the intended function of the TEA 21?

_____

Urban Blight

Economic and Ethnic Segregation

11.   What is the main, negative, societal impact of exurban migration?

_____

12.   Exurban migration

a. has profound economic and social impacts. (True or False)
b. segregates the population along economic and/or ethnic lines. (True or False)

13.   Indicate whether the following characteristics do [+] or do not [-] describe the people who are left behind after exurban migration.

[   ] They are the poor, elderly, handicapped, and minority groups.
[   ] They can obtain mortgage loans.
[   ] They reflect a **gentrification** of society.
[   ] They represent the economically depressed subset of society.

Vicious Cycle of Urban Blight

14.   Local governments

a. are responsible for providing societal services. (True or False)
b. operate from budgets derived from property taxes. (True or False)
c. are separate entities in the central city and suburbs. (True or False)

15.   Property values are generally (Higher, Lower) in the central city when compared to suburbs.

16.   Property taxes are generally (Higher, Lower) in the central city when compared to suburbs.

17.   An **eroding tax base** is characteristic of the (Suburbs, Central City).

18.   The quality and quantity of government services is markedly (Better, Worse) in the central city when compared to the suburbs.

19.   Exurban migration causes the central city infrastructure to (Deteriorate, Improve).

Economic Exclusion of the Inner City

20.   It is (True or False) that people will never return to the central city from the suburbs.

21.   It is (True or False) that some programs are being implemented to revitalize the urban core.

22.   Indicate whether the following events are likely to increase [+] or decrease [-] as a result of urban decay.

[   ] industrial development in the central city
[   ] job opportunities in the central city
[   ] the tax base of the central city
[   ] the unemployment rate of central city residents
[   ] social unrest and crime rates within the central city
[   ] the level of education offered to and achieved by central city residents
[   ] accessibility of jobs outside the central city by its residents

23.   Suburban sprawl is a (Sustainable, Nonsustainable) system.

What Makes Cities Livable?

24.   The most livable cities around the world have which of the following characteristics? (Check all that apply.)

[   ] reduced outward sprawl
[   ] reduced automobile traffic
[   ] improved access by foot or bicycle
[   ] mass transit
[   ] high population density
[   ] heterogeneity of residences and businesses
[   ] people meet people not cars

25.   List three changes used by Portland, Oregon to make this city more livable.

a. _____

b. _____

c. _____

Moving Toward Sustainable Communities

26. List two outcomes of National Environmental Action Plans.

    a. _____   b. _____

Sustainable Cities

27. It is (True or False) that cities can be sustainable.

28. Which of the following are characteristics of a sustainable city? (Check all that apply)

    [   ] proximity of people to residences, shops, and workplaces
    [   ] need for cars
    [   ] use of solar energy
    [   ] ability to utilize alternative energy resources
    [   ] self-sufficiency in provision of food
    [   ] stable population
    [   ] cluster development

Sustainable Communities

29. Which of the following were outcomes of the **Chattanooga Venture**? (Check all that apply)

    [   ] renovation and recycling
    [   ] greenways development
    [   ] modernized parking garages
    [   ] reclaimed waterways
    [   ] new industries

President's Council on Sustainable Development

30. What is the purpose of the President's Council on Sustainable Development?

    _____

31. Economic growth + environmental protection + social equity ➔ _____
    development.

32. Indicate whether the following statements are [+] or are not [-] Council recommendations.

    [   ] intergenerational equity
    [   ] national goals for achieving sustainable development
    [   ] maintained levels of conflict between business and environmental groups
    [   ] individual responsibility

Epilogue

33. Define or describe the three themes of how we must live on our planet.

    Sustainability: _____

243

Stewardship: _____

Sound Science: _____

Our Dilemma

34. It is (True or False) that all environmental problems can be addressed successfully.

35. It is (True or False) that the root of all environmental problems are humans.

36. List the basic values of stewardship.

    a. _____ b. _____ c. _____

    d. _____ e. _____ f. _____

Lifestyle Changes

37. Match the actions below with a level of personal change (lifestlye change, volunteerism, political involvement, career choice) required to achieve a sustainable society.

    a. recycle: _____

    b. be a joiner: _____

    c. do something for nothing: _____

    d. become something: _____

VOCABULARY DRILL					
Directions:	Match each term with the list of definitions and examples given in the tables below. Term definitions and examples might be used more than once.				

Term	Definition	Example(s)	Term	Definition	Example(s)
economic exclusion			gentrification		
empowerment zones			urban decay		
eroding tax base			urban blight		
exurban migration			urban sprawl		
exurbs					

VOCABULARY DEFINITIONS
a. residential areas, shopping malls, and other facilities laced together by multilane highways
b. relocation of people, residences, shopping areas, and workplaces in the city to outlying areas
c. communities farther from cities than suburbs
d. segregation of the population into groups sharing common economic, social, and cultural backgrounds
e. declining tax revenue resulting from declining property values
f. whole process of exurban migration and debilitating economic and social impacts on the inner city
g. lack of access by inner city residents to suburban jobs
h. coordinated efforts by government and business to revitalize designated inner city zones

VOCABULARY EXAMPLES		
a. 50% unemployment rates	d. suburbia	f. Fig. 24-9
b. Fig. 24-7	e. Fig. 24-8	g. Detroit
c. Fig. 24-3		

SELF-TEST

1. The prime factor that initiated suburban growth and has supported urban sprawl is

   a. farmers selling land to developers. b. widespread ownership of private cars.
   c. decline of the central city. d. developers buying farms.

2. Which of the following factors had no effect on the rate of urban sprawl?

   a. ability to purchase cars b. commuting time c. commuting distance d. rural tax base

3. An environmental consequence of urban sprawl was

   a. depletion of energy resources. b. air pollution, acid rain, and the greenhouse effect.
   c. loss of agricultural land. d. all of the above.

4. Which of the following is not a characteristic of the people left behind after exurban migration?

   a. They are the poor, elderly, handicapped, and minority groups.
   b. They have high levels of education and income.
   c. They are predominantly economically depressed minorities and whites.
   d. They reflect the gentrification of society.

5. The factor that led most directly to a decline of the central cities was because

   a. the most affluent people moved out. b. the rural poor moved into the cities.
   c. tax revenues of the cities declined. d. the welfare costs of the city increased.

Indicate whether the following events increased [+] or decreased [-] as a result of exurban migration?

6.   [   ] industrial development in the central city
7.   [   ] job opportunities in the central city
8.   [   ] social unrest and crime rates within the central city
9.   [   ] accessibility of jobs outside the central city to its residents
10.   [   ] level of education offered to and achieved by central city residents
11.   [   ] types and efficiency of local government services
12.   [   ] property tax base of the central city
13.   [   ] unemployment rate of central city residents
14.   [   ] urban decay
15.   [   ] gentrification
16.   [   ] urban sprawl

Use the terms MORE or LESS to complete the below statement.

More urban sprawl leads to 17. _____ exurban migration which leads to 18. _____ urban decay which leads to 19. _____ urban sprawl.

The Highway Trust Fund

20.   benefits both central city and suburb residents. (True or False)
21.   brings people closer together. (True or False)
22.   promotes the establishment of parks and other natural areas. (True or False)
23.   tends to disrupt existing communities. (True or False)

24.   One effective way to stem urban blight is to

        a. provide mechanisms that give residents ownership in their homes and community.
        b. give residents large sums of money to purchase goods and services.
        c. have volunteers come in and fix things up for residents.
        d. use government funds to build large multiple housing units.

25.   List one program that gives central city residents ownership of their homes and community.

        _____

ANSWERS TO STUDY GUIDE QUESTIONS

1. car ownership, low home mortgages, highways; 2. time; 3. +, +, +, -, -, +, +, -, +, +, +; 4. doubled, no change; 5. inner city → suburbs → exurbs; 6. e, d, c, b, a, d, a; 7. integrated living, working and recreational space; affordable housing; provision for pedestrian and bicycle traffic; 8. -, -, +, +; 9. provides funding from highway trust fund to construct modes of access for alternative forms of transportation; 10. provides transportation enhancements that include pedestrain and bicycle facilities; 11. urban decay; 12. all true; 13. +, -, +, +; 14. all true; 15. lower; 16. higher; 17. central city; 18. worse; 19. deteriorating; 20. false; 21. true; 22. -, -, -, +, +, -, -; 23. nonsustainable; 24. all checked; 25. reduced outward sprawl, reduced automobile traffic, mass transit; 26. Sustainable Cities Progam, U.S. Council on Sustainable Development; 27. true; 28. check, blank, all the rest checked; 29. check, check, blank, check, check; 30. guide the country toward greater sustainability; 31. sustainable; 32. +, +, -, +; 33. practical interaction between human and natural ecosystems, ethical and moral framework guiding our actions, basis for our understanding of how the world works; 34. ture; 35. true; 36. compassion, consern for justice, honesty, frugality, humility, neighborliness; 37. lifestyle change, political involvement, volunteerism, career choice.

ANSWERS TO SELF-TEST

1. b; 2. c; 3. d; 4. b; 5. c; 6. -; 7. -; 8. +; 9. -; 10. -; 11. -; 12. -; 13. +; 14. +; 15. +; 16. +; 17. more; 18. more; 19. more; 20. false; 21. false; 22. false; 23. true; 24. a; 25. empowerment zones or habitat for humanity.

		VOCABULARY DRILL ANSWERS			
Term	Definition	Example(s)	Term	Definition	Example(s)
economic exclusion	g	a	gentrification	d	d
empowerment zones	h	g	urban decay	f	f
eroding tax base	e	e	urban blight	f	f
exurban migration	b	b	urban sprawl	a	c
exurbs	c	f			

Glossary

abiotic. Pertaining to factors or things that are separate and independent from living things; nonliving.

abortion. The termination of a pregnancy by some form of surgical or medicinal intervention.

absolute poverty. The lack of sufficient income in cash or exchange items for meeting the most basic human needs for food, clothing, and shelter.

acid. Any compound that releases hydrogen ions when dissolved in water. Also, a water solution that contains a surplus of hydrogen ions.

acid deposition. Any form of acid precipitation and also fallout of dry acid particles. (See **acid precipitation**.)

acid precipitation. Includes acid rain, acid fog, acid snow, and any other form of precipitation that is more acidic than normal, i.e., less than pH 5.6. Excess acidity is derived from certain air pollutants, namely, sulfur dioxide and oxides of nitrogen.

activated sludge. Sludge made up of clumps of living organisms feeding on detritus that settles out and is recycled in the process of secondary wastewater treatment.

activated sludge system. A system for removing organic wastes from water. The system uses microorganisms and active aeration to decompose such wastes. The system is used most as a means of secondary sewage treatment following the primary settling of materials.

active safety features. Those safety features of nuclear reactors that rely on operator-controlled reactions, external power sources, and other features that are capable of failing. (See **passive safety features**.)

adaptation. An ecological or evolutionary change in structure or function that produces better adjustment of an organism to its environment and hence enhances its ability to survive and reproduce.

adiabatic cooling. The cooling that occurs when warm air rises and encounters lower atmospheric pressure. Adiabatic *warming* is the opposite process, where cool air descends into higher pressure.

adsorption. The process whereby chemicals (ions or molecules) stick to the surface of other materials.

aeration. *Soil:* The exchange within the soil of oxygen and carbon dioxide necessary for the respiration of roots. *Water:* The bubbling of air or oxygen through water to increase the dissolved oxygen.

aerosols. Microscopic liquid and solid particles originating from land and water surfaces and carried up into the atmosphere.

affluenza. A term used to describe a dysfunctional relationship with wealth or money.

Agenda 21. A definitive program to promote sustainable development that was produced by the 1992 Earth Summit in Rio de Janeiro, Brazil, and adopted by the U.N. General Assembly.

age structure. Within a population, proportions of people who are old, middle-aged, young adults, and children.

AID. The U.S. Agency for International Development, the development arm of the U.S. State Department responsible for administering international aid from the United States.

AIDS. Acquired immune deficiency syndrome, a fatal disease caused by the HIV (human immunodeficiency virus) and transmitted by sexual contact or the use of nonsterile needles (as in drug addiction).

air pollution disaster. Short-term situation in industrial cities in which intense industrial smog brings about a significant increase in human mortality.

air toxics. A category of air pollutants including radioactive materials and other toxic chemicals that are present at low concentrations but are of concern because they often are carcinogenic.

alga, pl. **algae.** Any of numerous kinds of photosynthetic plants that live and reproduce entirely immersed in water. Many species, the planktonic forms, exist as single or small groups of cells that float freely in the water. Other species, the "seaweeds," may be large and attached.

algal bloom. A relatively sudden development of a heavy growth of algae, especially planktonic forms. Algal blooms generally result from additions of nutrients, whose scarcity is normally limiting.

alkaline, alkalinity. A *basic* substance; chemically, a substance that absorbs hydrogen ions or releases hydroxyl ions; in reference to natural water, a measure of the base content of the water.

alleles. The two or more variations of a gene for any particular characteristic, e.g., blue and brown are alleles of the gene for eye color.

alley cropping. Agricultural cropping in which rows of shade-producing trees are alternated with rows of food crops to promote growth in dry climates.

ambient standards. Air-quality standards (set by the EPA) stating that outside average air should always maintain a level of purity. That is, certain levels of pollution should not be exceeded in order to maintain environmental and human health.

anaerobic. Oxygen-free.

anaerobic digestion. The breakdown of organic material by microorganisms in the absence of oxygen. The process results in the release of methane gas as a waste product.

anaerobic respiration. Respiration carried on by certain bacteria in the absence of oxygen. Methane, which can be used as fuel gas (it is the same as natural gas), may be a byproduct of the process.

animal testing. Procedures used to assess the toxicity of chemical substances using rats, mice and guinea pigs as surrogates for humans who might be exposed to the substances.

anthropogenic. Referring to pollutants and other forms of impacts on natural environments that can be traced to human activities.

appropriate technology. Technology that seeks to increase the efficiency and productivity of hand labor without displacing workers. That is, it seeks to enable people to improve their well-being without disrupting the existing social and economic system.

aquaculture. A propagation and/or rearing of any aquatic (water) organism in a more or less artificial system.

aquifer. An underground layer of porous rock, sand, or other material that allows the movement of water between layers of nonporous rock or clay. Aquifers are frequently tapped for wells.

artificial selection. Plant and animal breeders' practice of selecting individuals with the greatest expression of desired traits to be the parents of the next generation.

asbestos fibers. Crystals of asbestos, a natural mineral, that have the form of minute strands; asbestos is a serious health hazard in indoor spaces.

atom. The fundamental unit of all elements.

autotroph. Any organism that can synthesize all its organic substances from inorganic nutrients, using light or certain inorganic chemicals as a source of energy. Green plants are the principal autotrophs.

background radiation. Radioactive radiation that comes from natural sources apart from any human activity. We are all exposed to such radiation.

bacterium, pl. **bacteria.** Any of numerous kinds of microscopic organisms that exist as simple single cells that multiply by simple division. Along with fungi, they constitute the decomposer component of ecosystems. A few species cause disease.

balanced herbivory.   A diversified plant community held in balance by various herbivores specific to each plant species.

base.   Any compound that releases hydroxyl ions (OHs) when dissolved in water. A solution that contains a surplus of hydroxyl ions.

baseload power plant.   A power plant, usually large and fueled by coal or nuclear energy, that is kept operating continuously to supply electricity.

bedload.   The load of coarse sediment, mostly coarse silt and sand, that is gradually moved along the bottom of a riverbed by flowing water rather than being carried in suspension.

benefit-cost analysis.   (See **cost-benefit analysis** also see **cost-benefit-ratio.**)

benthic plants.   Plants that grow under water attached to or rooted in the bottom. For photosynthesis, they depend on light penetrating the water.

best management practice.   Farm management practices that serve best to reduce soil and nutrient runoff and subsequent pollution.

bioaccumulation.   The accumulation of higher and higher concentrations of potentially toxic chemicals in organisms. It occurs in the case of substances that are ingested but cannot be excreted or broken down (nonbiodegradable substances).

biochemical oxygen demand (BOD).   The amount of oxygen that will be absorbed or "demanded" as wastes are being digested or oxidized in both biological and chemical processes. Potential impacts of wastes are commonly measured in terms of the BOD.

biocide.   Applies to any pesticide or other chemical that is toxic to many, if not all, kinds of living organisms.

bioconversion.   The use of biomass as fuel. Burning materials such as wood, paper, and plant wastes directly to produce energy, or converting such materials into fuels such as alcohol and methane.

biodegradable.   Able to be consumed and broken down to natural substances such as carbon dioxide and water by biological organisms, particularly decomposers. Opposite: nonbiodegradable.

biodiversity.   The diversity of living things found in the natural world. The concept usually refers to the different species but also includes ecosystems and the genetic diversity within a given species.

biogas.   The mixture of gases—about two-thirds methane, one-third carbon dioxide, and small portions of foul-smelling compounds—resulting from the anaerobic (without air) digestion of organic matter. The methane content enables biogas to be used as a fuel gas.

biological control.   Control of a pest population by introduction of predatory, parasitic, or disease-causing organisms.

biological nutrient removal.   A process employed in sewage treatment to remove nitrogen and phosphorus from the effluent coming from secondary treatment.

biological species concept.   The concept that defines species as a group of interbreeding natural populations that does not breed with other groups because of various barriers.

biological wealth.   The life-sustaining combination of commercial, scientific, and aesthetic values imparted to a region by its biota.

biomagnification.   Bioaccumulation occurring through several levels of a food chain.

biomass.   Mass of biological material. Usually the total mass of a particular group or category; for example, biomass of producers.

biomass energy, biomass fuels.   Energy or fuels, such as alcohol and methane, produced from current photosynthetic production of biological materials; also, firewood. (See **bioconversion.**)

biomass pyramid.   Refers to the structure that is obtained when the respective biomasses of producers, herbivores, and carnivores in an ecosystem are compared. Producers have the largest biomass, followed by herbivores and then carnivores.

biome.   A group of ecosystems that are related by having a similar type of vegetation governed by similar climatic conditions.

Examples include prairies, deciduous forests, arctic tundra, deserts, and tropical rain forests.

bioremediation.   Refers to the use of microorganisms for the decontamination of soil or groundwater. Usually involves injecting organisms and/or oxygen into contaminated zones.

biosolids.   Organic material removed from sewage effluents in the course of treatment. Formerly referred to as sludge.

biosphere.   The overall ecosystem of Earth. It is the sum total of all the biomes and smaller ecosystems, which ultimately are all interconnected and interdependent through global processes such as water and atmospheric cycles.

biosphere reserves.   Natural areas in various parts of the world set aside to protect genetic biodiversity.

biota.   Refers to any and all living organisms, usually in the setting of natural ecosystems.

biotic.   Living or derived from living things.

biotechnology.   The use of genetic-engineering techniques, enabling researchers to introduce new genes into food crops and domestic animals to produce valuable products and more nutritious crops.

biotic community.   All the living organisms (plants, animals, and microorganisms) that live in a particular area.

biotic potential.   Reproductive capacity. The potential of a species for increasing its population and/or distribution. The biotic potential of every species is such that given optimum conditions, its population will increase. (Contrast environmental resistance.)

biotic structure.   The organization of living organisms in an ecosystem into groups such as producers, consumers, detritus feeders, and decomposers.

birth control.   Any means, natural or artificial, that may be used to reduce the number of live births.

BOD.   See **biochemical oxygen demand.**

bottle law (bottle bill).   A law that provides for the recycling or reuse of beverage containers, usually by requiring a returnable deposit at the purchase of the item.

breeder reactor.   A nuclear reactor that in the course of producing energy also converts nonfissionable uranium-238 into fissionable plutonium-239, which can be used as fuel. Hence, a reactor that produces as much nuclear fuel as it consumes or more.

broad-spectrum pesticides.   Chemical pesticides that kill a wide range of pests. They also kill a wide range of nonpest and beneficial species; therefore, they may lead to environmental upsets and resurgences. The opposite of narrow-spectrum pesticides and biorational pesticides.

brownfields.   Abandoned, idled, or underused industrial and commercial facilities where real or perceived chemical contamination inhibits further development.

BTU (British thermal unit).   A fundamental unit of energy in the English system. The amount of heat required to raise the temperature of 1 pound of water 1 degree Fahrenheit.

buffer.   A substance that will maintain the pH of a solution by reacting with the excess acid. Limestone is a natural buffer that helps to maintain water and soil at a pH near neutral.

buffering capacity.   Refers to the amount of acid that may be neutralized by a given amount of buffer.

calorie.   A fundamental unit of energy. The amount of heat required to raise the temperature of 1 gram of water 1 degree Celsius. All forms of energy can be converted to heat and measured in calories. Calories used in connection with food are kilocalories, (1000 calories) the amount of heat required to raise the temperature of 1 liter of water 1 degree Celsius.

cancer.   Any of a number of cellular changes that result in uncontrolled cellular growth.

capillary water.   Water that clings in small pores, cracks, and spaces against the pull of gravity, like water held in a sponge.

carbon monoxide.   A highly poisonous gas, the molecules of which consist of a carbon atom with one oxygen attached. Not to be confused with nonpoisonous carbon dioxide, a natural gas in the atmosphere.

carbon tax.   A tax levied on all fossil fuels in proportion to the amount of carbon dioxide that is released as they burn.

carcinogenic.   Having the property of causing cancer, at least in animals and by implication in humans.

carnivore.   An animal that feeds more or less exclusively on other animals.

carrying capacity.   The maximum population of a given species that an ecosystem can support without being degraded or destroyed in the long run. The carrying capacity may be exceeded, but not without lessening the system's ability to support life in the long run.

castings.   The humus-rich pellets resulting from earthworm activity.

catalyst.   A substance that promotes a given chemical reaction without itself being consumed or changed by the reaction. Enzymes are catalysts for biological reactions. Also, catalysts are used in some pollution control devices, e.g., the catalytic converter.

catalytic converter.   The device used by U.S. automobile manufacturers to reduce the amount of carbon monoxide and hydrocarbons in the exhaust. The converter contains a catalyst that oxidizes these compounds to carbon dioxide and water as the exhaust passes through.

cell.   The basic unit of life; the smallest unit that still maintains all the attributes of life. Many microscopic organisms consist of a single cell. Large organisms consist of trillions of specialized cells functioning together.

cell respiration.   The chemical process that occurs in all living cells wherein organic compounds are broken down to release energy required for life processes. Higher plants and animals require oxygen for the process as well and release carbon dioxide and water as waste products, but certain microorganisms do not require oxygen. (See **anaerobic respiration**.)

cellulose.   The organic macromolecule that is the prime constituent of plant cell walls and hence the major molecule in wood, wood products, and cotton. It is composed of glucose molecules, but because it cannot be digested by humans, its dietary value is only as fiber, bulk, or roughage.

center pivot irrigation.   An irrigation system consisting of a spray arm several hundred meters long supported by wheels pivoting around a central well from which water is pumped.

centrally planned economy.   An economic system in which a ruling class makes most of the basic decisions about how the economy will be structured; typical of communist countries.

CFCs.   See **chlorofluorocarbons**.

chain reaction.   Nuclear reaction wherein each atom that fissions (splits) causes one or more additional atoms to fission.

channelization/channelized.   The straightening and deepening of stream or river channels to speed water flow and reduce flooding. A waterway so treated is said to be channelized.

chemical barrier.   In reference to genetic pest control, a chemical aspect of the plant that makes it resist pest attack.

chemical energy.   The potential energy that is contained in certain chemicals; most importantly, the energy contained in organic compounds such as food and fuels, which may be released through respiration or burning.

chemical technology.   Applied to the control of agricultural pests, it refers to the use of pesticides and herbicides to control or eradicate the pests.

chemosynthesis.   The ability of some microorganisms to utilize the chemical energy contained in certain inorganic chemicals such as hydrogen sulfide for the production of organic material. Such organisms are producers.

chlorinated hydrocarbons.   Synthetic organic molecules in which one or more hydrogen atoms have been replaced by chlorine atoms. They are extremely hazardous compounds because they tend to be nonbiodegradable and therefore to bioaccumulate; many have been shown to be carcinogenic. Also called organochlorides.

chlorination.   The disinfection process of adding chlorine to drinking water or sewage water in order to kill microorganisms that may cause disease.

chlorine cycle.   In the stratosphere, a cyclical chemical process involving chlorine monoxide that breaks down ozone.

chlorofluorocarbons (CFCs).   Synthetic organic molecules that contain one or more of both chlorine and fluorine atoms, and are known to cause ozone destruction.

chlorophyll.   The green pigment in plants responsible for absorbing the light energy required for photosynthesis.

chromosome.   In a living organism, one of a number of structures on which genes are arranged in a linear fashion.

CITES (Convention on Trade in Endangered Species).   An international treaty conveying some protection to endangered and threatened species by restricting trade in those species or their products.

Clean Air Act of 1970.   Amended in 1977 and 1990, this is the foundation of U.S. air pollution control efforts.

Clean Water Act of 1972.   The cornerstone of federal legislation addressing water pollution.

clearcutting.   In forestry, cutting every tree, leaving the area completely clear.

climate.   A general description of the average temperature and rainfall conditions of a region over the course of a year.

climax ecosystem.   The last stage in ecological succession. An ecosystem in which populations of all organisms are in balance with each other and with existing abiotic factors.

clone.   A group of genetically identical individuals derived from the asexual propagation of a single individual.

cogeneration.   The joint production of useful heat and electricity. For example, furnaces may be replaced with gas turbogenerators that produce electricity while the hot exhaust still serves as a heat source. An important avenue of conservation, it effectively avoids the waste of heat that normally occurs at centralized power plants.

combined-cycle natural gas unit.   A recent technology for generating electricity, employing both a natural gas turbine and a steam turbine that uses excess heat coming from the gas turbine.

combustion/combustion facility.   The practice of disposing of wastes by incineration in a special facility designed to handle large amounts of waste; modern combustion facilities also capture some of the energy by generating electricity on site.

command-and-control strategy.   The basic strategy behind most air and water pollution public policy. It involves setting limits on pollutant levels and specifying control technologies that must be used to accomplish those limits.

commons, common pool resources.   Resources (usually natural ones) owned by many people in common or, as in the case of the air or the open oceans, owned by no one but open to exploitation.

compaction.   Packing down. *Soil:* Packing and pressing out air spaces present in the soil. Reduces soil aeration and infiltration and thus reduces the capacity of the soil to support plants. *Trash:* Packing down trash to reduce the space that it requires.

competitive exclusion principle.   The concept that when two species compete very directly for resources, one eventually excludes the other from the area.

composting/compost.   The process of letting organic wastes decompose in the presence of air. A nutrient-rich humus, or compost, is the resulting product.

composting toilet. A toilet that does not flush wastes away with water but deposits them in a chamber where they will compost. (See **composting**.)

compound. Any substance (gas, liquid, or solid) that is made up of two or more different kinds of atoms bonded together. (Contrast **element**.)

Comprehensive Environmental Response, Compensation, and Liability Act of 1980. See **Superfund**.

condensation. The collecting of molecules from the vapor state to form the liquid state, as, for example, water vapor condenses on a cold surface and forms droplets. Opposite: **evaporation**.

conditions. Abiotic factors that vary in time and space but are not used up by organisms; for example, temperature.

confusion technique. Pest control method in which a quantity of sex attractant is applied to an area so that males become confused and are unable to locate females. The actual quantities of pheromones applied are very small because of their extreme potency.

conservation. The management of a resource in such a way as to assure that it will continue to provide maximum benefit to humans over the long run. Conservation may include various degrees of use or protection, depending on what is necessary to maintain the resource over the long run. *Energy*: Saving energy. It entails not only cutting back on use of heating, air conditioning, lighting, transportation, and so on but also increasing the efficiency of energy use. That is, developing and instigating means of doing the same jobs, e.g., transporting people, with less energy.

consumers. In an ecosystem, those organisms that derive their energy from feeding on other organisms or their products.

consumptive use. In reference to the harvest of natural resources, the harvesting of those resources in order to provide for people's immediate needs for food, shelter, fuel, and clothing.

consumptive water use. Use of water for such things as irrigation, where the water does not remain available for potential purification and reuse.

containment building. Reinforced concrete building housing the nuclear reactor. Designed to contain an explosion should one occur.

contour farming. The practice of cultivating land along the contours across rather than up and down slopes. In combination with strip cropping, it reduces water erosion.

contraceptive. A device or method employed by couples during sexual intercourse to prevent the conception of a child.

control rods. Part of the core of a nuclear reactor; the rods of neutron-absorbing material that are inserted or removed as necessary to control the rate of nuclear fissioning.

convection. The vertical movement of air due to heating and cooling atmospheric events.

Convention on Biological Diversity. The Biodiversity Treaty signed by 158 nations at the Earth Summit in Rio de Janeiro in 1992 calling for various actions and cooperative steps between nations to protect the world's biodiversity.

conveyor system. The giant pattern of oceanic currents that moves water masses from the surface to the depths and back again, producing major effects on the climate.

cooling tower. A massive tower designed to dissipate waste heat from a power plant (or other industrial process) into the atmosphere.

corrosion. In a nuclear power plant, the deterioration of pipes receiving hot pressurized water in the circulation system that conveys heat from one part of the unit to another.

cosmetic spraying. Spraying of pesticides to control pests that damage only the surface appearance of fruits and vegetables.

cost-benefit analysis. An analysis and/or comparison of the value benefits in contrast to the costs of any particular action or project. (also see **benefit-cost analysis**.)

cost-benefit ratio/benefit-cost ratio. The value of the benefits to be gained from a project divided by the costs of the project. If the ratio is greater than 1, the project is economically justified; if the ratio is less than 1, the project is not economically justified.

cost-effective. Pertaining to a project or procedure that produces economic returns or benefits that are significantly greater than the costs.

covalent bond. A chemical bond between two atoms, formed by sharing a pair of electrons between the two atoms. Atoms of all organic compounds are joined by covalent bonds.

credit associations. Associated with microlending. Groups of poor people with no collateral to assure loans forming an association to assure each other's loans.

criteria pollutants. Certain pollutants the level of which is used as a gauge for the determination of air (or water) quality.

critical level. The level of one or more pollutants above which severe damage begins to occur and below which few if any ill effects are noted.

critical number. The minimum number of individuals of a given species that is required to maintain a healthy, viable population of the species. If a population falls below its critical number, its extinction will almost certainly occur.

crop rotation. The practice of alternating the crops grown on a piece of land. For example, corn one year, hay for two years, then back to corn. (Contrast **monocropping**)

crude birth rate. Number of births per 1000 individuals per year.

crude death rate. Number of deaths per 1000 individuals per year.

cultivar. A cultivated variety of a plant species. All individuals of the cultivar are genetically highly uniform.

cultural control. A change in the practice of growing, harvesting, storing, handling, or disposing of wastes that reduces the susceptibility or exposure to pests. For example, spraying the house with insecticides to kill flies is a chemical control; putting screens on the windows to keep flies out is a cultural control.

cultural eutrophication. The process of natural eutrophication accelerated by human activities. (See **eutrophication**.)

DDT (dichlorodiphenyltrichloroethane). The first and most widely used of the synthetic organic pesticides belonging to the chlorinated hydrocarbon class.

debt crisis. Refers to the fact that many less-developed nations are so heavily in debt that they may not be able to meet their financial obligations, e.g., interest payments. Their failure to meet such obligations could have severe economic impacts on the entire world.

declining tax base. The loss of tax revenues that occurs when affluent taxpayers and businesses leave an area and property values subsequently decline. Also referred to as eroding tax base.

decommissioning. Refers to the inevitable need to take nuclear power plants out of service after 25–35 years because the effects of radiation will gradually make them inoperable.

decomposers. Organisms whose feeding action results in decay or rotting of organic material. The primary decomposers are fungi and bacteria.

deep-well injection. A technique used for the disposal of liquid chemical wastes that involves putting them into deep dry wells where they permeate dry strata.

deforestation. The process of removing trees and other vegetation covering the soil, leading to erosion and loss of soil fertility.

Delaney clause. A provision of the Federal Food, Drug and Cosmetic Act of 1958 that prohibited the use of any food additive that induced cancer in people or animals.

demographic transition. The transition from a condition of high birth rate and high death rate through a period of declining death rate but continuing high birth rate finally to low birth rate and low death rate. This transition may result from economic development.

demography/demographer. The studies of population trends (growth, movement, development, and so on). People who perform such studies and make projections from them.

denitrification. The process of reducing oxidized nitrogen compounds present in soil or water back to nitrogen gas in the atmosphere. It is a natural process conducted by certain bacteria (see text discussion of the nitrogen cycle), and it is now utilized in the treatment of sewage effluents.

density-dependent. Refers to population balancing, factors such as parasitism that increase and decrease in intensity corresponding to population density.

Department of Transportation regulations (DOT Regs). Regulations intended to reduce the risk of spills, fires, and poisonous fumes by specifying the kinds of containers and methods of packing to be used in transporting hazardous materials.

deregulation. In the electric power industry, a policy that requires utility companies to divest themselves of the power generating facilities and allows consumers greater choice.

desalinization. Processes that purify seawater into high-quality drinking water via distillation or microfiltration.

desertification. Declining land productivity caused by mismanagement. Overgrazing and overcultivation allowing erosion and salinization are the major causes.

desertified. Land for which productivity has been significantly reduced (25% or more) because of human mismanagement. Erosion is the most common cause.

desert pavement. A covering of stones and coarse sand protecting desert soils from further wind erosion. The covering results from the differential erosion of finer material.

detritus. The dead organic matter, such as fallen leaves, twigs, and other plant and animal wastes, that exists in any ecosystem.

detritus feeders. Organisms such as termites, fungi, and bacteria that obtain their nutrients and energy mainly by feeding on dead organic matter.

deuterium (^{2}H). A stable, naturally occurring isotope of hydrogen. It contains one neutron in addition to the single proton normally in the nucleus.

developed countries. Industrialized countries—United States, Canada, Western European nations, Japan, Australia, and New Zealand—in which the gross domestic product exceeds $8200 per capita.

developing countries. All free-market countries in which the gross domestic product is less than $8200 per capita. Includes nations of Latin America, Africa, and Asia, except Japan.

differential reproduction. Refers to the fact that within a population, certain individuals reproduce more successfully than others.

dioxin. A synthetic organic chemical of the chlorinated hydrocarbon class. It is one of the most toxic compounds known to humans, having many harmful effects, including induction of cancer and birth defects, even in extremely minute concentrations. It has become a widespread environmental pollutant because of the use of certain herbicides that contain dioxin as a contaminant.

discharge permit. (Technically called NPDES permit.) A permit that allows a company to legally discharge certain amounts or levels of pollutants into air or water.

discount rate. In economics, a rate applied to some future benefit or cost in order to calculate its present real value.

disinfection. The killing (as opposed to removal) of microorganisms in water or other media where they might otherwise pose a health threat. For example, chlorine is commonly used to disinfect water supplies.

dissolved oxygen (DO). Oxygen gas molecules (O_2) dissolved in water. Fish and other aquatic organisms depend on dissolved oxygen for respiration. Therefore, concentration of dissolved oxygen is a measure of water quality.

distillation. A process of purifying water or other liquids by boiling the liquid and recondensing the vapor. Contaminants remain behind in the boiler.

distributive justice. The ethical process of making certain that all parties receive equal rights regardless of their economic status.

disturbance. A natural or human-induced event or process that interrupts ecological succession and creates new conditions on a site (for example, forest fire).

DNA (deoxyribonucleic acid). The natural organic macromolecule that carries the genetic or hereditary information for virtually all organisms.

dose. A consideration of the concentration of a hazardous material times the length of exposure to it. For any given material or radiation, effects correspond to the product of these two factors.

doubling time. The time it will take a population to double in size, assuming the continuation of current rate of growth.

drift-netting. The practice of harvesting marine fish and squid by laying down miles of gill nets across the open seas. The nets collapse around larger organisms and kill many whales, dolphins, seals, marine birds, and turtles.

drip irrigation. Supplying irrigation water through tubes that literally drip water onto the soil at the base of each plant.

drought. A local or regional lack of precipitation such that the ability to raise crops and water animals is seriously impaired.

earth-sheltered housing. Making use of the insulating properties of earth materials to site a house that is also oriented for passive solar heating; often involves earth berms built up against the building walls.

easement. In reference to land protection, an arrangement whereby a landowner gives up development rights into the future but retains ownership.

ecdysone. The hormone that promotes molting in insects.

ecological economist. An economist who thoroughly integrates ecological and economic concerns; part of a new breed of economist who disagrees with classical economic theory.

ecological pest management. Control of pest populations through understanding the various ecological factors that provide natural control and, so far as possible, utilizing these factors as opposed to using synthetic chemicals.

ecological restoration. See **restoration ecology**.

ecological succession. Process of gradual and orderly progression from one ecological community to another.

ecologists. Scientists who study ecology, i.e., the ways in which organisms interact with each other and with their environment.

ecology. The study of any and all aspects of how organisms interact with each other and with their environment.

economic exclusion. The cutting of access of certain ethnic or economic groups to jobs, quality education, and other opportunities and thus preventing them from entering the economic mainstream of society—a condition that prevails in poor areas of cities.

economic threshold. The level of pest damage that, to be reduced further, would require an application of pesticides that is more costly than the economic damage caused by the pests.

ecosystem. A grouping of plants, animals, and other organisms interacting with each other and with their environment in such a way as to perpetuate the grouping more or less indefinitely. Ecosystems have characteristic forms such as deserts, grasslands, tundra, deciduous forests, and tropical rain forests.

ecosystem management.   The management paradigm adopted by all federal agencies managing public lands, involving a long-term stewardship approach to maintaining the lands in their natural state.

ecotone.   A transitional region between two ecosystems that contains some of the species and characteristics of the two adjacent ecosystems and also certain species characteristic of the transitional region.

ecotourism.   The enterprises involved in promoting tourism of unusual or interesting ecological sites.

El Niño.   A major climatic phenomenon characterized by the movement of unusually warm surface water into the eastern equatorial Pacific Ocean. It results in extensive disruption of weather around the world.

electrolysis.   The use of electrical energy to split water molecules into their constituent hydrogen and oxygen atoms. Hydrogen gas and oxygen gas result.

electrons.   Fundamental atomic particles that have a negative electrical charge but virtually no mass. They surround the nuclei of atoms and thus balance the positive charge of protons in the nucleus. A flow of electrons in a wire is synonymous with an electrical current.

element.   A substance that is made up of one and only one distinct kind of atom. (Contrast **compound.**)

embrittlement.   Becoming brittle. Pertains especially to the reactor vessel of nuclear power plants gradually becoming prone to breakage or snapping as a result of continuous bombardment by radiation. It is the prime factor forcing the decommissioning of nuclear power plants.

emergent vegetation.   Aquatic plants whose lower parts are under water but whose upper parts emerge from the water.

emission allowance/standards.   See **discharge permit.**

endangered species.   A species of which the total population is declining to relatively low levels, a trend that if continued will result in extinction.

Endangered Species Act (ESA).   The federal legislation that mandates protection of species and their habitats that are determined to be in danger of extinction.

endocrine disrupters.   Any of a class of organic compounds, often pesticides, that are suspected of having the capacity of interfering with hormone activities in wild organisms.

energy.   The ability to do work. Common forms of energy are light, heat, electricity, motion, and chemical bond energy inherent in compounds such as sugar, gasoline, and other fuels.

energy flow.   The movement of energy through ecosystems starting from the capture of solar energy by primary producers and ending with the loss of heat energy.

enrichment.   With reference to nuclear power, it signifies the separation and concentration of uranium-235 so that, in suitable quantities, it will sustain a chain reaction.

entomologist.   A scientist who studies insects, their life cycles, physiology, behavior, and so on.

entropy.   Refers to the degree of disorder: increasing entropy means increasing disorder.

environment.   The combination of all things and factors external to the individual or population of organisms in question.

environmental accounting.   A process of keeping national accounts of economic activity that includes gains and losses of environmental assets.

environmental impact.   Effects on the natural environment caused by human actions. Includes indirect effects through pollution, for example, as well as direct effects such as cutting down trees.

environmental impact statement.   A study of the probable environmental impacts of a development project. The National Environmental Policy Act of 1968 (NEPA) requires such studies prior to proceeding with any project receiving federal funding.

environmental justice.   The fair treatment and meaningful involvement of all people regardless of race, color, national origin, or income with respect to the development, implementation, and enforcement of environmental laws and regulations.

environmental movement.   Refers to the upwelling of public awareness and citizen action regarding environmental issues that began during the 1960s.

environmental resistance.   The totality of factors such as adverse weather conditions, shortage of food or water, predators, and diseases that tend to cut back populations and keep them from growing or spreading. (Contrast **biotic potential.**)

Environmental Revolution.   In the view of some, a coming change in the adaptation of humans to the rising deterioration of the environment that brings about sustainable interactions with the environment.

environmental science.   The branch of science concerned with environmental issues.

environmentalism.   An attitude or movement involving concern about pollution, resource depletion, population pressures, biodiversity loss, and other environmental issues. Usually implies involvement in action to address those concerns.

environmentalist.   Any person who is concerned about the degradation of the natural world and is willing to act on that concern.

EPA (U.S. Environmental Protection Agency.)   The federal agency responsible for control of all forms of pollution and other kinds of environmental degradation.

epidemiology, epidemiological study.   Determination of causes of disease (e.g., lung cancer) through the study and comparison of large populations of people living in different locations or following different lifestyles and/or habits (e.g., smoking versus nonsmoking).

epidemiologic transition.   The pattern of change in mortality from high death rates to low death rates; contributes to the demographic transition.

epiphytes.   Air plants that are not parasitic but "perch" on tree branches, where they can get adequate light.

equilibrium theory.   The theory that ecosystems are systems maintained over time by natural checks and balances; it is challenged by many ecologists.

erosion.   The process of soil particles' being carried away by wind or water. Erosion moves the smaller soil particles first and hence degrades the soil to a coarser, sandier, stonier texture.

estimated reserves.   See **reserves.**

estuary.   A bay and/or river system open to the ocean at one end and receiving fresh water at the other. Hence, mixing of fresh and salt water occurs (brackish).

ETS (environmental tobacco smoke).   "Secondhand" tobacco smoke to which nonsmokers are exposed when in the presence of smokers.

euphotic zone.   In aquatic systems, the layer or depth of water through which there is adequate light penetration to support photosynthesis.

eutrophic.   Refers to a body of water characterized by nutrient-rich water supporting abundant growth of algae and/or other aquatic plants at the surface. Deep water has little or no dissolved oxygen.

eutrophication.   The process of becoming eutrophic.

evaporation.   Molecules leaving the liquid state and entering the vapor or gaseous state as, for example, water evaporates to form water vapor. Opposite: **condensation**.

evapotranspiration.   The combination of evaporation and transpiration.

evolution.   The theory that all species now on Earth descended from ancestral species through a process of gradual change brought about by natural selection.

evolutionary succession. The succession of different species that have inhabited Earth at different geological periods, as revealed through the fossil record. The process whereby new species come in through the process of speciation while other species pass into extinction.

experimental group. The group in an experiment that receives the experimental treatment in contrast to the control group, used for comparison, which does not receive the treatment. Synonym: **test group.**

exotic species. A species introduced to a geographical area where it is not native.

exponential increase. The growth produced when the base population increases by a given percentage (as opposed to a given amount) each year. It is characterized by doubling again and again, each doubling occurring in the same period of time. It produces a J-shaped curve.

externality/external cost. Any effect of a business process not included in the usual calculations of profit and loss. Pollution of air or water is an example of a *negative* externality—one that imposes a cost on society that is not paid for by the business itself.

extinction. The death of all individuals of a particular species. When this occurs, all the genes of that particular line are lost forever.

extractive reserves. As now established in Brazil, forest lands that are protected for native peoples and rubber tappers who harvest natural products of the forests, such as latex and Brazil nuts.

exurban migration. Refers to the pronounced trend since World War II of relocating homes and businesses from the central city and older suburbs to more outlying suburbs.

exurbs. New developments beyond the traditional suburbs but from which most residents still commute to the associated city for work.

family planning. Making contraceptives and other health and reproductive services available to couples in order to enable them to achieve a desired family size.

famine. A severe shortage of food accompanied by a significant increase in the local or regional death rate.

FAO. Food and Agriculture Organization of the United Nations.

farm cooperatives. An association of consumers who jointly own and manage a farm for the production of produce specifically for their own consumption.

Farmer-centered Agricultural Resource Management (FARM) Program. A U.N. FAO-administered program in Asian countries fostering sustainable agriculture.

FDA. Federal Food and Drug Administration, with jurisdiction over foods, drugs, and all products that come in contact with the skin or are ingested.

fecal coliform test. A test for the presence of *Escherichia coli*, the bacterium that normally inhabits the gut of humans and other mammals. A positive test indicates sewage contamination and the potential presence of disease-causing microorganisms carried by sewage.

Federal Agricultural Improvement and Reform (FAIR) Act of 1996. Major legislation removing many subsidies and controls from farming.

fermentation. A form of respiration carried on by yeast cells in the absence of oxygen. It involves a partial breakdown of glucose (sugar) that yields energy for the yeast and the release of alcohol as a byproduct.

fertility rate. See **total fertility rate.**

fertility transition. The pattern of change in birth rates in a society from high rates to low; a major component of the demographic transition.

fertilizer. Material applied to plants or soil to supply plant nutrients, most commonly nitrogen, phosphorus, and potassium but may include others. *Organic* fertilizer is natural organic material such as manure, which releases nutrients as it breaks down. *Inorganic* fertilizer, also called chemical fertilizer, is a mixture of one or more necessary nutrients in inorganic chemical form.

field scouts. Persons trained to survey crop fields and determine whether applications of pesticides or other pest-management procedures are actually necessary to avert significant economic loss.

FIFRA (Federal Insecticide, Fungicide, and Rodenticide Act.) The key U.S. legislation to control pesticides.

fire climax ecosystems. Ecosystems that depend on the recurrence of fire to maintain the existing balance.

first principle of ecosystem sustainability. Ecosystems use sunlight as their source of energy.

first-generation pesticides. Toxic inorganic chemicals that were first used to control insects, plant diseases, and other pests. Included mostly compounds of arsenic and cyanide and various heavy metals such as mercury and copper.

first law of thermodynamics. The fact based on irrefutable observations that energy is never created or destroyed but may be converted from one form to another, e.g., electricity to light. Also called the **law of conservation of energy.** (See also **second law of thermodynamics.**)

fishery. Fish species being exploited, or a limited marine area containing commercially valuable fish.

fish ladder. A stepwise series of pools on the side of a dam where the water flows in small falls that fish can negotiate.

fission. The splitting of a large atom into two atoms of lighter elements. When large atoms such as uranium or plutonium fission, tremendous amounts of energy are released.

fission products. Any and all atoms and subatomic particles resulting from splitting atoms in nuclear reactors. All or most such products are highly radioactive.

fitness. Features of an organism that adapt it to survive and reproduce in the environment.

flat-plate collector. A solar collector that consists of a stationary, flat, black surface oriented perpendicular to the average sun angle. Heat absorbed by the surface is removed and transported by air or water (or other liquid) flowing over or through the surface.

flood irrigation. Technique of irrigation in which water is diverted from rivers through canals and flooded through furrows in fields.

food aid. Food of various forms that is donated or sold below cost to needy people for humanitarian reasons.

food chain. The transfer of energy and material through a series of organisms as each one is fed upon by the next.

food guide pyramid. A graphic presentation of six basic food needs arranged in a pyramid to indicate the relative proportions of each food type needed for good nutrition.

food security. For families, the ability to meet the food needs of everyone in the family, providing freedom from hunger and malnutrition.

food web. The combination of all the feeding relationships that exist in an ecosystem.

fossil fuels. Energy sources, mainly crude oil, coal, and natural gas, that are derived from prehistoric photosynthetic production of organic matter on Earth.

fourth principle of ecosystem sustainability. Biodiversity is maintained.

FQPA (Food Quality Protection Act of 1996.) Legislation that amended the Delaney clause and replaced many provisions of FIFRA.

fragmentation. The division of a landscape into patches of habitat by road construction, agricultural lands, or residential areas.

Framework Convention on Climate Change.   A result of the 1992 Earth Summit, this international treaty was a start in negotiating agreements on steps to prevent future catastrophic climate change.

free market economy.   In its purest form, an economy where the market itself determines what goods will be exchanged and how. The system is wholly in private hands.

fresh water.   Water that has a salt content of less than 0.1% (1000 parts per million).

front.   The boundary where different air masses meet.

fuel assembly.   The assembly of many rods containing the nuclear fuel, usually uranium, positioned close together. The chain reaction generated in the fuel assembly is controlled by rods of neutron-absorbing material between the fuel rods.

fuel elements.   The pellets of uranium or other fissionable material that are placed in tubes, which, with the control rods, form the core of the nuclear reactor.

fuel rods.   See **fuel elements**.

fuelwood.   The use of wood for cooking and heating, the most common energy source for people in developing countries.

fungus, pl. fungi.   Any of numerous species of molds, mushrooms, brackets, and other forms of nonphotosynthetic plants. They derive energy and nutrients by consuming other organic material. Along with bacteria, they form the decomposer component of ecosystems.

fusion.   The joining together of two atoms to form a single atom of a heavier element. When light atoms such as hydrogen are fused, tremendous amounts of energy are released.

gasohol.   A blend of 90% gasoline and 10% alcohol, which can be substituted for straight gasoline. It serves to stretch gasoline supplies.

gene.   Segment of DNA that codes for one protein, which in turn determines a particular physical, physiological, or behavioral trait.

gene pool.   The sum total of all the genes that exist among all the individuals of a species.

genetic bank.   The concept that natural ecosystems with all their species serve as a tremendous repository of genes that is frequently drawn upon to improve domestic plants and animals and to develop new medicines, among other uses.

genetic control.   Selective breeding of the desired plant or animal to make it resistant to attack by pests. Also, attempting to introduce harmful genes—for example, those that cause sterility—into the pest populations.

genetic engineering.   The artificial transfer of specific genes from one organism to another.

genetics.   The study of heredity and the processes by which inherited characteristics are passed from one generation to the next.

genetic variation.   An expression of the range of genetic (DNA) differences that occur among individuals of the same species.

geothermal.   Refers to the naturally hot interior of Earth. The heat is maintained by naturally occurring nuclear reactions in Earth's interior.

geothermal energy.   Useful energy derived from the naturally hot interior of Earth.

global climate change.   The term given to the various impacts of rising levels of greenhouse gases on Earth's climate. These effects include global warming, weather changes, and a rising sea level.

glucose.   A simple sugar, the major product of photosynthesis. Serves as the basic building block for cellulose and starches and as the major "fuel" for the release of energy through cell respiration in both plants and animals.

grassroots movement.   An environmental movement that begins at the level of the populace, often in response to a perceived problem not being addressed by public policy.

gravitational water.   Water that is not held by capillary action in soil but percolates downward by the force of gravity.

graying.   The increasing average age in populations in developed and many developing countries that is occurring because of decreasing birth rates and increasing longevity.

gray water.   Wastewater, as from sinks and tubs, that does not contain human excrements. Such water can be reused without purification for some purposes.

greenhouse effect.   An increase in the atmospheric temperature caused by increasing amounts of carbon dioxide and certain other gases that absorb and trap heat radiation, which normally escapes from Earth.

greenhouse gases.   Gases in the atmosphere that absorb infrared energy and contribute to the air temperature. These gases are like a heat blanket and are important in insulating Earth's surface. They include carbon dioxide, water vapor, methane, nitrous oxide, and chlorofluorocarbons and other halocarbons.

green fee.   An added cost for a particular service or product that has the effect of internalizing the external costs and reflects the real environmental cost of the activity.

green manure.   A legume crop such as clover that is specifically grown to enrich the nitrogen and organic content of soil.

green revolution.   Refers to the development and introduction of new varieties of wheat and rice (mainly) that increased yields per acre dramatically in many countries.

green products.   Products that are more environmentally benign than their traditional counterparts.

grit chamber.   Part of preliminary treatment in wastewater-treatment plants; a swimming pool–like tank in which the velocity of the water is slowed enough to let sand and other gritty material settle.

grit-settling tank.   See **grit chamber**.

gross domestic (national) product per capita.   The total value of all goods and services exchanged in a year in a country, divided by its population. A common indicator for the average level of development and standard of living for a country.

gross primary production.   The rate at which primary producers in ecosystems are fixing organic matter.

groundwater.   Water that has accumulated in the ground, completely filling and saturating all pores and spaces in rock and/or soil. Groundwater is free to move more or less readily. It is the reservoir for springs and wells and is replenished by infiltration of surface water.

groundwater remediation.   The repurification of contaminated groundwater by any of a number of techniques.

gully erosion.   Gullies, large or small, resulting from water erosion.

habitat.   The specific environment (woods, desert, swamp) in which an organism lives.

Habitat Conservation Plan (HCP).   Under the Endangered Species Act, a plan that may be drafted by the U.S. Fish and Wildlife Service in working with landowners to help mitigate conflicts over the Act.

Hadley cell.   A system of vertical and horizontal air circulation predominating in tropical and subtropical regions and creating major weather patterns.

half-life.   The length of time it takes for half of an unstable isotope to decay. The length of time is the same regardless of the starting amount. Also refers to the amount of time it takes compounds to break down in the environment.

halogenated hydrocarbon.   Synthetic organic compound containing one or more atoms of the halogen group, which includes chlorine, fluorine, and bromine.

hard water.   Water that contains relatively large amounts of calcium and/or certain other minerals that cause soap to precipitate. (Contrast **soft water**.)

hazard. Anything that can cause (1) injury, disease, or death to humans; (2) damage to property; or (3) degradation of the environment. *Cultural* hazards refer to factors that are often a matter of choice, like smoking or sunbathing. *Biological* hazards are pathogens and parasites that infect humans. *Physical* hazards are natural disasters like earthquakes and tornadoes. *Chemical* hazards refer to the chemicals in use in different technologies and household products.

hazard assessment. The process of examining evidence linking a particular hazard to its harmful effects.

hazardous materials (HAZMAT). Any material having one or more of the following attributes: ignitability, corrosivity, reactivity, toxicity.

heavy metals. Any of the high atomic weight metals, such as lead, mercury, cadmium, and zinc. All may be serious pollutants in water or soil because they are toxic in relatively low concentrations and they tend to bioaccumulate.

herbicide. A chemical used to kill or inhibit the growth of undesired plants.

herbivore/herbivorous. An organism such as rabbit or deer that feeds primarily on green plants or plant products such as seeds or nuts. Such an organism is said to be herbivorous. Synonym: **primary consumer.**

herbivory. The feeding on plants that occurs in an ecosystem. The total feeding of all plant-eating organisms.

heterotroph/heterotrophic. Any organism that consumes organic matter as a source of energy. Such an organism is said to be heterotrophic.

HHW (household hazardous wastes). An important component of the hazardous waste problem in the United States.

Highway Trust Fund. The monies collected from the gasoline tax designated for construction of new highways.

hormones. Natural chemical substances that control development, physiology, and/or behavior of an organism. Hormones are produced internally and affect only that individual. Hormones are coming into use in pest control. (See also **pheromones.**)

host. In feeding relationships, particularly parasitism, refers to the organism that is being fed upon, i.e., supporting the feeder.

host-parasite relationship. The combination of a parasite and the organism upon which it feeds.

host-specific. Referring to insects, fungal diseases, and other parasites that are unable to attack species other than their particular host.

human resources. One component of the wealth of nations. Comprises the *human capital*, or the population and its attributes; and the *social capital*, or the social and political environment that people have created in a society.

human system. The entire system that humans have created for their own support, consisting of agriculture, industry, transportation, and communications networks, etc.

humidity. The amount of water vapor in the air. (See also **relative humidity.**)

humus. A dark brown or black, soft, spongy residue of organic matter that remains after the bulk of dead leaves, wood, or other organic matter has decomposed. Humus does oxidize, but relatively slowly. It is extremely valuable in enhancing physical and chemical properties of soil.

hunger. A general term referring to the lack of basic food required for meeting nutritional and energy needs, such that the individual is unable to lead a normal, healthy life.

hunter-gatherers. Humans surviving by hunting wild game and gathering seeds, nuts, berries, and other edible things from the natural environment.

hybrid. A plant or animal resulting from a cross between two closely related species that do not normally cross.

hybrid electric vehicle (HEV). An automobile with a small gasoline motor and batteries capable of getting over 60 mpg and producing only 1/10 the pollution of a comparable gasoline car.

hybridization. Cross-mating between two more or less closely related species.

hydrocarbon emissions. Exhaust of various hydrogen-carbon compounds due to incomplete combustion of fuel. They are a major contribution to photochemical smog.

hydrocarbons. *Chemistry:* Natural or synthetic organic substances that are composed mainly of carbon and hydrogen. Crude oil, fuels from crude oil, coal, animal fats, and vegetable oils are examples. *Pollution:* A wide variety of relatively small carbon-hydrogen molecules resulting from incomplete burning of fuel and emitted into the atmosphere. (See **volatile organic compounds.**)

hydroelectric dam. A dam and associated reservoir used to produce electrical power by letting the high-pressure water behind the dam flow through and drive a turbogenerator.

hydroelectric power. Electrical power that is produced from hydroelectric dams or, in some cases, natural waterfalls.

hydrogen bonding. A weak attractive force that occurs between a hydrogen atom of one molecule and, usually, an oxygen atom of another molecule. It is responsible for holding water molecules together to produce the liquid and solid states.

hydrogen ions. Hydrogen atoms that have lost their electrons. Chemical symbol, H^+.

hydrologic cycle. The movement of water from points of evaporation through the atmosphere, through precipitation, and through or over the ground, returning to points of evaporation.

hydroponics. The culture of plants without soil. The method uses water with the required nutrients in solution.

hydroxyl radical. The hydroxyl group (OH) missing the electron. It is a natural cleansing agent of the atmosphere. It is highly reactive and readily oxidizes many pollutants on contact and thus contributes to their removal.

hypothesis. An educated guess concerning the cause of an observed phenomenon that is then subjected to experimental tests to prove its accuracy or inaccuracy.

ICPD. (International Conference on Population and Development) A U.N.-sponsored conference held in 1994 in Cairo, Egypt, where the Program of Action was drafted and later adopted by the United Nations.

incremental value. Used in calculating the value of natural services performed by ecosystems, it is the value of some finite change in a natural service.

indicator organism. An organism, the presence or absence of which indicates certain conditions. For example, the presence of *Escherichia coli* indicates that water is contaminated with fecal wastes and pathogens may be present; the absence indicates that the water is free of pathogens.

individual transfer quota (ITQ). A system for fishery management where quotas are set and individual fishers are given or sold the right to harvest some proportion of the quota.

industrialized agriculture. Using fertilizer, irrigation, pesticides, and energy from fossil fuels to produce large quantities of crops and livestock with minimal labor for domestic and foreign sale.

industrialized countries. (See **developed countries.**)

Industrial Revolution. During the nineteenth century, the development of manufacturing processes using fossil fuels and based on applications of scientific knowledge.

industrial smog. The grayish mixture of moisture, soot, and sulfurous compounds that occurs in local areas where industries are concentrated and coal is the primary energy source.

infant mortality.   The number of babies that die before age 1, per 1000 babies born.

infiltration.   The process in which water soaks into soil as opposed to running off the surface.

infiltration-runoff ratio.   The ratio of the amount of water soaking into the soil to that running off the surface. The ratio is obtained by dividing the first amount by the second.

infrared radiation.   Radiation of somewhat longer wavelengths than red light, the longest wavelengths of the visible spectrum. Such radiation manifests itself as heat.

infrastructure.   The sewer and water systems, roadways, bridges, and other facilities that underlie the functioning of a city and that are owned, operated, and maintained by the city.

inherently safe reactor.   In theory, a nuclear reactor that is designed in such a way that any accident would be automatically corrected and no radioactivity released.

inorganic compounds/molecules.   *Classical definition:* All things such as air, water, minerals, and metals that are neither living organisms nor products uniquely produced by living things. *Chemical definition:* All chemical compounds or molecules that do not contain carbon atoms as an integral part of their molecular structure. (Contrast **organic compounds.**)

insecticide.   Any chemical used to kill insects.

instrumental value.   Based on the belief that living organisms or species are worthwhile if their existence or use benefits people; the degree to which they benefit humans. (Contrast **intrinsic value.**)

insurance spraying.   Spraying of pesticides when it is not really needed in the belief that it will ensure against loss due to pests.

integrated pest management (IPM).   Two or more methods of pest control carefully integrated into an overall program designed to avoid economic loss from pests. The objective is to minimize the use of environmentally hazardous, synthetic chemicals. Such chemicals may be used in IPM, but only as a last resort to prevent significant economic losses.

integrated waste management.   The approach to municipal solid waste that provides for several options for dealing with wastes, including recycling, composting, waste reduction, and landfilling and incineration where unavoidable.

International Whaling Commission (IWC).   The international body that regulates the harvesting of whales; the IWC placed a ban on all whaling in 1986.

intrinsic value.   Based on the belief that living organisms or species are worthwhile in their own right; they do not have to be useful to have value. (Contrast **instrumental value.**)

inversion.   See **temperature inversion**.

ion.   An atom or group of atoms that has lost or gained one or more electrons and consequently has acquired a positive or negative charge. Ions are designated by "+" or "−" superscripts following the chemical symbol.

ion-exchange capacity.   See **nutrient-holding capacity.**

ionic bond.   The bond formed by the attraction between a positive and a negative ion.

IPCC (Intergovernmental Panel on Climate Change).   The U.N.-sponsored organization delegated to the continual production of an assessment of the science of global climate change and the potential impacts and means of response to the threat.

IPM.   See **integrated pest management**.

irrigation.   Any method of artificially adding water to crops.

isotope.   A form of an element in which the atoms have more (or less) than the usual number of neutrons. Isotopes of a given element have identical chemical properties, but they differ in mass (weight) as a result of the additional (or lesser) neutrons. Many isotopes are unstable and radioactive. (See **radioactive decay, radioactive emissions,** and **radioactive materials.**)

ISTEA (Intermodal Surface Transportation Efficiency Act).   Legislation that provides for funding of alternative transportation (mass transit, cycling paths) using money from the Highway Trust Fund.

junk science.   Information presented as valid science but unsupported by peer-reviewed research. Often, politically motivated and biased results are selected to promote a particular point of view.

juvenile hormone.   The insect hormone that at sufficient levels preserves the larval state. Pupation requires diminished levels; hence, artificial applications of the hormone may block development.

keystone species.   A species whose role is essential for the survival of many other species in an ecosystem.

kinetic energy.   The energy inherent in motion or movement, including molecular movement (heat) and movement of waves, hence radiation including light.

Kyoto Protocol.   An international agreement among the developed nations to agree to curb greenhouse gas emissions, forged in December 1997.

Lacey Act.   The first national act, passed in 1900, that gave protection to wildlife by forbidding interstate commerce in illegally killed wildlife.

landfill.   A site where wastes (municipal, industrial, or chemical) are disposed of by burying them in the ground or placing them on the ground and covering them with earth. Also used as a verb meaning to dispose of a material in such a way.

landscape.   A group of interacting ecosystems occupying adjacent geographical areas.

land subsidence.   The phenomenon whereby land gradually sinks. It may result from removing groundwater or oil, which is frequently instrumental in supporting the overlying rock and soil.

land trust.   Land that is purchased and held by various private organizations specifically for the purpose of protecting its natural environment and biota that inhabit it.

larva, pl. **larvae.**   A free-living immature form that occurs in the life cycle of many organisms and that is structurally distinct from the adult. For example, caterpillars are the larval stage of moths and butterflies.

law of conservation of energy.   See **first law of thermodynamics.**

law of conservation of matter.   Law stating that in chemical reactions, atoms are neither created, changed, nor destroyed; they are only rearranged.

law of limiting factors.   Also known as Liebig's law of minimums. A system may be limited by the absence or minimum amount (in terms of that needed) of any required factor. (See **limiting factor**.)

leachate.   The mixture of water and materials that are leaching.

leaching.   The process in which materials in or on the soil gradually dissolve and are carried by water seeping through the soil. It may result in the removal of valuable nutrients from the soil, or it may carry buried wastes into groundwater, thereby contaminating it.

legumes.   The group of pod-bearing land plants that is virtually alone in its ability to fix nitrogen; includes such common plants as peas, beans, clovers, alfalfa, and locust trees but no major cereal grains. (See **nitrogen fixation**.)

Liebig's law of minimums.   See **law of limiting factors.**

lifeboat ethic.   An argument that suggests we should limit the amount of food aid to high-population countries in order to prevent further population growth.

limiting factor.   A factor primarily responsible for determining the growth and/or reproduction of an organism or a population. The limiting factor may be a physical factor such as temperature or light, a chemical factor such as a particular nutrient, or a biological factor such as a competing species. The limiting factor may differ at different times and places.

limits of tolerance. The extremes of any factor, e.g., temperature, that an organism or a population can tolerate and still survive and reproduce.

lipids. A class of natural organic molecules that includes animal fats, vegetable oils, and phospholipids, the last being an integral part of cellular membranes.

litter. In an ecosystem, the natural cover of dead leaves, twigs, and other dead plant material. This natural litter is subject to rapid decomposition and recycling in the ecosystem, whereas human litter, such as bottles, cans, and plastics, is not.

livability. A subjective index of how viable a city is for enjoyable living.

loam. A solid consisting of a mixture of about 40% sand, 40% silt, and 20% clay.

longevity. The average life span of individuals of a given population.

LULU. ("locally unwanted land use"). Expresses the difficulty in siting a facility that is necessary but which no one wants in the immediate locality.

macromolecules. Very large, organic molecules such as proteins and nucleic acids that constitute the structural and functional parts of cells.

MACT (maximum achievable control technology). The best technologies available for reducing the output of especially toxic industrial pollutants.

Magnuson Act. An act passed in 1976 which extended the limits of jurisdiction over coastal waters and fisheries of the United States to 200 miles offshore.

malnutrition. The lack of essential nutrients such as vitamins, minerals, and amino acids. Malnutrition ranges from mild to severe and life-threatening.

mass number. The number that accompanies the chemical name or symbol of an element or isotope. It represents the number of neutrons and protons in the nucleus of the atom.

material safety data sheets (MSDS). Documents containing information on the reactivity and toxicity of chemicals that must accompany the shipping, storage, and handling of over 600 chemicals.

materials recycling facility (MRF). A processing plant where regionalized recycling is facilitated. Recyclable municipal solid waste, usually presorted, is prepared in bulk for the recycling market.

matter. Anything that occupies space and has mass. Refers to any gas, liquid, or solid. (Contrast **energy**.)

maximum sustainable yield (MSY). The maximum amount of a renewable resource that can be taken year after year without depleting the resource. It is the maximum rate of use or harvest that will be balanced by the regenerative capacity of the system—for example, the maximum rate of tree cutting that can be balanced by tree regrowth.

meltdown. The event of a nuclear reactor's getting out of control or losing its cooling water so that it melts from its own production of heat. The melted reactor would continue to produce heat and could melt its way out of the reactor vessel and eventually down into groundwater, where it would cause a violent eruption of steam that could spread radioactive materials over a wide area.

metabolism. The sum of all the chemical reactions that occur in an organism.

methane. A gas, CH_4. It is the primary constituent of natural gas. It is also produced as a product of fermentation by microbes. Methane from ruminant animals is thought to be responsible for the rise in atmospheric methane, of concern because methane is one of the greenhouse gases.

microbe. A term used to refer to any microscopic organism, primarily bacteria, viruses, and protozoans.

microclimate. The actual conditions experienced by an organism in its particular location. Owing to numerous factors such as shading, drainage, and sheltering, the microclimate may be quite distinct from the overall climate.

microfiltration. A process for purifying water in which water is forced under very high pressure through a membrane that is fine enough to filter out ions and molecules in solution; used by small desalination plants to filter salt from seawater. Also called reverse osmosis.

microlending. The process of providing very small loans (usually $50–$500) to poor people to facilitate their starting a small enterprise to become economically self-sufficient.

microorganism. Any microscopic organism, particularly bacteria, viruses, and protozoans.

midnight dumping. The wanton illicit dumping of materials, particularly hazardous wastes, frequently under the cover of darkness.

Milankovitch cycle. Major oscillations in the Earth's orbit taking place over frequencies of thousands of years, known to influence the distribution of solar radiation and therefore the global weather patterns.

Minamata disease. A "disease" named for a fishing village in Japan where an "epidemic" was first observed. Symptoms, which included spastic movements, mental retardation, coma, death, and crippling birth defects in the next generation, were found to be the result of mercury poisoning.

mineral. Any hard, brittle, stonelike material that occurs naturally in Earth's crust. All consist of various combinations of positive and negative ions held together by ionic bonds. Pure minerals, or crystals, are one specific combination of elements. Common rocks are composed of mixtures of two or more minerals.

mineralization. The process of gradual oxidation of the organic matter (humus) present in soil that leaves just the gritty mineral component of the soil.

mixture. Forms when there is no chemical bonding between the molecules of the elements involved. For example, air contains (is a mixture of) oxygen, nitrogen, and carbon dioxide.

mobilization. In soil science, the bringing into solution of normally insoluble minerals. Presents a particular problem when the elements of such minerals have toxic effects.

moderator. In a nuclear reactor, the moderator is any material that slows down neutrons from fission reactions so that they are traveling at the right speed to trigger another fission. Water and graphite represent two types of moderators.

molecule. A specific union of two or more atoms. The smallest unit of a compound that still has the characteristics of that compound.

monoculture/monocropping. The practice of growing the same crop over very wide areas, for example, thousands of square kilometers of wheat, and only wheat, grown in the Midwest.

monsoon. The seasonal air flow created by major differences in cooling and heating between oceans and continents, usually bringing extensive rain.

Montreal Protocol. An agreement made in 1987 by a large group of nations to cut back the production of chlorofluorocarbons by 50% by the year 2000 in order to protect the ozone shield. A 1990 amendment calls for the complete phaseout of these chemicals by 2000 in developed nations and by 2010 in less-developed nations.

morbidity. The incidence of disease in a population.

mortality. The incidence of death in a population.

MRF. See **materials recycling facility**.

municipal solid waste (MSW). The entirety of refuse or trash generated by a residential and business community. The refuse that a municipality is responsible for collecting and disposing of, distinct from agricultural and industrial wastes.

mutagenic. Causing mutations.

mutation.   A random change in one or more genes of an organism. Mutations may occur spontaneously in nature, but their number and degree are vastly increased by exposure to radiation and/or certain chemicals. Mutations generally result in a physical deformity and/or metabolic malfunction.

mutualism.   Refers to a close relationship between two organisms in which both organisms benefit from the relationship.

mycelia.   The threadlike feeding filaments of fungi.

mycorrhiza, pl. mycorrhizae.   The mycelia of certain fungi that grow symbiotically with the roots of some plants and provide for additional nutrient uptake.

NASA.   National Aeronautics and Space Administration.

National Ambient Air Quality Standards (NAAQS).   The allowable levels of ambient air pollutants set by regulation.

National Emission Standards for Hazardous Air Pollutants (NESHAPS).   The standards for allowable emissions of certain toxic substances.

national forests.   Administered by the National Forest Service, these are public forest and woodlands that have been managed for multiple uses, such as logging, mineral exploitation, livestock grazing, and recreation.

national parks.   Administered by the National Park Service, national parks are lands and coastal areas of great scenic, ecological, or historical importance. They are managed with the dual goals of protection and providing public access.

national priorities list (NPL).   A list of the chemical waste sites presenting the most immediate and severe threats. Such sites are scheduled for cleanup ahead of other sites.

natural.   Describes a substance or factor that occurs or is produced as a normal part of nature apart from any activity or intervention of humans. Opposite of artificial, synthetic, human-made, or caused by humans.

natural capital.   The natural assets and the services they perform are referred to as natural capital; one form of the wealth of a nation is its complement of natural capital.

natural chemical control.   The use of one or more natural chemicals such as hormones or pheromones to control a pest.

natural control methods.   Any of many techniques of controlling a pest population without resorting to the use of synthetic organic or inorganic chemicals. (See **biological control, cultural control, genetic control, hormones,** and **pheromones.**)

natural enemies.   All the predators and/or parasites that may feed on a given organism. Organisms used to control a specific pest through predation or parasitism.

natural increase.   The number of births minus the number of deaths in a given population. It does not consider immigration and emigration.

natural laws.   Derivations from our observations that matter, energy, and certain other phenomena apparently always act (or react) according to certain "rules."

natural rate of change.   The percent of growth (or decline) of a given population during a year. It is found by subtracting the crude death rate from the crude birth rate and changing the result to a percent. It does not include immigration or emigration.

natural resources.   As applied to natural ecosystems and species, this term indicates that they are expected to be of economic value and may be exploited. Likewise, the term applies to particular segments of ecosystems such as air, water, soil, and minerals.

natural selection.   The process whereby the natural factors of environmental resistance tend to eliminate those members of a population that are least well adapted to cope and thus, in effect, select those best adapted for survival and reproduction.

natural services.   Functions performed free of charge by natural ecosystems such as control of runoff and erosion, absorption of nutrients, and assimilation of air pollutants.

Neolithic Revolution.   The development of agriculture by human societies around 10,000 years ago, leading to more permanent settlement and population increases.

net primary production.   The rate at which new organic matter is made available to consumers by primary producers (equals gross primary production minus plant respiration).

neutron.   A fundamental atomic particle found in the nuclei of atoms (except hydrogen) and having one unit of atomic mass but no electrical charge.

New Forestry.   A forestry management strategy that is now part of the Forest Service's management practice, placing priority on protecting the ecological health and diversity of forests rather than maximizing the harvest of logs.

niche (ecological).   The total of all the relationships that bear on how an organism copes with both biotic and abiotic factors it faces.

NIMBY. ("Not in my backyard").   A common attitude regarding undesirable facilities such as incinerators, nuclear facilities, and hazardous waste treatment plants, whereby people do everything possible to prevent the location of such facilities nearby.

NIMTOO. ("Not in my term of office").   Refers to the reluctance of office-holders to make unpopular decisions on matters like siting waste facilities.

nitric acid (HNO_3).   One of the acids in acid rain. Formed by reactions between nitrogen oxides and the water vapor in the atmosphere.

nitrogen fixation.   The process of chemically converting nitrogen gas (N_2) from the air into compounds such as nitrates (NO_3^-) or ammonia (NH_3) that can be used by plants in building amino acids and other nitrogen-containing organic molecules.

nitrogen oxides (NO_x).   A group of nitrogen-oxygen compounds formed when some of the nitrogen gas in air combines with oxygen during high-temperature combustion; they are a major category of air pollutants. Along with hydrocarbons, they are a primary factor in the production of ozone and other photochemical oxidants that are the most harmful components of photochemical smog. They also contribute to acid precipitation (see **nitric acid**). Major nitrogen oxides are nitric oxide, NO; nitrogen dioxide, NO_2; and nitrogen tetroxide, N_2O_2.

nitrous oxide.   A gas, N_2O. Nitrous oxide comes from biomass burning, fossil fuel burning, and the use of chemical fertilizers. It is of concern because in the troposphere it is a greenhouse gas and in the stratosphere it contributes to ozone destruction.

NOAA.   National Oceanic and Atmospheric Administration.

nonbiodegradable.   Not able to be consumed and/or broken down by biological organisms. Nonbiodegradable substances include plastics, aluminum, and many chemicals used in industry and agriculture. Particularly dangerous are human-made nonbiodegradable chemicals that are also toxic and tend to accumulate in organisms, i.e., nonbiodegradable synthetic organic compounds. (See **biodegradable** and **bioaccumulation**.)

nonconsumptive water use.   Use of water for such purposes as washing and rinsing where the water, albeit polluted, remains available for further uses. With suitable purification, such water may be recycled indefinitely.

nongovernmental organization (NGO).   Any of a number of private organizations involved in studying and/or advocating action on different environmental issues.

nonpersistent.   Refers to chemicals that break down readily to harmless compounds, as, for example, natural organic compounds break down to carbon dioxide and water.

nonpoint sources.   Sources of pollution such as general runoff of sediments, fertilizer, pesticides, and other materials from farms and urban areas as opposed to specific points of discharge such as factories. Also called diffuse sources. (Contrast **point sources.**)

nonrenewable resources. Resources such as ores of various metals, oil, and coal that exist as finite deposits in Earth's crust and that are not replenished by natural processes as they are mined. (Contrast **renewable resources.**)

no-till agriculture. The farming practice in which weeds are killed with chemicals (or other means) and seeds are planted and grown without resorting to plowing or cultivation. The practice is very effective in reducing soil erosion.

NRCS. Natural Resources Conservation Service, formerly the SCS (U.S. Soil Conservation Service).

nuclear power. Electrical power that is produced by using a nuclear reactor to boil water and produce steam, which, in turn, drives a turbogenerator.

Nuclear Regulatory Commission (NRC). The agency within the Department of Energy that sets and enforces safety standards for the operation and maintenance of nuclear power plants.

nucleic acids. The class of natural organic macromolecules that function in the storage and transfer of genetic information.

nucleus. *Biology:* The large body contained in most living cells that contains the genes or hereditary material, DNA. *Physics:* The central core of atoms, which is made up of neutrons and protons. Electrons surround the nucleus.

nutrient. *Animal:* Material such as protein, vitamins, and minerals required for growth, maintenance, and repair of the body and material such as carbohydrates required for energy. *Plant:* An essential element in a particular ion or molecule that can be absorbed and used by the plant. For example, carbon, hydrogen, nitrogen, and phosphorus are essential elements; carbon dioxide, water, nitrate (NO_3^-), and phosphate (PO_4^{3-}) are the respective nutrients.

nutrient cycle. The repeated pathway of particular nutrients or elements from the environment through one or more organisms back to the environment. Nutrient cycles include the carbon cycle, the nitrogen cycle, the phosphorus cycle, and so on.

nutrient-holding capacity. The capacity of a soil to bind and hold nutrients (fertilizer) against their tendency to be leached from the soil.

observations. Things or phenomena that are perceived through one or more of the basic five senses in their normal state. In addition, to be accepted as factual, the observations must be verifiable by others.

ocean thermal energy conversion (OTEC). The concept of harnessing the temperature difference between surface water heated by the sun and colder deep water to produce power.

oil field. The area in which exploitable oil is found.

oil sand. Sedimentary material containing bitumen, a tar-like hydrocarbon, subject to exploitation under favorable economic conditions.

oil shale. A natural sedimentary rock that contains a material, kerogen, that can be extracted and refined into oil and oil products.

oligotrophic. Refers to a lake in which the water is nutrient-poor. Therefore, it will not support much phytoplankton, but it will support submerged aquatic vegetation, which get nutrients from the bottom.

omnivore. An animal that feeds on both plant material and other animals.

OPEC. Organization of Petroleum Exporting Countries.

optimal range. With respect to any particular factor or combination of factors, the maximum variation that still supports optimal or near-optimal growth of the species in question.

optimum. The condition or amount of any factor or combination of factors that will produce the best result. For example, the amount of heat, light, moisture, nutrients, and so on that will produce the best growth. Either more or less than the optimum is not as good.

organically grown. Generally refers to produce grown without the use of hard chemical pesticides or inorganic fertilizer. However, as of yet there are no official standards defining the use of the term.

organic compounds/molecules. *Classical definition:* All living things and products that are uniquely produced by living things, such as wood, leather, and sugar. *Chemical definition:* All chemical compounds or molecules, natural or synthetic, that contain carbon atoms as an integral part of their molecular structure. Their structure is based on bonded carbon atoms with hydrogen atoms attached. They can be either biodegradable to non-biodegradable. (Contrast **inorganic compounds.**)

organic fertilizer. See **fertilizer.**

organic gardening/farming/food. Gardening or farming without the use of inorganic fertilizers, synthetic pesticides, or other human-made materials; food raised with organic farming techniques.

organic phosphate. Phosphate (PO_4^{-3}) bonded to an organic molecule.

organism. Any living thing—plant, animal, or microbe.

orphan site. The location of a hazardous waste site abandoned by former owners.

OSHA. (Occupational Safety and Health Administration). Promulgates regulations concerning measures that must be taken to protect workers.

osmosis. The phenomenon whereby water diffuses through a semipermeable membrane toward an area where there is more material in solution (where there is a relatively lower concentration of water). Has particular application regarding salinization of soils where plants are unable to grow because of osmotic water loss.

OTEC (ocean thermal energy conversion). Any energy conversion scheme that uses the thermal gradient of the ocean to generate power.

outbreak. A population explosion of a particular pest. Often caused by an application of pesticides that destroys the pest's natural enemies.

overcultivation. The practice of repeated cultivation and crop growing more rapidly than the soil can regenerate, leading to a decline in soil quality and productivity.

overgrazing. The phenomenon of animals' grazing in greater numbers than the land can support in the long term. There may be a temporary economic gain in the short term, but the grassland (or other ecosystem) is destroyed, and its ability to support life in the long term is vastly diminished.

oxidation. Chemical reaction process that generally involves breakdown through combining with oxygen. Both burning and cellular respiration are examples of oxidation. In both cases, organic matter is combined with oxygen and broken down to carbon dioxide and water.

ozone. A gas, O_3, that is a pollutant in the lower atmosphere but necessary to screen out ultraviolet radiation in the upper atmosphere. May also be used for disinfecting water.

ozone hole. First discovered over the Antarctic, this is a region of stratospheric air that is severely depleted of its normal levels of ozone during the Antarctic spring because of CFCs from anthropogenic (human-made) sources.

ozone shield. The layer of ozone gas (O_3) in the upper atmosphere that screens out harmful ultraviolet radiation from the Sun.

PANs (peroxyacetylnitrates). A group of compounds present in photochemical smog that are extremely toxic to plants and irritating to eyes, nose, and throat membranes of humans.

parasites.   Organisms (plant, animal, or microbial) that attach themselves to another organism, the host, and feed on it over a period of time without killing it immediately but usually doing harm to it. Commonly divided into ectoparasites, those that attach to the outside, and endoparasites, those that live inside their hosts.

parent material.   The rock material, the weathering and gradual breakdown of which is the source of the mineral portion of soil.

particulates.   (See **PM-2, PM-10** and **suspended particulate matter.**)

parts per million (ppm).   A frequently used expression of concentration. It is the number of units of one substance present in a million units of another. For example, 1 g of phosphate dissolved in 1 million grams (t 1 ton) of water would be a concentration of 1 ppm.

passive safety features.   Those safety features of nuclear facilities that involve processes that are not vulnerable to operator intrusion or electrical power failures. Passive safety features enhance the degree of safety of nuclear reactors. (See **active safety features.**)

passive solar heating system.   A solar heating system that does not use pumps or blowers to transfer heated air or water. Instead, natural convection currents are used or the interior of the building itself acts as the solar collector.

pasteurization.   The process of applying heat to kill pathogens.

pastoralist.   One involved in animal husbandry, usually in subsistence agriculture.

pathogen.   An organism, usually a microbe, that is capable of causing disease. Such an organism is said to be pathogenic.

PCBs (polychlorinated biphenyls).   A group of widely used industrial chemicals of the chlorinated hydrocarbon class. They have become serious and widespread pollutants, contaminating most food chains on Earth, because they are extremely resistant to breakdown and are subject to bioaccumulation. They are known to be carcinogenic.

percolation.   The process of water seeping through cracks and pores in soil or rock.

permafrost.   The ground of arctic regions that remains permanently frozen. Defines tundra, since only small herbaceous plants can be sustained on the thin layer of soil that thaws each summer.

persistent.   Refers to pesticides or other chemicals that are nonbiodegradable and very resistant to breakdown by other means. Such chemicals therefore remain present in the environment more or less indefinitely.

pest.   Any organism that is noxious, destructive, or troublesome; usually means agricultural pests.

pesticide.   A chemical used to kill pests. Pesticides are further categorized according to the pests they are designed to kill—for example, herbicides kill plants, insecticides kill insects, fungicides kill fungi, and so on.

pesticide treadmill.   Refers to the fact that use of chemical pesticides simply creates a vicious cycle of "needing more pesticides" to overcome developing resistance and secondary outbreaks caused by the pesticide applications.

pest-loss insurance.   Insurance that a grower can buy that will pay in the event of loss of crop due to pests.

petrochemical.   A chemical made from petroleum (crude oil) as a basic raw material. Petrochemicals include plastics, synthetic fibers, synthetic rubber, and most other synthetic organic chemicals.

pH.   Scale used to designate the acidity or basicity (alkalinity) of solutions or soil, expressed as the logarithm of the concentration of hydrogen ions (H^+). pH 7 is neutral; values decreasing from 7 indicate increasing acidity; values increasing from 7 indicate increasing basicity. Each unit from 7 indicates a tenfold increase over the preceding unit.

pheromone.   A chemical substance secreted externally by certain members of a species that affects the behavior of other members of the same species. The most common examples are sex attractants, which female insects secrete to attract males. Pheromones are coming into use in pest control. (See also **hormones.**)

phosphate.   An ion composed of a phosphorus atom with four oxygen atoms attached. PO_4^{3-}. It is an important plant nutrient. In natural waters it is frequently the limiting factor. Therefore, additions of phosphate to natural water are frequently responsible for algal blooms.

photochemical oxidants.   A major category of secondary air pollutants, including ozone, that are highly toxic and damaging especially to plants and forests. Formed as a result of interactions between nitrogen oxides and hydrocarbons driven by sunlight.

photochemical smog.   The brownish haze that frequently forms on otherwise clear sunny days over large cities with significant amounts of automobile traffic. It results largely from sunlight-driven chemical reactions among nitrogen oxides and hydrocarbons, both of which come primarily from auto exhausts.

photosynthesis.   The chemical process carried on by green plants through which light energy is used to produce glucose from carbon dioxide and water. Oxygen is released as a byproduct.

photovoltaic cells.   Devices that convert light energy into an electrical current.

physical barrier.   A genetic feature on a plant, such as sticky hairs, that physically blocks attack by pests.

phytoplankton.   Any of the many species of algae that consist of single cells or small groups of cells that live and grow freely suspended in the water near the surface.

phytoremediation.   The use of plants to accomplish the cleanup of some hazardous chemical wastes.

planetary albedo.   The reflection of solar radiation due to cloud cover, contributing to cooling of the atmosphere.

plankton.   Any and all living things that are found freely suspended in the water and that are carried by currents as opposed to being able to swim against currents. It includes both plant (phytoplankton) and animal (zooplankton) forms.

plant community.   The array of plant species, including numbers, ages, and distribution, that occupies a given area.

plate tectonics.   The theory that postulates the movement of sections of the Earth's crust, creating earthquakes, volcanic activity, and the buildup of continental masses over long periods of time.

PM-2, PM-10.   Standard criterion pollutants for suspended particulate matter. PM-10 refers to particles smaller than 10 micrometers in diameter, and PM-2 for particles smaller than 2 micrometers in diameter. The smaller particles are readily inhaled directly into the lungs.

point sources.   Specific points of origin of pollutants, such as factory drains or outlets from sewage-treatment plants. (Contrast **nonpoint sources.**)

policy life cycle.   A typical course of events occurring over time in the recognition, formulation, implementation, and control of an environmental problem requiring public policy action.

pollutant.   A substance that contaminates air, water, or soil.

pollution.   Contamination of air, water, or soil with undesirable amounts of material or heat. The material may be a natural substance, such as phosphate, in excessive quantities, or it may be very small quantities of a synthetic compound such as dioxin, which is exceedingly toxic.

pollution avoidance, pollution prevention.   A strategy of encouraging development of techniques that would not generate pollutants.

polyculture.   The growing of two or more species together. (Contrast **monoculture.**)

poor.   Economically unable to afford adequate food and/or housing.

population. A group within a single species, the individuals of which can and do freely interbreed. Breeding between populations of the same species is less common because of differences in location, culture, nationality, and so on.

population density. The numbers of individuals per unit of area.

population equilibrium. A state of balance between births and deaths in a population.

population explosion. The exponential increase observed to occur in a population when or if conditions are such that a large percentage of the offspring are able to survive and reproduce in turn. Frequently leads, in turn, to overexploitation, upset, and eventual collapse of the ecosystem.

population momentum. Refers to the fact that a rapidly growing human population may be expected to grow for 50–60 years after replacement fertility (2.1) is reached because of increasing numbers entering reproductive age.

population profile. A bar graph that shows the number of individuals at each age or in each five-year age group.

population structure. See **age structure.**

potential energy. The ability to do work that is stored in some chemical or physical state. For example, gasoline is a form of potential energy; the ability to do work is stored in the chemical state and is released as the fuel is burned in an engine.

power grid. The combination of central power plants, high-voltage transmission lines, poles, wires, and transformers that makes up an electricial power system.

ppm. See **parts per million.**

pOH. The negative logarithm of the concentration of hydroxyl ions (OHs). Like pH, the scale ranges from 0 to 14, each unit representing a tenfold increase over the preceding unit. The lower the pOH, the higher the concentration of hydroxyl ions.

precautionary principle. The principle that says that where there are threats of serious or irreversible damage, lack of full scientific certainty shall not be used as a reason for postponing cost-effective measures to prevent environmental degradation.

precipitation. Any form of moisture condensing in the air and depositing on the ground.

predator. An animal that feeds on another living organism, either plant or animal.

predator-prey relationship. A feeding relationship existing between two kinds of organisms. The predator is the animal feeding on the prey. Such relationships are frequently instrumental in controlling populations of herbivores.

preliminary treatment. The removal of debris and grit from wastewater by passing the water through a coarse screen and grit-settling chamber.

primary consumer. An organism such as a rabbit or deer that feeds more or less exclusively on green plants or their products, such as seeds and nuts. (Synonym: **herbivore**)

primary energy sources. Fossil fuels, radioactive material, solar, wind, water, and other energy sources that exist as natural resources subject to exploitation.

primary pollutants. The air pollutants that are emitted into the air as direct byproducts of combustion or other process as opposed to those (secondary air pollutants) that form as a result of various chemical reactions occurring in the atmosphere.

primary producers/production. Refers to photosynthetic organisms and their activities in creating new organic matter in ecosystems.

primary standard. The maximum tolerable level of a pollutant. The standard is intended to protect human health.

primary succession. See **succession.**

primary treatment. The process that follows preliminary sewage treatment. It consists of passing the water very slowly through a

large tank so that the particulate organic material in the water can settle out. The settled material is raw sludge.

prior informed consent (PIC). A U.N.-approved procedure whereby countries exporting pesticides must inform importing countries of actions they have taken to restrict the use of the pesticides.

produced assets. One component of the wealth of nations. Refers to the stock of buildings, machinery, vehicles, and other infrastructure essential to the production of economic goods and services.

producers. In an ecosystem, those organisms, mostly green plants, that use light energy to construct their organic constituents from inorganic compounds.

productive use. The exploitation of ecosystem resources for economic gain.

profligate growth. Growth characterized by extravagant and wasteful use of resources.

property taxes. Taxes that the local government levies on privately owned properties, generally a few dollars per hundred dollars of property value. This is the major source of revenue for local governments.

protein. The class of organic macromolecules that is the major structural component of all animal tissues and that functions as enzymes in both plants and animals.

proton. Fundamental atomic particle with a positive charge, found in the nuclei of atoms. The number of protons present equals the atomic number and is distinct for each element.

protozoan, pl. **protozoa.** Any of a large group of microscopic organisms that consist of a single, relatively large complex cell or, in some cases, small groups of cells. All have some means of movement. Amoebae and paramecia are examples.

proven reserves. See **reserves.**

punctuated equilibrium. "Step" model of evolution in which there is little change when an ecosystem is in a balanced state, but a shift will alter selective pressures and set into motion fairly rapid charges in almost all, if not all, species in the ecosystem until a new balance is reached.

RACT (reasonably available control technology). Applied to the goals of the Clean Air Act, EPA-approved forms of technology that will reduce the output of industrial air pollutants. (See **MACT**.)

radioactive decay. The reduction of radioactivity that occurs as an unstable isotope (radioactive substance) gives off radiation and becomes stable.

radioactive emissions. Any of various forms of radiation and/or particles that may be given off by unstable isotopes. Many such emissions have very high energy and can destroy biological tissues or cause mutations leading to cancer or birth defects.

radioactive materials. Substances that are or that contain unstable isotopes and that consequently give off radioactive emissions. (See **isotope** and **radioactive emissions**.)

radioactive wastes. Waste materials that contain, are contaminated with, or are radioactive substances. Many materials used in the nuclear industry become wastes because of their contamination with radioactive substances.

radioisotope. An isotope of an element that is unstable and may tend to gain stability by giving off radioactive emissions. (See **isotope** and **radioactive decay**).

radon. A radioactive gas produced by natural processes in Earth that is known to seep into buildings. It can be a major hazard within homes and is a known carcinogen.

rain shadow. The low-rainfall region that exists on the leeward (downwind) side of a mountain range. It is the result of the mountain range's causing the precipitation of moisture on the windward side.

range of tolerance. The range of conditions within which an organism or population can survive and reproduce, for example, the range from the highest to lowest temperature that can be tolerated. Within the range of tolerance is the optimum, or best, condition.

raw sludge. The untreated organic matter that is removed from sewage water by letting it settle. It consists of organic particles from feces, garbage, paper, and bacteria.

raw wastewater. (See **raw sludge**.)

reactor vessel. Steel-walled vessel that contains the nuclear reactor.

recharge area. With reference to groundwater, the area over which infiltration and resupply of a given aquifer occurs.

recruitment. With reference to populations, the maturation and entry of young into the adult breeding population.

recycling. Recovery of materials that would otherwise be buried in landfills or combusted. Primary recycling refers to remaking the same material from the waste; in secondary recycling, waste materials are made into different products.

relative humidity. The percentage of moisture in the air compared with how much the air can hold at the given temperature.

rem. An older unit of measurement of the ability of radioactive emissions to penetrate biological tissue. (See **sievert**.)

remediation. The return to the original uncontaminated state. (See **bioremediation** and **groundwater remediation**.)

renewable energy. Energy sources, namely solar, wind, and geothermal, that will not be depleted by use.

renewable resources. Biological resources such as trees that may be renewed by reproduction and regrowth. Conservation to prevent overcutting and protection of the environment are still required, however. (Contrast **nonrenewable resources**.)

replacement fertility/level. The fertility rate that will just sustain a stable population.

reproductive isolation. One of the processes of speciation, involving anything that keeps individuals or subpopulations from interbreeding.

reproductive strategy. The particular methodologies seen in nature to enhance the chance of subsequent generations—for example, producing massive numbers of young but offering no care or protection vs. producing few young and caring for them.

reserves. The amount of a mineral resource (including oil, coal, and natural gas) remaining in Earth that can be exploited using current technologies and at current prices. Usually given as *proven* reserves, those that have been positively identified, and *estimated* reserves, those that have not yet been discovered but that are presumed to exist.

resistance. In reference to pesticides, the development of more hardy pests by the effects of repeated use of pesticides in selecting against sensitive individuals in a pest population.

resource partitioning. The outcome of competition by a group of species, where natural selection favors division of a resource in time or space by specialization of the different species.

resources. Biotic and abiotic factors that are consumed by organisms.

Resources Conservation and Recovery Act of 1976 (RCRA). The cornerstone legislation to control indiscriminate land disposal of hazardous wastes.

respiration. See **cell respiration**.

restoration ecology. The branch of ecology devoted to restoring degraded and altered ecosystems to their natural state.

resurgence. The rapid comeback of a population especially of pests after a severe dieoff, usually caused by pesticides, and the return to even higher levels than before the treatment.

reuse. The practice of reusing items as opposed to throwing them away and producing new items, as, for example, bottles collected and refilled (recycling).

riparian woodlands. The strip of woods that grow along natural watercourses.

risk. The probability of suffering injury, disease, death, or other loss as a result of exposure to a hazard.

risk analysis. The process of evaluating the risks associated with a particular hazard before taking some action. Often called risk assessment.

risk characterization. The process of determining a risk and its accompanying uncertainties after hazard assessment, dose-response assessment, and exposure assessment have been accomplished.

risk management. The task of regulators whose job is to review risk data and make regulatory decisions based on the evidence. The process often is influenced by considerations of costs and benefits as well as by public perception.

risk perception. Nonexperts' intuitive judgments about risks, which often are not in agreement with the level of risk as judged by experts.

rooftop garden. A form of above-ground gardening on the top of a flat residential roof, using shallow soil beds and containers like old automobile tires or plastic wading pools.

rubber tapper. In tropical forests of South America, people who harvest the latex rubber and Brazil nuts from rain forest trees and are opposed to clearing the forests. Also called *serengueros*.

runoff. The portion of precipitation that runs off the surface as opposed to soaking in.

Safe Drinking Water Act of 1974. Legislation to protect the public from the risk that toxic chemicals will contaminate drinking water supplies. Mandates regular testing of municipal water supplies.

safety net. In a society, the availability of food and other necessities extended to people who are unable to meet their own needs for a variety of reasons.

salinization. The process whereby soil becomes saltier and saltier until finally the salt prevents the growth of plants. It is caused by irrigation because salts brought in with the water remain in the soil as the water evaporates.

saltwater intrusion, saltwater encroachment. The phenomenon of seawater's moving back into aquifers or estuaries. It occurs when the normal outflow of freshwater is diverted or removed for use.

sand. Mineral particles 0.2û2.0 mm in diameter.

sanitary sewer. Separate drainage system used to receive all the wastewater from sinks, tubs, and toilets.

SARA (Title III). A section of the Superfund Amendments and Reauthorization Act that promulgates community Right-to-Know requirements. Also known as Emergency Planning and Community Right-to-Know Act of 1986. (See **toxics release inventory**).

savanna. A type of grassland typical of subtropical regions, particularly in Africa, usually dotted with trees and supported by wet and dry seasons and frequent natural fires.

secondary air pollutants. Air pollutants resulting from reactions of primary air pollutants while resident in the atmosphere. These include ozone, other reactive organic compounds, and sulfuric and nitric acids. (See **ozone**, **PANs**, and **photochemical oxidants**.)

secondary consumer. An organism such as a fox or coyote that feeds more or less exclusively on other animals that feed on plants.

secondary energy source. A form of energy such as electricity that must be produced from a primary energy source such as coal or radioactive material.

secondary pest outbreak. The phenomenon of a small, and therefore harmless, population of a plant-eating insect suddenly exploding to become a serious pest problem. Often caused by the elimination of competitors through pesticide use.

secondary succession. See **succession**.

secondary treatment. Also called biological treatment. A sewage-treatment process that follows primary treatment. Any of a variety of systems that remove most of the remaining organic matter by enabling organisms to feed on it and oxidize it through their respiration. Trickling filters and activated-sludge systems are the most commonly used methods.

second principle of ecosystem sustainability. Whereby resources are supplied and wastes are disposed of by recycling all elements.

second-generation pesticides. Synthetic organic compounds used to kill insects and other pests. Started with the use of DDT in the 1940s.

second law of thermodynamics. The fact based on irrefutable observations that in every energy conversion (e.g., electricity to light), some of the energy is converted to heat and some heat always escapes from the system because it always moves toward a cooler place. Therefore, in every energy conversion, a portion of energy is lost. And, since energy cannot be created (first law), the functioning of any system requires an energy input.

secondary energy source. A source of energy, especially electricity, that depends on a primary energy source for its origin.

secure landfill. A landfill with suitable barriers, leachate drainage, and monitoring systems such that it is deemed secure against contaminating groundwater with hazardous wastes.

sediment. Soil particles—namely, sand, silt, and clay—carried by flowing water. The same material after it has been deposited. Because of different rates of settling, deposits generally are pure sand, silt, or clay.

sedimentation. The filling in of lakes, reservoirs, stream channels, and so on with soil particles, mainly sand and silt. The soil particles come from erosion, which generally results from poor or inadequate soil conservation practices in connection with agriculture, mining, and/or development. Also called siltation.

sediment trap. A device for trapping sediment and holding it on a development or mining site.

seep. Where groundwater seeps from the ground over some area as opposed to a spring, which is the exit as a single point.

selective breeding. The breeding of certain individuals because they bear certain traits and the exclusion from breeding of others.

selective pressure. A fundamental mechanism of evolution. An environmental factor that causes individuals with certain traits, which are not the norm for the population, to survive and reproduce more than the rest of the population. The result is a shift in the genetic makeup of the population. For example, the presence of insecticides provides a selective pressure to increase pesticide resistance in the pest population.

septic system. An on-site method of treating sewage widely used in suburban and rural areas. Requires a suitable amount of land and porous soil.

sex attractant. A natural chemical substance (pheromone) secreted by the female of many insect species that serves to attract males for the function of mating. Sex attractants may be used in traps or for the confusion technique to aid in the control of insect pests.

shadow pricing. In cost-benefit analysis, a technique used to estimate benefits where normal economic analysis is ineffective. For example, people could be asked how much they might be willing to pay monthly to achieve some improvement in their environment.

sheet erosion. The loss of a more or less even layer of soil from the surface due to the impact and runoff from a rainstorm.

shelterbelts. Rows of trees around cultivated fields for the purpose of reducing wind erosion.

sievert. A unit of measurement of the ability of radioactive emissions to penetrate biological tissues. 1 sievert = 100 rem.

silt. Soil particles between the size of sand particles and clay particles; namely, particles 0.002–0.2 mm in diameter.

siltation. See **sedimentation.**

sinkhole. A large hole resulting from the collapse of an underground cavern.

slash-and-burn agriculture. The practice, commonly exercised throughout tropical regions, of cutting and burning vegetation to make room for agriculture. The process is highly destructive of soil humus and may lead to rapid degradation of soil.

sludge cake. Treated sewage sludge that has been dewatered to make a moist solid.

sludge digesters. Large tanks in which raw sludge (removed from sewage) is treated through anaerobic digestion by bacteria.

Smart Growth. A movement that addresses urban sprawl by protecting sensitive lands and directing growth to limited areas.

smog. See **industrial smog** and **photochemical smog.**

social modernization. A process of sustainable development that leads developing countries through the demographic transition by promoting education, family planning, and health improvements rather than simply economic growth.

soft water. Water with little or no calcium, magnesium, or other ions in solution that will cause soap to precipitate (form a curd that makes a "ring" around the bathtub). (Contrast **hard water.**)

soil. A dynamic terrestrial system involving three components: mineral particles, detritus, and soil organisms feeding on the detritus.

soil classes. Major groupings of soil arranged like a taxonomy that define the properties of the soils of different regions.

soil erosion. The loss of soil caused by particles' being carried away by wind and/or water.

soil fertility. Soil's ability to support plant growth; often refers specifically to the presence of proper amounts of nutrients. The soil's ability to fulfill all the other needs of plants is also involved.

soil horizons. Distinct layers within a soil that convey different properties to the soil and derive from natural processes of soil formation.

soil profile. A description of the different, naturally formed layers, called horizons, within a soil.

soil structure. The composition of soil in terms of particles (sand, silt, and clay) stuck together to form clumps and aggregates, generally with considerable air spaces in between. Structure affects infiltration and aeration. It develops as organisms feed on organic matter in and on the soil.

soil texture. The relative size of the mineral particles that make up the soil. Generally defined in terms of the sand, silt, and clay content.

solar cells. See **photovoltaic cells.**

solar energy. Energy derived from the Sun. Includes direct solar energy (the use of sunlight directly for heating and/or production of electricity) and indirect solar energy (the use of wind, which results from the solar heating of the atmosphere, and biological materials such as wood, which result from photosynthesis).

solar-trough collectors. Reflectors in the shape of a parabolic trough, which reflect the sunlight onto a tube of oil at the focal point. The oil thus heated is used to boil water to drive a steam turbine.

solid waste. The total of materials discarded as "trash" and handled as solids, as opposed to those that are flushed down sewers and handled as liquids.

solubility. The degree to which a substance will dissolve and enter into solution.

solution. A mixture of molecules (or ions) of one material in another. Most commonly, molecules of air and/or ions of various minerals in water. For example, seawater contains salt in solution.

sound science. The results of scientific work based on peer-reviewed research. As one of the unifying themes of this text, the basis for our understanding of how the world works and how human systems interact with it.

special-interest group. In politics, people organized around a particular environmental issue who lobby and exert their influence on the political process to bring about changes they favor.

specialization. With reference to evolution, the phenomenon whereby species become increasingly adapted to exploit one particular niche but, thereby, are less able to exploit other niches.

speciation. The evolutionary process whereby populations of a single species separate and, through being exposed to different forces of natural selection, gradually develop into distinct species.

species. All the organisms (plant, animal, or microbe) of a single kind. The "single kind" is determined by similarity of appearance and/or by the fact that members do or potentially can mate and produce fertile offspring. Physical, chemical, or behavioral differences block breeding between species.

splash erosion. The compaction of soil that results when rainfall hits bare soil.

springs. Natural exits of groundwater.

standing crop biomass. The biomass of primary producers in an ecosystem at any given time.

starvation. The failure to get enough calories to meet energy needs over a prolonged period of time. It results in a wasting away of body tissues until death occurs.

Statement on Forest Principles. One of the treaties signed at the 1992 Earth Summit, where the signatory countries agreed to a number of nonbinding principles stressing sustainable management.

sterile male technique. Saturating an infested area with males of the pest species that have been artificially reared and sterilized by radiation. Matings between normal females and sterile males render the eggs infertile.

stewardship/steward. A steward is one to whom a trust has been given. In reference to natural lands, stewardship is an attitude of active care and concern for nature. As one of the unifying themes of this text, the ethical and moral framework that informs our public and private actions.

stoma, pl. stomata. Microscopic pores in leaves, mostly on the undersurface, that allow the passage of carbon dioxide and oxygen into and out of the leaf and that also permit the loss of water vapor from the leaf.

storm drains. Separate drainage systems used for collecting and draining runoff from precipitation in urban areas.

storm water. In cities, the water that results directly from rainfall, as opposed to municipal water and sewage water piped to and from homes, offices, and so on. The extensive hard surfacing in cities creates a vast amount of stormwater runoff, which presents a significant management problem.

stormwater management. Policies and procedures for handling stormwater in acceptable ways to reduce the problems of flooding and erosion of streambanks.

stormwater retention reservoirs. Reservoirs designed to hold stormwater temporarily and let it drain away slowly in order to reduce problems of flooding and streambank erosion.

stratosphere. The layer of Earth's atmosphere between 10 and 40 miles above the surface that contains the ozone shield. This layer mixes only slowly; pollutants that enter may remain for long periods of time. (See **troposphere**.)

strip cropping. The practice of growing crops in strips alternating with grass (hay) at right angles to prevailing winds or slopes in order to reduce erosion.

strip mining. The mining procedure in which all the earth covering a desired material such as coal is stripped away with huge power shovels in order to facilitate removal of the desired material.

subduction. In plate tectonics, where an oceanic plate slides under a continental plate.

submerged aquatic vegetation (SAV). Aquatic plants rooted in bottom sediments growing under water that depend on light's penetration through the water for photosynthesis.

subsistence agriculture farming. Farming that meets the food needs of the farmers and their families but little more. It involves hand labor and is practiced extensively in the developing world.

subsoil. In a natural situation, the soil beneath topsoil. In contrast to topsoil, subsoil is compacted and has little or no humus or other organic material, living or dead. In many cases, topsoil has been lost or destroyed as a result of erosion or development, and subsoil is at the surface.

succession. The gradual, or sometimes rapid, change in the species that occupy a given area, with some species invading and becoming more numerous while others decline in population and disappear. Succession is caused by a change in one or more abiotic or biotic factors that benefits some species at the expense of others. *Primary succession:* The gradual establishment, through a series of stages, of a climax ecosystem in an area that has not been occupied before, e.g., a rock face. *Secondary succession:* The reestablishment, through a series of stages, of a climax ecosystem in an area from which it was previously cleared.

sulfur dioxide (SO_2). A major air pollutant, this toxic gas is formed as a result of burning sulfur. The major sources are burning coal (coal-burning power plants) that contains some sulfur and refining metal ores (smelters) that contain sulfur.

sulfuric acid (H_2SO_4). The major constituent of acid precipitation. Formed when sulfur dioxide emissions react with water vapor in the atmosphere. (See **sulfur dioxide**.)

Superfund. The popular name for the Comprehensive Environmental Response, Compensation, and Liability Act of 1980. This act is the cornerstone legislation that provides the mechanism and funding for the cleanup of potentially dangerous hazardous waste sites to protect groundwater.

surface impoundments. Closed ponds formerly used to collect and hold liquid chemical wastes.

surface water. Includes all bodies of water, lakes, rivers, ponds, and so on that are on Earth's surface in contrast to groundwater, which lies below the surface.

surge flow irrigation. Microprocessor control that allows the periodic release of irrigation water, thus cutting water use substantially.

suspended particulate matter (SPM). A category of major air pollutants consisting of solid and liquid particles suspended in the air. (See **PM-2, PM-10**.)

suspension. With reference to materials contained in or being carried by water, materials kept "afloat" only by the water's agitation that settle as the water becomes quiet.

sustainability. Refers to whether a process can be continued indefinitely without depleting the energy or material resources on which it depends. As one of the unifying themes of the text, the practical goal toward which our interactions with the natural world should be working.

sustainable agriculture. Agriculture that maintains the integrity of soil and water resources such that it can be continued indefinitely.

sustainable development. Development that provides people with a better life without sacrificing or depleting resources or causing environmental impacts that will undercut the ability of future generations to meet their needs.

sustainable forest management. Management of forests as ecosystems where the primary objective of management is to maintain the biodiversity and function of the ecosystem.

sustainable society. A society that functions in a way so as not to deplete the energy or material resources on which it depends. It interacts with the natural world in ways that sustain existing species and ecosystems.

sustainable yield. The taking of a biological resource (e.g., fish or forests) that does not exceed the capacity of the resource to reproduce and replace itself.

symbiosis. The intimate living together or association of two kinds of organisms.

synergism. The phenomenon in which two factors acting together have an exceedingly greater effect than would be indicated by the sum of their effects separately—as, for example, the sometimes fatal mixture of modest doses of certain drugs in combination with modest doses of alcohol.

synfuels, synthetic fuels. Fuels similar or identical to those that come from crude oil and/or natural gas, produced from coal, oil shale, or tar sands.

tar sands. Sedimentary material containing bitumin that can be "melted out" using heat and then refined in the same way as crude oil.

taxonomy. The science of identification and classification of organisms according to evolutionary relationships.

TEA (total exposure assessment). The analysis of the impact of air pollutants on people based on the amount of time they spend in different spaces, especially indoors.

tectonic plates. Huge slabs of rock that make up Earth's crust.

temperature inversion. The weather phenomenon in which a layer of warm air overlies cooler air near the ground and prevents the rising and dispersion of air pollutants.

teratogenic. Causing birth defects.

terracing. The practice of grading sloping farmland into a series of steps and cultivating only the level portions in order to reduce erosion.

territoriality. The behavioral characteristic exhibited by many animal species, especially birds and mammalian carnivores, to mark and defend a given territory against other members of the same species.

theory. A conceptual formulation that provides a rational explanation or framework for numerous related observations.

thermal pollution. The addition of abnormal and undesirable amounts of heat to air or water. It is most significant with respect to discharging waste heat from electric generating plants, especially nuclear power plants, into bodies of water.

third principle of ecosystem sustainability. The size of consumer populations is maintained such that overgrazing and other forms of overuse do not occur.

third world. See **developing countries.**

threatened species. A species the population of which is declining precipitously because of direct or indirect human impacts.

threshold level. The maximum degree of exposure to a pollutant, drug, or other factors that can be tolerated with no ill effect. The threshold level will vary depending on the species, the sensitivity of the individual, the length of exposure, and the presence of other factors that may produce synergistic effects.

tidal wetlands. Areas of marsh grasses and reeds along coasts and estuaries where the ground is covered by high tides but drained at low tide.

topsoil. The surface layer of soil, which is rich in humus and other organic material, both living and dead. As a result of the activity of organisms living in the topsoil, it generally has a loose, crumbly structure as opposed to being a compact mass. In many cases, because of erosion, development, or mining activity, the topsoil layer may be absent.

total allowable catch (TAC). In fisheries management, a yearly quota set for the harvest of a species by fisheries managers.

total fertility rate. The average number of children that would be born alive to each woman during her total reproductive years if she followed the average fertility at each age.

total product life cycle. Consideration of all steps from the obtaining of raw materials through the manufacture, use, and finally disposal of a product. Consideration of byproducts and pollution resulting from each step.

Toxic Substances Control Act (TOSCA). A law requiring the assessment of the potential hazards of a chemical before the chemical is put on the market.

toxicology. The study of the impacts of toxic substances on human health and the pathways by which such substances reach humans.

toxics release inventory. An annual record of releases of toxic chemicals to the environment and the locations and quantities of toxic chemicals stored on site, required by the Emergency Planning and Community Right-to-Know Act (EPCRA) of 1986.

trace elements. Those essential elements that are needed in only very small amounts.

trade winds. The more or less constant winds blowing in horizontal directions over the surface as part of Hadley cells.

traditional agriculture. Farming methods as they were practiced before modern industrialized agriculture.

tragedy of the commons. The overuse or overharvesting and consequent depletion and/or destruction of a renewable resource that tends to occur when the resource is treated as a commons, that is, when it is open to be used or harvested by any and all with the means to do so.

trait. Any physical or behavioral characteristic or talent that an individual is born with.

transgenic organism. Any organism with genes introduced from other species via biotechnology to convey new characteristics.

transpiration. The loss of water vapor from plants. Water evaporates from cells within the leaves and exits through stomata.

trapping technique. The use of sex attractants to lure male insects into traps.

treated sludge. Solid organic material that has been removed from sewage and treated so that it is nonhazardous.

trickling filter system. System in which wastewater trickles over rocks or a framework coated with actively feeding microorganisms. The feeding action of the organisms in a well-aerated environment results in the decomposition of organic matter. Used in secondary or biological treatment of sewage.

tritium (^{3}H). An unstable isotope of hydrogen that contains two neutrons in addition to the usual single proton in the nucleus. It does not occur in significant amounts naturally but is human-made.

trophic level. Feeding level with respect to the primary source of energy. Green plants are at the first trophic level, primary consumers at the second, secondary consumers at the third, and so on.

trophic structure. The major feeding relationships between organisms within ecosystems, organized into trophic levels.

troposphere. The layer of Earth's atmosphere from the surface to about 10 miles in altitude. The tropopause is the boundary between the troposphere and the stratosphere above. This layer is well mixed and is the site and source of our weather, as well as the primary recipient of air pollutants. (See also **stratosphere**.)

turbid. Refers to water purity; means cloudy.

turbine. A sophisticated "paddle wheel" driven at a very high speed by steam, water, or exhaust gases from combustion.

turbogenerator. A turbine coupled to and driving an electric generator. Virtually all commercial electricity is produced by such devices. The turbine is driven by gas, steam, or water.

ultraviolet radiation. Radiation similar to light but with wavelengths slightly shorter than violet light and with more energy. The greater energy causes it to severely burn and otherwise damage biological tissues.

underground storage tanks. The tanks used to store petroleum products at service stations, now subject to mandated protection against leakage. See **UST legislation.**

undernutrition. A form of hunger in which there is a lack of adequate food energy as measured in calories. Starvation is the most severe form of undernutrition.

upwelling. In oceanic systems, the upward vertical movement of water masses, caused by diverging currents and offshore winds, bringing nutrient-rich water to the surface.

urban blight/decay. General deterioration of structures and facilities such as buildings and roadways, in addition to the decline in quality of services such as education, that has occurred in inner city areas where growth has been focused on suburbs and exurbs.

urban sprawl. The rapid expansion of metropolitan areas through building housing developments and shopping centers farther and farther from urban centers and lacing them together with more and more major highways. Widespread development that has occurred without any overall land-use plan.

UST legislation. Amendments to the Resources Conservation and Recovery Act of 1976, passed in 1984 to address the mounting problem of leaking underground storage tanks (USTs).

value. In ethical considerations, value means imparting a property to objects or species or individuals that implies a moral duty to the one assigning the value; intrinsic and instrumental value are two types of valuing.

vector. An agent like an insect or tick that carries a parasite from one host to another.

vigor. Applied to crop plants or animals, refers to traits for hardiness to disease, drought, cold, and other adverse factors or conditions.

vitamin. A specific organic molecule that is required by the body in small amounts but that cannot be made by the body and therefore must be present in the diet.

volatile organic compounds (VOCs). A category of major air pollutants present in the air in vapor state, including fragments of hydrocarbon fuels from incomplete combustion and evaporated organic compounds such as paint solvents, gasoline, and cleaning solutions. They are major factors in the formation of photochemical smog.

waste-to-energy (WTE). A process of combustion of solid wastes that also generates electrical energy.

water cycle. See **hydrologic cycle.**

water-holding capacity. The ability of a soil to hold water so that it will be available to plants.

waterlogging. The total saturation of soil with water. Results in plant roots' not being able to get air and dying as a result.

watershed. The total land area that drains directly or indirectly into a particular stream or river. The watershed is generally named from the stream or river into which it drains.

water table. The upper surface of groundwater. It rises and falls with the amount of groundwater.

water vapor. Water molecules in the gaseous state.

weather. The day-to-day variations in temperature, air pressure, wind, humidity, and precipitation mediated by the atmosphere in a given region.

weathering. The gradual breakdown of rock into smaller and smaller particles, caused by natural chemical, physical, and biological factors.

wet cleaning. A water-based alternative to dry cleaning that avoids the use of hazardous chemicals.

wetlands. Areas that are constantly wet and are flooded at more or less regular intervals. Especially, marshy areas along coasts that are regularly flooded by tide.

wetland systems. A biological aquatic system (usually a restored wetlands) to remove nutrients from treated sewage wastewater and return it, virtually pure, to a river or stream. Wetland systems are sometimes used when using treated wastewater for irrigation is not feasible.

Wilderness Act of 1964. Federal legislation that provides for the permanent protection of undeveloped and unexploited areas so that natural ecological processes can operate freely. Most uses are excluded from such areas, which now total 100 million acres in the United States.

wind farms. Arrays of numerous, modestly sized wind turbines for the purpose of producing electrical power.

windrows. Piles of organic material extended into long rows to facilitate turning and aeration to enhance composting.

wind turbines. "Windmills" designed for the purpose of producing electrical power.

WIPP (waste isolation pilot plant). A facility built by the Department of Energy in New Mexico to receive defense-related nuclear wastes.

Wise Use. A form of environmental backlash reacting against regulations and restrictions on the use of public lands for recreation and extractive activities like mining, forestry, and grazing.

work. Any change in motion or state of matter. Any such change requires the expenditure of energy.

workability. With reference to soils, the relative ease with which a soil can be cultivated.

World Bank. A branch of the United Nations that acts as a conduit to handle loans to developing countries.

world view. A set of assumptions that a person holds regarding the world and how it works.

xeroscaping. Landscaping with drought-resistant plants that need no watering.

yard wastes. Grass clippings and other organic wastes from lawn and garden maintenance.

Younger Dryas event. A rapid change in global climate between 10,000 and 12,000 years ago that brought on 1500 years of colder weather.

zones of stress. Regions where a species finds conditions tolerable but suboptimal. Where a species survives but under stress.

zooplankton. Any of a number of animal groups, including protozoa, crustaceans, worms, molluscs, and cnidarians, which live suspended in the water column and feed on phytoplankton and other zooplankton.

Photo Credits

Index